ИЗДАТЕЛЬСТВО МИР

А. И. АНСЕЛЬМ

# ВВЕДЕНИЕ В ТЕОРИЮ ПОЛУПРОВОДНИКОВ

Издательство «Наука» Москва

A. ANSELM

# INTRODUCTION to SEMICONDUCTOR THEORY

Translated from the Russian
by M.M. Samokhvalov
Cand. Sc.



MIR PUBLISHERS · MOSCOW

PRENTICE-HALL, INC.,
Englewood Cliffs, New Jersey 07632

*To the memory of Abram Fyodorovich Ioffe*

English language edition, for distribution throughout the world excluding the Union of Soviet Socialist Republics, Bulgaria, Czechoslovakia, Hungary, Poland, Romania, and India, published 1981 by Prentice-Hall, Inc., Englewood Cliffs, New Jersey 07632.

First published 1981
Revised from the 1978 Russian edition

ISBN 0-13-496034-3

Library of Congress Catalog Card Number 80-84675

PRENTICE-HALL INTERNATIONAL, INC., London
PRENTICE-HALL OF AUSTRALIA PTY. LIMITED., Sydney
PRENTICE-HALL OF CANADA, LTD., Toronto
PRENTICE-HALL OF JAPAN, INC., Tokyo
PRENTICE-HALL OF SOUTHEAST ASIA PTE. LTD., Singapore
WHITEHALL BOOKS LIMITED, Wellington. New Zealand





*Printed in the Union of Soviet Socialist Republics*

# From the Preface to the First Russian Edition

This book has been written mainly for the benefit of people engaged in experimental work in the field of semiconductor physics. It will probably prove useful to students specializing in physics.

Among the principal subjects treated in this book are crystal lattice vibrations, the laws of electron motion in an ideal and a perturbed periodic fields, the kinetic equation and transport phenomena (electric current).

The reader must be familiar with mathematics, quantum mechanics and physical statistics within the limits specified in the curricula of physical faculties of universities or physical and mathematical faculties of polytechnical colleges. He or she need not have a detailed knowledge of those courses but is expected to be able to find a way through the appropriate sections of textbooks referred to.

The special feature of the book is that those elementary facts are used to derive all the formulae. This, I hope, is done meticulously enough to make the book comprehensible for the abovementioned category of readers.

Some mathematical derivations of a more complex nature and less connected with the main text are presented in the end of the book in Appendices.

Quite naturally, such detailed derivation of the fundamental relations in a book of limited size precluded the discussion of some important problems and detailed comparison of theory with experiment.

I have dedicated this book to the noble memory of Abram Fyodorovich Ioffe on whose initiative it has been written. I express sincere gratitude to the editor G. E. Pikus for his numerous remarks which helped to improve the book.

A. I. Anselm

# Preface to the Second Russian Edition

In its new edition the book has been greatly augmented and revised. Yet the general aims and purpose of the book have remained the same. Like the first edition, the second is intended for the same category of readers: for students specializing in physics and for experimental physicists. This sets the standard of knowledge required for reading this book.

Two new chapters have been written for the 2nd edition: Elements of Group Theory and Crystal Symmetry (Chapter 2) and Optical Phenomena in Semiconductors (Chapter 7). Besides, old sections have been revised and new ones added: Sections 3.6-3.10, 4.8-4.15, 5.1, 6.9, 8.6, 9.2-9.6, 9.9, and 9.11.

With the extent of modern solid state physics having grown boundlessly, the choice of the additional material naturally had to be subjective. I have as before tried to present the material so that "the reader will derive some pleasure from the perusal of this book and from our refusal to invite prolonged confusion by the phrase 'it can readily be shown that'" (from P. T. Landsberg's preface to *Solid State Theory: Methods and Applications* (ed. by P. T. Landsberg, Wiley, London, 1969, p. V).

Although the book has grown in size almost one half as large as the first edition, many important sections of semiconductor theory (strong electric fields, disordered semiconductors, optics of impurity centres, semiconductor electronics, etc.) could not be included.

Being a textbook, the book contains almost no references to original papers. At appropriate places the reader is referred to monographs and textbooks; in part they have been used in the preparation of the book. Besides, I have used the lectures on semiconductor theory I have been reading at the Leningrad Polytechnical Institute and at the Leningrad State University.

I express my sincere gratitude to Grigori Evgen'ievich Pikus, who has carefully read the whole manuscript, for his numerous remarks that helped to improve the book appreciably. I am thankful to Yu. N. Obraztzov and to G. L. Bir for their part in discussions on several problems leading to improvement of the text, and to A. G. Aronov for Section 7.8.3.

A.I.A.

# Contents

# 1. The Geometry of Crystal Lattices and X-Ray Diffraction

## 1.1 Simple and Complex Crystal Lattices

**1.1.1** Most solid semiconductors and solid metals have a crystalline structure, i.e., they are collections of an enormous number of atoms regularly arranged in space. Regular arrangement is the property of periodicity in space, or *translational symmetry*, characteristic of crystal lattices. In other words, we assume the existence of three vectors, $\mathbf{a}_1$, $\mathbf{a}_2$, $\mathbf{a}_3$, not lying in one plane, such that when the crystal as a whole is displaced along any of those vectors it coincides with itself. Of course, we ignore the existence of thermal motion and the surface of the crystal. As we will show below, in the lattice the directions of the vectors $\mathbf{a}_i$ $(i = 1, 2, 3)$ may be chosen in a number of ways. It is also evident that the displacement of the crystal by an integral number of vectors $\mathbf{a}_i$ results in it coinciding with itself. In what follows we will assume $\mathbf{a}_i$ to be the vectors of minimum length and fixed direction.

The vectors $\mathbf{a}_i$ so chosen are called the *basis vectors* of the crystal lattice. A parallelepiped built on the three vectors $\mathbf{a}_i$ is called the *unit cell.* Let us agree to arrange the vectors $\mathbf{a}_1$, $\mathbf{a}_2$, and $\mathbf{a}_3$ in the same order as the positive axes $x$, $y$, $z$ of a right-handed coordinate frame. Making use of the usual definition of a triple scalar product in a right-handed coordinate frame [1.1, p. 214], one can demonstrate that the volume of the unit cell is

$$\Omega_0 = |\,\mathbf{a}_1\cdot(\mathbf{a}_2 \times \mathbf{a}_3)\,| = |\,\mathbf{a}_3\cdot(\mathbf{a}_1 \times \mathbf{a}_2)\,| = |\,\mathbf{a}_2\cdot(\mathbf{a}_3 \times \mathbf{a}_1)\,|. \quad (1.1)$$

**1.1.2** We shall begin the study of the geometry of crystal lattices with the discussion of a linear (one-dimensional) lattice, which is a collection of particles arranged periodically along an infinite straight line. Such a lattice can be obtained by successive displacements of an atom or a group of atoms by equal segments along a straight line. In this case we have only one basis vector $\mathbf{a}_1 = \mathbf{a}$, and the "volume" of the unit cell, $\Omega_0$, is equal to the length of the segment, $a$. Figure 1.1 depicts three linear lattices. White and black circles depict atoms of different types. Keeping in mind that the basis vector $\mathbf{a}$ is the least distance by which the lattice must be displaced to make it coincide with itself, we see that lattice $(a)$ contains one atom per unit cell $\Omega_0 = a$, and the lattices $(b)$ and $(c)$

two atoms each. The lattice (*a*) is termed *simple* or *primitive*, and the lattices (*b*) and (*c*) *complex*.

Figure 1.2*a* depicts a planar lattice with atoms arranged at the vertices of parallelograms. It can be obtained as a result of parallel

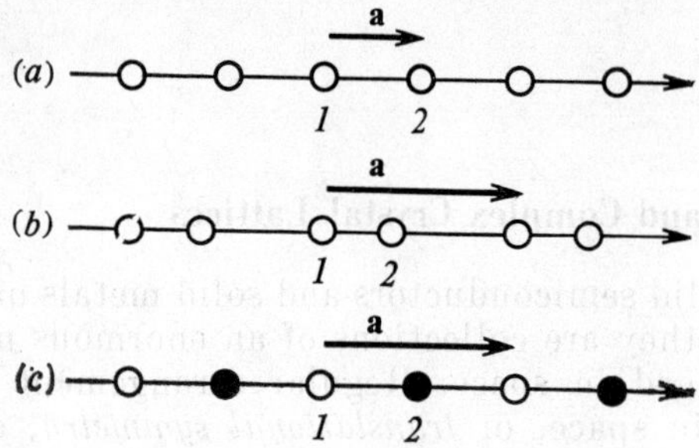


**Fig. 1.1**

displacements of a simple linear lattice (Fig. 1.1*a*) by equal distances in a plane. Figure 1.2*a* shows that the choice of basis vectors $\mathbf{a}_1$ and $\mathbf{a}_2$ is not unique. The unit cells *I* and *II* contain one atom each, and their "volumes" $\Omega_0 = |\mathbf{a}_1 \times \mathbf{a}_2|$ equal to the area of the shaded

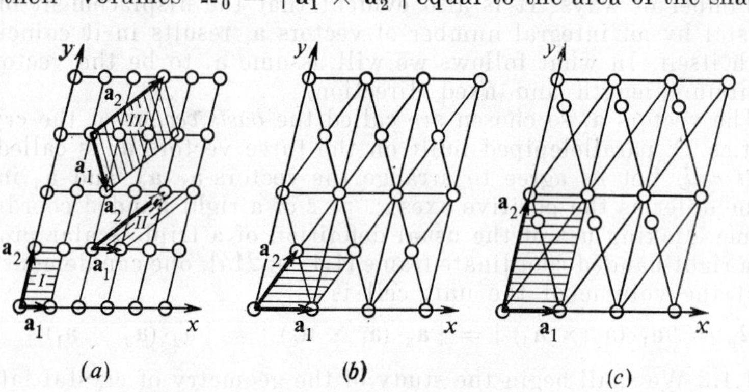


**Fig. 1.2**

parallelograms are equal. We see from the example of the unit cell *III* containing three atoms that a unit cell of more than one atom can be built even in a simple lattice. If the basis vectors $\mathbf{a}_i$ are chosen so that any translation of the lattice can be expressed in the form $\sum n_i \mathbf{a}_i$, with the $n_i$ being integers, the unit cell built on $\mathbf{a}_i$ is termed *primitive*. The unit cells *I* and *II* in Fig. 1.2*a* are primitive, and the cell *III* is not. Indeed, in the latter case a displacement along the $x$ axis by one (two) minimum translations is

$n_1\mathbf{a}_1 + n_2\mathbf{a}_2$, with $n_1 = n_2 = 1/3$ $(= 2/3)$. If the primitive cell of a lattice contains one atom, the lattice is termed *simple*, and if the number of the atoms exceeds one, it is termed *complex*. This defini-

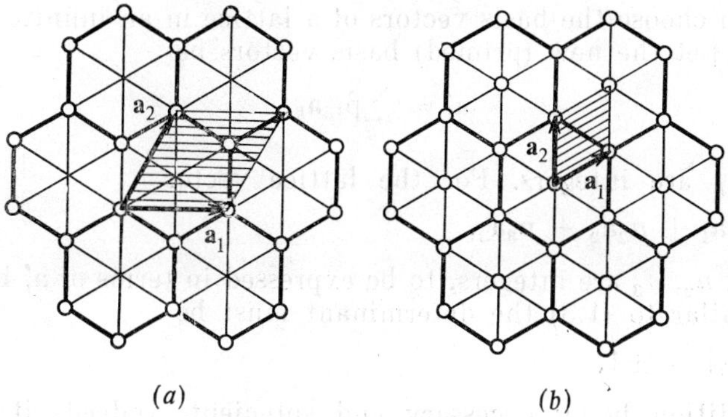


**Fig. 1.3**

tion holds also for a three-dimensional lattice. Hence, the lattice in Fig. 1.2*a* is simple.

Figure 1.2*b* depicts another simple lattice obtained from the lattice shown in Fig. 1.2*a* by placing atoms of the same type at the intersections of diagonals of the parallelograms. Now we can choose the primitive cell as shown in Fig. 1.2*b*. If we displace all the atoms at the intersections of the diagonals in a similar manner (see Fig. 1.2*c*), we obtain a complex lattice with two atoms per primitive cell, which we choose as shown in the figure. We can imagine this complex lattice as consisting of two simple lattices one inserted into the other. If we place atoms of different types at the intersections of the diagonals of the parallelograms (Fig. 1.2*b*), we obtain a complex lattice, because in this case the lattice sites are not equivalent.

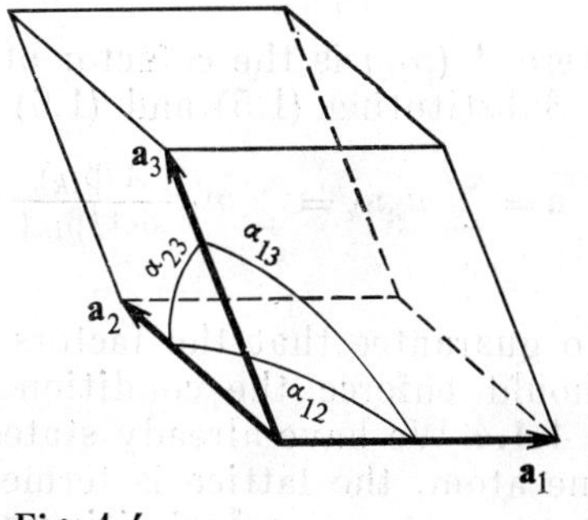


**Fig. 1.4**

Figure 1.3*a* depicts a very symmetrical lattice whose atoms are located at the vertices of octagons filling the plane. This lattice may easily be seen to be complex, since the primitive lattice shown in the figure by the vectors $\mathbf{a}_1$ and $\mathbf{a}_2$ contains two atoms. If we place an additional atom of the same kind in the centre of each octagon, we obtain a simple lattice (1.3*b*).

**1.1.3** A three-dimensional crystal lattice is based on a unit cell in the shape of a parallelepiped built on the basis vectors $\mathbf{a}_1$, $\mathbf{a}_2$, and $\mathbf{a}_3$ (Fig. 1.4; $\alpha_{12}$ denotes the angle between $\mathbf{a}_1$ and $\mathbf{a}_2$; the other angles are denoted in a similar manner).

We can choose the basis vectors of a lattice in an infinite number of ways. Let the new (primed) basis vectors be

$$\mathbf{a}'_i = \sum_k \beta_{ik}\mathbf{a}_k, \tag{1.2}$$

where $\beta_{ik}$ are integers. For the lattice vector

$$\mathbf{a} = n_1\mathbf{a}_1 + n_2\mathbf{a}_2 + n_3\mathbf{a}_3, \tag{1.3}$$

where $n_1$, $n_2$, $n_3$ are integers, to be expressed in terms of $\mathbf{a}'_i$ by a formula similar to (1.3) the determinant must be

$$\det[\beta_{ik}] = \pm 1, \tag{1.4}$$

this condition being necessary and sufficient. Indeed, it follows from (1.2) that

$$\mathbf{a}_k = \sum_i \beta^{-1}_{ki}\mathbf{a}'_i, \tag{1.5}$$

where the coefficients of the inverse transformation are [1.1, p. 266]

$$\beta^{-1}_{ki} = \frac{A(\beta_{ik})}{\det[\beta_{ik}]}. \tag{1.6}$$

Here $A(\beta_{ik})$ is the cofactor of the $\beta_{ik}$ element in det $\beta$.

Substituting (1.5) and (1.6) into (1.3), we obtain

$$\mathbf{a} = \sum_k n_k\mathbf{a}_k = \sum_{k,\,i} n_k \frac{A(\beta_{ik})}{\det[\beta_{ik}]}\mathbf{a}'_i.$$

To guarantee that the factors of $\mathbf{a}'_i$ remain integers for all $n_k$, we should enforce the condition (1.4).

**1.1.4** We have already stated that, if the primitive cell contains one atom, the lattice is termed simple. Complex lattices with two or more atoms per primitive cell can be made up both of atoms of one and of several types. Let us consider some of the more important simple lattices.

Figure 1.5 depicts three cubic lattices whose basis vectors $\mathbf{a}_1$, $\mathbf{a}_2$, and $\mathbf{a}_3$ are mutually perpendicular ($\alpha_{12} = \alpha_{23} = \alpha_{13} = 90°$) and of equal length. Such lattices are said to belong to the *cubic system*. The term for the lattice in Fig. 1.5*a* is *simple cubic*. The unit cells in Fig. 1.5*b* are termed *body-centred cubic* and *face-centred cubic*, respectively. In the former case the additional atom is located in the cube's centre and in the latter in the centres of each of the six faces of the cube.

Let us calculate the number of atoms per unit cell of a cubic lattice. To do this we must bear in mind that the number of unit cells (cubes) in contact at the vertex of the cube is eight, or that there is 1/8 of an atom per cell, the atom located on the cell's face

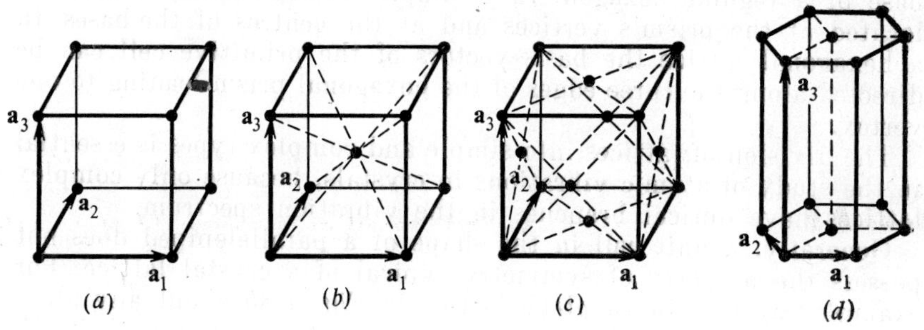


Fig. 1.5

contributes 1/2 of an atom (by the way, an atom located on the cell's edge contributes 1/4 of an atom). Obviously, in the case of a simple cubic lattice, a cell contains $1/8 \times 8 = 1$ atom; in the case of a body-centred cubic lattice $1/8 \times 8 + 1 = 2$ atoms, in

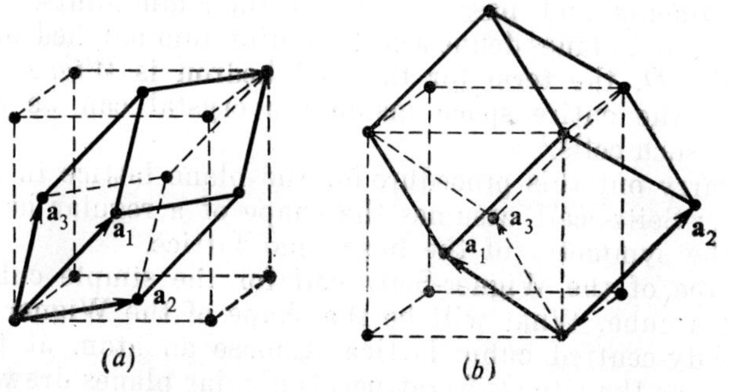


Fig. 1.6

the case of a face-centred cubic lattice $1/8 \times 8 + 1/2 \times 6 = 4$ atoms. However, the body-centred and face-centred cubic lattices are simple lattices, for their primitive cells contain one atom each. In case of the face-centred cubic lattice (when there are 4 atoms to the cubic cell), the basis vector can point from the cube's vertex to the centres of adjoining faces, as shown in Fig. 1.6*a*. In this case the primitive cell ($\alpha_{12} = \alpha_{23} = \alpha_{13} = 60°$) contains one atom. In the body-centred cubic lattice the basis vectors can be made to point from the cube's vertex to the centres of adjoining cubes, as depicted in Fig. 1.6*b*.

Figure 1.5*d* depicts the so-called *hexagonal cell* in the shape of a regular hexagonal prism whose edges are perpendicular to the base of a regular hexagon. In a simple lattice similar atoms are located at the prism's vertices and at the centres of the bases. In a hexagonal lattice the basis vectors of the primitive cell can be directed along the three edges of the hexagonal prism leading to one vertex.

The division of lattices into simple and complex types is essential in the study of atomic vibrations in crystals, because only complex lattices have optical branches in the vibration spectrum.

Generally, a unit cell in the shape of a parallelepiped does not possess the property of symmetry typical of a crystal lattice. For example, by rotating a plane lattice in Fig. 1.3*b* about any atom through an angle of 60° the lattice is made to coincide with itself, a symmetry not exhibited by the primitive cell depicted in Fig. 1.3*b*. Obviously the primitive cell of the face-centred cubic lattice shown in Fig. 1.6*a* does not possess the symmetry of a cube.

E. P. Wigner and F. Seitz demonstrated how to choose a primitive cell so that it possesses the symmetry of the crystal lattice. Let us take some atom $O$ belonging to the lattice and draw lines connecting it with the nearest atoms; draw planes normal to the segments and passing through their midpoints. The intersections of those planes define a certain minimum polyhedron containing the site $O$, the term for this polyhedron is *Wigner-Seitz cell*. Obviously, the entire space inside the crystal can be compactly filled with such cells.[1]

If we carry out this procedure for the plane lattice in Fig. 1.3*b*, the Wigner-Seitz cell assumes the shape of a regular hexagon possessing the symmetry of the hexagonal lattice.

The shape of the Wigner-Seitz cell for the simple cubic lattice is that of a cube. What will be the shape of the Wigner-Seitz cell for the body-centred cubic lattice? Choose an atom at the centre of the cube as the site $O$. Eight perpendicular planes drawn through the midpoints of the lines connecting $O$ with the atoms at the eight vertices of the cube constitute a regular octahedron. Six perpendicular planes drawn through the midpoints of the segments connecting $O$ with the central atoms of neighbouring cells will cut off six vertices from the octahedron, thus forming a polyhedron of fourteen faces depicted in Fig. 4.5*a*. Eight of its faces are regular

---

[1] A complex lattice consists of several simple lattices inserted into each other; in this case the above construction is carried out for one of the simple lattices.

hexagons, and six faces are squares. The Wigner-Seitz cell of fourteen faces has the symmetry of a cube. Wigner-Seitz cells for crystal lattices of other types may be constructed in a similar way.

## 1.2 Examples of Crystal Structures

**1.2.1** Let us consider some concrete crystal structures that serve to illustrate the propositions of the preceding section and are essential for further study of the theory.

X-ray structural analysis shows that most crystals of pure metals belong to the cubic or hexagonal system (Fig. 1.5*a*)[2]. Monovalent alkali metals Li, Na, K, Rb, Cs, bivalent Ba, the transition metals, $\alpha$-, $\beta$-, and $\delta$-modifications of iron and several other elements crystallize in the form of a body-centred cubic lattice (Fig. 1.5*b*). The metals Cu, Ag, Au, Al, Pb, $\gamma$-modification of iron, Ni, Ir, Pt and some other metals crystallize in the form of a face-centred cubic lattice (Fig. 1.5*c*). The elements Be, Mg, Zn, Cd and some other metals have a unit cell of a hexagonal structure (Fig. 1.5*d*). It was shown with the aid of X-ray structural analysis that in the latter case we are dealing with so-called *close hexagonal packing*. In this case a hexahedral prism contains in its volume three addi-

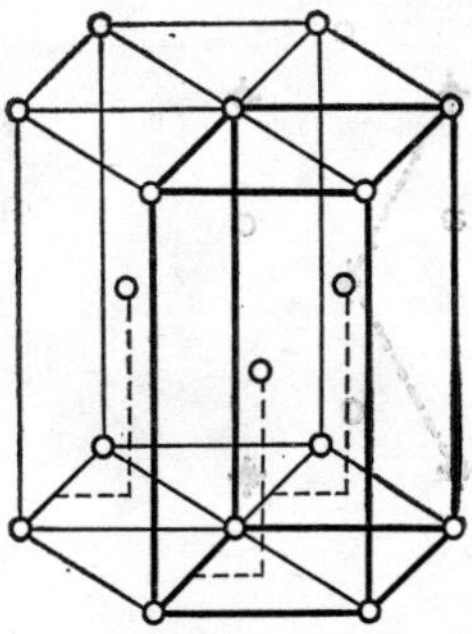

Fig. 1.7

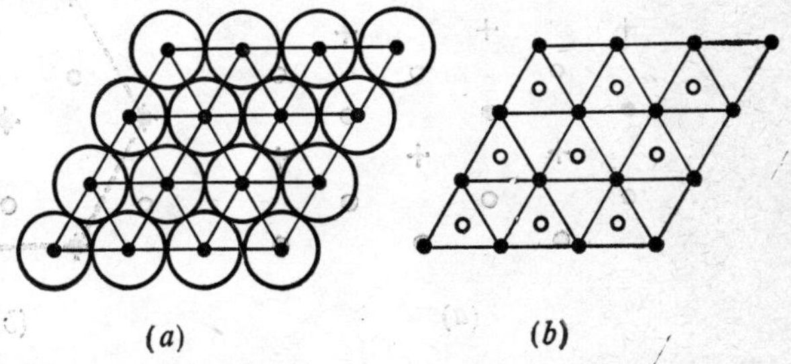


Fig. 1.8

tional atoms, as shown in Fig. 1.7. In the case of close hexagonal packing the lattice is no longer a simple one and contains two atoms per primitive cell.

**1.2.2** Consider the problem of the number of nearest atoms surrounding a given atom and located at the same distance *d* from it. Denote this number, termed *coordination number*, by the letter *z*. In any simple lattice this number is the same for all sites. In a simple cubic lattice $z = 6$, and the distance *d* is equal to the length of the

[2] A rigorous definition of a crystal system will be presented in Section 2.4.

cube's edge $a$. In a body-centred cubic lattice $z = 8$ and $d = (\sqrt{3}/2)\,a$; in a face-centred cubic lattice $z = 12$ and $d = (\sqrt{2}/2)\,a$. Indeed, for a face-centred cubic lattice the atoms nearest to the cube's vertex are those located in the centres of adjoining faces a distance $(\sqrt{2}/2)\,a$ away from the vertex. Obviously, in a three-dimensional lattice, twelve such faces meet at the vertex of each cube ($z = 12$).

**1.2.3** A problem of importance for X-ray structural analysis concerns different close packings of solid spheres of equal diameter. When spheres of equal diameter are closely packed on a horizontal plane, the centres of adjoining spheres are located at the vertices of equilateral triangles with sides equal to the diameter (Fig. 1.8*a*). The second horizontal layer of the spheres compactly laid on the first layer also forms a mesh of identical equilateral triangles. The arrangement of the centres of the spheres in the second layer with respect to the triangles of the first is indicated by circles in Fig. 1.8*b*. There are different ways in which the third

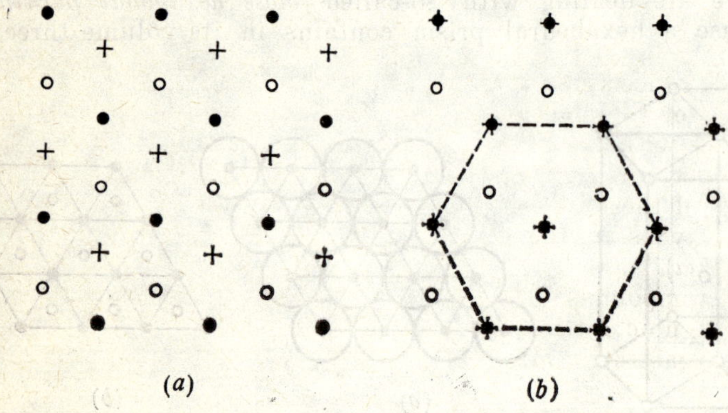

**Fig. 1.9**

layer can be laid, as depicted in Fig. 1.9, where ● are atoms of the first layer, ○ atoms of the second, and + atoms of the third. In Fig. 1.9*b* the centres of atoms of the third layer are directly above the centres of the atoms of the first. This is the case of close hexagonal packing. The cross section of the hexagonal prism in Fig. 1.9*b* is represented by a dashed line. It can be demonstrated that the structure corresponding to the packing in Fig. 1.9*a* is that of a face-centred cubic lattice (*close cubic packing*); [1.2, p. 283].

It should be noted, however, that in this case no face of the cube will be parallel to the horizontal plane.

**1.2.4** Alkali-halide compounds NaCl, LiF, NaI, KCl, etc., as well as binary compounds MgO, CaO, MgS, CaSe, BaTe, etc., crystallize in the shape of a simple cubic lattice whose sites are alternately occupied by atoms (by ions, to be more exact) of the elements making up the compound. The term for such a crystal lattice is rock salt structure (the name comes from a very abundant compound, NaCl). In this case each ion of, for example, $Na^+$ is surrounded by six $Cl^-$

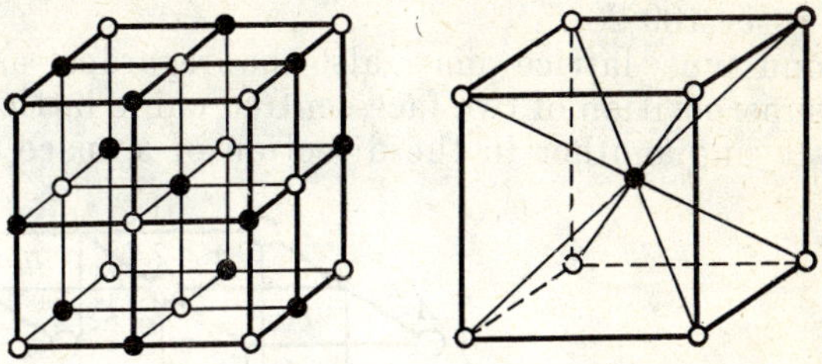

(*a*) (*b*)

**Fig. 1.10**

ions, and vice versa (Fig. 1.10*a*). It may easily be seen that lattices of $Na^+$ (or $Cl^-$) form the structure of a face-centred cubic lattice. Lattices with rock salt structure are complex lattices with two atoms per unit cell. In the case of the rock salt structure the basis vectors of the unit cell containing two atoms can be chosen in the same way as in the case of a simple face-centred cubic lattice (Fig. 1.6*a*).

The compounds CsCl, CsBr, CsI have the structure of a body-centred cubic lattice. In this case each $Cs^+$ ion is surrounded by eight negative halide ions and each halide ion is surrounded by eight $Cs^+$ ions (Fig. 1.10*b*).

In the crystals discussed above the number of positive ions is equal to the number of the negative ions. Therefore both the crystal as a whole and the unit cell are neutral.

It is easy to see that in all the above cases the crystal lattice is not only neutral but that its dipole electric moment is zero. For instance, in the case of the cell of the face-centred cubic lattice of CsCl, the zero electric moment of the cell is a geometrical sum of eight dipole moments directed along the four space diagonals of the cube (the magnitude of each moment is $(e/8a\sqrt{3}/2)$, where $e$ is the charge of the monovalent ions, and $a$ is the length of the cube's edge). Thermal motion violates the conditions in which the cell's dipole moment is zero. This, as we shall learn later, is the cause of conduction electron scattering in ionic crystals.

1.2.5 In recent years some materials with the diamond-type lattice (Ge, Si, InSb) became very important in technology. In a lattice of this type each atom located in the centre of a regular tetrahedron is surrounded by four atoms of the same (Ge or Si) or of another (InSb) type located at its vertices. Figure 1.11 depicts a site of the diamond lattice *O* with four neighbouring atoms *1*, *2*, *3*, *4* located at the vertices of a regular tetrahedron inscribed into a cube. The angle between two directions from the site *O* to the atoms surrounding it is 109°28′.

The diamond-type lattice may also be regarded as being the result of the superposition of two face-centred cubic lattices displaced with respect to one another in the direction of a space diagonal by

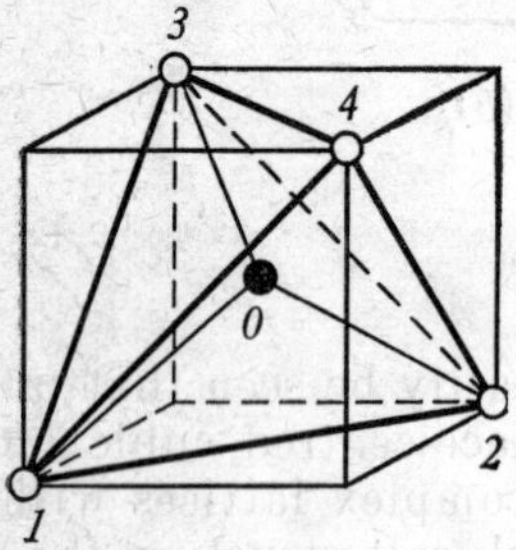


**Fig. 1.11**

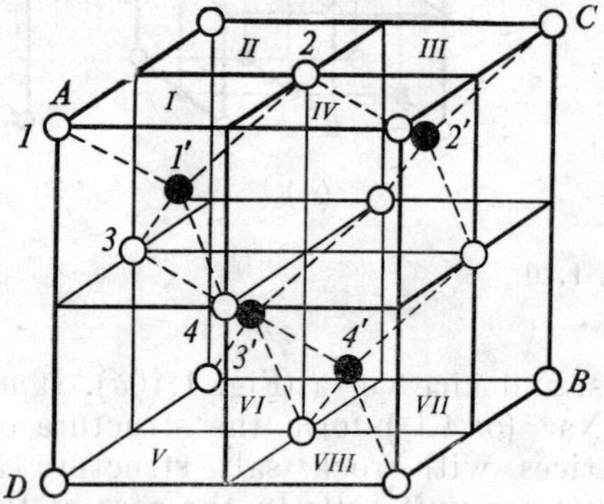


**Fig. 1.12**

one-fourth of its length. This is clearly seen in Fig. 1.12. Suppose initially we had a face-centred cubic lattice with atoms represented by open circles ○. If we displace it in the direction of the space diagonal *AB* by one-fourth of its length, the ○-atom *1* occupies the place of the ●-atom *1*′, the ○-atom *2* the place of the ●-atom *2*′, the ○-atom *3* the place of the ●-atom *3*′, and the ○-atom *4* the place of the ●-atom *4*′. Since the face-centred cubic lattice is simple, in the diamond lattice we can separate the unit cell containing two atoms.

In the case of germanium the open and black circles in Fig. 1.12 represent atoms of one sort, but in the case of the InSb compound they represent atoms of different types (for instance, atoms ○ are those of In and atoms ● those of Sb).

In a diamond lattice we can separate the group of 18 atoms forming the cubic lattice depicted in Fig. 1.13. The arrangement of atoms of this group may be visualized as follows. Divide the face-centred cube into eight identical cubes (*I*-*VIII*) (Fig. 1.12). Place atoms (black circles) at the centres of four of those cubes, as shown in the figure. We obtain the diamond-type lattice shown in Fig. 1.13. Count the number of atoms in such a cell. It follows from Fig. 1.12 that eight atoms are located at the cube's vertices, six on its faces, and four in its volume. Hence, the number of atoms in the cubic cell as a whole is

$$8 \times \frac{1}{8} + 6 \times \frac{1}{2} + 4 = 8.$$

It should not be inferred that the cubic cell (Figs. 1.12 and 1.13) we separated in the diamond lattice possesses all the properties of

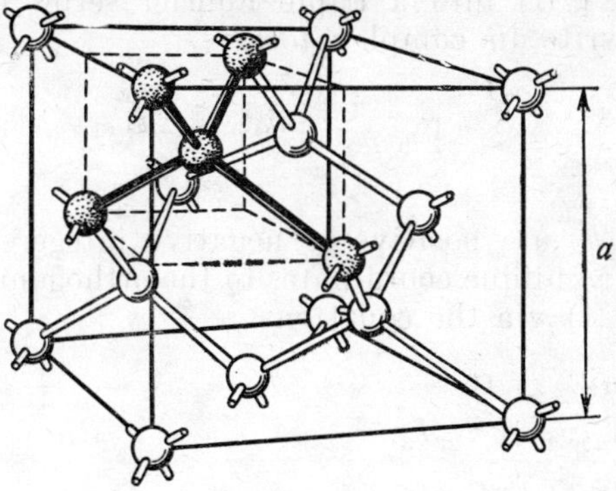


**Fig. 1.13**

symmetry of a cube. For instance, the rotation about a vertical axis passing through the cube's centre through an angle of 90° does not cause the atoms to coincide with themselves. It can, however, be demonstrated that with respect to its macroscopic properties the diamond crystal possesses cubic symmetry.

## 1.3 Direct and Reciprocal Crystal Lattices

**1.3.1** The property of paramount importance of an ideal crystal, as has already been stated, is the periodic arrangement of its atoms (or, to be more exact, of atomic nuclei) in space. This means that, when the crystal as a whole is displaced by a vector,

$$\mathbf{a} \equiv \mathbf{a_n} = n_1\mathbf{a}_1 + n_2\mathbf{a}_2 + n_3\mathbf{a}_3, \tag{3.1}$$

where $\mathbf{a}_i$ are the basis vectors of the lattice, and $n_i$ are integers ($i = 1, 2, 3$), the crystal coincides with itself.

Such quantities as electrostatic potential or electron density regarded at some point inside the crystal are obviously functions periodic in space (three-dimensional periodic functions). Indeed, some point inside the crystal defined by its position vector **r** and the point $\mathbf{r} + \mathbf{a}_n$ are physically equivalent. Therefore, for example, the electrostatic potential

$$V(\mathbf{r}) = V(\mathbf{r} + \mathbf{a}_n). \tag{3.2}$$

To expand the periodic function $V(\mathbf{r})$ into the Fourier series we introduce the coordinates $\xi_1$, $\xi_2$, and $\xi_3$ of an oblique coordinate frame with axes pointing in the directions of the vectors $\mathbf{a}_1$, $\mathbf{a}_2$, and $\mathbf{a}_3$.

Then the function $V(\mathbf{r})$ will be periodic in the variables $\xi_1$, $\xi_2$, and $\xi_3$ with the periods $\mathbf{a}_1$, $\mathbf{a}_2$, and $\mathbf{a}_3$, respectively. Expand the periodic function $V(\mathbf{r})$ into a triple Fourier series [1.3, p. 143], which we shall write in complex form:

$$V(\mathbf{r}) = \sum_{k_1=-\infty}^{+\infty} \sum_{k_2=-\infty}^{+\infty} \sum_{k_3=-\infty}^{+\infty} V_{k_1k_2k_3} e^{2\pi i(k_1\xi_1/a_1 + k_2\xi_2/a_2 + k_3\xi_3/a_3)}, \tag{3.3}$$

where $k_1$, $k_2$ and $k_3$ are positive or negative integers or zero.

Passing from the oblique coordinates to the orthogonal coordinates $x_i$ (see Appendix 1) via the equations

$$\begin{aligned} \xi_1 &= \alpha_{11}x_1 + \alpha_{12}x_2 + \alpha_{13}x_3, \\ \xi_2 &= \alpha_{21}x_1 + \alpha_{22}x_2 + \alpha_{23}x_3, \\ \xi_3 &= \alpha_{31}x_1 + \alpha_{32}x_2 + \alpha_{33}x_3, \end{aligned} \tag{3.4}$$

where $\alpha_{ik}$ are coefficients that depend on the angles between the axes of the oblique and the orthogonal frames, we obtain

$$V(\mathbf{r}) = \sum_{b_1} \sum_{b_2} \sum_{b_3} V_{b_1b_2b_3} e^{i(b_1x_1 + b_2x_2 + b_3x_3)}, \tag{3.5}$$

where $b_1$, $b_2$, $b_3$ are coefficients that depend on $\alpha_{ik}$, $k_i$, and $a_i$. The summation in (3.5) should be performed over all integral values of the indices $k_i$. Regarding $b_1$, $b_2$, and $b_3$ as orthogonal components of the vector $\mathbf{b}$, we write (3.5) in the form

$$V(\mathbf{r}) = \sum_{\mathbf{b}} V_{\mathbf{b}} e^{i(\mathbf{b}\cdot\mathbf{r})}. \tag{3.6}$$

The easiest way to determine $\mathbf{b}$ is from the requirement (3.2) of the periodicity of $V(\mathbf{r})$:

$$V(\mathbf{r} + \mathbf{a}_n) = \sum_{\mathbf{b}} V_{\mathbf{b}} e^{i\mathbf{b}\cdot(\mathbf{r}+\mathbf{a}_n)} = \sum_{\mathbf{b}} V_{\mathbf{b}} e^{i(\mathbf{b}\cdot\mathbf{r})} e^{i(\mathbf{b}\cdot\mathbf{a}_n)}.$$

Hence, it follows that $e^{i(\mathbf{b}\cdot\mathbf{a}_n)}$ should be unity, i.e., that $\mathbf{b}\cdot\mathbf{a}_n = n_1(\mathbf{b}\cdot\mathbf{a}_1) + n_2(\mathbf{b}\cdot\mathbf{a}_2) + n_3(\mathbf{b}\cdot\mathbf{a}_3) = 2\pi \times$ integer for all integral values of $n_1$, $n_2$, and $n_3$, which is possible only if

$$\mathbf{b}\cdot\mathbf{a}_1 = 2\pi g_1, \quad \mathbf{b}\cdot\mathbf{a}_2 = 2\pi g_2, \quad \mathbf{b}\cdot\mathbf{a}_3 = 2\pi g_3, \tag{3.7}$$

where $g_1$, $g_2$ and $g_3$ are arbitrary positive or negative integers or zero. Each vector is determined by its three components; therefore the three independent equations (3.7) suffice to determine $\mathbf{b}$. It can be demonstrated (Appendix 2) that

$$\mathbf{b}_g \equiv \mathbf{b} = g_1\mathbf{b}_1 + g_2\mathbf{b}_2 + g_3\mathbf{b}_3,$$

where

$$\mathbf{b}_1 = \frac{2\pi\mathbf{a}_2 \times \mathbf{a}_3}{\Omega_0}, \quad \mathbf{b}_2 = \frac{2\pi\mathbf{a}_3 \times \mathbf{a}_1}{\Omega_0}, \quad \mathbf{b}_3 = \frac{2\pi\mathbf{a}_1 \times \mathbf{a}_2}{\Omega_0}, \tag{3.8}$$

and $\Omega_0 = \mathbf{a}_1 \cdot (\mathbf{a}_2 \times \mathbf{a}_3)$ is the volume of the unit cell.

It may easily be checked that $\mathbf{b} \cdot \mathbf{a}_n = \mathbf{b}_g \cdot \mathbf{a}_n = 2\pi \times \text{integer} = 2\pi\,(n_1 \cdot g_1 + n_2 \cdot g_2 + n_3 \cdot g_3)$. It follows directly from the definition (3.8) of the $\mathbf{b}_k$ that

$$\mathbf{a}_i\mathbf{b}_k = 2\pi\delta_{ik} = \begin{cases} 0 \text{ for } i \neq k, \\ 2\pi \text{ for } i = k. \end{cases} \tag{3.9}$$

Conversely, from equations (3.9) we see that the $\mathbf{b}_k$ are determined by (3.8). The vectors $\mathbf{b}_k$ are called the *basis vectors of the reciprocal lattice.* The vectors $\mathbf{a}_n$ and $\mathbf{b}_g$ are termed *vectors of the direct and reciprocal lattices.* The dimensionality of the vectors $\mathbf{b}_k$ is $(\text{length})^{-1}$ as follows from (3.8). An infinite periodic lattice built on the basis vectors $\mathbf{b}_k$ is termed a *reciprocal lattice.* The space of the reciprocal lattice has the dimensionality of $(\text{length})^{-3}$. A parallelepiped built on the vectors $\mathbf{b}_k$ is termed a *unit cell of the reciprocal lattice.* It can be proved [1.4, pp. 11-12] that its "volume" is $\mathbf{b}_1 \cdot (\mathbf{b}_2 \times \mathbf{b}_3) = (2\pi)^3/\Omega_0$.

It follows from the condition (3.8) that the vector $\mathbf{b}_1$ is perpendicular to the vectors $\mathbf{a}_2$ and $\mathbf{a}_3$, the vector $\mathbf{b}_2$ to the vectors $\mathbf{a}_1$ and $\mathbf{a}_3$, and the vector $\mathbf{b}_3$ to the vectors $\mathbf{a}_1$ and $\mathbf{a}_2$. If the unit cell of the direct lattice is in the shape of a right parallelepiped, the vectors $\mathbf{b}_1$, $\mathbf{b}_2$, $\mathbf{b}_3$ are parallel to the vectors $\mathbf{a}_1$, $\mathbf{a}_2$, $\mathbf{a}_3$, respectively, and $\mathbf{b}_i = 2\pi/\mathbf{a}_i$. It is obvious that if the direct lattice is a simple cubic lattice, the reciprocal lattice is also a simple cubic lattice with $\mathbf{b}_1 = \mathbf{b}_2 = \mathbf{b}_3 = 2\pi/\mathbf{a}$.

The concept of the reciprocal lattice stemmed directly from the problem of expanding a function with the period of the direct lattice into a Fourier series. We shall see below how the concept of the reciprocal lattice is effectively used in the analysis of X-ray diffraction in crystals, in the study of atomic vibrations in crystals, and in the quantum mechanical study of electronic motion in a periodic field.

**1.3.2** Let us introduce an important concept of the Miller indices (*hkl*), which is closely connected with the reciprocal lattice concept.

Imagine a plane in the crystal passing through the centres of the atoms. Figure 1.14 depicts four such planes oriented in different ways with respect to the unit cell of a simple cubic lattice. We shall describe the position (the orientation) of the plane in the crystal passing through atomic centres in terms of the Miller indices $(hkl)$ which we shall define as follows. Let three integers, $s_1$, $s_2$ and $s_3$, be the number of units of $a_1$, $a_2$ and $a_3$ in the three segments that the plane cuts off the coordinate axes $\mathbf{a}_1$, $\mathbf{a}_2$, and $\mathbf{a}_3$, respectively. We form the ratio $1/s_1 \div 1/s_2 \div 1/s_3$ and express it as the ratio of three smallest integers. The latter are termed the *Miller indices* $(hkl)$. Hence $h \div k \div l = 1/s_1 \div 1/s_2 \div 1/s_3$. Figure 1.14*a* depicts the axes $\mathbf{a}_1$, $\mathbf{a}_2$, and $\mathbf{a}_3$. The Miller indices of the shaded plane in

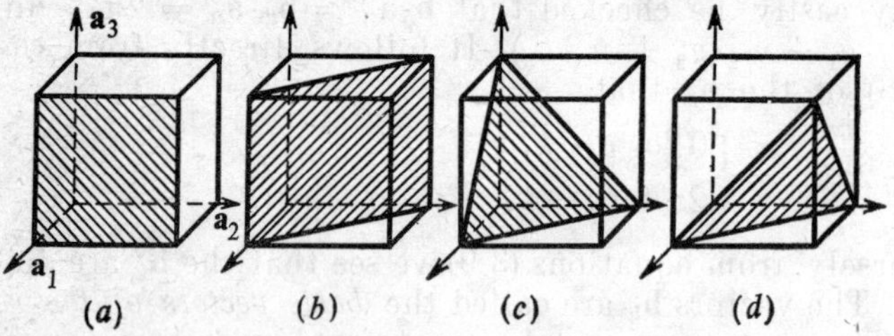


**Fig. 1.14**

Fig. 1.14*a* are obviously $1/s_1 \div 1/s_2 \div 1/s_3 = 1/1 \div 1/\infty \div 1/\infty = 1 \div 0 \div 0 = (100)$. Likewise, the Miller indices of the shaded planes in Fig. 1.14*b*, *c*, and *d* are (110), (111), $(11\bar{1})$, where $\bar{1}$ means minus one and expresses the fact that the shaded plane in Fig. 1.14*d* cuts a segment $a$ off the negative axis $\mathbf{a}_3$. It is obvious that the specific Miller indices $(hkl)$ define not a single plane but a whole family of parallel planes. It may easily be seen that $(\bar{h}\bar{k}\bar{l}) = (hkl)$.

The set of physically equivalent planes (e.g. all six faces of a cube) is denoted by a symbol in curly brackets (e.g. {100}). The notation for the direction of a straight line passing through the atomic centres is $[uvw]$, where $u$, $v$ and $w$ are the three least integers whose ratio $u \div v \div w$ is equal to the ratio of the lengths (in units of $a_1$, $a_2$, $a_3$) of the components along $\mathbf{a}_1$, $\mathbf{a}_2$, and $\mathbf{a}_3$ of the vector directed along the straight line. For example, the symbol for the $x$ axis in Fig. 1.14, which coincides with the vector $\mathbf{a}_1$, is [100]. The symbol corresponding to the direction of the space diagonal of the cube that passes through the origin is [111]. Physically equivalent directions in the crystal are designated by the symbol $\langle uvw \rangle$. In a cubic crystal the direction $\langle uvw \rangle$ is perpendicular to the plane $(uvw)$; this is not the case with crystals of lower symmetry.

Let us prove two important propositions. Construct a reciprocal lattice in the space of a direct lattice whose sites are determined by the vectors $\mathbf{a}_i$, the sites of the reciprocal lattice being determined by the vectors $\mathbf{b}_i$ oriented with respect to $\mathbf{a}_i$ in accordance with equations (3.8). Of course, the scale of the vectors $\mathbf{b}_i$ remains in this case arbitrary. We now prove that the vector of the reciprocal lattice, $\mathbf{b}_g = g_1\mathbf{b}_1 + g_2\mathbf{b}_2 + g_3\mathbf{b}_3$ is perpendicular to the plane with the Miller indices $(hkl)$ if $g_1 \div g_2 \div g_3 = h \div k \div l$. The terminations of the vectors $\mathbf{a}_1/h$, $\mathbf{a}_2/k$, $\mathbf{a}_3/l$ obviously lie in the plane $(hkl)$; therefore, the vectors $(\mathbf{a}_1/h - \mathbf{a}_2/k)$ and $(\mathbf{a}_1/h - \mathbf{a}_3/l)$ lie in the plane $(hkl)$.

Take the vector $\mathbf{b}_{hkl} = h\mathbf{b}_1 + k\mathbf{b}_2 + l\mathbf{b}_3$ parallel to the vector $\mathbf{b}_g$ and prove that $\mathbf{b}_{hkl}$ is perpendicular to the plane $(hkl)$. To this end it suffices to make sure that the scalar products vanish:

$$\mathbf{b}_{hkl}\cdot\left(\frac{\mathbf{a}_1}{h}-\frac{\mathbf{a}_2}{k}\right)=(h\mathbf{b}_1+k\mathbf{b}_2+l\mathbf{b}_3)\cdot\left(\frac{\mathbf{a}_1}{h}-\frac{\mathbf{a}_2}{k}\right)=0,$$

$$\mathbf{b}_{hkl}\cdot\left(\frac{\mathbf{a}_1}{h}-\frac{\mathbf{a}_3}{l}\right)=(h\mathbf{b}_1+k\mathbf{b}_2+l\mathbf{b}_3)\cdot\left(\frac{\mathbf{a}_1}{h}-\frac{\mathbf{a}_3}{l}\right)=0,$$

in accordance with (3.9).

Second, let us prove that the distance $d_{hkl}$ between neighbouring planes of the $(hkl)$ family is $2\pi/b_{hkl}$. Let $\mathbf{n} = \mathbf{b}_{hkl}/b_{hkl}$ be the unit vector of the normal to the $(hkl)$ planes. Then

$$d_{hkl}=\frac{\mathbf{a}_1}{h}\cdot\mathbf{n}=\frac{1}{h}\frac{1}{b_{hkl}}\,\mathbf{a}_1\cdot(h\mathbf{b}_1+k\mathbf{b}_2+l\mathbf{b}_3)=\frac{2\pi}{b_{hkl}}.$$

Let us see if the distance between the (100) planes in a cubic lattice is equal to the lattice constant $\mathbf{a}$. Indeed, in this case $b_{hkl} = b_{100} = b_1 = 2\pi/a$ and, consequently, $d_{100} = 2\pi/b_{100} = a$. It is just as easy to compute the distance between the (111) planes in a cubic lattice: $b_{111} = \sqrt{b_1^2 + b_2^2 + b_3^2} = 2\pi\sqrt{3}/a$, or $d_{111} = 2\pi/b_{111} = a/\sqrt{3}$.

## 1.4 The Laue and Bragg Equations for X-Ray Diffraction in Crystals. Atomic and Structural Scattering Factors

**1.4.1** Let us discuss, following Laue, the conditions for the diffraction of X rays in a crystal. Figure 1.15 depicts two atoms $O$ and $A$ separated by the basis vector $\mathbf{a}_1$. A bunch of X rays falling in the direction $\mathbf{n}$ ($|\mathbf{n}| = 1$) is scattered by the atoms in all directions. Consider a bunch $RS$ scattered in a direction (as determined by the unit vector $\mathbf{n}'$ ($|\mathbf{n}'| = 1$)) in which the rays $IAR$ and $KOS$ augment each other as the result of interference. To this end their geometrical path difference, $BO + OC$, should be equal to an integral number of wavelengths, i.e. $BO + OC = -(\mathbf{a}_1\cdot\mathbf{n}) + (\mathbf{a}_1\cdot\mathbf{n}') = \mathbf{a}_1\cdot(\mathbf{n}' - \mathbf{n}) = g_1\lambda$, where $g_1$ is an arbitrary integer. In a three-dimensional lattice the condition of augmentation as the result of interference should hold simultaneously for all atoms of the crystal corresponding to the basis vectors $\mathbf{a}_1$, $\mathbf{a}_2$, and $\mathbf{a}_3$. This leads to the following Laue conditions for interference:

$$\mathbf{a}_1\cdot(\mathbf{n}' - \mathbf{n}) = g_1\lambda,\ \mathbf{a}_2\cdot(\mathbf{n}' - \mathbf{n}) = g_2\lambda,$$
$$\mathbf{a}_3\cdot(\mathbf{n}' - \mathbf{n}) = g_3\lambda, \tag{4.1}$$

which can also be written in the more conventional form

$$a_1(\cos\alpha_1' - \cos\alpha_1) = g_1\lambda,\ a_2(\cos\alpha_2' - \cos\alpha_2) = g_2\lambda,$$
$$a_3(\cos\alpha_3' - \cos\alpha_3) = g_3\lambda, \tag{4.2}$$

if we introduce the angles $\alpha_i$ and $\alpha_i'$ between the directions of the incident and the reflected rays and the crystal axes $\mathbf{a}_i$, i.e., if we put $\mathbf{a}_i\cdot\mathbf{n} = a_i\cos\alpha_i$ and $\mathbf{a}_i\cdot\mathbf{n}' = a_i\cos\alpha_i'$.

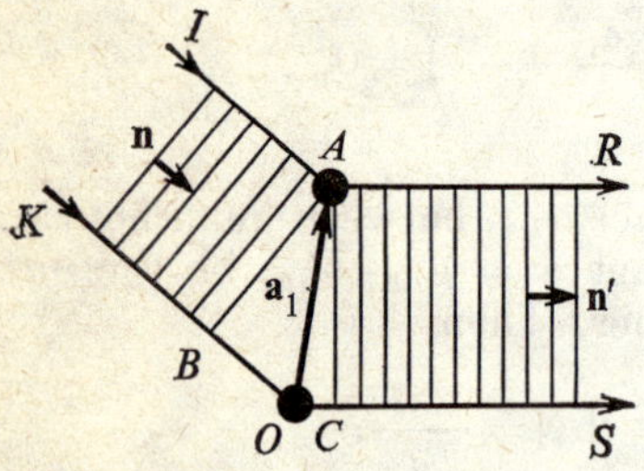

Fig. 1.15

It follows from (4.1) that the vector $2\pi(\mathbf{n}'/\lambda - \mathbf{n}/\lambda)$ satisfies the same equations (3.7) as the vector $\mathbf{b}$. Hence, it follows that

$$2\pi\left(\frac{\mathbf{n}'}{\lambda} - \frac{\mathbf{n}}{\lambda}\right) = \mathbf{b}_g, \tag{4.3}$$

where $\mathbf{b}_g = g_1\mathbf{b}_1 + g_2\mathbf{b}_2 + g_3\mathbf{b}_3$ is the vector of the reciprocal lattice. The condition (4.3) may be regarded as equivalent to the Laue conditions (4.1) and (4.2).

We now introduce the wave vector $\mathbf{k} = (2\pi/\lambda)\mathbf{n}$. The condition (4.3) then assumes the form

$$\mathbf{k}' - \mathbf{k} = \mathbf{b}_g. \tag{4.4}$$

Since $k' = k$, it follows from (4.4) that $\mathbf{k}^2 = \mathbf{k}'^2 = (\mathbf{b}_g + \mathbf{k})^2 = \mathbf{b}_g^2 + \mathbf{k}^2 + 2(\mathbf{b}_g\cdot\mathbf{k})$, whence

$$(1/2)\,\mathbf{b}_g^2 + \mathbf{b}_g\cdot\mathbf{k} = 0, \tag{4.5}$$

which, of course, is also equivalent to (4.1) and (4.3).

**1.4.2** We shall now demonstrate how the Laue conditions in the form (4.4) can be interpreted geometrically in the space of the reciprocal lattice. To make the discussion visualizable, a planar reciprocal lattice is shown in Fig. 1.16. Draw a vector $-\mathbf{k}$ from an arbitrary site $O$ of the reciprocal lattice and describe a circumference of radius $k = 2\pi/\lambda$ with the centre at the vector's origin $R$ (Ewald's sphere). If this circumference (sphere) passes quite close to another site $S$ of the reciprocal lattice, the vector $\overrightarrow{RS}$ will be equal to $\mathbf{k}'$, the wave vector of the scattered bunch augmented by the interference. Then the vector of the reciprocal lattice $\overrightarrow{OS}$ will be equal to $\mathbf{b}_g$, which satisfies equation (4.4). Thus, by building Ewald's sphere in the space of the reciprocal lattice we can graphically find all the directions in which interference maxima may be observed.

**1.4.3** The Laue conditions for the interference of X rays (4.1)-(4.5) can be formulated in another way. Draw a straight line (a plane) $PQ$ perpendicular to the vector of the reciprocal lattice $\mathbf{b_g} = \overrightarrow{OS}$ (Fig. 1.16). As demonstrated at the end of the preceding section, this plane in the space of the direct lattice determines an atomic plane with the Miller indices $(hkl)$, where $h \div k \div l = g_1 \div g_2 \div g_3$. Denoting the common multiplier of the numbers $g_1$, $g_2$, and $g_3$ by $n$, we have $b_g = nb_{hkl}$. On the other hand, the distance between neighbouring planes $(hkl)$ is $d = 2\pi/b_{hkl} = 2\pi n/b_g$. Consider now the triangle $ORS$. Denoting the angle $\measuredangle QRS$ by $\theta$, we have $OS = 2RS \sin\theta$, or $b_g = 2k \sin\theta = 2\,(2\pi/\lambda) \sin\theta$. Substituting the quantity $d$, we obtain the Bragg equation

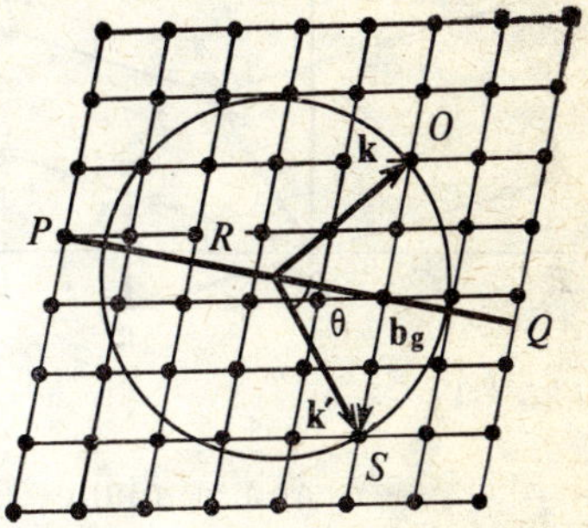

**Fig. 1.16**

$$2d \sin\theta = n\lambda. \qquad (4.6)$$

It is evident from Fig. 1.16 that the incident bunch $\mathbf{k}$ and the scattered $\mathbf{k}'$ can formally be treated as a reflection of the bunch from the atomic plane. Hence, the Bragg condition (4.6) which is equivalent to the Laue conditions (4.1) may formally be treated as a reflection of the rays from the crystal's atomic planes at angles $\theta$ satisfying (4.6).

**1.4.4** Let us introduce the concept of the atomic scattering factor $f$ which takes into account the distribution of the atom's electric charge in the scattering of X rays. Let $\rho(\mathbf{r})\,d\tau$ be the average number of electrons in the volume of an atom $d\tau$. Obviously, the total number of electrons in an atom is $\int \rho\, d\tau = Z$. Figure 1.17 depicts the centre of an atom $O$ and the volume element $d\tau$ whose position at point $A$ is determined by the position vector $\overrightarrow{OA} = \mathbf{r}$. Let $OS$ be the direction of the ray scattered by one electron which we shall imagine (for simplicity) to be located at the centre $O$ of the atom. The amplitude of this scattered radiation at some sufficiently distant point (compared with the atomic dimensions) at a given moment of time $t$ is $A(t)$. Assuming the scattering to be coherent and taking into account the fact that its amplitude is proportional to the number of scattered electrons, we obtain for the amplitude of the ray $AR$ at the same point of observation and at the same moment of time $t$ the expression $dA_\rho(t) = A(t)\rho\, d\tau\, e^{i\Phi}$. Here $\rho\, d\tau$ is the number of electrons in $d\tau$ and $\Phi$ is the phase difference of the rays $IAR$ and $KOS$. Obviously, $\Phi = (2\pi/\lambda)\, d$, where $d$ is the path difference of the rays. It follows from Fig. 1.15 that the path difference for the rays $IAR$ and $KOS$ scattered by the atoms $O$ and $A$, which must be calculated to find the Laue interference conditions, is identical to the path difference of the rays scattered by the centre of the atom and by $d\tau$. Hence $\mathbf{d} = \mathbf{r}\cdot(\mathbf{n}' - \mathbf{n})$, where $\mathbf{n}$ and $\mathbf{n}'$ are unit vectors in the directions of the incident and scattered bunches, respectively. It may be generally assumed that $\mathbf{r}$, $\mathbf{n}'$, and $\mathbf{n}$ lie in the same plane (that of the figure). Draw an axis parallel to the vector $\mathbf{s} = \mathbf{n}' - \mathbf{n}$ and a plane $POQ$ perpendicular to the $z$ axis (see Fig. 1.17). If the atom being considered belongs to the crystal, and if the directions of the incident and the scattered bunches correspond to an interference maximum, then $POQ$ is a planar atomic mesh from which Bragg reflection takes place. In this case the angle $\measuredangle\, POK = \measuredangle\, QOS = \theta$ satisfies the Bragg equation (4.6). It may be seen from Fig. 1.17 that $s = 2 \sin \theta$ $(n = n' = 1)$, therefore $\Phi = (2\pi/\lambda)\,(\mathbf{r}\cdot\mathbf{s}) = (2\pi/\lambda)\, sr \cos \vartheta = (4\pi/\lambda)\, r \sin \theta \cos \vartheta$, where $\vartheta$ is the angle between the direction of s (the $z$ axis) and the position vector $\mathbf{r}$.

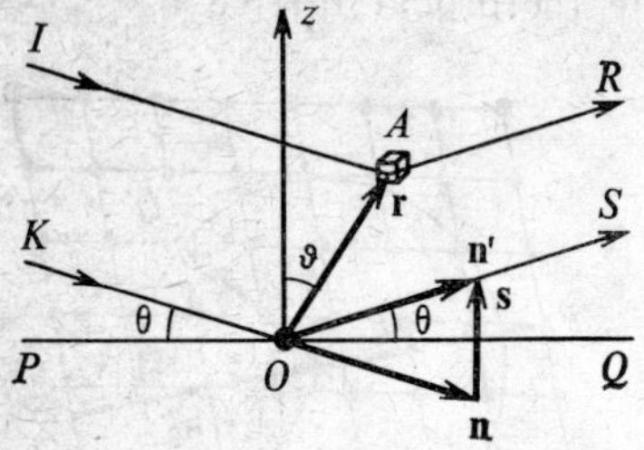

Fig. 1.17

The amplitude of scattering by all the electrons of an atom is

$$A_\rho(t) = A(t) \int \rho(\mathbf{r})\, e^{i\Phi}\, d\tau .$$

Let the distribution of the electrons in the atom be spherically symmetric: $\rho = \rho(r)$. Putting $d\tau = 2\pi r^2\, dr \sin \vartheta\, d\vartheta$, we obtain for the atomic scattering factor, by definition,

$$f = \frac{A_\rho(t)}{A(t)} = 2\pi \int_0^\infty \int_0^\pi \rho(r)\, r^2 e^{iar\cos\vartheta}\, dr \sin \vartheta\, d\vartheta,$$

where $a = (4\pi/\lambda) \sin \theta$. Integrating with respect to $\vartheta$, we obtain, as will be demonstrated below,

$$f = \int_0^\infty W(r) \frac{\sin(ar)}{ar}\, dr, \tag{4.7}$$

where $W(r)\, dr = 4\pi r^2 \rho(r)\, dr$ is the probable (average) number of electrons in the spherical layer between $r$ and $r + dr$ in an atom.

To evaluate the integral $\int_0^\pi e^{iar\cos\vartheta}\sin\vartheta\, d\vartheta$ we put $\cos\vartheta = x$; then $dx = -\sin\vartheta\, d\vartheta$ and the limits of integration are $+1$ and $-1$. Accordingly,

$$\int_{-1}^{+1} e^{iarx}\, dx = \frac{1}{iar}\left(e^{iar} - e^{-iar}\right) = \frac{2}{ar}\sin ar.$$

Thus, $f$ is the *atomic scattering factor*[3], which is equal to the ratio of the amplitude of radiation scattered by all the electrons of an atom in a certain direction to the amplitude of radiation scattered in the same direction by one electron.

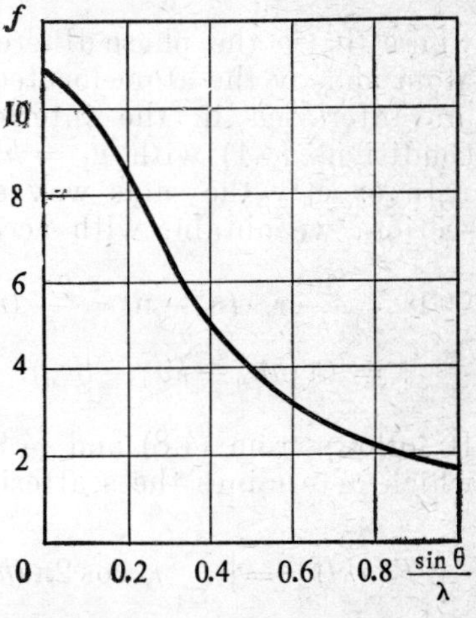

Fig. 1.18

The quantity $W(r)$ that enters the definition of $f$ can be calculated with the aid of the quantum mechanical (the Hartree-Fock) method or the statistical (the Thomas-Fermi) method. The results of calculating $f$ for sodium are presented in Fig. 1.18. At $\theta = 0$, equation (4.7) yields $f = Z$ (the number of electrons in the atom). Theoretical computations of the quantity $f$ have been proved to be very accurate by experimental studies of intensity distribution near interference maxima (with account taken of the temperature factor of scattering). The same investigations carried out at low temperatures also served as proof of the existence of zero-point energy of vibrations of the lattice atoms.

1.4.5 We deduced the Laue interference conditions (4.1), or the equivalent Bragg conditions (4.6), while considering the scattering from identical atoms of a simple lattice. How will the result change if we take a complex lattice with a basis consisting of $s$ atoms whose position with respect to the cell's origin is determined by the vectors

$$\mathbf{r}_n = u_n\mathbf{a}_1 + v_n\mathbf{a}_2 + w_n\mathbf{a}_3, \quad n = 1, 2, \ldots, s?$$

[3] This is also known as the atomic form factor or the atomic scattering power. Sometimes $f$ is called the atomic scattering amplitude; the term atomic scattering factor is reserved for $f^2$, to which the relative intensity of the scattered radiation is proportional.

Since the sublattice of every $n$th atom of the cell has the same basis vectors $\mathbf{a}_1$, $\mathbf{a}_2$, $\mathbf{a}_3$, it produces interference maxima determined by the same relation (4.1) as for the atom with $n = 1$ located at the site (at the cell's origin). However, since the radiation is coherent, we must take into account phase relationships arising from scattering by different sublattices. We introduce the *structural scattering amplitude* $F\,(hkl)$ defined as the ratio of the scattering amplitude at the interference maximum $(hkl)$ from all the sublattices to the scattering amplitude in the same direction from one electron.

If $f_n$ is the atomic scattering factor of the $n$th atom,

$$F\,(hkl) = \sum_{n=1}^{s} f_n e^{i\Phi_n}, \tag{4.8}$$

where $\Phi_n$ is the phase difference of the rays scattering by the $n$th atom and by the atom located at the cell's origin ($n = 1$). Here we are interested in the interference maximum determined by the conditions (4.1) with $g_1 = h$, $g_2 = k$, and $g_3 = l$.

In exactly the same way as $\Phi$ was calculated in the preceding section, we obtain, with account taken of (4.1),

$$\Phi_n = \frac{2\pi}{\lambda}\mathbf{r}_n\cdot(\mathbf{n}'-\mathbf{n}) = \frac{2\pi}{\lambda}(u_n\mathbf{a}_1 + v_n\mathbf{a}_2 + w_n\mathbf{a}_3)\cdot(\mathbf{n}'-\mathbf{n})$$
$$= 2\pi\,(hu_n + kv_n + lw_n). \tag{4.9}$$

It follows from (4.8) and (4.9) that the *structural scattering factor*, which determines the scattering intensity, is equal to

$$|\,F\,(hkl)\,|^2 = \Big|\sum_{n=1}^{s} f_n \cos 2\pi\,(hu_n + kv_n + lw_n)\Big|^2$$
$$+ \Big|\sum_{n=1}^{s} f_n \sin 2\pi\,(hu_n + kv_n + lw_n)\Big|^2. \tag{4.10}$$

If all the atoms of the crystal are of the same type, $f_1 = f_2 = \ldots$ $\ldots = f$ and

$$F\,(hkl) = f \sum_{n=1}^{s} \exp\,[2\pi i\,(hu_n + kv_n + lw_n)] = fS, \tag{4.11}$$

where

$$S = \sum_{n=1}^{s} \exp\,[2\pi i\,(hu_n + kv_n + lw_n)]. \tag{4.11a}$$

Considering the centred cubic lattice a complex lattice with a basis (0, 0, 0; 1/2, 1/2, 1/2), we obtain

$$S = 1 + \exp\,[\pi i\,(h + k + l)].$$

If $h + k + l$ is an odd number, then $\exp [i\pi (h + k + l)] = -1$ and $S = 0$, i.e. the intensity of the corresponding maxima is zero.

Hence, in a body-centred cubic lattice consisting of atoms of one type there will be, for example, no interference maxima of the types (100), (111), (210), (300), etc. Of course, in the case of a body-centred cubic lattice we can obtain the same result without bringing in the concept of the structural amplitude if we isolate the unit cell corresponding to a simple lattice.

# 2. Elements of Group Theory and Crystal Symmetry

## 2.1. Introduction

**2.1.1** Physics, and the theory of semiconductors in particular, is becoming increasingly permeated with the methods evolved in group theory.

We have chosen to devote a separate chapter to this theory, despite the fact that the theory itself and its applications to physics are treated in several monographs and textbooks.[1] There are two reasons for this. First, most books on the subject are too advanced and contain too much material, at least for the experimental physicist. Second, we do not intend to teach the reader how to apply group theory, as do most books, but desire only to make him familiar with the minimum material essential for understanding the following chapters and the current literature on semiconductor physics.

The goal of group theory is to answer several questions pertaining to the state of a physical system solely on the basis of consideration of its symmetry. With this in view, we shall be mainly interested in the symmetry of the potential field acting on the system (for instance, the symmetry of the periodic field acting on the conduction electrons in a crystal), as well as in symmetry involving the reversal of time: $t \rightarrow -t$.

Certain general methods are evolved in group theory that enable it to be used for (1) classification of crystals according to their symmetry, (2) classification of normal vibrations of crystals, (3) classification of quantum mechanical states of conduction electrons in crystals and of electrons at impurity atoms, (4) determination of selection rules for matrix elements in a system, etc.

**2.1.2** Consider two simple examples, in which the results following from the symmetry of the system can be obtained without resorting to the formalism of group theory.

A. Consider a helium atom that has two electrons. If we neglect weak interaction between the electron spins and their orbital motion,

---

[1] *Landsberg* [2.1, Chaps. IV and V] is a very consistent and comprehensible treatise, *Tinkham* [2.2] is one of the best books on the applications of group theory in physics (only Chap. 8 deals with the solid state as a separate subject), and *Bir and Pikus* [2.3] is a very extensive and thorough treatise on the subject.

we can represent the full wave function of the electrons as a product of the coordinate function $\psi(\mathbf{r}_1, \mathbf{r}_2)$, where $\mathbf{r}_1$ and $\mathbf{r}_2$ are the position vectors of the electrons, and the spin function $\nu(s_{z1}, s_{z2})$, where $s_{z1}$ and $s_{z2}$ are the projections of the electron spins on the $z$ axis). Since quantum mechanics regards the electrons as indistinguishable, their permutation can change the coordinate part of the wave function $\psi(\mathbf{r}_1, \mathbf{r}_2)$ only by a factor $\varepsilon$ of unit modulus. But as the result of a repetition of the electron permutation

$$\psi(\mathbf{r}_1, \mathbf{r}_2) = \varepsilon\psi(\mathbf{r}_2, \mathbf{r}_1) = \varepsilon^2\psi(\mathbf{r}_1, \mathbf{r}_2),$$

whence $\varepsilon^2 = 1$; therefore, $\varepsilon = +1$ or $\varepsilon = -1$. Thus, we are able to classify the electron states in the helium atom:

$\psi(\mathbf{r}_1, \mathbf{r}_2) = \psi(\mathbf{r}_2, \mathbf{r}_1)$ is parahelium with a symmetric coordinate wave function;

$\psi(\mathbf{r}_1, \mathbf{r}_2) = -\psi(\mathbf{r}_2, \mathbf{r}_1)$ is orthohelium with an antisymmetric coordinate wave function.

Since the full wave function equal to the product $\psi(\mathbf{r}_1, \mathbf{r}_2)\,\nu(s_{z1}, s_{z2})$ must, in compliance with Pauli's principle, be antisymmetric, the spin wave function $\nu(s_{z1}, s_{z2})$ is antisymmetric for parahelium and symmetric for orthohelium.

It follows from the symmetry connected with the permutation of indistinguishable electrons that the state corresponding to parahelium with its antisymmetric spin function is the singlet state with opposite spins and a zero total spin; similarly, the state corresponding to parahelium is the triplet state with parallel spins and a total spin equal to $\hbar$.

Indeed, let $\alpha(i)$ and $\beta(i)$ be the spin functions of the $i$th electron for spins directed "upward" and "downward". The three symmetrical spin functions for both electrons in that case will be

$$\alpha(1)\,\alpha(2), \qquad \beta(1)\,\beta(2), \qquad \frac{\alpha(1)\,\beta(2) + \alpha(2)\,\beta(1)}{\sqrt{2}},$$

where $\sqrt{2}$ was introduced as a normalization factor. It can be demonstrated that the total spin of each of these three states is $\hbar$ (for the first two this is obvious and for the third state see [2.4, § 72]). On the other hand, only one antisymmetric spin function $(1/\sqrt{2})\,[\alpha(1)\,\beta(2) - \alpha(2)\,\beta(1)]$ with zero total spin is possible.

We see that the electron states in a helium atom can be classified on the grounds of simple symmetry considerations based on the indistinguishability of the electrons.

B. As the second example, consider the matrix element of the dipole moment

$$M_x = \int\int\int_{-\infty}^{+\infty} \psi_2^*(\mathbf{r})\, x\, \psi_1(\mathbf{r})\, dx\, dy\, dz$$

appearing in an atom when the electric field of a light wave oscillates along the $x$ axis. Here $\psi_1(\mathbf{r})$ and $\psi_2(\mathbf{r})$ are wave functions of the valence electron in the atom. The probability of light being absorbed in the course of the transition $\psi_1 \rightarrow \psi_2$ in the dipole approximation is proportional to the square of the modulus of the matrix element, $|M_x|^2$. The electronic states in a spherically symmetrical atomic field can be classified by the orbital quantum number $l$:

$l = 0, \ldots,$ $s$ is a nondegenerate state,

$l = 1, \ldots,$ $p$ is a three-fold degenerate state,

$l = 2, \ldots,$ $d$ is a five-fold degenerate state,

and so on. The wave function corresponding to the $s$-state is the spherically symmetrical wave function $\psi_s(\mathbf{r})$ $(r = |\mathbf{r}|)$. There are three wave functions corresponding to the $p$-state, which can be written in the form: $x\varphi(r)$, $y\varphi(r)$, $z\varphi(r)$; etc. This form of the electron wave functions is a direct consequence of the spherical symmetry of the electric field acting on the electron.

It may easily be seen that, if the two electronic states $\psi_1$ and $\psi_2$ are $s$-states, the expression under the integral in $M_x$ will be an odd function of $x$, and therefore integration with respect to $x$ from $-\infty$ to $+\infty$ will yield zero. If, on the other hand, one state is an $s$-state and the other a $p$-state, the expression under the integral for the $p$-state $x\varphi(r)$ will be an even function in all the variables, and as a result the integral is nonzero.

We come to the following selection rules. In the dipole approximation, the probability of light being absorbed in a transition between $s$-states is zero. But if the transitions are between an $s$- and a $p$-state, the probability is nonzero.

Group theory enables answers to be obtained to questions similar to those discussed under A and B, as well as in much more complicated cases.

## 2.2 Elements of Abstract Group Theory

**2.2.1** The term *group* applies to a finite or infinite set of distinct elements: $A, B, C, \ldots, P, Q, R, \ldots$ that satisfies the following four requirements.

I. The operation of multiplication[2] is defined for each pair of elements chosen in a definite succession, the resulting product being a definite element of the same set. For instance, the product of the

[2] We shall abstain from writing the words "multiplication" and "product" in quotation marks although the appropriate operation may be quite unlike conventional algebraic multiplication.

elements $B$ and $A$ of our set is equal to the element $C$ of the same set, this being written as $AB = C$.[3]

*Note.* In general the product of the elements of the group is not commutative: $AB \neq BA$. A group all of whose elements commute with one another is termed *Abelian*.

II. The product of the group's elements obeys the associative law: $(AB)\,C = A\,(BC)$.

III. The set of the group's elements must include an element termed *unit* or *identity* element (we shall denote this by $E$) characterized by the property $ER = RE = R$, where $R$ is an arbitrary element of the group.

IV. For every element $Q$ of the group there must be among the elements of the group an *inverse* element (we shall denote this by $Q^{-1}$) such that $Q^{-1}Q = QQ^{-1} = E$.

*Note.* Demonstrate that $(ABC)^{-1} = C^{-1}B^{-1}A^{-1}$. Indeed, $(ABC)^{-1}ABC = E$. Multiplying both parts of the latter equality by $C^{-1}$, then by $B^{-1}$ and lastly by $A^{-1}$ and making use of the associative law, we obtain the desired result.

Note that the number of the elements of a finite group is its *order*.

2.2.2 Examples of Groups.

A. Positive and negative integers and zero form an infinite (Abelian) group if multiplication in the group is interpreted to mean algebraic addition. The unit element in this case is zero ($E = 0$), and the inverse element for the element (the number) $Q$ is $-Q$. Integers do not form a group with respect to conventional multiplication, because the inverse element $1/Q$ is not an integer, i.e., is not part of the group.

B. The totality of all positive rational numbers forms an infinite Abelian group with respect to conventional arithmetical multiplication. In this case the unit element $E = 1$, and if $Q = A/B$, then $Q^{-1} = B/A$.

C. The set of vectors **a** forms an infinite Abelian group if the multiplication in the group means geometrical addition of the vectors $\mathbf{a} + \mathbf{a}' = \mathbf{a}''$. The unit element $E = \mathbf{a} = 0$; the inverse element $(\mathbf{a})^{-1} = -\mathbf{a}$.

D. An instructive example of a group is the rotation of an equilateral triangle *123* (Fig. 2.1) about a fixed centre $O$, as a result of which the triangle comes into coincidence with itself (the operation of self-coincidence). The altitudes of the triangle, $1a$, $2b$ and $3c$, intersect in the centre of the triangle, $O$, and are at the same time its bisectors and medians.

Introduce the following notation for the operations of self-coincidence of the triangle: (1) rotation through 180° about the height $1a$, $A$; (2) rotation through 180° about the height $2b$, $B$; (3) rotation

[3] The multiplicands in a product are read from right to left.

through 180° about the height $3c$, $C$; (4) rotation through an angle of $120° = 2\pi/3$ clockwise about the axis perpendicular to the plane of the triangle and passing through its centre $O$, $D$; (5) rotation about the same axis and in the same direction through an angle $120° \times \times 2 = 4\pi/3$ (this motion can also be imagined as the rotation about the same axis in a counterclockwise direction through a negative angle $120° = = 2\pi/3$), $F$; (6) absence of motion of the triangle, or its rotation through an angle $2\pi$ about any axis, $E$.

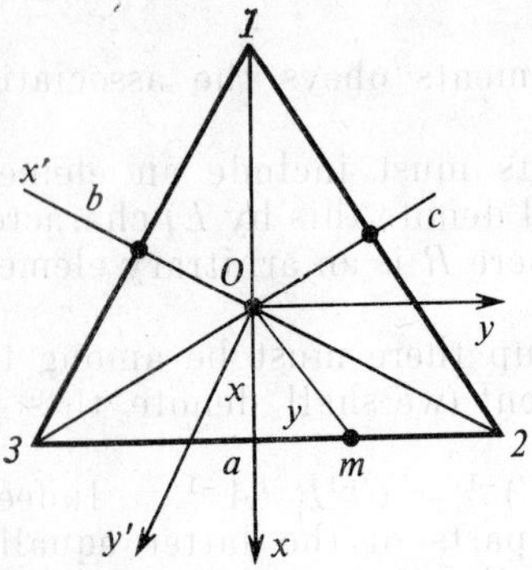


Fig. 2.1

In addition to the operations of rotation discussed above, there are also operations of symmetry involving the reflection in planes perpendicular to that of the triangle and passing through the altitudes $1a$, $2b$, $3c$. However, we shall consider the sixth order group consisting only of rotations $A$, $B$, $C$, $D$, $E$, $F$.

It should be noted that the operation of self-coincidence of a body can be considered from two equivalent points of view: (1) the coordinate frame $(xy)$ remains at rest but the body moves (from this point of view the operation $D$, for example, is the rotation of the triangle through 120° clockwise, so that its vertices turn into each other in the order $1 \to 2$, $2 \to 3$, $3 \to 1$), and (2) the body is at rest but the coordinate frame is in motion (from this point of view the operation $D$ is the rotation of the coordinate frame through 120° counterclockwise). Sometimes it will be expedient to apply the first point of view and sometimes the second. The important thing to keep in mind is the relative motion of the body and the coordinate frame.

The product $AB$ is interpreted as meaning the application of operation $B$ followed by the operation $A$ (similar to operators acting on a function). All the rotation axes retain their positions and do not move with the triangle. Let us represent the product $AB$ in the form of a diagram

$$AB\left\{\begin{matrix} & 1 & \\ & \triangle & \\ 3 & & 2\end{matrix}\right\} = A\left\{\begin{matrix} & 3 & \\ & \triangle & \\ 1 & & 2\end{matrix}\right\} = \begin{matrix} & 3 & \\ & \triangle & \\ 2 & & 1\end{matrix} = D\left\{\begin{matrix} & 1 & \\ & \triangle & \\ 3 & & 2\end{matrix}\right\},$$

whence it is evident that the product is $D$.

It may easily be checked that this set of six elements satisfies all the requirements I-IV of a group.

The product of two of the group's elements always produces a third element. For example, it can easily be checked that $AB = D$, $BA = F$, $AA = A^2 = E$, $DD = D^2 = F$. The application of the operations $B$ and $A$ in turn is equivalent to the operation $D$, the application of those operations in the reverse order being equivalent to $F$. Hence, the group is non-Abelian.

It follows directly from the geometrical meaning that $A^{-1} = A$, $B^{-1} = B$, $C^{-1} = C$, $D^{-1} = F$, $F^{-1} = D$. As a result we can write

Table 2.1

| 2nd multiplier (left) \ 1st multiplier (right) | $E$ | $A$ | $B$ | $C$ | $D$ | $F$ |
|---|---|---|---|---|---|---|
| $E$ | $E$ | $A$ | $B$ | $C$ | $D$ | $F$ |
| $A$ | $A$ | $E$ | $D$ | $F$ | $B$ | $C$ |
| $B$ | $B$ | $F$ | $E$ | $D$ | $C$ | $A$ |
| $C$ | $C$ | $D$ | $F$ | $E$ | $A$ | $B$ |
| $D$ | $D$ | $C$ | $A$ | $B$ | $F$ | $E$ |
| $F$ | $F$ | $B$ | $C$ | $A$ | $E$ | $D$ |

a multiplication table for the group (Table 2.1). This sixth order group is often denoted as $\boldsymbol{D}_3$.

**2.2.3** Note that in each row and in each column of Table 2.1 every element of the group appears only once. This is so for the following reason. If we have an $h$-order group $\mathfrak{A} = \{A_1 \equiv E, A_2, A_3, \ldots, A_h\}$ and form a product of every element of the group times one of the elements,

$$A_kE,\ A_kA_2,\ A_kA_3,\ \ldots,\ A_kA_h, \tag{2.1}$$

we shall obtain all the elements of the group, but in another order (the group shifts). To prove this, we must show only that all the products are different, so that there cannot be, for example,

$$A_kA_2 = A_kA_3.$$

Indeed, multiply both sides of this equality by $A_k^{-1}$. Then $A_k^{-1}A_kA_2 = A_k^{-1}A_kA_3$, or $EA_2 = EA_3$, or $A_2 = A_3$ and this is in contradiction to the definition of a group. If we raise some element $A_k$

to a power, we obtain other elements of the group until we reach a power $n$, termed *order* of the element $A_k$, such that $A_k^n = E$. Accordingly, we obtain a sequence of elements

$$A_k, A_k^2, A_k^3, \ldots, A_k^n = E. \tag{2.2}$$

If the element is raised to still higher powers, this sequence will be repeated ($A_k^{n+1} = A_k^n A_k = EA_k = A_k$); therefore, it is termed the *period* of the element $A_k$ and denoted $\{A_k\}$. For example, for the element $D$ of the group $\boldsymbol{D}_3$ we have $\{D\} = \{D, D^2 = F, D^3 = E\}$.

The period of the element $A_k$ obviously forms an Abelian group of the $n$th order, the term for this being *cyclic* group. In general, if within a given group we can separate a set of elements that also forms a group, the term for such a group is *subgroup*. Thus, $E$, $D$, $F$ form a subgroup of the group $\boldsymbol{D}_3$. A trivial subgroup of the first order (for every group) is the unit element $E$. Since every subgroup must include a unit element, only one subgroup can be selected out of a group, because the remaining elements do not include the unit element.

In the case of finite groups it can be proved that the order $g$ of a subgroup is an integral divisor of the group's order $h$, i.e., $h/g$ is an integer [2.5, p. 61].

For the cyclic subgroup $\{D\}$ of $\boldsymbol{D}_3$ we have: $6 \div 3 = 2$. Specifically, it follows that, if the group's order $h$ is a prime number, it has only one (trivial) subgroup $E$.

**2.2.4** The element $B$ is said to be *conjugate* to $A$ if

$$B = V^{-1}AV, \tag{2.3}$$

where $V$ is an element of the group. Premultiplying (2.3) by $V$ and multiplying it into $V^{-1}$, we obtain

$$A = VBV^{-1} = (V^{-1})^{-1}BV^{-1} = U^{-1}BU,$$

where $U = V^{-1}$ is also an element of the group. Hence, if $B$ is conjugate to $A$, then $A$ is also conjugate to $B$.

Let us show that if $B$ is conjugate to $A$ and $C$ is conjugate to $A$, then $B$ is conjugate to $C$. We have

$$B = V^{-1}AV, \; C = X^{-1}AX.$$

Then $A = XCX^{-1}$, and

$$B = V^{-1}XCX^{-1}V = (X^{-1}V)^{-1}CX^{-1}V = Z^{-1}CZ,$$

where $Z = X^{-1}V$ is also an element of the group. Hence, if instead of $V$ we substitute into (2.3) all the elements of the group in succession, we shall obtain the collection of all the mutually conjugate elements, the term for which is *class*. Define the class of the group $\boldsymbol{D}_3$ which includes the element $A$. The elements conjugate to $A$ are:

$$E^{-1}AE = A,\quad A^{-1}AA = A,\quad B^{-1}AB = B^{-1}D = C,$$
$$C^{-1}AC = B,\quad D^{-1}AD = C,\quad F^{-1}AF = B.$$

Here we resorted to Table 2.1. Hence, the corresponding class is $A, B, C$.

The unit element alone forms a class, since $V^{-1}EV = E$ for any $V$. It may easily be demonstrated that the elements $D$ and $F$ also form a class of the group $\boldsymbol{D}_3$. Hence, $\boldsymbol{D}_3$ can be subdivided into three classes:

$$1)\ E,\quad 2)\ A, B, C,\quad 3)\ F, D. \tag{2.4}$$

Any group can be subdivided into classes, so that no element is part of two different classes.

Every element of an Abelian group forms a class. In an Abelian (i.e., commutative) group: $V^{-1}AV = V^{-1}VA = A$ for any $V$, i.e., every element is conjugate only to itself.

The concept of a class is very important. It does not, of course, coincide with the subgroup concept. All classes, with the exception of $E$ (for instance, the classes 2) and 3) in (2.4)), lack the unit element without which there can be no subgroup.

A subgroup consisting of one or several complete classes is termed *invariant*, or *normal divisor*. (In general, a subgroup does not always consist of an integral number of classes.)

Therefore, if $a$ is an element of an invariant subgroup $\mathscr{H}$ of the group $G$, then all the elements $a' = g^{-1}ag$, where $g$ is an arbitrary element of the group $G$ ($g \in G$), belong to the invariant subgroup (this is because all the elements of the class are part of the subgroup $\mathscr{H}$).

All elements of one class are of the same order $n$; indeed, if $A^n = E$, then $B^n = (V^{-1}AV)^n = V^{-1}AVV^{-1}AV \ldots V^{-1}AV = V^{-1}A^nV = E$. It may easily be seen that there is a "similarity" between the elements of one class. Class 2) in (2.4) consists of rotations through 180°, and class 3) of rotations through 120°. Below we shall define the essence of the similarity of the elements of one class more precisely.

2.2.5 Let there be two (in general, non-Abelian) groups

$$\mathfrak{A} = \{A_1 \equiv E, A_2, A_3, \ldots, A_{h_a}\},\ \mathfrak{B} = \{B_1 \equiv E, B_2, B_3, \ldots, B_{h_b}\} \tag{2.5}$$

of the orders $h_a$ and $h_b$. Let all the elements of both groups (with the exception of the unit elements $A_1 = B_1 = E$) be different and commute with each other, i.e., $A_kB_l = B_lA_k$ (for all $k$ and $l$). Form $h_ah_b$ products $A_kB_l$ and demonstrate that this set is a group.

First of all we shall show that the product of two elements is an element of the same set:

$$(A_{k'}B_{l'})\,(A_kB_l) = (A_{k'}A_k)\,(B_{l'}B_l) = A_mB_n,$$

where $A_{k'}A_k = A_m$ and $B_{l'}B_l = B_n$. The product $A_mB_n$ is an element of the same set.

Next we shall show that the set $A_kB_l$ contains both a unit and an inverse element. It follows from (2.5) that the unit element of the set $A_kB_l$ is equal to $A_1B_1 = EE = E$. The inverse of $A_kB_l$ is equal to $(A_kB_l)^{-1} = B_l^{-1}A_k^{-1} = A_k^{-1}B_l^{-1}$ and also belongs to the same set.

The group of elements $A_kB_l$ is termed the *direct product* of the groups $\mathfrak{A}$ and $\mathfrak{B}$ and denoted $\mathfrak{A} \times \mathfrak{B}$. The order of the direct-product group $\mathfrak{A} \times \mathfrak{B}$ is equal to the product of the orders of $\mathfrak{A}$ and $\mathfrak{B}$.

The elements of a class of $\mathfrak{A} \times \mathfrak{B}$ determined by the element $A_kB_l$ are

$$(A_{k'}B_{l'})^{-1}\,(A_kB_l)\,(A_{k'}B_{l'}) = B_{l'}^{-1}A_{k'}^{-1}A_kB_lA_{k'}B_{l'}$$
$$= (A_{k'}^{-1}A_kA_{k'})\,(B_{l'}^{-1}B_lB_{l'}) = A_{k''}B_{l''}. \qquad (2.6)$$

Here we made use of the commutativity of the elements belonging to different groups $\mathfrak{A}$ and $\mathfrak{B}$. In (2.6) $A_{k'}$ and $B_{l'}$ are equal to all the elements in groups $\mathfrak{A}$ and $\mathfrak{B}$ in succession. If $A_k$ belongs to a certain class $\alpha$ of $\mathfrak{A}$ and $B_l$ to class $\beta$ of $\mathfrak{B}$, then $A_{k''}$ and $B_{l''}$ also belong to $\alpha$ and $\beta$. The element $A_{k''}B_{l''}$ will belong to the class of the group $\mathfrak{A} \times \mathfrak{B}$ derived from classes $\alpha$ and $\beta$ of the groups $\mathfrak{A}$ and $\mathfrak{B}$. Hence, there is a class of the group $\mathfrak{A} \times \mathfrak{B}$ to correspond to each pair of classes of $\mathfrak{A}$ and $\mathfrak{B}$; therefore, the number of classes of the direct product is equal to the product of the numbers of classes in each of them.

2.2.6 Let there be two groups, an "unprimed" and a "primed", $G = \{A, B, C, \ldots, Q, \ldots\}$ and $G' = \{A', B', C', \ldots, Q', \ldots\}$, consisting, in general, of elements of different nature and having a different meaning. If a one-to-one correspondence $A \leftrightarrow A'$, $B \leftrightarrow B'$, . . ., $Q \leftrightarrow Q'$, . . . can be established between the elements of both groups such that for $BA = Q$ we have $B'A' = Q'$, i.e., the same multiplication table for both groups is valid, then these groups are called *isomorphic*. For instance, the elements constituting the group $G$ may be the elements performing motions of a geometrical figure resulting in its coinciding with itself (see the equilateral triangle group $\boldsymbol{D}_3$), and the elements of group $G'$ may be matrices of the same rank whose multiplication is performed in compliance with the same multiplication table (Table 2.1)[4].

[4] We shall consider this case in detail in the following chapter.

From the point of view of the abstract group theory, isomorphic groups are identical.

But if several elements of group $G$ are the counterpart of one element of group $G'$ (for instance, $A \to A'$, $B \to A'$, . . ., $Q \to Q'$, . . .) and if it follows from $BA = Q$ that $A'A' = (A')^2 = Q'$, then group $G$ is called *homomorphic* to group $G'$. The homomorphic groups are of different orders (since all the elements in group $G'$ must be different). A trivial example of a homomorphism is the collation of all the elements of group $G$ with unity: $A \to 1$, $B \to 1$, . . ., $Q \to 1$ with the law of arithmetical multiplication acting in group $G'$. In this case the counterpart of any group will be a first order ($h = 1$) group.

If each symmetry operation on the triangle ($E$, $A$, $B$, . . .) in Fig. 2.1 is made to correspond to a permutation of the numbers *1*, *2*, and *3* marking the vertices of the triangle, then the set of the permutations (whose number is $3! = 6$) will form a group isomorphic to $\boldsymbol{D}_3$:

$$E \leftrightarrow (1,\ 2,\ 3), \quad A \leftrightarrow (1,\ 3,\ 2), \quad B \leftrightarrow (3,\ 2,\ 1), \ \ldots .$$

## 2.3 Point Groups[5]

**2.3.1** Displacements of a body of finite dimensions (e.g. a molecule) resulting in its coinciding with itself (in self-coincidence) form the so-called *point group*. For a body of finite dimensions, such displacements may involve rotations through certain angles about axes intersecting in a common point and reflections in planes containing this point.[6] The axis of rotation through an angle $2\pi/n$ is denoted by $C_n$, and through an angle $p\ (2\pi/n)$ by $C_n^p$; if $p$ is a multiplier of $n$, then $C_n^p = C_{n/p}$. Obviously, $C_n^n = E$, a unit element, since it corresponds to a rotation through an angle $n2\pi/n = 2\pi$ or to the absence of a rotation; therefore, $C_1 = E$.

Reflection in a plane is denoted by $\sigma$; obviously, $\sigma^2 = E$, a unit element. Usually the axis of the highest symmetry (with the greatest $n$) is chosen as the vertical axis. The reflection plane that passes through this vertical axis is denoted $\sigma_v$ (vertical), and the reflection plane perpendicular to this axis $\sigma_h$ (horizontal).

The rotary-reflection transformation (axis) $S_n$ is defined as a sequential application of two operations $C_n$ and $\sigma_h$:

$$\sigma_h C_n = C_n \sigma_h = S_n. \tag{3.1}$$

[5] An excellent and a more complete treatise on the subject may be found in *Landau and Lifshitz* [2.6, §§ 91 and 93].

[6] Indeed, if the rotation axes (or reflection planes) have no common points (or intersection lines), sequential rotation and reflection operations may result in a linear motion of the body, in the course of which it cannot coincide with itself.

A body possessing rotary-reflection symmetry $S_n$ coincides with itself if it is rotated about the symmetry axis through the angle $2\pi/n$ and then reflected in the plane perpendicular to the axis $C_n$ (or if such operations are performed in reverse order); see Fig. 2.2.

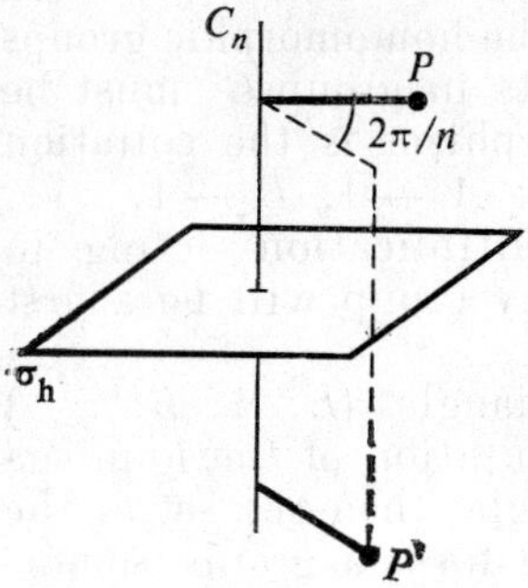


Fig. 2.2

It follows from (3.1) that $S_1 = \sigma_h$. An important specific case is

$$\sigma_h C_2 = C_2 \sigma_h = S_2 = J, \tag{3.2}$$

where $J$ is the operation of inversion. As the result of the inversion operation, every point of a body $P$ is transformed into a point $P'$ lying on the straight line that passes through $P$ and a stationary centre $O$, so that $\overrightarrow{OP} = -\overrightarrow{OP'}$. When an inversion operator is applied to the $x$, $y$, $z$ coordinates, they change sign: $\hat{J}\,\{x, y, z\} = \{-x, -y, -z\}$; therefore a right-handed coordinate frame changes to a left-handed one. It follows from (3.2) that $J\sigma_h = C_2$ and $JC_2 = \sigma_h$; hence, those three elements of symmetry $C_2$, $\sigma_h$ and $J$ are interconnected, so that the existence of two of them automatically leads to the existence of the third.

2.3.2 Here we present some geometrical properties inherent in rotations about an axis and reflections in a plane that will be helpful in the study of point groups.

It is a well-known fact from the kinematics of solid bodies that two rotations in succession about intersecting axes are equivalent to one rotation about an axis passing through the point of intersection, the resulting rotation being dependent on the order in which the former two have been performed (i.e., rotations through finite angles do not commute, and for this reason cannot be represented by vectors) [2.7, Sec. 22].

Successive reflections in two intersecting planes are equivalent to the rotation about an axis coinciding with the line of intersection of the planes through an angle twice the angle between the planes $\varphi$:

$$\sigma_v \sigma'_v = C\,(2\varphi). \tag{3.3}$$

The proof of this theorem follows directly from the analysis of Fig. 2.3 (note that reversal of the order of reflections changes the sign of the rotation).

Premultiplying (3.3) by $\sigma_v$ and taking into account that $\sigma_v^2 = E$, we obtain

$$\sigma'_v = \sigma_v C(2\varphi), \tag{3.3a}$$

i.e., the product of a rotation and a reflection in a plane passing through the rotation axis is equivalent to the reflection in another plane passing through the same axis and forming an angle with the first plane equal to one-half the rotation angle. Hence, it follows that if a plane $\sigma_v$ passes through an axis $C_n$, it automatically results in the appearance of $n - 1$ additional reflection planes passing through the same axis, so that the angles between them are $\pi/n$. Similarly, it may be demonstrated [2.6, p. 363] that the existence of a $C_2$ axis perpendicular to a $C_n$ axis automatically leads to the appearance of $n - 1$ additional $C_2$ axes perpendicular to $C_n$, so that the angles between them are $\pi/n$.

Although the result of two successive transformations generally depends on their order, in the following cases the operations are

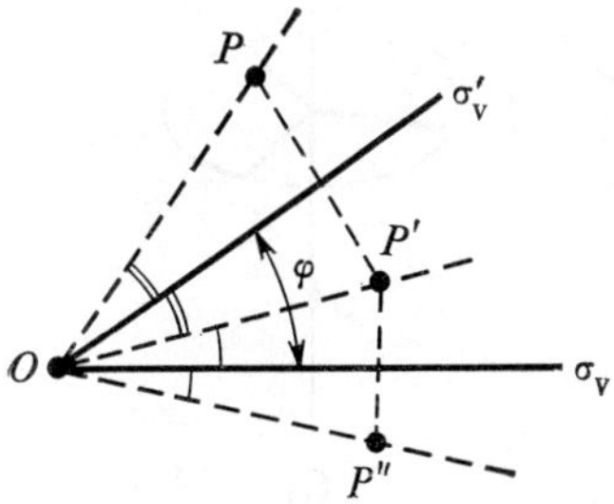


Fig. 2.3

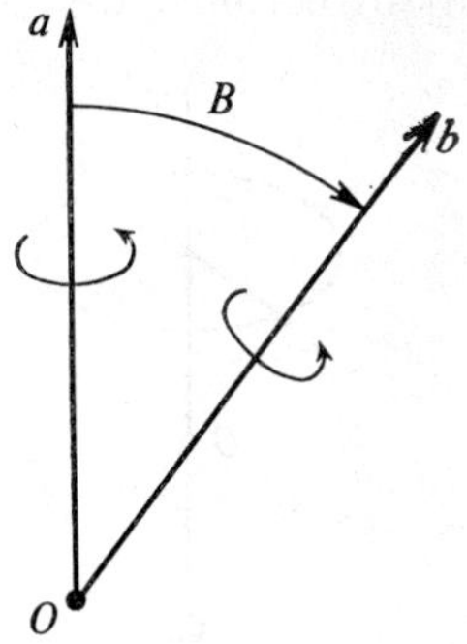


Fig. 2.4

commutative: (1) two rotations about the same axis; (2) two rotations through the angle $\pi$ about two mutually perpendicular axes; (3) two rotations in mutually perpendicular planes; (4) rotation and reflection in the plane perpendicular to the rotation axis (see (3.1)); (5) any rotation (or reflection) and inversion in a point lying on the rotation axis (or in the reflection plane).

We have already mentioned in Section 2.2.4 the "similarity" of elements belonging to the same class, using the group $\boldsymbol{D}_3$ as an example. To generalize this proposition, let us prove the following theorem: two rotations through an equal angle about two different axes (or two reflections in different planes) belong to one class of a point group if among the elements of the latter there is an operation that makes different rotation axes (or different reflection planes) coincide.

*Proof*. Figure 2.4 depicts the axes $Oa$ and $Ob$; an element of point group $B$ transforms the axis $Oa$ into the axis $Ob$; $C_n$ and $C'_n$ are rotations about the axes $Oa$ and $Ob$ through the angle $2\pi/n$. The product $B^{-1}C'_nB$ has the following geometrical meaning: the axis $Oa$ turns until it coincides with the $Ob$ axis; next a rotation is performed about the latter axis through an angle $2\pi/n$, after which the axis $Ob$ is rotated until it coincides with $Oa$. Evidently, the result is a rotation through an angle $2\pi/n$ about the $Oa$ axis:

$$C_n = B^{-1}C'_nB,$$

which means that $C_n$ and $C'_n$ are mutually conjugate elements, i.e., they belong to the same class.

The proof for reflections in two different planes is quite similar. Axes and planes that are made to coincide by the application of one of the group's elements are termed *equivalent*.

Two rotations about the same axis through equal angles but in different directions, i.e., the elements $C_n^k$ and $(C_n^k)^{-1}$, belong to the

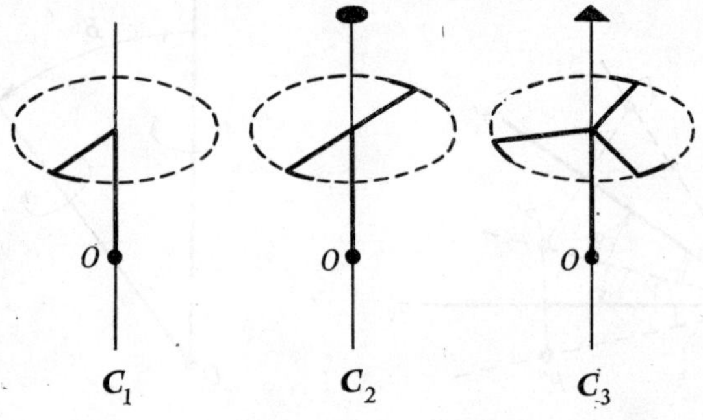


**Fig. 2.5**

same class if among the elements of the group there is an axis $C_2$ perpendicular to the rotation axis or a plane $\sigma_v$ passing through it. Indeed, in this case the operations $C_2$ and $\sigma_v$ reverse the direction of rotation. In this case the rotation axis $C_n$ is termed *bilateral.* Note that a reflection in a $\sigma_h$ plane perpendicular to a $C_n$ axis does not reverse the direction of rotation and hence does not make it bilateral.

Let us consider the major point groups. Here we shall make no attempt at a complete and thorough study of the problem. The symbols of the point groups will be printed in bold type.

**I. $C_n$ groups.** These groups consist of a symmetry axis of the $n$th order, i.e., the body coincides with itself if rotated about the axis through an angle $2\pi/n$. This is a cyclic group consisting of $n$ elements: $C_n$, $C_n^2$, . . ., $C_n^n = E$. Every element is a class by itself. The group $C_1$ consists of one element $C_1 = E$; the corresponding state is the absence of any symmetry. Figure 2.5 depicts bodies with symmetries $C_1$, $C_2$ and $C_3$.[7]

[7] The notation for the symmetry axes of the second, third, fourth, etc., orders is ●, ▲, ■, etc.

**II. $S_n$ groups.** A rotary-reflection axis $S_n$ for odd $n = 2p + 1$ reduces to a symmetry axis $C_{2p+1}$ and to a symmetry plane $\sigma_h$ perpendicular to it. Indeed, $S_{2p+1}^{2p+1} = \sigma_h$, since in this case the rotary-reflection axis corresponds to the type of symmetry discussed in III, the group $C_{nh}$.

For even $n = 2p$, the group $S_{2p}$ is a cyclic group consisting of $2p$ elements: $S_{2p}, S_{2p}^2, \ldots, S_{2p}^{2p} = E$, every one of which forms a separate class. The group $S_2$ consists of two elements: $S_2 = C_2\sigma_h = J$ (inversion) and $S_2^2 = E$ (a unit element); the notation for it is $C_i$.

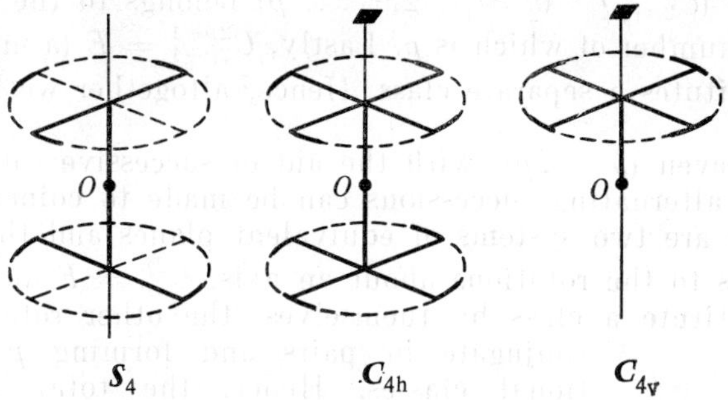


Fig. 2.6

Figure 2.6 depicts a body with the symmetry of $S_4$. Indeed, when the rigid framework (shown by solid lines) is rotated through an angle $2\pi/4 = 90°$ and reflected in the plane $\sigma_h$ passing through the centre $O$, it coincides with its original position (prior to rotation).

**III. $C_{nh}$ groups.** These groups are obtained when a plane $\sigma_h$ perpendicular to a $C_n$ axis is added to this axis. The group contains $2n$ elements: $n$ rotations about the $C_n$ axis and $n$ reflections[8]: $C_n^k\sigma_h = S_n^k$, $k = 1, 2, \ldots, n$ (including the reflection $C_n^n\sigma_h = \sigma_h$). All elements of the group are commutative, i.e., the group is Abelian; the number of classes is equal to the number of elements. In the case of an even $n$ ($n = 2p$) the group contains a symmetry centre (inversion) (i.e., $C_{2p}^p\sigma_h = C_2\sigma_h = J$). The simplest group $C_{1h}$ contains two elements: $E$ and $\sigma_h$. Another notation for it is $C_s$. Figure 2.6 depicts a body with a $C_{4h}$ symmetry.

Since for even $n = 2p$ a $C_{2ph}$ group contains an inversion, the $C_{2ph}$ group may be represented in the form of a direct product $C_{2p} \times C_i$, where $C_i = \{E, J\}$.

**IV. $C_{nv}$ group.** This group is realized if a plane $\sigma_v$ passing through the axis $C_n$ is added to it. According to the theorem proved

[8] A group must meet the following condition: a product of two elements must produce an element belonging to the group.

above, this automatically leads to the appearance of $n - 1$ additional planes passing through axis $C_n$, so that the angles between them are $\pi/n$. Thus, group $C_{nv}$ contains $2n$ elements: $n$ rotations about the $C_n$ axis and $n$ reflections in the $\sigma_v$ planes. The presence of the $\sigma_v$ planes makes the $C_n$ axis bilateral.

The division into classes is different for even and odd $n$'s. If $n$ is odd ($n = 2p + 1$), then the successive rotations make all the planes $\sigma_v$ coincide with one another, and because of that $n$ reflections belong to the same class. Since the $C_n$ axis is bilateral, each pair of rotations $C^k_{2p+1}$ and $(C^k_{2p+1})^{-1}$ ($k = 1, 2, \ldots, p$) belongs to the same class, the total number of which is $p$. Lastly, $C^{2p+1}_{2p+1} = E$ (a unit element) also constitutes a separate class. Hence, altogether we have $p + 2$ classes.

If $n$ is even ($n = 2p$), with the aid of successive rotations only planes in alternating successions can be made to coincide; in this case there are two systems of equivalent planes and therefore two classes. As to the rotations about an axis, $C^{2p}_{2p} = E$ and $C^{p}_{2p} = C_2$ each constitute a class by themselves, the other rotations being conjugate in pairs and forming $p - 1$ additional classes. Hence, the total number of classes is $2 + 2 + (p - 1) = p + 3$.

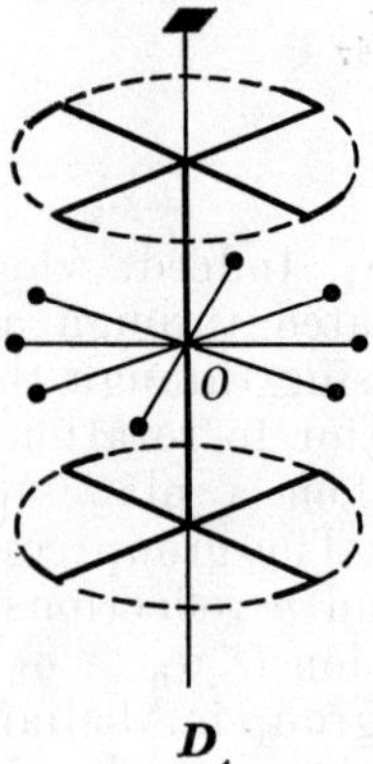
$D_4$

Fig. 2.7

Figure 2.6 depicts a body with the symmetry of $C_{4v}$ (two planes $\sigma_v$ pass through the symmetry axis and through solid diameters; two other planes bisect angles made by the former two).

**V. $D_n$ group.** If a horizontal axis $C_2$ perpendicular to a vertical symmetry axis $C_n$ is added to it, this will automatically create $n - 1$ additional horizontal axes $C_2$, so that the angles between them will be $\pi/n$ (see above). The number of elements is $2n$ ($n$ rotations about $C_n$ and $n$ rotations about the $C_2$ axes). The $C_n$ axis is bilateral. The division into classes is quite similar to that of the $C_{nv}$ groups. Figure 2.7 depicts a body with the symmetry of $D_4$ (note the difference between $D_4$ and $C_{4h}$); the four axes pass through the centre $O$. Note that the equilateral triangle group (Section 1.2) is a $D_3$ group.

**VI. $D_{nh}$ groups.** If a horizontal reflection plane is added to a system of $D_n$ axes, it will automatically lead to the appearance of $n$ vertical planes $\sigma_v$ passing through the $C_n$ axis and through one of the $C_2$ axes perpendicular to it [this follows from (3.3a) if we put $\sigma_v = \sigma_h$ and $2\varphi = 180°$]. The $D_{nh}$ group obtained in this way contains $4n$ elements: $2n$ elements of the $D_n$ group, $n$ reflections $\sigma_v$, and $n$ rotary-reflection transformations $C^k_n\sigma_h$.

The reflection $\sigma_h$ commutes with the rest of the group's elements; therefore, we can write $D_{nh}$ in the form of a direct product $D_n \times C_s$ (or $C_{nv} \times C_s$), where $C_s$ is a group of two elements $E$ and $\sigma_h$. Hence, it follows that the number of classes in a $D_{nh}$ group is equal to twice the number of classes in a $D_n$ group. One-half coincides with the classes of the $D_n$ group, and one-half is the result of their multiplication by $\sigma_h$. It is obvious that for an even $n = 2p$, when the group $D_{2ph}$ contains an inversion $J$, we can write $D_{2ph} = D_{2p} \times C_i$.

**VII. $D_{nd}$ groups.** If a plane of symmetry is drawn through a $C_n$ axis of a $D_n$ group so that it bisects the angle between adjacent $C_2$ axes, it will lead to the appearance of $n - 1$ additional planes passing through the $C_n$ axis.

The group $D_{nd}$ thus obtained contains $4n$ elements; $n$ reflections in the vertical planes are denoted by $\sigma_d$ (d for diagonal), and $n$ transformations of the type $C_2\sigma_d$ are added to the $2n$ elements of the $D_n$ group. It can be demonstrated [2.6, p. 364] that $C_2\sigma_d = S_{2n}^{2k+1}$, where $k = 1, 2, \ldots, (n - 1)$, i.e., are rotary-reflection transformations about a vertical axis which thus becomes not only a symmetry axis of the $n$th order but a rotary-reflection axis of order $2n$ as well.

Let us now turn to the discussion of very important point groups of higher symmetry (when there are several intersecting axes of the $n$th order) termed *cubic*. The justification for this term is that their elements of symmetry can be selected from among the axes and planes of symmetry of a cube.

**VIII. Tetrahedron axes group $T$.** This group is made up of three $C_2$ axes and four $C_3$ axes of a regular tetrahedron. Figure 2.8*a*

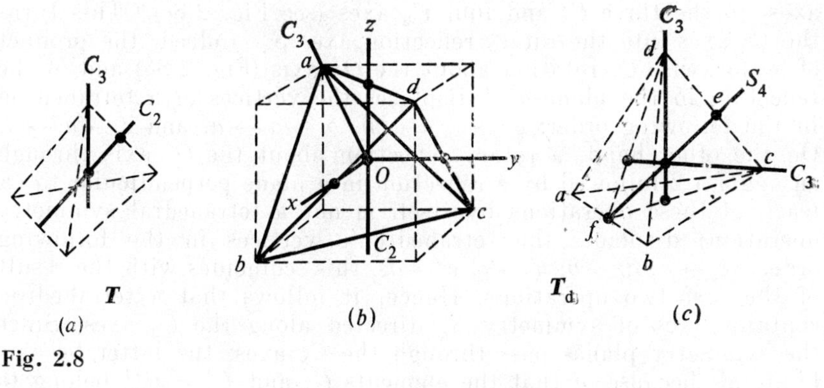


**Fig. 2.8**

and *b* depicts how those axes should be drawn in a tetrahedron and how they can be chosen in a cube (the $C_2$ axes are perpendicular to the cube's faces and the $C_3$ axes coincide with the cube's space diagonals); in Fig. 2.8*b* a tetrahedron with the vertex *a* and the base *bcd* is inscribed into a cube. Altogether there are twelve elements: $E$, $3C_2$, $4C_3$, $4C_3^2$ $(= 4C_3^{-1})$. The three $C_2$ axes are equivalent. The $C_3$ axes are also equivalent, but they are not bilateral, and for this reason the elements of a $T$ group are divided into four classes:

$$E,\ 3C_2,\ 4C_3,\ 4C_3^2.$$

The $T$ symmetry is not the full symmetry of a tetrahedron (see $T_d$ group). To obtain a body with the symmetry of $T$, it suffices to take a body without any symmetry and subject it to all twelve transformations of $T$; thus, we obtain a figure with the symmetry of $T$.

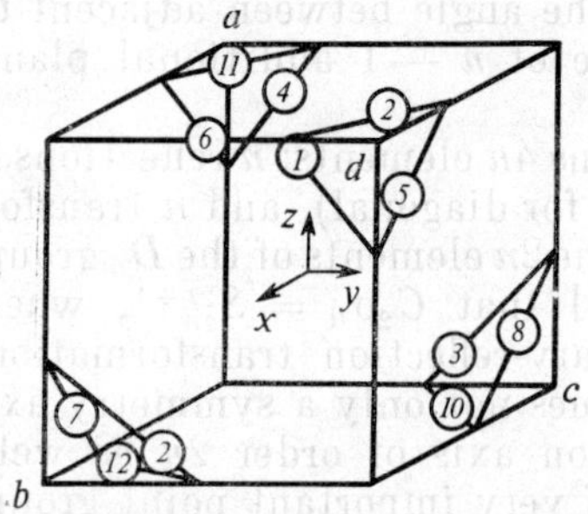

Fig. 2.9

Such a figure is shown in Fig. 2.9. The label *1* has been subjected to all the transformations of the group $T$ in turn. The $C_2$ rotations about the $x$, $y$, $z$ axes transform *1* into *2*, *3*, *4*; the $C_3$ and $C_3^2 = C_3^{-1}$ rotations about the axes passing through the vertices of the tetrahedron $a$, $b$, $c$, $d$ multiply the label so that it assumes the positions *5*, ..., *12*.

**IX. $T_d$ group.** This group contains all the symmetry transformations of a tetrahedron. To obtain it, we add six reflection planes, each of which passes through one $C_2$ axis and two $C_3$ axes to the three $C_2$ and four $C_3$ axes (see Fig. 2.8*c*).[9] This turns the $C_2$ axes into the rotary-reflection axes $S_4$. Indeed, the product of a clockwise $C_2$ rotation about the $Ox$ axis (Fig. 2.8*b*) and of the reflection in the plane $aOd$ displaces the vertices of a tetrahedron in the following order: $a \to c$, $d \to b$, $b \to d \to a$, and $c \to a \to d$. On the other hand, a rotary reflection about the $Oz$ axis through an angle $\pi/2$ followed by a reflection in a plane perpendicular to $Oz$ (each of these operations by itself is not a tetrahedral symmetry operation) displaces the tetrahedron's vertices in the following order: $b \to a$, $d \to b$, $a \to c$, $c \to d$; this coincides with the result of the first two operations. Hence, it follows that a tetrahedron contains axes of symmetry $S_4$ directed along the $C_2$ axes. Since the symmetry planes pass through the $C_3$ axes, the latter become bilateral; because of that the elements $C_3$ and $C_3^2 = C_3^{-1}$ belong to the same class. All the planes and axes of each type are equivalent.

[9] The second plane that passes through the same $C_2$ axis in Fig. 2.8*c* passes through the edge *afb* and through point *e*.

Therefore, the 24 elements of the group are divided into the following five classes:[10]

$$E,\quad 8\,(C_3,\ C_3^2),\quad 6\sigma,\quad 6\,(S_4,\ S_4^3),\quad 3C_2.$$

**X. $T_h$ group.** This group is obtained if a centre of symmetry (inversion) is added to the $T$ group, so that $T_h = T \times C_i$. The result is the doubling of the number of elements and classes as compared with the $T$ group (i.e., 24 elements distributed over 8 classes).

**XI. Cubic axes group $O$.** This is one of the most important point groups; it consists of all the cube's axes of symmetry—of three $C_4$ axes passing through the centres of the opposite faces, of four $C_3$ axes passing through the opposite vertices and of six $C_2$ axes passing

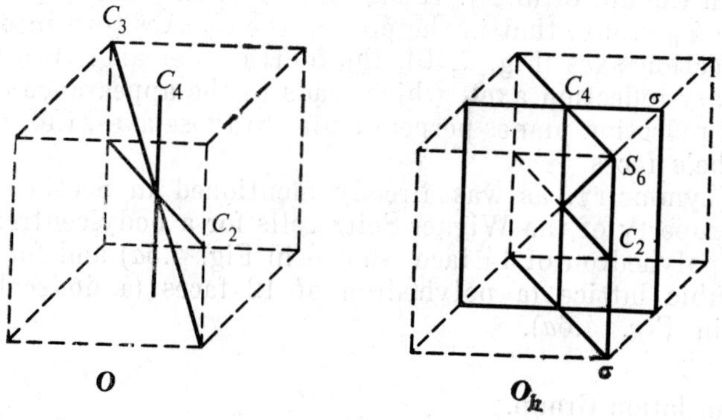


**Fig. 2.10**

through the midpoints of the opposite edges (Fig. 2.10). All the axes of the same order are equivalent and bilateral. Accordingly, the 24 elements of the group are distributed over five classes:

$$E,\quad 8\,(C_3,\ C_3^2),\quad 3\,(C_2 = C_4^2),\quad 6C_2,\quad 6\,(C_4,\ C_4^3).$$

We should like to point out that the $O$ group does not describe the full symmetry of a cube. A figure with the symmetry of the $O$ group can be obtained in the way similar to that described for the $T$ group, i.e., by "breeding" some asymmetrical figure as the result of 24 symmetry transformations of the $O$ group. It can easily be demonstrated that the $O$ group is isomorphic to the tetrahedral group $T_d$.

**XII. $O_h$ group.** This is the full symmetry group of a cube (octahedron). It contains, in addition to all the cube's symmetry

[10] Here and below the numerical coefficient is equal to the total number of elements in a class; e.g., $8\,(C_3,\ C_3^2) = 4C_3 + 4C_3^2$.

axes, a centre of symmetry (inversion). Since $E$ and the inversion $J$ commute with all the elements of group $\boldsymbol{O}$, the $\boldsymbol{O}_{\mathrm{h}}$ group is equal to the direct product of $\boldsymbol{O}$ and $C_{\mathrm{i}} = \{E, J\}$, i.e., $\boldsymbol{O}_{\mathrm{h}} = \boldsymbol{O} \times C_{\mathrm{i}}$. It can be demonstrated that the $\boldsymbol{O}_{\mathrm{h}}$ group can also be represented in the form of a direct product of $\boldsymbol{T}_{\mathrm{d}}$ and $C_{\mathrm{i}}$, i.e., $\boldsymbol{O}_{\mathrm{h}} = \boldsymbol{T}_{\mathrm{d}} \times C_{\mathrm{i}}$.[11] The elements of the $\boldsymbol{O}_{\mathrm{h}}$ group are 48 ($24 \times 2$) in number and the classes 10 ($5 \times 2$). All the elements and classes of the $\boldsymbol{O}_{\mathrm{h}}$ group can be obtained by multiplying the elements and classes of the $\boldsymbol{O}$ (or the $\boldsymbol{T}_{\mathrm{d}}$) group by $E$ and by $J$. (It follows that one half of the elements and classes of the $\boldsymbol{O}_{\mathrm{h}}$ group coincides with those of $\boldsymbol{O}$ or $\boldsymbol{T}_{\mathrm{d}}$ and the rest are obtained by multiplying by $J$.) The addition of a centre of inversion automatically results in the appearance of six reflection planes passing through the opposite edges [this may be shown with the aid of (3.2)]. It may also be easily shown, as in the case of the $\boldsymbol{T}_{\mathrm{d}}$ group, that in the process the $C_3$ axes turn into the $S_6$ rotary-reflection axes (Fig. 2.10), the fourth order axes turning into the $S_4$ rotary-reflection axes, which leads to the appearance of three additional reflection planes perpendicular to those axes, i.e., parallel to the cube's faces.

Such a symmetry, as was already mentioned in Section 1.1, is also the property of the Wigner-Seitz cells for a body-centred cubic lattice (a polyhedron of 14 faces shown in Fig. 4.5*a*) and for a face-centred cubic lattice (a polyhedron of 12 faces (a dodecahedron) depicted in Fig. 4.6*a*).

## 2.4 Translation Group. Crystal Systems and Bravais Lattices

**2.4.1** Consider a set of lattice vectors

$$\mathbf{a} \equiv \mathbf{a_n} = n_1\mathbf{a}_1 + n_2\mathbf{a}_2 + n_3\mathbf{a}_3 \tag{4.1}$$

corresponding to integral values of $n_1$, $n_2$, $n_3$. The set (4.1) will constitute a group if geometrical addition of the vectors **a** is accepted as the multiplication rule. The unit element of the group is

$$E = \mathbf{a_n} = 0. \tag{4.2}$$

The inverse of $\mathbf{a_n}$ is $-\mathbf{a_n}$. This group, which we shall denote by $\mathcal{T}$, is termed a *translation group*. The group is obviously an Abelian group, so that every element of it (4.1) is a class of the group.

For an ideal infinite crystal the translation group is infinite. We shall demonstrate in Section 2.8.1 how an infinite translation group can, through the introduction of certain conditions (cyclic conditions), be transformed into a finite group with a large but still finite number of elements.

---

[11] This will be demonstrated at length in Section 4.8.

If identical spherical atoms are placed into each site $\mathbf{n}$ $(n_1, n_2, n_3)$ (4.1), we shall obtain a *simple* (or a *vacant*) crystal lattice.

Simple lattices can meet not only the conditions of translational symmetry (4.1), but also the additional conditions of symmetry of some point group. For example, a simple cubic lattice meeting the conditions of translational symmetry with three mutually perpendicular basis vectors $\mathbf{a}_i$ $(i = 1, 2, 3)$ of equal length is symmetrical under transformations of the point group $\boldsymbol{O}_\mathrm{h}$ (the origin can be placed in one of the sites or in the centre of a cubic cell). We shall denote the point symmetry group of a simple lattice by $\mathscr{F}$. It is evident that any element $R$ of the group $\mathscr{F}$ transforms any lattice vector $\mathbf{a}_\mathrm{n}$ into another lattice vector $\mathbf{a}_{\mathrm{n}'}$, i.e., $R\mathbf{a}_\mathrm{n} = \mathbf{a}_{\mathrm{n}'}$.

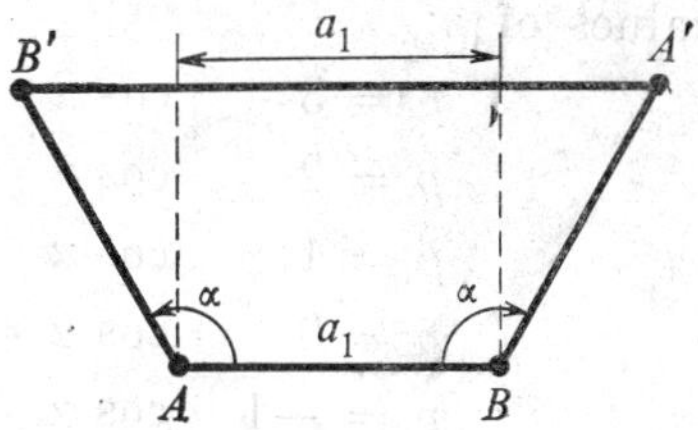


Fig. 2.11

We shall demonstrate that the translational symmetry (4.1) (for arbitrary $\mathbf{a}_i$) places limitations on the point symmetry groups $\mathscr{F}$ that a simple lattice should satisfy. As we intend to demonstrate now, for simple lattices there are only seven point groups compatible with translational symmetry.

First we note that, since in addition to the lattice vector $\mathbf{a}_\mathrm{n}$ there is also the vector $-\mathbf{a}_\mathrm{n}$ (to this end it suffices to reverse the signs of all the numbers $n_i$), a simple lattice is symmetrical (invariant) under inversion $J$. A complex lattice (e.g., see Fig. 1.2*c*) is generally not invariant under the inversion operation.

Second, let us demonstrate that only the symmetry axes of the second, third, fourth, and the sixth orders are compatible with translational symmetry. Axes of the fifth, seventh and of higher orders cannot exist in crystals.

Here is an elegant proof due to P. Niggli (1919).

Draw a plane in the crystal perpendicular to the $C_n$ symmetry axes ($n = 2\pi/\alpha$, where $\alpha$ is the minimum rotation angle about a $\boldsymbol{C_n}$ symmetry axis). Let this plane coincide with the plane of Fig. 2.11; $A$ and $B$ are the traces of the $C_n$ axes nearest to each other, so that $\overline{AB} = a_1 =$ the length of the basis lattice vector (if $\mathbf{a}_1$ does not lie in the plane being considered, $\overline{AB}$ is the projection of $\mathbf{a}_1$ on this plane). Rotating the crystal about the $A$ and $B$ axes through the angle $\alpha$ in opposite directions, we obtain an equilateral trapezoid $ABA'B'$ in which $\overline{BA'} = \overline{AB'} = a_1$. If the rotation through the angle $\alpha$ is a symmetry operation for the crystal, there will be $\boldsymbol{C_n}$ axes passing through the points $A'$ and $B'$; therefore, $\overline{B'A'} = pa_1$, where $p$ is an integer (including zero).

It follows from Fig. 2.11 that

$$\overline{B'A'} = a_1 + 2a_1 \sin(\alpha - 90°) = pa_1,$$

whence

$$\cos\alpha = \frac{1-p}{2}. \tag{4.3}$$

Since $|\cos\alpha| \leqslant 1$, only the following values of $p$ are possible: $p = 3, 2, 1, 0, -1$. The following angles $\alpha$ correspond to these values of $p$:

$$\begin{array}{lll} p = 3 & \cos\alpha = -1 & \alpha = 180° = 2\pi/2, \\ p = 2 & \cos\alpha = -1/2 & \alpha = 120° = 2\pi/3, \\ p = 1 & \cos\alpha = 0 & \alpha = 90° = 2\pi/4, \\ p = 0 & \cos\alpha = 1/2 & \alpha = 60° = 2\pi/6, \\ p = -1 & \cos\alpha = 1 & \alpha = 0° \text{ or } 2\pi. \end{array} \tag{4.4}$$

Hence, only the following symmetry axes can exist in a crystal: $C_2$, $C_3$, $C_4$ and $C_6$.

The third limitation placed on the point group $\mathcal{F}$ is that, if a point group of a simple lattice contains a $C_n$ axis (where $n > 2$), it must also satisfy the conditions of the $C_{nv}$ symmetry. (For proof see *Ljubarskii* [2.8].)

If we inspect all the point groups (Section 2.3), we will establish that only seven groups (namely, $\boldsymbol{S}_2$, $\boldsymbol{C}_{2h}$, $\boldsymbol{D}_{2h}$, $\boldsymbol{D}_{3d}$, $\boldsymbol{D}_{4h}$, $\boldsymbol{D}_{6h}$ and $\boldsymbol{O}_h$) meet all the three conditions stated above.

It may easily be seen that the groups $\boldsymbol{S}_2$, $\boldsymbol{C}_{2h}$ and $\boldsymbol{D}_{2h}$ meet the above conditions (in this case $n = 2$, and the third condition is not required). Next, the $\boldsymbol{C}_{nv}$ groups do not include an inversion centre, and the $\boldsymbol{C}_{nh}$ groups ($n > 2$) do not include the $\boldsymbol{C}_{nv}$ subgroup. The groups $\boldsymbol{D}_{4h}$ and $\boldsymbol{D}_{6h}$ satisfy all three of the above conditions. The $\boldsymbol{D}_{3h}$ group has no inversion centre and must be replaced by the $\boldsymbol{D}_{3d}$ group [2.6, p. 364]. Lastly, among the cubic groups only the $\boldsymbol{O}_h$ group meets the above requirements.

The seven point groups $\mathcal{F}$ specified above constitute seven *crystal systems*, the terms and designations for them being:

1) triclinic ($\boldsymbol{S}_2$) . . . *tr*,
2) monoclinic ($\boldsymbol{C}_{2h}$) . . . *m*,
3) orthorhombic ($\boldsymbol{D}_{2h}$) . . . *o*,
4) tetragonal or quadratic ($\boldsymbol{D}_{4h}$) . . . *t*,
5) rhombohedral or trigonal ($\boldsymbol{D}_{3d}$) . . . *rh*,
6) hexagonal ($\boldsymbol{D}_{6h}$) . . . *h*,
7) cubic ($\boldsymbol{O}_h$) . . . *c*.

It can be demonstrated that several different simple lattices may belong to one system. For example, the following lattices belong to the cubic system ($c$): the simple ($\Gamma_c$), the body-centred ($\Gamma_c^b$) and the face-centred ($\Gamma_c^f$) cubic lattices (see Figs. 1.5 and 2.13). It can be demonstrated that there are no other simple lattices with the symmetry of $\boldsymbol{O}_h$.

The triclinic system ($tr$) has the lowest symmetry $\boldsymbol{S}_2 = \boldsymbol{C}_i = \{E, J\}$—that of a simple lattice ($\Gamma_{tr}$) with basis vectors of arbitrary length ($a_1 \neq a_2 \neq a_3$) arbitrarily oriented with respect to each other.

The French physicist Auguste Bravais (1850) demonstrated that there are altogether 14 types of simple lattices (*Bravais lattices*) corresponding to the seven systems.

Let us demonstrate, using the monoclinic system ($m$) as an example, how the possible types of Bravais lattices can be determined in succession.

First of all, let us demonstrate that in the case of the monoclinic system, where the symmetry of the lattice is of the $\boldsymbol{C}_{2h}$ type, two basis vectors (for instance, $\mathbf{a}_1$ and $\mathbf{a}_2$) lie in the plane $\sigma_h$ perpendicular to the $C_2$ axis. Take two lattice vectors $\mathbf{a}$ and $\mathbf{a}'$ not lying in one plane passing through the $C_2$ axis. The vectors $\mathbf{a} + \hat{\sigma}_h\mathbf{a}$ and $\mathbf{a}' + \hat{\sigma}_h\mathbf{a}'$ are also lattice vectors ($\hat{\sigma}_h\mathbf{a}$ is the lattice vector obtained from $\mathbf{a}$ by reflection in the $\sigma_h$ plane) and, as may readily be seen, lie in the $\sigma_h$ plane. But this means that the two basis vectors $\mathbf{a}_1$ and $\mathbf{a}_2$ can be selected in the $\sigma_h$ plane.

Represent the third basis vector in the form

$$\mathbf{a}_3 = \boldsymbol{\alpha} + \boldsymbol{\beta},$$

where $\boldsymbol{\alpha} \parallel C_2$ and $\boldsymbol{\beta} \perp C_2$. The lattice vector

$$\hat{C}_2\mathbf{a}_3 - \mathbf{a}_3 = \hat{C}_2(\boldsymbol{\alpha} + \boldsymbol{\beta}) - \boldsymbol{\alpha} - \boldsymbol{\beta} = \boldsymbol{\alpha} - \boldsymbol{\beta} - \boldsymbol{\alpha} - \boldsymbol{\beta} = -2\boldsymbol{\beta}$$

lies in the $\sigma_h$ plane; therefore

$$2\boldsymbol{\beta} = m_1\mathbf{a}_1 + m_2\mathbf{a}_2$$

($m_1$ and $m_2$ are integers or zero). Hence

$$\mathbf{a}_3 = \boldsymbol{\alpha} + \boldsymbol{\beta} = \boldsymbol{\alpha} + \frac{m_1}{2}\mathbf{a}_1 + \frac{m_2}{2}\mathbf{a}_2. \tag{4.6}$$

It may easily be seen that if the vector $n_1\mathbf{a}_1 + n_2\mathbf{a}_2$ ($n_1$ and $n_2$ are integers) is subtracted from $\mathbf{a}_3$, then

$$\mathbf{a}_3' = \mathbf{a}_3 - n_1\mathbf{a}_1 + n_2\mathbf{a}_2$$

will also be a basis vector (for a planar lattice this is evident from Fig. 1.2$a$). Put the third basis vector equal to

$$\mathbf{a}_3' = \mathbf{a}_3 - n_1\mathbf{a}_1 - n_2\mathbf{a}_2 = \boldsymbol{\alpha} + \frac{m_1}{2}\mathbf{a}_1 + \frac{m_2}{2}\mathbf{a}_2 - n_1\mathbf{a}_1 - n_2\mathbf{a}_2$$
$$= \boldsymbol{\alpha} + \left(\frac{m_1}{2} - n_1\right)\mathbf{a}_1 + \left(\frac{m_2}{2} - n_2\right)\mathbf{a}_2$$

and choose the coefficients of $\mathbf{a}_1$ and $\mathbf{a}_2$ positive and not exceeding unity; then we shall obtain the following four cases, depending on whether $m_1$ and $m_2$ are even or odd:

$$\begin{aligned} &1)\ \mathbf{a}_3' = \boldsymbol{\alpha},\quad 2)\ \mathbf{a}_3' = \boldsymbol{\alpha} + \frac{1}{2}\mathbf{a}_1, \\ &3)\ \mathbf{a}_3' = \boldsymbol{\alpha} + \frac{1}{2}\mathbf{a}_1 + \frac{1}{2}\mathbf{a}_2, \\ &4)\ \mathbf{a}_3' = \boldsymbol{\alpha} + \frac{1}{2}\mathbf{a}_2. \end{aligned} \tag{4.7}$$

It may readily be seen that the only difference between (2) and (4) is in the designation of the vectors $\mathbf{a}_1$ and $\mathbf{a}_2$, and that (4) transforms into (3), if the basis vector $\mathbf{a}_2$ is replaced by $\mathbf{a}_1 + \mathbf{a}_2$. Essentially different simple lattices are (1) and one of the three remaining types [for instance, (4)]. It follows from (4.7) that the following mutual orientations of the basis vectors correspond to these cases (we omit the prime in $\mathbf{a}_3'$):

$$\mathbf{a}_1,\ \mathbf{a}_2 \perp \mathbf{a}_3; \quad \mathbf{a}_1,\ \mathbf{a}_2 \perp 2\mathbf{a}_3 - \mathbf{a}_2. \tag{4.8}$$

The terms for the Bravais lattices corresponding to the cases (4.8) (Fig. 2.12) are *simple monoclinic* ($\Gamma_m$) and *base-centred monoclinic*

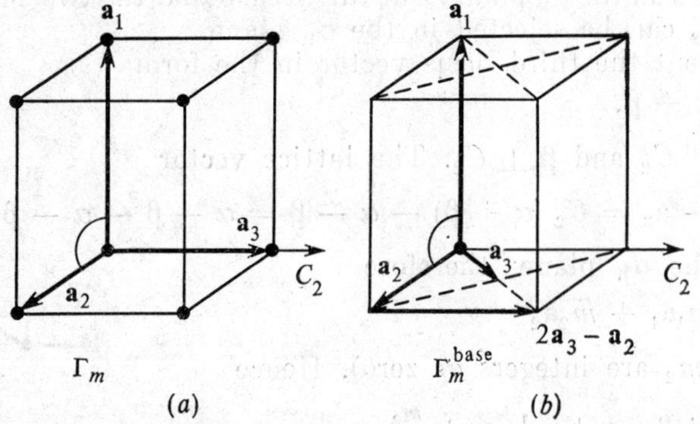


**Fig. 2.12**

($\Gamma_m^{\text{base}}$). We see that for the $\Gamma_m^{\text{base}}$ type the vector corresponding to the centre of the lower base is $\mathbf{a}_3$ and to the upper base $\mathbf{a}_3 + \mathbf{a}_1$. The centres of the other four faces of the parallelepiped considered here do not coincide with the ends of the vectors of the translation group.

Note that for the $\Gamma_m$ type six vectors $\pm\mathbf{a}_1$, $\pm\mathbf{a}_2$, $\pm\mathbf{a}_3$ are invariant under transformations of the $C_{2h}$ group. For the $\Gamma_m^{\text{base}}$ type the vectors invariant under those transformations are $\pm\mathbf{a}_1$, $\pm\mathbf{a}_2$ and $\pm(2\mathbf{a}_3 - \mathbf{a}_2)$.

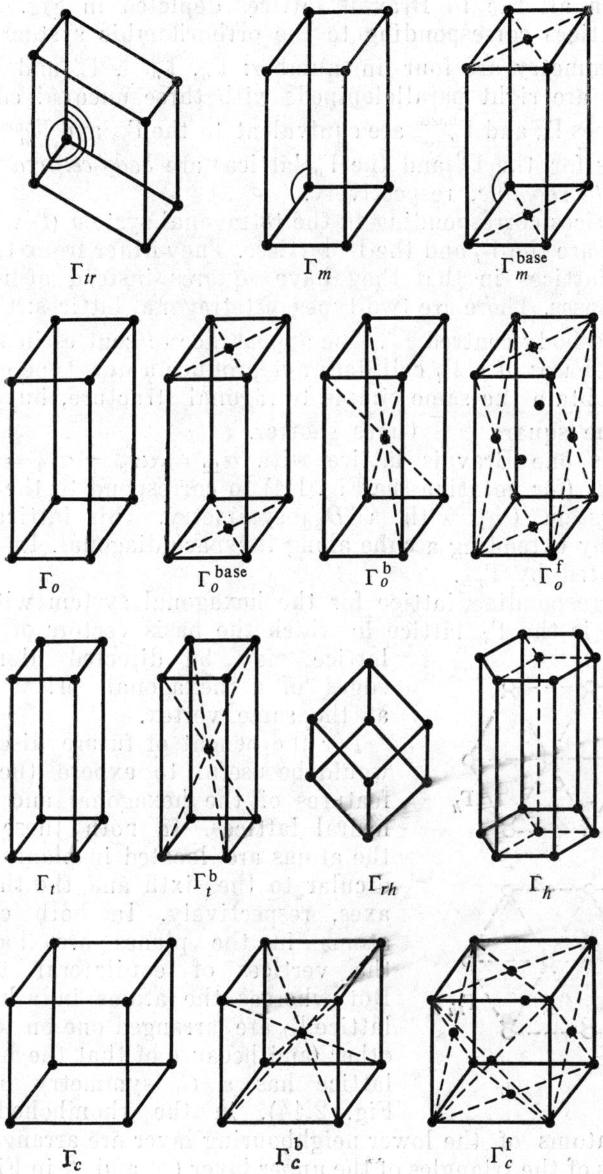
$\Gamma_{tr}$
$\Gamma_m$
$\Gamma_m^{\text{base}}$
$\Gamma_o$
$\Gamma_o^{\text{base}}$
$\Gamma_o^{\text{b}}$
$\Gamma_o^{\text{f}}$
$\Gamma_t$
$\Gamma_t^{\text{b}}$
$\Gamma_{rh}$
$\Gamma_h$
$\Gamma_c$
$\Gamma_c^{\text{b}}$
$\Gamma_c^{\text{f}}$

**Fig. 2.13**

By analyzing the five remaining systems in the same way we can obtain all the 14 Bravais lattices depicted in Fig. 2.13.

The lattices corresponding to the orthorhombic system ($o$) with a $\boldsymbol{D}_{2h}$ symmetry are four in number: $\Gamma_o$, $\Gamma_o^{\text{base}}$, $\Gamma_o^{\text{b}}$ and $\Gamma_o^f$. Their unit cells are right parallelepipeds with three unequal edges. The lattice types $\Gamma_o$ and $\Gamma_o^{\text{base}}$ are equivalent to the $\Gamma_m$ and $\Gamma_m^{\text{base}}$ lattices. The terms for the $\Gamma_o^{\text{b}}$ and the $\Gamma_o^{\text{f}}$ lattices are *body-centred* and *face-centred orthorhombic*, respectively.

The lattices corresponding to the tetragonal system ($t$) with a $\boldsymbol{D}_{4h}$ symmetry are the $\Gamma_t$ and the $\Gamma_t^b$ lattices. They differ from the orthorhombic lattices in that they have squares instead of rectangles for their bases. There are two types of tetragonal lattices: the simple $\Gamma_t$ and the body-centred $\Gamma_t^{\text{b}}$. The appearance of centres in the upper and lower bases of a $\Gamma_t$ cell does not produce a new type of lattice, since we obtain the same simple tetragonal structure, but with the side of the square $\sqrt{2}$ times shorter.

There is one Bravais lattice with $\alpha_{12} = \alpha_{23} = \alpha_{13} = 60°$ and $a_1 = a_2 = a_3$ (for notation see Fig. 1.4) to correspond to the rhombohedral system ($rh$) with a $\boldsymbol{D}_{3d}$ symmetry. This lattice can be obtained by extending a cube along its space diagonal. In Fig. 2.13 it is denoted by $\Gamma_{rh}$.

The corresponding lattice for the hexagonal system with a $\boldsymbol{D}_{6h}$ symmetry is the $\Gamma_h$ lattice in which the basis vectors of a simple lattice may be directed along three edges of a hexagonal prism meeting at the same vertex.

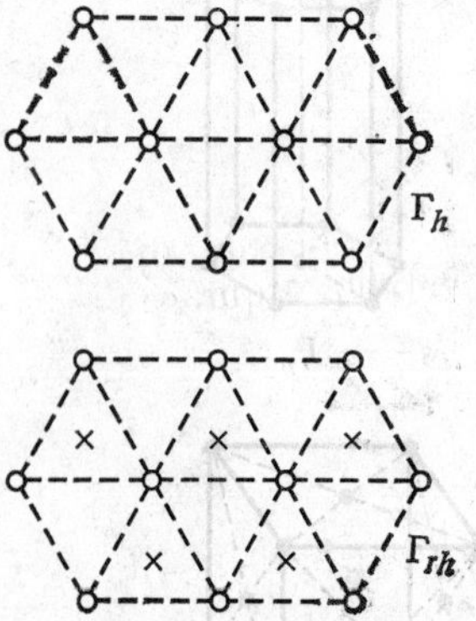

Fig. 2.14

For the benefit of future discussion it would be useful to expose the special features of the hexagonal and rhombohedral lattices. In both these lattices the atoms are located in planes perpendicular to the sixth and the third order axes, respectively. In both cases the atoms in the planes are located at the vertices of equilateral triangles. But whereas the atoms in a hexagonal lattice $\Gamma_h$ are arranged one on top of the other (and because of that the hexagonal lattice has a $C_6$ symmetry axis; see Fig. 2.14), in the rhombohedral lattice the atoms of the lower neighbouring layer are arranged under the centres of the triangles of the upper layer ($\times$ and $\circ$ in Fig. 2.14).

Lastly, as we have already mentioned, three Bravais lattices correspond to the cubic system with an $\boldsymbol{O}_h$ symmetry: the simple cubic $\Gamma_c$, the body-centred $\Gamma_c$, and the face-centred $\Gamma_c^f$ (Fig. 2.13).

In Section 1.1.3 we discussed the problem of how the basis vectors in body-centred and face-centred cubic lattices should be chosen for the primitive cells to contain one atom each.

## 2.5 Crystal Classes. Space Groups

2.5.1 In the preceding section we discussed the symmetry properties of simple lattices. We arrived at the conclusion that there are only 7 crystal systems compatible with the symmetry of simple lattices and 14 types of simple lattices (Bravais lattices). However, all the chemical compounds, as well as some atomic materials, crystallize in complex lattices. In this case the primitive cell of the crystal contains more than one atom, which inevitably reduces the symmetry of the crystal. The simplest pertinent examples have already been cited above. The planar complex lattice in Fig. 1.2*b* has no inversion centre characteristic of the simple lattice obtained from it when an atom located inside the cell is removed.

Consider from this point of view the complex diamond lattice (Ge, Si) depicted in Fig. 1.12. It is well known that this lattice can be obtained by displacing one face-centred cubic lattice with respect to another along the space diagonal *AB* (Fig. 1.12) by one-fourth of its length. The cubic lattice of diamond may be imagined as consisting of eight smaller cubes numbered *I*, ..., *VIII* in Fig. 1.12. As the result of the displacement described above (one-fourth of *AB*) the ● -atoms *1'*, *2'*, *3'* and *4'* take up positions in the centres of the smaller cubes *I*, *III*, *VI*, *VIII*. Choosing the primitive cell as shown in Fig. 1.6*a*, we shall ensure that it contains two atoms.

Let us demonstrate that the symmetry in the diamond lattice is lower than that of a simple face-centred cubic lattice. If the origin of the coordinate frame is placed in a site in the diamond lattice, it will not be invariant under inversion. The $C_4$ axes of the face-centred cubic lattice (Fig. 2.10) in the diamond lattice are transformed into $C_2$ axes; indeed, when the lattice is rotated about the vertical axis passing through the centre of the cube's face through an angle $\pi/2$, the atoms *1'* and *2'*, as well as *3'* and *4'* do not coincide with themselves, this being the case only if the rotation angle is $\pi$. The $C_3$ axes coinciding with the space diagonals of the cube are also the symmetry axes of the diamond lattice; when the lattice is rotated by an angle $2\pi/3$ about the $\overline{AB}$ diagonal the following atoms coincide: $2' \to 4'$, $3' \to 2'$, $4' \to 3'$. Finally, symmetry planes of the diamond lattice are the reflection planes passing through the $C_2$ axis and the two $C_3$ axes (Fig. 2.8*c*). If a reflection plane is drawn through the vertices of the cube *ABCD*, the atoms *1'*, *2'* will be located in this plane, the atoms *3'*, *4'* being arranged symmetrically with respect to it. For this reason this plane will be a plane of symmetry of the diamond lattice. Hence, the diamond lattice has the symmetry of the point group $\boldsymbol{T}_d$, a subgroup of the $\boldsymbol{O}_h$ group.

It is obvious that in all cases the point group obtained as the result of atoms filling the primitive cell of a simple lattice will be a subgroup of the system of the simple lattice (the Bravais lattice).

Because of this the complete set of point groups $\mathscr{F}$ of crystals, termed *crystal classes*, coincides with the set of all possible subgroups contained in the seven systems of the simple Bravais lattices.

We write out all the possible subgroups contained in the seven systems.

| *Crystal Systems* | *Crystal Classes* (*subgroups of systems*) |
|---|---|
| 1) triclinic $\boldsymbol{S}_2 = \boldsymbol{C}_i$ | $E$, $\boldsymbol{C}_i = \{E, J\}$, |
| 2) monoclinic $\boldsymbol{C}_{2h}$ | $\boldsymbol{C}_{1h} = \boldsymbol{C}_s = \{E, \sigma_h\}$, $\boldsymbol{C}_2$, $\boldsymbol{C}_{2h}$, |
| 3) orthorhombic $\boldsymbol{D}_{2h}$ | $\boldsymbol{C}_{2v}$, $\boldsymbol{D}_2$, $\boldsymbol{D}_{2h}$, |
| 4) tetragonal $\boldsymbol{D}_{4h}$ | $\boldsymbol{C}_4$, $\boldsymbol{C}_{4v}$, $\boldsymbol{C}_{4h}$, $\boldsymbol{S}_4$, $\boldsymbol{D}_{2d}$, $\boldsymbol{D}_4$, $\boldsymbol{D}_{4h}$, |
| 5) rhombohedral $\boldsymbol{D}_{3d}$ | $\boldsymbol{C}_3$, $\boldsymbol{S}_6$, $\boldsymbol{C}_{3v}$, $\boldsymbol{D}_3$, $\boldsymbol{D}_{3d}$, |
| 6) hexagonal $\boldsymbol{D}_{6h}$ | $\boldsymbol{C}_6$, $\boldsymbol{C}_{3h}$, $\boldsymbol{C}_{6h}$, $\boldsymbol{C}_{6v}$, $\boldsymbol{D}_{3h}$, $\boldsymbol{D}_6$, $\boldsymbol{D}_{6h}$, |
| 7) cubic $\boldsymbol{O}_h$ | $\boldsymbol{T}$, $\boldsymbol{T}_h$, $\boldsymbol{T}_d$, $\boldsymbol{O}$, $\boldsymbol{O}_h$. (5.1) |

32 crystal classes in all.

In determining crystal classes it usually turns out that each crystal class is a subgroup not of one but of several systems. In all such cases the crystal class is classified as belonging to the system with the lowest symmetry. For instance, the crystal class $\boldsymbol{C}_i = \{E, J\}$ is a subgroup of all the systems, since all the simple lattices have an inversion centre $J$, but we classify $\boldsymbol{C}_i$ under the triclinic system, which has no other symmetry elements. When determining, for example, the classes of the orthorhombic system $\boldsymbol{D}_{2h}$, we take into account that the $\boldsymbol{D}_{2h}$ group consists of eight elements: $E$, $C_2$, $2C_2'$ (horizontal axes of the second order), $2\sigma_v$, $\sigma_h$, $C_2\sigma_h = J$. The subgroups $E$, $\boldsymbol{C}_i$, $\boldsymbol{C}_{1h} = \{E, \sigma_h\}$, and $\boldsymbol{C}_{2h}$ must be classified as belonging to the triclinic and monoclinic systems; therefore, only the crystal classes $\boldsymbol{C}_{2v}$, $\boldsymbol{D}_2$, and $\boldsymbol{D}_{2h}$ are classified as belonging to the orthorhombic system.

The convention used to classify a class as belonging to the system with the lowest symmetry finds its justification in physical considerations. It is highly improbable that the atoms of a crystal belonging to a definite crystal class would constitute a Bravais lattice of higher symmetry than needed for the realization of this class (it may be reasoned that this is connected with the minimum of the crystal's thermodynamic potential). Should this improbable situation be realized, the crystal would be in a metastable state, and a small perturbation (for instance, heating) would cause it to return to the equilibrium state with a less symmetrical

Bravais lattice. Should, for example, a crystal of the $\boldsymbol{D}_4$ class, which can crystallize in the tetragonal system, crystallize in the more symmetrical cubic system, a negligible interaction would be able to extend or contract one of the edges of the cubic cell, turning it into a right prism with a square base.

It is evident from this example that for the principle of crystallization in a Bravais lattice of the lowest symmetry to be realized, there should exist the possibility of continuous transition by means of infinitesimal deformations from a Bravais lattice of higher symmetry to one of lower symmetry.

Comparing Bravais lattices of different symmetries, we can establish that there is one exception where such a continuous transformation is impossible. Namely, there is no such infinitesimal deformation that could transform a hexagonal lattice into a rhombohedral lattice with a lower symmetry. Indeed, the atoms of the neighbouring layers of a rhombohedral lattice are displaced with respect to each other by a finite distance, whereas in the hexagonal lattice they are arranged one on top of the other (see Fig. 2.14). In this case the classification of all the crystal classes under the rhombohedral system is a matter of convention, since all of them, being subgroups of the $\boldsymbol{D}_{6h}$ groups, can crystallize into the hexagonal structure.

Thus, we arrived at the conclusion that any complex crystal can be described in terms of a definite system, a Bravais lattice and a crystal class.

The symmetry of a crystal class involving definite rotations, reflections and inversion obviously determines physically equivalent directions in the crystal. Since this symmetry does not involve discrete translations, we may say that it determines the macroscopic symmetry of an anisotropic continuous medium. A symmetry of this kind determines such physical phenomena as the propagation of light in a crystal, its thermal expansion and its mechanical properties, when acted upon by external forces.

2.5.2 We have studied the connection between the types of simple lattices (Bravais lattices) and their point symmetries. We also discussed the symmetry of directions in crystals (classes) that determines their macroscopic properties. Let us now turn to the study of the full crystal symmetry, i.e., transformations resulting in the coincidence of all atoms of a definite kind. The set of such transformations to which a crystal can be subjected constitutes a group termed *space group*. This symmetry of the crystal, which depends on the arrangement of all its atoms and which it is natural to term *microscopic symmetry*, determines, for example, the scattering of $X$ rays by the crystal.

Every crystal is invariant (if we ignore its surface and defects) under translations by certain basis vectors $\mathbf{a}_1$, $\mathbf{a}_2$ and $\mathbf{a}_3$.

The translation group $\mathscr{T}$

$$\mathbf{a} \equiv \mathbf{a_n} = n_1\mathbf{a}_1 + n_2\mathbf{a}_2 + n_3\mathbf{a}_3, \tag{5.2}$$

where $n_1$, $n_2$, $n_3$ are integers (including zero), is an invariant Abelian subgroup of a space group.

In addition to translations $\mathbf{a_n}$, a space group generally also includes various point symmetry transformations: rotations about simple and rotary-reflection axes of the second, third, fourth and sixth orders, reflections in planes and inversion (which coincides with the rotary-reflection axis of the second order, $S_2$). However, in addition to such point symmetry transformations a space group (i.e., a crystal) can also have the following symmetry elements: *screw axis* and *glide plane*.

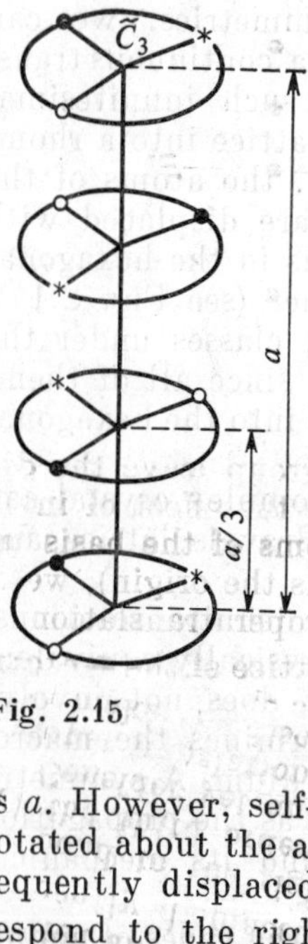


Fig. 2.15

The term screw axis applies to a symmetry transformation of a crystal consisting of two consecutive operations (performed in an arbitrary order): rotation about an axis through the angle $2\pi/n$ with a subsequent displacement along the axis by some integral part of the lattice vector termed *improper translation*.

Figure 2.15 depicts a screw axis of the third order, $\widetilde{C}_3$. Three atoms of different kinds, ●, ○ and *, are arranged on the circumference of a circle perpendicular to the axis at equal distances from one another. The displacement corresponding to proper translation along the axis is $a$. However, self-coincidence also takes place when the crystal is rotated about the axis through the angle $\alpha = 2\pi/3 = 120°$ and subsequently displaced by $a/3$. Since rotation and linear motion correspond to the right screw, the $\widetilde{C}_3$ axis is termed *right screw axis* of the third order.

Consider yet another lattice symmetry element involving an improper translation, the glide plane. Figure 2.16 depicts a planar lattice (to make the discussion more graphic) with a glide plane $PP'$. When the lattice is reflected in the $PP'$ plane, it does not coincide with itself (the $A$ layers coincide but the $B$ layers do not). However, if after reflection the lattice is displaced in the direction $PP'$ by $\mathbf{a}_1/2$, where $\mathbf{a}_1$ is the basis vector in the $PP'$ direction, the lattice will coincide with itself (we can first shift the lattice by $\mathbf{a}_1/2$ and then reflect it in the $PP'$ plane). It can be demonstrated that there can be no other symmetry elements in a crystal involving an improper translation except the two considered above.

Consider the diamond lattice (Fig. 1.12) from the point of view of the existence of *improper symmetry elements*, i.e., symmetry elements involving improper translations. We shall apply the term *basis lattice* to a face-centred cubic lattice consisting of ○-atoms and the term *displaced lattice* to one consisting of ●-atoms.

In Section 2.5.1 we demonstrated that the diamond lattice is invariant with respect to the transformations of the point group $T_d$. Let us investigate the operations $JT_d = T_dJ$ (where $J$ is inversion)

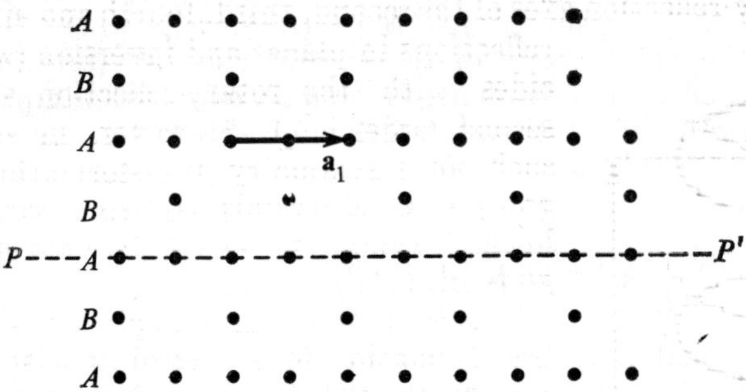


**Fig. 2.16**

for this group. Since the operations of the $T_d$ group leave the diamond lattice invariant, it suffices to investigate the effect of inversion $J$. When inversion is applied to the ○-atoms of the basis lattice (we choose the $A$ atom of the basis lattice as the origin), we reproduce the basis lattice, which after an improper translation by one-fourth of $\overline{AB}$ coincides with the displaced lattice of the ●-atoms. When, on the other hand, inversion is applied to the ●-atoms of the displaced lattice (with the origin remaining at $A$), they occupy "vacant places," but after an improper translation by one-fourth of $\overline{AB}$ they coincide with the ○-atoms of the basis lattice. Hence, the $JT_d$ operations followed by an improper translation by one-fourth of $\overline{AB}$ are also symmetry operations of the diamond lattice.

The reader may be perplexed by the above statement that there are only two improper symmetry operations for a crystal, a screw axis and a glide plane, whereas in the diamond lattice there appears to be a new improper symmetry element involving inversion.

It should, however, be pointed out that the definition of improper symmetry elements is, in a sense, a matter of convention. If we place the origin not at $A$ but between $A$ and a ●-atom $1'$, the diamond lattice is invariant under inversion (not followed by an improper translation). However, this would lead to the appearance of glide planes in the diamond lattice [2.9, Sec. 34].

Finally, we would like to point out that indium antimonide (InSb), which crystallizes in a diamond-like lattice, has no improper symmetry elements, because here the ○- and ●-atoms are of different kinds (In and Sb; see Fig. 1.12). The symmetry of InSb is that of the point group $T_d$.

2.5.3 All possible space groups of crystals are obtained when all the Bravais lattices are matched to all possible rotational symmetry axes, reflection planes, screw axes and glide planes. To gain insight into the peculiarities of such matching, we introduce an operator (element) of the space group $\{R \mid \alpha(R) + \mathbf{a_n}\}$, where $R$ is an element of the point group of the crystal class $\mathscr{F}$ (rotation, rotary-reflection, reflection in a plane, inversion), $\alpha(R)$ is an improper translation (always a rotational fraction of the lattice vector) corresponding to the element $R$, and $\mathbf{a_n}$ is a lattice vector. The action of this operator on the position vector $\mathbf{r}$ is described by the equation[12]

$$\{R \mid \alpha(R) + \mathbf{a_n}\}\,\mathbf{r} = \mathbf{Rr} + \alpha(R) + \mathbf{a_n}. \tag{5.3}$$

Here $\mathbf{R}$ is a matrix of orthogonal rotational transformation, inversion or reflection $R$. For the elements of $R$ for which the improper translation is zero, $\alpha(R) = 0$. For the operator

$$g = \{R \mid \alpha(R) + \mathbf{a_n}\} \tag{5.4}$$

to be an element of the crystal's space group, the set of all elements of type (5.4) must contain the unit element and the inverse elements, and the product of two elements of the set must belong to the same set.

The unit element of the set is obviously $\{E \mid 0\}$, where $E$ is the nuit element of the point group $\mathscr{F}$. Indeed,

$$\{E \mid 0\}\,\mathbf{r} = E\mathbf{r} = \mathbf{r}. \tag{5.5}$$

The inverse element of (5.4) is

$$g^{-1} = \{R \mid \alpha(R) + \mathbf{a_n}\}^{-1} = \{R^{-1} \mid -\mathbf{R}^{-1}\alpha(R) - \mathbf{R}^{-1}\mathbf{a_n}\}. \tag{5.6}$$

Indeed,

$$\begin{aligned} g^{-1}g\mathbf{r} &= \{R^{-1} \mid -\mathbf{R}^{-1}\alpha(R) - \mathbf{R}^{-1}\mathbf{a_n}\}\,\{\mathbf{Rr} + \alpha(R) + \mathbf{a_n}\} \\ &= \mathbf{R}^{-1}\mathbf{Rr} + \mathbf{R}^{-1}\alpha(R) + \mathbf{R}^{-1}\mathbf{a_n} - \mathbf{R}^{-1}\alpha(R) - \mathbf{R}^{-1}\mathbf{a_n} \\ &= \mathbf{r} = \{E \mid 0\}\,\mathbf{r}. \end{aligned} \tag{5.7}$$

---

[12] We use the common designation of the operator (element) of a space group in which $\mathbf{r}$ is operated on first with the operator $R$ to the left of the vertical line and then with the operator $\alpha(R) + \mathbf{a_n}$ (displacement) to the right of the line.

Finally, let us demonstrate under what conditions the product of two elements of the set of elements of type (5.4) belongs to the same set. If $R_1$ and $R_2$ are both elements of the crystal point group $\mathcal{F}$, then

$$\{R_2 \mid \alpha(R_2) + \mathbf{a}_m\} \{R_1 \mid \alpha(R_1) + \mathbf{a}_n\} \mathbf{r}$$
$$= \{R_2 \mid \alpha(R_2) + \mathbf{a}_m\} [(\mathbf{R}_1\mathbf{r} + \alpha(R_1) + \mathbf{a}_n]$$
$$= \mathbf{R}_2\mathbf{R}_1\mathbf{r} + \mathbf{R}_2\alpha(R_1) + \mathbf{R}_2\mathbf{a}_n + \alpha(R_2) + \mathbf{a}_m. \quad (5.8)$$

The first addend in the last line of (5.8) is of the same structure as that of the first addend in the right-hand side of (5.3), since $R_2R_1 = R$, which is an element of the point group $\mathcal{F}$. A strict condition for the elements of the space group (5.4) is the requirement that the remaining four addends in (5.8) be of the structure of (5.3), i.e.,

$$\mathbf{R}_2\alpha(R_1) + \mathbf{R}_2\mathbf{a}_n + \alpha(R_2) + \mathbf{a}_m = \alpha(R) + \mathbf{a}_p, \quad (5.9)$$

where $\alpha(R)$ is an improper translation corresponding to the element $R = R_2R_1$, and $\mathbf{a}_p$ is a lattice vector. Clearly, (5.9) can be satisfied only if there is an adequate matching of the operations of improper translations $\alpha(R_1)$, $\alpha(R_2)$ and $\alpha(R_2, R_1)$ and the basis lattice vectors $\mathbf{a}_1$, $\mathbf{a}_2$, $\mathbf{a}_3$. An accurate analysis carried out in 1891 independently by the Russian crystallographer E. S. Fedorov and the German scientist A. Schonflies demonstrated that altogether there are 230 space groups (see, e.g., [2.10] and [2.11]).

The space groups for which $\alpha(R) = 0$ for all $R$ are termed *simple*, or *symmorphic*; if, on the other hand, $\alpha(R) \neq 0$ even for a single $R$, the group is termed *nonsymmorphic*.

The condition (5.9) for symmorphic groups is simplified thus:

$$\mathbf{R}_2\mathbf{a}_n + \mathbf{a}_m = \mathbf{a}_p. \quad (5.10)$$

If $R_2$ is the operation of rotation about an axis, the condition (5.10) means that the only possible axes are those of the second, third, fourth and sixth orders (see Section 2.4.1).

We can obtain a notion about the number of space groups, if he takes into account that each crystal class in combination with a possible Bravais lattice forms a space group. For example, three classes [see (5.1)] and four Bravais lattices $\Gamma_o$, $\Gamma_o^{base}$, $\Gamma_o^b$, $\Gamma_o^f$ (Fig. 2.13) correspond to the orthorhombic system $D_{2h}$; therefore, for this case we have at least twelve symmorphic groups. Taking into account the fact that the classes of the rhombohedral system may also belong to the hexagonal system, we obtain for the 14 Bravais lattices and the 32 classes (5.1) 61 symmorphic space groups. If, in addition, we take into account that in some cases the symmetry elements of a crystal class can be differently arranged with respect to basis vectors (e.g., the horizontal two-fold axes of the $D_3$ class in a hexagonal structure

Table 2.2

| Crystal systems | Bravais lattices | Point groups | Number of space groups | Total number of space groups |
|---|---|---|---|---|
| Triclinic | Triclinic | $E$<br>$C_i$ | 1<br>1 | 2 |
| Monoclinic | Simple monoclinic, base-centred monoclinic | $C_{1h}$<br>$C_2$<br>$C_{2h}$ | 4<br>3<br>6 | 13 |
| Orthorhombic | Simple orthorhombic, base-centred orthorhombic, face-centred orthorhombic, body-centred orthorhombic | $C_{2v}$<br>$D_2$<br>$D_{2h}$ | 22<br>9<br>28 | 59 |
| Tetragonal | Simple tetragonal, body-centred tetragonal | $C_4$<br>$C_{4v}$<br>$C_{4h}$<br>$S_4$<br>$D_{2d}$<br>$D_4$<br>$D_{4h}$ | 6<br>12<br>6<br>2<br>12<br>10<br>20 | 68 |
| Rhombohedral or trigonal | Trigonal | $C_3$<br>$S_6$<br>$C_{3v}$<br>$D_3$<br>$D_{3d}$ | 4<br>2<br>6<br>7<br>6 | 25 |
| Hexagonal | Hexagonal | $C_6$<br>$C_{3h}$<br>$C_{6h}$<br>$C_{6v}$<br>$D_{3h}$<br>$D_6$<br>$D_{6h}$ | 6<br>1<br>2<br>4<br>4<br>6<br>4 | 27 |
| Cubic | Simple cubic, face-centred cubic, body-centred cubic | $T$<br>$T_h$<br>$T_d$<br>$O$<br>$O_h$ | 5<br>7<br>6<br>8<br>10 | 36 |
| Total | 14 | 32 | 230 | 230 |

can point either to the vertices of the hexagon or to the midpoints of its sides), then the total number of symmorphic groups rises to 73.

Screw axes and glide planes present extensive additional possibilities for the construction of space groups; the number of nonsymmorphic groups turns out to be 157, so that the total number of space groups will be $73 + 157 = 230$.

Table 2.2 shows the distribution of the space groups over crystal systems and crystal classes.

Let us represent the element of a space group (5.4) as a product of two operators:

$$g = \{R \mid \boldsymbol{\alpha}(R) + \mathbf{a_n}\} = \{E \mid \mathbf{a_n}\}\ \{R \mid \boldsymbol{\alpha}(R)\}, \tag{5.11}$$

where $E$ is the unit element of the point group $\mathscr{F}$ with the elements $R$. Although the transformations $\{R \mid \boldsymbol{\alpha}(R)\}$ are the symmetry elements of the crystal, they do not constitute a group. Indeed, their product may produce a lattice vector $\mathbf{a}$ belonging to the translation group $\mathscr{T}$.

For example, by performing three times a transformation corresponding to the screw axis depicted in Fig. 2.15, we obtain a displacement along the axis by the length of the basis vector $\mathbf{a}$. Similarly, applying twice the operator $\{R \mid a_1/2\}$, where $R$ is a reflection in the $PP'$ plane (Fig. 2.16), we obtain a translation by the basis vector $\mathbf{a}_1$.

At the same time the totality of all orthogonal transformations $R$ (including those for which $\boldsymbol{\alpha}(R) \neq 0$) constitutes a point group $\mathscr{F}$ that determines the class of the crystal. Indeed, it follows from (5.8) that, if $R_1$ and $R_2$ are orthogonal transformations of the crystal's symmetry (with zero or nonzero improper translations), $R_2R_1 = R$ will be an orthogonal transformation of the crystal's symmetry (with $\boldsymbol{\alpha}(R) = 0$ or $\boldsymbol{\alpha}(R) \neq 0$).

Hence, to determine the class of the crystal we have to take into account all the symmetry axes and planes, replacing the screw axes and glide planes by equivalent simple axes and reflection planes.

## 2.6 Irreducible Representations of Groups and the Theory of Characters

2.6.1 Applications of group theory in physics and, in particular, solid state physics is based on the theory of irreducible representations and on the theory of characters.[13]

Consider a group $G$ consisting of $h$ elements: $A \equiv E, B, C, \ldots, S, T, \ldots$. To be definite, let us imagine that this is a point symmetry group, every element of which corresponds to some transformation of the body's coordinates $x$, $y$, $z$. Take an arbitrary single-valued

[13] Information about matrices and their properties required for understanding this section is contained in Appendix 3.

function $\psi_1(x, y, z) = \psi_1(\mathbf{x})$ and operate on it with the operator $\hat{P}_R$ ($R$ is one of the group's elements) defined as follows:

$$\hat{P}_R\psi_1(x, y, z) = \psi_1(\mathbf{R}^{-1}\mathbf{x}) = \Phi_R(\mathbf{x}). \tag{6.1}$$

Here $R^{-1}$ is an element inverse to $R$; $\mathbf{R}^{-1}\mathbf{x}$ is an orthogonal transformation of the coordinates corresponding to the element $R^{-1}$. At first it would seem more natural to define the operator $\hat{P}_R$ as follows:

$$\hat{P}'_R\psi_1(\mathbf{x}) = \psi_1(\mathbf{R}\mathbf{x}). \tag{6.1a}$$

We shall demonstrate, however, that the operators $\hat{P}_R$ constitute a group isomorphic to the group **G**, i.e.,

$$\hat{P}_S\hat{P}_R = \hat{P}_{SR}, \tag{6.1b}$$

only if defined as in (6.1). Indeed, it follows from the definition of $\hat{P}_R$ (6.1) that

$$\hat{P}_S\hat{P}_R\psi_1(\mathbf{x}) = P_S\Phi_R(\mathbf{x}) = \Phi_R(\mathbf{S}^{-1}\mathbf{x}) = \psi_1(\mathbf{R}^{-1}\mathbf{S}^{-1}\mathbf{x})$$
$$= \psi_1[(\mathbf{SR})^{-1}\mathbf{x}] = \hat{P}_{SR}\psi_1(\mathbf{x}),$$

whence (6.1b) follows directly. In the case of the definition (6.1a) we would obtain $\hat{P}'_S\hat{P}'_R = \hat{P}'_{RS}$, and this is not always convenient.[14] If instead of $R$ in (6.1) we take the other elements of $G$, we shall not generally obtain $h$ linearly independent functions $\Phi_R(\mathbf{x})$. The number of linearly independent functions $r$ will, in general, be less or equal to $h$; we denote them by $\psi_1(\mathbf{x}), \psi_2(\mathbf{x}), \ldots, \psi_r(\mathbf{x})$, where $\psi_1(\mathbf{x})$ is one of such functions (since $R$ may be equal to $E$). It is always possible, by applying an appropriate linear transformation, to make the functions $\psi_i(\mathbf{x})$ $(i = 1, 2, \ldots, r)$ orthogonal and normalized, and in the future we shall assume them to be such. The terms for the functions $\psi_i$ are: *basis functions* or simply *basis*. For a different choice of $\psi_1(\mathbf{x})$, the number of basis functions $r$ generally changes. The application of $\hat{P}_S$ to one of the basis functions, e.g., $\psi_i(\mathbf{x})$, is expressed in the form of a linear superposition of basis functions (this follows from (6.1) if $\hat{P}_S$ is applied to both sides of the equality and if account is taken of the fact that $\Phi_R(\mathbf{x})$ is a linear combination of the functions $\psi_i(\mathbf{x})$:

$$\hat{P}_S\psi_i = \sum_{k=1}^{r}\Gamma(S)_{ki}\,\psi_k, \tag{6.2}$$

where the $\Gamma(S)_{ki}$ are the elements of the $r$-rank matrix $\Gamma(S)$[15].

---

[14] Some authors (e. g., see [2.1, p. 157]) use definition (6.1a).

[15] The conjugation of the number of the basis function with the first index of the matrix in the sum (6.2) makes it easier to establish the correspondence between the matrices and the group elements (see below).

Premultiplying both parts of (6.2) by $\psi_l^* (\mathbf{x})$, integrating with respect to $\mathbf{x}$ ($d\mathbf{x} = dx\, dy\, dz$) and making use of the orthonormality of the basis functions, we find that

$$\int \psi_l^* (\mathbf{x}) \hat{P}_S \psi_i (\mathbf{x})\, d\mathbf{x} = \sum_{k=1}^{r} \Gamma (S)_{ki} \int \psi_l^* (\mathbf{x}) \psi_k (\mathbf{x})\, d\mathbf{x}$$

$$= \sum_{k=1}^{r} \Gamma (S)_{ki}\, \delta_{lk} = \Gamma (S)_{li}. \qquad (6.3)$$

Hence, the matrix elements of $\Gamma (S)$ in (6.2) are the matrix elements of the operator $\hat{P}_S$ expressed in terms of the basis functions $\psi_i (\mathbf{x})$. If the operator $\hat{P}_T$ is applied to both sides of (6.2), then

$$\hat{P}_T \hat{P}_S \psi_i = \hat{P}_{TS} \psi_i = \sum_l \Gamma (TS)_{li} \psi_l = \sum_k \Gamma (S)_{ki} \hat{P}_T \psi_k$$

$$= \sum_k \Gamma (S)_{ki} \sum_l \Gamma (T)_{lk}\, \psi_l = \sum_l \sum_k \Gamma (T)_{lk}\, \Gamma (S)_{ki}\, \psi_l$$

$$= \sum_l [\Gamma (T)\, \Gamma (S)]_{li}\, \psi_l. \qquad (6.4)$$

Here we made use of the matrix multiplication rule. The first row has been obtained as the result of the application of $\hat{P}_T$ to the left-hand side of (6.2) and the second to the right-hand side of (6.2).

Comparing the third member in the chain with the last, we obtain

$$\Gamma (TS)_{li} = [\Gamma (T)\, \Gamma (S)]_{li}, \qquad (6.5)$$

or

$$\Gamma (TS) = \Gamma (T)\, \Gamma (S). \qquad (6.5a)$$

We see that the matrix corresponding to the product $TS$ is equal to the product of the matrices for $T$ and $S$. The matrices $\Gamma (S)$ in (6.2), where $S$ is an arbitrary element of the group, multiplied in accordance with the same multiplication table as the group's elements, are termed the *representation* of the group. If all the matrices of a representation are different, the representation is termed *true*, or *faithful*; otherwise it is termed *untrue*. In the first case the matrix representation constitutes a group isomorphic to the initial group; in the second the initial group is homomorphic to the matrix representation.

If we make use of the definition (6.1a) for the operator $\hat{P}'_R$, instead of (6.5a) we shall obtain

$$\Gamma (TS) = \Gamma (S)\, \Gamma (T), \qquad (6.5b)$$

i.e., the multiplication of the representation matrices takes place in the reverse order to that of the group's elements (this is because, as was noted above, in the case (6.1a) $\hat{P}'_{TS} = \hat{P}'_S \hat{P}'_T$). However, transposed representation matrices, even if $\hat{P}'_R$ is defined as in (6.1a), are multiplied in accordance with (6.5a). This follows from (6.5b):

$$\tilde{\Gamma}(TS) = \widetilde{\Gamma(S)\Gamma(T)} = \tilde{\Gamma}(T)\tilde{\Gamma}(S), \tag{6.5c}$$

where we made use of (A3.30). We shall see in Section 2.7 how representations and basis functions naturally appear in the study of the symmetry of quantum mechanical systems.

A unit matrix

$$\Gamma(E) = \begin{pmatrix} 1 & 0 & \dots & 0 \\ 0 & 1 & \dots & 0 \\ \cdot & \cdot & \cdot & \cdot \\ 0 & 0 & \dots & 1 \end{pmatrix} = [\delta_{ik}], \tag{6.6}$$

where $\delta_{ik} = 1$ for $i = k$ and $\delta_{ik} = 0$ for $i \neq k$, corresponds to the unit element $E$ of a group. Indeed,

$$\Gamma(E)\Gamma(S) = \Gamma(S), \tag{6.7}$$

which follows from

$$[\Gamma(E)\Gamma(S)]_{ik} = \sum_l \Gamma(E)_{il}\Gamma(S)_{lk} = \sum_l \delta_{il}\Gamma(S)_{lk} = \Gamma(S)_{ik},$$

this coinciding with (6.7), i.e., we have demonstrated that the matrix element of the product of matrices on the left-hand side of (6.7) is equal to the corresponding matrix element of the matrix $\Gamma(S)$.

If $Q^{-1}Q = E$, i.e., $Q^{-1}$ is an element inverse to $Q$, then

$$\Gamma(Q^{-1}) = \Gamma(Q)^{-1}, \tag{6.8}$$

i.e., the matrix of an inverse element is the inverse matrix of the direct element. Indeed, $\Gamma(Q^{-1})\Gamma(Q) = \Gamma(E)$; multiplying both parts by the inverse matrix $\Gamma(Q)^{-1}$, we obtain (6.8). It follows from (6.8) that a group can be represented only by matrices that have inverse matrices, i.e., only by nonsingular matrices (the determinant of such matrices must not be zero).

If all the matrices of a representation are subjected to a *similarity transformation* (see Appendix 3, item 4),

$$S^{-1}\Gamma(A)S = \Gamma(A)', \quad S^{-1}\Gamma(B)S = \Gamma(B)',$$
$$S^{-1}\Gamma(C)S = \Gamma(C)^1, \dots, \tag{6.9}$$

where $S$ is a (nonsingular) matrix, then the new (primed) matrices will also be a representation of the group. Indeed,

$$\Gamma(B)'\Gamma(C)' = S^{-1}\Gamma(B)SS^{-1}\Gamma(C)S$$
$$= S^{-1}\Gamma(B)\Gamma(C)S = S^{-1}\Gamma(BC)S = \Gamma(BC)', \tag{6.10}$$

where in the second equality the use is made of the fact that $SS^{-1} = E$. Since the choice of the matrix $S$ is arbitrary, it is obvious that there cannot be much difference between the representations (6.9) and that they cannot convey different information. We shall apply the term *equivalent* to all such representations obtained with the aid of a similarity transformation and shall regard them as one. It is important in this connection to note that the trace of a matrix does not change when a similarity results:

$$\begin{aligned} \operatorname{Tr} \Gamma (A)' &= \operatorname{Tr} [S^{-1}\Gamma (A) S] = \operatorname{Tr} [\Gamma (A) SS^{-1}] \\ &= \operatorname{Tr} [\Gamma (A) E] = \operatorname{Tr} \Gamma (A), \end{aligned} \tag{6.11}$$

where we made use of the fact that the trace of a product of two matrices is independent of the order in which they are multiplied. Hence, the matrices of all the equivalent representations for a given element of the group have an equal trace.

Let us demonstrate that, if the basis functions are orthonormal, i.e., if

$$(\psi_i, \psi_k) = \int \psi_i^*(\mathbf{x}) \psi_k(\mathbf{x}) d\mathbf{x} = \delta_{ik}, \tag{6.12}$$

then the matrices of the representation $\Gamma(R)$ are unitary matrices. Indeed,

$$\begin{aligned} \delta_{ik} &= (\psi_i, \psi_k) = (\hat{P}_R\psi_i, \hat{P}_R\psi_k) = (\sum_r \Gamma(R)_{ri}\psi_r, \sum_s \Gamma(R)_{sk}\psi_s) \\ &= \sum_{r,s} \Gamma^*(R)_{ri}\Gamma(R)_{sk}(\psi_r, \psi_s) = \sum_{r,s} \Gamma^*(R)_{ri}\Gamma(R)_{sk}\delta_{rs} \\ &= \sum_r \Gamma^*(R)_{ri}\Gamma(R)_{rk} = \sum_r \Gamma^+(R)_{ir}\Gamma(R)_{rk} = [\Gamma^+(R)\Gamma(R)]_{ik}. \end{aligned}$$

When effecting the transition from the second to the third equality of the chain we made use of the property of the scalar product $(\psi_i, \psi_k)$ remaining invariant under an orthogonal transformation $\hat{P}_R$ of the integration variables (the Jacobian of the transformation is unity); in going from the fifth to the sixth equality we profited by the orthonormality of the functions $\psi_i(\mathbf{x})$; finally, in the step from the seventh to the eighth equality we made use of the definition of a conjugate matrix. Comparing the first equation in the chain with the last, we obtain

$$\Gamma^+(R)\Gamma(R) = E, \tag{6.13}$$

i.e., the $\Gamma(R)$ matrices are unitary.

It can easily be demonstrated (we shall not go into it) that, if the old basis functions are subjected to an arbitrary linear transformation

$$\psi_i' = \sum_k S_{ki}\psi_k, \tag{6.14}$$

the new (primed) representation will be

$$\Gamma'(R) = \mathbf{S}^{-1}\Gamma(R)\,\mathbf{S}, \tag{6.15}$$

i.e., will be equivalent to the old.

It may easily be seen (by a direct check) that the group of the equilateral triangle $D_3$ (Section 2.2, Table 2.1) has the following (nonequivalent) representations. (1) The trivial unit representation, when each element of the group is related to unity; the group $D_3$ is homomorphic to this representation of dimension one (all groups have such a representation). (2) A representation of dimension one in the form

$$\Gamma(E) = 1, \quad \Gamma(A) = \Gamma(B) = \Gamma(C) = -1, \quad \Gamma(D) = \Gamma(F) = 1,$$

which can easily be checked with the aid of Table 2.1. (3) An isomorphic representation of dimension two:

$$\Gamma(E) = \begin{pmatrix} 1 & 0 \\ 0 & 1 \end{pmatrix}, \quad \Gamma(A) = \begin{pmatrix} 1 & 0 \\ 0 & -1 \end{pmatrix}, \quad \Gamma(B) = \begin{pmatrix} -1/2 & \sqrt{3}/2 \\ \sqrt{3}/2 & 1/2 \end{pmatrix},$$

$$\Gamma(C) = \begin{pmatrix} -1/2 & -\sqrt{3}/2 \\ -\sqrt{3}/2 & 1/2 \end{pmatrix}, \quad \Gamma(D) = \begin{pmatrix} -1/2 & \sqrt{3}/2 \\ -\sqrt{3}/2 & -1/2 \end{pmatrix},$$

$$\Gamma(F) = \begin{pmatrix} -1/2 & -\sqrt{3}/2 \\ \sqrt{3}/2 & -1/2 \end{pmatrix}. \tag{6.16}$$

Indeed, in accordance with the matrix multiplication rule and Table 2.1 we have, for example,

$$\Gamma(A)\,\Gamma(D) = \begin{pmatrix} 1 & 0 \\ 0 & -1 \end{pmatrix} \begin{pmatrix} -1/2 & \sqrt{3}/2 \\ -\sqrt{3}/2 & -1/2 \end{pmatrix} = \begin{pmatrix} -1/2 & \sqrt{3}/2 \\ \sqrt{3}/2 & 1/2 \end{pmatrix} = \Gamma(B),$$

$$\Gamma(F)\,\Gamma(D) = \begin{pmatrix} -1/2 & -\sqrt{3}/2 \\ \sqrt{3}/2 & -1/2 \end{pmatrix} \begin{pmatrix} -1/2 & \sqrt{3}/2 \\ -\sqrt{3}/2 & -1/2 \end{pmatrix} = \begin{pmatrix} 1 & 0 \\ 0 & 1 \end{pmatrix} = \Gamma(E).$$

Note that all the matrices (6.16) are unitary. Since all the matrices of the representation (6.16) are different, this representation is termed true.

Should we choose an arbitrary nonsingular matrix S of rank 2 and apply (6.9), we would be able to write the matrix representation (6.16) in a quite different (but actually equivalent) form.

The two-dimensional representation (6.16) can be built if use is made of the basis functions

$$\psi_1 = \sqrt{\frac{2}{\pi}}\, x \exp\left[-\frac{1}{2}(x^2+y^2)\right],$$

$$\psi_2 = \sqrt{\frac{2}{\pi}}\, y \exp\left[-\frac{1}{2}(x^2+y^2)\right].$$

To find the matrices $\Gamma(S)_{ik}$ in (6.2), we subject the coordinates $x$, $y$ to transformations of group $D_3$ (Sec. 2.2.2); note that the factor $\sqrt{2/\pi}\exp[-(x^2+y^2)/2]$ remains invariant under all the transformations of $D_3$. In the process we can either rotate the triangle assuming the coordinate axes $x$, $y$ to be at rest (so the operations $A$, $B$, $C$, . . . are defined), or rotate the frame (in the reverse direction) assuming the triangle to be at rest (this is more convenient in our case). It follows from Fig. 2.1 that the transformations corresponding to the symmetry operations of group $D_3$ are those of (6.2)

$$\hat{P}_A\psi_1\colon\quad \hat{A}^{-1}x = x, \qquad P_A\psi_2\colon\quad \hat{A}^{-1}y = -y,$$

or

$$\hat{P}_A\begin{pmatrix}\psi_1\\ \psi_2\end{pmatrix} = \begin{pmatrix}1 & 0\\ 0 & -1\end{pmatrix}\begin{pmatrix}\psi_1\\ \psi_2\end{pmatrix}, \qquad \Gamma(A) = \begin{pmatrix}1 & 0\\ 0 & -1\end{pmatrix},$$

$$\hat{P}_B\psi_1\colon\ \hat{B}^{-1}x = -\frac{1}{2}x + \frac{\sqrt{3}}{2}y, \qquad \hat{P}_B\psi_2\colon\ \hat{B}^{-1}y = \frac{\sqrt{3}}{2}x + \frac{1}{2}y,$$

or

$$\hat{P}_B\begin{pmatrix}\psi_1\\ \psi_2\end{pmatrix} = \begin{pmatrix}-1/2 & \sqrt{3}/2\\ \sqrt{3}/2 & 1/2\end{pmatrix}\begin{pmatrix}\psi_1\\ \psi_2\end{pmatrix}, \qquad \Gamma(B) = \begin{pmatrix}-1/2 & \sqrt{3}/2\\ \sqrt{3}/2 & 1/2\end{pmatrix}.$$

In the same way we can consider the transformations corresponding to the elements $C$, $D$, $F$ of the group, i.e., construct all the matrices (6.16).

**2.6.2** Suppose we have two non-Euclidean representations, generally of different dimensions $r$ and $s$:

$$\Gamma_1(A),\quad \Gamma_1(B),\quad \Gamma_1(C),\ \ldots,\quad \Gamma_1(P),\ \ldots,$$
$$\Gamma_2(A),\quad \Gamma_2(B),\quad \Gamma_2(C),\ \ldots,\quad \Gamma_2(P),\ \ldots \qquad (6.17)$$

Construct *block* or *quasi-diagonal*, matrices of rank $(r+s)$ from matrices $\Gamma_1$ of rank $r$ and matrices $\Gamma_2$ of rank $s$:

$$\Gamma(A)=\left(\begin{array}{c|c}\Gamma_1(A) & 0\\ \hline 0 & \Gamma_2(A)\end{array}\right),\quad \Gamma(B)=\left(\begin{array}{c|c}\Gamma_1(B) & 0\\ \hline 0 & \Gamma_2(B)\end{array}\right),$$
$$\Gamma(C)=\left(\begin{array}{c|c}\Gamma_1(C) & 0\\ \hline 0 & \Gamma_2(C)\end{array}\right),\quad \text{etc.,} \tag{6.18}$$

where the zeros in the upper right corner fill in a rectilinear block with $s$ columns and $r$ rows, and the zeros in the lower left corner a rectilinear block with $r$ columns and $s$ rows. Demonstrate that the matrices (6.18) are also a representation of the group, i.e., if $BA = C$, then

$$\Gamma(B)\,\Gamma(A) = \Gamma(BA) = \Gamma(C).$$

Indeed, it follows from the equation that determines the matrix elements of a matrix equal to the product of two other matrices that

$$[\Gamma(B)\,\Gamma(A)]_{ik}=\sum_l \left(\begin{array}{c|c}\Gamma_1(B) & 0\\ \hline 0 & \Gamma_2(B)\end{array}\right)_{il}\left(\begin{array}{c|c}\Gamma_1(A) & 0\\ \hline 0 & \Gamma_2(A)\end{array}\right)_{lk},$$

where the sum in the right-hand side is calculated in accordance with the diagram

$$[\quad]_{ik}=i\,(***)\overset{k}{\begin{pmatrix}*\\ *\\ *\end{pmatrix}},$$

i.e., so that the elements of the $i$-th row of the first matrix are multiplied by the elements of the $k$-th column of the second. Evidently, as long as $i$ changes from 1 to $r$, $k$ also changes from 1 to $r$ (since the values corresponding to greater $k$'s are zeros). Thus the upper left block $\Gamma_1(B)$ is multiplied by the upper left block $\Gamma_1(A)$; a similar situation exists in case of the lower right blocks. Hence

$$\Gamma(B)\,\Gamma(A)=\left(\begin{array}{c|c}\Gamma_1(B)\,\Gamma_1(A) & 0\\ \hline 0 & \Gamma_2(B)\,\Gamma_2(A)\end{array}\right)$$
$$=\left(\begin{array}{c|c}\Gamma_1(C) & 0\\ \hline 0 & \Gamma_2(C)\end{array}\right)=\Gamma(C),$$

which is what had to be proved.

If we now subject the $\Gamma$ matrices in the representation (6.18) to some similarity transformation (with the aid of a matrix of rank $(r+s)$), the $\Gamma$ matrices will lose their quasi-diagonal (block) appearance, although they will still be a representation of the group. By subjecting this equivalent representation to an inverse similarity transformation we return it to the quasi-diagonal (block) form.

Hence we arrive at the conclusion that there are cases when, by subjecting all the matrices to a similarity transformation, we can reduce them to a quasi-diagonal (block) form. In this case the representation is termed *reducible*. In some cases, however, no similarity transformation can reduce the matrices to a quasi-diagonal form. Such representations are termed *irreducible*.

The existence of irreducible representations in the form of matrices of rank 2 and higher appears quite natural for non-Abelian groups, because in this case noncommutativity of matrix multiplication may (but not necessarily) correspond to the noncommutativity of multiplication of the group's elements. The dimension of all the irreducible representations of Abelian groups should be unity.

Irreducible representations of groups play a major part in the application of group theory to physical problems. As will be explained below, the number of irreducible representations of a group is equal to the number of its classes.

2.6.3 The following *theorem of orthogonality* holds for irreducible, nonequivalent, unitary representations (e.g., see [2.12, Chap. 3, Sec. 8]):

$$\sum_{R} = \Gamma_i^* (R)_{\mu\nu}\, \Gamma_j (R)_{\alpha\beta} = \frac{h}{l_i}\, \delta_{ij}\delta_{\mu\alpha}\delta_{\nu\beta}. \tag{6.18}$$

Here the summation is performed over all the $h$ elements of the group ($R = A_1, A_2, A_3, \ldots, A_h$); $i$ and $j$ are the numbers of the irreducible representations $l_i$, is the dimension of the $i$-th representation (since the right-hand side of (6.19) is nonzero only if $i = j$, we can write $l_j$ instead of $l_i$); $\Gamma_i^* (R)_{\mu\nu}$ is the complex conjugate $\mu\nu$-th element of the matrix of the $i$-th irreducible representation for element $R$ of the group.

The right-hand side of (6.19) is nonzero only for $i = j$, $\mu = \alpha$, and $\nu = \beta$. In this case

$$\sum_{R} |\Gamma_i (R)_{\alpha\beta}|^2 = \frac{h}{l_i}. \tag{6.20}$$

We see that the left-hand side is independent of $\alpha$ and $\beta$. It follows from (6.19) that

$$\sum_{R} \Gamma_i^* (R)_{\mu\nu}\, \Gamma_j (R)_{\mu\nu} = 0 \quad \text{for } i \neq j$$

and

$$\sum_{R} \Gamma_i^* (R)_{\mu\nu}\, \Gamma_i (R)_{\alpha\beta} = 0 \quad \text{for } \mu \neq \alpha \text{ or } \nu \neq \beta \tag{6.21}$$

It is evident from (6.21) that the matrix elements $\Gamma_i(A_1)_{\mu\nu}$, $\Gamma_i(A_2)_{\mu\nu}, \ldots, \Gamma_i(A_h)_{\mu\nu}$ for all $h$ elements of the group may be treated as the components of an $h$-dimensional vector orthogonal to any of the vectors with other indices $\mu$ and $\nu$, as well as to any of the similar vectors of another $j$-th irreducible representation.

Similarly, in the usual three-dimensional space the orthogonality of three- and two-dimensional vectors is written in the form

$$\sum_{i=1}^{3} a_i b_i = \mathbf{a}\cdot\mathbf{b} = 0, \qquad i = x, y, z,$$

$$\sum_{i=1}^{2} a_i b_i = \mathbf{a}\cdot\mathbf{b} = 0, \qquad i = x, y.$$

If there are altogether $s$ such irreducible representations, the total number of mutually orthogonal vectors will be $\sum_{i=1}^{s} l_i^2$ since for the $i$-th representation the number of the matrix components is $l_i^2$. But it is impossible to draw more than $h$ mutually orthogonal vectors in an $h$-dimensional space, and, therefore, $\sum_{i=1}^{s} l_i^2 \leqslant h$. It can be demonstrated (we shall not dwell on this here) that in this respect a limiting equality (see [2.13, p. 107]) is valid:

$$\sum_{i=1}^{s} l_i^2 = h. \tag{6.22}$$

We introduce the important concept of the trace (the sum of the diagonal elements) of a representation matrix; in this case the trace is termed *character* and is denoted by $\chi_i(R)$. We have

$$\chi_i(R) = \operatorname{Tr}\Gamma_i(R) = \sum_{\mu=1}^{l_i} \Gamma_i(R)_{\mu\mu}. \tag{6.23}$$

Here $l_i$ is the dimension of the $i$-th representation. From (6.23) the character of the unit element is

$$\chi_i(E) = \sum_{\mu=1}^{l_i} \Gamma_i(E)_{\mu\mu} = \sum_{\mu=1}^{l_i} 1 = l_i, \tag{6.24}$$

i.e., the character of the irreducible representation of the unit element of a group is equal to the dimension of the representation.

It follows from (6.11) that the character (the trace of a representation) does not change in the process of a similarity transformation. For this reason the characters of equivalent representations coincide. Since the representations of the elements belonging to the same class of a group are related by a similarity transformation (see (2.3)), it follows that $\Gamma(B) = \Gamma(V)^{-1}\,\Gamma(A)\,\Gamma(V)$, and the characters of all elements of a class are identical. Therefore, we may write

$$\chi_i(R) = \chi_i(\mathbb{C}_k), \quad \text{if } R \in \mathbb{C}_k, \tag{6.25}$$

where $\mathbb{C}_k$ is the symbol for the $k$-th class.

The following theorem of the *orthogonality of characters* is important for applications:

$$\sum_R \chi_i^*(R)\,\chi_j(R) = h\delta_{ij}, \tag{6.26}$$

or, taking (6.25) into account,

$$\sum_k N_k \chi_i^*(\mathbb{C}_k)\,\chi_j(\mathbb{C}_k) = h\delta_{ij}, \tag{6.27}$$

where $N_k$ is the number of elements in the $\mathbb{C}_k$ class, and the summation is performed over all the $k$ classes. To prove (6.26), in (6.19) we set $\mu = \nu$ and $\alpha = \beta$; then $\delta_{\mu\alpha}\delta_{\nu\beta} = \delta_{\mu\alpha}\delta_{\mu\alpha} = \delta_{\mu\alpha}$ and

$$\sum_R \Gamma_i^*(R)_{\mu\mu}\Gamma_j(R)_{\alpha\alpha} = \frac{h}{l_i}\,\delta_{ij}\delta_{\mu\alpha}.$$

Taking the sum of both sides of the equality over $\mu$ and $\alpha$ we obtain

$$\sum_R \sum_\mu \Gamma_i^*(R)_{\mu\mu} \sum_\alpha \Gamma_j(R)_{\alpha\alpha} = \frac{h}{l_i}\,\delta_{ij} \sum_{\mu\alpha} \delta_{\mu\alpha}.$$

Making use of the definition of characters (6.23), bearing in mind that $\sum_{\mu\alpha} \delta_{\mu\alpha} = \sum_\mu 1 = l_i$ (or $\sum_{\mu\alpha} \delta_{\mu\alpha} = l_j$, which makes no difference), and canceling $l_i$ in the right-hand side, we obtain (6.26). From (6.27) it follows that

$$\sum_k N_k \chi_i^*(\mathbb{C}_k)\,\chi_j(\mathbb{C}_k) = \begin{cases} 0 & \text{if } i \neq j, \\ h & \text{if } i = j. \end{cases} \tag{6.28}$$

Note that this expression can serve as a test of irreducibility of a representation. A representation $\Gamma_i$ is irreducible if and only if

$$\sum_k N_k |\chi_i(\mathbb{C}_k)|^2 = h. \tag{6.29}$$

If a representation $\Gamma(R)$ contains, for example, two irreducible representations $I$ and $II$, then obviously for it $\chi(R) = \chi^{I}(R) + \chi^{II}(R)$ [see (6.32) below]. Therefore, the sum $\sum_R |\chi(R)|^2$ for it is $2h$ if the representations $I$ and $II$ are nonequivalent and their characters are orthogonal, and $4h$ if they are equivalent. This means that (6.29) is the necessary and sufficient condition of irreducibility.

Now we shall consider orthonormal characters $\sqrt{N_1}\chi_i(\mathbb{C}_1)$, $\sqrt{N_2}\chi_i(\mathbb{C}_2), \ldots, \sqrt{N_k}\chi_i(\mathbb{C}_k)$ of the classes as the components of a vector in the class space (in the same way as it was done for $\Gamma_i(R)_{\mu\nu}$). It follows then from the condition of mutual orthogonality of such vectors (6.27) that the number of irreducible representations is less or equal to the number of classes. In this case, too, it can be proved (see [2.2, Chap. 3]) that the limiting relation remains valid:

number of representations = number of classes. (6.30)

An expression similar to (6.27) can be deduced (see [2.2, Chap. 3]):

$$\sum_i \chi_i^*(\mathbb{C}_k)\,\chi_i(\mathbb{C}_l) = \frac{h}{N_k}\,\delta_{kl}, \tag{6.31}$$

where the orthogonality of characters exists not between different representations but between different classes.

Although the characters of irreducible representations provide less information than the representations themselves, in many cases they are adequate for the solution of physical problems. For this reason it seems to be a lucrative goal to obtain the characters of irreducible representations of various classes without determining the representations themselves in an explicit form. This may be done with the aid of the following relations: (6.30), (6.22), (6.27) and (6.31).

We shall compile the table of characters so that there is a definite irreducible representation to correspond to each row and a definite class to each column. It follows from (6.30) that the number of rows and columns should be equal, i.e., that the table of characters should be square. If the corresponding representation for the first row is a unit one, then its characters for all the classes must be unity. If the class corresponding to the first column is $E$, then the characters of the irreducible representations, according to (6.24), are equal to their dimensions $l_i$ which can be determined from (6.22) (usually for specified $h$ and $s$ the values of $l_i$ can be determined uniquely). If we know the first row and the first column of the table, we can find the other characters by trial, so as to satisfy the conditions of orthonormality of the characters by rows (6.27) and by columns (6.31). Usually there is an overabundance of such conditions; therefore some of them can serve to check the table[16].

By making use of the properties of the table formulated above, we can easily compile it for any group of the second order. In this case the group is Abelian, and both irreducible representations, in compliance with (6.22), are of unit dimension. Applying the condition of orthogonality of rows (or of columns), we obtain for the groups $\boldsymbol{C}_\mathrm{i} = \{E, J\}$ and $\boldsymbol{C}_\mathrm{s} = \boldsymbol{C}_\mathrm{1h} = \{E, \sigma_\mathrm{h}\}$ Table 2.3.

[16] There is no definite algorithm for compiling the table of characters; therefore some experience is required. It must be said, however, that the tables for groups that are of interest to the physicists have already been made.

Table 2.3

| $C_1$ | $E$ | $J$ |
|---|---|---|
| $C_s$ | $E$ | $\sigma_h$ |
| $\Gamma_1$ | 1 | 1 |
| $\Gamma_2$ | 1 | —1 |

Table 2.4

| $D_3$ | $E$ | $3C_2$ | $2(C_3, C_3^2)$ |
|---|---|---|---|
| $\Gamma_1$ | 1 | 1 | 1 |
| $\Gamma_2$ | 1 | —1 | 1 |
| $\Gamma_3$ | 2 | 0 | —1 |

Let us cite the example of group $\boldsymbol{D}_3$ (Sec. 2.2.2) to illustrate the above relations. This non-Abelian group of the sixth order ($h = 6$) consists of three classes:

$$\mathbb{C}_1 = E,\ \mathbb{C}_2 = \{A,\ B,\ C\} = 3C_2,\ \mathbb{C}_3 = \{D,\ F\} = 2\,(C_3;\ C^2).$$

Since the number of irreducible representations of this group should also be equal to three, the three nonequivalent representations (6.16) are irreducible representations. We find that for each representation the characters belonging to the same class are equal. Table 2.4 contains the characters of the $\boldsymbol{D}_3$ group (the factors in $\mathbb{C}_2$ and $\mathbb{C}_3$ remind of the number of elements in the class). It follows from the table that the sum of the squares of the dimensions of the representations is equal to the group's order (6.22): $1^2 + 1^2 + 2^2 = 6$. We can easily check the validity of (6.27). For example, the normalization condition for $\Gamma_3$ is of the form $2^2 + 3 \times 0 + 2 \times (-1)^2 = 6$; the orthogonality condition for $\Gamma_2$ and $\Gamma_3$ is $1 \times 2 + 3 \times (-1) \times 0 + 2 \times 1 \times (-1) = 0$. In the same way we can check the validity of (6.31). For example, the normalization condition for the column $\mathbb{C}_2$ is $1^2 + (-1)^2 + 0 = 6/3 = 2$, and the orthogonality condition for $\mathbb{C}_1$ and $\mathbb{C}_3$ is $1 \times 1 + 1 \times 1 + 2 \times (-1) = 0$. The other relations can also be verified.

2.6.4 Imagine some reducible representation $\Gamma$. With the aid of an appropriate similarity transformation all the matrices of a reducible representation $\Gamma(R)$ ($R$ is an element of the group) can be simultaneously reduced to quasi-diagonal matrices of a similar structure in which the irreducible representation $\Gamma_i(R)$ is contained $a_i$ times[17]. We shall write

$$\Gamma = a_1\Gamma_1 + a_2\Gamma_2 + \ldots + a_s\Gamma_s = \sum_i a_i\Gamma_i, \tag{6.32}$$

[17] The uniqueness of such an expansion can be proved; of course, some of the $a_i$ may turn out to be equal to unity or zero.

which is a formal statement (to the right we have a "sum" of matrices of different ranks; it is sometimes termed *direct* sum). Since such a similarity transformation does not change the character of the matrix, it follows that

$$\chi(R) = \sum_i a_i \chi_i(R) \tag{6.33}$$

(here the sum and the equality sign are no longer a mere formality). Making use of the orthogonality of characters of irreducible representations (6.26), we obtain from (6.33)

$$\sum_R \chi(R)\chi_j^*(R) = \sum_i a_i \sum_R \chi_i(R)\chi_j^*(R) = \sum_i a_i h\delta_{ij} = ha_j,$$

whence

$$a_j = \frac{1}{h}\sum_R \chi(R)\chi_j^*(R) = \frac{1}{h}\sum_{k=1}^{s} N_k \chi(\mathfrak{C}_k)\chi_j^*(\mathfrak{C}_k), \tag{6.34}$$

where $N_k$ is the number of elements in the class $\mathfrak{C}_k$.

This equation, which determines the number of times a specified irreducible representation $\Gamma_j$ is contained in a reducible representation $\Gamma$, plays an important part in all applications.

**2.6.5** In Section 2.2.5 we introduced the concept of a direct product of groups

$$\mathfrak{A} = \{A_1 \equiv E, A_2, A_3, \ldots, A_{h_a}\}, \mathfrak{B} = \{B_1 \equiv E, B_2, B_3, \ldots, B_{h_b}\}$$

and denoted it $\mathfrak{A} \times \mathfrak{B}$.

Let $\Gamma_1$ be a representation of the group $\mathfrak{A}$, and $\Gamma_2$ a representation of the group $\mathfrak{B}$, generally of a different dimension. Demonstrate that the direct product (see Appendix 3, Sec. 6) of the representations, $\Gamma_1(A_k) \times \Gamma_2(B_l)$ is a representation of the direct product $\mathfrak{A} \times \mathfrak{B}$. Let the multiplication rule for the elements of the direct product group be

$$(A_k B_l)(A_{k'} B_{l'}) = (A_k A_{k'})(B_l B_{l'}) = A_{k''} B_{l''}, \tag{6.35}$$

where

$$A_{k''} = A_k A_{k'} \quad \text{and } B_{l''} = B_l B_{l'}.$$

Demonstrate that the same multiplication rule holds for the direct product of the representations $\Gamma_1 \times \Gamma_2$. We have

$$\begin{aligned}
&[\Gamma_1(A_k) \times \Gamma_2(B_l)]\,[\Gamma_1(A_{k'}) \times \Gamma_2(B_{l'})] \\
&\quad = \Gamma_1(A_k)\,\Gamma_1(A_{k'}) \times \Gamma_2(B_{l'})\,\Gamma_2(B_{l'}) \\
&\quad = \Gamma_1(A_k A_{k'}) \times \Gamma_2(B_l B_{l'}) = \Gamma_1(A_{k''}) \times \Gamma_2(B_{l''}).
\end{aligned} \tag{6.36}$$

Here the first equality follows from (A.3.50) and the last from (6.35). Evidently, the direct products $\Gamma_1 \times \Gamma_2$ satisfy the multiplication rule (6.35).

It follows from (6.36) that the representation of the direct product of groups is equal to the direct product of their representations:

$$\Gamma(A_k B_l) = \Gamma_1(A_k) \times \Gamma_2(B_l). \tag{6.37}$$

Making use of (A.3.52) we obtain

$$\begin{aligned} \mathrm{Tr}\,\Gamma(A_k, B_l) &= \mathrm{Tr}\,[\Gamma_1(A_k) \times \Gamma_2(B_l)] \\ &= \mathrm{Tr}\,\Gamma_1(A_k) \times \mathrm{Tr}\,\Gamma_2(B_l), \end{aligned}$$

or

$$\chi(A_k B_l) = \chi_1(A_k)\,\chi_2(B_l). \tag{6.38}$$

i.e., the character of the representation of a direct product of the groups for an element $A_k B_l$ is equal to the product of the characters of the representations for the elements $A_k$ and $B_l$.

The fact that this relation connects the characters of irreducible representations can be explained as follows. From (6.38) and (6.26) we have

$$\sum_{k,l} |\chi(A_k B_l)|^2 = \sum_k |\chi_1(A_k)|^2 \sum_l |\chi_2(B_l)|^2 = h_a h_b = h,$$

where $h_a$, $h_b$ and $h$ are the numbers of the elements in the groups $\mathfrak{A}$ and $\mathfrak{B}$ and in their direct product $\mathfrak{A} \times \mathfrak{B}$, respectively. Hence, if $\Gamma_1$ and $\Gamma_2$ are irreducible representations of $\mathfrak{A}$ and $\mathfrak{B}$, then $\Gamma_1 \times \Gamma_2$ will also be an irreducible representation of the group $\mathfrak{A} \times \mathfrak{B}$ [since the relation (6.26) holds only for an irreducible representation].

Relation (6.38) facilitates the compilation of tables for such groups that can be regarded as direct products of simpler groups.

We have seen in Section 2.3.4 that the group $\boldsymbol{D}_{n\mathrm{h}} = \boldsymbol{D}_n \times \boldsymbol{C}_\mathrm{s}$, where the group $\boldsymbol{C}_\mathrm{s} = \{E, \sigma_\mathrm{h}\}$. Table 2.4 presents the characters of group $\boldsymbol{D}_3$. The group $\boldsymbol{D}_{3\mathrm{h}}$ contains twice as many elements, classes and representations as the group $\boldsymbol{D}_3$. Using equation (6.38), we can easily draw up a table of characters for the $\boldsymbol{D}_{3\mathrm{h}}$ group (Table 2.5). Generally, if group $\boldsymbol{G}$ is considered a direct product of some group $\mathcal{H}$ and a second-order group $\boldsymbol{C}_\mathrm{s}$ or $\boldsymbol{C}_\mathrm{i}$ with characters as shown in Table 2.3, then the table of characters for $\boldsymbol{G}$ should be compiled in accordance with Table 2.6, where $\chi$ are the characters of the group $\mathcal{H}$.

Table 2.7 presents the characters of the $\boldsymbol{O}$ group and of the $\boldsymbol{T}_\mathrm{d}$ group isomorphic to it. The notation in the first and the second columns is that used by L. P. Bouckaert, R. Smoluchowski and E. Wigner and in papers on molecular spectra, respectively [in the latter case the usual designations are $A$ for one-dimensional, $E$ for two-dimensional and $F$ (or $T$) for three-dimensional representations].

Since the full cubic symmetry group is $\boldsymbol{O}_\mathrm{h} = \boldsymbol{O} \times \boldsymbol{C}_\mathrm{i}$ where $\boldsymbol{C}_\mathrm{i} = \{E, J\}$, the table of characters for group $\boldsymbol{O}_\mathrm{h}$ can be obtained from Table 2.7 in conjunction with Table 2.6.

Table 2.5

| $D_{3h}$ | $E$ | $3C_2$ | $2(C_3;\ C_3^2)$ | $\sigma_h$ | $3\sigma_h C_2$ | $2\sigma_h(C_3;\ C_3^2)$ |
|---|---|---|---|---|---|---|
| $\Gamma_{1+}$ | 1 | 1 | 1 | 1 | 1 | 1 |
| $\Gamma_{2+}$ | 1 | −1 | 1 | 1 | −1 | 1 |
| $\Gamma_{3+}$ | 2 | 0 | −1 | 2 | 0 | −1 |
| $\Gamma_{1-}$ | 1 | 1 | 1 | −1 | −1 | −1 |
| $\Gamma_{2-}$ | 1 | −1 | 1 | −1 | 1 | −1 |
| $\Gamma_{3-}$ | 2 | 0 | −1 | −2 | 0 | 1 |

Table 2.6

| $G$ | Classes | Classes |
|---|---|---|
| Irreducible representations | $\chi$ | $\chi$ |
| Irreducible representations | $\chi$ | $-\chi$ |

If the elements $R$ and $R^{-1}$ of some point group belong to different classes (for instance, this is the case if the rotation axis for the element $R$ is not a bilateral one), then for some irreducible representation

$$\Gamma(R^{-1}) = \Gamma^{-1}(R) = \Gamma^{+}(R) = \tilde{\Gamma}^{*}(R),$$

because $\Gamma$ is a unitary matrix. Hence

$$\Gamma(R^{-1})_{ik} = \Gamma^{*}(R)_{ki},$$

and, consequently,

$$\chi(R^{-1}) = \sum_i \Gamma(R^{-1})_{ii} = \sum_i \Gamma^{*}(R)_{ii} = \chi^{*}(R),$$

or

$$\chi(R^{-1}) = \chi^{*}(R), \tag{6.39}$$

i.e., the characters of the elements $R$ and $R^{-1}$ are complex conjugate.

Table 2.7

| $O$ | | $T_d$ | | $E$ | $8C_3$ | $C_{24}^2$ | $6C_2$ | $6C_4$ |
|---|---|---|---|---|---|---|---|---|
| | | | | $E$ | $8C_3$ | $3C_4^2$ | $6\sigma$ | $6S_4$ |
| $\Gamma_1$ | $A_1$ | $\Gamma_1$ | $A_1$ | 1 | 1 | 1 | 1 | 1 |
| $\Gamma_2$ | $A_2$ | $\Gamma_2$ | $A_2$ | 1 | 1 | 1 | −1 | −1 |
| $\Gamma_{12}$ | $E$ | $\Gamma_{12}$ | $E$ | 2 | −1 | 2 | 0 | 0 |
| $\Gamma_{15}$ | $F_1$ $xyz$ | $\Gamma_{25}$ | $F_1$ | 3 | 0 | −1 | −1 | 1 |
| $\Gamma_{25}$ | $F_2$ | $\Gamma_{15}$ | $F_2$ $xyz$ | 3 | 0 | −1 | 1 | −1 |

Table 2.8

| $C_3$ | $E$ | $C_3$ | $C_3^2$ |
|---|---|---|---|
| $\Gamma_1$ | 1 | 1 | 1 |
| $\Gamma_2$ | 1 | $\omega$ | $\omega^2$ |
| $\Gamma_3$ | 1 | $\omega^2$ | $\omega$ |

$\omega = e^{2\pi i/3}$

Table 2.8 presents the characters of the group $C_3$, whose elements $C_3$ and $C_3^{-1}$ belong to different classes (for more detailed tables of characters of point groups see, e.g., *Landau and Lifshitz* [2.6, pp. 377-8]. We see that, indeed, for the representation $\Gamma_2$,

$$\chi(C_3) = \omega = e^{2\pi i/3} \text{ and}$$

$$\chi(C_3^{-1}) = \omega^2 = e^{4\pi i/3} = e^{-2\pi i/3},$$

i.e.,

$$\chi(C_3) = \chi^*(C_3^{-1}).$$

A similar situation exists in the case of the irreducible representation $\Gamma_3$.

## 2.7 Quantum Mechanics and Group Theory

**2.7.1** At the beginning of Section 2.6 we stated that matrix representations of groups are the natural outcome of the study of physical problems.

The most important example of this sort is the study of the solutions of the stationary-state Schrödinger equation.

Imagine a physical system in the configurational space of $n$ coordinates $x_1, x_2, \ldots, x_n$ forming an $n$-dimensional vector $\mathbf{x} = (x_1, x_2, \ldots, x_n)$. The wave function of the system, $\psi(\mathbf{x})$, in the stationary state satisfies the Schrödinger equation

$$\hat{\mathscr{H}}(\mathbf{x})\,\psi(\mathbf{x}) = \mathscr{E}\psi(\mathbf{x}), \tag{7.1}$$

where $\hat{\mathscr{H}}(\mathbf{x})$ is the Hamiltonian of the system, and $\mathscr{E}$ the energy eigenvalue.

If the state with the energy $\mathscr{E}$ is $l$-fold degenerate, there are $l$ linearly independent eigenfunctions corresponding to it:

$$\psi_1(\mathbf{x}), \quad \psi_2(\mathbf{x}), \quad \ldots, \quad \psi_i(\mathbf{x}), \quad \ldots, \quad \psi_l(\mathbf{x}). \tag{7.2}$$

We know from quantum mechanics that these functions can always be assumed to be orthonormal,[18] i.e., their scalar products can be assumed to be equal to

$$(\psi_i, \psi_k) = \int \psi_i^*(\mathbf{x})\,\psi_k(\mathbf{x})\,d\mathbf{x} = \delta_{ik}, \tag{7.3}$$

where $d\mathbf{x} = dx_1 dx_2 \ldots dx_n$.

Any solution of the equation (7.1) corresponding to the eigenvalue $\mathscr{E}$ can be represented as a linear combination of the functions (7.2).

Imagine now that the system being considered possesses a symmetry of some sort, for instance, a space symmetry of some point group or a symmetry with respect to permutations of identical particles.

Every operation of such a symmetry (rotation, reflection, inversion, permutation of particles) involves some linear transformation of the configurational coordinates of the system, which can be written in the form

$$\mathbf{x}' = \mathbf{R}\mathbf{x}, \tag{7.4}$$

where $\mathbf{R}$ is a real orthogonal matrix of the linear transformation. The inverse of transformation (7.4) is

$$\mathbf{x} = \mathbf{R}^{-1}\mathbf{x}', \tag{7.5}$$

---

[18] See in Appendices 3 and 5 the discussion of a similar problem of orthogonalization of eigenvectors corresponding to the same eigenvalue of hermitian matrices.

where $\mathbf{R}^{-1}$ is the inverse matrix. For real orthogonal matrices, $\mathbf{R}^{-1} = \tilde{\mathbf{R}}$, i.e., $(\mathbf{R}^{-1})_{ij} = \mathbf{R}_{ji}$ (see A.3.28). For example, an equilateral triangle (Fig. 2.1) can be described with the aid of six coordinates $x_1$, $x_2$, $x_3$, $x_4$, $x_5$, $x_6$ which, in turn, are equal to the $x$ and $y$ coordinates of the vertices *1*, *2* and *3*.

The transformation corresponding to operation $D$ (clockwise rotation through 120° about the axis perpendicular to the plane of the triangle and passing through its centre) is[19]

$$\begin{aligned}
x_1' &= x_3 = 0\cdot x_1 + 0\cdot x_2 + 1\cdot x_3 + 0\cdot x_4 + 0\cdot x_5 + 0\cdot x_6,\\
x_2' &= x_4 = 0\cdot x_1 + 0\cdot x_2 + 0\cdot x_3 + 1\cdot x_4 + 0\cdot x_5 + 0\cdot x_6,\\
x_3' &= x_5 = \ldots\ldots\ldots\ldots\ldots\ldots 1\cdot x_5 + 0\cdot x_6,\\
x_4' &= x_6 = \ldots\ldots\ldots\ldots\ldots\ldots + 1\cdot x_6,\\
x_5' &= x_1 = 1\cdot x_1 + \ldots\ldots\ldots\ldots\ldots\ldots,\\
x_6' &= x_2 = 0\cdot x_1 + 1\cdot x_2 \ldots\ldots\ldots\ldots\ldots\ldots,
\end{aligned} \tag{7.6}$$

so that the matrix of orthogonal transformation becomes

$$\mathbf{R} = \begin{pmatrix} 0 & 0 & 1 & 0 & 0 & 0\\ 0 & 0 & 0 & 1 & 0 & 0\\ 0 & 0 & 0 & 0 & 1 & 0\\ 0 & 0 & 0 & 0 & 0 & 1\\ 1 & 0 & 0 & 0 & 0 & 0\\ 0 & 1 & 0 & 0 & 0 & 0 \end{pmatrix}. \tag{7.7}$$

It may easily be seen that the inverse transformation is

$$\mathbf{R}^{-1} = \begin{pmatrix} 0 & 0 & 0 & 0 & 1 & 0\\ 0 & 0 & 0 & 0 & 0 & 1\\ 1 & 0 & 0 & 0 & 0 & 0\\ 0 & 1 & 0 & 0 & 0 & 0\\ 0 & 0 & 1 & 0 & 0 & 0\\ 0 & 0 & 0 & 1 & 0 & 0 \end{pmatrix} \tag{7.8}$$

and that, indeed,

$$\mathbf{R}^{-1} = \tilde{\mathbf{R}}. \tag{7.9}$$

If the system possesses a certain symmetry, so that it coincides with itself as the result of transformation (7.5), the potential energy

---

[19] In the process the triangle's vertex ***1*** becomes vertex *2*, *2* becomes *3*, and *3* becomes *1*.

of the system, $V(\mathbf{x})$, which enters the Hamiltonian $\hat{\mathscr{H}}(\mathbf{x})$ in (7.1), satisfies the condition

$$V(\mathbf{x}) = V(\mathbf{R}^{-1}\mathbf{x}') = V(\mathbf{x}'), \tag{7.10}$$

where we performed the transformation (7.5) from $\mathbf{x}$ to $\mathbf{x}'$ in $V(\mathbf{x})$. For instance, the potential energy of the electron in the field of an atomic nucleus, $Ze^2/r$, is spherically symmetric, i.e., does not change with any rotations of the atom (or of the coordinate frame) about the origin coinciding with the nucleus. Indeed, performing the orthogonal transformation (7.5) with respect to the electron's coordinates $x_1 = x$, $x_2 = y$, $x_3 = z$ we obtain

$$\frac{Ze^2}{\sqrt{x^2+y^2+z^2}} = \frac{Ze^2}{\sqrt{x'^2+y'^2+z'^2}}. \tag{7.11}$$

Since the Laplace operator is invariant under orthogonal transformations (7.5), it follows that

$$\hat{\mathscr{H}}(\mathbf{x}) = \hat{\mathscr{H}}(\mathbf{R}^{-1}\mathbf{x}') = \hat{\mathscr{H}}(\mathbf{x}'). \tag{7.12}$$

Changing the coordinates as in (7.5) throughout equation (7.1), we obtain

$$\hat{\mathscr{H}}(\mathbf{x}')\,\psi(\mathbf{R}^{-1}\mathbf{x}') = \mathscr{E}\psi(\mathbf{R}^{-1}\mathbf{x}').$$

We can, of course, denote $\mathbf{x}'$ by $\mathbf{x}$ and obtain

$$\hat{\mathscr{H}}(\mathbf{x})\,\psi(\mathbf{R}^{-1}\mathbf{x}) = \mathscr{E}\psi(\mathbf{R}^{-1}\mathbf{x}). \tag{7.13}$$

The term for all symmetry transformations that leave the Hamiltonian invariant is the *Schrödinger equation group*.

Comparing (7.13) with (7.1), we see that the function $\psi(\mathbf{R}^{-1}\mathbf{x}) \equiv \varphi(\mathbf{x})$ satisfies the same equation as the function $\psi(\mathbf{x})$ for the same energy eigenvalue $\mathscr{E}$.

If the eigenvalue of $\mathscr{E}$ is nondegenerate, $\varphi(\mathbf{x})$ can differ from $\psi(\mathbf{x})$ only by a constant factor, i.e.,

$$\varphi(\mathbf{x}) = c\psi(\mathbf{x}). \tag{7.14}$$

If $R^2 = E$, then $c^2 = 1$ and, consequently, $c = \pm 1$. Hence, in the case of $R^2 = E$ there are two solutions to equation (7.1): one that does not change sign as the result of a symmetry transformation (the symmetric solution), and one that changes sign as the result of a symmetry transformation (the antisymmetric solution)[20].

If, on the other hand, the eigenvalue of $\mathscr{E}$ is $l$-fold degenerate, so that $\psi(\mathbf{x})$ is equal to one of the functions (7.2), then $\psi_i(\mathbf{R}^{-1}\mathbf{x})$ (see (7.13)) is a solution of the Schrödinger equation (7.1) for the

[20] See the solution for the helium atom in Section 2.1.

same energy eigenvalue $\mathscr{E}$ and because of that should be expressed linearly in terms of the wave functions (7.2):

$$\psi_i(\mathbf{R}^{-1}\mathbf{x}) = \hat{P}_R\psi_i(\mathbf{x}) = \sum_{k=1}^{l} \Gamma(R)_{ki}\psi_k(\mathbf{x}). \tag{7.15}$$

This expression coincides with (6.2) for the transformation of the basis functions $\psi_i$. It follows from (6.3) that the elements of the matrix $\Gamma(R)_{ki}$ are the matrix elements of the operator $\hat{P}_R$ constructed on the wave (basis) functions (7.2). It follows from (6.5a) that the unitary matrices $\Gamma(R)$ in (7.15) represent the Schrödinger equation group (7.1).

We can draw an important conclusion that there is a definite (to within a similarity transformation) representation of the Schrödinger equation group to correspond to every energy eigenvalue $\mathscr{E}$.

In what follows we shall assume that, in the absence of *accidental degeneracy*, the representation of the Schrödinger equation group corresponding to a definite energy eigenvalue is irreducible.

Indeed, if the representation were reducible, $\Gamma(R)$ could be reduced, with the aid of an appropriate similarity transformation [or by an appropriate choice of basis functions (7.2)], to the quasi-diagonal form, for example, to two blocks of ranks $n$ and $l - n$ arranged along the principal diagonal of $\Gamma(R)$. In this case, $n$ functions from (7.2) would in all symmetry transformations $R$ be transformed only into each other, the same being true for the remaining $l - n$ functions. Hence, each of the two groups of functions would behave as if it belonged to some definite energy level. The probability of such a coincidence at both levels (accidental degeneracy) would be quite small; in any case it would have nothing to do with the symmetry of the Hamiltonian $\mathscr{H}(\mathbf{x})$.

Thus, we assume that a definite irreducible representation of the Schrödinger equation group corresponds to every energy level. Of course, one irreducible representation can correspond to different energy levels. For example, we shall learn below that the irreducible representations of the electron in an atom in the single-particle approximation are characterized by the orbital quantum number $l$, so that the $s$-state ($l = 0$), the $p$-state ($l = 1$), the $d$-state ($l = 2$), etc. are different irreducible representations of a continuous rotation group in which a definite irreducible representation with a specified $l$ corresponds to each energy level of the electron. Still, of course, the same irreducible representations with the same $l$ can correspond to different energy levels with different principal quantum numbers $n$. Consider an atom in a site of a cubic crystal lattice. Let the resultant field acting on the atom possess the symmetry of the $\boldsymbol{O}$ group (in this approximation an electric field with the symmetry of the lattice can be substituted for the effect of all the other electrons and of all the nuclei of the lattice), which has five irreducible representations (Table 2.7). In that case all the energy levels of the electron in an atom located in a site of a cubic crystal must belong to one of the five irreducible representations of the $\boldsymbol{O}$ group. Hence, we may assert that in a field with the symmetry of $\boldsymbol{O}$ there cannot be any energy levels of greater than three-fold degeneracy.

In Section 2.8 we shall consider the problem of splitting energy levels of a valence electron acted upon by the crystal field with an $\boldsymbol{O}$ or an $\boldsymbol{O}_{\mathrm{h}}$ symmetry[21].

**2.7.2** Consider an important case in which the Hamiltonian in (7.1) can be represented as a sum

$$\hat{\mathscr{H}} = \hat{\mathscr{H}}_0 + \hat{\mathscr{H}}', \tag{7.16}$$

with $\hat{\mathscr{H}}'$ being of lower symmetry than $\hat{\mathscr{H}}_0$, so that the symmetry group of $\hat{\mathscr{H}}'$ is a subgroup of the symmetry group of $\hat{\mathscr{H}}_0$. Such a situation is brought about, for example, by the application of an electric field $\mathbf{E}_0$ to an atom, with the result that the spherical symmetry of the field acting on its electrons by the perturbation $\hat{\mathscr{H}}' = -eE_0 z$ ($\mathbf{E}_0 \parallel z$ axis) is violated.

A similar reduction in symmetry takes place if the atom is placed in one of the sites of a crystal lattice and it is assumed that in addition to the central forces with which the atomic nucleus and the other electrons act on it there is also the averaged electric field determined by the symmetry of the lattice. The symmetry of the total Hamiltonian will obviously be determined by its least symmetrical part $\hat{\mathscr{H}}'$.

Consider what will happen to some (in general, degenerate) energy level $\mathscr{E}_0$ corresponding to the Hamiltonian $\hat{\mathscr{H}}_0$ when $\hat{\mathscr{H}}'$ is "switched on". Since the group[22] of $\hat{\mathscr{H}}'$ is a subgroup of the group of $\hat{\mathscr{H}}_0$, the irreducible representation of the level $\mathscr{E}_0$ will, in general, be a reducible representation of the group of $\hat{\mathscr{H}}'$. We can decompose this reducible representation in terms of the irreducible representations of the group of $\hat{\mathscr{H}}'$; in that case, if the energy corresponding to $\hat{\mathscr{H}}'$ is much less than the spacing between the energy terms of the unperturbed Hamiltonian, $\hat{\mathscr{H}}_0$, we will be able to say into what number of levels of what degeneracy the level $\mathscr{E}_0$ will split as the result of the application of $\hat{\mathscr{H}}'$. However, we shall not be able to say anything about

---

[21] As we shall demonstrate below, in this case group $\boldsymbol{O}$ has the same properties, from the point of view of the nature of term splitting, as the group $\boldsymbol{O}_{\mathrm{h}} = \boldsymbol{O} \times \boldsymbol{C}_i$.

[22] We say "group of $\hat{\mathscr{H}}'$" instead of "symmetry group of $\hat{\mathscr{H}}'$" for the sake of brevity.

the magnitude of this splitting, or about the order in which the split levels are arranged.

Let the unperturbed system, i.e., the Hamiltonian $\hat{\mathscr{H}}_0$, have the symmetry of group $\boldsymbol{O}$ (see Table 2.7). We are interested in what will happen to the irreducible representation $F_2$ (three-fold degenerate level $F_2$) when a field $\mathscr{H}'$ with a $\boldsymbol{D}_3$ symmetry (see Table 2.4) is applied, if its $C_3$ axis coincides with one of the $C_3$ axes of group $\boldsymbol{O}$. We shall see that the elements of group $\boldsymbol{O}$, i.e., $E$, 2 $(C_3, C_3^2)$, $3C_2$ form a subgroup of $\boldsymbol{D}_3$. The representation $F_2$ is a reducible representation of this subgroup (this is evident from the fact that a subgroup of $\boldsymbol{D}_3$ has no irreducible representations of dimension three). Thus, we have to decompose the reducible (for $\boldsymbol{D}_3$) representation $F_2$ with the characters 3, 0, 1 (for the elements $E$, 8 $(C_3, C_3^2)$ and $6C_2$) in terms of the irreducible representations $\Gamma_1$, $\Gamma_2$ and $\Gamma_3$ of group $\boldsymbol{D}_3$. In accordance with (6.32) and (6.34) we obtain

$$F_2 = a_1\Gamma_1 + a_2\Gamma_2 + a_3\Gamma_3$$

and

$$a_1 = \frac{1}{6}[3\times 1+2\times 0\times 1+3\times 1\times 1]=1,$$

$$a_2 = \frac{1}{6}[3\times 1+2\times 0\times 1+3\times 1\times(-1)]=0,$$

$$a_3 = \frac{1}{6}[3\times 2+2\times 0\times(-1)+3\times 1\times 0]=1.$$

Hence

$$F_2 = \Gamma_1 + \Gamma_3,$$

i.e., the three-fold degenerate level $F_2$ splits under the influence of a field with a $\boldsymbol{D}_3$ symmetry into a nondegenerate level $\Gamma_1$ and a two-fold degenerate level $\Gamma_3$.

**2.7.3** In addition to the direct product of two groups, we introduce the important concept of a direct product of irreducible representations of one group.

Let there be two irreducible representations of the group $\Gamma_1$ and $\Gamma_2$, of dimensions $l$ and $m$, respectively. The basis functions for these irreducible representations are

$$\Gamma_1 : \psi_1, \psi_2, \ldots, \psi_i, \ldots, \psi_l; \; \Gamma_2 : \varphi_1, \varphi_2, \ldots, \varphi_k, \ldots, \varphi_m. \tag{7.17}$$

It may easily be seen that the products $\psi_i\varphi_k$ ($i = 1, 2, \ldots, l$, $k = 1, 2, \ldots, m$) also form the basis of a representation of a group of dimension $l \times m$. Indeed,

$$\hat{P}_R(\psi_i\varphi_k) = \hat{P}_R\psi_i \times \hat{P}_R\varphi_k = \sum_j \Gamma_i(R)_{ji}\psi_j \sum_p \Gamma_2(R)_{pk}\varphi_p$$
$$= \sum_{j,p} [\Gamma_1(R)_{ji}\,\Gamma_2(R)_{pk}]\,\psi_j\varphi_p = \sum_{j,p} \Gamma(R)_{jp,\,ik}\psi_j\varphi_p.$$

where the matrix $\Gamma(R) = \Gamma_1(R) \times \Gamma_2(R)$, i.e., equal to the direct product of the matrices $\Gamma_1$ and $\Gamma_2$ (see Appendix 3, Sec. 6).

Hence it follows, in exactly the same way as in Section 2.6.5, that the direct product of the matrices of irreducible representations of a group is also its representation. However, whereas a direct product of irreducible representations of two different groups is also an irreducible representation of the direct product of the groups, a direct product of two irreducible representations of one group is, in general, a reducible representation of the group.

As we did in (6.38), we can demonstrate that the character $\chi(R)$ of the representation $\Gamma$ is equal to the product of characters $\chi_1(R)$ and $\chi_2(R)$ of irreducible representations $\Gamma_1$ and $\Gamma_2$:

$$\chi(R) = \chi_1(R)\,\chi_2(R). \tag{7.18}$$

The reducible representation $\Gamma = \Gamma_1 \times \Gamma_2$ can be decomposed in accordance with (6.32) and (6.34) in terms of irreducible representations of the group.

## 2.8 Application of Group Theory to the Study of Splitting Energy Levels of Impurity Atoms in Crystals and to the Classification of Normal Vibrations in Polyatomic Molecules

Consider two examples of applications of group theory in physics: (1) splitting energy levels of an atom (or ion) in the crystal field and (2) the study of natural vibrations of a molecule.

The first problem is directly related to the study of impurity atoms in solids; the second problem not only provides an instructive example of applications of group theory in physics but is also of major importance for the study of lattice vibrations.

**2.8.1** As an approach to the first problem, consider a continuous group of rotations.

In addition to the finite point groups discussed in Section 2.3, there are also the so-called continuous point groups with an infinite number of elements, namely, the groups of *axial* and *spherical symmetries*. We shall deal only with the spherical symmetry group $\mathscr{R}_i$, which consists of an infinite number of rotations through arbitrary angles about arbitrary axes passing through a static point $O$ and of reflections in any plane containing this point. Since $\sigma_h C_2 = J$ is inversion, we may say that the group of spherical symmetry consists of arbitrary rotations about all axes passing through the centre and of inversion. The subgroup of the spherical group that consists only of arbitrary rotations became known as the *proper rotation* group $\mathscr{R}$ or simply as the rotation group (see A.3.33), in contrast to the spherical group (which includes inversion), another term for which is *improper rotation* group. Obviously, $\mathscr{R}_i = \mathscr{R} \times \boldsymbol{C}_i$, where $\boldsymbol{C}_i = \{E, J\}$.

It is obvious that in the rotation group $\mathscr{R}$ all the axes are equivalent and bilateral; therefore the classes of this group consist of rotations through a specified angle $\alpha$ about any axis. The classes of the group $\mathscr{R}_i$ are obtained immediately from the direct product $\mathscr{R} \times C_i$.

To determine the characters of the continuous rotation group $\mathscr{R}$, consider the solutions of the Schrödinger equation with a Hamiltonian invariant with respect to the group $\mathscr{R}$. Consider an electron in a spherically symmetric field for which the Hamiltonian is

$$\hat{\mathscr{H}}(\mathbf{x}) = -\frac{\hbar^2}{2m}\nabla^2 + V(r), \tag{8.1}$$

where the potential energy of the electron $V(r) = V(\sqrt{x^2 + y^2 + z^2})$ is spherically symmetric[23].

The wave functions of the Schrödinger equation with a Hamiltonian of the type (8.1) are of the form (e.g., see [2.4, § 34])

$$\psi_{nlm}(r, \vartheta, \varphi) = R_{nl}(r)\, Y_{lm}(\vartheta, \varphi). \tag{8.2}$$

Here $r$, $\vartheta$, $\varphi$ are polar coordinates of the electron; $n$, $l$, $m$ are the principal, orbital and magnetic quantum numbers, respectively; $R_{nl}(r)$ is the radial part of the wave function that depends on the specific form of the potential $V(r)$; and

$$Y_{lm}(\vartheta, \varphi) = P_{lm}(\cos\vartheta)\, e^{im\varphi} \tag{8.3}$$

is a spherical harmonic, where $P_{lm}(\cos\vartheta)$ is a polynomial of degree $l$ in $\cos\vartheta$ expressed explicitly in terms of the Legendre polynomial $P_l(\cos\vartheta)$. The energy eigenvalues $\mathscr{E}_{nl}$ are $(2l + 1)$-fold degenerate with respect to the magnetic quantum number $m$ ($m = -l, -l + 1, \ldots, 1, \ldots, l$).

To determine the irreducible representation corresponding to the rotation about any axis through the angle $\alpha$, choose the axis $z$ as the rotation axis to obtain, in accordance with (7.15),

$$\begin{aligned} \hat{P}_\alpha \psi_{nlm} &= \psi_{nlm}(\alpha^{-1}\mathbf{x}) = R_{nl}(r)\, P_{lm}(\cos\vartheta)\, e^{im(\varphi-\alpha)} \\ &= \sum_{m'=-l}^{l} D_l(\alpha)_{m'm} \psi_{nlm'} = \sum_{m'=-l}^{l} D_l(\alpha)_{m'm} R_{nl}(r)\, P_{lm'} \\ &\times (\cos\vartheta)\, e^{im'\varphi}, \end{aligned} \tag{8.4}$$

where $D_l(\alpha)$ denotes the matrix of rank $(2l + 1)$ of the irreducible representation corresponding to the rotation through the angle $\alpha$ (Wigner's notation derived from the German word *Darstellung* for

---

[23] We consider the general case, in which $V(r)$ is not just the Coulomb energy.

representation). Comparing the third equation in the chain (8.4) with the last, we easily see that

$$D_l(\alpha)_{m'm} = \delta_{m'm} e^{-im'\alpha}, \tag{8.5}$$

where $\delta_{m'm}$ is the Kronecker delta.

We see that irreducible representations of the rotation group are characterized by the number $l$ (the dimension of the representation is $2l+1$). Hence, the *s*-state ($l=0$), the *p*-state ($l=1$), the *d*-state ($l=2$), etc. are irreducible representations of the electron in a spherically symmetric field. We see that both the number of irreducible representations and the number of elements in the group are infinite.

The character of the irreducible representation $D_l(\alpha)$ is

$$\chi_l(\alpha) = \sum_{m=-l}^{l} D_l(\alpha)_{mm} = \sum_{m=-l}^{l} e^{-im\alpha}$$
$$= e^{il\alpha} + e^{i(l-1)\alpha} + \ldots + 1 + \ldots + e^{-i(l-1)\alpha} + e^{-il\alpha}.$$

We have a geometric progression of $2l+1$ terms with the first term $e^{il\alpha}$ and the ratio $e^{-i\alpha}$, its sum being

$$\chi_l(\alpha) = \frac{e^{il\alpha}\,[(e^{-i\alpha})^{2l+1} - 1]}{e^{-i\alpha} - 1} = \frac{e^{-i(l+1)\alpha} - e^{il\alpha}}{e^{-i\alpha} - 1}.$$

Multiplying the numerator and the denominator by $e^{i\alpha/2}$, we obtain for the character of the irreducible representation $D_l(\alpha)$

$$\chi_l(\alpha) = \frac{\sin(2l+1)\dfrac{\alpha}{2}}{\sin\dfrac{\alpha}{2}}. \tag{8.6}$$

**2.8.2** If a foreign atom (an impurity atom) is placed in a lattice site, then the spherical symmetry of the field acting on the electron in a free atom is reduced to the symmetry determined by that of the crystal. Such a reduction of the symmetry of the field should, generally, cause a splitting of degenerate electron energy levels to occur (see Section 2.7.2).

If the spherically symmetric field is replaced by a field of lower symmetry, for instance, of cubic symmetry $\boldsymbol{O}$, the basis functions $Y_{lm}(\vartheta, \varphi)$ (8.3) will still represent group $\boldsymbol{O}$ (i.e., the matrices $D_l(R)$, (8.4), where $R$ is an element of group $\boldsymbol{O}$, will represent group $\boldsymbol{O}$ of dimension $(2l+1)$. However, in general, this representation will be a reducible one; this follows at least from the fact that the maximum dimension of the irreducible representations of group $\boldsymbol{O}$ is three, therefore for $(2l+1) > 3$ the representation $D_l(R)$ must be reducible. Since we assume that different energy levels of the electron correspond to irreducible representations, we determine the nature of splitting energy levels with different $l$'s in a field of cubic symmetry by decomposing the reducible representation $D_l(R)$ into irreducible representations of group $\boldsymbol{O}$.

Making use of expression (8.6), let us determine the characters of the classes of group $\boldsymbol{O}$ for different $l$'s. For instance, at $l = 1$ ($p$-state) we have:

$$\chi_1(\alpha) = \frac{\sin(3\alpha/2)}{\sin(\alpha/2)},$$

$$\chi_1(E) = \chi_1(0) = \left[\frac{\sin(3\alpha/2)}{\sin(\alpha/2)}\right]_{\alpha \to 0} = 3,$$

$$\chi_1(C_3) = \chi_1(120^\circ) = \frac{\sin 180^\circ}{\sin 60^\circ} = 0,$$

$$\chi_1(C_2) = \chi_1(180^\circ) = \frac{\sin 270^\circ}{\sin 90^\circ} = -1,$$

$$\chi_1(C_4) = \chi_1(90^\circ) = \frac{\sin 135^\circ}{\sin 45^\circ} = 1.$$

In the same way we can find the characters $\chi_l$ for other values of $l$. The result will be Table 2.9 for reducible representations $D_l$ corre-

Table 2.9

| $O$ | $E$ | $8C_3$ | $3C_4^2$ | $6C_2$ | $6C_4$ |
|---|---|---|---|---|---|
| $D_0$ | 1 | 1 | 1 | 1 | 1 |
| $D_1$ | 3 | 0 | −1 | −1 | 1 |
| $D_2$ | 5 | −1 | 1 | 1 | −1 |
| $D_3$ | 7 | 1 | −1 | −1 | −1 |
| $D_4$ | 9 | 0 | 1 | 1 | 1 |

sponding to the classes of group $\boldsymbol{O}$. Making use of Table 2.7 and equation (6.34) we can find what irreducible representations of cubic group $\boldsymbol{O}$ are contained in representations $D_l$.

$D_0 = A_1$, i.e., the nondegenerate level of the $s$-state, of course, cannot split and becomes the unit representation of group $\boldsymbol{O}$.

$D_1 = F_1$, i. e., the three-fold degenerate level of the $d$-state does not split in a field of cubic symmetry.

$D_2 = E + F_2$, the five-fold degenerate $d$-state must split, because the maximum dimension of the irreducible representations of group $\boldsymbol{O}$ is three; the $d$-level splits into two levels; into a two-fold degenerate and a three-fold degenerate.

$D_3 = A_2 + F_1 + F_2$, the seven-fold degenerate $f$-state splits into one nondegenerate and two three-fold degenerate.

$D_4 = A_1 + E + F_1 + F_2$, the nine-fold degenerate $g$-state splits into one nondegenerate, one two-fold degenerate and two three-fold degenerate states. (8.7)

Consider now splitting the terms of an atom in a crystal field of cubic symmetry with an inversion centre $J$, the point group for which is $\boldsymbol{O}_{\mathrm{h}} = \boldsymbol{O} \times \boldsymbol{C}_{\mathrm{i}}$, where the second order group $\boldsymbol{C}_{\mathrm{i}} = \{E, J\}$. In this case the effect of the inversion operator $\hat{J}$ on the wave function of the system is either to leave it unchanged or to change its sign [2.6, § 30]:

$$\hat{J}\psi(\mathbf{x}) = \psi(-\mathbf{x}) = \pm\psi(\mathbf{x}).$$

This follows immediately from the fact that $J^2 = E$ [see (7.14)] and that, accordingly,

$$\chi(J) = \pm 1, \tag{8.8}$$

in compliance with Table 2.2.

Hence, the character of inversion for even states of the system is equal to $+1$ and that for odd states $-1$. The term for the wave function (and the corresponding state) is *even* in the first case and *odd* in the second. Since for the systems being considered the operator $\hat{J}$ commutes with the Hamiltonian $\hat{\mathscr{H}}$, the fact that the wave function is either even or odd, is retained in time; this rule became known as the *parity conservation law*. It can be demonstrated that the parity of the wave function of a many-electron atom in the single-particle approximation is $\prod_k (-1)^{l_k}$, where $l_k$ is the orbital quantum number of the $k$-th electron.

Imagine now the characters $D_l(\alpha)$ to be determined not only for proper rotations $\alpha$ but for improper rotations $J\alpha$ as well. In this case five classes should be added to Table 2.7: $JE = J$, $8JC_3$, $3JC_4^2$, etc. The characters of the representations of these additional classes either coincide with the characters in Table 2.7 (for even states) or are of opposite signs (for odd states).

Since the representation $\Gamma_1$ in Table 2.3 corresponds to even states and $\Gamma_2$ to odd states, the table of characters of irreducible representations of the group (which includes inversion $J$) is of the form shown in Table 2.6, where $\chi$ coincides with that in Table 2.7.

Consider now the contributions of the individual quadrants of Table 2.6 to the sum, when the coefficients $a_j$ are calculated with the aid of equation (6.34). Quadrants *3* and *4* cancel each other in the sum; the contributions of quadrants *1* and *2* are equal and have to be divided by $2h$. The resulting value of $a_j$ is the same as for the single quadrant *1*, i.e., as for group $\boldsymbol{O}$.

2.8.3 Demonstrate now how group theory should be applied to the classification of normal vibrations of polyatomic molecules. We shall consider the problem of small vibrations of the atomic nuclei of the molecule only within the limits of classical mechanics.

It has been established in classical mechanics that a system of $N$ particles possessing $3N$ degrees of freedom and performing small oscillations about the equilibrium positions of the particles $\mathbf{r}_i$ ($i=1, 2, \ldots$ $\ldots, N$) has the energy

$$\mathscr{E}=\frac{1}{2}\sum_{i,\,\alpha}\frac{\dot{u}_{i\alpha}^2}{m_i}+\frac{1}{2}\sum_{i,\,k,\,\alpha,\,\beta}\varkappa_{i\alpha,\,k\beta}u_{i\alpha}u_{k\beta}. \tag{8.9}$$

Here $u_{i\alpha}$ ($i = 1, 2, \ldots, N$; $\alpha = x, y, z$) is the $\alpha$-th orthogonal (Cartesian) coordinate of the displacement of the $i$-th atomic nucleus in the molecule from its equilibrium position $\mathbf{r}_i$ and $\varkappa_{i\alpha,k\beta}$ is the quasi-elastic force constant for displacements $u_{i\alpha}$ and $u_{k\beta}$. From the total of $3N$ degrees of freedom of the molecule three correspond to its linear motion as a whole and three to its rotational motion as a whole. Hence, the number of vibrational degrees of freedom in the molecule is $3N - 6$. By means of an appropriate linear transformation it is possible to eliminate from (8.9) those six degrees of freedom. Next, by means of an appropriate linear transformation we can introduce *normal coordinates* $Q_{i\alpha}$ ($i, \alpha = 1, 2, 3, \ldots, (3N - 6)$; e.g., see [2.14, § 23]. Then the energy $\mathscr{E}$ will assume the form

$$\mathscr{E}=\frac{1}{2}\sum_{i,\,\alpha}\dot{Q}_{i\alpha}^2+\frac{1}{2}\sum_{i}\omega_i^2\sum_{\alpha=1}^{f_i}Q_{i\alpha}^2, \tag{8.10}$$

where $\omega_i$ is the frequency of normal vibrations. We have introduced a double label $i\alpha$ for the normal coordinates $Q_{i\alpha}$, with $i$ assuming values corresponding to different frequencies $\omega_i$ and $\alpha$ assuming $f_i$ values corresponding to $f_i$ linearly independent normal coordinates vibrating with the specified frequency $\omega_i$; $f_i$ is called the *degree of degeneracy of frequency* $\omega_i$ (obviously, $\sum_i f_i = 3N - 6$).

Usually the structure of a molecule corresponds to some point symmetry group. For example, the ammonia molecule $NH_3$ has the shape of a regular triangular pyramid whose one vertex is occupied by the nitrogen atom N, with the three hydrogen atoms H occupying the vertices of the equilateral base triangle. The $NH_3$ molecule has a vertical axis of symmetry $C_3$, which passes through the N atom, and three planes of symmetry $\sigma_v$, each of which passes through $C_3$

and one H atom. The symmetry group of the ammonia molecule $C_{3v}$ consists of six elements: $E$, $2C_3$, $3\sigma_v$. Transformations of the symmetry group of the molecule leave its dynamical coefficients, $\varkappa_{i\alpha,k\beta}$ and $m_i$, invariant; because of that the normal vibration frequencies $\omega_i$ remain unchanged.

Apply operator $\hat{P}_R$ (6.1), where $R$ is a symmetry element of the molecule, to the potential energy in (8.10):

$$\hat{P}_R\left\{\frac{1}{2}\sum_i \omega_i^2 \sum_{\alpha=1}^{f_i} Q_{i\alpha}^2\right\} = \frac{1}{2}\sum_i \omega_i^2 \sum_{\alpha=1}^{f_i} \hat{P}_R Q_{i\alpha}^2$$

$$= \frac{1}{2}\sum_i \omega_i^2 \sum_{\alpha=1}^{f_i} (\hat{P}_R Q_{i\alpha})(\hat{P}_R Q_{i\alpha}). \qquad (8.11)$$

We see that for the new normal coordinates $\hat{P}_R Q_{i\alpha}$ to retain the meaning of normal coordinates at the original frequencies $\omega_i$,

$$\hat{P}_R Q_{i\alpha} = \sum_\beta \Gamma(R)^{(i)}_{\beta\alpha} Q_{R^{-1}(i)\,\beta}, \qquad (8.11a)$$

where $R^{-1}(i) = k$ is the position of an identical atomic nucleus of the molecule, and $\Gamma(R)^{(i)}_{\beta\alpha}$ is the matrix of a linear orthogonal transformation. Exactly in the same way as it was done in Section 2.6, we can demonstrate that the matrices $\Gamma(R)^{(i)}$ constitute a representation of the molecule's symmetry group with the basis $Q_{i\alpha}$ ($\alpha = 1, 2, \ldots, f_i$). In the absence of accidental degeneracy of the frequencies $\omega_i$, this representation will be irreducible. The dimension of this irreducible representation is equal to the degree of degeneracy of the vibration $f_i$.

The transformation (8.11a) is similar to the transformation (7.15) in quantum mechanics. The frequency of normal vibrations $\omega_i$ plays the part of the degenerate energy level $\mathcal{E}$, the part of the wave functions $\psi_i(\mathbf{x})$ ($i = 1, 2, \ldots, l$) being played by the normal vibrations of $Q_{i\alpha}$ ($\alpha = 1, 2, \ldots, f_i$).

Determining the natural frequencies and the normal coordinates of a polyatomic molecule is a complicated dynamical problem. Group theory enables us to find the number of natural frequencies $\omega_i$, the degree of their degeneracy and even the pattern of corresponding vibrations without solving this problem.

To carry out this program, we shall determine the common reducible representation realized by all the vibrational coordinates at a time, and then decompose it into irreducible representations of the molecule's symmetry group. To find the full representation we make use of the fact that the characters of representations are invariant under a similarity transformation. Therefore, for computing them we need not use normal coordinates, but can use displacements of the nuclei in orthogonal coordinates $u_{i\alpha}$.

Determine the effect of the operator $\hat{P}_R$ on the field of displacements $u_{i\alpha}$, when $R$ is an element of the molecule's point symmetry group that displaces a nucleus $i$ into the site of an identical nucleus $k = R^{-1}(i)$.

Figure 2.17 depicts a flat molecule consisting of four identical atoms *1*, *2*, *3*, *4* located at the ends of an equilateral cross. Let $R^{-1}$ be the clockwise rotation of the molecule about its centre $O$ through the angle $\pi/2$. In the course of it atom *1* takes the place of atom *2*,

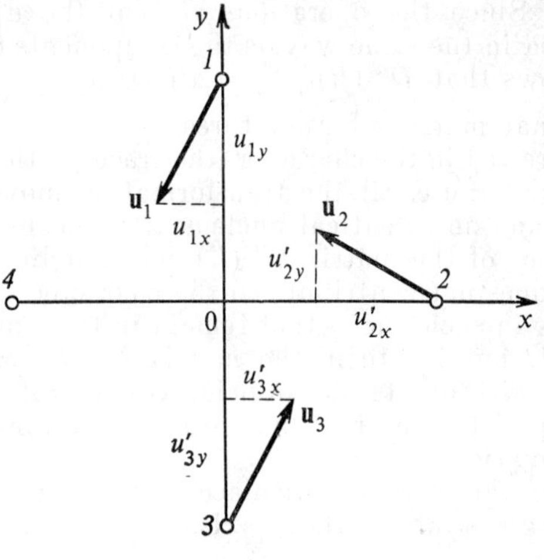


Fig. 2.17

atom *2* the place of atom *3*, etc. If we were to permute atoms *1*, *2*, etc. together with their displacements, we would gain no new information, because this is equivalent to the rotation of the molecule as a whole in space. We can extract some information if we attribute the displacement $\mathbf{u}_1$ to atom *2*; the displacement $\mathbf{u}_2$ to atom *3*, etc. Such a transformation of the displacement field $\mathbf{u}_i$ will, in view of the identity of the atoms, correspond to the same frequencies $\omega_i$ of molecular vibrations.

What transformations do the displacements $u_{1x}$ and $u_{1y}$ undergo in the course of rotation $C_4$? It may be seen from Fig. 2.17 that

$$\hat{P}C_4^{-1}u_{1x} = u'_{2y} = -u_{1x}, \quad \hat{P}C_4^{1}u_{1y} = u'_{2x} = u_{1y}. \tag{8.12}$$

If $R^{-1}$ is the rotation through the angle $\pi$, such that atom *1* moves to the site of atom *3*, then

$$\hat{P}C_2^{-1}u_{1x} = u'_{3x} = -u_{1x}, \quad \hat{P}C_2^{-1}u_{1y} = u'_{3y} = -u_{1y}. \tag{8.12a}$$

Generalizing (8.12) and (8.12a) to include the case of the more complex molecular structure, we express $\hat{P}_R u_{i\alpha}$ in the form of a linear combination of the quantities $u_{k\beta}$, i.e.,

$$\hat{P}_R u_{i\alpha} = \sum_{k\beta}^{1,\,3N} D^u(R)_{k\beta,\,i\alpha} u_{k\beta} = \sum_{k\beta} A(R)_{\beta\alpha} \delta_{k,\,R^{-1}(i)} u_{k\beta}, \tag{8.13}$$

where $R^{-1}(i) = k$, i.e., the nucleus $i$ by the transformation $R^{-1}$ is transformed into an identical nucleus $k$. The matrix $D^u$ of rank $3N$ yields the reducible representation realized by all the vibrations of the molecule. Since the operations $R^{-1}$ of the displacement $u_{i\alpha}$ are transformed in the same way as the components of a polar vector (8.12), it follows that $D^u(R)_{k\beta,\,i\alpha} = A(R)_{\beta\alpha}\delta_{k,\,R^{-1}(i)}$, where $A(R)$ is an orthogonal matrix of rank three.

We are interested in the character (the trace) of the matrix $D^u(R)$; therefore, if $k \neq i$, i.e., if the transformation moves the nucleus $i$ to the site of another identical nucleus $k$, it means that the corresponding element of the matrix $D^u(R)$ does not lie on the principal diagonal and does not contribute to the character. Therefore we are interested only in such nuclei that remain in their places after transformation $R^{-1}$; but for them the matrix $D^u(R)$ assumes a quasi-diagonal form with all its blocks being equal to $A(R)$ (the number of blocks is equal to the number of atoms remaining in their places after transformation $R^{-1}$).

If $R^{-1}$ is a rotation of the displacement $\mathbf{u}$ (we omit the label $i$) through an angle $\varphi$ about the $z$ axis,

$$\hat{P}_{C(\varphi)} u_\alpha = \sum_{\beta=1}^{3} A(C)_{\beta\alpha} u_\beta = \begin{cases} u_x \cos\varphi + u_y \sin\varphi, \\ -u_x \sin\varphi + u_y \cos\varphi, \\ u_z. \end{cases} \tag{8.14}$$

The trace of the matrix $A(C)$ is equal to $1 + 2\cos\varphi$. If there are $N_C$ nuclei on the rotation axis, then the trace of $D^u$ in (8.13) is

$$N_C(1 + 2\cos\varphi). \tag{8.15}$$

However, this trace is connected with all the $3N$ degrees of freedom of the molecule, including the rectilinear and rotational displacements of the molecule as a whole. The rectilinear displacement of the molecule is determined by the polar displacement vector of its centre of gravity $\mathbf{U}$, for which the character (the trace) of the transformation is equal to $(1+2\cos\varphi)$, the same as for the vibrations $\mathbf{u}$. The rotation of the molecule as a whole through a small angle $\delta\Omega$ is characterized by the axial vector $\boldsymbol{\delta\Omega}$ directed along the rotation axis (in accordance with the right-hand screw rule). In the course of proper rotations of the coordinate frame the axial vector $\boldsymbol{\delta\Omega}$ behaves as a polar vector with a corresponding trace also equal to $(1 + 2\cos\varphi)$. Therefore, in order to determine the character of the full representation corresponding to the vibrational degrees of freedom, we have to subtract $2(1 + 2\cos\varphi)$ from the trace (8.15), so that

$$\chi^u(C) = (N_C - 2)(1 + 2\cos\varphi). \tag{8.16}$$

The character of the unit element ($N_C = N$, $\varphi = 0$) is equal to $\chi^u(E) = (N - 2) \times 3 = 3N - 6$. The character of the full vibrational representation for the rotary-reflection transformation $S(\varphi)$, i.e., for the rotation through the angle $\varphi$ about the $z$ axis and reflection in the $xy$ plane, is computed in the same way. If this transformation does not displace the atomic nucleus, then the displacement is transformed, by analogy with (8.14), in accordance with the equations

$$\hat{P}_{S(\varphi)}u_\alpha = \sum_{\beta=1}^{3} A(S)_{\beta\alpha}u_\beta = \begin{cases} u_x\cos\varphi + u_y\sin\varphi, \\ -u_x\sin\varphi + u_y\cos\varphi, \\ -u_z, \end{cases} \tag{8.17}$$

the corresponding trace being equal to $(-1 + 2\cos\varphi)$. The character of the representation realized by all the $3N$ degrees of freedom is equal to

$$N_S(-1 + 2\cos\varphi), \tag{8.18}$$

where $N_S$ is the number of nuclei remaining stationary in the course of the transformation $S(\varphi)$. The trace corresponding to the vector of displacement of the centre of gravity, $\mathbf{U}$, in the transformation $S(\varphi)$ is $(-1 + 2\cos\varphi)$, and that corresponding to the axial vector $\boldsymbol{\delta\Omega}$ in the same transformation is $(1 - 2\cos\varphi)$; the latter follows from the fact that a reflection in the $xy$ plane transforms a right coordinate frame into a left one with the axial vector $\boldsymbol{\delta\Omega}$ reversing its direction, i.e., in the right-hand side of (8.17) $-u_\alpha$ should be substituted for $u_\alpha$. We see that the total trace of $\mathbf{U}$ and $\boldsymbol{\delta\Omega}$ under the transformation $S(\varphi)$ is $-1 + 2\cos\varphi + 1 - 2\cos\varphi = 0$, and therefore (8.18) is the character of the full representation of all the vibrational degrees of freedom under the transformation $S(\varphi)$:

$$\chi^u(S) = N_S(-1 + 2\cos\varphi). \tag{8.19}$$

If $S(\varphi)$ consists only of reflection, i.e., if $S = \sigma$ ($\varphi = 0$), then $\chi^u(\sigma) = N_S$.

To be able to classify normal vibrations of a molecule, it suffices now to decompose the full representation (8.16), (8.19) into the irreducible representations of the point group of the molecule's symmetry.

Consider now the $NH_3$ molecule mentioned above. As has already been mentioned, the point group of its symmetry, $C_{3v}$, consists of six elements distributed over three classes: $E$, $2C_3$, $3\sigma_v$ (see Sec. 2.3, Group IV $C_{nv}$).

Using equations (8.16) and (8.19), we can determine the characters of the full representation $D^u$ for the $NH_3$ molecule:

$$\begin{aligned}\chi^u(E) &= (N-2)(1+2\cos 0°) = \\ &= (N-2)\times 3 = (4-2)\times 3 = 6, \\ \chi^u(C_3) &= (N_C - 2)(1 + 2\cos 120°) = \\ &= -(1+2\cos 120°) = 0, \\ \chi^u(\sigma) &= N_6 = 2. \end{aligned} \tag{8.20}$$

We can easily show, making use of equations (6.34), (8.20) and Table 2.3 for the characters of group $D_3$ isomorphic to group $C_{3v}$, that the reducible representation $D^u$ splits into the following irreducible representations of group $C_{3v}$:

$$D^u = 2\Gamma_1 + 2\Gamma_3 = 2A_1 + 2E, \tag{8.21}$$

where we, in accordance with the convention existing in molecular physics, denoted the unit representation $\Gamma_1$ by $A_1$ and the double degenerate representation $\Gamma_3$ by $E$. Hence, the $NH_3$ molecule, which possesses six vibrational degrees of freedom, has two nondegenerate normal vibrations of the $A_1$-type and two double degenerate vibrations of the $E$-type. The full symmetry of the molecule (a regular triangular pyramid) is retained in the vibrations of the $A_1$-type corresponding to the unit representation. We shall not dwell on the nature of atomic displacements in the vibrations of the $E$-type.

## 2.9 Application of Group Theory to Translational Symmetry of Crystals

**2.9.1** The most important kind of symmetry of a solid is its translational symmetry, thanks to which the crystal lattice coincides with itself when the crystal as a whole is displaced by a lattice vector

$$\mathbf{a_n} = n_1\mathbf{a}_1 + n_2\mathbf{a}_2 + n_3\mathbf{a}_3, \tag{9.1}$$

where $\mathbf{a}_1$, $\mathbf{a}_2$, $\mathbf{a}_3$ are the basis vectors and $n_i$ are integers (see Section 1.3.1). The elements of the crystal's translation group $\mathscr{T}$, an invariant subgroup of the crystal's space group, can be written in the form (5.4):

$$\begin{aligned} t_\mathbf{n} &= \{E \mid \mathbf{a_n}\} = \{E \mid n_1\mathbf{a}_1 + n_2\mathbf{a}_2 + n_3\mathbf{a}_3\} \\ &= \{E \mid n_1\mathbf{a}_1\}\{E \mid n_2\mathbf{a}_2\}\{E \mid n_3\mathbf{a}_3\} = t_{n_1}t_{n_2}t_{n_3}, \end{aligned} \tag{9.2}$$

where $R = E$ is the unit element of the crystal's point group.

It may be seen from (9.2) that the group $\mathcal{T}$ can be represented in the form of a direct product of three one-dimensional translation groups with the elements

$$t_{n_i} = \{E \mid n_i \mathbf{a}_i\} \quad (i = 1, 2, 3).$$

The translation group is Abelian (and cyclic); therefore each of its elements is a class, and the dimension of its irreducible representations is unity, i.e., they are numbers (in general, complex). Indeed, for Abelian groups the number of classes $s$, equal to the number of irreducible representations, is equal to the group's order $h$. It then follows from (6.22) that the dimension of all irreducible representations $l_i = 1$.

To avoid difficulties which, in case of a finite crystal, are connected with the fixing of boundary conditions, we divide the crystal into identical parallelepipeds with edges $G\mathbf{a}_1$, $G\mathbf{a}_2$, $G\mathbf{a}_3$, where $G$ is a large integer (for convenience, $G$ is assumed to be an odd number). We shall substitute the *Born-von Kármán cyclic conditions* for the boundary conditions. These new conditions specify that all physical properties and functions (the wave function, quasi-elastic force constants, etc.) attain the same values at point $\mathbf{r}$ and $\mathbf{r} + G\mathbf{a}_i$ ($i = 1, 2, 3$), i.e., are periodically repeated in all the parallelepipeds into which we have divided the infinite crystal.

Thus, using cyclic conditions enables us to consider all the phenomena and all the properties of the crystal within a single parallelepiped isolated in it (the *principal region*), of the volume $V = G^3\Omega_0$, where $\Omega_0 = \mathbf{a}_1 \cdot (\mathbf{a}_2 \times \mathbf{a}_3)$ is the volume of the crystal's unit cell.

In the case of a three-dimensional crystal it is impossible to attribute to the Born-von Kármán cyclic conditions such graphic meaning as in the case of a one-dimensional chain of equally spaced atoms (see Section 3.2.3). However, it can be demonstrated that in the mathematical sense the cyclic conditions for a three-dimensional crystal are equivalent to any boundary conditions on the surface of the principal region that do not affect the volume properties of the crystal [2.15, p. 391].

The cyclic conditions reduce the cyclic translation group to a finite group, since the element corresponding to the displacement $G\mathbf{a}_i$ is the unit displacement element $\{E \mid G\mathbf{a}_i\} = \{E \mid 0\}$; hence the order of the translation group in $\mathcal{T}$ is $G^3$.

Consider the irreducible representations of the translation group $\mathcal{T}$. Determine first the irreducible representations of the group of one-dimensional translations along the vector $\mathbf{a}_1$ made up of the elements $t_{n_1} = \{E \mid n_1\mathbf{a}_1\}$.

Let the irreducible representation (a number) corresponding to the element $\{E \mid \mathbf{a}_1\}$ be $s$:

$$\{E \mid \mathbf{a}_1\} \to s,$$

which yields

$$\{E \mid 2\mathbf{a}_1\} = \{E \mid \mathbf{a}_1\}\ \{E \mid \mathbf{a}_1\} \rightarrow s \times s = s^2,$$
$$\{E \mid n_1\mathbf{a}_1\} = \{E \mid \mathbf{a}_1\} \ldots \{E \mid \mathbf{a}_1\} \rightarrow s \times s \times \ldots \times s = s^{n_1},$$
$$\{E \mid G\mathbf{a}_1\} = \{E \mid 0\} \rightarrow s^G = 1. \tag{9.3}$$

The latter equality follows from the fact that the irreducible representation (a number) corresponding to the unit translation element $\{E \mid 0\}$ is unity. Hence [2.16, p. 480].

$$s = \sqrt[G]{1} = \exp\left(\frac{2\pi i}{G} g_1\right), \tag{9.4}$$

where

$$g_1 = 0, 1, 2, \ldots, G-1.$$

It follows from (9.3) and (9.4) that there are $G$ irreducible representations to correspond to the element $t_{n_1} = \{E \mid n_1\mathbf{a}_1\}$:

$$\Gamma_{g_1}(t_{n_1}) = s^{n_1} = \exp\left(\frac{2\pi i}{G} g_1 n_1\right), \quad g_1 = 0, 1, 2, \ldots, G-1. \tag{9.5}$$

Thus, in agreement with the general theory, the number of irreducible representations is equal to the number of classes (elements) [see (6.30)]. The irreducible elements for the other two groups of one-dimensional translations with the elements $t_{n_2} = \{E \mid n_2\mathbf{a}_2\}$ and $t_{n_3} = \{E \mid n_3\mathbf{a}_3\}$ are, by analogy, equal to

$$\Gamma_{g_2}(t_{n_2}) = \exp\left(\frac{2\pi i}{G} g_2 n_2\right), \quad g_2 = 0, 1, 2, \ldots, G-1,$$
$$\Gamma_{g_3}(t_{n_3}) = \exp\left(\frac{2\pi i}{G} g_3 n_3\right), \quad g_3 = 0, 1, 2, \ldots, G-1. \tag{9.6}$$

It was demonstrated in Section 2.6.5 that the irreducible representation of a direct product of groups is equal to the direct product of their irreducible representations. As has already been noted above, the three-dimensional group of translations $\mathcal{T}$ is equal to the direct product of three one-dimensional translation groups with the elements $t_{n_1}, t_{n_2}, t_{n_3}$; therefore, the irreducible representation of the group $\mathcal{T}$ is

$$\Gamma_{g_1 g_2 g_3}(t_n) = \Gamma_{g_1}(t_{n_1})\,\Gamma_{g_2}(t_{n_2})\,\Gamma_{g_3}(t_{n_3})$$
$$= \exp\frac{2\pi i}{G}(g_1 n_1 + g_2 n_2 + g_3 n_3), \tag{9.7}$$

in accordance with (9.5) and (9.6).

We introduce the *wave vector*

$$\mathbf{k} = \frac{g_1}{G}\mathbf{b}_1 + \frac{g_2}{G}\mathbf{b}_2 + \frac{g_3}{G}\mathbf{b}_3, \tag{9.8}$$

where $\mathbf{b}_1$, $\mathbf{b}_2$ and $\mathbf{b}_3$ are the basis vectors of the reciprocal lattice (1.3.8). The dimension of $\mathbf{k}$ is obviously that of $\mathbf{b}_i$, i.e., inverse length ($cm^{-1}$). Three numbers $g_1$, $g_2$, $g_3$ determine the vector $\mathbf{k}$. It follows from (1.3.9) that the exponent in (9.7) can be represented in the form $i\mathbf{k}\cdot\mathbf{a_n}$; hence (9.7) can be written in the form

$$\Gamma_{\mathbf{k}}(t_{\mathbf{n}}) = \exp(i\mathbf{k}\cdot\mathbf{a_n}), \tag{9.9}$$

where the wave vector $\mathbf{k}$ numbers the irreducible representation. If the vector $\mathbf{k}' = \mathbf{k} + \mathbf{b_m}$, where $\mathbf{b_m} = m_1\mathbf{b}_1 + m_2\mathbf{b}_2 + m_3\mathbf{b}_3$ is an arbitrary vector of the reciprocal lattice, is substituted for $\mathbf{k}$, then

$$\begin{aligned}\exp(i\mathbf{k}'\cdot\mathbf{a_n}) &= \exp(i\mathbf{k}\cdot\mathbf{a_n})\exp(i\mathbf{b_m}\cdot\mathbf{a_n})\\ &= \exp(i\mathbf{k}\cdot\mathbf{a_n})\exp(2\pi i \times \text{integer}) = \exp(i\mathbf{k}\cdot\mathbf{a_n}),\end{aligned}$$

where we made use of equation (1.3.9).

Hence, $\mathbf{k}$ and $\mathbf{k}'$ correspond to the same irreducible representation, i.e., they are equivalent. For this reason only the values of $\mathbf{k}$ for which the $\mathbf{k}\cdot\mathbf{a}_i$ ($i = 1, 2, 3$) lie inside the $2\pi$ interval need be considered (because if $\mathbf{k}\cdot\mathbf{a}_i$ lies outside this interval, it can be reduced to it by excluding from $\mathbf{k}$ the appropriate vector of the reciprocal lattice), we may, for instance, put

$$-\pi < \mathbf{k}\cdot\mathbf{a}_i < \pi \quad (i = 1, 2, 3). \tag{9.10}$$

Substituting the value of $\mathbf{k}$ from (9.8), we obtain

$$-G/2 < g_i < G/2, \tag{9.10a}$$

i.e. we arrive at the same conditions (9.5) and (9.6), which specify that $g_i$ assumes $G$ values [(9.10a) makes it clear why it is convenient to choose an odd number for $G$].

We can choose the unit cell of the reciprocal lattice with the "volume" $(2\pi)^3/\Omega_0$ as the region of nonequivalent values of the vector (9.10). Such a choice of the region (9.10) is not always convenient since the parallelepiped built on the vectors $\mathbf{b}_1$, $\mathbf{b}_2$, $\mathbf{b}_3$, in general, does not possess the symmetry of the lattice itself (of the direct or the reciprocal lattice). In the same way as we have built the Wigner-Seitz symmetrical cell (see Section 1.1.4), in the case of the reciprocal lattice we can also always isolate a region (9.10) possessing the full symmetry of the reciprocal (and the direct) lattice. To this end draw the vectors $\mathbf{b_g}$ from some site $O$ of the reciprocal lattice to all its other sites. Draw planes perpendicular to the vectors $\mathbf{b_g}$ through their midpoints. The equations for these planes are of the form

$$\mathbf{b}_{\mathbf{g}}^2/2 - (\mathbf{b_g}\cdot\mathbf{k}) = 0, \tag{9.11}$$

where $\mathbf{k}$ is a position vector drawn from the origin $O$ to some point of the plane. Note that (9.11) coincides with the conditions of X-ray diffraction in the crystal (1.4.5) [the difference in signs of the scalar products in (1.4.5) and (9.11) is because **k** in (1.4.5) points from the plane to the origin $O$; see Fig. 1.16]. Those intersecting planes define certain polyhedrons with a common centre at point $O$. The smallest polyhedron with its centre at point $O$ is termed *first*, or *reduced*, *Brillouin zone* (sometimes simply *Brillouin zone*); its volume is also equal to $(2\pi)^3/\Omega_0$ (see below). The "volume" contained between the surface of the first Brillouin zone and the surface of the next polyhedron is termed *second Brillouin zone*. And so on. It can be demonstrated that the volumes of all the Brillouin zones are equal to $(2\pi)^3/\Omega_0$.

As has already been mentioned in Section 1.3, the reciprocal lattice of a simple cubic lattice is also a simple cubic lattice; hence it follows immediately that in this case the first Brillouin zone has the shape of a cube with an edge $2\pi/a$ ($a$ is the edge of the cube of the direct lattice); the "volume" of the Brillouin zone is $(2\pi/a)^3 = (2\pi)^3/a^3 = (2\pi)^3/\Omega_0$.

It follows from the method of construction of the first Brillouin zone that the difference between the wave vectors **k** whose ends lie inside it is less than the vector of the reciprocal lattice. If, on the other hand, the end of the vector **k** lies on the boundary of the Brillouin zone, there is always at least one vector $\mathbf{k}' = \mathbf{k} \pm \mathbf{b}_i$ equivalent to it whose end also lies on the boundary.

The first Brillouin zone may also be defined as a set of points; the distance from this set of points to a specified site of the reciprocal lattice is less than the distance to all other sites (only the points lying on the boundary are equidistant from two sites of the reciprocal lattice). Thus, all the space of the reciprocal lattice can be subdivided into reduced Brillouin zones built around each site.

Consider the reciprocal lattice that corresponds to the principal region of a crystal containing $G^3$ unit cells. The total volume of such a reciprocal lattice is equal to the volume of the reciprocal lattice unit cell, $(2\pi)^3/\Omega_0$, multiplied by $G^3$. On the other hand, it is equal to the volume of the Brillouin zone multiplied by $G^3$; hence it follows that the volume of the Brillouin zone is equal to that of the unit cell of the reciprocal lattice, i.e., to $(2\pi)^3/\Omega_0$, as stated above.

A more thorough investigation of the Brillouin zones will be undertaken in Chapter 4 in conjunction with the study of the motion of the electron in an ideal crystal.

We have introduced the wave vector **k** only from translational symmetry considerations. And although in the following chapter, for instance, the wave vector will make its appearance in the solution of the mechanical problem of vibrations of atoms in a crystal, we must be aware that the more fundamental cause for its appearance is the crystal's translational symmetry.

2.9.2 Consider the effect of the operator $\hat{P}_R$ (6.1), where $R$ is a translation by the vector $-\mathbf{a}_n$, on the wave function $\psi_{\mathbf{k}}(\mathbf{r})$ of an electron moving in the periodic field of a crystal[24]. The label $\mathbf{k}$ of the wave function means that we are considering a basis function of the k-th irreducible representation.

Since the dimension of the irreducible representations of the translation group is unity and since these representations are equal to (9.9), it follows from (6.1) that

$$\hat{P}_R\psi_{\mathbf{k}}(\mathbf{r}) = \psi_{\mathbf{k}}(R^{-1}\mathbf{r}) = \psi_{\mathbf{k}}(\mathbf{r} + \mathbf{a}_n)$$
$$= \Gamma_{\mathbf{k}}\psi_n(\mathbf{r}) = e^{i\mathbf{k}\cdot\mathbf{a}_n}\psi_{\mathbf{k}}(\mathbf{r}),$$

i.e.,

$$\psi_{\mathbf{k}}(\mathbf{r} + \mathbf{a}_n) = e^{i\mathbf{k}\cdot\mathbf{a}_n}\psi_{\mathbf{k}}(\mathbf{r}). \tag{9.12}$$

It may easily be shown that for the electron wave function $\psi_{\mathbf{k}}(\mathbf{r})$ to satisfy (9.12), it is necessary that

$$\psi_{\mathbf{k}}(\mathbf{r}) = u_{\mathbf{k}}(\mathbf{r})\, e^{i\mathbf{k}\cdot\mathbf{r}}, \tag{9.13}$$

where the function $u_{\mathbf{k}}(\mathbf{r})$ has the periodicity of the lattice, i.e.,

$$u_{\mathbf{k}}(\mathbf{r} + \mathbf{a}_n) = u_{\mathbf{k}}(\mathbf{r}). \tag{9.13a}$$

Indeed, in this case it follows from (9.13) that

$$\psi_{\mathbf{k}}(\mathbf{r} + \mathbf{a}_n) = u_{\mathbf{k}}(\mathbf{r} + \mathbf{a}_n)\, e^{i\mathbf{k}\cdot(\mathbf{r}+\mathbf{a}_n)}$$
$$= u_{\mathbf{k}}(\mathbf{r})\, e^{i\mathbf{k}\cdot\mathbf{r}} e^{i\mathbf{k}\cdot\mathbf{a}_n} = e^{i\mathbf{k}\cdot\mathbf{a}_n}\psi_{\mathbf{k}}(\mathbf{r}),$$

in agreement with (9.12).

The term for the electron wave function in a crystal, (9.13), is *Bloch's wave function* (F. Bloch, 1928); its form is entirely due to the crystal's translational symmetry. The specific form of the function $u_{\mathbf{k}}(\mathbf{r})$ depends on the form of the periodic potential acting on the electron.

2.9.3 The stationary states of the electron in the crystal's periodic field are described with the aid of the Schrödinger equation

$$\hat{\mathscr{H}}(\mathbf{r})\,\psi_{n\mathbf{k}j}(\mathbf{r}) \equiv \left[-\frac{\hbar^2}{2m}\nabla^2 + V(\mathbf{r})\right]\psi_{n\mathbf{k}j}(\mathbf{r}) = \mathscr{E}_n(\mathbf{k})\,\psi_{n\mathbf{k}j}(\mathbf{r}) \tag{9.14}$$

where the Hamiltonian $\hat{\mathscr{H}}(\mathbf{r})$ is invariant under all symmetry transformations of the crystal.

Bloch's wave function of the electron, (9.13),

$$\psi_{n\mathbf{k}j}(\mathbf{r}) = u_{n\mathbf{k}j}(\mathbf{r})\, e^{i\mathbf{k}\cdot\mathbf{r}} \tag{9.15}$$

---

[24] Of course, we could have defined the operation $R$ as a translation by the vector $+\mathbf{a}_n$, but this would have resulted in less customary relations.

depends not only on the wave vector **k**, but also on the number of the energy band, $n$; in the general case, it can also depend on the additional quantum numbers $j$ characterizing different degenerate states of the electron for specified $n$ and **k**. The $\mathscr{E}_n(\mathbf{k})$ in (9.14) are eigenvalues of the electron's energy. Let the space group with the elements (Sec. 2.5.3)

$$g = \{ R \mid \boldsymbol{\alpha}(R) + \mathbf{a_n} \} \tag{9.16}$$

and with the reciprocal elements (see Section 2.5.3)

$$g^{-1} = \{R^{-1} \mid -R^{-1}\boldsymbol{\alpha}(R) - R^{-1}\mathbf{a_n}\} \tag{9.17}$$

be the crystal's symmetry group.

Operate with the operator $\hat{P}_g$ (6.1) on both sides of (9.14); since the crystal's periodic field $V(\mathbf{r})$ and the kinetic energy operator $(-\hbar^2/2m)\nabla^2$ remain unchanged under all transformations of the space group (9.16), it follows that

$$\mathscr{H}(\mathbf{r})\hat{P}_g\psi_{n\mathbf{k}j}(\mathbf{r}) = \mathscr{E}_n(\mathbf{k})\hat{P}_g\psi_{n\mathbf{k}j}(\mathbf{r}), \tag{9.18}$$

i.e., the function $\hat{P}_g\psi_{n\mathbf{k}j}(\mathbf{r})$ must also be Bloch's function corresponding to the same energy $\mathscr{E}_n(\mathbf{k})$.

It follows from (6.1), (9.17), and (9.15) that

$$\begin{aligned}\hat{P}_g\psi_{n\mathbf{k}j}(\mathbf{r}) &= \hat{P}_g[u_{n\mathbf{k}j}(\mathbf{r})e^{i\mathbf{k}\cdot\mathbf{r}}] = u_{n\mathbf{k}j}(g^{-1}\mathbf{r})e^{i\mathbf{k}\cdot g^{-1}\mathbf{r}}\\ &= u_{n\mathbf{k}j}(R^{-1}\mathbf{r} - R^{-1}\boldsymbol{\alpha}(R) - R^{-1}\mathbf{a_n})\\ &\times \exp[-i\mathbf{k}\cdot(R^{-1}\boldsymbol{\alpha} + R^{-1}\mathbf{a_n})]\, e^{i\mathbf{k}\cdot R^{-1}\mathbf{r}}.\end{aligned} \tag{9.19}$$

Vector $R^{-1}\mathbf{a_n}$ is equal to the lattice vector (this also holds for symmorphic groups, when $R^{-1}$ can be other than a symmetry element of a complex lattice). If we also take into account that the scalar product of two vectors does not change when the transformation $R$ is applied to each of them, i.e., when

$$\mathbf{k}\cdot R^{-1}\mathbf{r} = R\mathbf{k}\cdot RR^{-1}\mathbf{r} = R\mathbf{k}\cdot E\mathbf{r} = R\mathbf{k}\cdot\mathbf{r}, \tag{9.20}$$

then (9.19) can be written in the form

$$\hat{P}_g\psi_{n\mathbf{k}j}(\mathbf{r}) = \tilde{u}_{nR\mathbf{k}j}(\mathbf{r})e^{iR\mathbf{k}\cdot\mathbf{r}}, \tag{9.21}$$

where

$$\begin{aligned}\tilde{u}_{nR\mathbf{k}j}(\mathbf{r}) &= u_{n\mathbf{k}j}(R^{-1}\mathbf{r} - R^{-1}\boldsymbol{\alpha}(R)) \times\\ &\times \exp[-iR\mathbf{k}\cdot(\boldsymbol{\alpha}(R) + \mathbf{a_n})]\end{aligned} \tag{9.21a}$$

is a periodic function with the periods equal to that of the lattice (this follows from the fact that $R^{-1}\mathbf{a_m}$ is equal to the lattice period). Bloch's function (9.21) with the wave vectors $R\mathbf{k}$ satisfies the same Schrödinger equation (9.14) with energy eigenvalues $\mathscr{E}_n(\mathbf{k})$; in other words,

$$\mathscr{E}_n(\mathbf{k}) = \mathscr{E}_n(R\mathbf{k}). \tag{9.22}$$

Hence, the energy surfaces $\mathscr{E}_n(\mathbf{k}) = \text{const.}$ in a Brillouin zone for all energy bands have the symmetry of the crystal's point group.

The set of the vectors $R\mathbf{k}$, where $R$ is an element of the crystal's point group $\mathscr{F}$, is termed *wave vector star*.

Demonstrate that the invariance of the Schrödinger equation with respect to the reversal of time ($t \to -t$) in some cases leads to additional symmetry of the energy $\mathscr{E}_n(\mathbf{k})$.

Consider the time-dependent Schrödinger equation

$$i\hbar \frac{\partial \psi}{\partial t} = \hat{\mathscr{H}} \psi \tag{9.23}$$

and take the equation that is complex conjugate to it:

$$i\hbar \frac{\partial \psi^*}{\partial(-t)} = \hat{\mathscr{H}} \psi^* \tag{9.24}$$

(we assume the Hamiltonian $\hat{\mathscr{H}}$ to be real). We see that the evolution of the state $\psi^*$ in the direction of $-t$ is the same as that of the state $\psi$ in the direction of $t$. Since the probability density is proportional to $|\psi|^2$, it is not affected by time reversal. At the same time the velocities proportional to $(1/i)\,\psi^*\nabla\psi$ change their direction.

For a stationary state

$$\hat{\mathscr{H}} \psi_n = \mathscr{E}_n \psi_n, \tag{9.24}$$

whence

$$\hat{\mathscr{H}} \psi_n^* = \mathscr{E}_n \psi_n^*, \tag{9.24a}$$

and because of that $\psi_n$ and $\psi_n^*$ correspond to the same energy $\mathscr{E}_n$. If $\psi_n$ and $\psi_n^*$ are linearly independent, this results in additional degeneracy. It follows from (9.23) and (9.24) that this additional degeneracy is connected with the symmetry with respect to time reversal.

The wave function for the electron in a periodic field is of the form (9.15), therefore,

$$\psi_{n\mathbf{k}j}^*(\mathbf{r}) = u_{n\mathbf{k}j}^*(\mathbf{r})\, e^{-i\mathbf{k}\cdot\mathbf{r}}, \tag{9.25}$$

the differences, as compared to (9.15), being the substitution of $-\mathbf{k}$ for $\mathbf{k}$. Since both (9.15) and (9.25) satisfy the same equation (9.14), it follows that

$$\mathscr{E}_n(-\mathbf{k}) = \mathscr{E}_n(\mathbf{k}). \tag{9.26}$$

Hence, the constant-energy surfaces $\mathscr{E}_n(\mathbf{k}) = \text{const.}$ have a centre of symmetry, irrespective of whether the direct (reciprocal) lattice has such a centre.

If **k** is located inside the Brillouin zone and is directed along one of its axes or planes of symmetry, then there will be among the elements of the group $\mathscr{F}$ such that

$$R\mathbf{k} = \mathbf{k}. \tag{9.27}$$

If, on the other hand, the end of the vector **k** lies on the Brillouin zone boundary, then the symmetry transformation may transform **k** into an equivalent vector, i.e.,

$$R\mathbf{k} = \mathbf{k} \pm \mathbf{b}_i, \tag{9.27a}$$

where $\mathbf{b}_i$ is a basis vector of the reciprocal lattice.

The elements of the crystal's space group (9.16) whose point transformations $R$ satisfy relations (9.27) or (9.27a) constitute its subgroup and are termed *wave vector group* $G_\mathbf{k}$. The group $G_\mathbf{k}$ determines the degeneracy and the symmetry of Bloch's functions at point **k** of the Brillouin zone in all the bands of the energy, $n$. This means that every element of the group $G_\mathbf{k}$ transforms Bloch's function with a wave vector **k** and energy $\mathscr{E}_n(\mathbf{k})$ into other Bloch's functions with the same (or equivalent) wave vector and with the same energy. In order to apply group theory to the determination of the degeneracy and the symmetry of Bloch's functions at specified points of the **k**-space, we must know the irreducible representations of the group $G_\mathbf{k}$ at those points, or at least their characters.

It is easy to determine directly the characters of a point group $\mathscr{F}_\mathbf{k}$ whose elements $R$ leave the vector **k** unchanged or transform it into an equivalent one. Below we shall establish the relation between the irreducible representations $\Gamma(g)$ of the wave vector group $G_\mathbf{k}$ and the irreducible representations $\Gamma(R)$ of the point group $\mathscr{F}_\mathbf{k}$. As we shall see later, those representations coincide for symmorphic groups.

For the purpose of illustration, let us consider the group $\mathscr{F}_\mathbf{k}$ for a planar square lattice for different positions of the wave vector **k**. If the side of the square for the direct lattice is $a$, the Brillouin zone will also be in the shape of a square with the side $2\pi/a$ (Fig. 2.18). In this case the crystal's point group is $C_{4v}$ with eight elements: $E$, $2C_4$, $C_2$, $2\sigma_v$, $2\sigma_d$ ($\sigma_v$ are reflection planes parallel to the sides of the square, and $\sigma_d$ are reflection planes passing through its diagonals).

In the most general case (Fig. 2.18a) the wave vector star consists of eight different vectors **k** obtained as the result of the application of all eight elements of $C_{4v}$; the group $\mathscr{F}_\mathbf{k}$ consists in this case of one element $E$. In the case (*b*) the vector lies in the plane $\sigma_v$ ($AA'$), and the wave vector star consists of four vectors; in this case the group $\mathscr{F}_{\mathbf{k}}$ consists of two elements: $E$, $\sigma_v$. In case (*c*) the wave vector star consists also of four vectors; the group $\mathscr{F}_{\mathbf{k}}$ for one of them consists of the elements $E$ and $\sigma_d$. In cases (*b*), (*e*) and (*f*) the ends of the vectors **k** in the star are located on the Brillouin zone boundary. Since the vectors $k$ and $k' = k + 2\pi/a$ are equivalent, in case (*d*) the **k**-star consists of four vectors, and the group $\mathscr{F}_{\mathbf{k}}$ consists of two elements: $E$ and $\sigma_v$. In case (*e*) the wave vector star consists of two vectors, and the group $\mathscr{F}_{\mathbf{k}}$ of four elements $E$, $C_2$, $2\sigma_v$. Finally, in case (*f*) all four vectors illustrated are equivalent, and the group $\mathscr{F}$ coincides with the crystal's point group $C_{4v}$.

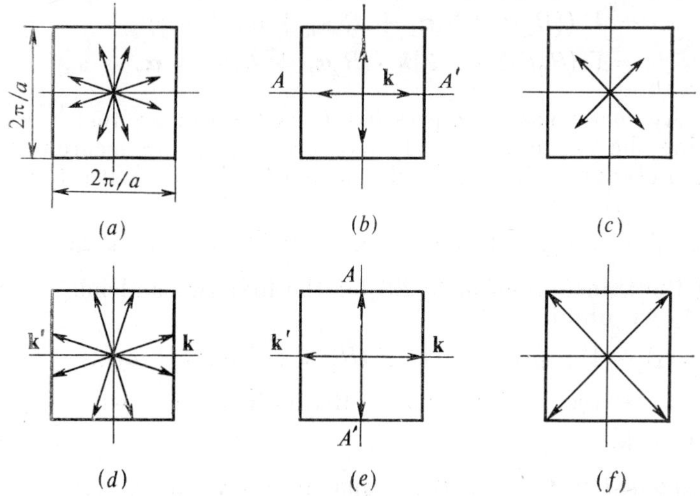


**Fig. 2.18**

**2.9.4** Demonstrate that if $\Gamma(R)$ is an irreducible representation of the point group $\mathscr{F}_{\mathbf{k}}$ then the irreducible representation of the wave vector group $G_{\mathbf{k}}$ is equal to (limitations are discussed below)

$$\Gamma(g) = \Gamma(\{R \mid \boldsymbol{\alpha}(R) + \mathbf{a}\}) = \Gamma(R) \exp i [\mathbf{k} \cdot (\boldsymbol{\alpha} + \mathbf{a})]. \qquad (9.28)$$

Here $g = \{R \mid \boldsymbol{\alpha}(R) + \mathbf{a}\}$ is an element of the group $G_{\mathbf{k}}$, $\mathbf{a} \equiv \mathbf{a}_n$ is a lattice vector and $\boldsymbol{\alpha}(R) \equiv \boldsymbol{\alpha}$ is an improper translation corresponding to the element $R$.

Since $\Gamma(R)$ is an irreducible representation of the group $\mathscr{F}_{\mathbf{k}}$, it follows that

$$\Gamma(R_2 R_1) = \Gamma(R_2)\, \Gamma(R_1). \qquad (9.29)$$

It follows from (9.28) that

$$\Gamma(g_2)\Gamma(g_1) = \Gamma(R_2)\Gamma(R_1)\exp i[\mathbf{k}\cdot(\boldsymbol{\alpha}_2 + \mathbf{a}_2) + \mathbf{k}\cdot(\boldsymbol{\alpha}_1 + \mathbf{a}_1)]. \tag{9.30}$$

Here $\mathbf{a}_1$ and $\boldsymbol{\alpha}_1 \equiv \boldsymbol{\alpha}(R_1)$ are a lattice vector and the improper translation corresponding to the element $R_1$ for the element $g_1 \in G_{\mathbf{k}}$; the meaning of $\mathbf{a}_2$ and $\boldsymbol{\alpha}_2$ is similar.

From (9.28) for the element $g_2g_1$ of the group $G_{\mathbf{k}}$ it follows that

$$\begin{aligned}\Gamma(g_2g_1) &= \Gamma(\{R_2 \mid \boldsymbol{\alpha}_2 + \mathbf{a}_2\}\{R_1 \mid \boldsymbol{\alpha}_1 + \mathbf{a}_1\})\\ &= \Gamma(\{R_2R_1 \mid R_2\boldsymbol{\alpha}_1 + R_2\mathbf{a}_1 + \boldsymbol{\alpha}_2 + \mathbf{a}_2\})\\ &= \Gamma(R_2R_1)\exp i[\mathbf{k}\cdot(R_2\boldsymbol{\alpha}_1 + R_2\mathbf{a}_1 + \boldsymbol{\alpha}_2 + \mathbf{a}_2)],\end{aligned} \tag{9.31}$$

where we made use of expression (5.8) for the product $g_2g_1$. Substituting the product $\Gamma(R_2)\Gamma(R_1)$ for $\Gamma(R_2R_1)$ in accordance with (9.29) and expressing each of the multiplicands with the aid of (9.28), we obtain

$$\Gamma(g_2g_1) = \Gamma(g_2)\Gamma(g_1)\exp i[\mathbf{k}\cdot(R_2\boldsymbol{\alpha}_1 + R_2\mathbf{a}_1) - \mathbf{k}\cdot(\boldsymbol{\alpha}_1 + \mathbf{a}_1)].$$

Using the transformation (9.20) for the first two addends in the exponent, we obtain

$$\Gamma(g_2g_1) = \Gamma(g_2)\Gamma(g_1)\exp[i(R_2^{-1}\mathbf{k} - \mathbf{k})\cdot(\boldsymbol{\alpha}_1 + \mathbf{a}_1)]. \tag{9.32}$$

If the wave vector lies inside a Brillouin zone, then

$$R_2^{-1}\mathbf{k} = \mathbf{k}. \tag{9.33}$$

The exponent in (9.32) is then equal to unity and therefore

$$\Gamma(g_2g_1) = \Gamma(g_2)\Gamma(g_1), \tag{9.34}$$

i.e., $\Gamma(g)$ as determined by equality (9.28) is an irreducible representation of the wave vector group $G_{\mathbf{k}}$.

If the vector $\mathbf{k}$ terminates at some point of a Brillouin's zone surface, then for some $R_2^{-1}$ the wave vector equivalent to it is

$$R_2^{-1}\mathbf{k} = \mathbf{k} + \mathbf{b}, \tag{9.35}$$

where $\mathbf{b}$ is a vector of the reciprocal lattice. For a symmorphic space group $\boldsymbol{\alpha}_1 = \boldsymbol{\alpha}(R_1) = 0$, and the exponent in (9.32) is also equal to unity; indeed,

$$e^{i(R_2^{-1}\mathbf{k}-\mathbf{k})\cdot\mathbf{a}_1} = e^{i\mathbf{b}\cdot\mathbf{a}_1} = e^{2\pi i \times \text{integer}} = 1. \tag{9.36}$$

Hence, in this case, too $\Gamma(g)$ of (9.29) is an irreducible representation of the wave vector group $G_{\mathbf{k}}$. We encounter a more complicated case when the end of the vector $\mathbf{k}$ is located on a Brillouin zone boundary but the space group is not symmorphic. We shall not consider this case here (see Section 4.9).

## 2.10 Selection Rules

**2.10.1** The transition probability of a system acted upon by a perturbation $\hat{Q}$ from a state with the wave function $\psi_{i\alpha}$ ($\alpha$-th wave (basis) function of the $i$-th energy level of the irreducible representation $\Gamma_i$) to a state with the wave function $\psi_{k\beta}$ is proportional to the square of the modulus of the matrix element

$$M = \int \psi_{k\beta}^* \hat{Q} \psi_{i\alpha} \, d\tau, \tag{10.1}$$

where $d\tau$ is the product of the differentials of the coordinates of the system's configurational space.

In numerous cases we are ignorant of the system's wave functions, and because of this we are unable to calculate the matrix element (10.1). However, often it is enough to know whether the matrix element is zero or not. We shall now demonstrate how group theory provides an answer to this question, i.e., enables a *selection rule* to be formulated.

For the simple case of an optical electron in a free atom, when the electric field of a light wave acts as the perturbation, such rules are known from quantum mechanics. For dipole transitions the variation of the orbital quantum number should be equal to $\pm 1$ (i.e., $\Delta l = \pm 1$) and that of the magnetic quantum number to $\pm 1$ or 0 (i.e., $\Delta m = \pm 1$ or $\Delta m = 0$) [2.4, § 95]. Before considering a more general case we shall prove that

$$\int \psi_{i\alpha} \, d\tau = 0 \tag{10.2}$$

if the irreducible representation $\Gamma_i$ is not a unit one (an identity).

Subjecting the function under the integral sign in (10.2) to an orthogonal symmetry transformation $\hat{P}_R$, where $R$ is an element of the Schrödinger group, we obtain

$$\int \psi_{i\alpha} \, d\tau = \int \hat{P}_R \psi_{i\alpha} \, d\tau = \int \sum_{\beta} \Gamma_i (R)_{\beta\alpha} \, \psi_{i\beta} \, d\tau. \tag{10.3}$$

Here the first equality is based on the fact that the Jacobian of an orthogonal transformation $\mathbf{x}' = \mathbf{R}\mathbf{x}$ is equal to unity, and the second equality is based on (7.15).

Performing the summation on both sides of (10.3) over all the elements of the group, we obtain

$$h \int \psi_{i\alpha} \, d\tau = \sum_{\beta} \int \psi_{i\beta} \, d\tau \sum_{R} \Gamma_i (R)_{\beta\alpha}, \tag{10.4}$$

where $h$ is the group's order.

If $\Gamma_i$ is not a unit representation, then

$$\sum_R \Gamma_i(R)_{\beta\alpha} = 0.$$

This follows from (6.19) if it is assumed that the representation $j$ is a unit representation and $i$ is not. Hence, the right-hand side of (10.4) is zero, and thereby (10.2) has been proved.

Let individual multiplicands (functions) in the integrand of the matrix element (10.1) be transformed with the aid of irreducible representations $\Gamma_k$, $\Gamma_Q$, and $\Gamma_i$ of the Hamiltonian group. In that case the product of these multiplicands is a basis function of a reducible, in general, representation of the direct product $\Gamma_k \times \Gamma_Q \times \times \Gamma_i$ (see Section 2.7.3). If this direct product does not include a unit representation, there will be no basis function of the unit representation among its basis functions; in that case, in compliance with (10.2), the matrix element (10.1) will vanish.

Thus, to determine the selection rules it suffices to see whether an identity is contained in the direct product $\Gamma_k \times \Gamma_Q \times \Gamma_i$, the number of unit representations contained in the product determining the number of linearly independent matrix elements. This procedure can be simplified with the aid of the following theorem.

The direct product of two different irreducible representations, $\Gamma_l \times \Gamma_m$, does not contain a unit representation and the direct product of two identical irreducible representations $\Gamma_l \times \Gamma_l$ contains only one unit representation. Here is the proof of the theorem.

To determine the number of unit representations contained in a reducible representations we make use of (6.34). Substituting $a_j = = a_1$ and $\chi_j(R)^* = 1$ (the characters of a unit representation for all $R$'s are equal to unity), we obtain

$$a_1 = \frac{1}{h} \sum_R \chi(R).$$

For the direct product $\Gamma_l \times \Gamma_m$ the character is, according to (7.18),

$$\chi(R) = \chi_l(R)\,\chi_m(R),$$

whence

$$a_1 = \frac{1}{h} \sum_R \chi_l(R)\,\chi_m(R).$$

Hence, applying the theorem about the orthogonality of characters (6.26), we obtain

$$a_1 = 0 \text{ if } l \neq m,$$

$$a_1 = 1 \text{ if } l = m.$$

Thus, if the expansions of $\Gamma_k$ and $\Gamma_Q \times \Gamma_i$ in irreducible representations contain common irreducible representations, the matrix element $M$ (10.1) will be nonzero; otherwise it will be zero.

**2.10.2** Consider, by way of an example, the selection rules for dipole transitions (in this case $\hat{Q}$ = position vector = r = $\{x, y, z\}$) in a field of cubic symmetry $\boldsymbol{O}$.

As will be demonstrated in Section 4.8, the position vector **r**, i.e., the coordinates $x$, $y$, $z$, are transformed with the aid of the irreducible representation $\Gamma_{15}$ or $F_1$ of the $\boldsymbol{O}$ group.

Table 2.10

| $F_1 \times \Gamma_i$ | $E$ | $8(C_3; C_3^2)$ | $3C_4^2$ | $6C_2$ | $6C_4$ |
|---|---|---|---|---|---|
| $F_1 \times A_1$ | 3 | 0 | −1 | −1 | 1 |
| $F_1 \times A_2$ | 3 | 0 | −1 | 1 | −1 |
| $F_1 \times E$ | 6 | 0 | −2 | 0 | 0 |
| $F_1 \times F_1$ | 9 | 0 | 1 | 1 | 1 |
| $F_1 \times F_2$ | 9 | 0 | 1 | −1 | −1 |

Making use of equation (7.18), compile the table of characters (Table 2.10) of the direct product $F_1 \times \Gamma_i$, where $\Gamma_i$ is one of the irreducible representations of the $\boldsymbol{O}$ group corresponding to the basis functions $\psi_{i\alpha}$.

Decompose the direct products $F_1 \times \{A_1, A_2, E, F_1, F_2\}$ in the irreducible representations of the $\boldsymbol{O}$ group; making use of equation (6.34), we obtain

$$F_1 \times A_1 = F_1, \quad F_1 \times A_2 = F_2, \quad F_1 \times E = F_1 + F_2,$$
$$F_1 \times F_1 = A_1 + E + F_1 + F_2, \quad F_1 \times F_2 =$$
$$= A_2 + E + E_1 + F_2. \qquad (10.5)$$

From the theorems proved above we learn that the nonzero matrix elements $M$ correspond to allowed dipole transitions between states corresponding to the second representation in the direct product on the left-hand side of the equality and states corresponding to the irreducible representations on the right-hand side. This means that the allowed transitions are

$$F_1 \leftrightarrow A_1, E, F_1, F_2; \quad F_2 \leftrightarrow A_2, E, F_1, F_2.$$

This follows explicitly from the last two equalities in (10.5) (the first three equalities convey nothing new). At the same time the transitions $F_1 \rightleftarrows A_2$, $F_2 \rightleftarrows A_1$, $A_1 \rightleftarrows A_2$ and $E \rightleftarrows A_2$ are forbidden.

# 3. Vibrations of Atoms in Crystal Lattices

## 3.1 The Interaction of Atoms in Crystals

**3.1.1** The fact that under certain conditions atoms can form stable molecules and crystals proves that forces of attraction can act between them, being compensated at distances of the order of $10^{-8}$ cm by forces of repulsion. Almost in every case it appears more convenient to operate not with forces but with the potential energy of interatomic interaction $\mathcal{U}(R)$, which we assume to be dependent only on the distance $R$ between the atoms.

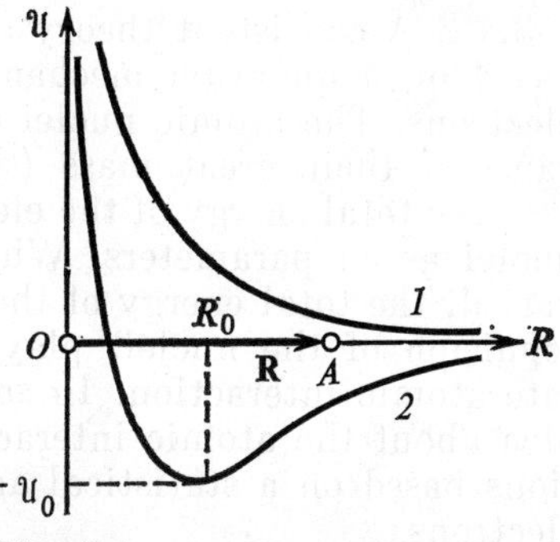


Fig. 3.1

The curves *1* and *2* in Fig. 3.1 depict possible cases of interaction between two atoms, one of which is placed in the origin $O$, and the other, $A$, can move along the $R$ axis. Since the potential energy is determined to within a constant term, we can always put $\mathcal{U} = 0$ for $R \to \infty$. The force acting on atom $A$ is $\mathbf{F} = -\operatorname{grad} \mathcal{U}(R) = -(d\mathcal{U}/dR)\mathbf{R}/R$, where $\mathbf{R} = \overrightarrow{OA}$ is the position vector drawn from $O$ to $A$. Hence, the forces of attraction act at those points where $d\mathcal{U}/dR > 0$ (**F** is antiparallel to **R**), and the forces of repulsion where $d\mathcal{U}/dR < 0$ (**F** is parallel to **R**). It may be seen from Fig. 3.1 that curve *1* corresponds to the case where the atoms repulse each other at all distances $R$. Curve *2* corresponds to a more complex case where the atoms attract each other at $R > R_0$, and repulse each other at $R < R_0$. The derivative $(d\mathcal{U}/dR)_{R_0} = 0$ at $R = R_0$, i.e., $\mathbf{F} = -(d\mathcal{U}/dR)_{R_0}\mathbf{R}/R = 0$, and the system of two atoms is in a state of stable equilibrium. In this case, as we immediately infer from the shape of curve *2*, $(d^2\mathcal{U}/dR^2)_{R_0} \equiv \beta > 0$. For small deflections of atom $A$ from the equilibrium position,

$$\mathcal{U}(R) = \mathcal{U}(R_0) + \left(\frac{d\mathcal{U}}{dR}\right)_{R_0}(R - R_0) + \frac{1}{2}\left(\frac{d^2\mathcal{U}}{dR^2}\right)_{R_0}(R - R_0)^2 + \frac{1}{6}\left(\frac{d^3\mathcal{U}}{dR^3}\right)_{R_0}(R - R_0)^3 + \ldots \quad (1.1)$$

to within terms of the third order of smallness. If, as is actually the case, the forces of repulsion close to the point $R = R_0$ grow more rapidly with a decrease in $R$ than the forces of attraction fall, then $(d^3\mathcal{U}/dR^3)_{R_0} \equiv -2\gamma < 0$. Denoting $\mathcal{U}(R_0)$ by $-\mathcal{U}_0$ and the deflection of atom $A$ from its equilibrium position $R - R_0$ by $x$, we obtain

$$\mathcal{U}(x) + \mathcal{U}_0 = \frac{1}{2}\beta x^2 - \frac{1}{3}\gamma x^3. \tag{1.2}$$

The force acting on atom $A$ as it moves near its equilibrium position along the $R$ axis is

$$F = -\frac{d\mathcal{U}}{dR} = -\beta x + \gamma x^2. \tag{1.3}$$

Note that the term for the force in the approximation $F = -\beta x$ is quasi-elastic. It may be reasoned that the order of magnitude of $\beta$ is $\gamma R_0$.

3.1.2 A consistent theory of atomic (ionic) interaction should be based on a quantum mechanical treatment of the motion of their electrons. The atomic nuclei can be presumed to remain at rest because of their great mass (the adiabatic approximation). In this case the total energy of the electrons depends on the position of the nuclei as on parameters. When the distance between the atoms is varied, the total energy of the electrons, together with the Coulomb repulsion of the nuclei, plays the part of the potential energy of interatomic interaction. In some cases it is possible to get a rough idea about the atomic interaction from more elementary considerations based on a statistical analysis of the behaviour of the atomic electrons.

Consider from this point of view the nature of the repulsion forces acting between the atoms (ions). Except in the trivial case of the Coulomb interaction between ions of like charge, which makes itself felt at great distances between them, the repulsion between atoms (ions) at short distances is the result of mutual penetration of their electron shells. This repulsion is due mainly to the rise in the kinetic energy of the atomic electrons on account of Pauli's principle.

To establish the nature of those forces we shall consider the electrons of the atoms (ions) as a degenerate Fermi gas at the absolute zero temperature. A simple calculation shows (Appendix 4) that the density of the kinetic energy of the electrons (kinetic energy per cubic centimetre) is $\mathcal{E}_k = (3^{5/3}\pi^{4/3}\hbar^2/10m)\, n^{5/3}$, where $n$ is the electron concentration (at a specified point), $2\pi\hbar$ Planck's constant, and $m$ the electron mass. As a zero approximation, we shall presume that the densities of electrons $n_a$ and $n_b$ in free atoms (ions) $a$ and $b$ (Fig. 3.2) do not change as the result of mutual penetration of their electron shells. This assumption corresponds to the use in quantum mechanical calculations of unperturbed wave functions. In this case the variation of the kinetic energy density in the region of overlapping electron shells (shaded area in Fig. 3.2) is

$$\Delta\mathscr{E}_k = \frac{3^{5/3}\pi^{4/3}\hbar^2}{10m}\left[(n_a+n_b)^{5/3} - n_a^{5/3} - n_b^{5/3}\right]. \qquad (1.4)$$

It may easily be seen that the effect of the overlapping of electron shells of atoms $a$ and $b$ is to increase the kinetic energy of the electrons in the system, i.e., to establish repulsive forces (the greater the smaller the distance between $a$ and $b$). We should add to them the repulsive forces of a purely Coulomb origin between the atomic nuclei $a$ and $b$, which appear when the nucleus of one atom penetrates the atomic shell of the other. A more detailed study shows that the inclusion of exchange effects into the statistical theory results in the appearance, as the result of mutual penetration, of certain attractive forces which, however, are unable to change the qualitative picture described above.

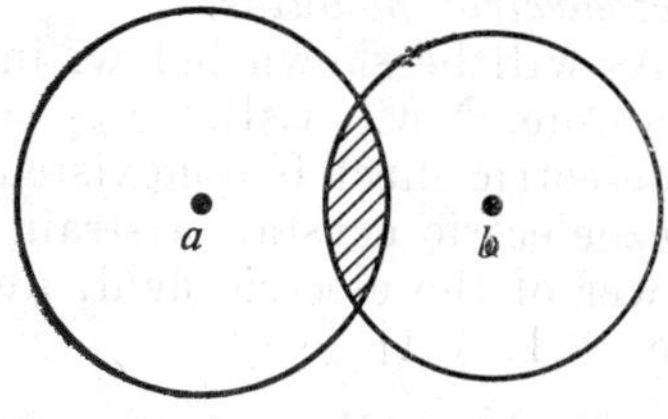

Fig. 3.2

The quantum mechanical theory of interatomic repulsive forces yields for the potential energy an expression of the form $A \exp(-R/a)$, where $A$ and $a$ are positive constants.

**3.1.3** For the formation of stable crystals there should be, in addition to the forces of repulsion between the atoms (ions), also forces of attraction acting between them. Usually four principal types of bonds in crystals are considered: (a) ionic or heteropolar, (b) covalent, (c) van der Waals or dispersion, and (d) metallic. It should be noted that in the majority of cases the bonds in crystals are of mixed character, therefore one often hears statements that a bond is covalent so many percent and ionic so many percent. When we speak of one type of bond, we mean that this type is prevalent.

The ionic bond in its purest form is realized in ionic crystals, for example, in alkali-halide compounds NaCl, KCl, CsCl. The interaction between the ions, in the first approximation, is considered as the interaction between point charges located at lattice sites. Since the ions of the first coordination group are always charged oppositely to the central ion, the Coulomb interaction of all the lattice ions results in some attraction, which provides for the stability of the lattice.

In a quantitative theory, because of the slow decrease of Coulomb forces with distance, we should take into account the interaction of the central ion with more distant ions of both signs. The next approximation takes into account the mutual polarization of the ions

In Section 1.2.4 we considered the lattices of some ionic crystals. In them the number of positive ions is equal to that of negative, and because of that both the crystal as a whole and its unit cell are neutral. When mechanical stress is applied to ionic crystals, they behave in different ways.

In some crystals, termed *piezoelectric*, the electric polarization $P_i$ ($i = x, y, z$) linearly depends on the components of the stress tensor $\sigma_{kl}$ [3.1, § 2]

$$P_i = \sum_{k,l} \gamma_{ikl}\sigma_{kl}. \tag{1.5}$$

Here $\gamma_{ikl}$ is the piezoelectric tensor of rank 3 (see Appendix 2) or the *piezoelectric modulus.*

As will be shown below, in crystals with an inversion centre (for example, NaCl, CsBr) $\gamma_{ikl} = 0$; therefore, in such crystals the piezoelectric effect is nonexistent. If an electric field $E_i$ is applied to a piezoelectric crystal, a strain appears in it proportional to the first power of the electric field, i.e., the components of the strain tensor are [3.1, § 1]

$$u_{kl} = \sum_i \gamma_{ikl}E_i. \tag{1.5a}$$

With the aid of thermodynamics one can demonstrate that the tensors $\gamma_{ikl}$ in (1.5) and (1.5a) coincide [3.2, Chap. X]. The strain of crystals in an electric field (1.5a) became known as the *converse piezoelectric effect*. Because of the symmetry of the strain tensor $u_{kl}$, the piezoelectric tensor $\gamma_{ikl}$ is symmetrical in $k$ and $l$, i.e., $\gamma_{ikl} = \gamma_{ilk}$. The latter condition brings the maximum number of independent components of the tensor $\gamma_{ikl}$ down to $3^3$ $3^2 = 27 - 9 = 18$.

Let us demonstrate that for a crystal with an inversion centre all the components of the $\gamma_{ikl}$ tensor are zeros, i.e., that such a crystal is not piezoelectric.

The components of the $\gamma_{ikl}$ tensor are transformed in the same way as a product of orthogonal coordinates, $x_i x_k x_l$. New, primed coordinates resulting from an inversion transformation are

$$x_1' = -x_1, \quad x_2' = -x_2, \quad x_3' = -x_3, \tag{1.6}$$

i.e., the cosines of the angles between the axes, $\alpha_{ik}$, are equal to $-\delta_{ik}$, where $\delta_{ik}$ is the Kronecker delta. Hence, an inversion transformation yields

$$\gamma_{ikl}' = \sum_{mnp} \alpha_{im}\alpha_{kn}\alpha_{lp}\gamma_{mnp} = -\sum_{mnp} \delta_{im}\delta_{kn}\delta_{lp}\gamma_{mnp} = -\gamma_{ikl}. \tag{1.6a}$$

But an inversion transformation is a symmetry transformation of the crystal and therefore the material tensor $\gamma_{ikl}$ cannot change in the process, i.e., $\gamma'_{ikl} = \gamma_{ikl}$; accordingly, it follows from (1.6a) that $\gamma_{ikl} = 0$.

Consider now the piezoelectric properties of the zinc blende crystal ZnS, which crystallizes in a lattice resembling that of germanium (Fig. 1.13) devoid of an inversion centre. The point symmetry group of the ZnS crystal is the tetrahedral group $T_d$. Making use of Table 4.2, which shows how the elements of the $T_d$ group transform the coordinates $x = x_1$, $y = x_2$, $z = x_3$, we apply the transformation[1] $R_2 (x_1, \bar{x}_2, \bar{x}_3)$ to the component $\gamma_{112}$, which transforms as $x_1^2 x_2$. We have $\gamma'_{112} = -\gamma_{112}$ and $\gamma'_{112} = \gamma_{112}$, whence $\gamma_{112} = 0$. By performing the same transformation $R_2 (x_1, \bar{x}_2, \bar{x}_3)$ we can show that

$$\gamma_{211} = \gamma_{113} = \gamma_{311} = \gamma_{222} = \gamma_{333} = 0,$$

too. Using the transformation $R_3 (\bar{x}_1, x_2, \bar{x}_3)$, we can demonstrate that

$$\gamma_{122} = \gamma_{212} = \gamma_{223} = \gamma_{322} = \gamma_{111} = 0;$$

a similar result is

$$\gamma_{133} = \gamma_{331} = \gamma_{233} = \gamma_{323} = 0,$$

which is obtained by applying $R_4(\bar{x}_1, \bar{x}_2, \bar{x}_3)$. Hence, out of eighteen independent components of the tensor $\gamma_{ikl}$ only three are nonzero: $\gamma_{123}$, $\gamma_{213}$, and $\gamma_{312}$. Making use of the transformations $R_9 (x_3, x_1, x_2)$ and $R_{15} (\bar{x}_3, \bar{x}_2, \bar{x}_1)$, we can show that

$$\gamma_{123} = \gamma_{213} = \gamma_{312} \equiv \gamma_0. \tag{1.7}$$

Hence, all the nonzero components of the tensor $\gamma_{ikl}$ are equal. We see that the piezoelectric properties of the ZnS crystal are characterized by one material constant $\gamma_0$.

**3.1.4** The covalent bond occurs between closely spaced ($\sim 10^{-8}$ cm) neutral atoms if certain conditions are fulfilled. In its simplest form, the covalent bond is realized between two hydrogen atoms in a hydrogen molecule $H_2$ (W. Heitler and F. London, 1927). The covalent bond cannot be interpreted in termsof classical physics. Special quantum mechanical features in the behaviour of a system of identical particles (electrons) are essential for the explanation of the covalent bond [3.3. § 131]. Classical physics was quite powerless to explain the properties of the saturated covalent bond, for example, the inability of a hydrogen atom to become attached to more than one other hydrogen atom. This property is characterized in chemistry by the concept of *valency*; it stems from the pairing of the electrons belonging to both atoms and the formation of a singlet state in which the electron spins are antiparallel (the curve of atomic

---

[1] The meaning of the symbols $R_i$ (...) is made clear in Section 4.8.2.

interaction for the triplet state in which the electron spins are parallel is of the form *1* in Fig. 3.1, i.e., the atoms repulse each other at all distances).

The covalent bond may occur not only between two hydrogen atoms but between other atoms possessing electrons capable of forming pairs with opposite spins. For instance, the nitrogen atom N has two electrons in the 1*s*-, two electrons in the 2*s*- and three electrons in the 2*p*-states, i.e., it has an electron structure denoted by $(1s)^2\ (2s)^2\ (2p)^3$. Spectroscopic data show that the spins of three electrons in the 2*p*-state are all parallel, i.e., there is no spin saturation, and, consequently, they are able to form three covalent bonds; hence, nitrogen is trivalent.

This is confirmed by experiment. Thus, for example, when nitrogen reacts with hydrogen, $NH_3$ is produced. The diatomic nitrogen molecule $N_2$ in which the atoms are bonded by three pairs of electrons with antiparallel spins is formed in the same way. The magnetic quantum numbers of the three 2*p* electrons in the nitrogen atom are $m = +1, -1, 0$, and the corresponding wave functions are of the form $\psi_{+1} = xf(r)$, $\psi_{-1} = yf(r)$ and $\psi_0 = zf(r)$, i.e., the electron clouds of the three valence electrons are elongated in three mutually orthogonal directions $x$, $y$ and $z$ [3.4, p. 32]. The gain in energy accompanying the formation of a covalent bond depends to a great degree on the overlapping of the wave functions of the electrons forming the appropriate pair with antiparallel spins. Thus, it may be reasoned that the hydrogen atoms in an $NH_3$ molecule will be arranged in three mutually perpendicular directions with respect to the nitrogen atom (*directed valencies*). Experiment confirms that the $NH_3$ molecule, indeed, is of the shape of a pyramid with the $\widehat{HNH}$ angle close to 90° (109°, to be precise). A somewhat greater angle between the directed valencies in the $NH_3$ molecule can be explained by the mutual repulsion of the hydrogen atoms.

The electron structure of the carbon atom C is $(1s)^2\ (2s)^2\ (2p)^2$. Since the spins in the *s*-states are saturated (antiparallel), the carbon atom should be bivalent. However, this conclusion is in contradiction with the data obtained in organic chemistry, according to which the valency of carbon is four. A more scrupulous theoretical and experimental investigation of the problem shows that the carbon atom takes part in the reactions not in its ground state but in an excited state: $(1s)^2\ (2s)^1\ (2p)^3$. In this case the spins of all four electrons are not saturated (the 2*s* electron has no partner, and the spins of the 2*p* electrons are parallel) and can participate in the formation of a covalent bond. More precisely, the carbon atom forms covalent bonds in the electron state, which is a superposition of one 2*s*- and three 2*p*-states. The coefficients (the weights) of each of those states in the linear combination and the directions of the four valence bonds are determined by the condition that the free energy of the molecule be minimal. Mathematical analysis, which we are not in a position to carry out here, shows the directions of the valence bonds to coincide with the *01*, *02*, *03*, and *04* directions in a tetrahedron (Fig. 1.11), and we know those directions to be 109,5°. Experiment shows that the methane molecule $CH_4$ does, indeed, have such a tetrahedral structure. The directed four valency of carbon atoms manifests itself in the formation of the diamond crystal, in which every carbon atom is located in the centre of a tetrahedron formed by four other carbon atoms (Fig. 1.13). The silicon (Si) atom has four electrons in its *M*-shell in the $(3s)^2$ $(3p)^2$-states, and, because the spins in its *K*- and *L*-shells are saturated, is expected to behave in a similar way to the carbon atom. The properties of the germanium (Ge) atom with four electrons in its *N*-shell in the $(4s)^2$ $(4p)^2$-states, of the tin (Sn) atom with four electrons in its *O*-shell in the $(5s)^2$ $(5p)^2$-states and of the lead (Pb) atom with four electrons in its *P*-shell in the $(6s)^2(6p)^2$-states are similar. Actually, silicon, germanium and grey tin all crystallize in the diamond-type lattice and belong to typical covalent crystals. As for normal (white) tin and lead, the covalent nature of atomic bonding in them does not make itself felt, because it is suppressed by the metallic properties of the material (see below).

Experimental studies in recent years have proved the chemical compounds of the $A^{III}B^{V}$ type, i.e., of the elements of groups III and V of the Periodic Table, to possess numerous properties (crystal lattice, electron band structure) typical for the elements of Group IV, germanium and silicon. InSb and GaAs belong to compounds of this type. Indium has three electrons in its *O*-shell in the $(5s)^2$ $(5p)^1$-states, and antimony five electrons in the $(5s)^2$ $(5p)^3$-states. Hence, just as is the case with Ge or Si, there are four electrons in the *s*-state and four in the *p*-state. If one of the *p* electrons of Sb partly goes over to In, a covalent bond can be formed similar to that formed in Si and Ge crystals. The same is true of the GaAs compound. Of course, the bond InSb and GaAs is not purely covalent, being partly ionic.

**3.1.5** The covalent bond occurs only if certain conditions are fulfilled. First, the atoms should have electrons capable of forming pairs with opposite spins (singlet state). Second, the spacing between the atoms should be small enough for the quantum mechanical properties based on the indistinguishability of identical particles constituting the system to make themselves manifest. Calculation shows the covalent forces to decrease rapidly (exponentially) with the distance. At larger distances all atomic systems begin to display certain universal attraction forces. Those forces are termed *van der Waals*, or *dispersion*, *forces* because, on the one hand, they are the cause of the divergence in behaviour of real gases from the ideal and, on the other, their parameters determine the dispersion of light by atoms. The van der Waals interaction provides bonding between particles in solids (argon, crypton, xenon and molecular crystals) in such cases where for some reason (closed electron shells, saturated valence bonds in the interacting molecules, etc.) there are no covalent, ionic, or metallic bonds (see below).

If the distance between the atomic systems $R \gg a$ ($a$ is the atomic system's dimension), the van der Waals (dispersion) forces can be calculated in the second approximation of the quantum mechanical perturbation theory. It can be demonstrated [3.5, § 89] that the van der Waals interaction energy of two atomic systems is

$$\mathcal{U}(R) = -W_0/R^6. \tag{1.8}$$

Here $R$ is the distance between the systems, and

$$W_0 \approx \frac{3}{2} \frac{J_a J_b}{J_a + J_b} \alpha_a \alpha_b, \tag{1.9}$$

where $J_a$, $J_b$ are ionization potentials and $\alpha_a$, $\alpha_b$ the polarizations of the atoms (molecules) $a$ and $b$, respectively.

The van der Waals forces, like the ionic forces, do not exhibit saturation effects characteristic for the covalent bond.

In addition to the ionic, covalent, and van der Waals forces discussed above, there are also *dipole* and *induced forces* generated by the permanent electric dipole moment of some molecules. They may be of importance in the case of complex molecular lattices and shall not be considered here.

The condition $R \gg a$ is not fulfilled in crystals in which the atoms or molecules are bonded by van der Waals forces (noble gases at low temperatures, molecular crystals) and for this reason equation (1.6) can serve at best for qualitative estimates.

3.1.6 Typical metals such as, for example, Li, Na, K, Cu, Ag, Fe, Ni, have some characteristic electrical, optical and mechanical properties. They all feature a relatively high electrical conductivity and light absorption coefficient and high plasticity and malleability. These properties unambiguously point to the fact that the metals contain a large number (of the order of the number of atoms) of free electrons, i.e., electrons that can travel over macroscopic distances in the crystal already in weak external electric fields. The simplest model of a metal, proposed by Paul Drude, is made up of positively charged ions located at the sites of a crystal lattice and of an ideal gas of "free electrons" moving between the ions. Despite all the changes and complexities introduced into the modern electron theory of metals (which we shall in part touch on below), such a straightforward model has not lost its importance; we need only take into account that the ideal gas of free electrons is strongly degenerate at all practically attainable temperatures of a metal.

As was first demonstrated by Ya. I. Frenkel (we do not consider subsequent perfections of his idea to be essential), the forces of cohesion in a metal crystal can be explained on the basis of the original simple free electron model [3.6. Chap. 2]. Frenkel considers the total bonding energy in a metal as consisting of two parts: the negative energy of Coulomb interaction between the free electrons and the positively charged ions (it is proportional to $1/a$, where $a$ is the lattice parameter) and the positive kinetic energy of the degenerate gas of "free" electrons (it is proportional to $n^{2/3} \sim 1/a^2$, where $n$ is the free electron concentration; see Appendix 4).

The total bonding energy will obviously have a minimum value for some value of $a$ shown by simple calculation to be of the order of $10^{-8}$ cm.

## 3.2 Vibrations and Waves in Simple One-Dimensional (Linear) Lattices

**3.2.1** The atoms of a crystal are at rest in the lattice sites only at the temperature of absolute zero[2]. As the temperature rises, the atoms begin to vibrate about their stable equilibrium positions; therefore, we must consider the dynamics of atomic motion in a crystal.

We shall start our investigation of the problem with the simplest case of identical atoms vibrating in a one-dimensional (linear) lattice in compliance with the laws of classical mechanics. As we shall subsequently learn, all the regularities obtained for this artificial one-dimensional model prove to be true for three-dimensional lattices as well. It will furthermore turn out that at sufficiently elevated temperatures the atomic motion in crystals does in fact conform to the laws of classical mechanics.

Figure 3.3 depicts a linear chain of identical atoms of mass $m$ deflected from their equilibrium sites $(n-1)$, $n$, $(n+1)$ by the amounts

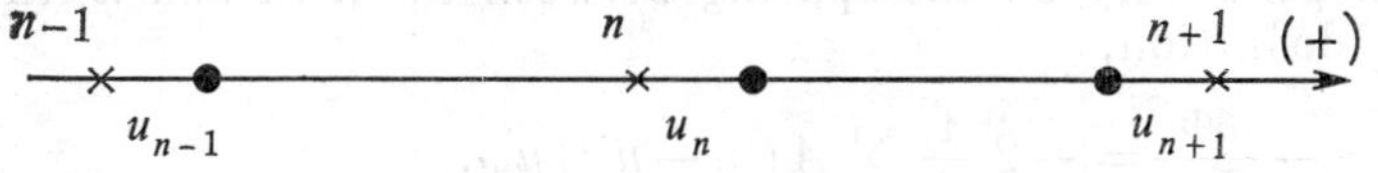


Fig. 3.3

$u_{n-1} > 0$, $u_n > 0$, $u_{n+1} < 0$. In the one-dimensional case we shall take into account only the interaction between the nearest neighbours, an assumption that little affects the final results. The deflec-

[2] We ignore here the existence of quantum effects, which are actually quite essential at low temperatures.

tions $u_n$ and the forces acting on the atoms are assumed to be positive if they coincide in direction with the direction of the positive axis and negative in the opposite case.

As was demonstrated in Section 3.1.1, for small deflections of the atoms from their equilibrium positions ($|u_n| \ll a$, here the distance between the sites) the interaction forces may be considered quasi-elastic, i.e., proportional to the variation of the interionic distance. Hence, the forces with which the $(n-1)$-st and $(n+1)$-st atoms act on the $n$-th atom are equal to $f_{n,n-1} = -\beta(u_n - u_{n-1})$ and $f_{n,n+1} = -\beta(u_n - u_{n+1})$, respectively, where $\beta > 0$ is the quasi-elastic force constant. The resultant force acting on the $n$-th atom is $f_n = f_{n,n-1} + f_{n,n+1} = -\beta(2u_n - u_{n-1} - u_{n+1})$. The equation of motion (mass × acceleration = force) for the $n$-th atom is of the form

$$m\ddot{u}_n + \beta(2u_n - u_{n-1} - u_{n+1}) = 0, \tag{2.1}$$

where

$$\ddot{u}_n \equiv \frac{d^2u_n}{dt^2}.$$

The expression for the force $f_n$ can be obtained otherwise. The potential energy of the lattice, $\Phi$, is a function of atomic displacements $u_n$. Expanding $\Phi$ in a power-series in the small displacements $u_n$, we obtain

$$\Phi = \Phi(0) + \sum_n \left(\frac{\partial\Phi}{\partial u_n}\right)_0 u_n + \frac{1}{2}\sum_n\sum_{n'}\left(\frac{\partial^2\Phi}{\partial u_n\,\partial u_{n'}}\right)_0 u_n u_{n'} + \ldots, \tag{2.2}$$

where the subscript 0 shows that all the $u_n$'s are assumed to vanish. Without loss of generality we can assume the potential energy of the ground state to vanish: $\Phi(0) = 0$. Since $u_n = 0$ corresponds to the equilibrium of the system, it follows that $(\partial\Phi/\partial u_n)_0 = 0$. For an infinite atomic chain the coefficients $(\partial^2\Phi/\partial u_n\,\partial u_{n'})_0 = A(|n-n'|)$, i.e., depend only on the spacing between the $n$-th and $n'$-th sites.

By definition,

$$f_n = -\frac{\partial\Phi}{\partial u_n} = -2\,\frac{1}{2}\sum_{n'} A(|n-n'|)\,u_{n'}. \tag{2.3}$$

If all the $u_{n'}$ = const. $= u_0$, the force $f_n = 0 = -u_0\sum_{n'} A(|n-n'|)$, where $n'$ runs through all the values. If only the nearest neighbours of the $n$-th atom are taken into account, then $n' = n, n+1, n-1$, and we obtain $A(0) + A(1) + A(-1) = 0$. It may easily be seen that (2.3) yields the same expression for the force $f_n$ as (2.1) if we make $A(1) = -\frac{1}{2}A(0) = \beta$.

**3.2.2** At first glance the solution to an infinite system of coupled equations (2.1), in which there are the unknown functions $u_{n-1}(t)$ and $u_{n+1}(t)$ presents considerable difficulties. Therefore, it is surprising what an elementary trick can be used to solve the system (2.1). An infinite atomic chain of atoms interacting with a quasi-elastic force resembles a tensioned string. It is well-known that for an endless string there is a simple type of motion in the form of a traveling monochromatic wave for which the deflection $u$ of the string from its equilibrium position at points $x$ at the moment $t$ is $u(x, t) = A \sin 2\pi \times \times (x/\lambda - \nu t)$, where $A$ is the amplitude, $\lambda$ the wave length, and $\nu$ the frequency. Introducing the cyclic frequency $\omega = 2\pi\nu$ and the wave number $q = 2\pi/\lambda$, we obtain $u(x, t) = A \sin(qx - \omega t)$. If we agree to consider not only the positive but also the negative values of $q$, we obtain in addition to the wave propagating in the positive direction of the $x$ axis ($q > 0$) also waves traveling in the opposite direction ($q < 0$). Finally, if we take into account that the equation of string vibrations is linear[3], so that the sum of the solutions is also a solution of the equation, we shall in many cases find it more convenient from the point of view of mathematics to use the complex form of the solution:

$$u(x, t) = A[\cos(qx - \omega t) + i \sin(qx - \omega t)] = Ae^{i(qx-\omega t)},$$

where the amplitude $A$ may be a complex number.

As we shall learn below, a traveling wave in a continuous string has two peculiar features that distinguish it from waves propagating in discrete atomic chains. First, the absolute value of the wave number $q$ can assume any values from 0 to $\infty$ with a definite shape of wave corresponding to each value of $q$. Second, the frequency $\omega = v_0 |q|$ can also change from 0 to $\infty$.

Try to solve the system (2.1) by substitution

$$u_n = Ae^{i(qan-\omega t)}, \tag{2.4}$$

where instead of the continuous coordinate $x$ there is a discrete quantity $an$ ($a$ is the spacing between the sites). The amplitude $A$ is independent of the site number $n$. Substituting (2.4) into (2.1), we obtain after canceling out $Ae^{i(qan-\omega t)}$:

$$-m\omega^2 = -\beta(2 - e^{-iqa} - e^{-iqa}).$$

---

[3] The equation for natural vibration of a string is of the form $\partial^2 u/\partial t^2 = v_0^2 (\partial^2 u/\partial x^2)$, where $v_0 = (T_0/\rho)^{1/2}$, $T_0$ is the tension, and $\rho$ is the linear density of the string; see [3.7, pp. 417-9].

Using the identity $\cos\alpha = \frac{1}{2}(e^{-i\alpha} + e^{i\alpha})$, we obtain

$$\left.\begin{aligned} &\omega^2 = 2\frac{\beta}{m}(1-\cos qa) = 4\frac{\beta}{m}\sin^2\frac{qa}{2} \\ &\text{or} \\ &\omega = \omega_{max}\left|\sin\frac{qa}{2}\right| = \omega_{max}\left|\sin\frac{\pi a}{\lambda}\right|, \end{aligned}\right\} \tag{2.5}$$

where

$$\omega_{max} = 2\sqrt{\beta/m}. \tag{2.5a}$$

We see that the solutions (2.4) of the traveling-wave type satisfy (2.1) for any $n$ if a *dispersion relation* (2.5) is established between the frequency $\omega$ and the wave number $q$ (or the wave length $\lambda$). Hence, a discrete atomic chain exhibits dispersion, i.e., the frequency $\omega$ is not proportional to the wave number $q$, as was the case for a continuous string.

We substitute in (2.4) $q' = q + 2\pi g/a$, where $g$ is a positive or a negative integer, for $q$. The new wave will be

$$u'_n = Ae^{i(q'an-\omega t)} = Ae^{i(qan-\omega t)}e^{i2\pi gn} = u_n,$$

since $\exp(i2\pi gn) = \exp(i2\pi \times \text{integer}) = 1$. Hence, the wave $u'_n$ is an identical (at all the points and at all moments of time) copy of the wave $u_n$. This means that $q$ and $q'$ are physically indistinguishable. In other words, only the variations of $q$ inside a $2\pi/a$ interval need be considered. All the physical properties of our one-dimensional crystal that depend on the wave number must be periodic with the period $2\pi/a$. We shall choose as the basic interval of variation of $q$ the region

$$-\pi/a \leqslant q \leqslant \pi/a. \tag{2.6}$$

Figure 3.4 depicts the dependence of $\omega$ on $q$, which in accordance with the aforesaid is periodic with a period $2\pi/a$. In most cases only the positive values of $q$ from 0 to $q_{max} = \pi/a$ need be considered, since the curve for $q < 0$ is symmetrical. A minimum wave length $\lambda$ corresponds to the maximum $q$. It follows from the condition $q_{max} = 2\pi/\lambda_{min} = \pi/a$ that $\lambda_{min} = 2a$. It is quite easy to see why there can be no waves with $\lambda/2$ shorter than $a$ in a discrete atomic chain. There is a maximum frequency $\omega_{max}$ (2.5a) to correspond to the wave with the shortest wave length $\lambda = \lambda_{min} = 2a$. The existence of a maximum frequency is also a characteristic feature of vibrations of discrete atomic structures.

**3.2.3** Macroscopic crystal specimens consist of a very large, but finite, number of atoms. If the number of atoms $G$ in an atomic chain is very great [4] and if the forces of atomic interaction are effective at distances up to one or several lattice parameters, the condition of the boundary atoms on the "surface" does not affect the motion of other atoms inside the chain. Specifically, if we were to arrange the $G$ atoms along a circumference of a very great radius, so that the $G$-th and the

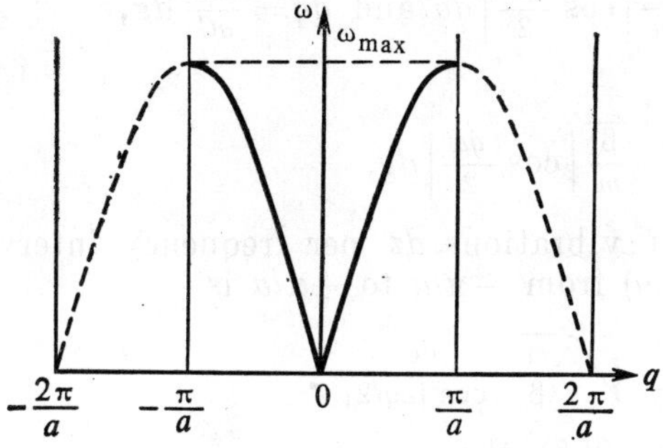


**Fig. 3.4**

first atoms would be in equilibrium at a distance $a$, we could substitute the Born-von Kármán cyclic conditions

$$u_{n\pm G} = u_n \tag{2.7}$$

for the boundary conditions, since in this case the $(n \pm G)$-th atom would coincide with the $n$-th atom.

It follows from the cyclic conditions (2.7) and expression (2.4) that $\exp(\pm iqaG) = 1$, i.e., $qaG = 2\pi g$, where $g$ is an integer. Hence and from (2.6) it follows that

$$q = \frac{2\pi}{a}\frac{g}{G}, \tag{2.8}$$

where

$$-G/2 < g < +G/2. \tag{2.8a}$$

Thus, for a finite atomic chain with $G$ degrees of freedom, the wave number $q$ varying inside the interval from $-\pi/a$ to $+\pi/a$ assumes $G$ discrete values determined by (2.8a). Of course, we always can (and must) choose $G$ so large that the variation of $q$ in (2.8) could be considered quasi-continuous. Such a method of counting $q$ appears quite convenient for solving problems in statistics and kinetics. The final equations to be compared with the experiment should not contain the arbitrary number $G$ of the atoms in the principal region.

---

[4] As we shall see below, $G$ is conveniently assumed to be a large odd number (which, of course, makes no difference).

Since, according to (2.5), $\omega$ is a function of $q$, and since $q$ varies discretely, a question may be asked about the number of vibrations (with distinct $q$'s) in the frequency interval $\omega$, $\omega + d\omega$.

It follows from equations (2.5) and (2.8) that

$$d\omega = a\sqrt{\frac{\beta}{m}}\left|\cos\frac{qa}{2}\right| dq \text{ and } dq = \frac{2\pi}{aG}\,dg,$$

whence

$$d\omega = \frac{2\pi}{G}\sqrt{\frac{\beta}{m}}\left|\cos\frac{qa}{2}\right| dg.$$

The number of vibrations $dz$ per frequency interval $d\omega$ in both branches of $\omega$ ($q$) from $-\pi/a$ to $+\pi/a$ is

$$dz = 2dg = \frac{G}{\pi}\sqrt{\frac{m}{\beta}}\,\frac{d\omega}{\cos|aq/2|}. \tag{2.9}$$

It follows from (2.5) that

$$\cos\frac{aq}{2} = \sqrt{1-\sin^2\frac{aq}{2}} = \sqrt{1-\frac{\omega^2}{4\beta/m}}.$$

Hence and from (2.9) we obtain for the density of vibrations per unit frequency interval

$$\frac{dz}{d\omega} = \frac{2G}{\pi}\,\frac{1}{\sqrt{\omega_{max}^2-\omega^2}}. \tag{2.10}$$

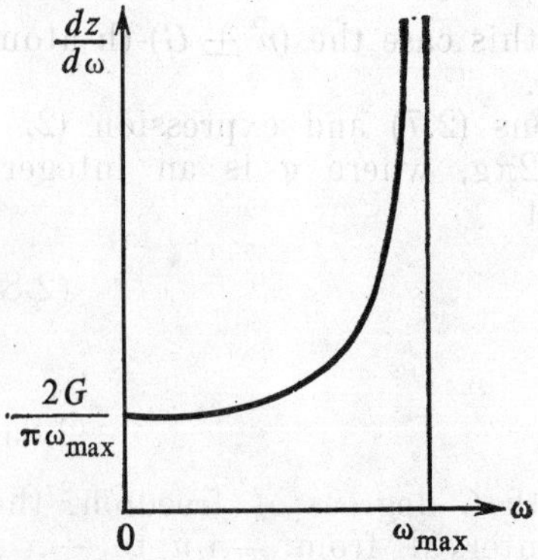


Fig. 3.5

Figure 3.5 depicts the dependence (2.10). As we shall see below, for a three-dimensional crystal in the continuous medium approximation,

$$\frac{dz}{d\omega} \propto \omega^2.$$

3.2.4 We know (e.g., see [3.8, Sec. 20-2]) that the propagation velocity of sound in a rigid bar is $v_0 = \sqrt{E/\rho}$, where $E$ is Young's modulus, and $\rho$ is the density. In the case of a linear atomic chain $\rho = m/a$,

$$E = \frac{\text{force}}{\text{relative elongation}} = \frac{|f_{n,\,n-1}|}{|u_n - u_{n-1}|/a} = \beta a,$$

whence

$$v_0 = \sqrt{E/\rho} = a\sqrt{\beta/m}. \tag{2.11}$$

For long waves and small $q$'s from (2.5) it follows that

$$\omega = 2\sqrt{\frac{\beta}{m}}\,\frac{aq}{2} = v_0 q. \tag{2.12}$$

In general in the presence of dispersion, i.e., when $\omega$ depends on the wave number $q$, we should distinguish between the *phase velocity*, $v_{\text{ph}}$ (the propagation velocity of the phase of a monochromatic

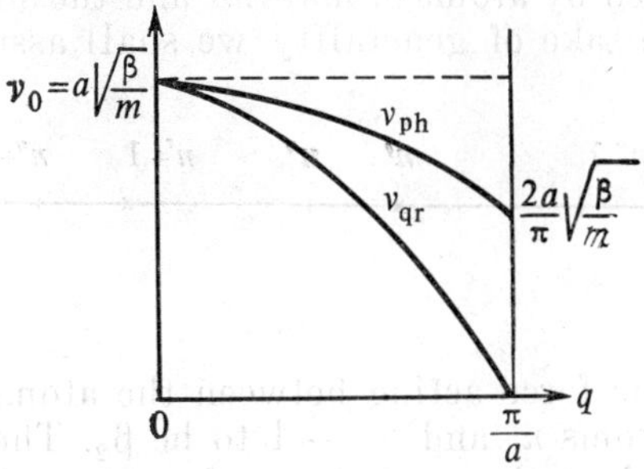


**Fig. 3.6**

wave), and the *group velocity*, $v_{\text{gr}}$ (the velocity of a wave packet and therefore of the wave energy).

In was established that [3.9. pp. 537-8]

$$v_{\text{ph}} = \omega / |\, q \,|, \tag{2.13}$$

$$v_{\text{gr}} = |\, d\omega/dq \,|. \tag{2.14}$$

It follows from expression (2.12) that for long waves

$$v_{\text{ph}} = v_{\text{gr}} = v_0 = \text{sound velocity}.$$

In the general case not limited to small $q$'s we have, according to 2.5),

$$v_{\text{ph}} = v_0 \left| \frac{\sin(aq/2)}{aq/2} \right|, \tag{2.15}$$

$$v_{\text{gr}} = v_0\, |\cos(aq/2)|. \tag{2.16}$$

Figure 3.6 depicts the dependence of $v_{\text{ph}}$ and of $v_{\text{gr}}$ on the wave number $q$. It should be noted that the group velocity with which the vibration energy is transported drops to zero for the shortest waves.

## 3.3 Vibrations and Waves in Complex One-Dimensional (Linear) Lattices

**3.3.1** In the preceding section we discussed the motion of atoms in a one-dimensional model of a simple lattice for which the unit cell could be chosen so that it would contain one atom. Three-dimensional analogues of such lattices are, for instance, the body-centred and face-centred cubic lattices into which most metals crystallize. Consider now vibrations in a one-dimensional model of a complex lattice whose unit cell contains two atoms. The ionic crystals NaCl, CsCl and the atomic crystals Si and Ge are examples of lattices whose unit cells contain two atoms each.

**3.3.2** Figure 3.7 depicts the sites of a complex linear lattice. The sites $n'$ are occupied by atoms of mass $m'$ and the sites $n''$ by atoms of mass $m''$. For the sake of generality we shall assume the constant

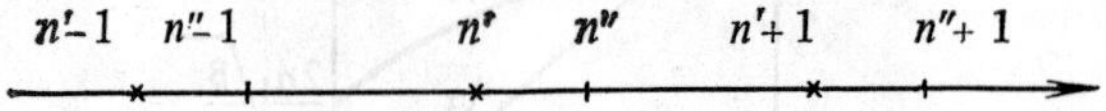


**Fig. 3.7**

of the quasi-elastic force acting between the atoms $n'$ and $n''$ to be $\beta_1$ and between atoms $n'$ and $n''-1$ to be $\beta_2$. The latter can occur even if the lattice is made up of atoms of one type (in this case $m' = m''$). The length of the lattice scale vector **a** is equal to the spacing between $n'-1$, $n'$, $n'+1$ or between $n''-1$, $n''$, $n''+1$. That is why the unit cell of "volume" $\Omega_0 = a$ contains two atoms. Denoting the displacements of the $n'$-th and the $n''$-th atoms by $u'_n$ and $u''_n$, respectively, we obtain in the approximation of the interaction only between nearest neighbours and of the quasi-elastic force (see Section 3.2.1):

$$\begin{aligned} m'\ddot{u}'_n &= -\beta_1(u'_n - u''_n) - \beta_2(u'_n - u''_{n-1}), \\ m''\ddot{u}''_n &= -\beta_1(u''_n - u'_n) - \beta_2(u''_n - u'_{n+1}). \end{aligned} \tag{3.1}$$

We try to satisfy this system of equations by putting

$$u'_n = A'e^{i(qan-\omega t)}, \quad u''_n = A''e^{i(qan-\omega t)}, \tag{3.2}$$

where $q$ and $\omega$ are the wave number and the frequency identical in both expressions, and $A' \neq A''$ are amplitudes.

Substituting (3.2) into equations (3.1), we obtain after some simple transformations and after factoring out $e^{i(qan-\omega t)}$

$$\left(\omega^2 - \frac{\beta_1+\beta_2}{m'}\right)A' + \left(\frac{\beta_1+\beta_2 e^{-iaq}}{m'}\right)A'' = 0, \tag{3.3}$$

$$\left(\frac{\beta_1+\beta_2 e^{iaq}}{m''}\right)A' + \left(\omega^2 - \frac{\beta_1+\beta_2}{m''}\right)A'' = 0. \tag{3.3a}$$

We have thus obtained a system of homogeneous linear equations for the unknown and, in general, complex amplitudes $A'$ and $A''$. It is obvious that the solution of such a system can yield only the ratio $A'/A''$, since $cA'$ and $cA''$ ($c$ = const.) also satisfy (3.3) and

(3.3a). For simultaneous solution of (3.3) and (3.3a), the ratio $A'/A''$ obtained from each equation should be the same:

$$\frac{A'}{A''}=\frac{\beta_1+\beta_2 e^{-iaq}}{(\beta_1+\beta_2)-m'\omega^2}=\frac{(\beta_1+\beta_2)-m''\omega^2}{\beta_1+\beta_2 e^{iaq}}, \tag{3.4}$$

which leads to a quadratic equation with respect to $w^2$. The roots of this equation are[5]

$$\omega_1^2=\frac{1}{2}\omega_0^2\,[1-\sqrt{1-\gamma^2\sin^2(aq/2)}],$$

$$\omega_2^2=\frac{1}{2}\omega_0^2\,[1+\sqrt{1-\gamma^2\sin^2(aq/2)}], \tag{3.5}$$

where $\omega_0^2=\frac{(\beta_1+\beta_2)(m'+m'')}{m'm''}$ and $\gamma^2=16\frac{\beta_1\beta_2}{(\beta_1+\beta_2)^2}\frac{m'm''}{(m'+m'')^2}$. The quantity $\gamma^2$ attains its maximum value equal to unity at $\beta_1=\beta_2$ and $m'=m''$. Hence, it follows that $\gamma^2\sin^2(aq/2)\leqslant 1$, which guarantees that $\omega_1$ and $\omega_2$ are real.

We see that, as in the case of the simple linear lattice (Section 3.2), the solution (3.2) satisfies the equations of motion (3.1) if $\omega$ and $q$ are related by the dispersion law (3.5). However, there is an important difference here. The expressions (3.5) define two dispersion branches, one of which $\omega_1 = \omega_{ac}$, for reasons to be specified below, shall be termed *acoustic* and the other $\omega_2 = \omega_{op}$ *optical*.

Using the same reasoning as in Section 3.2.2, we can demonstrate that the nature of (3.2) makes it possible in the analysis of the variations of $q$ to limit ourselves to the interval $(0, \pi/a)$. It follows from (3.5) that, for $q = 0, \pi/a$,

$$\omega_{ac}(0)=0,\quad \omega_{ac}(\pi/a)=\frac{\omega_0}{\sqrt{2}}\sqrt{1-\sqrt{1-\gamma^2}},$$

$$\omega_{op}(0)=\omega_0,\quad \omega_{op}(\pi/a)=\frac{\omega_0}{\sqrt{2}}\sqrt{1+\sqrt{1-\gamma^2}}. \tag{3.6}$$

Hence,

$$\omega_{op}(0) = \omega_0 > \omega_{op}(\pi/a) > \omega_{ac}(\pi/a) > \omega_{ac}(0) = 0. \tag{3.6a}$$

For long waves, $\lambda \gg a$ or $aq \ll 1$, it follows from (3.5) that in the approximation $\sin(aq/2) \approx aq/2$, if the roots are expanded into a power series,

$$\omega_{ac}\approx\frac{1}{4}\omega_0\gamma aq,\qquad \omega_{op}\approx\omega_0\left(1-\frac{\gamma^2a^2}{32}q^2\right). \tag{3.7}$$

---

5 In algebraic transformations we made use of the relations $e^{iaq}+e^{-iaq}=2\cos aq$ and $1-\cos aq=2\sin^2(aq/2)$.

Taking the derivative of $\omega_{1,2}$ with respect to $q$ [see (3.5)], we can easily show that

$$\left(\frac{d\omega_{\text{ac}}}{dq}\right)_{\pi/a}=\left(\frac{d\omega_{\text{op}}}{dq}\right)_{\pi/a}=0$$

if $\gamma^2 \neq 1$. Making use of this fact and of (3.6) and (3.7), we obtain a uniqueness of the dispersion branches depicted in Fig. 3.8*a*. Such

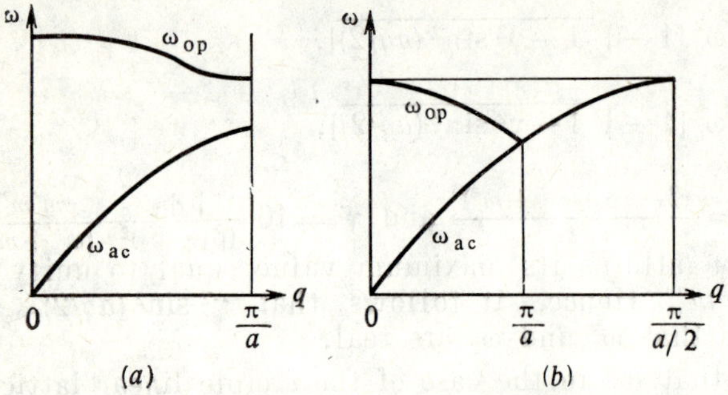


**Fig. 3.8**

a picture is true when $\gamma^2 < 1$, the adequate conditions for this being $m' \neq m''$ or $\beta_1 \neq \beta_2$. We would once again like to stress the essential properties of the acoustic and the optical branches: for $\lambda \to \infty$ the frequency $\omega_{\text{ac}} \to 0$ and $\omega_{\text{op}} \to \omega_0 \neq 0$.

Consider the "degenerate" case (Fig. 3.8*b*), when $m' = m'' = m$ and $\beta_1 = \beta_2 = \beta$, i.e., $\gamma^2 = 1$. It follows from expressions (3.5) that

$$\omega_{\text{ac}}=2\sqrt{\frac{\beta}{m}}\left|\sin\frac{aq}{4}\right|, \quad \omega_{\text{op}}=2\sqrt{\frac{\beta}{m}}\left|\cos\frac{aq}{4}\right|. \tag{3.8}$$

In this case all the atoms are identical, as are the bonds, and because of this all interatomic distances must be equal. Therefore the lattice parameter in our notation should be $a/2$, i.e., one-half that of the previous value. We should now consider the variation of the wave number $q$ in the interval $\left(0, \frac{\pi}{a/2}\right)$. It may easily be seen that $\omega(q)$ in (2.5) for the interval $\left(0, \frac{1}{2}\frac{\pi}{a/2}\right)$ corresponds to $\omega_{\text{ac}}$ in (3.8), and the same $\omega(q)$ in (2.5) for the interval $\left(\frac{1}{2}\frac{\pi}{a/2}, \frac{\pi}{a/2}\right)$ corresponds to $\omega_{\text{op}}$ in (3.8). In the latter case the equivalent value of $q$ is $q' = 2\pi/a - q$; indeed,

$$\cos\frac{aq'}{4}=\cos\left(\frac{\pi}{2}-\frac{aq}{4}\right)=\sin\frac{aq}{4}.$$

**3.3.3** Let us consider the nature of atomic vibrations in the acoustic and the optical branches. It follows from relations (3.2) and (3.4) that

$$\frac{u'_n}{u''_n}=\frac{A'}{A''}=\frac{\beta_1+\beta_2 e^{-iqa}}{(\beta_1+\beta_2)-m'\omega^2}. \tag{3.9}$$

Note that for $q=0$ and $q=\pi/a$ the factor $e^{-iqa}=1$ and $e^{-iqa}=-1$, respectively. In both cases, as may be seen from (3.9), the amplitudes $A'$ and $A''$ and the displacements $u'_n$ and $u''_n$ can be assumed to be real.

Consider first the important case of long (to be exact, infinitely long) waves:

$$\lambda=\infty,\quad q=\frac{2\pi}{\lambda}=0.$$

In this case $e^{-iqa}=e^0=1$. Making use of (3.6), i.e., of the extreme values of $\omega$ for $q=0$, we obtain from (3.9)

$$\left(\frac{u'_n}{u''_n}\right)_{\mathrm{ac}}=1,\quad \left(\frac{u'_n}{u''_n}\right)_{\mathrm{op}}=-\frac{m''}{m'}. \tag{3.10}$$

Hence, in an infinitely long wave in the acoustic branch of vibrations, different atoms move in step, their deflections being the same at any moment of time ($u'_n=u''_n$). In the optical branch different atoms of a cell in an infinitely long wave move in opposite phases, so that their centre of gravity remains at rest ($m'u'_n+m''u''_n=0$). Atomic motion in the first case corresponds to the mechanism of propagation of acoustic waves, and this explains the term for the corresponding branch of vibrations. If the cell of a complex crystal consists of oppositely charged ions, then the vibrations of the second type involve the vibrations of the electric dipole moment of the cell, and they frequently turn out to be optically active, i.e., result in absorption or emission of infrared radiation. This explains the term optical for the second branch.

Consider now the case of short waves:

$$\lambda=2a,\quad q=2\pi/\lambda=\pi/a.$$

In this case $e^{-iaq}=e^{-i\pi}=-1$. Making use of the expressions for $\omega(\pi/a)$ [see (3.6)] we obtain from equality (3.9)

$$\frac{u'_n}{u''_n}=\frac{\dfrac{\beta_1-\beta_2}{\beta_1+\beta_2}}{1-\dfrac{m'+m''}{2m''}\left[1\mp\sqrt{1-\dfrac{16\beta_1\beta_2}{(\beta_1+\beta_2)^2}\,\dfrac{m'm''}{(m'+m'')^2}}\right]}, \tag{3.11}$$

where the upper sign in front of the radical corresponds to the acoustic branch and the lower to the optical branch of vibrations. It appears expedient at this place to consider several specific cases.

If $\beta_1 = \beta_2$, the numerator in (3.11) vanishes, the denominator being nonzero and equal to $(m'' - m')/m''$ for the acoustic branch [when $m' < m''$, since the root in (3.11) is always meant to be positive], the same being the case for the optical branch if $m' > m''$. Hence we conclude that

$$u_n' = 0,\ u_n'' \neq 0 \text{ for the acoustic branch if } m' < m'',$$
$$u_n' = 0,\ u_n'' \neq 0 \text{ for the optical branch if } m' > m''. \quad (3.12)$$

For the shortest waves $\lambda = 2a$ in the acoustic branch the lighter atoms $m'$ remain at rest and the heavier $m''$ vibrate. In the optical branch the situation is reversed.

We have discussed the case $m' < m''$ for the acoustic branch and the case $m' > m''$ for the optical. We have done so only for the sake of convenience. Had we presumed that in the acoustic branch [the top sign before the radical in (3.11)] $m' > m''$ the denominator would have vanished and $(u_n'/u_n'') = 0/0$. Evaluating the indeterminate form, we find that $(u_n'/u_n'') = \infty$, i.e., $u_n' \neq 0$ and $u_n'' = 0$. But this would lead us to the conclusion arrived at previously that in the acoustic branch the atoms remaining at rest are the lighter ones $m''$, and the vibrating atoms are the heavier ones $m'$.

Consider now the second particular case: $m' = m''$ and $\beta_1 > \beta_2$. The denominator in (3.11) is equal to

$$1 - 1 \pm \sqrt{(\beta_1 - \beta_2)^2/(\beta_1 + \beta_2)} = \pm \sqrt{(\beta_1 - \beta_2)^2/(\beta_1 + \beta_2)},$$

i.e., to $(\beta_1 - \beta_2)/(\beta_1 + \beta_2)$ for the acoustic branch and to $-(\beta_1 - \beta_2)/(\beta_1 + \beta_2)$ for the optical branch. Hence

$$\left(\frac{u_n'}{u_n''}\right)_{ac} = 1 \text{ and } \left(\frac{u_n'}{u_n''}\right)_{op} = -1, \quad (3.13)$$

i.e., in this case the atoms in the shortest acoustic wave vibrate in phase and in the shortest optical wave in opposite phases. The case $\beta_1 < \beta_2$ can be considered in the same way, with the result that the right-hand sides in (3.13) change places.

**3.3.4** In the same way as it was done in Section 3.2.3, we can introduce the Born-von Kármán cycle conditions (2.7) for a complex lattice, in which case $G$ will be equal to the number of cells in the principal region. The density of vibrations, $dz/d\omega$, for both the acoustic and the optical branches can be computed in the same way, as it was done in (2.10).

### 3.4 Normal Coordinates for Simple One-Dimensional Lattices

In Section 3.2 we considered vibrations and waves in a linear chain made up of identical atoms and demonstrated that expression (2.4) satisfies the equations of motion (2.1), provided that the dispersion relations (2.5) are fulfilled. The harmonic waves (2.4), obviously, do not describe the more general case of atomic motion in a chain. An arbitrary motion of the atoms can be represented by means of a linear superposition (sum) of miscellaneous waves of the type (2.4). The individual waves will have different wave numbers $q$, frequencies $\omega(q) = \omega_q$ corresponding to it and an amplitude $A_g$, usually different for different waves. Make the actual real displacements of the atoms of the chain equal to

$$u_n = A_q e^{i(qan-\omega_q t)} + A_q^* e^{-i(qan-\omega_q t)} = 2\,|A_q|\cos(qan - \omega_q t + \alpha_q),$$

where $A_q = |A_q|\, e^{i\alpha_q}$.

Hence, in the most general case of atomic motion

$$u_n = \sum_q \left[A_q e^{i(qan-\omega_q t)} + A_q^* e^{-i(qan-\omega_q t)}\right] =$$

$$= \frac{1}{\sqrt{G}} \sum_q \{a_q e^{iqan} + a_q^* e^{-iqan}\}, \qquad (4.1)$$

where $a_q \equiv \sqrt{G}\, A_q e^{-i\omega_q t}$. Here we presume the cyclic boundary conditions to be enforced for the lattice, so that the summation in (4.1) is performed over $G$ discrete values of $q = (2\pi/aG)\, g$ [see 2.8)].

As we shall see below, the quantities $a_q$ or, to be more precise, certain simple combinations of them are *normal coordinates* and *generalized momenta* of our linear atomic chain (Appendix 5). Hence, (4.1) can be considered to be the transformation of the coordinates $u_n$ to normal coordinates and momenta $a_q$ (the quantites $a_q$ are complex-valued and therefore there are $2G$ real quantities corresponding to them).

This transformation is of a somewhat more general character than the one considered in Appendix 5, since in (4.1) the original coordinates $u_n$ are expressed simultaneously in terms of new coordinates and momenta corresponding to them (contact transformation [3.10, Chap. 7]). The kinetic energy $\mathcal{T}$ and the potential energy $\Phi$ for an atomic chain are equal to

$$\mathcal{T} = \frac{m}{2} \sum_{n=1}^{G} \dot{u}_n^2, \qquad (4.2)$$

$$\Phi = \frac{\beta}{2} \sum_{n=1}^{G} (u_n - u_{n-1})^2, \qquad (4.3)$$

respectively. Indeed, it follows from (4.3) that the force acting on the $n$-th particle,

$$f_n = -\frac{\partial \Phi}{\partial u_n} = -\beta\,(2u_n - u_{n-1} - u_{n+1}),$$

coincides with (2.1).

Making use of (4.1) and taking into account that $\dot{a}_q = -i\omega_q a_q$ and $\dot{a}_q^* = i\omega_q a_q^*$, we see that

$$\mathscr{T} = \frac{m}{2}\sum_{n=1}^{G} \dot{u}_n \dot{u}_n = \frac{m}{2}\sum_{n=1}^{G}\frac{1}{G}\sum_{q}\{\dot{a}_q e^{iqan} + \dot{a}_q^* e^{-iqan}\}$$

$$\times \frac{1}{\sqrt{G}}\sum_{q'}\{\dot{a}_{q'} e^{iq'an} + \dot{a}_{q'}^* e^{-iq'an}\} = -\frac{m}{2G}\sum_{qq'}\sum_{n=1}^{G}\omega_q\omega_{q'}\{a_q a_{q'} e^{i(q+q')an}$$

$$-a_q a_{q'}^* e^{i(q-q')an} - a_q^* a_{q'} e^{-i(q-q')an} + a_q^* a_{q'}^* e^{-i(q+q')an}\}. \qquad (4.4)$$

It may easily be shown (see Appendix 6) that

$$\sum_{n=1}^{G} e^{iqan} = \sum_{n=1}^{G} e^{i\frac{2\pi}{G}gn} = \begin{cases} 0 \text{ at } q \neq 0 \text{ (or } g \neq 0), \\ G \text{ at } q = 0 \left(\text{or } \frac{2\pi}{a} \times \text{integer}\right). \end{cases} \qquad (4.5)$$

Hence, the summation with respect to $n$ in the first addend in the braces in (4.4) yields

$$\sum_{n=1}^{G} e^{i(q+q')an} = \begin{cases} 0 \text{ for } q+q' \neq 0, \\ G \text{ for } q+q' = 0 \text{ (or } q' = -q). \end{cases} \qquad (4.6)$$

The summation with respect to $n$ in the second, third and fourth addends in the braces in (4.4) is quite similar. Taking into account that according to (2.5) $\omega_{-q} = \omega_q$, we obtain

$$\mathscr{T} = \frac{m}{2}\sum_{q}\omega_q^2 (2a_q a_q^* - a_q a_{-q} - a_q^* a_{-q}^*). \qquad (4.7)$$

Let us now express the potential energy (4.3) in terms of the quantities $a_q$ and $a_q^*$. Substituting $u_n$ and $u_{n-1}$ from (4.1) into (4.3), we obtain

$$\Phi = \frac{\beta}{2}\sum_{n=1}^{G}(u_n - u_{n-1})(u_n - u_{n-1}) = \frac{\beta}{2G}\sum_{qq'}\sum_{n=1}^{G}[a_q e^{iqan} + a_q^* e^{-iqan}$$

$$- a_q e^{-iqa} e^{iqan} - a_q^* e^{iqa} e^{-iqan}] \times [a_{q'} e^{iq'an} + a_{q'}^* e^{-iq'an} - a_{q'} e^{-iq'a} a^{-iq'an}$$

$$- a_{q'}^* e^{iq'a} e^{-iq'an}].$$

Multiplying the brackets and collecting like addends, we obtain

$$\sum_{n=1}^{G}\{a_q a_{q'}[1 + e^{-i(q+q')a} - e^{-iqa} - e^{-iq'a}]\, e^{i(q+q')an}$$

$$+ a_q a_{q'}^*[1 + e^{i(q'-q)a} - e^{iq'a} - e^{-iqa}]\, e^{i(q-q')an}$$

$$+ a_q^* a_{q'}[1 + e^{i(q-q')a} - e^{-iq'a} - e^{iqa}]\, e^{i(q'-q)an}$$

$$+ a_q^* a_{q'}^*[1 + e^{i(q+q')a} - e^{iqa} - e^{iq'a}]\, e^{-i(q+q')an}\}.$$

Taking into account the conditions (4.6), we see that summation over $n$ yields a nonzero result equal to $G$ only when $q' = -q$ or $q' = +q$. Then all the brackets become equal to

$$2 - e^{-iqa} - e^{iqa} = 2(1 - \cos qa) = 4\sin^2 \frac{aq}{2} = \frac{m\omega_q^2}{\beta},$$

as stipulated in (2.5). Hence

$$\Phi = \frac{m}{2} \sum_q \omega_q^2 (2a_q a_q^* + a_q a_{-q} + a_q^* a_{-q}^*). \tag{4.8}$$

The total energy is

$$\mathscr{E} = \mathscr{T} + \Phi = 2m \sum \omega_q^2 a_q a_q^*. \tag{4.9}$$

Define the quantities $x_q$ and $p_q$ with the aid of equalities

$$x_q = a_q + a_q^* = 2\mathrm{Re}\,\{a_q\},$$

$$p_q = \frac{m\omega_q}{i}(a_q - a_q^*) = \frac{m\omega_q}{i}\, 2\,\mathrm{Im}\,\{a_q\}. \tag{4.10}$$

Hence

$$a_q = \frac{1}{2}\left(x_q + i\frac{p_q}{m\omega_q}\right),$$

$$a_q^* = \frac{1}{2}\left(x_q - i\frac{p_q}{m\omega_q}\right). \tag{4.10a}$$

Substituting these values into expression (4.9) we obtain

$$\mathscr{E} = \sum_q \left\{\frac{1}{2m} p_q^2 + \frac{1}{2} m\omega_q^2 x_q^2\right\} = \mathscr{H}(x_q, p_q). \tag{4.11}$$

We see that the quantities $x_q$ and $p_q = m\dot{x}_q$ play the part of normal coordinates and momenta conjugate to them. Expression (4.11) is the Hamiltonian function expressed in those coordinates. Hence, the total energy $\mathscr{E}$ of the most general motion of atoms in a one-dimensional crystal can be represented as a sum of the normal vibration energies, which behave like linear harmonic oscillators with frequencies $\omega_q$.

## 3.5 Atomic Vibrations in Complex Three-Dimensional Lattices

3.5.1 In this section we intend to demonstrate that many properties of vibrations in a one-dimensional (linear) atomic chain also apply to three-dimensional crystal lattices.

Consider a complex crystal with $s$ (in general) different atoms in a unit cell, each having a mass $m_k$ ($k = 1, 2, \ldots, s$). In the future we shall consider the principal region of the crystal with a volume $V$ and in the shape of a parallelepiped built on the vectors $Ga_i$ ($i = 1, 2, 3$), where $\mathbf{a}_i$ are the basis vectors of the direct lattice, and $G$ is a large odd number. Obviously, $V = G^3\mathbf{a}_1\cdot(\mathbf{a}_2 \times \mathbf{a}_3) = G^3\Omega_0 = N\Omega_0$, where $\Omega_0$ is the volume of a unit cell, and $N = G^3$. The position of the atom of the $k$-th type in the $n$-th unit cell is determined by its position vector

$$\mathbf{r}_{\mathbf{n}}^{k} = \mathbf{a}_{\mathbf{n}} + \mathbf{r}^{k}, \tag{5.1}$$

where $\mathbf{a}_{\mathbf{n}} = \mathbf{n}_1\mathbf{a}_1 + \mathbf{n}_2\mathbf{a}_2 + \mathbf{n}_3\mathbf{a}_3$ is a vector of the direct lattice, and $\mathbf{r}^k$ is the position vector that determines the position of the $k$-th atom inside the crystal lattice.

Denote the displacement of the $k$-th atom in the $\mathbf{n}$-th cell from its equilibrium position by $\mathbf{u}_{\mathbf{n}}^{k}$, and denote the orthogonal projections of this displacement by $\mathbf{u}_{\mathbf{n}\alpha}^{k}$ ($\alpha = x, y, z$).

The potential energy $\Phi$ of the principal region of the crystal is a function of $3sN$ displacement $\mathbf{u}_{\mathbf{n}\alpha}^{k}$ and, evidently, has a minimum at $\mathbf{u}_{\mathbf{n}\alpha}^{k} = 0$, so that $(\partial\Phi/\partial\mathbf{u}_{\mathbf{n}\alpha}^{k})_0 = 0$. Expand $\Phi$ into a power series in the projections of displacements $u_{\mathbf{n}\alpha}^{k}$:

$$\Phi = \frac{1}{2}\sum_{\mathbf{n}\mathbf{n}'kk'\alpha\beta} \Phi_{\alpha\beta}\begin{pmatrix} kk' \\ \mathbf{n}\mathbf{n}' \end{pmatrix} u_{\mathbf{n}\alpha}^{k} u_{\mathbf{n}'\beta}^{k'} + \frac{1}{6}\sum_{\mathbf{n}\mathbf{n}'\mathbf{n}''kk'k''\alpha\beta\gamma} \Phi_{\alpha\beta\gamma}\begin{pmatrix} kk'k'' \\ \mathbf{n}\mathbf{n}'\mathbf{n}'' \end{pmatrix} u_{\mathbf{n}\alpha}^{k} u_{\mathbf{n}'\beta}^{k'} u_{\mathbf{n}''\gamma}^{k''}. \tag{5.2}$$

We have put the minimum potential energy equal to zero and introduced the following notation:

$$\Phi_{\alpha\beta}\begin{pmatrix} kk' \\ \mathbf{n}\mathbf{n}' \end{pmatrix} = \left(\frac{\partial^2\Phi}{\partial u_{\mathbf{n}\alpha}^{k}\,\partial u_{\mathbf{n}'\beta}^{k'}}\right)_{00},$$

$$\Phi_{\alpha\beta\gamma}\begin{pmatrix} kk'k'' \\ \mathbf{n}\mathbf{n}'\mathbf{n}'' \end{pmatrix} = \left(\frac{\partial^3\Phi}{\partial u_{\mathbf{n}\alpha}^{k}\,\partial u_{\mathbf{n}'\beta}^{k'}\,\partial u_{\mathbf{n}''\gamma}^{k''}}\right)_{000}. \tag{5.2a}$$

It can be demonstrated that the coefficients of the expansion (5.2a) satisfy the relations

$$\Phi_{\alpha\beta}\begin{pmatrix} kk' \\ \mathbf{n}\mathbf{n}' \end{pmatrix} = \Phi_{\alpha\beta}\begin{pmatrix} kk' \\ \mathbf{n}-\mathbf{n}' \end{pmatrix}, \quad \Phi_{\alpha\beta\gamma}\begin{pmatrix} kk'k'' \\ \mathbf{n}\mathbf{n}'\mathbf{n}'' \end{pmatrix} = \Phi_{\alpha\beta\gamma}\begin{pmatrix} kk'k'' \\ \mathbf{n}-\mathbf{n}', \mathbf{n}-\mathbf{n}'' \end{pmatrix}, \tag{5.3}$$

$$\Phi_{\alpha\beta}\begin{pmatrix} kk' \\ \mathbf{n}\mathbf{n}' \end{pmatrix} = \Phi_{\beta\alpha}\begin{pmatrix} k'k \\ \mathbf{n}'\mathbf{n} \end{pmatrix}, \tag{5.3a}$$

$$\sum_{\mathbf{n}'k'} \Phi_{\alpha\beta}\begin{pmatrix} kk' \\ \mathbf{n}\mathbf{n}' \end{pmatrix} = 0, \quad \sum_{\mathbf{n}'\mathbf{n}''k'k''} \Phi_{\alpha\beta\gamma}\begin{pmatrix} kk'k'' \\ \mathbf{n}\mathbf{n}'\mathbf{n}'' \end{pmatrix} = 0. \tag{5.3b}$$

The equalities (5.3) reflect the fact that the coefficients determining the atomic interaction depend only on the type of atoms, $(k, k')$ and on the spacing between the appropriate cells $(\mathbf{n}\text{-}\mathbf{n}', \mathbf{n}\text{-}\mathbf{n}'')$. Equality (5.3a) is a direct consequence of the definition (5.2a). To prove (5.3b) we expand the function $(\partial\Phi/\partial u^k_{\mathbf{n}\alpha})$ into a power series in atomic displacements:

$$\frac{\partial\Phi}{\partial u^k_{\mathbf{n}\alpha}}=\left(\frac{\partial\Phi}{\partial u^k_{\mathbf{n}\alpha}}\right)_0+\sum_{\mathbf{n}'k'\beta}\left(\frac{\partial^2\Phi}{\partial u^k_{\mathbf{n}\alpha}\,\partial u^{k'}_{\mathbf{n}'\beta}}\right)_{00}u^{k'}_{\mathbf{n}'\beta}+\frac{1}{2}\sum_{\substack{\mathbf{n}'k'\beta\\ \mathbf{n}''k''\gamma}}\left(\frac{\partial^3\Phi}{\partial u^k_{\mathbf{n}\alpha}\,\partial u^{k'}_{\mathbf{n}'\beta}\,\partial u^{k''}_{\mathbf{n}''\gamma}}\right)_{000}u^{k'}_{\mathbf{n}'\beta}u^{k''}_{\mathbf{n}''\gamma}. \quad (5.4)$$

Let all the atoms experience an equal displacement equal to $u^0$, in this case the crystal will move in space as a whole, and the variation of all the functions that depend only on the mutual arrangement of the atoms should be zero, i. e., $\partial\Phi/\partial u^k_{\mathbf{n}\alpha}=0$ [from the conditions of equilibrium $(\partial\Phi/\partial u^k_{\mathbf{n}\alpha})_0=0$]. Hence, it follows from (5.4) that

$$\sum_{\beta}u^0_\beta\sum_{\mathbf{n}'k'}\Phi_{\alpha\beta}\begin{pmatrix}kk'\\ \mathbf{nn}'\end{pmatrix}+\frac{1}{2}\sum_{\beta\gamma}u^0_\beta u^0_\gamma\sum_{\mathbf{n}'\mathbf{n}''k'k''}\Phi_{\alpha\beta\gamma}\begin{pmatrix}kk'k''\\ \mathbf{nn}'\mathbf{n}''\end{pmatrix}=0.$$

Since the projections $u^0_\beta$ and $u^0_\gamma$ are independent of one another, this leads to (5.3b). The kinetic energy of atomic motion is

$$\mathscr{T}=\frac{1}{2}\sum_{\mathbf{n}k}m_k(\dot{\mathbf{u}}^k_{\mathbf{n}})^2=\frac{1}{2}\sum_{\mathbf{n}k\alpha}m_k(\dot{u}^k_{\mathbf{n}\alpha})^2. \quad (5.5)$$

**3.5.2** The classical equations of motion of atoms in the harmonic approximation are of the form

$$m_k\ddot{u}^k_{\mathbf{n}\alpha}=-\frac{\partial\Phi}{\partial u^k_{\mathbf{n}\alpha}}=-\sum_{\mathbf{n}'k'\beta}\Phi_{\alpha\beta}\begin{pmatrix}kk'\\ \mathbf{nn}'\end{pmatrix}u^{k'}_{\mathbf{n}'\beta} \quad (5.6)$$

$(\mathbf{n}=1, 2, 3, \ldots, N,\ k=1, 2, 3, \ldots, s,\ \alpha=x, y, z)$, i.e., of the form of a system of $3sN$ differential equations for $3sN$ unkown functions $u^k_{\mathbf{n}\alpha}(t)$.

The solution of the system (5.6) will be sought, as in the case of a complex linear lattice [see (3.2)], in the form of traveling waves:

$$\tilde{u}^k_{\mathbf{n}\alpha}=\frac{1}{\sqrt{m_k}}A^k_\alpha(q)\,e^{i(q\cdot\mathbf{a_n}-\omega t)}, \quad (5.7)$$

where $(m_k)^{-1/2}A^k_\alpha$ $(\alpha=x, y, z)$ is the projection of the complex amplitude $(m_k)^{-1/2}\mathbf{A}^k$ different for different types of atoms $k$ $(k=1, 2, 3, \ldots, s)$, $\mathbf{q}=\frac{2\pi}{\lambda}\mathbf{v}$ is the wave vector ($\mathbf{v}$ is a unit vector normal to a plane wave of wavelength $\lambda$), and $\omega = \omega(\mathbf{q}) \equiv \omega_q$ is the cyclic frequency. The quantities $\tilde{u}^k_{\mathbf{n}\alpha}$ are complex, so that actual displacements are $\tilde{u}^k_{n\alpha} = \mathrm{Re}\,(\tilde{u}^k_{\mathbf{n}\alpha})$. We can look for the solution of equations (5.6) in complex form (5.7), in which case it, because of the linearity of (5.6), consists of a sum of cosines and sines. As in the one-dimensional case, the wave vector $\mathbf{q}$ possesses properties stemming from the fact that the plane wave (5.7) propagates in a discrete system (in a lattice). Substitute for $\mathbf{q}$ in (5.7) $\mathbf{q}' = \mathbf{q} + \mathbf{b}_g$, where $\mathbf{b}_g = g_1\mathbf{b}_1 + g_2\mathbf{b}_2 + g_3\mathbf{b}_3$ is a vector of the reciprocal lattice; then $\mathbf{q}'\mathbf{a}_\mathbf{n} = \mathbf{q}\cdot\mathbf{a}_\mathbf{n} + 2\pi\,(n_1g_1 + n_2g_2 + n_3g_3) = \mathbf{q}\cdot\mathbf{a}_n + 2\pi k$ ($k$ is an integer). Hence, it follows from (5.7) that

$$(\tilde{u}^k_{\mathbf{n}\alpha})' = \frac{1}{\sqrt{m_k}} A^k_\alpha e^{i(\mathbf{q}'\cdot\mathbf{a}_\mathbf{n} - \omega t)} = \frac{1}{\sqrt{m_k}} A^k_\alpha e^{i(\mathbf{q}\cdot\mathbf{a}_\mathbf{n} - \omega t)} e^{i2\pi k} = \tilde{u}^k_{\mathbf{n}\alpha}, \quad (5.8)$$

since $e^{i2\pi k} = 1$. It will be seen from (5.8) that the wave with $\mathbf{q}'$ coincides with that with $\mathbf{q}$. But this means that $\mathbf{q}'$ is physically equivalent to $\mathbf{q}$. This makes it possible to consider the variations of the wave vector $\mathbf{q}$ in a limited region, just as in the one-dimensional case. Making $\mathbf{a}_\mathbf{n} = \mathbf{a}_i$ and $\mathbf{b}_g = \mathbf{b}_i$, we obtain $\mathbf{q}'\cdot\mathbf{a}_i = (\mathbf{q} + \mathbf{b}_i)\cdot\mathbf{a}_i = \mathbf{q}\cdot\mathbf{a}_i + 2\pi$. Hence, it is always possible to choose the quantity $\mathbf{q}\cdot\mathbf{a}_i$ inside the interval $2\pi$. Make[6].

$$-\pi \leqslant \mathbf{q}\cdot\mathbf{a}_i \leqslant +\pi \quad (i = 1, 2, 3). \quad (5.9)$$

For a cubic crystal whose edges are directed along the axes of an orthogonal system of coordinates,

$$-\frac{\pi}{a} \leqslant q_\alpha \leqslant +\frac{\pi}{a} \quad (\alpha = x, y, z). \quad (5.9a)$$

For an arbitrary lattice the region of variation of $\mathbf{q}\cdot\mathbf{a}_i$ equal to $2\pi$ can be chosen in the form of a unit cell of the reciprocal lattice, i.e., in the form of a parallelepiped with the edges $\mathbf{b}_1$, $\mathbf{b}_2$, $\mathbf{b}_3$, or in a symmetrical form of the first Brillouin zone (Section 2.9.1). Substituting (5.7) into the system (5.6) and dividing both sides of the equation by $e^{i(\mathbf{q}\cdot\mathbf{a}_\mathbf{n} - \omega t)}$, we obtain

$$\omega^2 A^k_\alpha = \sum_{k'\beta} D^{kk'}_{\alpha\beta} A^{k'}_\beta, \quad (5.10)$$

where the elements of the dynamical matrix of the crystal are

$$D^{kk'}_{\alpha\beta}(\mathbf{q}) = \sum_{\mathbf{n}'} \frac{1}{\sqrt{m_k m_{k'}}} \Phi_{\alpha\beta}\begin{pmatrix} kk' \\ \mathbf{nn}' \end{pmatrix} e^{i\mathbf{q}(\mathbf{a}'_\mathbf{n} - \mathbf{a}_\mathbf{n})}. \quad (5.10a)$$

---

[6] We remind the reader that the more fundamental cause of the appearance of the wave vector $\mathbf{q}$ and of its properties is the crystal's translational symmetry (see Section 2.9).

In this matrix of rank $3s$ the double index $\binom{k}{\alpha}$ numbers the rows and the double index $\binom{k'}{\beta}$ numbers the columns. We have obtained a homogeneous linear system of $3s$ equations [instead of $3sN$ equations (5.6)] for $3s$ unknown complex quantities $A_\alpha^k$ ($k = 1, 2, \ldots, s$; $\alpha = x, y, z$). The system (5.10) can be rewritten in a more compact form:

$$\sum_{k'\beta} \{D_{\alpha\beta}^{kk'} - \omega^2 \delta_{kk'} \delta_{\alpha\beta}\} A_\beta^{k'} = 0 \quad (k = 1, 2, \ldots, s; \quad \alpha = x, y, z), \tag{5.10b}$$

where the symbols $\delta_{kk'}$ and $\delta_{\alpha\beta}$ that are nonzero and equal to unity only if $k = k'$ and $\alpha = \beta$ automatically provide for the coincidence of the system (5.10b) with (5.10).

It is an established fact [3.7, p. 258, Theorem 3] that the system (5.10b) has solutions other than the zero (trivial) solutions only if the system determinant is zero:

$$\det [D_{\alpha\beta}^{kk'} - \omega^2 \delta_{kk'} \delta_{\alpha\beta}] = 0. \tag{5.11}$$

This *characteristic* (or *secular*) *equation* of the $3s$-st degree for the frequency $\omega^2$ can be written out in more detail:

$$\begin{vmatrix} D_{xx}^{11} - \omega^2 & D_{xy}^{11} & D_{xz}^{11} & D_{xx}^{12} & \ldots & D_{xz}^{1s} \\ D_{yx}^{11} & D_{yy}^{11} - \omega^2 & D_{yz}^{11} & D_{yx}^{12} & \ldots & D_{yz}^{1s} \\ \cdot & \cdot & \cdot & \cdot & \cdot & \cdot \\ D_{zx}^{s1} & D_{zy}^{s1} & D_{zz}^{s1} & D_{zx}^{s2} & \ldots & D_{zz}^{ss} - \omega^2 \end{vmatrix} = 0. \tag{5.11a}$$

It follows from (5.10a) and (5.3a) that

$$D_{\alpha\beta}^{kk'} = D_{\beta\alpha}^{k'k*}, \tag{5.11b}$$

where the * denotes a complex conjugate quantity. A matrix $D_{\alpha\beta'}^{kk'}$ with the property (5.11b) is termed *hermitian*; the eigenvalues of an hermitian matrix, $\omega^2$, obtained from (5.11a) are real (see Appendix 3, Sec. 5). It is clear from physical considerations that they should be positive. Indeed, in the case of negative eigenvalues $\omega^2 = -\gamma^2 < 0$ there appears in the expression for the wave (5.7) a factor $\exp(\pm \gamma t)$, which assumes in the past or in the future an infinite value, i.e., causes the destruction of the lattice. The positive nature of the values of $\omega^2$ can also be demonstrated mathematically from the conditions imposed on the force constants $\Phi_{\alpha\beta}\binom{kk'}{\mathbf{nn}'}$ which, in their turn, originate from the requirement of the minimum lattice potential energy in the equilibrium position.

Hence, in general, the characteristic equation (5.11a) of the $3s$-st degree for $\omega^2$ yields $3s$ different real roots $\omega_j(\mathbf{q})$ ($j = 1, 2, 3, \ldots, 3s$), which determine $3s$ different vibration branches. The symmetry of the crystal cell sometimes results in the coincidence of some of the $\omega_j$s, so that the number of different vibration branches may turn out to be less than $3s$. Substituting $3s$ roots $\omega_j(\mathbf{q})$ into the system of homogenous equations (5.10b), we obtain (to within a constant factor) $3s$ different solutions for the complex amplitudes $A_{j\alpha}^{k}$.

**3.5.3** Consider an important question: In what circumstances are the amplitudes $A_{j\alpha}^{k}$ real? The amplitudes $A_{j\alpha}^{k}$ will obviously be real (to within an inessential common factor that may be complex) if the coefficients $D_{\alpha\beta}^{kk'}$ of the homogeneous system (5.10b) are real. For $\mathbf{q}=0$ and for $\mathbf{q}\cdot\mathbf{a}_i=\pm\pi$, the factor $\exp[i\mathbf{q}\cdot\mathbf{a}_{\mathbf{n}'}-\mathbf{a}_{\mathbf{n}}]$ is equal to $\pm 1$; but then it follows from (5.10a) that $D_{\alpha\beta}^{kk'}$ are real.

Hence, for the infinitely long and shortest waves the amplitudes $A_{j\alpha}^{k}$ in (5.7) may be assumed to be real.

It can easily be demonstrated that the amplitudes $A_{j\alpha}^{k}$ may be assumed to be real for a simple lattice. In this case the lattice is made up of identical atoms located in sites $\mathbf{a}_{\mathbf{n}}$, and there always is an identical atom in the site $-\mathbf{a}_{\mathbf{n}}$ to match one in $\mathbf{a}_{\mathbf{n}}$. The summation in (5.10a) can be performed over atomic pairs in $\mathbf{a}_{\mathbf{n}}'$ and $-\mathbf{a}_{\mathbf{n}}'$. Assuming that $\mathbf{a}_{\mathbf{n}}=0$ and taking into account that $\Phi_{\alpha\beta}(\mathbf{n})=$ $=\Phi_{\alpha\beta}(-\mathbf{n})$, we obtain from (5.10a)

$$D_{\alpha\beta}(\mathbf{q})=\frac{1}{m}\sum_{\mathbf{n}'}\Phi_{\alpha\beta}(\mathbf{n}')\,e^{-i\mathbf{q}\cdot\mathbf{a}_{\mathbf{n}}'}+\Phi_{\alpha\beta}(-\mathbf{n}')\,e^{i\mathbf{q}\cdot\mathbf{a}_{\mathbf{n}}'}$$

$$=\sum_{\mathbf{n}'}\frac{1}{m}\Phi_{\alpha\beta}(\mathbf{n}')\,e^{-i\mathbf{q}\cdot\mathbf{a}_{\mathbf{n}}'}+e^{i\mathbf{q}\cdot\mathbf{a}_{\mathbf{n}}'}=\sum_{\mathbf{n}'}\frac{2}{m}\Phi_{\alpha\beta}(\mathbf{n}')\cos(\mathbf{q}\cdot\mathbf{a}_{\mathbf{n}}').\quad(5.11\text{c})$$

Hence, the elements of the rank $3s$ matrix $D_{\alpha\beta}(\mathbf{q})$ are real, and consequently real amplitudes $A_{j\alpha}^{k}$ can be chosen.

It follows from the definition (5.10a) that

$$D_{\alpha\beta}^{kk'}(-\mathbf{q})=D_{\alpha\beta}^{kk'}(\mathbf{q})^{*}.\quad(5.12)$$

Hence, and from the fact that the matrix (5.11b) is hermitian, it follows that the characteristic equation (5.11a) does not change when $-\mathbf{q}$ is substituted for $\mathbf{q}$:

$$\omega_j(-\mathbf{q})=\omega_j(\mathbf{q}).\quad(5.13)$$

But then it follows from the system (5.10b) that

$$A_{j\alpha}^{k}(-\mathbf{q})=A_{j\alpha}^{k}(\mathbf{q})^{*}.\quad(5.14)$$

Relation (5.13) is a more profound consequence of the invariance of the equations of mechanics with respect to time reversal ($t\rightarrow -t$).

**3.5.4** Since $\omega_j$ is a function of $\mathbf{q}$, we can construct in a Brillouin zone a family of surfaces $\omega_j(\mathbf{q})=$ const. for each vibration branch $j$.

The structure of those constant frequency or constant energy ($\hbar\omega_j(\mathbf{q}) = \mathscr{E}_j(\mathbf{q})$) surfaces depends substantially on the symmetry of the direct crystal lattice. Specifically, the symmetry determines to a great extent (although not entirely) the values of $\mathbf{q}$ at which contacts and intersections of constant-energy surfaces (of different vibration branches) take place.

It follows from the equivalence of wave vectors that differ by $\mathbf{b}_g$ that the physical quantities dependent on $\mathbf{q}$ should be three-dimensional periodic functions in the $\mathbf{q}$-space with the periods equal to $\mathbf{b}_i$, where the $\mathbf{b}_i$ are the basis vectors of the reciprocal lattice. Specifically,

$$\omega_j(\mathbf{q} + \mathbf{b}_g) = \omega_j(\mathbf{q}). \tag{5.15}$$

Hence, the pattern of constant energy surfaces $\omega_j(\mathbf{q}) = \text{const.}$ will be periodically repeated in the unit cells of the reciprocal lattice and in the Brillouin zones.

Condition (5.13) shows that the constant energy surfaces $\omega_j(\mathbf{q}) = \text{const.}$ have an *inversion*, or *symmetry*, *centre* in the $\mathbf{q}$-space. Of course, (5.13) does not necessarily mean that in general $\omega_j(q_x) = \omega_j(-q_x)$. For this to be true the reciprocal lattice must have a symmetry (a reflection) plane $q_x = 0$, and this is the case only if the direct lattice has an appropriate symmetry plane $x = 0$. It can be demonstrated that the reciprocal lattice has all the symmetry elements of the direct lattice.

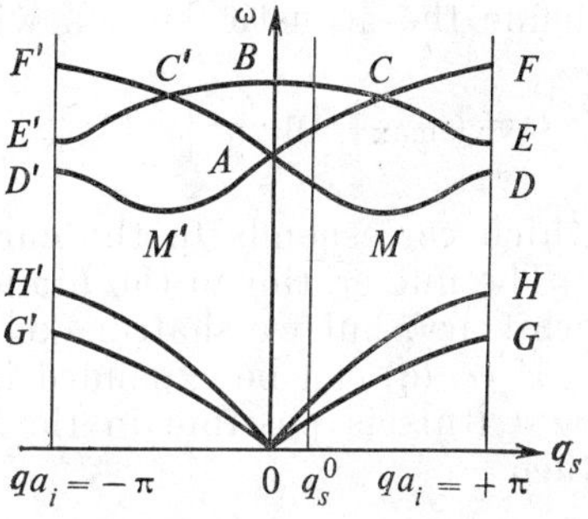


Fig. 3.9

Choose such a direction s in the $\mathbf{q}$-space (this direction may in a particular case coincide with that of the vector $\mathbf{q}$ itself) so that $\omega_j(q_s) = \omega_j(-q_s)$. Figure 3.9 depicts five vibration branches $\omega_j(q_s)$ for variations of $q_s$ corresponding to the limits of inequality (5.9). We see that the points $0, A, C$ and $C'$ are points of degeneracy, i.e., of coincidence (intersection) of several vibration branches. But how can we determine a specified vibration branch in the entire range of variation of $q_s$? Indeed, we can define the specified branch of vibrations $\omega_j$, for instance, alternatively as $DMAC'F'$, $DMAC'E'$, or $DMAM'D'$. Let us adopt the following procedure: number the vibration branches for a specified $q_s = q_s^0$, e.g.,

$$\omega_1(q_s^0) < \omega_2(q_s^0) < \omega_3(q_s^0) < \omega_4(q_s^0) < \omega_5(q_s^0)$$

(see Fig. 3.9). Now let the following inequality hold for all the $q_s$:

$$\omega_1(q_s) \leqslant \omega_2(q_s) \leqslant \omega_3(q_s) \leqslant \omega_4(q_s) \leqslant \omega_5(q_s); \tag{5.16}$$

this inequality determines the numeration of the branch in the entire range of variation of $q_s$. It is easily seen that the curve corresponding to the branch $j = 1$ in this case is $GOG'$, to $j = 3$ it is $DMAM'D'$, and to $j = 4$ it is $ECAC'E'$. Only such a choice of the branches satisfies the condition $\omega_j(q) = \omega_j(-q)$. However, the tangent $(\partial\omega_j/\partial q_s)_{q_s=0}$ for a specified branch generally experiences a jump (see the branches $GOG'$ or $DAD'$). For example, for the acoustic branch we had in the one-dimensional case the relation (2.5):

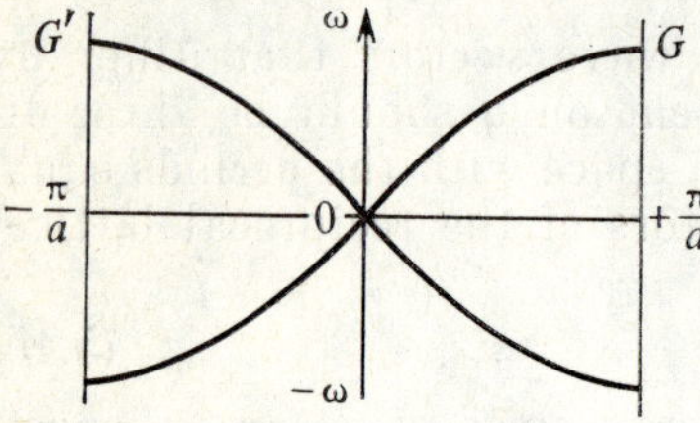


Fig. 3.10

$$\omega^2 = \omega_{\max}^2 \sin^2\frac{qa}{2}, \quad \text{or}$$

$$\omega = \pm\,\omega_{\max} \sin\frac{qa}{2},$$

i.e., the point $q = 0$ is a branch point (Fig. 3.10). Rejecting the parts of the branches corresponding to the negative frequencies $\omega$, we define the acoustic branch with the aid of the equality

$$\omega = \omega_{\max}\left|\sin\frac{qa}{2}\right|$$

which corresponds to the curve $GOG'$ in Figs. 3.9 and 3.10.

The numeration of the branches at point $C$ (Fig. 3.9) is, of course, arbitrary, but we shall make it comply with the general rule.

If $\omega_j(\mathbf{q})$ can be expanded in a power series in $q_\alpha$ $(x, y, z)$ for $\mathbf{q} = 0$ (this is possible in the absence of degeneracy at this point), then

$$\omega_j(\mathbf{q}) = \omega_j(0) + \sum_\alpha r_\alpha q_\alpha + \sum_{\alpha\beta} R_{\alpha\beta} q_\alpha q_\beta, \tag{5.17}$$

where

$$r_\alpha = \left(\frac{\partial\omega_j}{\partial q_\alpha}\right)_0, \quad R_{\alpha\beta} = \frac{1}{2}\left(\frac{\partial^2\omega_j}{\partial q_\alpha\,\partial q_\beta}\right)_{00}.$$

Substituting $-\mathbf{q}$ for $\mathbf{q}$ and making use of (5.13), we obtain $\sum_\alpha r_\alpha q_\alpha = 0$. Since $q_\alpha$ is arbitrary, this yields

$$r_\alpha = \left(\frac{\partial\omega_j}{\partial q_\alpha}\right) = 0.$$

Reducing the tensor $R_{\alpha\beta}$ to the principal axes, we obtain

$$\omega_j(\mathbf{q}) = \omega_j^0 + \sum_\alpha R_\alpha q_\alpha^2, \tag{5.18}$$

i.e., in the absence of degeneracy at point $q = 0$ the expansion of $\omega_j - \omega_j^0$ starts with quadratic terms. This case is depicted in Fig. 3.9 by point $B$.

For a cubic crystal the components of the tensor $R_\alpha$ reduce to a scalar $R$, so that

$$\omega_j(\mathbf{q}) = \omega_j^0 + Rq^2, \tag{5.18a}$$

i.e., the constant energy surfaces turn into spheres.

At point $A$ the expansion of $\omega_j$ in a power series is impossible (the derivative of $\omega_j$ is not unique); because of this at this point for every branch we have, in general,

$$\left(\frac{\partial \omega_j}{\partial q}\right)_0 \neq 0.$$

Finally, note that generally the extrema of $\omega_j(q_s)$ can exist both in the centre and on the boundaries of a Brillouin zone and at some points inside it (for instance, $M$ and $M'$ in Fig. 3.9).

**3.5.5** Consider the nature of the solutions of equations (5.10) in the limiting case of ultralong waves ($\mathbf{q} \to 0$), when, as we pointed out above, the amplitudes $\overset{k}{A_\alpha}(0)$ are real and, consequently, represent actual displacements $\overset{k}{u_{\mathbf{n}\alpha}}$ of atoms from their equilibrium positions.

It follows from (5.10) and (5.10a) that, for $\mathbf{q} = 0$,

$$\omega^2(0)\overset{k}{A_\alpha}(0) = \sum_{k'\beta\mathbf{n}'} \frac{1}{\sqrt{m_k m_{k'}}} \Phi_{\alpha\beta}\begin{pmatrix} kk' \\ \mathbf{nn}' \end{pmatrix} \overset{k'}{A_\beta}(0). \tag{5.19}$$

For the $j$-th branch of the vibration amplitude [see (5.7)] we put

$$\overset{k'}{A_{j\beta}}(0)/\sqrt{m_{k'}} = \text{const.}_{(k')}, \tag{5.20}$$

i.e., is independent of the atom's number $k'$. In this case the right-hand side of (5.19) due to (5.3b) is

$$\frac{1}{\sqrt{m_k}} \sum_\beta \frac{\overset{k'}{A_{j\beta}}(0)}{\sqrt{m_{k'}}} \sum_{k'\mathbf{n}'} \Phi_{\alpha\beta}\begin{pmatrix} kk' \\ \mathbf{nn}' \end{pmatrix} = 0.$$

But then it follows from (5.10) that

$$\omega_j(0) = 0 \tag{5.21}$$

(since we are not interested in the case $\overset{k}{A_{j\alpha}}(0) = 0$ for all three values of $\alpha = x, y, z$). On the other hand, for (5.21) to be valid it is sufficient that $\overset{k}{A_{j\alpha}}(0) \neq 0$ for one $\alpha$; therefore, it is natural to assume the existence of three vibration branches ($j = 1, 2, 3$) for which $\omega_j(0) = 0$, i.e., for which the frequency tends to zero when $\mathbf{q} \to 0$. Those three vibration branches for which the pattern of atomic motion is described by (5.20) are termed *acoustic*. They are similar to the acoustic branch considered in the one-dimensional case (Section 3.3).

Consider now the case when $\mathbf{q} = 0$ but (5.20) and the condition (5.21) which follows from it do not hold. We rewrite (5.19) in the form

$$\omega^2(0)\, m_k \frac{A_\alpha^k(0)}{\sqrt{m_k}} = \sum_{k'\beta\mathbf{n}'} \frac{1}{\sqrt{m_{k'}}}\, \Phi_{\alpha\beta}\begin{pmatrix} kk' \\ \mathbf{nn}' \end{pmatrix} A_\beta^{k'}(0)$$

and perform summation in both sides over $k$:

$$\omega^2(0) \sum_k m_k \frac{A_\alpha^k(0)}{\sqrt{m_k}} = \sum_{k'\beta} \frac{A_\beta^{k'}(0)}{\sqrt{m_{k'}}} \sum_{\mathbf{n}'k'} \Phi_{\alpha\beta}\begin{pmatrix} kk' \\ \mathbf{nn}' \end{pmatrix}$$

The last sum in the right-hand side, according to (5.3b), is zero, and since we have assumed that $\omega^2(0) \neq 0$, it follows that the sum over $k$ in the left-hand side vanishes. But then, making use of (5.7), we obtain

$$\sum_k m_k u_{\mathbf{n}\alpha}^k(0) = 0, \tag{5.22}$$

i.e., the atomic motion leaves the centre of gravity of the cell in place. Such vibration branches became known as *optical*. Their one-dimensional analogue was discussed in Section 3.3.

We see that in addition to the three acoustic branches there should be $3s$-3 optical branches for which $\omega_j(0) \neq 0$ $(j = 4, 5, \ldots, 3s)$.

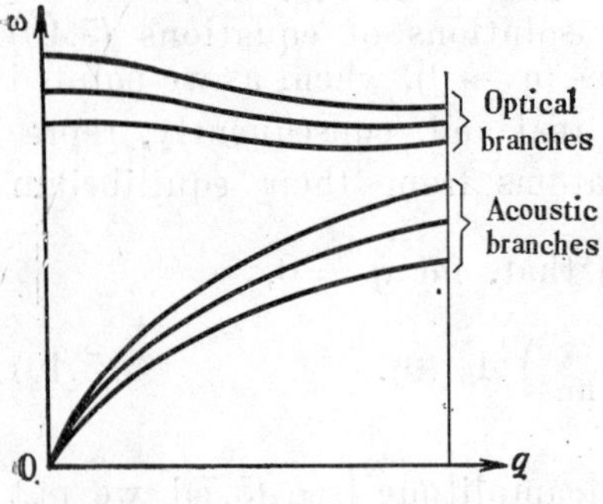


Fig. 3.11

Just as in the one-dimensional case, we can demonstrate that for $\mathbf{q}\cdot\mathbf{a}_i \ll 1$ the frequencies of the acoustic branches $\omega_j = v_{0j}q$ $(j = 1, 2, 3)$, where the wave propagation velocity $v_{0j}$ depends not only on the branch $j$ but also on the direction of $\mathbf{q}$. For optical branches the $\omega_j$ versus $\mathbf{q}$ dependence in the vicinity of zero in the absence of degeneracy is given by expression (5.18), i.e., it displays an extremum (usually a maximum). At the boundaries a Brillouin zone and in the absence of degeneracy of the vibration branches, the frequencies $\omega_j(\mathbf{q})$ also attain extremal values. Figure 3.11 is a schematic diagram of the acoustic and optical branches (in the absence of degeneracy) for some direction of a crystal with two atoms per unit cell.

To obtain a complete picture of atomic vibrations in a crystal, we should solve (5.11) for all values of $\mathbf{q}$ inside the interval $-\pi \leqslant \mathbf{q}\cdot\mathbf{a}_i \leqslant +\pi$ or for all the points of a Brillouin zone. Even if the force constants $\Phi_{\alpha\beta}\begin{pmatrix} kk' \\ \mathbf{nn}' \end{pmatrix}$ are known, this necessarily involves numerical calculations. Such calculations show that in a three-dimensional crystal the acoustic and optical branches are approximately of the form shown in Fig. 3.11.

**3.5.6** In exactly the same way as it was done in the one-dimensional case (Section 3.2.3), substitute the Born-von Kármán cyclic conditions for the boundary conditions on the surface of the crystal's principal region. In the case of a three-dimensional crystal their meaning is not so clear as in the one-dimensional case, when we could arrange the whole atomic chain in a circle joining the first and the $G$-th atoms. However, it can be demonstrated [3.11, p. 391] that in a three-dimensional crystal the cyclic conditions are equivalent to the boundary conditions on the free-surface of the crystal's principal region in determining its volume properties.

The Born-von Kármán cyclic conditions in a three-dimensional crystal are tantamount to the requirement that the wave field $u^{k}_{\mathbf{n}\alpha}$ proportional to $e^{i\mathbf{q}\cdot\mathbf{a_n}}$ should remain invariant as the result of a displacement by any one of the vectors $G\mathbf{a}_i$ $(i = 1, 2, 3)$. The necessary condition for this is $e^{i\mathbf{q}\cdot G\mathbf{a}_i} = 1$, i.e., $G\mathbf{q}\cdot\mathbf{a}_i = 2\pi g_i$, where $g_i$ is an integer. Hence,

$$\varphi_i \equiv \mathbf{q}\cdot\mathbf{a}_i = \frac{2\pi}{G} g_i, \quad (i = 1, 2, 3). \tag{5.23}$$

Comparing this relation with (1.3.7), we see that (5.23) holds if

$$\mathbf{q} = \frac{1}{G}\mathbf{b}_{\mathbf{g}} = \frac{1}{G}(g_1\mathbf{b}_1 + g_2\mathbf{b}_2 + g_3\mathbf{b}_3), \tag{5.24}$$

where $\mathbf{b}_i$ are the basis vectors of the reciprocal lattice. This representation of the vector $\mathbf{q}$ similar to (2.8) is quite formal, because $G$ is an arbitrary large number (for convenience, assumed to be odd) which should not affect the physically observable quantities.

Making use of the inequality (5.9), we obtain in the same way as (2.8a)

$$-G/2 < g_i < G/2. \tag{5.25}$$

We see that the projection of $\mathbf{q}$ on each of the vectors $\mathbf{a}_i$ can assume $G$ quasi-discrete values; therefore, there are altogether $G^3 = N$ distinct values of $\mathbf{q}$. How many distinct vibrations (possible values of $\mathbf{q}$) are there for a single branch in volume $d\tau_{\mathbf{q}} = dq_x dq_y dq_z$ of the wave vector space? This number is obviously

$$dz = dg_1\, dg_2\, dg_3 = \left(\frac{G}{2\pi}\right)^3 d\varphi_1\, d\varphi_2\, d\varphi_3, \tag{5.26}$$

where

$$\varphi_i = q_x a_{ix} + q_y a_{iy} + q_z a_{iz}.$$

Change over from the variables $\varphi_i$ to the variables $q_x$, $q_y$, $q_z$. To this end we calculate the Jacobian of the transformation [3.12, Sec. 4.5-6]:

$$\begin{vmatrix} \frac{\partial \varphi_1}{\partial q_x} & \frac{\partial \varphi_2}{\partial q_x} & \frac{\partial \varphi_3}{\partial q_x} \\ \frac{\partial \varphi_1}{\partial q_y} & \frac{\partial \varphi_2}{\partial q_y} & \frac{\partial \varphi_3}{\partial q_y} \\ \frac{\partial \varphi_1}{\partial q_z} & \frac{\partial \varphi_2}{\partial q_z} & \frac{\partial \varphi_3}{\partial q_z} \end{vmatrix} = \begin{vmatrix} a_{1x} & a_{2x} & a_{3x} \\ a_{1y} & a_{2y} & a_{3y} \\ a_{1z} & a_{2z} & a_{3z} \end{vmatrix} = \mathbf{a}_1 \cdot (\mathbf{a}_2 \times \mathbf{a}_3) = \Omega_0.$$

Hence,

$$dz = \left(\frac{G}{2\pi}\right)^3 d\varphi_1\, d\varphi_2\, d\varphi_3 = \left(\frac{G}{2\pi}\right)^3 \Omega_0\, dq_x\, dq_y\, dq_z = \frac{V}{(2\pi)^3}\, d\tau_{\mathbf{q}}. \quad (5.27)$$

The total number of vibrations for one branch is

$$z = \int dz = \frac{V}{(2\pi)^3} \int\int\int d\tau_{\mathbf{q}} = \frac{V}{(2\pi)^3} \frac{(2\pi)^3}{\Omega_0} = N.$$

For all vibration branches of a complex crystal this number is always $3sN$, i.e., is equal to the number of degrees of freedom of the atoms in the principal region.

In many cases, for instance, in the study of statistics, it is desirable to know the number of vibrations of the $j$-th branch inside the $[\omega, \omega + d\omega]$ interval. Consider the surfaces $\omega_j(\mathbf{q}) = \text{const.}$ in the $\mathbf{q}$-space and make

$$d\tau_{\mathbf{q}} = d\sigma\, dq_\perp, \quad (5.28)$$

where $d\sigma$ is an element of this surface, and $dq_\perp$ is an infinitesimal increment along a normal to $d\sigma$. Obviously,

$$d\omega = |\nabla_{\mathbf{q}} \omega_j|\, dq_\perp, \quad (5.29)$$

where $\nabla_{\mathbf{q}}$ is a gradient in the $\mathbf{q}$-space. Substituting (5.28) and (5.29) into (5.27) and integrating over the constant frequency surface, we obtain an expression for the number of vibrations in the $j$-th branch inside $[\omega, \omega + d\omega]$:

$$dz_\omega = \frac{V}{(2\pi)^3} \int \frac{d\sigma}{|\nabla_{\mathbf{q}} \omega_j|}\, d\omega = V g_j(\omega)\, d\omega, \quad (5.30)$$

where

$$g_j(\omega) = \frac{1}{(2\pi)^3} \int \frac{d\sigma}{|\nabla_{\mathbf{q}} \omega_j|} \quad (5.31)$$

is the *frequency* or the *normal vibration, distribution function for the j-th branch*. We see that to determine it we must know the dependence $\omega_j = \omega_j(\mathbf{q})$, i.e., the *dispersion law*.

If we are interested in the total number of vibrations in all dispersion branches, we must use the full frequency distribution function

$$g(\omega)=\sum_{j=1}^{3s} g_j(\omega)=\frac{1}{(2\pi)^3}\sum_{j=1}^{3s}\int\frac{d\sigma}{|\nabla_{\mathbf{q}}\omega_j|}. \qquad (5.32)$$

A simple example of a frequency distribution function was cited in (2.10) for a one-dimensional lattice.

In some cases it is more convenient to use another frequency distribution function, $\tilde{g}(\omega^2)$; $\tilde{g}(\omega^2)\,d\omega^2$ is the number of vibrations in the $[\omega^2, \omega^2+d\omega^2]$ interval. It is obvious that

$$g(\omega)=2\omega\tilde{g}(\omega^2), \qquad (5.33)$$

since $\tilde{g}(\omega^2)\,d\omega^2=\tilde{g}(\omega^2)2\omega\,d\omega$.

## 3.6 Normal Coordinates of Crystal Lattice Vibrations

**3.6.1** In Section 3.4 we considered the normal coordinates for a simple one-dimensional lattice.

We started by introducing complex normal coordinates $a_q$ and then transformed them into real normal coordinates $x_q$ and generalized momenta $p_q$.

The normal coordinates are distinguished by the property that each depends harmonically on time, i.e., is proportional to $\exp(i\omega_q t)$. This is tantamount to the assertion that the Hamiltonian is a sum of squares (with appropriate coefficients) of coordinates and momenta and is of the form (4.11).

The atomic motion in a complex three-dimensional crystal lattice in the most general form consists, obviously, of a sum of expressions (5.7) with different wave vectors $\mathbf{q}$ and frequencies $\omega_{\mathbf{q}}$. Indeed, such a sum will also satisfy the equations of motion (5.6) in view of their linearity.

To introduce normal coordinates for a three-dimensional lattice, we define *eigenvectors* $\mathbf{e}_{jk}(\mathbf{q})$ (in general, complex) of the dynamical matrix $D_{\alpha\beta}^{kk'}(\mathbf{q})$ [see (5.10a)] in the following way (see Appendix 3, Sec. 5):

$$\sum_{k'\beta} D_{\alpha\beta}^{kk'}(\mathbf{q})\,e_{jk'\beta}(\mathbf{q})=\omega_j^2(\mathbf{q})\,e_{jk\alpha}(\mathbf{q}), \qquad (6.1)$$

where $e_{jk\alpha}$ ($\alpha=x, y, z$) are the components of an eigenvector, and $\omega_j^2(\mathbf{q})$ is the $j$-th eigenvalue of the dynamical matrix. Since the dynamical matrix is hermitian, its eigenvectors are orthogonal (see Appendix 3, Sec. 5), and since they are determined from a homogenous system (6.1) to within an arbitrary factor, they can be normalized to unity. Hence,

$$\sum_{\alpha k} e_{jk\alpha} e^*_{j'k\alpha} = \delta_{jj'}, \tag{6.2}$$

and

$$\sum_{j} e_{jk\alpha} e^*_{jk'\beta} = \delta_{kk'} \delta_{\alpha\beta}. \tag{6.2a}$$

The only difference between the eigenvectors $\mathbf{e}_{jk\alpha}(\mathbf{q})$ and the complex amplitudes $A^k_{j\alpha}(\mathbf{q})$ satisfying (5.10) is in the normalizing conditions.

We see that (5.14) follows from (5.12):

$$e_{jk\alpha}(\mathbf{q}) = e^*_{jk\alpha}(-\mathbf{q}), \tag{6.3}$$

since substitution of $-\mathbf{q}$ for $\mathbf{q}$ changes all the coefficients of equations (6.1) to complex conjugate.

3.6.2 Introduce complex normal coordinates $a_j(\mathbf{q}, t)$ with the aid of the relation

$$u^k_{\mathbf{n}\alpha} = \frac{1}{\sqrt{Nm_k}} \sum_{\mathbf{q}j} e_{jk\alpha}(\mathbf{q})\, a_j(\mathbf{q}, t)\, e^{i\mathbf{q}\cdot\mathbf{a_n}}, \tag{6.4}$$

where $u^k_{\mathbf{n}\alpha}$ is the $\alpha$-th component of the displacement of the $(\mathbf{n}, k)$-th atom from its equilibrium position, $N$ is the number of unit cells in the crystal's principal region, and $m_k$ is the mass of the atom of type $k$.

It may easily be demonstrated that

$$a_j(\mathbf{q}) = a^*_j(-\mathbf{q}). \tag{6.5}$$

Indeed, only in this case are the displacements $u^k_{\mathbf{n}\alpha}$ real; actually, it follows from (6.4), (6.3) and (6.5) that

$$(u^k_{\mathbf{n}\alpha})^* = \frac{1}{\sqrt{Nm_k}} \sum_{\mathbf{q}j} e^*_{jk\alpha}(\mathbf{q})\, a^*_j(\mathbf{q})\, e^{-i\mathbf{q}\cdot\mathbf{a_n}}$$

$$= \frac{1}{\sqrt{Nm_k}} \sum_{\mathbf{q}j} e_{jk\alpha}(-\mathbf{q})\, a_j(-\mathbf{q})\, e^{-i\mathbf{q}\cdot\mathbf{a_n}} = u^k_{\mathbf{n}\alpha},$$

since summation over $-\mathbf{q}$ is equivalent to summation over $\mathbf{q}$. Taking into account that $\mathbf{q}$ runs through $N$ quasi-discrete values and $j$ through $3s$ values, we conclude that there are altogether $3sN$ complex values $a_j(\mathbf{q})$. Because of (6.5), this is equivalent to specifying $3sN$ real coordinates ($3sN$ is the number of degrees of freedom of the system).

Expression (6.4) for atomic displacements can be written in a somewhat different form which is more convenient for certain applications.

Draw a plane through the origin in the **q**-space; the vectors **q** and —**q** will be separated by this plane. Write expression (6.4) in terms of sums of **q**'s taken on each side of the plane:

$$u_{\mathbf{n}\alpha}^{k}=\frac{1}{\sqrt{Nm_k}}\Big\{\sum_{\mathbf{q}>0,\ j} e_{jk\alpha}(\mathbf{q})\,a_j(\mathbf{q},\ t)\,e^{i\mathbf{q}\cdot\mathbf{a}_\mathbf{n}}$$

$$+\sum_{\mathbf{q}<0,\ j} e_{jk\alpha}(\mathbf{q})\,a_j(\mathbf{q},\ t)\,e^{i\mathbf{q}\cdot\mathbf{a}_\mathbf{n}}\Big\}.$$

Substituting in the second sum —**q** for **q** and making use of (6.3) and (6.5), we obtain

$$u_{\mathbf{n}\alpha}^{k}=\frac{1}{\sqrt{Nm_k}}\sum_{\mathbf{q}j}{}'\{e_{jk\alpha}(\mathbf{q})\,a_j(\mathbf{q},\ t)\,e^{i\mathbf{q}\cdot\mathbf{a}_\mathbf{n}}$$

$$+e_{jk\alpha}^{*}(\mathbf{q})\,a_j^{*}(\mathbf{q},\ t)\,e^{-i\mathbf{q}\cdot\mathbf{a}_\mathbf{n}}\}, \tag{6.4a}$$

where the prime at the sum means that the summation over **q** is performed in one-half of a Brillouin zone.

In order to express the complex coordinates $a_j(\mathbf{q},\ t)$ in terms of real displacements $u_{\mathbf{n}\alpha}^{k}$, we multiply both sides of equality (6.4) by $\sqrt{m_k}e_{j'k\alpha}^{*}(\mathbf{q}')\times\exp(-i\mathbf{q}'\cdot\mathbf{a}_\mathbf{n})$ and sum them over **n**, $k$, and $\alpha$. The summation of the multiplicand $\exp[i(\mathbf{q}-\mathbf{q}')\cdot\mathbf{a}_\mathbf{n}]$ over **n** in the right-hand side yields, according to (A. 6.4), the quantity $N\delta_{\mathbf{q}\mathbf{q}'}$, and the summation of the multiplicand $e_{jk\alpha}(\mathbf{q})\,e_{j'k\alpha}^{*}$ over $k$ and $\alpha$ yields, according to (6.2), the quantity $\delta_{jj'}$. Finally, we obtain, after summation over **q** and $j$, for the right-hand side $\sqrt{N}a_{j'}(\mathbf{q}')$. Changing the notation of $j'$ and $\mathbf{q}'$ to $j$ and **q**, we obtain

$$a_j(\mathbf{q})=\frac{1}{\sqrt{N}}\sum_{\mathbf{n}k\alpha}\sqrt{m_k}\,u_{\mathbf{n}\alpha}^{k}e_{jk\alpha}(\mathbf{q})\,e^{-i\mathbf{q}\cdot\mathbf{a}_\mathbf{n}}. \tag{6.6}$$

To ensure that $a_j(\mathbf{q})$ are normal coordinates of the system, we express in terms of them the kinetic and potential energies of the crystal.

The kinetic energy of the atoms in the crystal's principal region is

$$\mathscr{T}=\frac{1}{2}\sum_{\mathbf{n}k\alpha}m_k(\dot{u}_{\mathbf{n}\alpha}^{k})^2=\frac{1}{2N}\sum_{\mathbf{n}k\alpha}\Big[\sum_{\mathbf{q}j}e_{jk\alpha}(\mathbf{q})\,\dot{a}_j(\mathbf{q},\ t)\,e^{i\mathbf{q}\cdot\mathbf{a}_\mathbf{n}}\Big]^2, \tag{6.7}$$

where we made use of relation (6.4).

Since the expression in the square bracket on the right-hand side is real, its square can be formally represented as the product of the expression in the brackets times the complex conjugate of it. Hence,

$$\mathcal{T} = \frac{1}{2N} \sum_{\mathbf{n}k\alpha} \sum_{\mathbf{q}j\mathbf{q}'j'} [e_{jk\alpha}(\mathbf{q})\, \dot{a}_j(\mathbf{q}, t)\, e^{i\mathbf{q}\cdot\mathbf{a_n}} e^*_{j'k\alpha}(\mathbf{q}')\, \dot{a}^*_{j'}(\mathbf{q}', t)\, e^{-i\mathbf{q}'\cdot\mathbf{a_n}}]. \tag{6.8}$$

It follows from (A.6.4) that

$$\sum_{\mathbf{n}} e^{i(\mathbf{q}-\mathbf{q}')\cdot\mathbf{a_n}} = N\delta_{\mathbf{q}\mathbf{q}'};$$

therefore, in the course of summation over $\mathbf{q}'$ we may make $\mathbf{q}' = \mathbf{q}$ in all the multiplicands, after which the summation over $k$, $\alpha$ of the products $e_{jk\alpha}e^*_{j'k\alpha}$ yields $\delta_{jj'}$ [see (6.2)]. Lastly, summing over $j'$, we obtain from (6.8)

$$\mathcal{T} = \frac{1}{2} \sum_{\mathbf{q}j} |\dot{a}_j(\mathbf{q}, t)|^2. \tag{6.9}$$

The potential energy of the crystal's principal region in the quadratic approximation with respect to atomic displacements is, according to (5.2), equal to

$$\Phi = \frac{1}{2} \sum_{\mathbf{nn}'k\alpha k'\beta} \Phi_{\alpha\beta}\begin{pmatrix} kk' \\ \mathbf{nn}' \end{pmatrix} u^k_{\mathbf{n}\alpha} u^{k'}_{\mathbf{n}'\beta}$$

$$= \frac{1}{2N} \sum_{\mathbf{nn}'k\alpha k'\beta} \Phi_{\alpha\beta}\begin{pmatrix} kk' \\ \mathbf{nn}' \end{pmatrix} \frac{1}{\sqrt{m_k}} \sum_{\mathbf{q},\, j} e^*_{jk\alpha}(\mathbf{q})\, a^*_j(\mathbf{q})\, e^{-i\mathbf{q}\cdot\mathbf{a_n}}$$

$$\times \frac{1}{\sqrt{m_{k'}}} \sum_{\mathbf{q}'j'} e_{j'k'\beta}(\mathbf{q}')\, a_{j'}(\mathbf{q}')\, e^{i\mathbf{q}'\cdot\mathbf{a_n}}, \tag{6.10}$$

where we expressed the atomic displacements using (6.4) and for convenience wrote the real quantity $u^k_{\mathbf{n}\alpha}$ in the form $(u^k_{\mathbf{n}\alpha})^*$.

Transforming the sum over $\mathbf{n}$ and $\mathbf{n}'$, we get

$$\sum_{\mathbf{nn}'} \frac{1}{\sqrt{m_k m_{k'}}} \Phi_{\alpha\beta}\begin{pmatrix} kk' \\ \mathbf{nn}' \end{pmatrix} e^{i(\mathbf{q}'\cdot\mathbf{a}'_\mathbf{n} - \mathbf{q}\cdot\mathbf{a_n})}$$

$$= \sum_{\mathbf{n}} e^{i(\mathbf{q}'-\mathbf{q})\cdot\mathbf{a_n}} \sum_{\mathbf{n}'} \frac{1}{\sqrt{m_k m_{k'}}} \Phi_{\alpha\beta}\begin{pmatrix} kk' \\ \mathbf{nn}' \end{pmatrix} e^{i\mathbf{q}'\cdot(\mathbf{a}'_\mathbf{n}\cdot\mathbf{a_n})} = N\delta_{\mathbf{q}\mathbf{q}'} D^{kk'}_{\alpha\beta}(\mathbf{q}'),$$

where we made use of (A.6.4) and of (5.10a).[7] Substituting this ex-

[7] It should be taken into account that the sum over $\mathbf{n}'$ is independent of $\mathbf{n}$, since the expressions under the summation sign depend only on the difference $'$ — $\mathbf{n}$.

pression into (6.10), summing over $\mathbf{q}'$ and making use of (6.1), we obtain

$$\Phi = \frac{1}{2} \sum_{jk\alpha j'k'\beta \mathbf{q}\mathbf{q}'} D_{\alpha\beta}^{kk'}(\mathbf{q}')\, \delta_{\mathbf{q}\mathbf{q}'} e_{jk\alpha}^{*}(\mathbf{q})\, a_j^{*}(\mathbf{q})\, e_{j'k'\beta}(\mathbf{q}')\, a_{j'}(\mathbf{q}')$$

$$= \frac{1}{2} \sum_{k\alpha jj'\mathbf{q}} \omega_{j'}^{2}(\mathbf{q})\, e_{j'k\alpha}(\mathbf{q})\, e_{jk\alpha}^{*}(\mathbf{q})\, a_{j'}(\mathbf{q})\, a_j^{*}(\mathbf{q}).$$

Summing over $k$, $\alpha$, making use of (6.2), and summing over $j'$, we finally obtain

$$\Phi = \frac{1}{2} \sum_{\mathbf{q}j} \omega_j^2(\mathbf{q})\, |a_j(\mathbf{q})|^2. \tag{6.11}$$

Hence, we obtain for the total energy of the crystal's principal region

$$\mathscr{E} = \mathscr{T} + \Phi = \frac{1}{2} \sum_{\mathbf{q}j} [|\dot{a}_j(\mathbf{q})|^2 + \omega_j^2(\mathbf{q})\, |a_j(\mathbf{q})|^2]. \tag{6.12}$$

It is evident from the structure of this expression that the vibration energy of the crystal's atoms is equal to the sum of vibration energies of individual independent "oscillators" with a "kinetic energy" of $|\dot{a}_j(\mathbf{q})|^2/2$ and a potential energy of $\omega_j^2(\mathbf{q})\,|a_j(\mathbf{q})|^2/2$ each; for this reason it is proper to term the quantities $a_j(\mathbf{q})$ *complex normal coordinates.*

**3.6.3** Since the laws of mechanics have been formulated for real coordinates and velocities, it is desirable to transform expression (6.12) to real normal coordinates. The most immediate way to introduce them is by making

$$a_j(\mathbf{q}) = \frac{1}{\sqrt{2}} [Q_{1j}(\mathbf{q}) + iQ_{2j}(\mathbf{q})], \tag{6.13}$$

where $Q_{1j}(\mathbf{q})$ and $Q_{2j}(\mathbf{q})$ are real. At first it appears that their number is twice the number of the degrees of freedom; however, from (6.5) it follows that

$$Q_{1j}(\mathbf{q}) = Q_{1j}(-\mathbf{q}),\quad Q_{2j}(\mathbf{q}) = -Q_{2j}(-q). \tag{6.14}$$

Because of this the number of independent real normal coordinates is equal to $3sN$, i.e., the number of the degrees of freedom [as was already stated in conjunction with (6.5)].

In order to express the energy $\mathscr{E}$ (6.12) in terms of the independent coordinates $Q_{1j}$ and $Q_{2j}$, we shall draw a plane through the origin in the $\mathbf{q}$-space and, after substituting (6.13) into (6.12), sum only over the **q**'s that lie to one side of the plane, i.e., in one-half of a Brillouin zone. Then

$$\mathscr{E} = \frac{1}{4} \sum_{\mathbf{q}j}{}' \{[\dot{Q}_{1j}^2(\mathbf{q}) + \omega_j^2(\mathbf{q})\, Q_{1j}^2(\mathbf{q})] + [\dot{Q}_{2j}^2(\mathbf{q}) + \omega_j^2(\mathbf{q})\, Q_{2j}^2(\mathbf{q})]\}. \tag{6.15}$$

It can be demonstrated that there are standing waves in a crystal shifted by one-fourth of a wavelength with respect to each other to correspond to the real normal coordinates (6.13) [3.11, Sec. 28]. Traveling waves are more useful for applications. Real normal coordinates and velocities (momenta) were first introduced by R. Peierls with the aid of the *canonical transformation* [3.10, § 45]

$$a_j(\mathbf{q}) = \widetilde{a}_j(\mathbf{q}) + \widetilde{a}_j^*(-\mathbf{q}). \tag{6.16}$$

Here

$$\widetilde{a}_j(\mathbf{q}) = \frac{1}{2}\left[Q_j(\mathbf{q}) + \frac{i}{\omega_j(\mathbf{q})}\dot{Q}_j(\mathbf{q})\right], \tag{6.16a}$$

where $Q_j(\mathbf{q})$ are real normal coordinates.

It is obvious that condition (6.5) is fulfilled automatically. It will be demonstrated below that $Q_j(\pm\mathbf{q})$ are in fact normal coordinates, and therefore,

$$\ddot{Q}_j(\pm\mathbf{q}) = -\omega_j^2(\mathbf{q})\, Q_j(\pm\mathbf{q}), \tag{6.17}$$

where we have taken into account that $\omega_j(-\mathbf{q}) = \omega_j(\mathbf{q})$ [see (5.13)].

From (6.16) and (6.17) we obtain

$$\dot{a}_j(\mathbf{q}) = \frac{1}{2}[\dot{Q}_j(\mathbf{q}) - i\omega_j(\mathbf{q})\, Q_j(\mathbf{q})] + \frac{1}{2}[\dot{Q}_j(-\mathbf{q}) + i\omega_j(\mathbf{q})\, Q_j(-\mathbf{q})]. \tag{6.18}$$

Substituting (6.16) and (6.18) into (6.12), we obtain

$$\mathscr{E} = \frac{1}{4}\sum_{\mathbf{q}j}\{[\dot{Q}_j^2(\mathbf{q}) + \omega_j^2(\mathbf{q})\, Q_j^2(\mathbf{q})] + [\dot{Q}_j^2(-\mathbf{q}) + \omega_j^2(\mathbf{q})\, Q_j^2(-\mathbf{q})]\}$$

$$= \frac{1}{2}\sum_{\mathbf{q}j}[\dot{Q}_j^2(\mathbf{q}) + \omega_j^2(\mathbf{q})\, Q_j^2(\mathbf{q})], \tag{6.19}$$

since the summation over $-\mathbf{q}$ is equivalent to that over $\mathbf{q}$ and since, according to (5.13), $\omega_j(-\mathbf{q}) = \omega_j(\mathbf{q})$. It follows from (6.19) that $Q_j(\mathbf{q})$ are indeed normal coordinates; this justifies expression (6.17) and confirms that (6.16) is a canonical transformation.

Introducing real generalized momenta

$$P_j(\mathbf{q}) = \frac{\partial \mathscr{E}}{\partial \dot{Q}_j} = \dot{Q}_j(\mathbf{q}), \tag{6.20}$$

conjugate to the coordinates $Q_j(\mathbf{q})$, we can write (6.19) in the form

$$\mathscr{E} = \sum_{\mathbf{q}j} \left[ \frac{1}{2} P_j^2(\mathbf{q}) + \frac{1}{2} \omega_j^2(\mathbf{q}) Q_j^2(\mathbf{q}) \right] = \mathscr{H}(Q, P), \tag{6.21}$$

where $\mathscr{H}(Q, P)$ is the system's Hamiltonian.

Substituting (6.16) into (6.4), we obtain for the atomic displacements

$$u_{\mathbf{n}\alpha}^k = \frac{1}{\sqrt{Nm_k}} \sum_{\mathbf{q}j} e_{jk\alpha}(\mathbf{q}) [\tilde{a}_j(\mathbf{q}) + \tilde{a}_j^*(-\mathbf{q})] e^{i\mathbf{q}\cdot\mathbf{a_n}}. \tag{6.22}$$

If the summation in the second addend that contains the multiplicand $\tilde{a}_j^*(-\mathbf{q})$ is performed over $-\mathbf{q}$ instead of $\mathbf{q}$ (this is obviously equivalent) and if use is made of (6.16), (6.20) and (6.3), we obtain

$$u_{\mathbf{n}\alpha}^k = \frac{1}{\sqrt{Nm_k}} \sum_{\mathbf{q}j} \mathrm{Re} \left\{ e_{jk\alpha}(\mathbf{q}) \left[ Q_j(\mathbf{q}) + \frac{i}{\omega_j(\mathbf{q})} P_j(\mathbf{q}) \right] e^{i\mathbf{q}\cdot\mathbf{a_n}} \right\} \tag{6.23}$$

(Re {. . .} is the real part of the expression in the braces).

Let us show that a traveling wave with a wave vector $\mathbf{q}$ and frequency $\omega_j(\mathbf{q})$ corresponds to the excitation of one normal coordinate $Q_j(\mathbf{q})$.

If only one normal coordinate $Q_j(\mathbf{q})$ is nonzero, then the sum in (6.22) reduces to one term

$$u_{\mathbf{n}\alpha}^k = \frac{1}{\sqrt{Nm_k}} \mathrm{Re} \left\{ e_{j\alpha k}(\mathbf{q}) \left[ Q_j(\mathbf{q}) + \frac{i}{\omega_j(\mathbf{q})} P_j(\mathbf{q}) \right] e^{i\mathbf{q}\cdot\mathbf{a_n}} \right\}. \tag{6.24}$$

Since $Q_j(\mathbf{q})$ is a normal coordinate, $Q_j(\mathbf{q})$ and $(i/\omega_j(\mathbf{q}))\, P_j(\mathbf{q})$ feature a harmonic time dependence, i.e., they are proportional to $\exp[-i\omega_j(\mathbf{q})\, t]$. Denoting the complex amplitude in the braces by $C_\alpha(\mathbf{q})\, e^{i\varphi}$ [the factor in front of Re {. . .} has been included in $C_\alpha(\mathbf{q})$], we obtain

$$u_{\mathbf{n}\alpha}^k = \mathrm{Re}\, \{C_\alpha(\mathbf{q})\, e^{i[\mathbf{q}\cdot\mathbf{a_n} - \omega_j(\mathbf{q})t] + \varphi}\}, \tag{6.25}$$

whence it follows that

$$u_{\mathbf{n}\alpha}^k = C_\alpha(\mathbf{q}) \cos[\mathbf{q}\cdot\mathbf{a_n} - \omega_j(\mathbf{q})\, t + \varphi], \tag{6.26}$$

where $C_\alpha(\mathbf{q})$ is the amplitude of the plane wave propagating in the direction of the wave vector $\mathbf{q}$ with a frequency $\omega_j(\mathbf{q})$, and $\varphi$ is the phase of this wave. We see that traveling waves (6.26) correspond to the normal coordinates $Q_j(\mathbf{q})$ introduced with the aid of the canonical transformation.

**3.6.4** It has been demonstrated in the preceding section that for a simple lattice (a Bravais lattice) the amplitudes $A_{j\alpha}$ are real. Obviously, the same is true for the eigenvectors $e_{j\alpha}$.[8]

It follows from (6.4a) that for a simple lattice

$$u_{\mathrm{n}\alpha}=\frac{1}{\sqrt{NM}}\sum_{\mathbf{q}j}{}' e_{j\alpha}(\mathbf{q})\{a_j(\mathbf{q},t)e^{i\mathbf{q}\cdot\mathbf{a}_\mathrm{n}}+a_j^*(\mathbf{q},t)e^{-i\mathbf{q}\cdot\mathbf{a}_\mathrm{n}}\}. \quad (6.27)$$

Here $M$ is the atomic mass, and the prime at the summation sign means that summation is performed over one half of a Brillouin zone.

Finally, from (6.23) it follows that for a simple lattice

$$u_{\mathrm{n}\alpha}=\frac{1}{\sqrt{NM}}\sum_{\mathbf{q}j}e_{j\alpha}(\mathbf{q})\left[Q_j(\mathbf{q})\cos(\mathbf{q}\cdot\mathbf{a}_\mathrm{n})-\frac{P_j(\mathbf{q})}{\omega_j(\mathbf{q})}\sin(\mathbf{q}\cdot\mathbf{a}_\mathrm{n})\right]. \quad (6.28)$$

## 3.7 Vibrations in Simple Cubic Lattice

**3.7.1** Consider a simple cubic lattice with an atom of mass $m$ in every site.

Figure 3.12 depicts a simple cubic lattice and an orthogonal coordinate frame $x$, $y$, $z$ with its origin at the atom $O$. The six nearest ●-

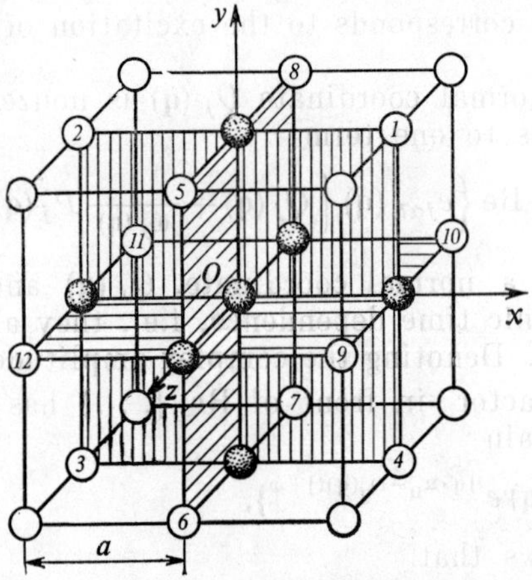


**Fig. 3.12**

atoms are located at a distance $a$ from the central $O$-atom (the *first coordination group*). The next 12 atoms nearest to $O$ (the *second*

[8] When performing the summation in (6.12) inside one half of a Brillouin zone, the factor 1/2 should be dropped.

*coordination group*) are located at a distance $a\sqrt{2}$ from $O$. Out of these, atoms 1, 2, 3, 4 lie in the plane $xy$, atoms 5, 6, 7, 8 in the plane $yz$, and atoms 9, 10, 11, 12 in the plane $xz$.[9]

To simplify notation, we denote the vector of the direct lattice by

$$\mathbf{a_n} = n_1\mathbf{a}_1 + n_2\mathbf{a}_2 + n_3\mathbf{a}_3 \equiv \mathbf{n} \qquad (a_1 = a_2 = a_3 = a). \tag{7.1}$$

It follows from (5.2) that the force acting on the $(\mathbf{n}, k)$-th atom in the direction $\alpha$, when the $(\mathbf{n}', k')$-th atom is displaced in the direction $\beta$ by the amount $u^{k'}_{\mathbf{n}'\beta}$, in the linear approximation is equal to

$$-\frac{\partial \Phi}{\partial u^{k}_{\mathbf{n}\alpha}} = -\Phi_{\alpha\beta}\begin{pmatrix} kk' \\ \mathbf{nn}' \end{pmatrix} u^{k'}_{\mathbf{n}'\beta}. \tag{7.2}$$

In the case of a simple lattice the indices $k$ and $k'$ are absent, and the force acting on the atom at the origin ($\mathbf{n} = 0$) in the direction of the $\alpha$ axis, when the atom $\mathbf{n}'$ ($\equiv \mathbf{n}$) is displaced in the direction $\beta$ by an amount $u_{\mathbf{n}\beta}$, is equal to

$$-\frac{\partial \Phi}{\partial u_{0\alpha}} = -\Phi^{0\mathbf{n}}_{\alpha\beta} u_{\mathbf{n}\beta}. \tag{7.3}$$

The force (7.3) can be expressed in terms of the quasi-elastic bonding constant $\alpha_{\mathbf{n}}$, which generally depends on the interatomic spacing $n = |\mathbf{n}|$. We have

$$-\Phi^{0\mathbf{n}}_{\alpha\beta} u_{\mathbf{n}\beta} = \alpha_{\mathbf{n}} u_{\mathbf{n}\beta} \frac{n_\beta}{n} \frac{n_\alpha}{n}, \tag{7.4}$$

where $n_\alpha$ and $n_\beta$ are projections of the spacing $n$ on the $\alpha$ and $\beta$ axes. Hence,

$$\Phi^{0\mathbf{n}}_{\alpha\beta} = -\alpha_n \frac{n_\alpha n_\beta}{n^2}. \tag{7.5}$$

When calculating $D_{\alpha\beta}(\mathbf{q})$ via (5.10a), we must perform summation over all the $\mathbf{n}'$, including $\mathbf{n}' = 0$, at the same time (7.4) and (7.5) are applicable only if $\mathbf{n} \neq 0$ (we have substituted $\mathbf{n}$ for $\mathbf{n}'$).

To find $\Phi^{00}_{\alpha\beta}$, we make use of (5.3b), whence

$$\Phi^{00}_{\alpha\beta} = -\sum_{(\mathbf{n}\neq 0)} \Phi^{0\mathbf{n}}_{\alpha\beta}. \tag{7.6}$$

It follows from (5.10a), (7.5) and (7.6) that for the dynamical matrix

$$D_{\alpha\beta}(\mathbf{q}) = \frac{1}{m} \sum_{(\mathbf{n}\neq 0)} \alpha_n \frac{n_\alpha n_\beta}{n^2} (1 - e^{i\mathbf{q}\cdot\mathbf{n}}), \tag{7.7}$$

where $m$ is the mass of an atom.

---

[9] The interaction with the second coordination group is vital for the stability of the lattice.

Let us calculate $D_{xx}(\mathbf{q})$. It may easily be seen that the product $n_x n_x = n_x^2$ is nonzero only for two ●-atoms lying on the $x$ axis (Fig. 3.12); therefore, for the first coordination group

$$D_{xx}^{(1)}(\mathbf{q}) = \frac{1}{m} \sum_{(\mathbf{n}\neq 0)} \alpha_1 \frac{n_x^2}{n^2} (1 - e^{i\mathbf{q}\cdot\mathbf{n}})$$

$$= \frac{\alpha_1}{m} [1 - e^{iq_x a} + 1 - e^{-iq_x a}] = \frac{2\alpha_1}{m} (1 - \cos q_x a)$$

$$= \frac{4\alpha_1}{m} \sin^2 \frac{q_x a}{2}. \qquad (7.8)$$

The four atoms of the second coordination group (5, 6, 7, 8; Fig. 3.12) contribute nothing to $D_{xx}^{(1)}$, since for them $n_x = 0$. Taking into account the contribution of the atoms belonging to the second coordination group, we obtain

$$D_{xx}^{(2)}(\mathbf{q}) = \frac{1}{m} \sum_{(\mathbf{n}\neq 0)} \alpha_2 \frac{n_x^2}{n^2} (1 - e^{i\mathbf{q}\cdot\mathbf{n}})$$

$$= \frac{\alpha_2}{m} \left(\frac{1}{\sqrt{2}}\right)^2 [8 - 2\cos(q_x a + q_y a)$$

$$- 2\cos(q_x a - q_y a) - 2\cos(q_x a + q_z a) - 2\cos(q_x a - q_z a)]$$

$$= \frac{2\alpha_2}{m} [2 - \cos(q_x a)\cos(q_y a) - \cos(q_x a)\cos(q_z a)]. \qquad (7.9)$$

Calculating this sum, we obtain, for example, for atoms *1* and *3* (Fig. 3.12),

$$(1 - e^{i(q_x a + q_y a)}) + (1 - e^{-i(q_x a + q_y a)}) = 2 - 2\cos(q_x a + q_y a).$$

Similar results are obtained for other atomic pairs.

It follows from (7.8) and (7.9) that

$$D_{xx}(\mathbf{q}) = D_{xx}^{(1)}(\mathbf{q}) + D_{xx}^{(2)}(\mathbf{q})$$

$$= \frac{4\alpha_1}{m} \sin^2 \frac{q_x a}{2} + \frac{2\alpha_2}{m} [2 - \cos(q_x a)(\cos q_y a + \cos q_z a)]. \qquad (7.10)$$

The other diagonal members of the dynamical matrix $D_{yy}(\mathbf{q})$ and $D_{zz}(\mathbf{q})$ are obtained from (7.10) with the aid of a cyclic permutation of the labels $x$, $y$, $z$.

The nondiagonal members are, for instance,

$$D_{xy}(\mathbf{q}) = \frac{1}{m} \sum_{(\mathbf{n}\neq 0)} \alpha_n \frac{n_x n_y}{n^2} (1 - e^{i\mathbf{q}\cdot\mathbf{n}}). \qquad (7.11)$$

The atoms of the first coordination group do not contribute to $D_{xy}(\mathbf{q})$, because for them either $n_x$ or $n_y$ vanish. From the second coordination group only four atoms lying in the $xy$ plane contribute:

$$D_{xy}(\mathbf{q}) = \frac{\alpha_2}{m}\left(\frac{1}{\sqrt{2}}\right)^2 [(1-e^{i(q_xa+q_ya)}) - (1-e^{i(q_ya-q_xa)}) + (1-e^{-i(q_xa+q_ya)}) - (1-e^{-i(q_ya-q_xa)})]$$

$$= \frac{\alpha_2}{2m}[2\cos(q_ya-q_xa) - 2\cos(q_xa+q_ya)]$$

$$= \frac{2\alpha_2}{m}\sin(q_xa)\sin(q_ya), \qquad (7.12)$$

$$D_{xy}(\mathbf{q}) = \frac{2\alpha_2}{m}\sin(q_xa)\sin(q_ya). \qquad (7.13)$$

The other nondiagonal members, $D_{xz}$, $D_{yz}$, etc. can be obtained via a cyclic permutation.

We see from (7.10) and (7.13) that the elements of the dynamical matrix are real and symmetric, i.e., $D_{\alpha\beta} = D_{\beta\alpha}$, as it should be

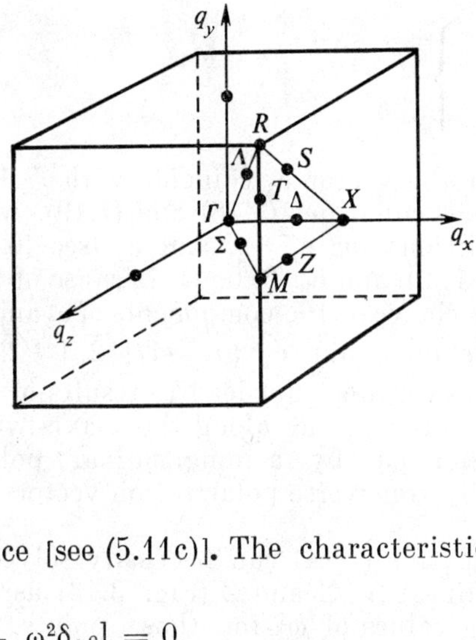


Fig. 3.13

for a simple lattice [see (5.11c)]. The characteristic equation (5.11) has the form

$$\det[D_{\alpha\beta}(\mathbf{q}) - \omega^2\delta_{\alpha\beta}] = 0, \qquad (7.14)$$

where $D_{\alpha\beta}(\mathbf{q})$ are given by (7.10) and (7.13).

The equation (7.14) is a degree three equation in $\omega^2$, and for this reason its solution is, in general, rather difficult. We shall solve it for selected symmetrical points (lines) of a Brillouin zone.

For a simple cubic lattice a Brillouin zone is in the shape of a cube, as depicted in Fig. 3.13.

Let us determine the form of the spectrum (dispersion) and the direction of polarization vectors $\mathbf{e}_j$ ($j = 1, 2, 3$) for points $\Delta$, i.e., for the axis of symmetry [100]. In this case $q_x = q$, $q_y = q_z = 0$, and from (7.10) and (7.13) it follows that

$$D_{xx} = \frac{4(\alpha_1 + 2\alpha_2)}{m} \sin^2 \frac{qa}{2}, \quad D_{yy} = D_{zz} = \frac{4\alpha_2}{m} \sin^2 \frac{qa}{2}. \tag{7.15}$$

All nondiagonal members $D_{xy}$, $D_{xz}$, etc. vanish; for this reason the characteristic equation is of the form

$$(D_{xx} - \omega^2)(D_{yy} - \omega^2)^2 = 0, \tag{7.16}$$

whence we obtain the three ($j = 1, 2, 3$) roots:

$$\omega_1^2 = D_{xx} = \frac{4(\alpha_1 + 2\alpha_2)}{m} \sin^2 \frac{qa}{2}, \quad \omega_2^2 = \omega_3^2 = \frac{4\alpha_2}{m} \sin^2 \frac{qa}{2}. \tag{7.17}$$

The equations for the polarization vectors are of the form

$$\left. \begin{aligned} D_{xx} e_{1x} &= \omega_1^2 e_{1x}, \\ D_{yy} e_{1y} &= \omega_1^2 e_{1y}, \\ D_{zz} e_{1z} &= \omega_1^2 e_{1z}. \end{aligned} \right\} \tag{7.18}$$

$$\left. \begin{aligned} D_{xx} e_{2x} &= \omega_2^2 e_{2x}, \\ D_{yy} e_{2y} &= \omega_2^2 e_{2y}, \\ D_{zz} e_{2z} &= \omega_2^2 e_{2z}. \end{aligned} \right\} \tag{7.19}$$

The equations for the vector $\mathbf{e}_3$ coincide with (7.19). The solutions of the homogeneous equations (7.18) and (7.19), which include the normalizing conditions for $\mathbf{e}_1$, $\mathbf{e}_2$, and $\mathbf{e}_3$ [see (6.2) and (6.3)] are $\mathbf{e}_1$ (1, 0, 0), $\mathbf{e}_2$ (0, 1, 0), and $\mathbf{e}_3$ (0, 0, 1). Because of frequency degeneracy $\omega_2 = \omega_3$, the choice of the components of $\mathbf{e}_2$ and $\mathbf{e}_3$ is not unique; one may, for instance, put $\mathbf{e}_2$ $(0, -1/\sqrt{2}, 1/\sqrt{2})$ and $\mathbf{e}_3$ $(0, 1/\sqrt{2}, 1/\sqrt{2})$, which, however, does not lead to results of physical interest.

Thus, the wave propagating along the $x$ axis with a wave vector $q = q_x$ is characterized by a longitudinal polarization vector $\mathbf{e}_L$ (1, 0, 0) and by transverse polarization vectors $\mathbf{e}_T^{(1)}$ (0, 1, 0) and $\mathbf{e}_T^{(2)}$ (0, 0, 1).

The secular equation (7.14) can be easily solved for other symmetrical points (lines) $\Lambda$, $\Sigma$ and $Z$ (Fig. 3.13) as well. If we find the corresponding values of $\omega_j^2$ for those points, we can find the polarization vectors from equations (6.1). Since this procedure is similar to the one we performed for the point $\Delta$, we do not give the calculations for other points; instead, we confine ourselves to the presentation of results in Table 3.1.

Looking at Table 3.1, we see that the polarization vectors are normalized to unity in cases (a) and (d). In the two other cases we deemed it more instructive to make all the nonzero components of

Table 3.1

(a) Line Δ: $q_x = q, \quad q_y = q_z = 0,$

$$[D_{\alpha\beta}] = \begin{pmatrix} A & 0 & 0 \\ 0 & B & 0 \\ 0 & 0 & B \end{pmatrix}, \qquad A = \frac{4\alpha_1 + 8\alpha_2}{m} \sin^2 \frac{qa}{2},$$

$$B = \frac{4\alpha_2}{m} \sin^2 \frac{qa}{2},$$

$\omega_1^2 = A, \qquad \mathbf{e}_L\ (1,\ 0,\ 0),$

$\omega_2^2 = \omega_3^2 = B, \qquad \mathbf{e}_T^{(1)}\ (0,\ 1,\ 0), \quad \mathbf{e}_T^{(2)}\ (0,\ 0,\ 1);$

(b) line Λ: $q_x = q_y = q_z = q' = \dfrac{q}{\sqrt{3}},$

$$[D_{\alpha\beta}] = \begin{pmatrix} A & B & B \\ B & A & B \\ B & B & A \end{pmatrix}, \qquad A = \frac{4\alpha_1}{m} \sin^2 \frac{q'a}{2} + \frac{4\alpha_2}{m} \sin^2 q'a,$$

$$B = \frac{2\alpha_2}{m} \sin^2 q'a,$$

$\omega_1^2 = A + 2B, \qquad \mathbf{e}_L\ (1,\ 1,\ 1),$

$\omega_2^2 = \omega_3^2 = A - B, \qquad \mathbf{e}_T^{(1)}\ (1,\ -1,\ 0), \quad \mathbf{e}_T^{(2)}\ (1,\ 0,\ -1);$

(c) line Σ: $q_x = q', \quad q_y = 0, \quad q_z = q' = \dfrac{q}{\sqrt{2}},$

$$[D_{\alpha\beta}] = \begin{pmatrix} A & 0 & C \\ 0 & B & 0 \\ C & 0 & A \end{pmatrix}, \qquad A = \frac{4(\alpha_1 + \alpha_2)}{m} \sin^2 \frac{q'a}{2} + \frac{2\alpha_2}{m} \sin^2 q'a,$$

$$B = \frac{8\alpha_2}{m} \sin^2 \frac{q'a}{2}, \qquad C = \frac{2\alpha_2}{m} \sin^2 q'a,$$

$\omega_1^2 = A + C, \qquad \mathbf{e}_L\ (1,\ 0,\ 1),$

$\omega_2^2 = B, \qquad \mathbf{e}_T^{(1)}\ (0,\ 1,\ 0),$

$\omega_3^2 = A - C, \qquad \mathbf{e}_T^{(2)}\ (1,\ 0,\ -1);$

(d) line Z: $q_x = \dfrac{\pi}{a}, \quad q_y = 0, \quad q_z = q' = \sqrt{q^2 - \left(\dfrac{\pi}{a}\right)^2},$

$$[D_{\alpha\beta}] = \begin{pmatrix} A & 0 & 0 \\ 0 & B & 0 \\ 0 & 0 & C \end{pmatrix}, \qquad A = \frac{4\alpha_1}{m} + \frac{2\alpha_2}{m} (3 + \cos q'a)$$

$$B = \frac{2\alpha_2}{m} (3 - \cos q'a)$$

$$C = \frac{4\alpha_1}{m} \sin^2 \frac{q'a}{2} + \frac{4\alpha_2}{m}$$

$\omega_1^2 = A, \qquad \mathbf{e}_1\ (1,\ 0,\ 0),$

$\omega_2^2 = B, \qquad \mathbf{e}_2\ (0,\ 1,\ 0) \equiv \mathbf{e}_T,$

$\omega_3^2 = C, \qquad \mathbf{e}_3\ (0,\ 0,\ 1).$

$\mathbf{e}_j$ equal to $\pm 1$. In case (b) the vectors $\mathbf{e}_T^{(1)}$ and $\mathbf{e}_T^{(2)}$ are not mutually orthogonal (they make an angle of 80°, a fact that the reader can establish by computing their scalar product). This does not contradict the general theorem proved in Appendix 3, Sec. 5, because the eigenvectors $\mathbf{e}_T^{(1)}$ and $\mathbf{e}_T^{(2)}$ belong to the same eigenvalue $\omega_2^2 = \omega_3^2$. Of course, in this case, too, with the aid of a linear transformation, mutually orthogonal vectors can be introduced. In case (*d*) the vibration branches are not separated into the transverse and the longitudinal (there is only $\mathbf{e}_2$ (0, 1, 0), a transverse vibration).

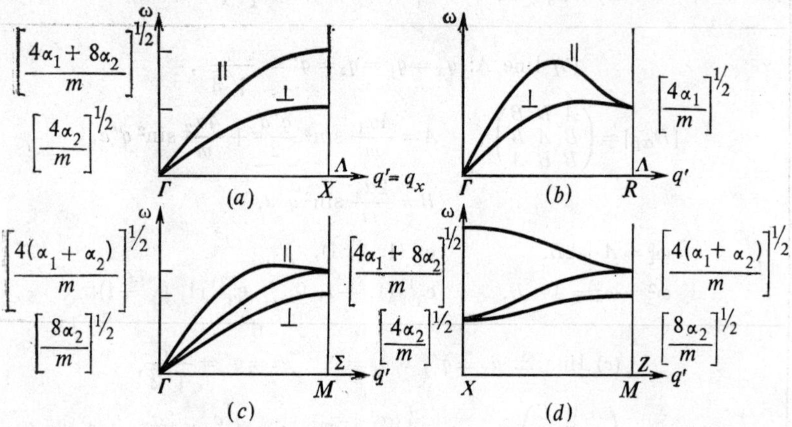


**Fig. 3.14**

Figure 3.14 is a schematic representation of the dependence of the frequencies $\omega_j$ on $q'$ for cases (a)-(d). In cases (*a*), (*b*), (*c*) the symmetry lines originate from point Γ, the centre of a Brillouin zone for which $\mathbf{q} = 0$; because of that the acoustic branches, as it should be, start from $\omega_j = 0$. As can be seen from Table 3.1, in those cases, for small values of $\mathbf{q}$, $\omega = v_0 q$, where $v_0$ is the sound velocity (the propagation velocity of long waves: $qa = 2\pi a/\lambda \ll 1$, i.e., the wave length $\lambda \gg a$ = lattice const.).

The velocities of the longitudinal and transverse sound waves for all three cases (*a*, *b*, *c*) can be easily determined from Table 3.1. For example, in case (*a*)

$$v_{0L} = a\sqrt{\frac{\alpha_1+2\alpha_2}{m}}, \quad v_{0T} = a\sqrt{\frac{\alpha_2}{m}},$$

whence it follows that $v_{0L} > v_{0T}$; this is true for all three cases [in cases (*b*) and (*c*) we should make $q' = q/\sqrt{3}$ and $q' = q/\sqrt{2}$].

The symmetry line *Z* does not pass through the centre of the Brillouin zone (Fig. 3.13); therefore, the dispersion curves $\omega_j = \omega_j(q)$

do not start from zero. Note that the same point $X$ inside the Brillouin zone corresponds to $q_x = q = \pi/a$ in case ($a$) and to $q_z = q' = 0$ in case ($d$); therefore the extreme values of the frequencies in both cases coincide.

## 3.8 Application of Group Theory to Normal Vibrations in Crystal Lattices

**3.8.1** In Section 2.7 we studied the general aspects of the application of group theory to a quantum-mechanical system whose Hamiltonian displays certain symmetry. We shall now deal with a similar problem as applied to the normal vibrations of a crystal lattice. Then the results obtained will be applied to the simple cubic lattice studied in the preceding section.

We shall base our discussion on the concept of the wave vector group $G_{\mathbf{q}}$ introduced in Section 2.8.3. The wave vector group $G_{\mathbf{q}}$ is a subgroup of the crystal's space group $G$ with the elements $g = \{R \mid \boldsymbol{\alpha}(R) + \mathbf{a}_{\mathbf{n}}\}$ such that $R\mathbf{q} = \mathbf{q}$ or $R\mathbf{q} = \mathbf{q} + \mathbf{b}_i$, where $\mathbf{b}_i$ is a vector of the reciprocal lattice.

Operate with the operator $\hat{P}_g$ (2.6.1), where $g \in G_{\mathbf{q}}$, on both sides of (6.1). The operation with $\hat{P}_g$ on the dynamical matrix $D_{\alpha\beta}^{kk'}(\mathbf{q})$ (5.10a) leaves it invariant. Indeed, since the elements $g$ are the crystal's symmetry elements ($g \in G_{\mathbf{q}} \in G$) and the force constants $\Phi_{\alpha\beta}\binom{kk'}{\mathbf{nn}'}$ are those for equilibrium positions of the atoms in the lattice, $\hat{P}_g$ can at most change the numeration of the lattice sites $\mathbf{n}$, $\mathbf{n}'$ over which the summation to determine $D_{\alpha\beta}^{kk'}(\mathbf{q})$ is performed ($\mathbf{n}$ can always be equated to zero). As to the wave vector $\mathbf{q}$, it experiences no change in compliance with the definition of the vector group $G_{\mathbf{q}}$. We obtain accordingly from (6.1)

$$\sum_{k'\beta} D_{\alpha\beta}^{kk'}(\mathbf{q})\,\hat{P}_g e_{jk'\beta}(\mathbf{q}) = \omega_j^2(\mathbf{q})\,\hat{P}_g e_{jk\alpha}(\mathbf{q}). \tag{8.1}$$

This equation is, in a sense, similar to the Schrödinger equation (2.7.13). The part of the Hamiltonian $\mathscr{H}(\mathbf{x})$ is played by the dynamical matrix $D_{\alpha\beta}^{kk'}$, the part of the eigenfunctions $\psi(\mathbf{x})$ by the eigenvectors $e_{jk\alpha}(\mathbf{q})$, and the part of the energy eigenvalues by the normal frequencies $\omega_j^2(\mathbf{q})$.

We see that the eigenvectors $e_{jk\alpha}(\mathbf{q})$ play the part of the basis functions of an irreducible representation of $G_{\mathbf{q}}$. If the eigenvalue $\omega_j^2(\mathbf{q})$ of (6.1) is degenerate, i.e., if there are several linearly independent eigenvectors $e_{jk\alpha}(\mathbf{q})$ corresponding to the same natural frequency $\omega_j(\mathbf{q})$, then the eigenvector $\hat{P}_g e_{jk\alpha}(\mathbf{q})$ can be represented as a linear combination of the eigenvectors corresponding to the frequency $\omega_j(\mathbf{q})$. The matrices of this linear representation, by analogy with (2.7.15), realize an irreducible representation of the wave vector group $G_{\mathbf{q}}$. The dimension of the irreducible representation is equal to the degree of degeneracy of the frequency $\omega_j(\mathbf{q})$.

To apply group theory to the classification of the vibrations of a crystal at different points of a Brillouin zone, we must find the full (reducible) representation corresponding to the vibrational degrees of freedom for a definite $\mathbf{q}$, just as it was done for molecules (Section 3.8.3). For a complex crystal the total number of vibrational degrees of freedom $u^k_{\mathbf{n}\alpha}$ is equal to $3Ns - 6 \approx 3Ns$ ($N$ is the number of crystal cells in the principal region, and $s$ is the number of atoms in a cell), i.e., it is very great. We are, however, interested in the classification of vibrations at a specified point of a Brillouin zone, i.e., for a definite value of the wave vector $\mathbf{q}$, which (for a vibration branch) itself assumes $N$ quasi-discrete values (5.24). Therefore the number of vibrational degrees of freedom for a definite $\mathbf{q}$ is $3s$, i.e., is equal to the number of degrees of freedom of the atoms in the unit cell.

We must effect the transition from $3sN$ atomic displacements $u^k_{\mathbf{n}\alpha}$ to their Fourier transforms $u^k_{\mathbf{n}\alpha}(\mathbf{q})$ (as it is done below) and find what representations of the wave vector space group $G_{\mathbf{q}}$ transform them.

We shall start with expression (6.4), which we write in the form

$$u^k_{\mathbf{n}\alpha} = \sum_{\mathbf{q}} u_{k\alpha}(\mathbf{q})\, e^{i\mathbf{q}\cdot\mathbf{a}_{\mathbf{n}}}, \tag{8.2}$$

where

$$u_{k\alpha}(\mathbf{q}) = \frac{1}{\sqrt{Nm_k}} \sum_j e_{jk\alpha}(\mathbf{q})\, a_j(\mathbf{q}, t). \tag{8.2a}$$

Here $\mathbf{e}_{jk}$ are polarization vectors and $a_j$ are complex normal coordinates.

Multiplying both sides of (8.2) by $\exp(-i\mathbf{q}'\cdot\mathbf{a}_{\mathbf{n}})$, summing over $\mathbf{n}$, and making use of (A.6.4), we obtain

$$u_{k\alpha}(\mathbf{q}) = \frac{1}{N} \sum_{\mathbf{n}} e^{-i\mathbf{q}\cdot\mathbf{a}_{\mathbf{n}}} u^k_{\mathbf{n}\alpha} \tag{8.3}$$

(in the last expression we denoted $\mathbf{q}'$ by $\mathbf{q}$).

Operate with the operator $\hat{P}_{g^{-1}}$ (2.6.1) on $3s$ quantities $u_{k\alpha}(\mathbf{q})$ (for a specified value of $\mathbf{q}$), where

$$g = \{R \mid \mathbf{a}_{\mathbf{m}} + \boldsymbol{\alpha}\} \tag{8.4}$$

is an element of the wave vector group $G_{\mathbf{q}}$ ($g^{-1}$ is obviously also an element of the wave vector group $G_{\mathbf{q}}$). It follows from (8.3) that

$$\hat{P}_{g^{-1}} u_{k\alpha}(\mathbf{q}) = \frac{1}{N} \sum_{\mathbf{n}} e^{-i\mathbf{q}\cdot\mathbf{a}_{\mathbf{n}}} \hat{P}_{g^{-1}} u^k_{\mathbf{n}\alpha}. \tag{8.5}$$

The operator $\hat{P}_{g^{-1}}$ on the right-hand side of (8.3) affects only the displacements $u_{\mathbf{n}\alpha}^{k}$, since the factors $\exp(-i\mathbf{q}\cdot\mathbf{a_n})$ are the expansion coefficients of the quantity $u_{k\alpha}(\mathbf{q})$ in the displacements $u_{\mathbf{n}\alpha}^{k}$ (the same as $\hat{P}_R$ in Section 2.8.3 transforms the displacements $u_{i\alpha}$ of the atoms in a molecule, not affecting their numeration).

Since $u_{\mathbf{n}\alpha}^{k}$ is a component of the displacement vector, by analogy with (2.8.13) we have

$$\hat{P}_{g^{-1}}u_{\mathbf{n}\alpha}^{k} = \sum_{\mathbf{n}'k'\alpha'} D^{u}(g^{-1})_{\mathbf{n}'k'\alpha',\,\mathbf{n}k\alpha}u_{\mathbf{n}'\alpha'}^{k'}. \tag{8.6}$$

The matrix $D^u$ of dimension $3sN$ represents the displacements of all the crystal's atoms. From (2.8.13) it follows that

$$D^{u}(g^{-1})_{\mathbf{n}'k'\alpha',\,\mathbf{n}k\alpha} = A(g^{-1})_{\alpha'\alpha}\,\delta_{\mathbf{n}',\,g(\mathbf{n})}\delta_{k',\,g(k)}. \tag{8.7}$$

Here $A(g^{-1})$ is the transformation matrix of the polar vector $u_{\mathbf{n}\alpha}^{k}$ which operates in the $g^{-1}$ transformation of the coordinate frame; the Kronecker deltas $\delta_{\mathbf{n}',\,g(\mathbf{n})}$ and $\delta_{k',\,g(k)}$ account for the fact that, as a result of the operation $g$, the atom $(\mathbf{n}, k)$ occupies the position of the atom $(\mathbf{n}', k')$ of the same type.

Substituting (8.7) and (8.6) into (8.5) and performing an identical transformation of the exponential, we obtain

$$\hat{P}_{g^{-1}}u_{k\alpha}(\mathbf{q}) = \frac{1}{N}\sum_{\mathbf{n},\,\mathbf{n}',\,k',\,\alpha'} e^{-i\mathbf{q}\cdot\mathbf{a}_{\mathbf{n}}'}u_{\mathbf{n}'\alpha'}^{k'}A(g^{-1})_{\alpha'\alpha} \times e^{i\mathbf{q}\cdot(\mathbf{a}_{\mathbf{n}}'-\mathbf{a_n})}\delta_{\mathbf{n}',\,g(\mathbf{n})}\delta_{k',\,g(k)}. \tag{8.8}$$

Since the operation $g$ transforms $\mathbf{a_n}$ into $\mathbf{a}_{\mathbf{n}}'$, it follows that

$$\begin{aligned} i\mathbf{q}\cdot(\mathbf{a}_{\mathbf{n}}' - \mathbf{a_n}) &= i\mathbf{q}\cdot(g\mathbf{a_n} - \mathbf{a_n}) = i\mathbf{q}\cdot(R\mathbf{a_n} + \mathbf{a_m} + \boldsymbol{\alpha} - \mathbf{a_n}) \\ &= i\,[(R^{-1}\mathbf{q} - \mathbf{q})\cdot\mathbf{a_n} + \mathbf{q}\cdot(\mathbf{a_m} + \boldsymbol{\alpha})]. \end{aligned} \tag{8.9}$$

If $g^{-1}$ is an element of the wave vector group $G_{\mathbf{q}}$, then

$$R^{-1}\mathbf{q} = \mathbf{q} \text{ or } R^{-1}\mathbf{q} = \mathbf{q} \pm \mathbf{b}_i \tag{8.10}$$

($\mathbf{b}_i$ is a vector of the reciprocal lattice). In both cases $\exp[i(R^{-1}\mathbf{q} - \mathbf{q})\cdot\mathbf{a_n}] = 1$; therefore

$$\hat{P}_{g^{-1}}u_{k\alpha}(\mathbf{q}) = \frac{1}{N}\sum_{\mathbf{n},\,\mathbf{n}',\,k',\,\alpha'} e^{-i\mathbf{q}\cdot\mathbf{a}_{\mathbf{n}}'}u_{\mathbf{n}'\alpha'}^{k'}A(g^{-1})_{\alpha'\alpha} \times e^{i\mathbf{q}\cdot(\mathbf{a_m}+\boldsymbol{\alpha})}\delta_{\mathbf{n}',\,g(\mathbf{n})}\delta_{k',\,g(k)}. \tag{8.11}$$

If we add up the terms on the right-hand side over $\mathbf{n}'$ and make use of (8.3), we obtain

$$\hat{P}_{g^{-1}}u_{k\alpha}(\mathbf{q}) = \sum_{k'\alpha'} A(g^{-1})_{\alpha'\alpha}\,e^{i\mathbf{q}\cdot(\mathbf{a_m}+\boldsymbol{\alpha})}\delta_{k',\,g(k)}u_{k'\alpha'}(\mathbf{q}) \tag{8.12}$$

(the summation over **n** is performed automatically, thanks to the presence of $\delta_{\mathbf{n}', g(\mathbf{n})}$).

Let us demonstrate that for symmorphic groups of the wave vector $G_{\mathbf{q}}$, as well as for internal points of a Brillouin zone of any group $G_{\mathbf{q}}$, the matrices $A(g^{-1}) \exp[i\mathbf{q}\cdot(\boldsymbol{\alpha} + \mathbf{a_m})]\,\delta_{k', g(k)}$ serve as $3s$-dimensional representations of the wave vector groups $G_{\mathbf{q}}$ of the basis functions $u_{k\alpha}(\mathbf{q})$. Indeed, since these matrices include the factor $\exp[i\mathbf{q}(\boldsymbol{\alpha} + \mathbf{a_m})]$, the product of the matrices $g_1 = \{R_1 \mid \boldsymbol{\alpha}_1 + \mathbf{a}_1\}$ and $g_2 = \{R_2 \mid \boldsymbol{\alpha}_2 + \mathbf{a}_2\}$ for the elements of the group $G_{\mathbf{q}}$ will include the factor $\exp[i\mathbf{q}\cdot(\boldsymbol{\alpha}_1 + \mathbf{a}_1 + \boldsymbol{\alpha}_2 + \mathbf{a}_2)]$.

In Section 2.9.4 we demonstrated that the irreducible representation $\Gamma(g_2g_1)$ in such cases contains $\exp[i\mathbf{q}\cdot(R_2\boldsymbol{\alpha}_1 + R_2\mathbf{a}_1 + \boldsymbol{\alpha}_2 + \mathbf{a}_2)]$ (2.9.31); both exponentials coincide if $\exp[i\mathbf{q}\cdot(\boldsymbol{\alpha}_1 + \mathbf{a}_1)] = \exp[i\mathbf{q}\cdot(R_2\alpha_1 + R_2\mathbf{a}_1)]$, i.e., if $\exp[i(R^{-1}\mathbf{q} - \mathbf{q})\cdot(\boldsymbol{\alpha}_1 + \mathbf{a}_1)] = 1$. Because of (8.10) this condition is fulfilled either for symmorphic groups ($\alpha_1 = 0$) or for internal points of a Brillouin zone ($R_2^{-1}\mathbf{q} = \mathbf{q}$).

Decomposing this representation of dimension $3s$ in irreducible representations of the $G_{\mathbf{q}}$ group, we are able to classify the crystal's vibrations at point **q** of its Brillouin zone. Note the close analogy between (8.12) and (2.8.13) obtained in the course of the analysis of vibrations of polyatomic molecules.

Since only characters of an irreducible representation need be known to perform the decomposition of the representation (8.7) in the representations of the wave vector group $G_{\mathbf{q}}$, we find the characters of (8.7). Making use of (2.8.15), we see that the character of the representation $D^u$ of (8.7) for a proper rotation through the angle $\varphi$ is

$$\chi^u(C(\varphi) \mid \boldsymbol{\alpha}) = (1 + 2\cos\varphi)\exp(i\mathbf{q}\cdot\boldsymbol{\alpha})\, n_C, \tag{8.13}$$

where $n_C = \sum_k \delta_{k,\,C(k)}$ is the number of atoms in the cell remaining in their places in the course of the transformation $C(\varphi) \equiv C$.

In the same way we obtain for a rotary-reflection transformation $S(\varphi) \equiv S$ (2.8.18):

$$\chi^u(S(\varphi) \mid \alpha) = (-1 + 2\cos\varphi)\exp(i\mathbf{q}\cdot\boldsymbol{\alpha})\, n_S, \tag{8.14}$$

where $n_S$ is the number of atoms in a cell unaffected by the transformation $S(\varphi)$.

**3.8.2** Apply now the results obtained to the study of vibrations in a simple cubic crystal discussed in the preceding section.

Determine the irreducible representations of the wave vector group $G_{\mathbf{q}}$ in the centre of a Brillouin zone ($\mathbf{q} = 0$), i.e., at point $\Gamma$ (Fig. 3.13). The point $\Gamma$ has the symmetry of a cube $\boldsymbol{O}_h = \boldsymbol{O} \times \boldsymbol{C}_i$; here $\boldsymbol{O}$ is the symmetry group of the cube's axes, and $\boldsymbol{C}_i = \{E, J\}$, where $J$ is inversion (Section 2.3). Table 3.2 of the characters of the

$\boldsymbol{O}_h$ group can be obtained in accordance with the procedure in Table 2.6 with reference to Table 2.7.[10]

To find the characters of the representation $D^u_\Gamma$ (8.7) at point $\Gamma$ we make use of (8.13) in which we put $\boldsymbol{\alpha} = 0$ and $n_C = 1$. In addition, we must bear in mind that for the elements containing inversion $J$,

$$\chi^u [JC(\varphi) \mid \boldsymbol{\alpha}] = -(1 + 2\cos\varphi)\exp(i\mathbf{q}\cdot\boldsymbol{\alpha})\, n_{CJ}. \qquad (8.15)$$

This follows from the preceding result and from (2.8.14) if all the $-u\alpha$'s are substituted for $u_\alpha$.

We finally obtain

$$\chi^u(E) = 3, \quad \chi^u(C_3) = 0, \quad \chi^u(C_4^2) = -1, \quad \chi^u(C_2) = -1,$$
$$\chi^u C_4) = 1, \quad \chi^u(J) = -3, \quad \chi^u(JC_3) = 0, \quad \chi^u(JC_4^2) = 1,$$
$$\chi^u(JC_2) = 1, \quad \chi^u(JC_4) = -1. \qquad (8.16$$

We see that the characters of the representation $D^u_\Gamma$ coincide with the characters of the irreducible representation $\Gamma_{15}$ (Table 3.2), but in this case

$$D^u_\Gamma = \Gamma_{15}, \qquad (8.17)$$

i.e., the representation corresponding to the wave vector group in the centre of a Brillouin zone is the three-fold degenerate irreducible representation $\Gamma_{15}$.

To determine the wave vector group at point $\Delta$ (see Fig. 3.13) consider the elements of the group that leave the vector $\mathbf{q}_\Delta$ invariant. It may easily be seen that such elements are the elements of the group $\boldsymbol{C}_{4v}$: $E$, $C_4^2$ (the $q_x$ axis), $2C_4$ ($q_x$), $2JC_4^2$ (the $q_y$ and $q_z$ axes), and $2JC_2$. Note that $JC_4^2$ and $JC_2$ are reflections in planes passing through the axis $q_x$ and perpendicular to the axes $C_4$ and $C_2$. The point $T$ of a Brillouin zone satisfies the same point symmetry group $\boldsymbol{C}_{4v}$ (note that there are four equivalent points $T$ on the four vertical edges of the cube). Now employ the general procedure to build a table of characters of groups $\boldsymbol{G}_{\mathbf{q}_\Delta}$ and $\boldsymbol{G}_{\mathbf{q}_T}$ (see Table 3.3). The tables of characters of groups of the wave vectors pointing in the directions of points $\Lambda$ and $\Sigma$ (the symmetry of point $S$ is the same as that of point $\Sigma$) can be built in a similar way (Table 3.4).

The irreducible representations of the wave vector $\mathbf{q}_\Delta$ group can be found with the aid of two methods. First, we can find the characters of the representation $D^u_\Delta$ (8.7) for the $\boldsymbol{C}_{4v}$ group:

$$\chi^u(E) = 3, \quad \chi^u(C_4^2) = -1, \quad \chi^u(C_4) = 1,$$
$$\chi^u(JC_4^2) = 1, \; \chi^u(JC_2) = 1. \qquad (8.18)$$

[10] The notations for irreducible representations in Table 3.2 are those used in the paper by L. Bouckaert, R. Smoluchowski and E. Wigner [3.13]. In Jones' opinion, such notation is justified "because it has established itself in literature so that its use appears inevitable" [3.

Table 3.2

| $O_h$ | $E$ | $8C_3$ | $3C_4^2$ | $6C_2$ | $6C_4$ | $J$ | $8JC_3$ | $3JC_4^2$ | $6JC_2$ | $6JC_4$ |
|---|---|---|---|---|---|---|---|---|---|---|
| $\Gamma_1$ | 1 | 1 | 1 | 1 | 1 | 1 | 1 | 1 | 1 | 1 |
| $\Gamma_2$ | 1 | 1 | 1 | −1 | −1 | 1 | 1 | 1 | −1 | −1 |
| $\Gamma_{12}$ | 2 | −1 | 2 | 0 | 0 | 2 | −1 | 2 | 0 | 0 |
| $\Gamma'_{15}$ | 3 | 0 | −1 | −1 | 1 | 3 | 0 | −1 | −1 | 1 |
| $\Gamma'_{25}$ | 3 | 0 | −1 | 1 | −1 | 3 | 0 | −1 | 1 | −1 |
| $\Gamma'_1$ | 1 | 1 | 1 | 1 | 1 | −1 | −1 | −1 | −1 | −1 |
| $\Gamma'_2$ | 1 | 1 | 1 | −1 | −1 | −1 | −1 | −1 | 1 | 1 |
| $\Gamma'_{12}$ | 2 | −1 | 2 | 0 | 0 | −2 | 1 | −2 | 0 | 0 |
| $\Gamma_{15}$ | 3 | 0 | −1 | −1 | 1 | −3 | 0 | 1 | 1 | −1 |
| $\Gamma_{25}$ | 3 | 0 | −1 | 1 | −1 | −3 | 0 | 1 | −1 | 1 |

These characters are written out in the last row of Table 3.3. Decomposing the representation $D^u_\Delta$ with the characters (8.18) in irreducible representations of the $C_{4v}$ group, we obtain

$$D^u_\Delta = \Delta_1 + \Delta_5. \tag{8.19}$$

The same result can be obtained if we bear in mind that the transition from the centre of a Brillouin zone $\Gamma$ to the line $\Delta$ involves a reduction in symmetry from $O_h$ to $C_{4v}$. Such a reduction in symmetry, as was established in Section 2.7.2, causes the splitting of the level $\Gamma_{15}$. Writing out the characters of $\Gamma_{15}$ corresponding to the classes of the group $C_{4v}$, we find them to coincide with the characters $\chi^u$ (8.18). Therefore, proceeding in compliance with the general rule, we obtain

$$\Gamma_{15} = \Delta_1 + \Delta_5, \tag{8.19a}$$

Table 3.3

| | $\Delta, T$ | $E$ | $C_4^2$ | $2JC_2$ | $2JC_4^2$ | $2C_4$ |
|---|---|---|---|---|---|---|
| | $\Delta_1$ | 1 | 1 | 1 | 1 | 1 |
| | $\Delta_1'$ | 1 | 1 | −1 | −1 | 1 |
| | $\Delta_2$ | 1 | 1 | −1 | 1 | −1 |
| | $\Delta_2'$ | 1 | 1 | 1 | −1 | −1 |
| | $\Delta_5$ | 2 | −2 | 0 | 0 | 0 |
| $\chi^u$ | $\Gamma_{15}$ | 3 | −1 | 1 | 1 | 1 |

Table 3.4

| $\Lambda$ | $E$ | $2C_3$ | $3JC_2$ |
|---|---|---|---|
| $\Lambda_1$ | 1 | 1 | 1 |
| $\Lambda_2$ | 1 | 1 | −1 |
| $\Lambda_3$ | 2 | −1 | 0 |
| $\Gamma_{15}$ | 3 | 0 | 1 |

(*a*)

| $\Sigma, S$ | $E$ | $C_2$ | $JC_4^2$ | $JC_2$ |
|---|---|---|---|---|
| $\Sigma_1$ | 1 | 1 | 1 | 1 |
| $\Sigma_2$ | 1 | 1 | −1 | −1 |
| $\Sigma_3$ | 1 | −1 | −1 | 1 |
| $\Sigma_4$ | 1 | −1 | 1 | −1 |
| $\Gamma_{15}$ | 3 | −1 | 1 | 1 |

(*b*)

i.e., the state $\Gamma_{15}$, three-fold degenerate in the centre, splits along the line $\Delta$ into a nondegenerate state $\Delta_1$ and a two-fold degenerate state $\Delta_5$.[11]

However, the advantage of the first method employing (8.18) is that it is applicable to the point $T$ (see Fig. 3.13), which is not in contact with the centre $\Gamma$, but has the same symmetry $C_{4v}$ as point $\Delta$.

---

[11] This is the reason for attributing the index "15" for the irreducible representation $\Gamma_{15}$.

Comparing (8.19a) with the analytical solution of mechanical equations of the preceding section (Table 3.1 and Fig. 3.14*a*), we see that group theory solely on the basis of considerations involving the symmetry of the cubic lattice predicts the existence of two vibration branches along the $\Delta$-line of a Brillouin zone: of a nondegenerate longitudinal branch ($\Delta_1$) and of a two-fold degenerate transverse branch ($\Delta_5$). By analogy, we obtain from Table 3.4 for the $\Lambda$-line

$$\Gamma_{15} = \Lambda_1 + \Lambda_3, \tag{8.20}$$

i.e., the three-fold degenerate state $\Gamma_{15}$ splits along the line $\Lambda$ (see Fig. 3.13) into a unit representation $\Lambda_1$ and a two-fold degenerate representation $\Lambda_3$.

We have already noted that the symmetry of point $R$ coincides with that of point $\Gamma$, and for this reason it is obvious that at point $R$ the states $\Lambda_1$ and $\Lambda_3$ merge into a three-fold degenerate state. This is a qualitative description of the dispersion pattern shown in Fig. 3.14*b*. Finally, from Table 3.4 we obtain on the line $\Sigma$

$$\Gamma_{15} = \Sigma_1 + \Sigma_3 + \Sigma_4. \tag{8.21}$$

Hence, along the $\Sigma$-line a complete splitting of the state $\Gamma_{15}$ into three nondegenerate states takes place, in agreement with the analytical solution (Table 3.1*c* and Fig. 3.14*c*).

We see that the state $\Gamma_{15}$ in the centre of a Brillouin zone, when displaced along the lines $\Delta$, $\Lambda$, and $\Sigma$ is uniquely decomposed in states (8.19a), (8.20) and (8.21). In general, if the symmetry elements are in contact, the irreducible representations of a group of higher symmetry can be uniquely decomposed in irreducible representations of groups of lower symmetry. These conditions are known as *compatibility relations.* Table 3.5 presents the compatibility relations be-

Table 3.5

| $\Gamma_1$ | $\Gamma_2$ | $\Gamma_{12}$ | $\Gamma'_{15}$ | $\Gamma'_{25}$ |
|---|---|---|---|---|
| $\Delta_1$ | $\Delta_2$ | $\Delta_1\Delta_2$ | $\Delta'_1\Delta_5$ | $\Delta'_2\Delta_5$ |
| $\Lambda_1$ | $\Lambda_2$ | $\Lambda_3$ | $\Lambda_2\Lambda_3$ | $\Lambda_1\Lambda_3$ |
| $\Sigma_1$ | $\Sigma_4$ | $\Sigma_1\Sigma_4$ | $\Sigma_2\Sigma_3\Sigma_4$ | $\Sigma_1\Sigma_2\Sigma_3$ |
| $\Gamma'_1$ | $\Gamma'_2$ | $\Gamma'_{12}$ | $\Gamma_{15}$ | $\Gamma_{25}$ |
| $\Delta'_1$ | $\Delta'_2$ | $\Delta'_1\Delta'_2$ | $\Delta_1\Delta_5$ | $\Delta_2\Delta_5$ |
| $\Lambda_2$ | $\Lambda_1$ | $\Lambda_3$ | $\Lambda_1\Lambda_3$ | $\Lambda_2\Lambda_3$ |
| $\Sigma_2$ | $\Sigma_3$ | $\Sigma_2\Sigma_3$ | $\Sigma_1\Sigma_3\Sigma_4$ | $\Sigma_1\Sigma_2\Sigma_4$ |

tween the irreducible representations at point $\Gamma$ and at points $\Delta$, $\Lambda$, and $\Sigma$.

To complete the group analysis of the vibration branches of a simple cubic crystal, we still have to compile a table of characters of wave vector groups for points $X$ and $M$ (with the same symmetry as point $X$), and for $Z$.

When determining the wave vector group for point $X$ (and $M$), we must bear in mind that the point $X$ gains additional symmetry elements connected with the inversion $J$ of point $X$, as compared with point $\Delta$; this gives us an equivalent vector $\mathbf{q}$ differing from the original one by the reciprocal lattice vector $2\pi/a$. Hence, the wave vector group $G_{\mathbf{q}_X} = G_{\mathbf{q}_\Delta} \times C_i$, and its characters may be obtained from Table 3.3 in compliance with the plan set out in Table 2.6. If we subsequently build the tables of compatibility relations for $M$ and $\Sigma$, $Z$, $T$ and for $X$ and $\Delta$, $Z$, $S$, we shall be able to predict the nature of the merging of branches at point $M$ in case (*c*) and the nature of the splitting branches in case (*d*).

Finally, let us consider the irreducible representations of the group of the wave vector directed toward point $Z$ (see Fig. 3.13). Since the line $Z$ does not originate from the centre $\Gamma$, we cannot make use of the compatibility relations but must start from the representation $D_Z^u$ (8.7). The following operations leave the wave vector directed toward point $Z$ invariant (or transform it into an equivalent one): $E$, $C_4^2\ (q_z)$, $JC_4^2\ (q_x)$, $JC_4^2\ (q_y)$ (the corresponding $C_4$ axes are shown in parentheses). The table of characters of this fourth order group concides with the table of characters of the group for points $\Sigma$ and $S$ (see Table 3.4).[12] It can easily be demonstrated that the characters of the representation $D_Z^u$ coincide with the values in the last row of $\Gamma_{15}$. Accordingly, the decomposition of $D_Z^u$ in $Z_1$, $Z_2$, ... is of a form similar to (8.21):

$$D_Z^u = Z_1 + Z_3 + Z_4. \qquad (8.22)$$

We see that there are three nondegenerate vibration branches along the $Z$-line, as was previously demonstrated analytically (see Table 3.1 and Fig. 3.14*d*).

Because of the relative simplicity of the solution for a simple cubic lattice obtained in the preceding section, the reader may doubt the effectiveness of the application of group theory to the normal crystal vibrations. In this connection the following remarks are appropriate: first, the solution in the preceding section was obtained for a model of a cubic crystal, which takes into account only the interaction of every atom with the atoms of the first and second coordination groups, while the results of group theory analysis are

[12] We should just substitute the irreducible representations $Z_1$, $Z_2$, $Z_3$, $Z_4$ for the first column and $E$, $C_4^2\ (q_Z)$, $JC_4^2\ (q_X)$, $JC_4^2\ (q_y)$ for the first row.

generally valid for a cubic crystal. Moreover, group theory can be applied in the analysis of normal vibrations also in the case of complex lattices, when it is impossible to obtain an analytical solution in a closed form. Group theory makes it possible to reject everything that contradicts the system's symmetry, and thus to essentially reduce the number of alternative solutions that have to be analyzed by computation methods.

## 3.9 Vibrations and Waves in Crystals in the Approximation of an Isotropic Continuous Medium

Consider harmonic waves in a crystal lattice in the approximation of a continuous medium (P. Debye, 1912). This approximation is, obviously, good when the wavelength is much longer than the lattice constant, since in this case the discrete (atomic) structure of the crystal should not make itself felt. As we shall see, the approximation of a continuous medium assumes quite different forms for long acoustic waves and for long optical waves in an ionic (heteropolar) crystal.

**3.9.1** In the case of long acoustic waves the continuous approximation is equivalent to the application of the theory of elasticity. The equations of motion of a homogeneous isotropic continuous medium in the absence of forces acting in the volume are of the form [3.1, § 22]

$$\rho \frac{\partial^2 \mathbf{u}}{\partial t^2} = (M + \Lambda)\,\mathrm{grad}\,\mathrm{div}\,\mathbf{u} + M\nabla^2 \mathbf{u}. \tag{9.1}$$

Here $\mathbf{u}(\mathbf{r}, t)$ is the displacement vector of the medium at point $\mathbf{r}$ at the moment $t$; $M$ and $\Lambda$ are the Lamé coefficients; and $\rho$ is the constant density of the homogeneous continuous medium.

The theory of elasticity shows [3.1, § 22] that $\vartheta \equiv \mathrm{div}\,\mathbf{u}$ is the relative variation of volume $\Delta V/V$ at point $\mathbf{r}$, and that $\varphi = (1/2) \times \times\,\mathrm{curl}\,\mathbf{u}$ is the angle of rotation of the volume element (as a whole) at point $\mathbf{r}$. Taking the divergence of both sides of (9.1), we obtain the wave equation for compression $\vartheta$

$$\frac{\partial^2 \vartheta}{\partial t^2} = v_l^2 \nabla^2 \vartheta, \tag{9.2}$$

where $v_l = \sqrt{(2M + \Lambda)/\rho}$ is the velocity of propagation of the compression waves. In obtaining (9.2) we made use of relations

$$\mathrm{div}\,\frac{\partial^2 \mathbf{u}}{\partial t^2} = \frac{\partial^2}{\partial t^2}(\mathrm{div}\,\mathbf{u}), \quad \mathrm{div}\,\mathrm{grad} \equiv \nabla^2, \quad \mathrm{div}\,\nabla^2 \mathbf{u} = \nabla^2\,\mathrm{div}\,\mathbf{u}.$$

Similarly, taking the curl of both sides of (9.1), we obtain the wave equation for the torsional angle $\boldsymbol{\varphi}$:

$$\frac{\partial^2 \boldsymbol{\varphi}}{\partial t^2} = v_t \nabla^2 \boldsymbol{\varphi}, \tag{9.3}$$

where $v_t = \sqrt{M/\rho}$ is the velocity of propagation of the torsional waves (use was made of the equality curl grad $\equiv 0$). It may easily be seen that $v_l > v_t$, the explanation being that elastic resistance to compression exceeds that to twisting.

Let us demonstrate that compression waves are longitudinal and torsional waves are transverse. Consider a plane wave propagating in the direction of the $x$ axis (this, obviously, does not restrict generality):

$$\mathbf{u} = \mathbf{A} \sin 2\pi \left( \nu t - \frac{x}{\lambda} \right), \tag{9.4}$$

where $\mathbf{A}$ is a constant amplitude, $\nu$ is the frequency, and $\lambda$ is the wavelength. Hence it follows that

$$\vartheta = \operatorname{div} \mathbf{u} = \frac{\partial u_x}{\partial x} + \frac{\partial u_y}{\partial y} + \frac{\partial u_z}{\partial z} = -A_x \left( \frac{2\pi}{\lambda} \right) \cos 2\pi \left( \nu t - \frac{x}{\lambda} \right) \tag{9.5}$$

and

$$\boldsymbol{\varphi} = \frac{1}{2} \operatorname{curl} \mathbf{u} = -A_y \mathbf{j}_0 \left( \frac{\pi}{\lambda} \right) \cos 2\pi \left( \nu t - \frac{x}{\lambda} \right) + A_z \mathbf{k}_0 \left( \frac{\pi}{\lambda} \right) \cos 2\pi \left( \nu t - \frac{x}{\lambda} \right), \tag{9.6}$$

where $\mathbf{j}_0$ and $\mathbf{k}_0$ are unit vectors directed along the $y$ and $z$ axes, respectively.

It follows directly from (9.5) and (9.6) that the compression waves $\vartheta$ are longitudinal ($A_y = A_z = 0$) and the torsional waves $\boldsymbol{\varphi}$ transverse ($A_x = 0$).

The longitudinal compression waves $\vartheta$ (9.5) and the transverse torsional waves $\varphi_y$ and $\varphi_z$ (9.6) are continuous medium analogues of the three acoustic branches of atomic vibrations in complex crystals (Section 3.5.5); $v_l$ and $v_t$ may be said to be the longitudinal and the transverse sound velocities. For the case being considered we shall find the frequency distribution function $g(\omega)$ (5.33), which specifies the number of vibrations per unit interval of the frequency $\omega$.

The wave equation for the scalar $\vartheta$ (9.2) is similar to the wave equation (9.3) for every component of the torsional vector $\boldsymbol{\varphi}$; therefore, it suffices to consider equation (9.2). Consider the waves $\vartheta$ in a cube with an edge $L$. Direct the orthogonal coordinates $x$, $y$, $z$ along the cube's edges. Choose as the boundary conditions $\vartheta = 0$ on all the faces of the cube $x = y = z = 0$ and $x = y = z = L$.

The choice of the boundary conditions cannot be of much importance if the wavelength is small as compared with $L$. Specifically, nothing changes if we require that elastic stresses vanish on the cube's faces instead of the compression $\vartheta$ (free surfaces).

We shall look for the solution of (9.2) in the form

$$\vartheta = A \sin \omega t \sin ax \sin by \sin cz, \tag{9.7}$$

where $A$ is the amplitude, $\omega$ is the cyclic frequency, and $a$, $b$, $c$ are constants. Substituting (9.7) into (9.2), we obtain after canceling out $\vartheta$ on both sides:

$$\omega = v_l \sqrt{a^2 + b^2 + c^2}. \tag{9.8}$$

Hence, (9.7) satisfies equation (9.2) if the frequency is related to $a$, $b$, $c$ via (9.8). To satisfy the boundary conditions we must make

$$aL = n_1\pi, \quad bL = n_2\pi, \quad cL = n_3\pi, \tag{9.9}$$

where $n_1$, $n_2$, $n_3$ are positive integers or zero (negative numbers yield the same vibration but with a phase shift of $\pi$).

Substituting (9.9) into (9.8), we obtain

$$\omega = \frac{\pi v_l}{L} \sqrt{n_1^2 + n_2^2 + n_3^2}. \tag{9.10}$$

There is a distinct normal vibration with a definite frequency (9.10) to correspond to each set of three numbers $n_i$. If $n_1$, $n_2$, $n_3$ are large, i.e., if the wavelength is much shorter than $L$, then the $\omega$'s dependence on the numbers $n_i$ is quasi-continuous. In this case we can speak about the number of vibrations in a frequency interval $[\omega, \omega + d\omega]$.

We introduce the quantity

$$R^2 = n_1^2 + n_2^2 + n_3^2. \tag{9.11}$$

Then the frequency will be

$$\omega = \frac{\pi v_l}{L} R. \tag{9.12}$$

Figure 3.15 depicts a Cartesian coordinate frame with positive integers $n_1$, $n_2$, $n_3$ marked on the axes; there are integral coordinates $n_1$, $n_2$, $n_3$ and a definite position vector $\mathbf{R}$ (9.11) to correspond to each site of the cubic lattice depicted in Fig. 3.15.[13] On the other hand, there is a distinct normal vibration (9.7) with a frequency (9.10) to correspond to each site.

[13] To avoid complicating the figure, only the lattice sites in the $n_2n_3$ plane are shown.

Find the number of vibrations per frequency interval $[\omega, \omega + d\omega]$ for large $n_i$. It follows from (9.12) and (9.11) that this number is equal to the number of lattice sites inside a spherical layer $[R, R + dR]$ in a coordinate frame octant. Since the volume of a cubic cell is unity, this number is simply equal to the volume of the corresponding spherical layer.

Hence, the number of longitudinal vibrations per frequency interval $[\omega, \omega + d\omega]$ is

$$g_l(\omega)\,d\omega = \frac{4\pi R^2\,dR}{8} = \frac{V}{2\pi^2 v_l^3}\,\omega^2\,d\omega, \qquad (9.13)$$

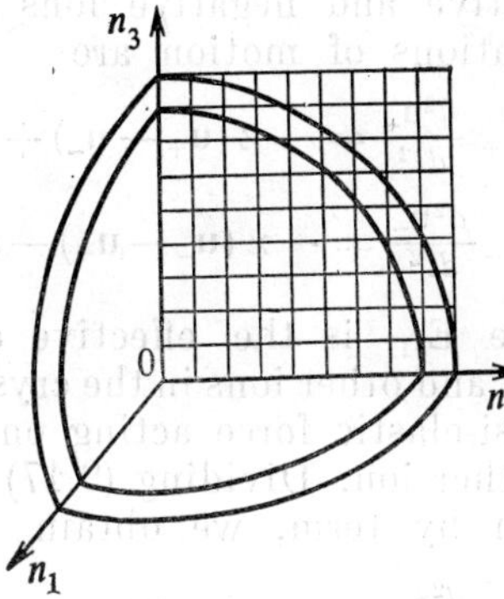

Fig. 3.15

in accordance with (9.12); here $V = L^3$ is the body's volume. Similar considerations hold for each of the two components of the vector $\varphi$ which satisfy equation (9.3); therefore the number of transverse vibrations in the interval $[\omega, \omega + d\omega]$ is

$$g_t(\omega)\,d\omega = \frac{2V}{2\pi^2 v_t^3}\,\omega^2\,d\omega, \qquad (9.14)$$

where two in the numerator reflects the existence of two components in a transverse wave (9.6).

The full frequency distribution function (5.32) is

$$g(\omega) = g_l(\omega) + g_t(\omega) = \frac{3V}{2\pi^2 v_0^3}\,\omega^2, \qquad (9.15)$$

where the velocity $v_0$ is determined by

$$\frac{1}{v_0^3} = \frac{1}{3}\left(\frac{1}{v_l^3} + \frac{2}{v_t^3}\right); \qquad (9.16)$$

$v_0^3$ may be said to represent the average of inverse cubic longitudinal and inverse cubic transverse sound velocities.

Note that in the continuous (Debye) approximation the frequency distribution function for acoustic vibrations $g(\omega)$ (5.32) is proportional to $\omega^2$.

**3.9.2** In the preceding section we saw how long acoustic waves are described in the approximation of a continuous medium.

Now we turn to the study of the continuous approximation for long optical waves in a cubic ionic crystal (K. Huang, 1950). Consider an ionic crystal whose every cell consists of two oppositely charged ions with the effective charges $\pm e^*$ and masses $m_+$ and $m_-$. In the process of long-wave optical vibrations the ions in all the cells vibrate in step (Section 3.5.5); therefore, only the ionic motion in one cell need be considered. If $\mathbf{u}_+$ and $\mathbf{u}_-$ are the displacements of the positive and negative ions from their equilibrium positions, the equations of motion are

$$m_+ \frac{d^2\mathbf{u}_+}{dt^2} = -\varkappa(\mathbf{u}_+ - \mathbf{u}_-) + e^*\mathbf{E}_{\text{eff}}, \tag{9.17}$$

$$m_- \frac{d^2\mathbf{u}_-}{dt^2} = -\varkappa(\mathbf{u}_- - \mathbf{u}_+) - e^*\mathbf{E}_{\text{eff}}. \tag{9.17a}$$

Here $\mathbf{E}_{\text{eff}}$ is the effective electric field with which the external field and other ions in the crystal act on the ion, and $\varkappa$ is the constant quasi-elastic force acting on the ion as it is displaced relative to another ion. Dividing (9.17) by $m_+$, (9.17a) by $m_-$, and subtracting term by term, we obtain

$$m_{\text{r}} \frac{d^2\mathbf{s}}{dt^2} = -\varkappa\mathbf{s} + e^*\mathbf{E}_{\text{eff}}, \tag{9.18}$$

where $m_{\text{r}}$ is the reduced mass ($m_{\text{r}}^{-1} = m_+^{-1} + m_-^{-1}$), $\mathbf{s} = \mathbf{u}_+ - \mathbf{u}_-$ is the displacement of the positive ion relative to the negative ion.

The effective field in a cubic crystal is [3.15, Sec. 2.9, equation (2.60)]

$$\mathbf{E}_{\text{eff}} = \mathbf{E} + \frac{4\pi}{3}\mathbf{P}, \tag{9.19}$$

where $\mathbf{E}$ is the average field in the dielectric, and the polarization vector is equal to

$$\mathbf{P} = N_0[e^*(\mathbf{u}_+ - \mathbf{u}_-) + \alpha_+\mathbf{E}_{\text{eff}} + \alpha_-\mathbf{E}_{\text{eff}}] = N_0(e^*\mathbf{s} + \alpha\mathbf{E}_{\text{eff}}). \tag{9.20}$$

Here $N_0$ is the number of cells per unit volume of the crystal, $\alpha_+$ and $\alpha_-$ are electronic polarizabilities of the positive and negative ions, and $\alpha = \alpha_+ + \alpha_-$. This is a continuous-medium approach, partly because it makes use of the macroscopic concept of the polarization vector $\mathbf{P}$.

Excluding the effective field $\mathbf{E}_{\text{eff}}$ from (9.19) and (9.20), we obtain

$$\mathbf{P} = N_0 \frac{e^*\mathbf{s} + \alpha\mathbf{E}}{1 - \frac{4\pi N_0}{3}\alpha}. \tag{9.21}$$

We shall exclude from the equations the polarizability $\alpha$, which is not measured directly in experiments, and turn to the expression for the induction vector [3.13, Sec. 2.3]

$$\mathbf{D} = \mathbf{E} + 4\pi\mathbf{P} = \varepsilon\mathbf{E}, \tag{9.22}$$

with $\varepsilon$ the dielectric constant[14]; hence

$$\mathbf{P} = \frac{\varepsilon - 1}{4\pi} \mathbf{E}. \tag{9.23}$$

In a high-frequency ($\omega \to \infty$) electric field the ions are unable to follow the field and therefore $\mathbf{s} \to 0$. By making $\mathbf{s} = 0$ in (9.21) and $\varepsilon = \varepsilon_\infty$ in (9.23), we exclude $\mathbf{P}$ from (9.21). Then the polarizability will be

$$\alpha = \frac{\varepsilon_\infty - 1}{\frac{4\pi N_0}{3} (\varepsilon_\infty + 2)}. \tag{9.24}$$

Substituting this value of $\alpha$ into (9.21), we obtain

$$\mathbf{P} = N_0 \frac{e^* (\varepsilon_\infty + 2)}{3} \mathbf{s} + \frac{\varepsilon_\infty - 1}{4\pi} \mathbf{E}. \tag{9.25}$$

Substituting into (9.19) this value of $\mathbf{P}$ and then excluding $\mathbf{E}_{\text{eff}}$ from (9.18), we obtain

$$m_r \frac{d^2 \mathbf{s}}{dt^2} = -m_r \omega_0^2 \mathbf{s} + \frac{e (\varepsilon_\infty + 2)}{3} \mathbf{E}, \tag{9.26}$$

where

$$\omega_0^2 = \frac{\varkappa}{m_r} - \frac{4\pi N_0 e^{*2} (\varepsilon_\infty + 2)}{9 m_r}. \tag{9.26a}$$

We introduce the "normalized" displacement

$$\mathbf{w} = \sqrt{N_0 m_r}\, \mathbf{s} \tag{9.27}$$

and substitute it into (9.26) and (9.25). Then

$$\frac{d^2 \mathbf{w}}{dt^2} = -\omega_0^2 \mathbf{w} + \sqrt{\frac{N_0}{m_r}}\, e^* \frac{\varepsilon_\infty + 2}{3} \mathbf{E} \tag{9.28}$$

and

$$\mathbf{P} = \sqrt{\frac{N_0}{m_r}}\, e^* \frac{\varepsilon_\infty + 2}{3} \mathbf{w} + \frac{\varepsilon_\infty - 1}{4\pi} \mathbf{E}. \tag{9.29}$$

If we were to introduce the static value of the dielectric constant $\varepsilon_0$ (for a constant electric field $\omega \to 0$), we would obtain from (9.23) and (9.29)

$$\varepsilon_0 - \varepsilon_\infty = \frac{N_0}{m_r} e^{*2} \frac{4\pi (\varepsilon_\infty + 2)^2}{9 \omega_0^2}. \tag{9.30}$$

---

[14] Here we make use of the customary notation for the dielectric constant without fear that it might be mistaken for energy.

From this we find the quantity $(N_0/m_r)\,e^{*2}$ and substitute it into (9.28) and (9.29) to obtain

$$\frac{d^2\mathbf{w}}{dt^2} = -\omega_0^2\mathbf{w} + \omega_0 \sqrt{\frac{\varepsilon_0 - \varepsilon_\infty}{4\pi}}\,\mathbf{E} \tag{9.31}$$

and

$$\mathbf{P} = \omega_0 \sqrt{\frac{\varepsilon_0 - \varepsilon_\infty}{4\pi}}\,\mathbf{w} + \frac{\varepsilon_\infty - 1}{4\pi}\,\mathbf{E}. \tag{9.32}$$

To determine the type of motion of the ions we make

$$\mathbf{w} = \mathbf{w}_t + \mathbf{w}_l, \tag{9.33}$$

where

$$\operatorname{div} \mathbf{w}_t = 0, \quad \operatorname{curl} \mathbf{w}_l = 0. \tag{9.34}$$

It is an established fact that such a representation of an arbitrary vector ($\mathbf{w}$) in the form of a sum of a solenoidal ($\mathbf{w}_t$) and an irrotational ($\mathbf{w}_l$) vector is always possible and unique [3.16].

In the absence of free charges

$$\operatorname{div} \mathbf{D} = \operatorname{div} \mathbf{E} + 4\pi \operatorname{div} \mathbf{P} = 0. \tag{9.35}$$

If we substitute $\mathbf{P}$ from (9.32) and make use of (9.33) and (9.34), we obtain

$$\operatorname{div} \mathbf{E} + \frac{\omega_0}{\varepsilon_\infty}\sqrt{4\pi(\varepsilon_0 - \varepsilon_\infty)}\operatorname{div} \mathbf{w}_l = 0, \tag{9.36}$$

whence

$$\mathbf{E} = -\frac{\omega_0}{\varepsilon_\infty}\sqrt{4\pi(\varepsilon_0 - \varepsilon_\infty)}\,\mathbf{w}_l. \tag{9.37}$$

Substituting (9.33) and (9.37) into (9.31), we obtain

$$\frac{d^2}{dt^2}(\mathbf{w}_t + \mathbf{w}_l) = -\omega_0^2\mathbf{w}_t - \omega_0^2\frac{\varepsilon_0}{\varepsilon_\infty}\mathbf{w}_l. \tag{9.38}$$

Separating in this equation the solenoidal part from the irrotational, we obtain

$$\frac{d^2\mathbf{w}_t}{dt^2} = -\omega_0^2\mathbf{w}_t, \tag{9.39}$$

$$\frac{d^2\mathbf{w}_l}{dt^2} = -\omega_0^2\frac{\varepsilon_0}{\varepsilon_\infty}\mathbf{w}_l. \tag{9.40}$$

If we represent $\mathbf{w}_t$ and $\mathbf{w}_l$ in the form of a plane wave $\mathbf{A}\exp[i(\mathbf{q}\cdot\mathbf{r} - \omega t)]$, we obtain from (9.39) and (9.40) for corresponding frequencies $\omega_t = \omega_0$ and $\omega_l = (\varepsilon_0/\varepsilon_\infty)^{1/2}\omega_0$.

On the other hand, substituting into (9.34) the expressions for a plane wave, we obtain

$$\text{div } \mathbf{w}_t = \text{div } \{\mathrm{A}_t \exp [i (\mathbf{q}\cdot\mathbf{r} - \omega_t t)\} \propto \mathbf{A}_t\cdot\mathbf{q} = 0,$$
$$\text{curl } \mathbf{w}_l = \text{curl } \{\mathbf{A}_l \exp [i (\mathbf{q}\cdot\mathbf{r} - \omega_l t)]\} \propto \mathbf{A}_l \times \mathbf{q} = 0, \tag{9.41}$$

whence it follows that $\mathbf{A}_t \perp \mathbf{q}$ and $\mathbf{A}_l \parallel \mathbf{q}$, i.e., the solenoidal wave $\mathbf{w}_t$ is transverse and the irrotational wave $\mathbf{w}_l$ is longitudinal.

The expressions obtained above lead to the Lyddane-Sachs-Teller relation

$$\omega_l/\omega_t = \sqrt{\varepsilon_0/\varepsilon_\infty}. \tag{9.42}$$

Since $\varepsilon_0 > \varepsilon_\infty$ [see (9.30)], it follows that the frequency of the longitudinal waves $\omega_l$ is higher than the frequency of the transverse waves $\omega_t$, which is an analogue of the relation $v_l > v_t$ for the acoustic waves (see the preceding section). Since it is easier to measure $\omega_t$ in experiments than $\omega_l$, equation (9.42) can serve to determine $\omega_l$.

## 3.10 Quantization of Crystal Lattice Vibrations. Phonons

**3.10.1** The quantum mechanical Hamiltonian in the $Q$ representation corresponding to the Hamiltonian for normal lattice vibrations (6.21) is obtained from the latter by substituting the operators for the momenta $P_j(\mathbf{q})$:

$$\dot{Q}_j(\mathbf{q}) = P_j(\mathbf{q}) \to \frac{\hbar}{i}\frac{\partial}{\partial Q_j(\mathbf{q})}, \tag{10.1}$$

where $\hbar$ is the Planck constant divided by $2\pi$, and $i = \sqrt{-1}$.

The Hamiltonian obtained from (6.21) is of the form

$$\hat{\mathscr{H}}(Q, P) = \sum_{j,\mathbf{q}} \left\{ -\frac{\hbar^2}{2}\frac{\partial^2}{\partial Q_j^2(\mathbf{q})} + \frac{1}{2}\omega_j^2(\mathbf{q})\, Q_j^2(\mathbf{q}) \right\}, \tag{10.2}$$

i.e., it splits up into a sum whose each addend is in the form of a Hamiltonian of a linear harmonic oscillator with the coordinate $Q_j(\mathbf{q})$, the frequency $\omega_j(\mathbf{q})$, and a unit mass. It has been established in quantum mechanics that the wave function of a system with a Hamiltonian in the form of a sum whose every addend depends only on one coordinate and one momentum conjugate to it is equal to the product of the wave functions corresponding to each addend, the energy being equal to the sum of the corresponding energies.

Consider an individual addend of the Hamiltonian (10.2) omitting, to simplify the notation, the symbols $j$ and $\mathbf{q}$ accompanying the coordinate $Q_j(\mathbf{q})$. The Schrödinger equation for such a linear oscillator assumes in this case the form

$$-\frac{\hbar^2}{2}\frac{\partial^2\psi}{\partial Q^2}+\frac{1}{2}\omega^2Q^2\psi=\varepsilon\psi. \tag{10.3}$$

It has also been established in quantum mechanics [3.5, § 23] that the eigenvalues and the normalized eigenfunctions of this equation are of the form

$$\varepsilon\equiv\varepsilon_N=\hbar\omega\,(N+1/2), \tag{10.4a}$$

$$\psi\equiv\psi_N(Q)=\left(\frac{\omega}{\pi\hbar}\right)^{1/4}\frac{1}{\sqrt{2^N N!}}\,e^{-\omega Q^2/2\hbar}\,H_N\left[\left(\frac{\omega}{\hbar}\right)^{1/2}Q\right]. \tag{10.4b}$$

Here $N$ is the oscillator's quantum number, $H_N[\xi]$ is a Hermite polynomial of the dimensionless coordinate $\xi=(\omega/\hbar)^{1/2}Q$.

As we shall see below, the matrix elements of the coordinate $Q$ and of the momentum $P$ (10.1) on the wave functions (10.4b) are of major importance for the study of the interaction of conduction electrons with the lattice vibrations. It can be demonstrated that they are equal to [3.5, § 23]

$$\langle N'|\hat{Q}|N\rangle=\int_{-\infty}^{\infty}\psi_{N'}^{*}Q\psi_N\,dQ$$

$$=\sqrt{\frac{\hbar}{2\omega}}\times\begin{cases}\sqrt{N} & \text{if } N'=N-1,\\ \sqrt{N+1} & \text{if } N'=N+1.\\ 0 & \text{in all other cases;}\end{cases} \tag{10.5}$$

$$\langle N'|\hat{P}|N\rangle=\frac{\hbar}{i}\int_{-\infty}^{\infty}\psi_{N'}^{*}\frac{\partial\psi_N}{\partial Q}\,dQ$$

$$=i\sqrt{\frac{\hbar\omega}{2}}\times\begin{cases}-\sqrt{N} & \text{if } N'=N-1,\\ \sqrt{N+1} & \text{if } N'=N+1,\\ 0 & \text{in all other cases.}\end{cases} \tag{10.6}$$

**3.10.2** Introduce new complex normal coordinates $\alpha_j(\mathbf{q})$ instead of $a_j(\mathbf{q})$, making

$$a_j(\mathbf{q})=\sqrt{\frac{\hbar}{2\,\omega_j(\mathbf{q})}}\,[\alpha_j^{*}(-\mathbf{q})+\alpha_j(\mathbf{q})], \tag{10.7}$$

$$\dot{a}_j(\mathbf{q})=i\sqrt{\frac{\hbar\omega_j(\mathbf{q})}{2}}\,[\alpha_j^{*}(-\mathbf{q})-\alpha_j(\mathbf{q})]. \tag{10.8}$$

The quantities $a_j(\mathbf{q})$ defined in this way automatically satisfy the condition (6.5). Solving the system (10.7) and (10.8) to find $\alpha_j(\mathbf{q})$, and $\alpha_j(-\mathbf{q})$, we obtain

$$\alpha_j(\mathbf{q}) = \sqrt{\frac{\omega_j(\mathbf{q})}{2\hbar}}\, a_j(\mathbf{q}) + i\sqrt{\frac{1}{2\hbar\omega_j(\mathbf{q})}}\, \dot{a}_j(\mathbf{q}). \tag{10.9}$$

Expressing $a_j(\mathbf{q})$ and $\dot{a}_j(\mathbf{q})$ with the aid of (6.16) in terms of $Q_j(\mathbf{q})$ and $\dot{Q}_j(\mathbf{q}) = P_j(\mathbf{q})$, we obtain

$$\alpha_j(\mathbf{q}) = \sqrt{\frac{\omega_j(\mathbf{q})}{2\hbar}}\, Q_j(\mathbf{q}) + i\sqrt{\frac{1}{2\hbar\omega_j(\mathbf{q})}}\, P_j(\mathbf{q}). \tag{10.10}$$

**3.10.3** Introduce the operators corresponding to the quantities $\alpha_j(\mathbf{q})$ and $\alpha_j^*(\mathbf{q})$. Omitting the label $j$ and the argument $\mathbf{q}$ to simplify the notation, we obtain from (10.10) and (10.1)

$$a = \left(\frac{\omega}{2\hbar}\right)^{1/2} Q + \left(\frac{\hbar}{2\omega}\right)^{1/2} \frac{\partial}{\partial Q}, \tag{10.11}$$

$$a^+ = \left(\frac{\omega}{2\hbar}\right)^{1/2} Q - \left(\frac{\hbar}{2\omega}\right)^{1/2} \frac{\partial}{\partial Q} \tag{10.12}$$

[$a^+$ is the operator of $\alpha_j^*(\mathbf{q})$].

It can easily be shown that those are not hermitian operators. Operating with them on the wave function (10.4b), we obtain

$$a\psi_N = \sqrt{N}\psi_{N-1}, \tag{10.13}$$

$$a^+\psi_N = \sqrt{N+1}\,\psi_{N+1}. \tag{10.14}$$

Indeed, calculating

$$a\psi_N = \left[\left(\frac{\omega}{2\hbar}\right)^{1/2} Q + \left(\frac{\hbar}{2\omega}\right)^{1/2} \frac{\partial}{\partial Q}\right] \times$$
$$\times \frac{\omega \quad \exp(-\omega Q^2/2\hbar)\, H_N\left[\left(\frac{\omega}{\hbar}\right)^{1/2} Q\right]}{(\pi\hbar)^{1/4}\,(2^N N!)^{1/2}},$$

we obtain (10.13). Here the product of the first addend in the square bracket times the wave function $\psi_N$ and the derivative $\partial/\partial Q$ of the exponential factor in $\psi_N$ cancel out. In addition, we should make use of the relation [3.7, p. 145, Problem 19]

$$H_N'(\xi) = 2NH_{N-1}(\xi); \tag{10.15}$$

(10.14) is obtained in a similar way.

As we shall see below, the normal lattice vibrations in interactions with the conduction electrons and with other normal lattice vibrations behave like particles with an energy $\hbar\omega_j(\mathbf{q})$ and a *quasi-momentum*[15] $\hbar\mathbf{q}$. Such quasi-particles are known as *phonons*.

The state of a crystal with the Hamiltonian (10.2) is specified in the $Q$ representation with the aid of a symmetrical product of oscillator functions (10.4b) [3.5, § 61], each of which is characterized by a quantum number $N \equiv N_{j\mathbf{q}}$. Because of the quantum indistinguishability of phonons of one type, the crystal's state is fully described by the numbers of phonons $N_{j\mathbf{q}}$. Such a method of describing systems of particles identical in the quantum mechanical sense is known as the method of *second quantization*. The operators $a$ and $a^+$ introduced above are operators in the second quantization representation, since they act directly on the *occupation numbers* $N$. Indeed, the operator $a$, as can be seen from (10.13), decreases the number of phonons $N$ by one (for this reason it is termed *annihilation operator*), and the operator $a^+$, as can be seen from (10.14), increases the number of phonons by one (for this reason it is termed *creation operator*).

Consider certain properties of the operators $a$ and $a^+$. From quantum mechanics it is known that for a conjugate pair of coordinate $Q \equiv Q_j(\mathbf{q})$ and momentum $P \equiv P_j(\mathbf{q})$ there is the commutation relations [3.5, § 61]

$$QP - PQ = [Q, P] = i\hbar. \tag{10.16}$$

Substituting herein the expressions for $Q$ and $P = \dot{Q}$ in terms of $a$ and $a^+$ [see (10.11) and (10.12)], we obtain

$$aa^+ - a^+a = [a, a^+] = 1. \tag{10.17}$$

Since the operators $Q_j(\mathbf{q})$ and $P_{j'}(\mathbf{q}')$ for $j \neq j'$ or $\mathbf{q} \neq \mathbf{q}'$ commute with one another, (10.17) can be rewritten in the form

$$[a_j(\mathbf{q}), \quad a_{j'}^+(\mathbf{q}')] = \delta_{jj'}\delta_{\mathbf{q}\mathbf{q}'}. \tag{10.18}$$

Quasi-particles (particles) whose creation and annihilation operators satisfy the commutation rule (10.18) are termed *bosons*. In the state of thermal equilibrium they are described by the *Bose-Einstein* statistics [3.17].

We express the Hamiltonian (10.2) in terms of the operators $a$ and $a^+$. Making use of (10.11) and (10.12), we obtain

$$\hat{\mathscr{H}}(a) = \sum_{j,\mathbf{q}} \frac{\hbar\omega}{2}[aa^+ + a^+a]. \tag{10.19}$$

---

[15] We use the term quasi-momentum for the quantity $\hbar\mathbf{q}$ instead of momentum because it displays certain properties not peculiar to the momentum of a free particle; for instance, in collisions the sum of quasi-momenta is conserved to within an arbitrary vector of the reciprocal lattice.

This, with reference to (10.18), may be rewritten as

$$\mathcal{H}(a) = \sum_{j,\mathbf{q}} \hbar\omega_j [a^+a + 1/2]. \qquad (10.20)$$

Operating with $a^+a$ on the wave function (10.4b), we obtain, with the help of (10.13) and (10.14),

$$a^+a\psi_N = a^+\sqrt{N}\psi_{N-1} = N\psi_N. \qquad (10.21)$$

We see that the eigenvalues of the operator $a^+a$ are equal to the number of particles (phonons) $N_{j\mathbf{q}}$ in the state $(j, \mathbf{q})$.

Operating with the Hamiltonian (10.20) on the eigenfunction $\psi_N$, we obtain, making use of (10.21), for the energy eigenvalues

$$\mathscr{E} = \sum_{j,\mathbf{q}} \hbar\omega_j(\mathbf{q}) [N_{j\mathbf{q}} + 1/2], \qquad (10.22)$$

which agrees with (10.4a).

Making use of (10.13), we obtain for the matrix elements of the annihilation and creation operators

$$\langle N' \mid a \mid N\rangle = \begin{cases} \sqrt{N} & \text{if } N' = N-1, \\ 0 & \text{in all other cases;} \end{cases} \qquad (10.23)$$

$$\langle N' \mid a^+ \mid N\rangle = \begin{cases} \sqrt{N+1} & \text{if } N' = N+1, \\ 0 & \text{in all other cases.} \end{cases} \qquad (10.24)$$

The matrix elements of the operators corresponding to the quantities $a_j(\mathbf{q})$, according to (10.7), are

$$\langle N'_{\mathbf{q}j} \mid a_j(\mathbf{q}) \mid N_{\mathbf{q}j}\rangle = \begin{cases} \sqrt{\hbar N_{\mathbf{q}j}/2\omega_j(\mathbf{q})} & \text{if } N'_{\mathbf{q}j} = N_{\mathbf{q}j} - 1, \\ 0 & \text{in all other cases;} \end{cases} \qquad (10.25)$$

$$\langle N'_{\mathbf{q}j} \mid a_j^*(\mathbf{q}) \mid N_{\mathbf{q}j}\rangle = \begin{cases} \sqrt{\hbar(N_{\mathbf{q}j}+1)/2\omega_j(\mathbf{q})} & \text{if } N'_{\mathbf{q}j} = N_{\mathbf{q}j} + 1, \\ 0 & \text{in all other cases.} \end{cases} \qquad (10.26)$$

Here we take into account that the matrix elements of the operators $a_j^*(-\mathbf{q})$ for $N'_{\mathbf{q}j} = N_{\mathbf{q}j} \mp 1$ are zero's; indeed, in the $Q$ representation (10.10) these operators depend on $Q_j(-\mathbf{q})$, whereas the wave functions depend on $Q_j(\mathbf{q})$.

The method of second quantization is especially expedient for the quantum mechanical treatment of systems with a very large (even varying) number of identical particles (phonons, photons, conduction electrons and holes in crystals, electrons and positrons).

## 3.11 Specific Heat of Crystal Lattices

The theory of specific heat of crystal lattices is one of the important applications of the theory of crystal lattice vibrations.

**3.11.1** A very simple theory is the theory of specific heat of crystals in the classical region, where the motion of the lattice atoms obeys the laws of classical mechanics. As was demonstrated by L. Boltzmann, the average kinetic energy per degree of freedom for systems in the state of statistical equilibrium is

$$\bar{\varepsilon}_{\text{kin}} = \frac{1}{2}\frac{R}{N_0} T = \frac{1}{2} k_0 T, \tag{11.1}$$

where $R = 1.987$ cal/K-mol is the universal gas constant, $N_0 = 6.023 \times 10^{23}$ is the Avogadro number (the number of particles per mole of substance), $k_0 = R/N_0 = 1.38 \times 10^{-16}$ erg/K is the Boltzmann constant, and $T$ is the absolute temperature [3.17, Chap. V, § 3].

It can easily be demonstrated that for a linear harmonic oscillator the average potential energy $\bar{\varepsilon}_{\text{pot}}$ is equal to the average kinetic energy $\bar{\varepsilon}_{\text{kin}}$; therefore, the average total energy is

$$\bar{\varepsilon} = \bar{\varepsilon}_{\text{kin}} + \bar{\varepsilon}_{\text{pot}} = 2\bar{\varepsilon}_{\text{kin}}. \tag{11.2}$$

For a mole of crystal of an element the internal energy in the state of statistical equilibrium is equal to the energy of $3N_0$ normal vibrations:

$$\mathscr{E} = 3N_0\bar{\varepsilon} = 3N_0 2\bar{\varepsilon}_{\text{kin}} = 3N_0 \frac{R}{N_0} T = 3RT. \tag{11.3}$$

The (molar) heat of such a crystal at constant volume is equal to

$$c_V = \partial\mathscr{E}/\partial T = 3R = 5.96 \text{ cal/K-mol}, \tag{11.4}$$

i.e., is independent of temperature and is equal to approximately 6 cal/K-mol (*Dulong-Petit's law*).

The great simplicity of the classical theory is due to two factors: the possibility of representing the atomic motion in the crystal (in the harmonic approximation) in the form of normal vibrations and to the universal character of the law of equidistribution of energy over the degrees of freedom (11.1). Relation (11.4) is in good agreement with experiment (metals are no exception although the first impression would be that this should not be the case). Indeed, the number of free electrons in metals is of the order of that of atoms. Every electron from the classical point of view has an average kinetic energy $(3/2)\,k_0T$, the corresponding additional contribution to the specific heat (11.4) is $\frac{\partial}{\partial T}\left(N_0 \frac{3}{2} k_0 T\right) = \frac{3}{2} R$, i. e., the increase in $c_V$ being 1.5. This is one of the major inconsistencies of the classical electron theory of metals, and it will be discussed in Section 6.3.

**3.11.2** Studies of the specific heats of solids in the low temperature range indicate that Dulong and Petit's law (11.4) is an asymptotical law valid only in the high temperature range. As the temperature is decreased, the specific heat, starting from a definite characteristic temperature $T_D$ (the so-called *Debye temperature*), begins to decrease rapidly and tends to zero as $T \to 0$. The Debye temperature varies for different substances, but for most solids it is of the order of 100-400 K. The explanation for the failure of Dulong and Petit's law for solids at low temperatures is that a reduction in temperature brings about a decrease in the average velocities of the crystal's atoms and consequently an increase in the corresponding de Broglie wavelength. It is shown in quantum mechanics that when the de Broglie wavelength becomes comparable to or exceeds the linear dimensions of the effective region of motion the motion no longer obeys the laws of classical mechanics. Therefore, when we consider the normal crystal vibrations from the quantum mechanics point of view, we must attribute to them, in compliance with (10.4), the energy

$$\varepsilon_N = \hbar\omega\,(N + 1/2), \tag{11.5}$$

where we omitted the labels **q** and $j$, which describe the particular oscillator.

The probability of an oscillator in the state of statistical equilibrium to be in the $N$-th quantum state with the energy $\varepsilon_N$ is, according to Boltzmann, equal to $w_N = ce^{-\varepsilon_N/k_0T}$. The constant $c$ is determined from the normalization condition $\sum_N w_N = c\sum_N e^{-\varepsilon_N/k_0T} = 1$, where $c = (\sum_N e^{-\varepsilon_N/k_0T})^{-1}$. The average energy of an oscillator is equal to the sum of the energies multiplied by the corresponding probabilities $w_N$:

$$\bar{\varepsilon} = \sum_{N=0}^{\infty} \varepsilon_N w_N = c\sum_{N=0}^{\infty} \varepsilon_N e^{-\varepsilon_N/k_0T} = \frac{\sum_{N=0}^{\infty} \varepsilon_N e^{-\varepsilon_N/k_0T}}{\sum_{N=0}^{\infty} e^{-\varepsilon_N/k_0T}}. \tag{11.6}$$

To calculate (11.6), we introduce the so-called *partition function*

$$Z = \sum_{N=0}^{\infty} e^{-\varepsilon_N/k_0T}. \tag{11.7}$$

Then

$$\bar{\varepsilon} = \frac{\sum_{N=0}^{\infty} \varepsilon_N e^{-\varepsilon_N/k_0T}}{\sum_{N=0}^{\infty} e^{-\varepsilon_N/k_0T}} = -\frac{d}{d\left(\frac{1}{k_0T}\right)} \ln Z, \tag{11.8}$$

which is possible to verify directly by finding the derivative ot (11.7).

On the other hand, making use of (11.5), we obtain

$$Z=\sum_{N=0}^{\infty} e^{-\varepsilon_N/k_0T}=e^{-\hbar\omega/2k_0T}\sum_{N=0}^{\infty} e^{-N\hbar\omega/k_0T}$$
$$=e^{-\hbar\omega/2k_0T}\,[1+e^{-\hbar\omega/k_0T}+e^{-2\hbar\omega/k_0T}+\dots]$$
$$=\frac{e^{-\hbar\omega/2k_0T}}{1-e^{-\hbar\omega/k_0T}}, \tag{11.9}$$

in accordance with the equation for an infinite decreasing geometrical progression. Substituting the result of (11.9) into expression (11.8) and differentiating with respect to the argument $(k_0T)^{-1}$, we obtain

$$\bar{\varepsilon}=\frac{\hbar\omega}{2}+\frac{\hbar\omega}{e^{\hbar\omega/k_0T}-1}, \tag{11.10}$$

where the term $(1/2)\,\hbar\omega$ independent of temperature is termed *oscillator's zero-point energy*. Expression (11.10) (to be precise, the second addend on the right-hand side of that expression) can be obtained, if the normal lattice vibrations are treated like quasi-particles, phonons.

Since the phonon creation and annihilation operators satisfy the commutation rules (10.18), the phonons are bosons, i.e., are described by the Bose-Einstein statistics.

In some respects the phonons behave differently from a gas of normal particles, and this is why they are termed quasi-particles. First, in interactions with electrons or other phonons the phonons are created or annihilated. Second, the average number of phonons in a definite volume (their concentration) depends on temperature. For a gas of normal particles (atoms, electrons) the variables $V$, $T$, and $N$ (the number of particles) are independent. In the case of phonons their number for specified $V$ and $T$ is determined from the conditions of equilibrium, i.e., from the conditions that the free energy $\mathcal{F}$ $(T, V, N)$ be minimal.

Hence

$$\left(\frac{\partial\mathcal{F}}{\partial N}\right)_{T,V}=\zeta=0,$$

where $\zeta$ by definition is the *chemical potential*.

The equilibrium number of phonons $\bar{N}$ per quantum state, i.e., in a cell of the phase space of volume 1 cm$^3$ $\times$ 1 $h^3$, with the energy $\hbar\omega$ and the chemical potential $\zeta=0$ is [3.17, Chap. IX, equation (2.9)]

$$\bar{N}=\frac{1}{e^{\hbar\omega/k_0T}-1}. \tag{11.10a}$$

Their average energy is obviously $\bar{\varepsilon} = \hbar\omega \times \bar{N}$, which coincides with the second addend on the right-hand side of (11.10).

Hence it follows that the phonons are elementary excitations of a crystal above its zero energy level $\mathscr{E}_0 = \sum_{j,\mathbf{q}} \frac{\hbar\omega_j(\mathbf{q})}{2}$.

Only such excitations interact with the conduction electrons, i.e., the zero-point vibrations set up a sort of permanent background (vacuum) in the crystal.

However, if the amplitude of the zero-point vibrations becomes comparable to the lattice constant, then such *quantum crystals* begin to exhibit some interesting properties; under certain conditions, these properties are exhibited by solid helium crystals.

Making use of expression (11.10), we obtain for the total internal energy of a crystal lattice in the state of thermodynamical equilibrium the following relationship:

$$\mathscr{E} = \mathscr{E}_0 + \sum_{j=1}^{3} \sum_{\mathbf{q}} \frac{\hbar\omega_{\mathbf{q}j}}{e^{\hbar\omega_{\mathbf{q}j}/k_0T} - 1} + \sum_{j=4}^{3s} \sum_{\mathbf{q}} \frac{\hbar\omega_{\mathbf{q}j}}{e^{\hbar\omega_{\mathbf{q}j}/k_0T} - 1}, \quad (11.11)$$

where $\mathscr{E}_0 = \sum_{\mathbf{q},j} \frac{\hbar\omega_{\mathbf{q}j}}{2}$ is the temperature independent zero-point energy, and the sum $\sum_{j=1}^{3}$ over the three acoustic branches has been written as a separate term.

**3.11.3** We shall assume (as a rough approximation) that the frequencies of the optical branches are independent of $\mathbf{q}$ and equal to their extreme values $\omega_j^0 = 0$[16]. In this case summation over $\mathbf{q}$ for every optical branch is equivalent to multiplication by $N$. The energy of acoustic vibrations (the second addend in (11.11) will be calculated in the approximation of a continuous medium (Section 3.9); this approximation is the better the longer the wavelength as compared with the lattice constant.

Making use of the frequency distribution function $g(\omega)$ (9.15), we represent the energy of the acoustic vibrations in the form

$$\mathscr{E}_{\text{ac}} = \int_0^{\omega_{\max}} \frac{\hbar\omega}{e^{\hbar\omega/k_0T} - 1} g(\omega)d\omega = \frac{3V\hbar}{2\pi^2 v_0^3} \int_0^{\omega_{\max}} \frac{\omega^3 \, d\omega}{e^{\hbar\omega/k_0T} - 1}, \quad (11.12)$$

where $v_0^3$ is the average of the inverse cubic longitudinal and inverse cubic transverse velocities of acoustic waves (9.16).

[16] For instance, making use of (3.6) we can demonstrate that the width of the optical branch, i.e., $\omega_{\text{op}}(0) - \omega_{\text{op}}(\pi/a)$, is small when the mass of one atom greatly exceeds the mass of another atom.

The maximum frequency $\omega_{max}$ is determined from the condition that the total number of vibrations is equal to the total number (3$N$) of normal vibrations in all three acoustic branches. Hence

$$\int_0^{\omega_{max}} g(\omega)\,d\omega = \frac{3V}{2\pi^2 v_0^3}\int_0^{\omega_{max}} \omega^2\,d\omega = \frac{V\omega_{max}^3}{2\pi^2 v_0^3} = 3N,$$

whence

$$\omega_{max} = v_0\left(\frac{6\pi^2}{\Omega_0}\right)^{1/3} \quad \text{and} \quad q_{max} = \frac{\omega_{max}}{v_0} = \left(\frac{6\pi^2}{\Omega_0}\right)^{1/3}. \qquad (11.13)$$

Here $\Omega_0 = V/N$ is the volume of a unit cell, and $q_{max}$ is the maximum value of the wave vector. Define the "lattice constant" with the aid of the equality $\Omega_0 = a^3$. The order of magnitude of $\omega_{max}$ is $v_0/a$; consequently, the maximum value of the wave vector and the minimum value of the wave length are $q_{max} = \omega_{max}/v_0 \sim 1/a$ and $\lambda_{min} = 2\pi/q_{max} \sim a$. A Brillouin zone of a cubic crystal has the shape of a cube with the edge equal to $2\pi/a$, so that the maximum values of the orthogonal components of $\mathbf{q}$ are equal to $\pi/a$ (5.9a).

In the approximation of a continuous medium employed in Debye's theory, the region of possible values of $\mathbf{q}$ is enclosed inside a sphere of radius $q_{max}$ (11.13).

Define the characteristic temperature of a solid (the *Debye temperature*) as

$$T_D = \frac{\hbar\omega_{max}}{k_0} = \left(\frac{6\pi^2}{\Omega_0}\right)^{1/3}\frac{\hbar}{k_0}v_0. \qquad (11.14)$$

Since $\omega_{max} \sim v_0/a \sim 10^5$ cm/sec/$10^{-8}$ cm $\sim 10^{13}$ Hz and $k_0 \sim \sim 10^{-16}$ erg/K, $T_D \sim 100$ K. We can introduce Debye temperatures corresponding to the extreme frequencies of the optical branches as

$$T_{Dj} = \hbar\omega_j^0/k_0, \qquad (11.14a)$$

their orders of magnitude being $10^2$-$10^3$ K, although in general $T_{Dj} > T_D$.

Introducing into (11.12) the integration variable $x = \hbar\omega/k_0T$ and making use of the definitions of characteristic temperatures (11.14) and (11.14a), we obtain

$$\mathscr{E} = \mathscr{E}_0 + Nk_0T\left\{3D\left(\frac{T_D}{T}\right) + \sum_{j=4}^{3s}\frac{T_{Dj}/T}{e^{T_{Dj}/T}-1}\right\} \qquad (11.15)$$

where the *Debye function*

$$D(t) = \frac{3}{t^3}\int_0^t \frac{x^3\,dx}{e^x-1}. \qquad (11.15a)$$

Consider first the case of high temperatures, where $T \gg T_{Dj}$ and the more so $T \gg T_D$, since $T_D < T_{Dj}$. Then the argument of the Debye function $t = T_D/T \ll 1$. Substituting $e^x \approx 1 + x$ in the integrand of (11.15a), we may easily see that $D(t) \approx 1$. Expanding the exponential in terms corresponding to optical branches, we obtain

$$\sum_{j=4}^{3s} \frac{T_{Dj}/T}{e^{T_{Dj}/T}-1} \approx \sum_{j=4}^{3s} 1 = 3s-3.$$

Hence, $\mathscr{E} = \mathscr{E}_0 + 3sNk_0T$, and the specific heat $c_V = \partial\mathscr{E}/\partial T = 3sNk_0$, in accordance with Dulong and Petit's law (11.4).

Consider now the low-temperature case, where $T \ll T_D$ and the more so $T \ll T_{Dj}$. Neglecting the quantities of the order of $e^{-T_D/T}$ in comparison with unity, we drop in (11.15) the terms corresponding to the optical branches and substitute $\infty$ for the upper limit in the integral of the function $D(t)$. Since [3.18, § 66]

$$\int_0^\infty \frac{x^3\,dx}{e^x-1} = \frac{\pi^4}{15},$$

it follows that

$$\mathscr{E} = \mathscr{E}_0 + \frac{3\pi^4 N k_0 T^4}{5T_D^3} = \mathscr{E}_0 + \frac{\pi^2 V (k_0 T)^4}{10\hbar^4 v_0^3}, \tag{11.16}$$

i.e., is the temperature-dependent part of the energy is proportional to $T^4$. The specific heat

$$c_V = \frac{\partial\mathscr{E}}{\partial T} = \frac{12\pi^4 k_0}{5}\left(\frac{T}{T_D}\right)^3, \tag{11.17}$$

i.e., proportional to $T^3$. The $T^3$-*law* is in good agreement with the experiment in the temperature range of 20-50 K. Debye's theory of specific heat based on expressions for $g(\omega)$ (9.15) and for the extreme frequency (11.13) should be true for low temperatures, when only the long waves are excited, and, consequently, the approximation of a continuous medium is valid. On the other hand, in the high-temperature range, when the heat capacity is determined solely by the number of degrees of freedom of the lattice, Debye's theory produces a correct result, Dulong and Petit's law. In the intermediate-temperature range the expression for the specific heat $c_V$ following from Debye's theory can be regarded only as a more or less successful interpolation. Differentiating expression (11.15) with respect to $T$, we obtain

$$c_V = \frac{\partial \mathscr{E}}{\partial T} = 3Nk_0 \left\{ 4D\left(\frac{T_D}{T}\right) - \frac{3(T_D/T)}{e^{T_D/T}-1} \right\}$$

$$+ Nk_0 \sum_{j=4}^{3s} \frac{e^{T_{Dj}/T}(T_{Dj}/T)^2}{(e^{T_{Dj}/T}-1)^2}. \qquad (11.18)$$

The temperature dependence of the heat of acoustic branches is characterized by the quantity

$$\frac{c_V^{(D)}}{3k_0 N} = \left\{ 4D\left(\frac{T_D}{T}\right) - \frac{3(T_D/T)}{e^{T_D/T}-1} \right\} = 3\left(\frac{T}{T_D}\right)^3 \int_0^{T_D/T} \frac{x^4\,dx}{(e^x-1)^2} \qquad (11.18a)$$

where the right-hand side can be obtained by integration by parts of the integral determining the function $D(T_D/T)$. Figure 3.16 depicts the graph of the quantity (11.18a) as a function of $T/T_D$. In the low-temperature range it is proportional to $(T/T_D)^3$, in the high-temperature range it tends to unity. The difference between the value of expression (11.18a) and unity is the measure of the deviation of the specific heat of acoustic branches from the classical value $3Nk_0$. The curve shows that the characteristic temperature that separates the classical region from the quantum one is not $T_D$ but rather $T_D/3$. The temperature range $T \geqslant T_D$ is the range of validity of classical expressions for specific heat, and the range $T < T_D/10$ is the range for quantum mechanics laws.

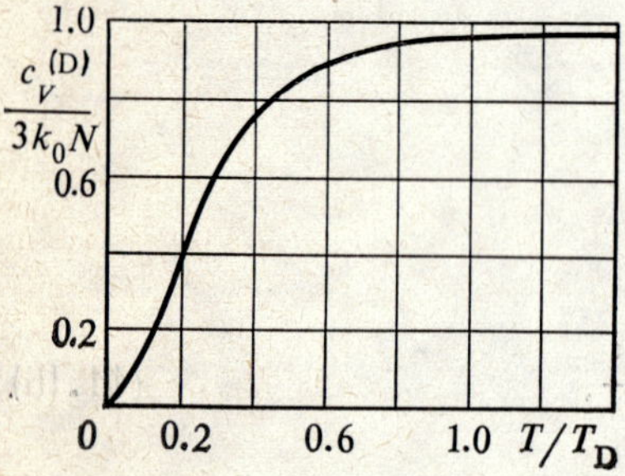


Fig. 3.16

The specific heats corresponding to the optical branches are frequently termed *Einstein's terms*.[17] As $T \to 0$ their decrease does not obey the $T^3$-law, but proceeds more rapidly, as $T^{-2}e^{-T_{Dj}/T}$. In cases where $T_{Dj} \gg T_D$ (molecular lattices) the situation might arise when $T \geqslant T_D$ and $T \ll T_{Dj}$, so that the acoustic branches are all excited (the heat of the optical branches being negligible). In this case the

[17] A. Einstein was the first to consider from quantum mechanics viewpoint the specific heat of a monatomic solid in the simplifying assumption that all atoms vibrate with a single definite frequency. This led him to the expression for specific heat which coincides with that for specific heat of a single optical branch in (11.18) (multiplied by 3).

lattice behaves as a classical nonatomic one with "atomic" masses $M = \sum_{k=1}^{s} m_k$.

**3.11.4** A more profound experimental and theoretical investigation of the specific heat of solids in the low-temperature range demonstrated that the true region of the $T^3$-law extends only up to several degrees above absolute zero. The impression that the $T^3$-law holds in the range 20-50 K is created by other factors discussed below. It was demonstrated by calculations that even in the low-temperature range ($T < T_D/10$) the specific heat remains sensitive to the discrete nature of the lattice structure.

Imagine that we succeeded experimentally or theoretically in determining the precise temperature dependence of $c_V$. Making use of Debye's theory of specific heat based on the concept of a continuous medium, we assume that

$$c_V (T) = c_V^{(D)} (T_D/T), \tag{11.19}$$

where $c_V^{(D)}$ is the Debye specific heat (11.18a), which depends on the temperature and on a single parameter $T_D$.[18] The difference between $c_V (T)$ and the expression for it given in Debye's theory of specific heat can be formally described by a characteristic temperature assumed to be dependent on the temperature $T$.

For example, E. W. Kellermann ([3.19] and [3.20]) discussed in detail the vibrations of NaCl. He calculated the coefficients $D_{\alpha\beta}^{kk'}$ (**q**) with the aid of (5.10a) and solved the characteristic equation (5.11) for the frequency $\omega^2$. In determining the Coulomb interaction forces, he treated the ions as point charges. The forces of interionic repulsion were determined from data on compressibility.

In general the number of vibrations in all $3s$ branches in a small frequency interval $\Delta\omega$ is equal to[19]

$$g(\omega)\, \Delta\omega = \frac{V}{(2\pi)^3} \sum_{j=1}^{3s} \iiint\limits_{\omega \leqslant \omega_j(q_x, q_y, q_z) \leqslant \omega + \Delta\omega} dq_x\, dq_y\, dq_z. \tag{11.20}$$

To determine the vibration frequency distribution function $g(\omega)$, we should know the function $\omega_j (q_x, q_y, q_z)$, which necessarily requires numerical calculations. The expression (11.20) may be used to determine the crystal's internal energy (11.12) instead of the approximate expression (9.15).

---

18 We consider here only the acoustic branches and presume that the optical branches not to be excited are in the temperature range of interest to us.

19 Here we must bear in mind that not all vibration branches in the specified interval $[\omega, \omega + \Delta\omega]$ necessarily contribute.

Figures 3.17 and 3.18 (after Kellermann) depict the vibration frequency distribution function $g(\omega)$ for the acoustic branches and the temperature dependence of $T_D$, which follows from it in accordance with Debye's theory.[20] We see that $g(\omega)$ differs very appre-

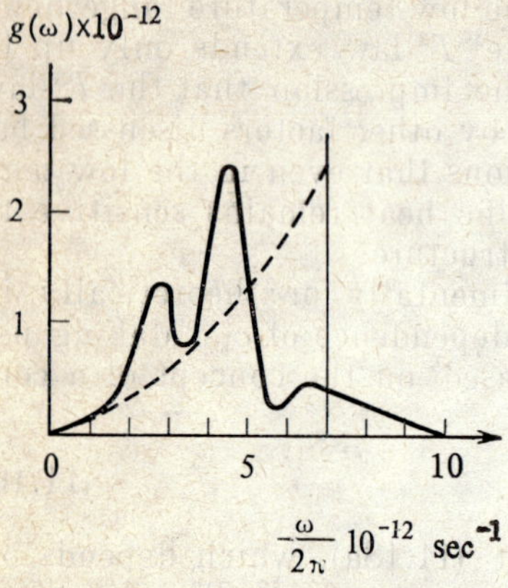


Fig. 3.17

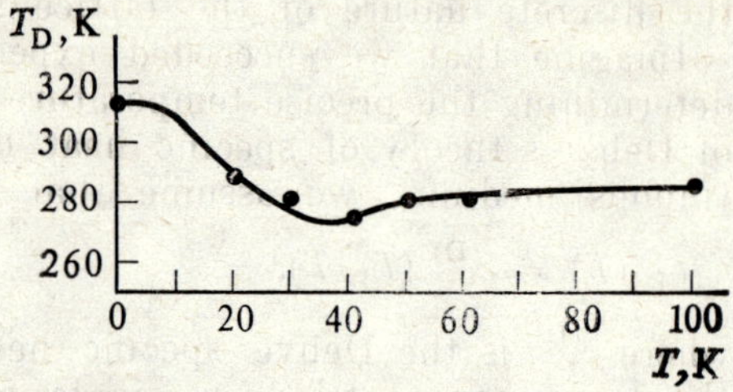


Fig. 3.18

ciably from the parabolic law (9.15) depicted in Fig. 3.17 by a dashed curve. It follows from Debye's equation (11.17) that the $T^3$-law is realized only when $T_D$ attains in the low-temperature range a constant value, i.e., as can be seen from Fig. 3.18, below 5-7 K.

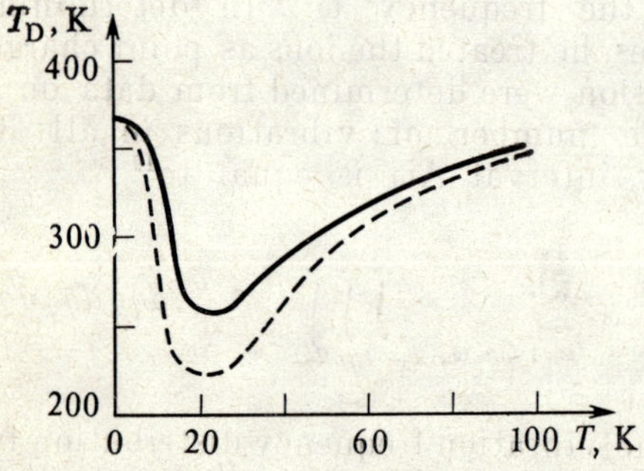


Fig. 3.19

On the other hand, a sloping minimum of $T_D$ is observed in the range 20-50 K close to which $T_D$ may be regarded as constant, and therefore the $T^3$-law may be expected to be valid.

---

[20] The circles in Fig. 3.18 are experimental data.

In his calculations of the heat of NaCl crystals, Kellermann employed empirical data on the compressibility of NaCl crystals. A more straightforward approach is the direct use of the frequency distribution function $g(\omega)$ obtained from experiments on neutron scattering in solids.

F. A. Johnson and W. Cochran [3.21] determined the function $g(\omega)$ for Ge from dispersion curves $\omega_j = \omega_j(\mathbf{q})$ obtained from neutron scattering experiments. The temperature dependence of the Debye temperature $T_D$ calculated by them (dashed curve in Fig. 3.19) can be compared with the same dependence obtained from the specific heat (solid curve). The coincidence, as we see, is not perfect, but still quite satisfactory. We can attribute the discrepancy to the inaccuracy in determining the frequency distribution function $g(\omega)$.

## 3.12 Equation of State for a Solid

**3.12.1** Let us deduce the equation of state, i.e., find the relation between the pressure $P$, the volume $V$, and the temperature $T$ of a solid. For the sake of simplicity, we shall consider a monatomic body in Debye's approximation.

The most direct method of obtaining the equation of state is to calculate the free energy of the system [3.17, pp. 53, 111]

$$\mathcal{F} = -k_0 T \ln Z, \tag{12.1}$$

where the partition function

$$Z = \sum_n e^{-\varepsilon_n/k_0 T} \Omega(\varepsilon_n). \tag{12.1a}$$

Here $\Omega(\varepsilon_n)$ is the statistical weight or the number of different states of the system with the same energy $\varepsilon_n$; in other words, $\Omega(\varepsilon_n)$ is the degree of degeneracy of the energy level $\varepsilon_n$. For a linear harmonic oscillator $\Omega(\varepsilon_n) = 1$, and $Z$ is given by (11.9). In the case of a continuous variation of energy $\varepsilon$, the partition function is replaced by the configurational integral.

It is known from thermodynamics [3.17] that

$$P = -\left(\frac{\partial \mathcal{F}}{\partial V}\right)_T, \tag{12.2}$$

where the derivative with respect to $V$ is at constant $T$.

The free energy of the oscillator at the frequency $\omega$ is

$$\mathcal{F}_{\text{osc}} = \frac{\hbar\omega}{2} + k_0 T \ln\left(1 - e^{-\hbar\omega/k_0 T}\right), \tag{12.3}$$

as stipulated by (12.1) and (11.9).

The free energy of a solid described by $3G^3 = 3N$ independent normal vibrations (oscillators) is equal to the sum of the free energies of the oscillators:

$$\mathcal{F} = \sum \mathcal{F}_{osc} = \sum_{\mathbf{q}j} \frac{\hbar\omega_{\mathbf{q}j}}{2} + k_0 T \sum_{\mathbf{q}j} \ln\left(1 - e^{-\hbar\omega_{\mathbf{q}j}/k_0 T}\right). \tag{12.4}$$

In Debye's approximation for a continuous medium,

$$\mathcal{F} = \mathcal{E}_0 + k_0 T \frac{3V}{2\pi^2 v_0^3} \int_0^{\omega_{max}} \ln\left(1 - e^{-\hbar\omega/k_0 T}\right) \omega^2 \, d\omega$$

$$= \mathcal{E}_0 + 9Nk_0 T \left(\frac{T}{T_D}\right)^3 \int_0^{T_D/T} \ln\left(1 - e^{-x}\right) x^2 \, dx \tag{12.4a}$$

if use is made of (9.15) and (11.14).

When calculating pressure with the aid of equation (12.2), we should differentiate (12.4a) with respect to the volume $V$ at constant temperature $T$. It can be demonstrated that the characteristic temperature $T_D$, or the maximum frequency $\omega_{max}$ proportional to it, depends in the anharmonic approximation on the volume $V$. To illustrate this point, we take the example of a simple one-dimensional lattice whose maximum vibration frequency, according to (2.5a), is $\omega_{max}^2 = (4/m)\,\beta$, where $\beta$ is the quasi-elastic force constant. According to (1.3), the force of atomic interaction in the anharmonic approximation is

$$F = -\beta x + \gamma x^2.$$

Here $x = R - R_0$ ($R_0$ is the equilibrium interatomic spacing), $\beta = \mathcal{U}''(R_0) > 0$ and $\gamma = -(1/2)\,\mathcal{U}'''(R_0) > 0$; $\mathcal{U}(R)$ is the potential energy of atomic interaction.

Uniform extension or compression of an atomic chain results in an additional uniform field $A = \text{const.}$ acting on every atom, so that the atom's energy in it is $AR$. Since the force in the new equilibrium position $R_0 + \Delta R_0$ is zero, it follows that $\mathcal{U}'(R_0 + \Delta R_0) + A = 0$. Expanding $\mathcal{U}'(R_0 + \Delta R_0)$ in a series in powers of $\Delta R_0$ and taking account of the fact that $\mathcal{U}'(R_0) = 0$, we obtain $\mathcal{U}''(R_0)\,\Delta R_0 + A = 0$, or $A = -\beta\Delta R_0$. On the other hand, the modified maximum frequency $\omega_{max} + \Delta\omega_{max}$ satisfies the relation

$$(\omega_{max} + \Delta\omega_{max})^2 = \frac{4}{m}\,\mathcal{U}''(R_0 + \Delta R_0)$$

(the second derivative of $R$ with respect to the energy $AR$ is zero). Expanding the right-hand side in powers of $\Delta R_0$, neglecting $(\Delta\omega_{max})^2$ on the left-hand side, and making use of the value of $\omega^2_{max}$, we obtain

$$\frac{\Delta\omega_{max}}{\omega_{max}} = -\frac{\gamma \Delta R_0}{\beta}.$$

We see that the maximum frequency, and therefore $T_D$, vary with the variation of the "volume" $\Delta R_0$ only in the anharmonic approximation ($\gamma \neq 0$). Since both $\beta$ and $\gamma$ are positive, an increase in the "volume" $\Delta R_0$ results in a decrease in the maximum frequency ($\Delta\omega_{max} < 0$), and vice versa.

Hence, the pressure is

$$P = -\left(\frac{\partial \mathcal{F}}{\partial V}\right)_T = -\frac{\partial \mathcal{E}_0}{\partial V} - 3Nk_0TD\left(\frac{T_D}{T}\right)\frac{1}{T_D}\frac{\partial T_D}{\partial V} \tag{12.5}$$

if we transform the integral in (12.4a) by parts and make use of the designation (11.15a).

Introduce the *Grüneisen constant*

$$\gamma_G \equiv -\frac{V}{T_D}\frac{dT_D}{dV} = -\frac{d\omega_{max}/\omega_{max}}{dV/V} = -\frac{d\ln\omega_{max}}{d\ln V} > 0, \tag{12.6}$$

which is independent of temperature. It follows from the definition (12.6) that it is directly connected with the anharmonism of atomic interaction. In the simple one-dimensional case,

$$\gamma_G = -\frac{\Delta\omega_{max}/\omega_{max}}{\Delta R_0/R_0} = \frac{\gamma R_0}{\beta}. \tag{12.6a}$$

In the harmonic approximation ($\gamma = 0$) the Grüneisen constant $\gamma_G = 0$. Denoting the temperature-dependent part of the internal energy by $\mathcal{E}_T = 3Nk_0TD\,(T_D/T)$, we obtain the *equation of state for a solid*

$$P = -\frac{\partial \mathcal{E}_0}{\partial V} + \frac{\gamma_G \mathcal{E}_T}{V}, \tag{12.7}$$

where the first term is independent of $T$.

**3.12.2** Deduce the so-called *Grüneisen relation* from the equation of state. Differentiating expression (12.7) with respect to $T$ and taking into account that $(\partial\mathcal{E}_T/\partial T)_V = c_V$, we obtain

$$\left(\frac{\partial P}{\partial T}\right)_V = \frac{\gamma_G c_V}{V}. \tag{12.8}$$

We make use of the thermodynamic identity [3.17, p. 117]

$$\left(\frac{\partial P}{\partial V}\right)_T\left(\frac{\partial V}{\partial T}\right)_P\left(\frac{\partial T}{\partial P}\right)_V = -1 \tag{12.9}$$

and write it in the form

$$\left(\frac{\partial V}{\partial T}\right)_P = -\left(\frac{\partial V}{\partial P}\right)_T\left(\frac{\partial P}{\partial T}\right)_V, \tag{12.9a}$$

which is possible because $(\partial P/\partial V)_T = 1 \div (\partial V/\partial P)_T$, etc.

Introducing the coefficient of linear expansion at constant pressure

$$\alpha = \frac{1}{3V}\left(\frac{\partial V}{\partial T}\right)_P \tag{12.10}$$

and isothermic compressibility

$$k = -\frac{1}{V}\left(\frac{\partial V}{\partial P}\right)_T, \tag{12.11}$$

we obtain from (12.9a), (12.10), (12.11), and (12.8) the Grüneisen relation

$$3V\alpha = \gamma_G k c_V. \tag{12.12}$$

Experiment proves that, indeed, the temperature dependence of the right-hand and the left-hand sides of this equality is similar if $\gamma_G$ is assumed to be independent of temperature. With the aid of experimental studies of the variations of the compressibility $k$ at high pressures $P$, the Grüneisen constant $\gamma_G$ can be determined independently, the result being in good agreement with $\gamma_G$ calculated via (12.12).

3.12.3 Consider a gas of equilibrium phonons in the approximation of a continuous medium (Debye). The number of phonons per frequency interval $[\omega, \omega + d\omega]$ is

$$dN_\omega = \bar{N} g(\varepsilon)\, d\omega = \frac{3V}{2\pi^2 v_0^3} \frac{\omega^2\, d\omega}{e^{\hbar\omega/k_0T} - 1}. \tag{12.13}$$

Here $\bar{N}$ is the average number of phonons per unit cell of the phase space (11.10a), and $g(\omega)$ is the frequency distribution function in the approximation of a continuous medium (9.15). The energy of these phonons

$$d\mathscr{E}_\omega = \hbar\omega\, dN_\omega = \frac{3V\hbar}{2\pi^2 v_0^3} \frac{\omega^3\, d\omega}{e^{\hbar\omega/k_0T} - 1}. \tag{12.14}$$

If we substitute in this expression the factors 2 and 3, to account for the fact that only two polarizations (transverse) are possible for a photon, and the velocity of light $c$ for the sound velocity $v_0$, we obtain the well-known Planck equation for the energy distribution in the radiation spectrum of blackbody. The total number of phonons in a volume $V$ is

$$N = \int dN_\omega = \frac{3V}{2\pi^2 v_0^3} \int_0^{\omega_{max}} \frac{\omega^2\, d\omega}{e^{\hbar\omega/k_0T} - 1} = \frac{3V}{2\pi^2 v_0^3} \left(\frac{k_0 T}{\hbar}\right)^3 \int_0^{T_D/T} \frac{x^2\, dx}{e^x - 1}, \tag{12.15}$$

where $T_D$ is the Debye temperature.

From (12.15) it follows that for $T \ll T_D$, when $\infty$ can, with an accuracy up to the order of $e^{-T_D/T}$, be substituted for the integral's upper limit, the number of phonons $N \propto VT^3$. For high temperatures, when $T \gg T_D$ $e^x - 1 \approx x$, the number of phonons $N \propto VT$. Figure 3.20 depicts the dependence of phonon concentration $N/V$ on $T$.

The total energy of the phonons in a volume $V$ is

$$\mathscr{E} = \int d\mathscr{E}_\omega = \frac{3V\hbar}{2\pi^2 v_0^3} \int\limits_0^{\omega_{max}} \frac{\omega^3 \, d\omega}{e^{\hbar\omega/k_0T} - 1}, \qquad (12.16)$$

which coincides with the energy of normal vibrations (11.12). Hence, the energy of normal lattice vibrations coincides with the energy of phonons distributed in accordance with the Bose-Einstein law.

In thermodynamics it is shown that

$$\mathscr{F} = -T \int \frac{\mathscr{E}(T)\, dT}{T^2}, \qquad (12.17)$$

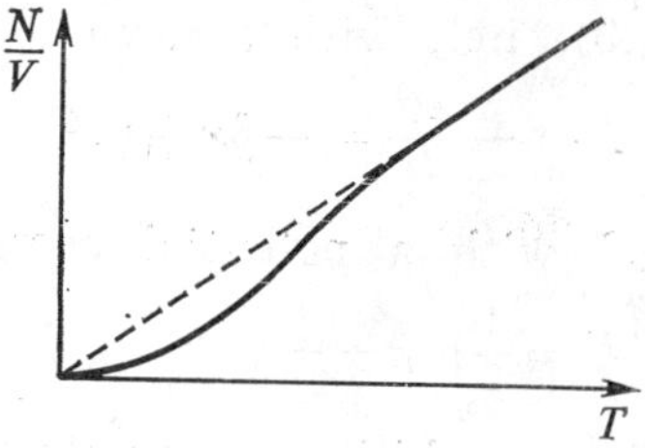

Fig. 3.20

where $\mathscr{E}(T)$ is the system's internal energy [3.17, p. 110]. The use of an indefinite integral in (12.17) is due to the presence of an indefinite additive constant in the expression for the free energy. If we substitute $\mathscr{E}(T)$ from (12.16) into (12.17) and perform integration with respect to $T$, we obtain free energy coinciding with $\mathscr{F} - \mathscr{E}_0 \equiv \mathscr{F}_T$ in (12.4a). We can regard $\mathscr{F}_T$ in (12.4a) as the free energy of a phonon gas, and equations (12.5) and (12.7) without the term $\partial\mathscr{E}_0/\partial V$ as the equations of state of a phonon gas. As we have indicated above, in the harmonic approximation the Grüneisen constant $\gamma_G = 0$; therefore, in this approximation the pressure of the phonon gas is zero.

In the anharmonic approximation the pressure of a phonon gas, according to expression (12.7), is

$$P_{phon} = \gamma_G \frac{\mathscr{E}_T}{V}. \qquad (12.18)$$

Analysis of the problem demonstrates that the momentum flux connected with a plane harmonic traveling wave with a definite quasi-momentum $\hbar\mathbf{q}$ is zero, i.e., a phonon has no momentum. This is the essential difference between the phonon and the photon which has a momentum $\hbar\mathbf{k}$ ($\mathbf{k}$ is the photon's wave vector).

### 3.13 Thermal Expansion and Heat Conductivity of Solids

We have united in one section the phenomena of thermal expansion and heat conductivity of solids because both are determined by the anharmonic part of the forces of atomic interaction. Both the thermal expansion and thermal resistance[21] (equal to $1/\varkappa$, where $\varkappa$ is the heat conductivity) vanish if the coefficient of anharmonicity is made equal to zero: $\gamma = 0$.

We shall consider thermal expansion for a simple model of two interacting atoms. This model not only enables the main points of the phenomenon to be clarified, but it also yields a correct order of magnitude for the expansion coefficient.

In the problem of heat conduction we shall limit ourselves to some general ideas and determine the heat conductivity in the high-temperature range on dimensional grounds.

**3.13.1** Consider two atoms that for small displacements from the equilibrium position, $x = R - R_0$, interact according to the law (1.3), i.e., with a force

$$F = -\frac{dU}{dx} = -\beta x + \gamma x^2 \tag{13.1}$$

and with a potential energy

$$U = \frac{1}{2}\beta x^2 - \frac{1}{3}\gamma x^3. \tag{13.1a}$$

The probability that an atom will be displaced from its equilibrium position by the distance $x$ is, according to Boltzmann, equal to

$$f(x) = A\exp\left(-\frac{U}{k_0T}\right) \approx Ae^{-\beta x^2/2k_0T}\left(1+\frac{\gamma x^3}{3k_0T}\right), \tag{13.2}$$

where the exponential corresponding to the anharmonic term has been expanded in a series:

$$\exp\left(\frac{\gamma x^3}{3k_0T}\right) \approx 1+\frac{\gamma x^3}{3k_0T}.$$

The constant $A$ in (13.2) is determined from the normalization condition

$$\int_{-\infty}^{+\infty} f(x)\,dx = A\int_{-\infty}^{+\infty} e^{-\beta x^2/2k_0T}\left(1+\frac{\gamma x^2}{3k_0T}\right)dx = 1.$$

The integral of the second addend, which is proportional to $\gamma$, is zero, because the integrand is an odd function; therefore (Appendix 7)

$$A = \left(\frac{\beta}{2\pi k_0T}\right)^{1/2}.$$

[21] Determined by phonon-phonon interaction.

The average displacement of an atom from its equilibrium position is

$$\bar{x} = \int_{-\infty}^{+\infty} x f(x)\, dx$$

$$= \left(\frac{\beta}{2\pi k_0 T}\right)^{1/2} \int_{-\infty}^{+\infty} e^{-\beta x^2/2k_0T} \left(x + \frac{\gamma x^4}{3k_0T}\right) dx = \frac{\gamma k_0 T}{\beta^2}, \quad (13.3)$$

where the integral of the first addend term, which contains the multiplicand $x$, is again zero because the integrand is an odd function. The integral of the second addend is easily calculated (Appendix 7).

By definition, the coefficient of linear expansion is the extension per unit length per kelvin; hence,

$$\alpha = \frac{\bar{x}}{aT} = \frac{\gamma k_0}{a\beta^2}, \quad (13.4)$$

where $a = R_0$ is the lattice constant. We see that the expansion coefficient is proportional to the coefficient of anharmonicity and vanishes for $\gamma = 0$.

By way of an example, consider a monovalent ionic crystal. Here

$$F = -\frac{e^2}{R^2} + \frac{B}{R^{10}}, \quad (13.5)$$

where $-e^2/R^2$ is the Coulomb attraction between neighboring oppositely charged undeformable ions, and $B/R^{10}$ is the repulsive force between those ions. This force increases rapidly as the distance between the ions $R$ is reduced, the proportionality of it to $R^{-10}$ being a good approximation of the exponential dependence obtained in quantum mechanics calculations. In equilibrium, $F = 0 = -e^2/a^2 + B/a^{10}$, where $a = R_0$ is the equilibrium spacing between the nearest ions. Hence, $B = e^2a^8$.

Since $R = a + x$, for small $x$

$$F = -\frac{e^2}{(a+x)^2} + \frac{e^2a^8}{(a+x)^{10}} \approx -\frac{8e^2}{a^3}x + \frac{52e^2}{a^4}x^2. \quad (13.6)$$

It follows from comparison with (13.1) that

$$\beta = 8e^2/a^3, \quad \gamma = 52e^2/a^4. \quad (13.6a)$$

Substituting the result (13.6a) into (13.4), we obtain

$$\alpha = 52ak_0/64e^2. \quad (13.6b)$$

For $a = 3 \times 10^{-8}$ cm, $k_0 = 1.38 \times 10^{-16}$ erg/K, and $e = 4.8 \times 10^{-10}$ esu, we obtain $\alpha = 1.5 \times 10^{-5}$ K$^{-1}$, which has a correct order of magnitude.

**3.13.2** Debye was the first to demonstrate (in 1914) that thermal resistance in a solid is due to the anharmonicity of atomic vibrations, and that if the vibrations are treated only in the harmonic approximation the thermal resistance will be zero. This statement appears obvious: for harmonic waves the principle of linear superposition is valid (according to this principle the waves propagate independently in the crystal without scattering on each other). In such a model the thermal resistance is zero because the heat flux propagates with the velocity of sound. Since there is a phonon with a quasi-momentum $\hbar \mathbf{q}$ and an energy $\hbar\omega_{\mathbf{q}}$ to correspond to a plane wave with a definite wave vector $\mathbf{q}$, we may say that in the harmonic approximation the phonons do not interact, i.e., they do not collide with one another. In general, the anharmonicity in a crystal lattice is described by terms in the expansion of the potential energy $\Phi(u)$ [see (5.2)] containing the atomic displacements $u^k_{n\alpha}$ in the third power.

It follows from the theory that if the anharmonic terms in the potential energy $\Phi(u)$ are treated as a small perturbation, there is a possibility of a simultaneous "collision" of three phonons, the results of such collision processes being either the transformation of two phonons into one or the decomposition of one phonon into two. Hence, in the process of collision the quasi-particles, phonons, are created or annihilated. A consistent theory of heat conductivity of crystals based on the kinetic equation for phonons was developed by R. Peierls (1929).

In this section we shall limit ourselves to the determination of the heat conductivity $\varkappa$ in the high-temperature range $T \gg T_D$ based on considerations of dimensionality. Debye's theory together with other more consistent theories lead to the conclusion that at high temperatures ($T \gg T_D$) the heat conductivity $\varkappa$ is inversely proportional to the absolute temperature $T$. On the other hand, the heat conductivity is proportional to the phonon's mean free path $l$, the latter being inversely proportional to the phonon scattering probability which, in its turn, is proportional to the square of the matrix element of the perturbation energy. Since the energy of an anharmonic perturbation is proportional to $\gamma$ and, accordingly, the square of the matrix element is proportional to $\gamma^2$, it follows that $\varkappa$ is inversely proportional to $\gamma^2$. Hence

$$\varkappa \propto \frac{1}{\gamma^2 T}. \tag{13.7}$$

At high temperatures, when the quantum effects are unessential, the heat conductivity $\varkappa$ may, according to Debye, depend, in addition, on the following quantities: the Boltzmann constant $k_0$, the lattice constant $a$, the crystal's atomic mass $M$, and sound velocity $v_0$.

Since $v_0 = \text{const.} \times a\sqrt{\beta/M}$, where $\beta$ is the quasi-elastic force constant, $\beta$ cannot simultaneously play the part of an independent parameter.

Hence, the problem is reduced to the determination of powers in the dimensionality equation

$$\varkappa = \text{const.} \times \frac{k_0^l a^m M^n v_0^p}{\gamma^2 T}, \tag{13.8}$$

where the constant is dimensionless.

Applying Bridgeman's Π-theorem [3.22] (or simply matching), we can demonstrate that for the dimensionalities on the right-hand and left-hand sides of (13.8) to be the same, we must make sure that

$$l = 0, \quad m = -8, \quad n = +3, \quad p = +7. \tag{13.8a}$$

Hence

$$\varkappa = \text{const.} \times \frac{M^3 v_0^7}{a^8 \gamma^2 T}. \tag{13.9}$$

Making use of (13.4) and of the relations between the quantities $\beta$ and $v_0$, and $a$ and $M$, we can exclude from equation (13.9) the unknown coefficient $\gamma^2$:

$$\varkappa = \text{const.} \times \frac{k_0^2}{a^2 M v_0 \alpha^2 T}, \tag{13.9a}$$

where the dimensionless constant is not the same as in (13.9).

Equation (13.9a) can be compared with experiment after the constant has been determined for some substance. The Debye temperature $T_D$ can be introduced into (13.9) and (13.9a) with the aid of (11.4) instead of the sound velocity $v_0$.

It should be kept in mind that the expressions (13.9) and (13.9a) are approximate, because the potential energy in this case is determined only by two force constants $\beta$ and $\gamma$ (or $v_0$ and $\gamma$). However, within the scope of this model, equations (13.9) and (13.9a) are precise, and it would be useless to try to obtain a more precise expression for the heat conductivity with the same set of parameters determining the properties of a crystal.

# 4. Electrons in an Ideal Crystal

## 4.1 General Formulation of the Problem. The Adiabatic Approximation

**4.1.1** Any solid of macroscopic dimensions is a collection of an enormous number of atomic nuclei and electrons. Normally the atomic nuclei of an element are a natural mixture of its isotopes, however this fact is of little importance for most of the solid's properties. An important peculiarity of solids established empirically is the fact that its atomic nuclei occupy more or less fixed positions in space. In an ideal crystal these positions form a three-dimensional periodic lattice. In a neutral solid the total positive charge of the nuclei is equal to the magnitude of the combined charge of all its electrons.

In some cases, especially when crystal boundaries (surfaces), where the value of the potential energy of conduction electrons is not the same as in the body, are being studied, the need arises for the consideration of charged regions of a body, in which the space charge and the electric potential are self-consistent with each other.

If we ignore such processes in a solid that involve nuclear transmutations and in which relativistic effects and electron and nuclear spins become important[1], then the stationary states of a system are described by the Schrödinger equation

$$\hat{\mathscr{H}}\Psi = W\Psi \tag{1.1}$$

with the Hamiltonian

$$\hat{\mathscr{H}} = -\frac{\hbar^2}{2m}\sum_i \nabla^2_{\mathbf{r}_i} - \frac{\hbar^2}{2}\sum_J \frac{1}{M_J}\nabla^2_{\mathbf{R}_J} + V(\mathbf{r}, R), \tag{1.1a}$$

where the Coulomb energy of interaction of the electrons and the nuclei is

$$V(\mathbf{r}, R) = \sum_{J<K} \frac{Z_J Z_K e^2}{R_{JK}} + \sum_{i<k} \frac{e^2}{r_{ik}} - \sum_{iJ} \frac{Z_J e^2}{r_{iJ}}. \tag{1.1b}$$

[1] It is an established fact that some phenomena in crystals (spin-orbital splitting of the electron spectrum, paramagnetic and nuclear resonance, etc.) essentially depend on the electron and nuclear spins.

Here $m$ is the electron mass, $M_J$ is the mass of the $J$-th nucleus, $\mathbf{r}_i$ and $\mathbf{R}_J$ are position vectors of the $i$-th electron and the $J$-th nucleus, $R_{JK}$, $r_{ik}$ and $r_{iJ}$ are the spacings between the corresponding nuclei and electrons, $Z_J$ is the atomic number of the $J$-th nucleus.

The eigenvalues spectrum of $\hat{\mathcal{H}}$ is generally of a complex type. Since the behaviour of a physical system is completely determined by its wave function, the solution of equation (1.1) would in principle answer to all questions pertaining to the solid's properties, specifically, to the questions: why does the collection of particular atoms and electrons constitute a crystal lattice of some sort or another, what are the thermal, the electric, the magnetic and the optical properties of the particular solid, etc.

Since a macroscopic specimen of a solid contains about $10^{23}$ particles, and for this reason the wave function of equation (1.1) depends on an equal number of variables, it is, of course, practically impossible to obtain (or even to write down) the solution of equation (1.1). Moreover, even if we were to find a method for writing down (specifying) a wave function of such a number of variables, it would be of no avail, since its application for the calculation of quantities observed in experiment would have met with difficulties that would have to be acknowledged as principal.

The task of a physical theory lies specifically in the establishment of reasonable approximations that would enable the quantities observed in the experiment to be interpreted and calculated.

**4.1.2** Let us try to simplify equation (1.1) making use of the fact that the electron mass $m$ is much less than the nuclear masses $M_J$ (adiabatic approximation). In considering the motion of the electrons we shall at first assume the heavy nuclei to be static, and this will enable us to neglect in the Hamiltonian (1.1) the addend $\left(-\frac{\hbar^2}{2}\right)\sum\limits_J \frac{1}{M_J}\nabla^2_{\mathbf{R}_J}$ related to the kinetic energy of the nuclei. The wave function of the electrons $\varphi$ moving in the field of static nuclei satisfies the equation

$$\left\{-\frac{\hbar^2}{2m}\sum_i \nabla^2_{\mathbf{r}_i}+V(r, R)\right\}\varphi=\mathcal{E}\varphi. \tag{1.2}$$

Now $\mathbf{R}_J$ are no longer variables of a differential equation, but parameters of the potential field of the nuclei. The eigenfunction and the eigenvalues of equation (1.2) will obviously depend on the $\mathbf{R}_J$ as on parameters: $\varphi$ ($\mathbf{r}$, $R_J$) and $\mathcal{E}$ ($R_J$).[2]

---

[2] From the appearance of $V$ ($r$, $R$) it will be clear that $\mathcal{E}$ ($R$) also includes the Coulomb energy of nuclear interaction.

Try to represent the full wave function of the system of electrons and nuclei in the form

$$\Psi(\mathbf{r}, R) = \Phi(R)\,\varphi(\mathbf{r}, R). \tag{1.3}$$

As we shall learn below, such a representation of the wave function $\Psi$ is an approximation, since, strictly speaking, the ratio $\frac{\Psi(\mathbf{r}, R)}{\varphi(\mathbf{r}, R)}$ to some extent depends on $\mathbf{r}_i$. The function $\Phi(R)$ will turn out to be a wave function that describes the motion of the nuclei in an average field set up by the electrons. Substituting expression (1.3) into equation (1.1) we obtain, taking into account (1.2),

$$-\frac{\hbar^2}{2}\sum_J \frac{1}{M_J}\left[\varphi\nabla^2_{\mathbf{R}_J}\Phi + 2\left(\nabla_{\mathbf{R}_J}\varphi\nabla_{\mathbf{R}_J}\Phi\right) + \Phi\nabla^2_{\mathbf{R}_J}\varphi\right]$$
$$+\mathscr{E}(R)\,\varphi\Phi = W\varphi\Phi. \tag{1.4}$$

Indeed, $\Phi$ is independent of the $\mathbf{r}_i$ and therefore, when the electron kinetic energy operator plus $V$ acts on $\Phi\varphi$, this immediately yields the term $\mathscr{E}\varphi\Phi$ in accordance with (1.2).

On the other hand, when the operator $\nabla^2_{\mathbf{R}_J}$ acts on the product $\Phi\varphi$ in which every multiplicand depends on $\mathbf{R}_J$, this immediately yields

$$\frac{\partial^2}{\partial X^2}(\Phi, \varphi) = \varphi\frac{\partial^2\Phi}{\partial X^2} + 2\frac{\partial\varphi}{\partial X}\frac{\partial\Phi}{\partial X} + \Phi\frac{\partial^2\varphi}{\partial X^2}$$

where $X$ is an orthogonal coordinate of $\mathbf{R}_J$.

For all three projections of $\mathbf{R}_J$ we obtain the expression in the square brackets of equation (1.4).

Without loss of generality the electron wave function $\varphi$ may be assumed to be real (we exclude the presence of macroscopic currents in the crystal); in that case the normalizing condition takes the orm

$$\int \varphi^2\, d\tau = 1, \tag{1.5}$$

where the integration is performed with respect to all electron coordinates over the entire volume of the crystal.

It follows from condition (1.5) that

$$\nabla_{\mathbf{R}_J}\int \varphi^2\, d\tau = 2\int \varphi\nabla_{\mathbf{R}_J}\varphi\, d\tau = 0. \tag{1.5a}$$

Multiply equation (1.4) by $\varphi$ and integrate with respect to $\tau$. Making use of (1.5) and (1.5a), we obtain

$$-\frac{\hbar^2}{2}\sum_J \frac{1}{M_J}\nabla^2_{\mathbf{R}_J}\Phi + \left[\mathscr{E}(R) - \frac{\hbar^2}{2}\sum_J \frac{1}{M_J}\int \varphi\nabla^2_{\mathbf{R}_J}\varphi\, d\tau\right]\Phi = W\Phi. \tag{1.6}$$

A more accurate analysis shows that $\sum_J$ in the square brackets of the latter expression can be neglected. The reason is that because of the small values of the ratios $m/M_J$, the wave function $\varphi(\mathbf{r}, R)$ is little dependent on the $\mathbf{R}_J$, so that the terms containing $\nabla^2_{\mathbf{R}_J}\varphi$ can be neglected. It can be demonstrated that the inclusion of these terms in (1.6) results in corrections of the order of $(m/M_J)^{1/4}$ to the values of various physical quantities.

Hence, instead of (1.6), we obtain

$$\left[-\frac{\hbar^2}{2}\sum_J \frac{1}{M_J}\nabla^2_{\mathbf{R}_J} + \mathscr{E}(R)\right]\Phi = W\Phi. \qquad (1.7)$$

We see that the function $\Phi$, determined by the differential equation (1.7) does, indeed, depend on the variables $\mathbf{R}_J$. It follows from (1.7) that $\Phi(R)$ is the wave function of the nuclei moving in a field with the potential energy $\mathscr{E}(R)$ equal to the energy eigenvalue of the electron system for a specific configuration of the nuclei and for a specific value of the Coulomb energy of the nuclei interaction.

We see that in the adiabatic approximation the precise quantum mechanics problem of the behaviour of a system of electrons and nuclei disintegrates into two simpler problems: (1) the problem of the motion of the electrons in the field of static nuclei (1.2) and (2) the problem of the motion of nuclei in an averaged field $\mathscr{E}(R)$ set up by the electrons.[3] It should be kept in mind that in some phenomena, for instance in multiphonon electron transitions in impurity centers, the second addend in the square brackets of equation (1.6), the so-called nonadiabatic term, plays a dominant part, being the cause of such transitions.

## 4.2 The Hartree-Fock Method

**4.2.1** In the preceding paragraph we saw, how the general quantum mechanics problem of the motion of electrons and nuclei in a crystal can be simplified on the basis of the so-called adiabatic approximation. The adiabatic approximation based on the smallness of the ratio $m/M_J$ makes it possible to reduce the problem of the behaviour of an electron-nuclei system of a solid to the problem of electron motion in the field of static nuclei. However, in this case, too, the problem of the motion of the collection of all electrons in a crystal remains extremely intricate and requires for its solution some or other approximate methods. One of such extremely effective methods that won prominence in the electron theory of crystals

---

[3] As was pointed out above $\mathscr{E}(R)$ also includes the Coulomb energy of interaction of the nuclei.

is the *Hartree-Fock method* of reducing the multielectron problem to that for a single electron.

Consider a system of $N$ interacting electrons in the field of static nuclei. The Schrödinger equation for stationary states of this system is of the form

$$\hat{\mathcal{H}}\psi(\mathbf{r}_1, \mathbf{r}_2, \ldots, \mathbf{r}_N) = \mathcal{E}\psi(\mathbf{r}_1, \mathbf{r}_2, \ldots, \mathbf{r}_N) \tag{2.1}$$

with the Hamiltonian

$$\hat{\mathcal{H}} = \sum_{i}^{1, N} \hat{\mathcal{H}}_i + \frac{1}{2} {\sum_{i, j}^{1, N}}' \frac{e^2}{r_{i, j}}, \tag{2.1a}$$

where

$$\hat{\mathcal{H}}_i = -\frac{\hbar^2}{2m} \nabla_i^2 + V(\mathbf{r}_i). \tag{2.1b}$$

Here $\nabla_i^2 = \frac{\partial^2}{\partial x_i^2} + \frac{\partial^2}{\partial y_i^2} + \frac{\partial^2}{\partial z_i^2}$; $V(\mathbf{r}_i)$ is the potential energy of the $i$-th electron in the field of the nuclei, the prime after the summation sign in expression (2.1a) indicating that the terms with $i=j$ should be omitted from the potential energy of the Coulomb interaction of the electrons.

The Hartree-Fock method is based on the idea of the substitution in the Hamiltonian $\hat{\mathcal{H}}$ of some effective external field $\mathcal{U}_{\text{eff}}(\mathbf{r})$ in which every electron moves independently for the potential energy. The field $\mathcal{U}_{\text{eff}}$ should provide the best description of the *averaged* action of all electrons on the specified electron. The Hamiltonian of the system will now be equal to a sum of Hamiltonians, everyone of which depends only on the coordinates of a single electron, i.e.,

$$\hat{\mathcal{H}} = \sum_{i=1}^{N} \hat{\mathcal{H}}'_i, \tag{2.2}$$

where

$$\hat{\mathcal{H}}'_i = -\frac{\hbar^2}{2m} \nabla_i^2 + V(\mathbf{r}_i) + \mathcal{U}_{\text{eff}}(\mathbf{r}_i). \tag{2.2a}$$

It will be demonstrated below, how the field $\mathcal{U}_{\text{eff}}$ should best be chosen. It may easily be shown that the solution of equation (2.1) with the Hamiltonian (2.2) is

$$\psi(\mathbf{r}_1, \mathbf{r}_2, \ldots, \mathbf{r}_N) = \psi_{n_1}(\mathbf{r}_1)\ \psi_{n_2}(\mathbf{r}_2) \ldots \psi_{n_N}(\mathbf{r}_N). \tag{2.3}$$

The index $n_i$ of the function $\psi_{n_i}$ means three quantum numbers characterizing the quantum state of the $i$-th electron as determined by the equation

$$\hat{\mathcal{H}}'_i \psi_{n_i}(\mathbf{r}_i) = \mathcal{E}_{n_i} \psi_{n_i}(\mathbf{r}_i), \tag{2.3a}$$

where $\mathscr{E}_{n_i}$ is the corresponding energy eigenvalue. The total energy of the system in the approximation (2.2) is equal to

$$\mathscr{E} = \sum_{i=1}^{N} \mathscr{E}_{n_i}. \tag{2.3b}$$

Indeed, substitute expression (2.3) into equation (2.1) with $\hat{\mathscr{H}}$ being equal to the sum (2.2). We have

$$\sum_i \hat{\mathscr{H}}'_i [\psi_{n_1}(\mathbf{r}_1) \ldots \psi_{n_i}(\mathbf{r}_i) \ldots \psi_{n_N}(\mathbf{r}_N)] = \mathscr{E} [\psi_{n_1}(\mathbf{r}_1) \ldots \psi_{n_N}(\mathbf{r}_N)]$$

or

$$\sum_i [\psi_{n_1}(\mathbf{r}_1) \ldots \psi_{n_{i-1}}(\mathbf{r}_{i-1}) \psi_{n_{i+1}}(\mathbf{r}_{i+1}) \ldots \psi_{n_N}(\mathbf{r}_N)] \hat{\mathscr{H}}'_i \psi_{n_i}(\mathbf{r}_i)$$
$$= \mathscr{E} [\psi_{n_1}(\mathbf{r}_1) \ldots \psi_{n_N}(\mathbf{r}_N)],$$

since $\hat{\mathscr{H}}'_i$ acts only on the coordinates of the $i$-th electron. Making use of equation (2.3a), we get

$$\sum_i [\psi_{n_1}(\mathbf{r}_1) \ldots \psi_{n_N}(\mathbf{r}_N)] \mathscr{E}_{n_i} = \mathscr{E} [\psi_{n_1}(\mathbf{r}_1) \ldots \psi_{n_N}(\mathbf{r}_N)].$$

Canceling out $[\psi_{n_1}(\mathbf{r}_1) \ldots \psi_{n_N}(\mathbf{r}_N)]$ in both sides of the equality, we obtain expression (2.3b).

Since $|\psi_{n_i}(\mathbf{r}_i)^2|$ is the probability density of the $i$-th electron located in space at point $\mathbf{r}_i$, the multiplicative nature of the solution (2.3) from the point of view of the probabilities multiplication theorem characterizes the independence of motion of noninteracting electrons in an external field $V + \mathscr{U}_{\text{eff}}$.

**4.2.2** Now let us set ourselves the task of finding the best method of determining $\mathscr{U}_{\text{eff}}(\mathbf{r})$ in the approximate Hamiltonian (2.2a). As we shall see, this can be done on the basis of a certain self-consistent procedure.

To make the following calculations less voluminous let us perform them for a system of $N = 2$ electrons, subsequently generalizing the results to embrace an arbitrary number of electrons $N$.

For two electrons expression (2.3) is of the form

$$\psi(\mathbf{r}_1, \mathbf{r}_2) = \psi_{n_1}(\mathbf{r}_1) \psi_{n_2}(\mathbf{r}_2).$$

Now we must take into account that the state of the electron in addition to the three space coordinates $x$, $y$, $z$ ($\equiv \mathbf{r}$) is characterized by the value of the projection of its own (internal) momentum (spin) $S_z$ on a specified direction (e.g., the $z$ axis). Theory and experiment prove that for an electron $S_z$ assumes only two values $+\hbar/2$ and $-\hbar/2$, if we put $S_z = s\hbar$, the spin coordinate will be either $s = +1/2$ or $s = -1/2$. Accordingly, we shall introduce spin functions $\nu_i(s)$ $(i = 1, 2)$, where the index $i$ describes the spin state: $i = 1$ for $S_z = \hbar/2$ and $i = 2$ for $S_z = -\hbar/2$, so that

$$\nu_1(1/2) = 1, \ \nu_1(-1/2) = 0, \quad \nu_2(1/2) = 0, \ \nu_2(-1/2) = 1. \quad (2.4)$$

With such a definition the spin functions are orthonormal, i.e.,

$$\sum_{s=\pm 1/2} \nu_i^*(s)\,\nu_k(s) = \delta_{ik}. \quad (2.5)$$

If the interaction of the electron magnetic moment (connected with the spin) with the magnetic field set up by its orbital motion is ignored, the full single-electron wave function of the $l$-th electron in the $(n_i, k) \equiv j$-th quantum state is equal to

$$\psi_{n_i}(\mathbf{r}_l)\,\nu_k(s_l) = \varphi_j(l). \quad (2.6)$$

The argument $l$ of the function $\varphi_j$ denotes a set of four coordinates of the $l$-th electron, three space coordinates $x_l$, $y_l$, $z_l$ and the spin coordinate $s_l$. Presuming the wave functions $\psi_{n_i}$ to be orthonormal we obtain

$$\int \varphi_j^*(l)\,\varphi_{j'}(l)\,d\tau_l = \int d\mathbf{r}_l \sum_s \psi_{n_i}^*(\mathbf{r}_l)\,\psi_{n_i'}(\mathbf{r}_l)\,\nu_k^*(s)\,\nu_{k'}(s)$$

$$= \delta_{n_i n_i'}\delta_{kk'} = \delta_{jj'}. \quad (2.7)$$

Here the symbol $\int d\tau_l$ denotes integration with respect to the space coordinates $x_l$, $y_l$, $z_l$, and summation over the spin coordinate $s$.

According to the *Pauli principle*, the full wave function for an electron system should always be *antisymmetric*, i.e., it should change sign as a result of two electrons changing places (of transmutation of their four coordinates). For two electrons ($N = 2$) such a wave function is of the form

$$\Phi(1, 2) = \frac{1}{\sqrt{2}}\{\varphi_1(1)\,\varphi_2(2) - \varphi_1(2)\,\varphi_2(1)\} = \frac{1}{\sqrt{2!}}\begin{vmatrix} \varphi_1(1) & \varphi_1(2) \\ \varphi_2(1) & \varphi_2(2) \end{vmatrix}. \quad (2.8)$$

Indeed, $\Phi(2, 1) = -\Phi(1, 2)$. The factor $1/\sqrt{2}$ was introduced to normalize the system's wave function. Making use of relation (2.7), we can easily demonstrate that

$$\int \Phi^*(1, 2)\,\Phi(1, 2)\,d\tau_1\,d\tau_2 = 1. \quad (2.8a)$$

The antisymmetrical form of expression (2.8) automatically satisfies the Pauli principle, according to which there can be no more than one electron in each quantum state. Indeed, if $\varphi_1 = \varphi_2$, then $\Phi = 0$.

Generalizing the determinant (2.8) for the case of $N$ electrons we obtain a correct antisymmetric wave function for a system of $N$ electrons:

$$\Phi(1, 2, 3, \ldots, N) = \frac{1}{\sqrt{N!}} \begin{vmatrix} \varphi_1(1) & \varphi_1(2) & \ldots & \varphi_1(N) \\ \varphi_2(1) & \varphi_2(2) & \ldots & \varphi_2(N) \\ \ldots & \ldots & \ldots & \ldots \\ \varphi_N(1) & \varphi_N(2) & \ldots & \varphi_N(N) \end{vmatrix}, \quad (2.8b)$$

which satisfies the Pauli principle.

If we were to exchange in this expression the coordinates of two electrons, for example $1 \rightleftarrows 2$, it would be equivalent to the transposition of two columns of the determinant resulting in its changing sign, i.e., $\Phi(1, 2, 3, \ldots, N) = -\Phi(2, 1, 3, \ldots, N)$. If two quantum states coincide, for example $\varphi_1 = \varphi_2$, two rows of the determinant will be equal, and in this case $\Phi = 0$. The factor $1/\sqrt{N!}$ normalizes the function $\Phi(1, 2, \ldots, N)$, if condition (2.7) is valid.

Making use of the precise Hamiltonian of the system (2.1a) and wave function (2.8b) calculate the total energy of the system (Appendix 8, Section 1):

$$\begin{aligned} \mathscr{E} &= \int \Phi^* \hat{\mathscr{H}} \Phi \, d\tau_1 \, d\tau_2 \ldots d\tau_N \\ &= \sum_{i}^{1, N} \int \varphi_i^*(1) \hat{\mathscr{H}}_1 \varphi_i(1) \, d\tau_1 + \frac{1}{2} \sum_{i, j}^{1, N}{}' \int |\varphi_i(1)|^2 \frac{e^2}{r_{12}} |\varphi_j(2)|^2 \, d\tau_1 \, d\tau_2 \\ &\quad - \frac{1}{2} \sum_{i, j}^{1, N}{}' \int \varphi_i^*(1) \varphi_j(1) \frac{e^2}{r_{12}} \varphi_i(2) \varphi_j^*(2) \, d\tau_1 \, d\tau_2. \end{aligned} \quad (2.9)$$

Note that the primes after the second and the third sums may be dropped, since the terms with $i = j$ cancel out.

Since the operator $\hat{\mathscr{H}}_1$ and $e^2/r_{12}$ do not depend on the spin variable $s_1$, the summation over it can be performed independently from the integration with respect to the space coordinates. In cases, when summation over $s_1$ is performed for a pair of functions $\varphi_i$ with equal indices, the result, according to condition (2.5), will obviously be unity. For the integrals in the last sum of expression (2.9) the result is unity, only if summation is performed over electrons with parallel spins, otherwise it is zero.

Hence, expression (2.9) can be written down in the form

$$\mathscr{E} = \sum_{i}^{1,\,N} \int \psi_{n_i}^{*}(\mathbf{r}_1)\, \hat{\mathscr{H}}_1 \psi_{n_i}(\mathbf{r}_1)\, d\mathbf{r}_1$$

$$+\frac{1}{2} \sum_{i,\,j}^{1,\,N} \int |\psi_{n_i}(\mathbf{r}_1)|^2 \frac{e^2}{r_{12}} |\psi_{n_j}(\mathbf{r}_2)|^2\, d\mathbf{r}_1\, d\mathbf{r}_2$$

$$-\frac{1}{2} \sum_{i,\,j}^{1,\,N} \int \psi_{n_i}^{*}(\mathbf{r}_1)\, \psi_{n_j}(\mathbf{r}_1) \frac{e^2}{r_{12}} \psi_{n_i}(\mathbf{r}_2)\, \psi_{n_j}^{*}(\mathbf{r}_2)\, d\mathbf{r}_1\, d\mathbf{r}_2, \tag{2.9a}$$

where in the last sum the summation is performed over electron states with parallel spins.

Since the electron functions $\psi_{n_i}$ are not eigenfunctions of the Hamiltonian $\hat{\mathscr{H}}_1$, the first sum is not the sum of energies of noninteracting electrons. The integrals under the second summation sign (with a plus) are termed *Coulomb* integrals—they are equal to the potential energy of interaction of two charges distributed with the densities: $-e\,|\psi_{n_i}(\mathbf{r}_1)|^2$ and $-e\,|\psi_{n_j}(\mathbf{r}_2)|^2$. The integrals under the last summation sign are termed *exchange* integrals—they have no classical electrostatic counterpart, the corresponding interaction for them being a Coulomb interaction with "complex charge densities":

$$e\psi_{n_i}^{*}(\mathbf{r}_1)\, \psi_{n_j}(\mathbf{r}_1) \text{ and } -e\psi_{n_j}^{*}(\mathbf{r}_2)\, \psi_{n_i}(\mathbf{r}_2).$$

Both electrons with parallel spins exist partly in the $n_i$-th state, and partly in the $n_j$-th state, they sort of "change places".

One question remains unanswered: How should the single-electron wave functions $\psi_{n_i}$ be chosen? To determine the best single-electron wave functions $\psi_{n_i}(\mathbf{r})$, the system energy $\mathscr{E}$ must be minimum with respect to small variations of the functions $\psi_{n_i} \to \psi_{n_i} + \delta\psi_{n_i}$. Since the functions $\psi_{n_i}$ are complex (i.e. equivalent to two real functions each), $\psi_{n_i}$ and $\psi_{n_i}^{*}$ should be varied independently. However, the equations obtained as a result of variation $\psi_{n_i}^{*}$ are complex conjugate to those obtained as a result of variation of $\psi_{n_i}$, therefore we shall limit ourselves to the variations of one of the functions, i.e. $\psi_{n_i}^{*} \to \psi_{n_i}^{*} + \delta\psi_{n_i}^{*}$. We shall require the orthonormality conditions to be satisfied for the varied functions, as well:

$$\int (\psi_{n_i}^{*} + \delta\psi_{n_i}^{*})\, \psi_{n_j}^{*}\, d\mathbf{r} = \delta_{n_i n_j}. \tag{2.10}$$

Taking into account condition (2.7), we obtain

$$\int \delta\psi_{n_i}^{*} \psi_{n_j}\, d\mathbf{r} = 0 \tag{2.10a}$$

for all $i$ and $j$.

Varying the functions $\psi^*_{n_i}$ on the right-hand side of equation (2.9a), calculate the corresponding variation of the system energy $\delta\mathscr{E}$. Equating $\delta\mathscr{E}$ to zero with the additional condition (2.10a), which is accounted for by the method of indeterminate Lagrange multipliers, we obtain the following equation for the determination of the function $\psi_{n_i}$ (Appendix 8, Sec. 2):

$$\hat{\mathscr{H}}_1\psi_{n_i}(\mathbf{r}_1)+\Big[\sum_j{}' \int \frac{e^2|\psi_{n_j}(\mathbf{r}_2)|^2}{r_{12}}\,d\mathbf{r}_2\Big]\psi_{n_i}(\mathbf{r}_1)$$
$$-\sum_j{}' \int \frac{e^2\psi_{n_i}(\mathbf{r}_2)\,\psi^*_{n_j}(\mathbf{r}_2)}{r_{12}}\,d\mathbf{r}_2\psi_{n_j}(\mathbf{r}_1)-\mathscr{E}_{n_i}\psi_{n_i}(\mathbf{r}_1)=0 \quad (i=1,\ 2,\ \ldots). \tag{2.11}$$

Here $\mathscr{E}_{n_i}$ is an indeterminate Lagrange multiplier that plays the part of an energy eigenvalue of a single-electron state. The second sum in expression (2.11) can be formally represented in the form of an operator operating on $\psi_{n_i}$, and equation (2.11) can be written as follows

$$\Big[\hat{\mathscr{H}}_1+\sum_j{}' \int \frac{e^2|\psi_{n_j}(\mathbf{r}_2)|^2}{r_{12}}\,d\mathbf{r}_2$$
$$-\sum_j{}' \frac{\psi_{n_j}(\mathbf{r}_1)}{\psi_{n_i}(\mathbf{r}_1)}\int \frac{e^2\psi_{n_i}(\mathbf{r}_2)\,\psi^*_{n_j}(\mathbf{r}_2)\,d\mathbf{r}_2}{r_{12}}\Big]\psi_{n_i}(\mathbf{r}_1)=\mathscr{E}_{n_i}\psi_{n_i}(\mathbf{r}_1). \tag{2.11a}$$

Comparing this expression with (2.3a) we see that

$$\mathscr{U}_{\text{eff}}(\mathbf{r}_1)=\sum_j{}' \int \frac{e^2|\psi_{n_j}(\mathbf{r}_2)|^2}{r_{12}}\,d\mathbf{r}_2$$
$$-\sum_j{}' \frac{\psi_{n_j}(\mathbf{r}_1)}{\psi_{n_i}(\mathbf{r}_1)}\int \frac{e^2\psi_{n_i}(\mathbf{r}_2)\,\psi_{n_j}(\mathbf{r}_2)}{r_{12}}\,d\mathbf{r}_2, \tag{2.11b}$$

where in the second sum the summation is performed over the electrons having spins parallel to that of the electron in the $n_i$ state.

The term for equations (2.11a) is *Hartree-Fock self-consistent field equations.* Hartree wrote the system's wave function in the form (2.3) which took no account of the Pauli principle; he formulated equation (2.11a) in a form in which the second sum in the square brackets was absent, thus neglecting the energy of exchange interaction. In calculations of electron cloud densities of multielectron atoms this leads to errors of about 20%. A rigorous theory of self-consistent field of multielectron systems was developed by Fock (1930).

Since $\mathcal{U}_{\text{eff}}(\mathbf{r}_1)$ itself depends on unknown functions $\psi_{n_i}$, equation (2.11a) represents a system of nonlinear integro-differential equations for the functions $\psi_{n_i}$. We can imagine the following procedure leading to their solution. Specify in the zero approximation some single-electron wave functions $\psi_{n_i}$ and using these functions calculate the effective field $\mathcal{U}_{\text{eff}}$. It will then be possible to determine from the system of equations (2.11a) the functions $\psi_{n_i}$ in the next approximation. Having determined with the aid of the new $\psi_n$ the field $\mathcal{U}_{\text{eff}}$ in the first approximation, we can solve equations (2.11a) and determine $\psi_{n_i}$ in the second approximation, etc. Should the functions substituted into $\mathcal{U}_{\text{eff}}$ coincide with those obtained from solving (2.11a), the task of finding self-consistent solutions of the Hartree-Fock equations could be regarded as completed. Of course, in practice such an operation presents considerable difficulties.

## 4.3 Electron in a Periodic Field

**4.3.1** As has been demonstrated theoretically and experimentally, a multielectron problem for a crystal can with adequate accuracy in many cases be considered a single-electron one. In other words, the electrons in a crystal can be to a good approximation described by the Hartree-Fock equations. How should the effective field $\mathcal{U}_{\text{eff}}(\mathbf{r})$ in (2.2a) be chosen in this case?

The crystal's symmetry suggests that $\mathcal{U}_{\text{eff}}(\mathbf{r})$ should have the periodicity of the crystal.

In Section 2.9.2 it was demonstrated that the wave function of an electron in a periodic field is of the form (2.9.13)

$$\psi_{\mathbf{k}}(\mathbf{r}) = u_{\mathbf{k}}(\mathbf{r})\, e^{i\mathbf{k}\cdot\mathbf{r}}, \tag{3.1}$$

where $\mathbf{k}$ is the wave vector of the electron, and the amplitude function

$$u_{\mathbf{k}}(\mathbf{r} + \mathbf{a}_n) = u_{\mathbf{k}}(\mathbf{r}), \tag{3.1a}$$

i.e., displays the periodicity of the crystal lattice. It can be demonstrated (Appendix 9) that the Bloch wave functions (3.1) substituted into expression (2.11b) do, indeed, lead to an effective field $\mathcal{U}_{\text{eff}}(\mathbf{r})$ having the periodicity of the lattice, i.e. that the solution (3.1) is self-consistent.

The electron wave vector can be represented in the form (2.9.8):

$$\mathbf{k} = \frac{g_1}{G}\mathbf{b}_1 + \frac{g_2}{G}\mathbf{b}_2 + \frac{g_3}{G}\mathbf{b}_3 \quad (g_i = 0,\ 1,\ 2,\ \ldots,\ G-1), \tag{3.2}$$

where $G$ is a large (odd) number, and $\mathbf{b}_1$, $\mathbf{b}_2$ and $\mathbf{b}_3$ are the basis vectors of the reciprocal lattice. Hence, $\mathbf{k}$ assumes $G^3$ quasi-discrete values.

We shall arrive at the same expression for $\mathbf{k}$ if we separate the principal region of the crystal in the shape of a parallelepiped with the edges $G\mathbf{a}_1$, $G\mathbf{a}_2$, $G\mathbf{a}_3$ and the volume $V = G^3\Omega_0$ ($\Omega_0 = |\mathbf{a}_1 \cdot (\mathbf{a}_2 \times \mathbf{a}_3)|$ is the volume of the crystal's unit cell) and demand that the wave function (3.1) should not change when its coordinate is displaced by a vector $G\mathbf{a}_i$ ($i = 1, 2, 3$) (*Born-Kármán cyclic conditions*; see Section 3.5.6).

Physically distinct (nonequivalent) values of $\mathbf{k}$ lie inside the interval (2.9.10)

$$-\pi < \mathbf{k}\cdot\mathbf{a}_i < \pi \quad (i = 1, 2, 3). \tag{3.3}$$

Substituting herein (3.2), we obtain (2.9.10a)

$$-\frac{G}{2} < g_i < \frac{G}{2}, \tag{3.3a}$$

so that in its region of variation the vector $\mathbf{k}$ (3.3) assumes $G^3$ quasi-discrete values.

We shall choose a Brillouin zone, as defined in Section 2.9.1 as the most convenient region of nonequivalent (distinct) values of the wave vector $\mathbf{k}$.

The conditions (3.2), (3.3) and (3.3a) coincide with the conditions (3.5.9), (3.5.24), (3.5.25) for the wave vector $\mathbf{q}$, which is quite natural, since the very existence of the wave vector and its properties are due to the existence of the translational symmetry of crystal lattices.

If the Bloch wave function (3.1) is normalized inside the crystal's principal region, then

$$\int_V \psi_{\mathbf{k}}^*(\mathbf{r})\,\psi_{\mathbf{k}}(\mathbf{r})\,d^3r = \int_{G^3\Omega_0} |u_{\mathbf{k}}(\mathbf{r})|^2\,d^3r = 1. \tag{3.4}$$

Since $|u_{\mathbf{k}}(\mathbf{r})|^2$ is periodic with the periods of the basic lattice vectors, it follows from (3.4) that

$$G^3 \int_{\Omega_0} |u_{\mathbf{k}}(\mathbf{r})|^2\,d^3r = 1, \tag{3.4a}$$

where the integration is performed over the crystal's unit cell. In some cases it appears expedient to introduce into the Bloch function the factor $G^{-3/2}$ in an explicit form, i.e., to make

$$\psi_{\mathbf{k}}(\mathbf{r}) = \frac{1}{G^{3/2}}\,u_{\mathbf{k}}(\mathbf{r})\,e^{i\mathbf{k}\mathbf{r}}. \tag{3.5}$$

Then we obtain instead of (3.4a)

$$\int_{\Omega_0} |u_{\mathbf{k}}(\mathbf{r})|^2 \, d^3r = 1, \tag{3.6}$$

and in this case the average value is

$$\overline{|u_{\mathbf{k}}|} = 1/\sqrt{\Omega_0}. \tag{3.6a}$$

Substituting the Bloch function (3.1) into the Schrödinger equation

$$-\frac{\hbar^2}{2m}\nabla^2\psi_{\mathbf{k}} + V(\mathbf{r})\,\psi_{\mathbf{k}} = \varepsilon_{\mathbf{k}}\psi_{\mathbf{k}}, \tag{3.7}$$

where $V(\mathbf{r})$ is the periodic potential acting on the electron in the crystal, we obtain after canceling out the multiplier $\exp(i\mathbf{k}\cdot\mathbf{r})$ the following equation for $u_{\mathbf{k}}(\mathbf{r})$ (Appendix 10):

$$-\frac{\hbar^2}{2m}\nabla^2 u_{\mathbf{k}} + V(\mathbf{r})\,u_{\mathbf{k}} - \frac{i\hbar^2}{m}(\mathbf{k}\cdot\nabla u_{\mathbf{k}}) = \left(\varepsilon_{\mathbf{k}} - \frac{\hbar^2k^2}{2m}\right)u_{\mathbf{k}}. \tag{3.8}$$

Here $\nabla u_{\mathbf{k}} \equiv \text{grad}\, u_{\mathbf{k}}$.

For $\mathbf{k} = 0$ equation (3.8) reduces to the equation

$$-\frac{\hbar^2}{2m}\nabla^2 u_0 + V(\mathbf{r})\,u_0 = \varepsilon_0 u_0, \tag{3.8a}$$

coinciding with equation (3.7) for $\psi_{\mathbf{k}}$.

**4.3.2** The study of different types of periodic fields (unidimensional models, cases of weak and strong bonds), as well as general considerations, points to the conclusion that not all values of the electron energy $\varepsilon_{\mathbf{k}}$ in (3.7) are allowed. Generally, the energy spectrum of the electron in a periodic fields splits up into *allowed* and *forbidden bands* (*energy bands*; another term for the forbidden band is *band gap*).

In other words, the dependence of the electron energy in a periodic field on the wave vector $\mathbf{k}$ (within the bounds of the first Brillouin zone) is not a unique function, but consists of a set of $\varepsilon_n(\mathbf{k})$, where $n$ is the number of an allowed energy band. The situation is similar to that which led to the formation of vibration branches $\omega_j(\mathbf{q})$ ($j = 1, 2, \ldots, 3s$) (see Section 3.5.2), the only difference being that the number of allowed energy bands $\varepsilon_n(\mathbf{k})$ is infinite.

The electron wave function (3.1) depends not only on the wave vector $\mathbf{k}$, but on the band number $n$, as well, i.e.,

$$\psi_{n\mathbf{k}}(\mathbf{r}) = u_{n\mathbf{k}}(\mathbf{r})\exp(i\mathbf{k}\cdot\mathbf{r}).$$

It follows from the physical equivalence of the wave vectors $\mathbf{k}$ and $\mathbf{k}' = \mathbf{k} + \mathbf{b}_g$ ($\mathbf{b}_g$ is a vector of the reciprocal lattice) that all the quantities that depend on $\mathbf{k}$ should be periodic with the periods of the basis vectors of the reciprocal lattice $\mathbf{b}_i$ ($i = 1, 2, 3$).

Specifically, for the energy $\varepsilon_n(\mathbf{k})$ we have

$$\varepsilon_n(\mathbf{k} + \mathbf{b}_g) = \varepsilon_n(\mathbf{k}), \tag{3.9}$$

which means that $\varepsilon_n(\mathbf{k})$ can be expanded in a three-dimensional Fourier series in the $\mathbf{k}$-space:

$$\varepsilon_n(\mathbf{k}) = \sum_{\mathbf{l}} c_{\mathbf{l}} e^{i\mathbf{k}\mathbf{a}_{\mathbf{l}}}, \tag{3.10}$$

where $\mathbf{l} \equiv \{l_1, l_2, l_3\}$ are integral indices of a direct lattice vector (1.3.1); indeed, substituting in (3.10) $\mathbf{k} + \mathbf{b}_g$ for $\mathbf{k}$ we confirm (3.9) ($\exp(i\mathbf{b}_g\mathbf{a}_{\mathbf{l}}) = \exp(2\pi i \times \text{integer}) = 1$). The expansion (3.10) is similar to the expansion (1.3.6) in the $\mathbf{r}$-space.

Calculate the result of the operator $\exp(\mathbf{a}_{\mathbf{l}}\cdot\nabla)$ acting on the Bloch wave function $\psi_{n\mathbf{k}}(\mathbf{r})$. Expanding the exponent into a series, we obtain

$$\exp(\mathbf{a}_{\mathbf{l}}\cdot\nabla)\,\psi_{n\mathbf{k}}(\mathbf{r}) = [1 + \mathbf{a}_{\mathbf{l}}\nabla + 1/2\,(\mathbf{a}_{\mathbf{l}}\cdot\nabla)^2 + \ldots]\,\psi_{n\mathbf{k}}(\mathbf{r})$$

$$= \psi_{n\mathbf{k}}(\mathbf{r}) + \mathbf{a}_{\mathbf{l}}\nabla\psi_{n\mathbf{k}}(\mathbf{r}) + \frac{1}{2}\sum_{\alpha,\,\beta}^{1,\,2,\,3} \mathbf{a}_{\mathbf{l}\alpha}\mathbf{a}_{\mathbf{l}\beta}\frac{\partial}{\partial x_\alpha}\frac{\partial}{\partial x_\beta}[\psi_{n\mathbf{k}}(\mathbf{r})] + \ldots$$

$$= \psi_{n\mathbf{k}}(\mathbf{r} + \mathbf{a}_{\mathbf{l}}), \tag{3.11}$$

where the series obtained may be regarded as a Taylor series in the powers of the orthogonal components of the vector $\mathbf{a}_{\mathbf{l}}$.

Substituting into (3.10) $-i\nabla$ for $\mathbf{k}$ and making use of (3.11), we obtain

$$\varepsilon_n(-i\nabla)\,\psi_{n\mathbf{k}}(\mathbf{r}) = \sum_{\mathbf{l}} c_{\mathbf{l}} e^{\mathbf{a}_{\mathbf{l}}\cdot\nabla}\psi_{n\mathbf{k}}(\mathbf{r}) = \sum_{\mathbf{l}} c_{\mathbf{l}}\psi_{n\mathbf{k}}(\mathbf{r} + \mathbf{a}_{\mathbf{l}})$$

$$= \sum_{\mathbf{l}} c_{\mathbf{l}} e^{i\mathbf{k}\mathbf{a}_{\mathbf{l}}} \psi_{n\mathbf{k}}(\mathbf{r}) = \varepsilon_n\,\psi_{n\mathbf{k}}(\mathbf{r}). \tag{3.12}$$

Relation (3.12) became known as the *Wannier theorem*, in its deduction the energy $\varepsilon_n(\mathbf{k})$ was presumed to be nondegenerate for the given $\mathbf{k}$.

Let there be, in addition to the periodic field $V(\mathbf{r})$, a field $\mathcal{U}(\mathbf{r})$ acting on a conduction electron in a crystal (for instance, an external electric field on the potential of an impurity ion). In that case the Schrödinger equation for an electron assumes the form

$$\hat{\mathcal{H}}\psi(\mathbf{r}) = \left[-\frac{\hbar^2}{2m}\nabla^2 + V(\mathbf{r}) + \mathcal{U}(\mathbf{r})\right]\psi(\mathbf{r}) = \varepsilon\psi(\mathbf{r}). \tag{3.13}$$

Expand the wave function $\psi(\mathbf{r})$ in a closed system of the Bloch functions $\psi_{n\mathbf{k}}(\mathbf{r})$ satisfying equation (3.7):

$$\psi(\mathbf{r}) = \sum_n \sum_{\mathbf{k}} c_{n\mathbf{k}}\psi_{n\mathbf{k}}(\mathbf{r}). \tag{3.14}$$

Here $c_{n\mathbf{k}}$ are expansion coefficients.

Substituting (3.14) into (3.13) and taking into account that $\psi_{n\mathbf{k}}(\mathbf{r})$ is an eigenfunction of equation (3.7), we obtain

$$\hat{\mathcal{H}}\psi(\mathbf{r})=\sum_{n}\sum_{\mathbf{k}} c_{n\mathbf{k}}[\varepsilon_n(\mathbf{k})+\mathcal{U}(\mathbf{r})]\,\psi_{n\mathbf{k}}=\varepsilon\psi(\mathbf{r}). \tag{3.15}$$

Making use of (3.12), we obtain

$$\hat{\mathcal{H}}\psi(\mathbf{r})=\sum_{n}[\varepsilon_n(-i\nabla)+\mathcal{U}(\mathbf{r})]\sum_{\mathbf{k}} c_{n\mathbf{k}}\psi_{n\mathbf{k}}=\varepsilon\psi(\mathbf{r}). \tag{3.16}$$

If the Bloch functions of a single band, for instance of the $n$-th, may be used to obtain a fair approximation, then

$$[\varepsilon_n(-i\nabla)+\mathcal{U}(\mathbf{r})]\,\psi(\mathbf{r})=\varepsilon\psi(\mathbf{r}). \tag{3.17}$$

In this approximation the Schrödinger equation (3.13) is replaced by the equation (3.17), which no longer contains the periodic potential $V(\mathbf{r})$. At first glance, this is of little use, since in equation (3.17) we should know $\varepsilon_n(\mathbf{k})$ inside the entire Brillouin zone, and to this end we must find the energy eigenvalues for equation (3.13). However, we may try to use approximate expressions for $\varepsilon_n(\mathbf{k})$.

**4.3.3** If there is a minimum or a maximum of the energy $\varepsilon_n(\mathbf{k})$ at point $\mathbf{k}=\mathbf{k}_0$, it can be expanded in a series in the vicinity of that point:

$$\varepsilon_n(\mathbf{k})=\varepsilon_n(\mathbf{k}_0)+\frac{1}{2}\sum_{\alpha,\beta}\left(\frac{\partial^2\varepsilon_n(\mathbf{k})}{\partial k_\alpha\,\partial k_\beta}\right)_{\mathbf{k}=\mathbf{k}_0}(k_\alpha-k_{\alpha 0})(k_\beta-k_{\beta 0}), \tag{3.18}$$

where $k_\alpha$ and $k_\beta$ are orthogonal components of the vector $\mathbf{k}$. In (3.18) we have cut short the expansion at quadratic terms and taken into account that at the point of an extremum the first derivatives $(\partial\varepsilon_n/\partial k_\alpha)_{\mathbf{k}=\mathbf{k}_0}=0$. Since the energy $\varepsilon_n$ is a scalar, and $\mathbf{k}$ is a vector, the quantities $(\partial^2\varepsilon_n/\partial k_\alpha\partial k_\beta)_{\mathbf{k}_0}$ are components of a rank 2 tensor (Appendix 11). Reducing this tensor to the principal axes and placing the origins of the energy and of the wave vector at the point of extremum ($\varepsilon_n(\mathbf{k}_0)=0$, $\mathbf{k}_0=0$), we obtain

$$\varepsilon_n(\mathbf{k})=\frac{1}{2}\sum_{\alpha}\left(\frac{\partial^2\varepsilon_n}{\partial k_\alpha^2}\right)_0 k_\alpha^2. \tag{3.18a}$$

To bring the description of the electron motion in a periodic field as much as possible in line with that of free electron motion, we introduce the *inverse effective mass tensor*

$$\frac{1}{m_{\alpha\beta}}=\frac{1}{\hbar^2}\left(\frac{\partial^2\varepsilon_n}{\partial k_\alpha\,\partial k_\beta}\right)_0, \tag{3.19}$$

whose principal axes components are

$$\frac{1}{m_\alpha}=\frac{1}{\hbar^2}\left(\frac{\partial^2\varepsilon_n}{\partial k_\alpha^2}\right)_0. \tag{3.19a}$$

Introduce the *quasi-momentum*

$$\mathbf{p} = \hbar \mathbf{k}. \tag{3.20}$$

Its only difference from the wave vector **k** is the presence of a constant factor $\hbar$, and for this reason it has the properties (3.2)-(3.3a).

Note that the dimensionality of the inverse effective mass tensor is $(\text{mass})^{-1}$, and that of the quasi-momentum is the same as of a momentum. It follows from expressions (3.18a), (3.19a) and (3.20) that

$$\varepsilon_n(\mathbf{k}) = \sum_\alpha \frac{\hbar^2 k_\alpha^2}{2m_\alpha} = \sum_\alpha \frac{p_\alpha^2}{2m_\alpha}, \tag{3.21}$$

i.e., the expression is of the form of that for the kinetic energy of an electron with different masses along the $x$, $y$, and $z$ axes. The quantities $m_\alpha$ with the dimensionality of mass are not components of a tensor (since the quantities inverse to tensor components generally do not constitute a tensor); however, for the sake of brevity they are called the *effective mass tensor*. It should be kept in mind that equation (3.18) holds only for small $p_\alpha$, i.e., in cases when $p_\alpha \ll \hbar/a$ ($a$ is the lattice parameter), this condition being usually well satisfied in semiconductors because of the low concentrations of conduction electrons in them. Close to the points $\mathbf{k}_0$, where $\varepsilon_n(\mathbf{k}_0)$ attains an extremum, the constant-energy surfaces $\varepsilon_n(\mathbf{k}) = \text{const.}$ have the shape of ellipsoids.

Finally, in the case when all tensor components $1/m_\alpha$ are equal, a scalar effective mass $m^*$ can be introduced:

$$\frac{1}{m^*} = \frac{1}{m_\alpha} = \frac{1}{\hbar^2}\left(\frac{\partial^2 \varepsilon_n}{\partial k_x^2}\right)_{\mathbf{k}_0} = \frac{1}{\hbar^2}\left(\frac{\partial^2 \varepsilon_n}{\partial k_y^2}\right)_{\mathbf{k}_0} = \frac{1}{\hbar^2}\left(\frac{\partial^2 \varepsilon_n}{\partial k_z^2}\right)_{\mathbf{k}_0}. \tag{3.22}$$

The electron energy is in this case equal to

$$\varepsilon_n(\mathbf{k}) = \hbar^2 k^2/2m^* = p^2/2m^*, \tag{3.22a}$$

i.e., is equal to the kinetic energy of a free electron with the mass $m^*$ and the momentum **p**.

Near the lower band edge, where $\varepsilon_n(\mathbf{k})$ attains its minimum $\frac{1}{m_\alpha} = \frac{1}{\hbar^2}\left(\frac{\partial^2 \varepsilon_n}{\partial k_\alpha^2}\right)_{k_0} > 0$, i.e., the effective mass $m_\alpha$ is positive. On the contrary, near the maximum of $\varepsilon_n(\mathbf{k})$ the second derivatives $\left(\frac{\partial^2 \varepsilon_n}{\partial k_\alpha^2}\right)_{k_0} < 0$, and the effective masses are negative. We shall see below that the effective masses, just like the normal masses, determine the force-to-acceleration ratio. For $m_\alpha < 0$ the electron's acceleration is opposite in direction to the direction of the force acting on it. This should not perplex us, since the effective mass method accounts in the simplest way for the effect of the periodic crystal field on the electron, the combined action of the periodic field and an external force on the electron, which possesses wave properties, being able to cause the above phenomenon.

**4.3.4** If we were to substitute the expression for the energy (3.21) into (3.17), we would obtain the equation

$$-\left[\frac{\hbar^2}{2m_x}\frac{\partial^2}{\partial x^2}+\frac{\hbar^2}{2m_y}\frac{\partial^2}{\partial y^2}+\frac{\hbar^2}{2m_z}\frac{\partial^2}{\partial z^2}+\mathcal{U}(\mathbf{r})\right]\psi(\mathbf{r})=\varepsilon\psi(\mathbf{r}), \quad (3.23)$$

which describes the motion of an electron with unequal masses $m_x$, $m_y$, $m_z$ along the axes in the field $\mathcal{U}(\mathbf{r})$.[4] Equation (3.23) assumes an especially simple form, when the tensor $m_{\alpha}^{-1}$ degenerates into a scalar $m_x = m_y = m_z = m^*$, and the field $\mathcal{U}(\mathbf{r}) = 0$. Then

$$-\frac{\hbar^2}{2m^*}\nabla^2\psi(\mathbf{r})=\varepsilon\psi(\mathbf{r}). \quad (3.23a)$$

This corresponds to a free particle with a renormalized mass ($m \to m^*$). Such a situation can come about, when the minimum of the energy $\varepsilon_n(\mathbf{k})$ is located in the centre of a Brillouin zone of a cubic crystal.

It should, however, be kept in mind that equation (3.23) was obtained under the assumption that (3.21) could be substituted for $\varepsilon_n(\mathbf{k})$ inside the entire Brillouin zone and this is generally incorrect, even when the electrons move close to an energy extremum.

A rigorous theory developed by Kohn and Luttinger (1955) generalized for degenerate bands demonstrates that equation (3.17) describes only the smoothly varying part $F(\mathbf{r})$ of the wave function, the full electron wave function being

$$\psi(\mathbf{r}) = F(\mathbf{r})\, u_{n\mathbf{k}_0}(\mathbf{r})\, e^{i\mathbf{k}_0\cdot\mathbf{r}}, \quad (3.24)$$

where $u_{n\mathbf{k}_0}(\mathbf{r})$ is the amplitude multiplier of the Bloch function (3.1) at point $\mathbf{k} = \mathbf{k}_0$. On the other hand, in certain cases the smoothly varying part of the wave function $F(\mathbf{r})$ can be sufficiently used as the function itself, the eigenvalues of the electron energy being determined from equation (3.23).

**4.3.5** It was demonstrated in Section 2.8.3 that the electron energy surface in a crystal field $\varepsilon_n(\mathbf{k}) = \text{const.}$ has the symmetry of the lattice point group $\mathcal{F}$; it was also noted that in all cases $\varepsilon_n(-\mathbf{k}) = \varepsilon_n(\mathbf{k})$, i.e., that the surface $\varepsilon_n(\mathbf{k}) = \text{const.}$ has a centre of symmetry (no matter whether the crystal lattice has such a centre). Hence, if, for example in the cubic lattice some arbitrary point $\mathbf{k}_0$ of the reciprocal space not located on any of the symmetry elements there is an extremum $\varepsilon_n(\mathbf{k}_0)$, then in a Brillouin zone there should be 48 additional symmetrically arranged extrema. If $\mathbf{k}_0$ coincides

[4] It should be remembered that the axes $x$, $y$ and $z$ coincide with the principal axes of the tensor $m_{\alpha\beta}^{-1}$, i.e., are oriented in a definite way with respect to the crystal.

with the [100] direction, the total number of extrema will be six, and the rotational ellipsoids (3.15) with axes coinciding with the [100] direction will serve as constant energy surfaces. If the centres of those ellipsoids, i.e., in other words, the minima of the energy $\varepsilon_n$, lie in the centres of a Brillouin zone faces, then one-half of each ellipsoid will be located outside the first zone, the latter's share being six half-ellipsoids, the equivalent of three whole ellipsoids.

A similar situation for a planar square lattice is depicted in Fig. 4.1. Figure 4.1*a* and *b* shows "constant energy ellipses" in the **k**-space for the cases, when the energy minima are located at four equivalent (symmetric) points inside a Brillouin zone and on its faces. Figure 4.1*c* depicts the periodic repetition of the structure shown in Fig. 4.1*b*. Similar and still more intricate energy band structures are not only theoretically possible, but they are actually realized in numerous substances (silicon, germanium, etc.). Below we shall treat this subject at length.

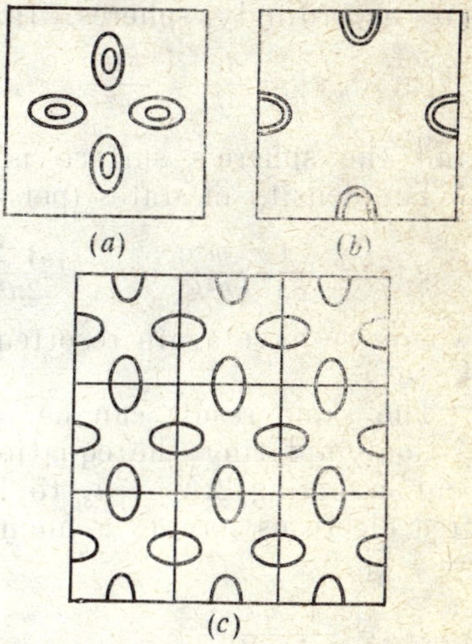

Fig. 4.1

**4.3.6** The Born-Kármán cyclic conditions imposed on the Bloch electron wave function or on the vibrations of lattice atoms cause the wave vectors **k** and **q** to assume $G^3$ quasi-discrete values (Section 2.9.1). Hence, we can speak of the number of electron states in a volume $V$ in a region $d\tau_{\mathbf{k}}$ of the **k**-space or in an energy interval $(\varepsilon, \varepsilon + d\varepsilon)$.

Since the expression for **q** (3.5.24) coincides with that for **k** (3.2), the number of electron states in the energy interval $(\varepsilon, \varepsilon + d\varepsilon)$ will, with reference to (3.5.27), be equal to

$$g(\varepsilon)\, d\varepsilon = \frac{V}{(2\pi)^3} \sum_n \quad \int\int\int_{\varepsilon \leqslant \varepsilon_n(\mathbf{k}) \leqslant \varepsilon + d\varepsilon} d\tau_{\mathbf{k}}, \tag{3.25}$$

and the density of states, by analogy with (3.5.32), will be

$$g(\varepsilon) = \frac{V}{(2\pi)^3} \sum_n \int \frac{d\sigma}{|\nabla_{\mathbf{k}}\varepsilon_n|}, \tag{3.26}$$

where $\nabla_{\mathbf{k}}\varepsilon_n \equiv \mathrm{grad}_{\mathbf{k}}\, \varepsilon_n(\mathbf{k})$, and the integration is performed over the surface $\varepsilon_n(\mathbf{k}) = \mathrm{const}$. Expressions (3.25) and (3.26) are deduced exactly in the same way as the frequency distribution function in Section 3.5.6; we only have to substitute $\mathbf{q} \rightarrow \mathbf{k}$ and $\omega_j \rightarrow \varepsilon_n$.

Apply this equation to the simple case of one band when the energy is determined by expression (3.22a) and the constant energy surfaces are accordingly spheres. In this case

$$|\operatorname{grad}_{\mathbf{k}} \varepsilon| = \frac{d\varepsilon}{dk} = \frac{\hbar^2 k}{m^*}$$

and the sphere's surface is $\sigma = 4\pi k^2$.

The density of states (per 1 cm$^3$) is equal to

$$g(\varepsilon) = \frac{1}{8\pi^3} \frac{m^*}{\hbar^2 k} 4\pi k^2 = \frac{\sqrt{2}}{2\pi^2} \frac{m^{*3/2}}{\hbar^3} \sqrt{\varepsilon} \propto \varepsilon^{1/2}, \tag{3.27}$$

where we have again resorted to expression (3.22a) to go over from $\mathbf{k}$ to $\varepsilon$.

The same result can be obtained directly from Appendix 4.

Contemplating the equation for a variable momentum $p$ (A.4.1) and assuming $\Delta N = n_s$ to be the number of quantum states for free electrons having a momentum less than $p$, we obtain ($\Delta V = 1$ cm$^3$)

$$dn_s = \frac{1}{\pi^2} \frac{p^2\, dp}{\hbar^3}.$$

Going over to the variable $\varepsilon = p^2/2m^*$ we obtain for the density of states $\frac{dn_s}{d\varepsilon} = 2g(\varepsilon)$ equation (3.27).

**4.3.7** Calculate the quantum mechanical mean velocity $\langle \mathbf{v} \rangle = \frac{1}{m} \langle \mathbf{p} \rangle$ of the electron in a state with the wave vector $\mathbf{k}$ in a periodic field. As we shall see below, in general, it is nonzero. Hence, the electron can freely move around the entire crystal, even if its total energy is less than its maximum potential energy in the crystal. There is a mean velocity $\mathbf{v}(\mathbf{k})$ and, consequently, a nonvanishing current $\mathbf{j} = e\mathbf{v}$, where $e$ is the electron charge, to correspond to an electron in the $\mathbf{k}$ state. The finite electrical resistance of a crystal is due not to the potential barriers of the periodic field that the electron passes by "tunneling" but to violations of the strict periodicity of the crystal field caused by thermal vibrations or by static lattice defects.

In compliance with the general equation for mean values in quantum mechanics [4.1, § 6]

$$\langle \mathbf{v} \rangle = \frac{1}{m} \langle \mathbf{p} \rangle = \frac{1}{m} \frac{\hbar}{i} \int\limits_V \psi_{\mathbf{k}}^* \nabla \psi_{\mathbf{k}}\, d\tau, \tag{3.28}$$

where $\frac{\hbar}{i} \nabla \equiv \frac{\hbar}{i} \operatorname{grad}$ is the momentum operator; the integration is performed over the crystal's principal region.

For a free electron the wave function normalized inside the principal region is equal to

$$\psi_{\mathbf{k}}^{(0)} = \frac{1}{\sqrt{V}} e^{i\mathbf{k}\cdot\mathbf{r}}, \tag{3.29}$$

i.e., it is in the form of a plane wave with a constant amplitude. Here the wave vector $\mathbf{k} = \mathbf{p}/\hbar$, where $\mathbf{p}$ is the electron's momentum (not a quasi-momentum!). Substituting (3.29) into (3.28) and noting that $\nabla e^{i\mathbf{kr}} = i\mathbf{k}e^{i\mathbf{kr}}$ we obtain (omitting the brackets around $\langle \mathbf{v} \rangle$)

$$\mathbf{v} = \frac{\hbar}{mi} \int \psi_{\mathbf{k}}^{(0)*} i\mathbf{k} \psi_{\mathbf{k}}^{(0)} \, d\tau = \frac{\hbar \mathbf{k}}{m} = \frac{\mathbf{p}}{m}, \tag{3.30}$$

i.e., the same relation as in classical mechanics.

If the mean quantum mechanical electron velocity in a crystal $\mathbf{v}$ is identified with the group velocity of a wave packet made up of Bloch functions $\mathbf{v}_{gr}$ (this is not the obvious operation), then [4.1, § 3]

$$\mathbf{v}_{gr} = \mathrm{grad}_{\mathbf{k}}\, \omega = \frac{\partial \omega}{\partial \mathbf{k}} = \frac{1}{\hbar} \frac{\partial \hbar \omega}{\partial \mathbf{k}} = \frac{1}{\hbar} \frac{\partial \varepsilon}{\partial \mathbf{k}} \tag{3.31}$$

and, consequently,

$$\mathbf{v} = \frac{1}{\hbar} \frac{\partial \varepsilon}{\partial \mathbf{k}} = \frac{1}{\hbar} \mathrm{grad}_{\mathbf{k}}\, \varepsilon(\mathbf{k}). \tag{3.32}$$

In the case of a free electron the energy is

$$\varepsilon(\mathbf{k}) = \frac{\hbar^2 k^2}{2m} = \frac{p^2}{2m}, \tag{3.33}$$

and equation (3.32) yields (3.30).

At the points of energy extrema $(\partial \varepsilon / \partial \mathbf{k}) = 0$, and therefore, $\mathbf{v} = 0$.

A rigorous deduction of equation (3.32) is presented in Appendix 12.

**4.3.8** Consider an electron in a periodic field with an additional external field $\mathbf{F}$ acting on it. If the force $\mathbf{F}$ is small enough, so that $F \cdot a \ll \mathscr{E}_g$, where $a$ is the lattice parameter, and $\mathscr{E}_g$ is the width of the corresponding forbidden band (band gap), the force $F$ will be unable to cause electron transitions between different energy bands and will only change the electron's wave vector $\mathbf{k}$. Since the classical equations of motion remain valid for the mean quantum mechanical values, it may naturally be presumed that the energy conservation law takes the form

$$\frac{d\varepsilon(\mathbf{k})}{dt} = \mathbf{v} \cdot \mathbf{F}, \tag{3.34}$$

where $\mathbf{v}$ is the velocity as determined by equality (3.32), and $\varepsilon(\mathbf{k})$ is the electron energy in the band in which it moves. Equality

(3.34) states that the work of the force **F** applied to the electron in one second is equal to the rate of variation of its energy $\varepsilon(\mathbf{k})$. Since

$$\frac{d\varepsilon(\mathbf{k})}{dt}=\sum_{\alpha}\frac{\partial\varepsilon}{\partial k_\alpha}\frac{dk_\alpha}{dt}=\mathrm{grad}_{\mathbf{k}}\,\varepsilon\cdot\frac{d\mathbf{k}}{dt},$$

we obtain from equation (3.32), making use of equation (3.34)

$$\frac{d(\hbar\mathbf{k})}{dt}=\frac{d\mathbf{p}}{dt}=\mathbf{F}. \tag{3.35}$$

Specifically, it follows from here that in a periodic field the electron's quasi-momentum $\mathbf{p}=\hbar\mathbf{k}$, plays in the equation of motion (3.35) the part of the momentum of a free electron.

The electron's acceleration, the term being understood as meaning the rate of variation of its mean quantum mechanical velocity **v**, is equal to

$$\frac{d\mathbf{v}}{dt}=\frac{d}{dt}\left(\frac{1}{\hbar}\,\mathrm{grad}_{\mathbf{k}}\,\varepsilon(\mathbf{k})\right)=\frac{1}{\hbar}\frac{d}{dt}\left(\frac{\partial\varepsilon(\mathbf{k})}{\partial\mathbf{k}}\right).$$

Since $\frac{\partial\varepsilon}{\partial\mathbf{k}}$ depends on time only via **k**, it follows that

$$\frac{dv_\alpha}{dt}=\frac{1}{\hbar}\frac{d}{dt}\left(\frac{\partial\varepsilon}{\partial k_\alpha}\right)=\frac{1}{\hbar}\sum_{\beta}\left(\frac{\partial^2\varepsilon}{\partial k_\alpha\,\partial k_\beta}\right)\frac{dk_\beta}{dt}=\sum_{\beta}\frac{1}{\hbar^2}\left(\frac{\partial\varepsilon}{\partial k_\alpha\,\partial k_\beta}\right)\frac{dp_\beta}{dt}. \tag{3.36}$$

We shall term the set of quantities

$$\frac{1}{\hbar^2}\left(\frac{\partial^2\varepsilon}{\partial k_\alpha\,\partial k_\beta}\right)\equiv\tilde{m}^{-1}_{\alpha\beta} \tag{3.37}$$

*generalized inverse effective mass tensor*. It differs from the tensor $m^{-1}_{\alpha\beta}$ (3.19) in that it is dependent on **k**. If the electron energy is equal to (3.18), i.e. is calculated in the quadratic approximation, then $\tilde{m}^{-1}_{\alpha\beta}=m^{-1}_{\alpha\beta}$. It follows from equations (3.35) and (3.36) reduced to principal axes of the tensor $\tilde{m}^{-1}_{\alpha\beta}$ that

$$\frac{dv_\alpha}{dt}=\tilde{m}^{-1}_\alpha F_\alpha. \tag{3.38}$$

In the case of $\tilde{m}^{-1}_{\alpha\beta}=m^{-1}_\alpha=\mathrm{const.}$ it is equivalent to classical equations of motion for an "anisotropic" mass.

Finally, in the approximation of a scalar effective mass (3.22) equation (3.38) assumes the form

$$m^*\frac{d\mathbf{v}}{dt}=\mathbf{F}, \tag{3.39}$$

coinciding with the usual equation of classical mechanics.

## 4.4 Concept of Positive Holes in an Almost Completely Filled Valence Band

**4.4.1** By analogy with the classical expression for the current set up by the charge $-e$, let us write down the quantum mechanics expression for current set up by an electron in the **k**-state:

$$\mathbf{j}_\mathbf{k} = -e\mathbf{v}(\mathbf{k}) = -\frac{e\hbar}{mi}\int \psi_\mathbf{k}^* \nabla \psi_\mathbf{k}\, d\tau = -\frac{e\hbar}{2mi}\int (\psi_\mathbf{k}^* \nabla \psi_\mathbf{k} - \psi_\mathbf{k} \nabla \psi_\mathbf{k}^*)\, d\tau. \quad (4.1)$$

The last integral may be obtained on the grounds that formally $\mathbf{j}_\mathbf{k} + \mathbf{j}_\mathbf{k}^* = 2\mathbf{j}_\mathbf{k}$ ($\mathbf{j}_\mathbf{k}$ is real). We see that the right-hand side of expression (4.1), is, indeed, the usual quantum mechanics expression for the current density averaged over the crystal's principal region [strictly speaking, for the purpose of averaging the integral in (4.1) should have been divided by $V$]. This averaging excludes circular currents inside the individual crystal cells. Such currents are due to the presence of a periodic multiplicand $u_\mathbf{k}(\mathbf{r})$ in the Bloch wave function and have nothing to do with the linear motion of the electron [4.2, p. 196].

We know that in a crystal of any symmetry (A.9.26)

$$\varepsilon(-\mathbf{k}) = \varepsilon(\mathbf{k}), \quad (4.2)$$

and that, therefore,

$$\mathbf{v}(-\mathbf{k}) = -\mathbf{v}(\mathbf{k}). \quad (4.3)$$

The latter expression follows immediately from the expression for the mean velocity (3.32).

Making use of equality (4.3), we can easily demonstrate that

$$\sum_\mathbf{k} \mathbf{v}(\mathbf{k}) = 0, \quad (4.4)$$

where the summation embraces all the values of **k** inside a Brillouin zone. According to the Pauli principle, there can be no more than two electrons with opposite spins $s_z$ in every **k**-state. Hence, the current in a completely filled valence band is

$$\mathbf{j} = \sum_{\mathbf{k}s} \mathbf{j}_\mathbf{k} = 2\sum_\mathbf{k} \mathbf{j}_\mathbf{k} = -e \times 2\sum_\mathbf{k} \mathbf{v}(\mathbf{k}) = 0. \quad (4.5)$$

Should an electric field $\mathbf{E} = -\text{grad}\,\varphi$, where $\varphi$ is the potential, be applied to the crystal, the current in its completely filled band would remain zero. Indeed, if the field is not too high, the electrons will not be thrown over to the conduction band, and the valence band will remain filled in the electric field as well.

If the band is not completely filled with electrons, but the external electric field is zero, the current will be

$$\mathbf{j} = \sum_{\mathbf{k}s}{}' \mathbf{j}_\mathbf{k} = 0 \quad (4.6)$$

owing to the symmetry of electron distribution in the k-space (the prime after the summation sign means that the summation is performed not in the entire Brillouin zone). But if the electrons do not fill the band completely, and an external field is applied to the crystal, the current (4.6) will not be zero, because there will be an increase in the number of $k$-states filled by electrons moving against the field.

We now introduce the symbol $\nu_n$ $(\mathbf{k}, s)$ equal to unity, if the state $(\mathbf{k}, s)$ is occupied by an electron, and zero otherwise. The probability of the state $(\mathbf{k}, s)$ being unoccupied by an electron (or a hole) is obviously equal to $\nu_p(\mathbf{k}, s) = 1 - \nu_n(\mathbf{k}, s)$. Now the current (4.6) can be written as follows

$$\mathbf{j} = -e \sum_{\mathbf{k}s} \nu_n(\mathbf{k}, s)\, \mathbf{v}(\mathbf{k}), \tag{4.6a}$$

where the summation is performed over the entire band, or

$$\mathbf{j} = -e \sum_{\mathbf{k}s} [1 - \nu_p(\mathbf{k}, s)]\, \mathbf{v}(\mathbf{k}) = -e \sum_{\mathbf{k}s} \mathbf{v}(\mathbf{k}) + \sum_{\mathbf{k}s} \nu_p(\mathbf{k}, s)\, \mathbf{v}(\mathbf{k})$$

$$= e \sum_{\mathbf{k}s} \nu_p(\mathbf{k}, s)\, \mathbf{v}(\mathbf{k}). \tag{4.6b}$$

We see that the electric current in a band incompletely filled with electrons can be described as a current of positively charged $+e$ quasi-particles, or holes, corresponding to the $\mathbf{k}$, $s$-states unoccupied by the electrons and moving at speeds $\mathbf{v}(\mathbf{k})$. The hole concentration (their number per 1 cm³) is $\sum_{\mathbf{k}s} \overline{\nu_p}(\mathbf{k}, s)$ if the number of quantum states is calculated per 1 cm³. The statement that the electric field brings about an increase in the number of $\mathbf{k}$-states occupied by electrons whose velocity $\mathbf{v}(\mathbf{k})$ is directed against the field is equivalent to the statement that there is an increase in the number of $\mathbf{k}$-states occupied by holes whose velocity is directed along the field. Calculate the energy flux $\mathbf{w}$ transported by the electrons of the partially filled band if there is a predominant direction of their motion (due to the presence of a gradient of temperature, of concentration, or of potential $\varphi$):

$$\mathbf{w} = \sum_{\mathbf{k}s} \nu_n(k, s)\, [\varepsilon(\mathbf{k}) - e\varphi]\, \mathbf{v}(\mathbf{k}) = \sum_{\mathbf{k}s} [1 - \nu_p(\mathbf{k}, s)]\, [\varepsilon(\mathbf{k}) - e\varphi]\, \mathbf{v}(\mathbf{k})$$

$$= \sum_{\mathbf{k}s} [\varepsilon(\mathbf{k}) - e\varphi]\, \mathbf{v}(\mathbf{k}) + \sum_{\mathbf{k}s} \nu_p(\mathbf{k}, s)\, [-\varepsilon(\mathbf{k}) + e\varphi]\, \mathbf{v}(\mathbf{k}). \tag{4.7}$$

The first sum, in accordance with relations (4.2) and (4.3), is zero, the second sum can be interpreted as the energy flux transported by holes with the charge $+e$ and the energy $-\varepsilon(\mathbf{k})$. If the band is almost filled so that the hole energy for $\mathbf{k} \approx \mathbf{k}_0$ is close to the band's upper edge, then in the effective mass approximation (3.21) we have

$$-\varepsilon(\mathbf{k}) = -\varepsilon(\mathbf{k}_0) + \sum_{\alpha} \frac{\hbar^2 k_\alpha^2}{2(-m_\alpha)}, \tag{4.7a}$$

where the effective mass of the hole $-m_\alpha$ is positive, because the effective mass of the electron $m_\alpha$ near the upper band edge is negative; hence,

$$m_\alpha^{(p)} = -m_\alpha^{(n)} > 0. \tag{4.7b}$$

In the presence of a magnetic field **H** the electron is acted upon by the Lorentz force $-(e/c)\,\mathbf{v} \times \mathbf{H}$ which, like the force of the electric field, is proportional to the charge $-e$; therefore, as can be demonstrated, in this case, too, the concept of a hole remains valid. It will be demonstrated below that the statistical behaviour of the electrons of an almost filled band is equivalent to the statistical behaviour of holes whose energy levels are determined by expression (4.7a). Thus, the hole concept can be employed both in the optical phenomena and in the transport phenomena caused by electric or magnetic fields or by temperature or concentration gradients.

**4.4.2** In general, in semiconductors the charge carriers are both the electrons of the lower part of the conduction band and the electrons of an almost completely filled valence band that can be interpreted as quasi-particles with a charge $+e$ and with a positive mass (4.7b). Inside a definite temperature range it may often be assumed that the transport phenomena are the work either of the conduction band electrons alone (electron- or $n$-type semiconductor), or solely of the holes of the valence band (hole- or $p$-type semiconductor). Studies of the semiconductor's electrical conductivity do not obviously enable us to determine whether it belongs to the $n$- or to the $p$-type. As we shall see below, this can be established in the course of studies of the Hall effect and of the other galvano- and thermomagnetic phenomena.

It should be pointed out that the concept of electrons and holes with definite effective masses can be substantiated only in the case of an external electromagnetic field acting on the conduction electrons in a crystal [4.3]. If, on the other hand, the conduction electrons in a crystal are acted upon by a gravitational field or by forces of inertia, the concept of holes and electrons with effective masses is no longer applicable. The most direct method used to discover the effect of forces of inertia acting on the conduction electrons is the Stuart and Talman experiment (1916). The idea of the experiment is as follows: a cylindrical coil with a large number of turns is made to rotate at high speed about its axis. When the coil is braked suddenly, the electrons continue to move inertially, and a short current pulse appears in the coil.

It can easily be demonstrated [4.4, p. 229] that the total charge $q$ flowing in the circuit while the coil is being slowed down is equal to

$$q = \frac{m}{e} \frac{v_0 l}{R}, \tag{4.8}$$

where $v_0$ is the initial linear velocity of the wire, $l$ is its length, $R$ is the circuit resistance, and $e$ and $m$ are the charge and the mass of the particles acted upon by the forces of inertia. If we measure the charge $q$ with the aid of a ballistic galvanometer, we can easily find the ratio of $e/m$ (for specified $l$, $R$ and $v_0$).

Experiment proved to be in agreement with the theory that the sign and the magnitude of the ratio $e/m$ always coincide with that for free electrons. For this reason the attempts to discover in experiments of the Stuart-Talman type the motion of holes with effective masses $m^*$ were doomed to failure right from the outset [4.5].

**4.4.3** The above discussion may create a false impression that the hole concept is perfectly equivalent to the idea of a band partially filled with electrons, no matter to what extent. It should, however, be pointed out that the holes appear as the result of a formal substitution of the symbol $\nu_n(\mathbf{k}, s)$ for $1 - \nu_p(\mathbf{k}, s)$ under the summation sign $\sum_{\mathbf{k}s}$, i.e., physically as the result of the replacement of a collection of electrons partially filling the band by a collection of electrons completely filling the band plus holes.

Both these collections are, indeed, equivalent from the point of view of balance of transport phenomena, but there is no reason to presume their equivalence from the viewpoint of the self-consistent periodic potential of the single-electron problem.

In the latter case the equivalence will be the better, the less the concentration of holes, i.e., the more completely is the valence band filled with electrons.

## 4.5 The Approximation of Almost Free (Weakly Bound) Electrons

Consider the motion of an electron in a weak periodic field $V(\mathbf{r})$, i.e., presume $V(\mathbf{r})$ to be a small perturbation of the free electron motion (R. E. Peierls).[5]

---

[5] In other words, the variations of the electron's potential energy $V(\mathbf{r})$ are small as compared with its kinetic energy.

Expand the periodic potential in a Fourier series (1.3.6):

$$V(\mathbf{r}) = \sum_{\mathbf{g} \neq 0} V_{\mathbf{g}} e^{i(\mathbf{b}_{\mathbf{g}} \cdot \mathbf{r})}, \tag{5.1}$$

where $\mathbf{b}_{\mathbf{g}}$ is a reciprocal lattice vector. Without impairing generality we may put the zeroth term in the expansion (5.1), i.e., the average value of the potential $V_0 = 0$. For the right-hand side of (5.1) to be real it should be $V_{-\mathbf{g}} = V_{\mathbf{g}}^*$. Presume the amplitudes $V_{\mathbf{g}}$ to be infinitesimals of the first order. Expand in a Fourier series the periodic multiplier of the Bloch wave function:

$$u_{\mathbf{k}}(\mathbf{r}) = \sum_{\mathbf{h}} a_{\mathbf{h}} e^{i(\mathbf{b}_{\mathbf{h}} \cdot \mathbf{r})}, \tag{5.2}$$

where $\mathbf{b}_{\mathbf{h}}$ is a reciprocal lattice vector.

Substitute expressions (5.1) and (5.2) into equation (3.8) for $u_{\mathbf{k}}(\mathbf{r})$ and obtain

$$\sum_{\mathbf{h}} \frac{\hbar^2}{2m} b_{\mathbf{h}}^2 a_{\mathbf{h}} e^{i(\mathbf{b}_{\mathbf{h}} \cdot \mathbf{r})} + \sum_{\mathbf{h}} \sum_{\mathbf{g} \neq 0} V_{\mathbf{g}} a_{\mathbf{h}} e^{i(\mathbf{b}_{\mathbf{g}} + \mathbf{b}_{\mathbf{h}}) \cdot \mathbf{r}}$$

$$+ \sum_{\mathbf{h}} \frac{\hbar^2}{m} (\mathbf{b}_{\mathbf{h}} \cdot \mathbf{k}) a_{\mathbf{h}} e^{i(\mathbf{b}_{\mathbf{h}} \cdot \mathbf{r})} = \sum_{\mathbf{h}} \left( \varepsilon_{\mathbf{k}} - \frac{\hbar^2 \mathbf{k}^2}{2m} \right) a_{\mathbf{h}} e^{i(\mathbf{b}_{\mathbf{h}} \cdot \mathbf{r})}. \tag{5.3}$$

Replace the summation over **h** and **g** in the double sum by summation over **h** — **g** and **g**. In this case the vector $\mathbf{b}_{\mathbf{h}}$ in the exponent should be substituted for $\mathbf{b}_{\mathbf{g}} + \mathbf{b}_{\mathbf{h}}$, so that the double sum assumes the form

$$\sum_{\mathbf{h}} \sum_{\mathbf{g}} V_{\mathbf{g}} a_{\mathbf{h}-\mathbf{g}} e^{i(\mathbf{b}_{\mathbf{h}} \cdot \mathbf{r})}. \tag{5.3a}$$

For the equality (5.3) to become an identity for all the **r** the sum of coefficients of all $\exp i(\mathbf{b}_{\mathbf{h}} \cdot \mathbf{r})$ should be zero.

Taking into account (5.3a) and performing a simple algebraic transformation, we obtain

$$\left[ \varepsilon_{\mathbf{k}} - \frac{\hbar^2}{2m} (\mathbf{k} + \mathbf{b}_{\mathbf{h}})^2 \right] a_{\mathbf{h}} - \sum_{\mathbf{g} \neq 0} V_{\mathbf{g}} a_{\mathbf{h}-\mathbf{g}} = 0$$

$$(h_1, h_2, h_3 = \pm 1, \pm 2, \pm 3, \ldots). \tag{5.4}$$

We have obtained an infinite homogeneous system of algebraic equations for the unknown coefficients $a_{\mathbf{h}}$. Dealing with the system as with a finite system requires its infinite determinant to turn zero. In this way we shall obtain an equation from which we can, in principle, determine the electron energy $\varepsilon_{\mathbf{k}}$ eigenvalue spectrum and subsequently, using equations (5.4), determine the corresponding system of the coefficients $a_{\mathbf{h}}$, i.e., the electron wave function. Of course, even with the known coefficients $V_{\mathbf{g}}$, i.e., with a specified periodic potential, the practical solution is hopelessly complicated.

Consider a free electron, in which all $V_g = 0$. In this case it follows from equations (5.4) that either

$$\varepsilon_{\mathbf{k}} - \frac{\hbar^2}{2m}(\mathbf{k}+\mathbf{b_h})^2 = 0, \quad \text{or} \quad a_{\mathbf{h}} = 0. \tag{5.5}$$

Besides, for a free electron

$$\varepsilon_{\mathbf{k}} = \varepsilon_0(\mathbf{k}) = \hbar^2\mathbf{k}^2/2m. \tag{5.5a}$$

It follows from the conditions (5.5) and (5.5a) that

$$a_0 \neq 0, \quad a_{\mathbf{h}} = 0 \quad (\mathbf{h} \neq 0). \tag{5.5b}$$

Make $a_0 = 1$, this corresponding to the free electron wave function normalized in a unit volume.

Solve now the system (5.4) for the case of a weak periodic field, in which all coefficients $V_g$ may be regarded as infinitesimals of the first order. It would be natural to presume (this will be confirmed by calculations) that in this case the $a_{\mathbf{h}\neq 0} \neq 0$ and will also be infinitesimals of the first order. For calculations up to the first order, retain on the left-hand side of the system (5.4) one addend with $\mathbf{g} = \mathbf{h}$ (proportional to $a_0 = 1$) and substitute for $\varepsilon_{\mathbf{k}}$ its value (5.5a); then

$$a_{\mathbf{h}} = -\frac{2m}{\hbar^2}\frac{V_{\mathbf{h}}}{\mathbf{b}_{\mathbf{h}}^2 + 2(\mathbf{b}_{\mathbf{h}}\cdot\mathbf{k})} \quad (\mathbf{h} \neq 0). \tag{5.6}$$

The set of coefficients $a_{\mathbf{h}\neq 0}$ (5.6) determines in the first approximation the correction to the electron wave function.

To find the perturbation energy make

$$\varepsilon_{\mathbf{k}} = \varepsilon_0 + \varepsilon' = \frac{\hbar^2\mathbf{k}}{2m} + \varepsilon'. \tag{5.7}$$

Substituting expressions (5.7) and (5.6) into equation (5.4) we obtain

$$\varepsilon' a_{\mathbf{h}} - V_{\mathbf{h}} = \sum_{\mathbf{g}\neq 0} V_{\mathbf{g}} a_{\mathbf{h}-\mathbf{g}}. \tag{5.8}$$

As we shall see, $\varepsilon'$ is an infinitesimal of the second order; for this reason the equation obtained above should be considered only for $\mathbf{h} = 0$ (the $\varepsilon' a_{\mathbf{h}\neq 0}$ are infinitesimals of the third order). In this case

$$\varepsilon' = \sum_{\mathbf{g}\neq 0} V_{\mathbf{g}} a_{-\mathbf{g}}. \tag{5.9}$$

Taking the sum over $-\mathbf{g}$ instead of over $\mathbf{g}$, substituting $a_{\mathbf{g}}$ and noting that $V_{-\mathbf{g}} = V_{\mathbf{g}}^*$, we obtain

$$\varepsilon' = -\frac{2m}{\hbar^2}\sum_{\mathbf{g}\neq 0}\frac{|V_{\mathbf{g}}|^2}{\mathbf{b}_{\mathbf{g}}^2 + 2(\mathbf{b}_{\mathbf{g}}\cdot\mathbf{k})}. \tag{5.9a}$$

We see that the correction to the energy $\varepsilon'$ is proportional to $|V_{\mathbf{g}}|^2$, i.e., is an infinitesimal of the second order, as was stated above.

Consider an important case, in which the denominators in expressions (5.6) and (5.9a) tend to zero, i.e.,

$$1/2\ \mathbf{b}_g^2 + (\mathbf{b}_g \cdot \mathbf{k}) = \mathbf{b}_g \cdot (\mathbf{b}_g/2 + \mathbf{k}) \approx 0. \tag{5.10}$$

The corresponding coefficient, $a_g$, and the correction to the energy $\varepsilon'$, will no longer be small, and consequently, equations (5.6) and (5.9a) will themselves be inapplicable. We can imagine that, if the electron's wave vector **k** satisfies the condition (5.10) for some vector $\mathbf{b}_g$, the almost free motion of the electron experiences great perturbation. This perturbation is of the nature of a mirror reflection of the Bragg type of the electron wave from an atomic plane perpendicular to the vector $\mathbf{b}_g$, since the condition (5.10) precisely coincides with (1.4.5). In this phenomenon the wave nature of the electron moving in a periodic crystal field is especially evident.

On the other hand, condition (5.10) coincides with the equation of a plane bounding the Brillouin zones (2.9.11). Hence, if the end of the wave vector **k** lies close to a Brillouin zone boundary, the electron experiences a strong Bragg reflection in the crystal.

Consider an electron with a wave vector **k** satisfying condition (5.10) for a specified $\mathbf{b}_g$. If the corresponding amplitude $V_g$ in equation (5.6) does not tend to zero, the coefficient $a_g$ will be large (together with the coefficient $a_0$). In this case the system of equations (5.4) reduces to two equations:

$$\left[\varepsilon_{\mathbf{k}} - \frac{\hbar^2\mathbf{k}^2}{2m}\right] a_0 = V_{-g} a_g,$$

$$\left[\varepsilon_{\mathbf{k}} - \frac{\hbar^2}{2m}(\mathbf{k}+\mathbf{b}_g)^2\right] a_g = V_g a_0. \tag{5.11}$$

The system of linear homogeneous equations (5.11) for the coefficients $a_0$ and $a_g$ has nonzero solutions, if its determinant is equal to zero, i.e.,

$$\left[\varepsilon_{\mathbf{k}} - \frac{\hbar^2\mathbf{k}^2}{2m}\right]\left[\varepsilon_{\mathbf{k}} - \frac{\hbar^2}{2m}(\mathbf{k}+\mathbf{b}_g)^2\right] - |V_g|^2 = 0. \tag{5.11a}$$

Solving this quadratic equation for $\varepsilon_{\mathbf{k}}$ and substituting (5.10) into the solution we obtain

$$\varepsilon_{\mathbf{k}} = \frac{\hbar^2\mathbf{k}^2}{2m} \pm |V_g|. \tag{5.12}$$

Hence, for **k** satisfying the condition of interference (5.10) there is a discontinuity in energy equal to $2\,|V_g|$.

To illustrate the point, consider a unidimensional case in which $b_g = 2\pi g/a$, and condition (5.10) assumes the form

$$1/2\ \frac{4\pi^2 g^2}{a^2} \pm \frac{k 2\pi}{a} = 0.$$

Hence,

$$k = k_g = \pm\frac{\pi}{a} g \quad (g = 1, 2, 3, \ldots). \tag{5.13}$$

Figure 4.2 depicts the dependence of $\varepsilon_k$ vs $k$ for this case. For $k$ removed from $k_g$ the correction to the energy is small (quadratic in $|V_g|$), i.e., it can be assumed that $\varepsilon_k = \frac{\hbar^2 k^2}{2m}$, where $m$ is the free electron mass. For $k = k_g = \pm\pi/a, \pm 2\pi/a, \pm 3\pi/a$, forbidden bands of widths $2|V_1|$, $2|V_2|$, $2|V_3|$, etc., appear in the energy spectrum of the electron. The electron energy spectrum acquires a band-type structure, i.e., allowed bands alternate with the forbidden ones. We see that the energy spectrum of almost free electrons is of

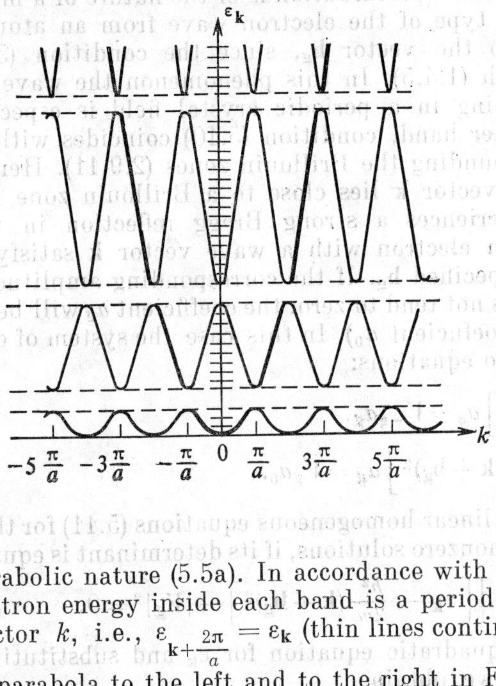


**Fig. 4.2**

an almost parabolic nature (5.5a). In accordance with the condition (3.9), the electron energy inside each band is a periodic function of the wave vector $k$, i.e., $\varepsilon_{k+\frac{2\pi}{a}} = \varepsilon_k$ (thin lines continuing the segments of the parabola to the left and to the right in Fig. 4.2). This makes it possible to consider all the energy bands inside the first, or reduced, Brillouin zone, i.e., for $-\frac{\pi}{a} \leqslant k \leqslant \frac{\pi}{a}$. We see that in the case of a weakly bonded electron the energy in the successive energy bands at $k = 0$ alternately assumes minimum and maximum values. One of the peculiarities of the unidimensional case depicted in Fig. 4.2 is that the successive allowed energy bands are always separated by the forbidden energy bands $2|V_1|$, $2|V_2|$, etc.

In the two- and three-dimensional cases this is not always so. Theory and experiment demonstrate that in such cases the electron energy of the upper energy band at some point of a Brillouin zone may prove to be lower than the electron energy of the lower energy band at some (in general, another) point of the zone. In such a case we speak of overlapping energy bands. This overlapping plays an important part in metals, where the occupation of the upper energy band by the electrons may start before the lower band is completely filled.

## 4.6 Brillouin Zones

**4.6.1** In Section 2.9.1 we considered the geometrical construction of Brillouin zones. To this end an arbitrary site of the reciprocal lattice $O$ is joined with the aid of segments with the other sites; next, through the midpoints of the segments planes normal to them are drawn. The Brillouin zones are formed by the polyhedrons bounded by those planes.

The reciprocal lattice for a planar square lattice with the side $a$ is also a square one with the side $2\pi/a$. Figure 4.3 depicts the first 10 Brillouin zones for a planar square lattice.

The reciprocal lattice of a three-dimensional simple cubic lattice is also a simple cubic one. The first Brillouin zone in the shape of a cube is obtained as the result of the intersection of the six planes passing through the midpoints of the segments joining the origin with the six nearest sites of the reciprocal cubic lattice. Twelve planes perpendicular to the segments joining the origin with the next twelve sites of the reciprocal lattice take part in shaping the second zone. Figure 4.4 depicts the first four Brillouin zones of a simple cubic lattice.

In Section 2.9.1 we noted that all Brillouin zones have the same volume equal to $(2\pi)^3/\Omega_0$, where $\Omega_0$ is the volume of the crystal unit cell. On the other hand, it follows from (3.5.27) that the number of states in the principal region $V$ per unit volume of the wave vector space is $V/(2\pi)^3$. Hence, the number of electron quantum states (spins are not taken into account) in all Brillouin zones is the same and is equal to $\frac{V}{(2\pi)^3}\frac{(2\pi)^3}{\Omega_0}=\frac{V}{\Omega_0}=G^3=N$, i.e., to the number of unit cells contained in the crystal's principal region.

By way of an example to be used in the following, consider the Brillouin zones for a face-centred and a body-centred cubic lattice.

It can be demonstrated (Appendix 13) that the reciprocal lattice for a face-centred cubic lattice is a body-centred cubic lattice, and vice versa, the reciprocal lattice for a body-centred cubic lattice is a face-centred cubic lattice. Find the shape of the first Brillouin zone for a face-centred cubic lattice. The reciprocal lattice in this case is a body-centred cube, and, consequently, each of its sites is surrounded by eight nearest sites (Section 1.2.2). Eight planes drawn perpendicul-

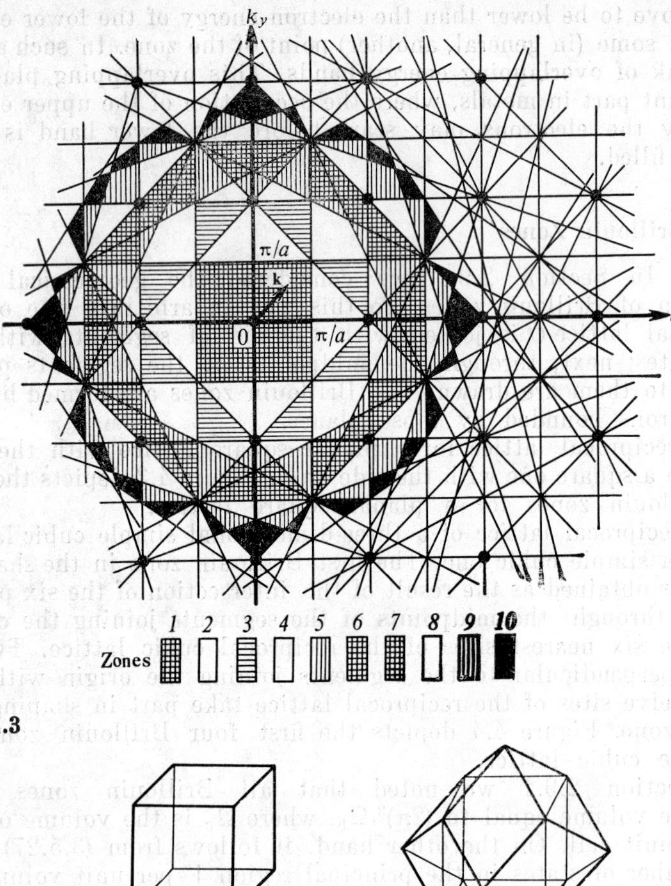


**Fig. 4.3**

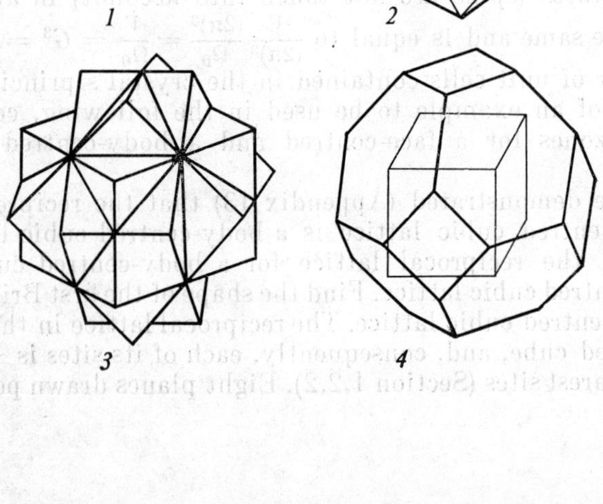


**Fig. 4.4**

arly through the midpoints of segments joining the origin with the eight nearest sites intersect to shape a regular octahedron with six vertices and with regular triangles forming its faces. The sites of the next coordination group are the six sites arranged in pairs along the $x$, $y$ and $z$ axes at a distance $a$ (the cube edge) from each other. They determine the six planes that cut off the octahedron's six vertices. As a result, we obtain for the first Brillouin zone a polyhedron with fourteen faces, six of them squares and eight hexa-

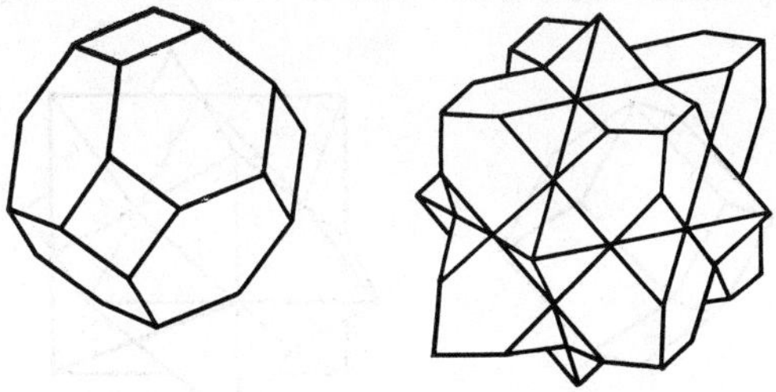

(a) (b)

Fig. 4.5

gons, as depicted in Fig. 4.5*a*. The geometrical shape of the second Brillouin zone is in this case very intricate, as depicted in Fig. 4.5*b*. Equations (5.10) for the faces of the first Brillouin zone can be easily written. Making, for example, $\mathbf{b}_g = \mathbf{b}_1$, we obtain from expressions (5.10) and (A. 13.3)

$$\frac{2\pi^2}{a^2}(\mathbf{i}_0-\mathbf{j}_0+\mathbf{k}_0)^2+\mathbf{k}\cdot\frac{2\pi}{a}(\mathbf{i}_0-\mathbf{j}_0+\mathbf{k}_0)=0$$

or

$$-k_x+k_y-k_z=3\pi/a.$$

This equation determines one of the eight hexagonal faces. If we make $\mathbf{b}_g=\mathbf{b}_1+\mathbf{b}_2=\frac{4\pi}{a}\mathbf{i}_0$, as stipulated by (A.13.3), we shall obtain from condition (5.10)

$$k_x=-2\pi/a,$$

thus determining one of the six square faces.

The shape of the Brillouin zones for a direct body-centred cubic lattice can be determined in a similar way. Since in this case the reciprocal lattice is of the face-centred cubic structure, the first coordination group consists of 12 sites (Section 1.2.2). The twelve planes drawn in accordance with equation (5.10) completely determine in this case the first Brillouin zone in the shape of the polyhedron with twelve faces (dodecahedron) depicted in Fig. 4.6*a*. The second Brillouin zone for this case is as shown in Fig. 4.6*b*.

**4.6.2** The planes in the **k** space determined by equation (5.10) form a band edge (the boundary of discontinuity of the electron energy $\varepsilon_{\mathbf{k}}$) only if $V_{\mathbf{g}} \neq 0$, since only in this case the coefficient $a_{\mathbf{h}}$ and the correction to the energy $\varepsilon'$, as stipulated by equations (5.6) and (5.9a), become large. In a complex lattice the conditions of interference of electron waves scattered by the atoms of individual sublattices may cause the extinction of some of the $V_{\mathbf{g}}$. In this case

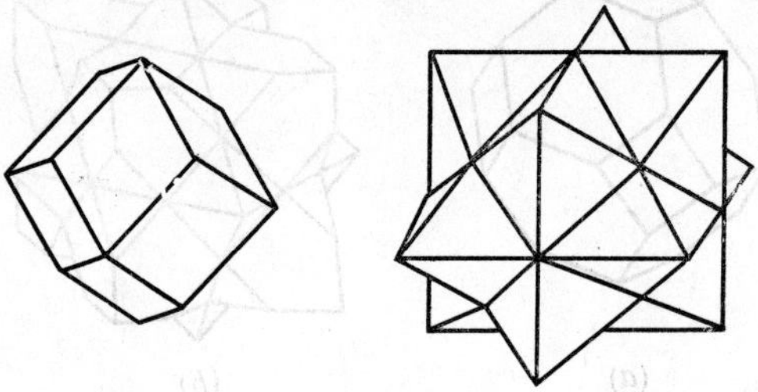

(*a*) (*b*)

**Fig. 4.6**

the energy $\varepsilon_{\mathbf{k}}$ for the corresponding face of the Brillouin zone experiences no discontinuity, i.e., the appropriate face is not actually a boundary of an energy band. As we shall now see, the situation here perfectly coincides with the one we encountered, when determining the structural factor of $X$-ray scattering (Section 1.4.5).

Consider a complex lattice with the main periods $\mathbf{a}_1$, $\mathbf{a}_2$, $\mathbf{a}_3$ and let the position of the $s$ atoms in a cell be determined with respect to its origin by $s$ vectors $\mathbf{r}_n = u_n\mathbf{a}_1 + v_n\mathbf{a}_2 + w_n\mathbf{a}_3$ ($n = 1, 2, \ldots, s$), so that the lattice basis is ($u_n$, $v_n$, $w_n$). If all the $s$ atoms are identical, the electron potential energy is the sum of the energies of its interaction with each of the $s$ sublattices:

$$V(\mathbf{r}) = \sum_n V_1(\mathbf{r} - \mathbf{r}_n), \tag{6.1}$$

where $|\mathbf{r} - \mathbf{r}_n|$ is the distance from the electron to the $n$-th atom in the zeroth cell, and $V_1(\mathbf{r})$ is the periodic potential of one of the sublattices. Hence,

$$V_1(\mathbf{r} - \mathbf{r}_n) = \sum_{\mathbf{g}} V_{1\mathbf{g}} e^{i\mathbf{b}_{\mathbf{g}} \cdot (\mathbf{r} - \mathbf{r}_n)}, \tag{6.1a}$$

and, therefore,

$$V(\mathbf{r}) = \sum_{\mathbf{g}} \sum_n V_{1\mathbf{g}} e^{-i\mathbf{b}_{\mathbf{g}} \cdot \mathbf{r}_n} e^{i\mathbf{b}_{\mathbf{g}} \cdot \mathbf{r}}. \tag{6.1b}$$

This may be written down in the following form:

$$V(\mathbf{r}) = \sum_{\mathbf{g}} V_{\mathbf{g}} e^{i\mathbf{b}_{\mathbf{g}}\cdot\mathbf{r}}, \tag{6.1c}$$

where

$$V_{\mathbf{g}} = V_{1\mathbf{g}} S_{\mathbf{g}}, \tag{6.1d}$$

and the structural factor

$$S_{\mathbf{g}} = \sum_{n=1}^{s} e^{-i\mathbf{b}_{\mathbf{g}}\cdot\mathbf{r}_n} = \sum_{n=1}^{s} e^{-2\pi i(g_1u_n+g_2v_n+g_3w_n)}. \tag{6.1e}$$

The expression for the structural factor for electrons (6.1e) coincides with that for $X$-rays (1.4.11a). Hence, if $S_{\mathbf{g}} = S_{g_1g_2g_3}$ is zero, all the corresponding physically equivalent planes $\{g_1g_2g_3\}$ do not effectively scatter $X$-rays or perturb the motion of almost free electrons.

Consider a simple cubic lattice for which the reciprocal lattice is also a simple cubic one with the basic vectors $b_i = 2\pi/a$ $(i = 1, 2, 3)$. In this case the distance from the plane $\{g_1g_2g_3\}$ to the origin, as can be seen from the equation for this plane (5.10), is equal to

$$h_{g_1g_2g_3} = \frac{1}{2} b_{g_1g_2g_3} = \frac{\pi}{a}\sqrt{g_1^2+g_2^2+g_3^2}.$$

Table 4.1 gives the numbers of some of the $\{g_1g_2g_3\}$ planes and their distances to the origin.

Table 4.1

| Plane | {100} | {110} | {111} | {200} | {220} |
|---|---|---|---|---|---|
| Number of planes | 6 | 12 | 8 | 6 | 12 |
| $h_{g_1g_2g_3}\times\frac{a}{\pi}$ | 1 | $\sqrt{2}$ | $\sqrt{3}$ | 2 | $2\sqrt{2}$ |

As the first example of a complex lattice, consider the face-centred cube already considered in the preceding section as a simple lattice. Should the cubic edges be chosen as the $a_i$, the unit cell of a face-centred cube would contain four atoms with the basis $(u_nv_nw_n) = (000, 0\,{}^1/_2, {}^1/_2, {}^1/_2\,0\,{}^1/_2, {}^1/_2\,{}^1/_2\,0)$. The structural factor, according to condition (6.1e), would be

$$S_{g_1g_2g_3} = 1 + e^{-\pi i(g_2+g_3)} + e^{-\pi i(g_1+g_3)} + e^{-\pi i(g_1+g_2)}. \tag{6.2}$$

Hence, it follows directly that

$$S_{100} = 0, \quad S_{110} = 0, \quad S_{111} = 4, \quad S_{200} = 4. \tag{6.2a}$$

Thus, the first Brillouin zone is bounded by eight {111} planes and by six {200} planes. The intersections of those planes form a truncated octahedron or a polyhedron of 14 faces depicted in Fig. 4.5*a*. This result coincides with that obtained in the preceding item.

Consider now a really complex lattice of the diamond type (see Fig. 1.13) that cannot be reduced to a simple Bravais lattice (Section 1.2, 5). Since the diamond lattice can be represented as two face-centred cubic lattices displaced with respect to one another along a cube's space diagonal by 1/4 of its length, it follows that if one chooses the unit cell in the shape of a cube (Fig. 1.13), it will contain eight atoms with the basis (000, $0\,{}^1/_2\,{}^1/_2$, ${}^1/_2\,0\,{}^1/_2$, ${}^1/_2\,{}^1/_2 0$, ${}^1/_4\,{}^1/_4\,{}^1/_4$, ${}^1/_4\,{}^3/_4\,{}^3/_4$, ${}^3/_4\,{}^1/_4\,{}^3/_4$, ${}^3/_4\,{}^3/_4\,{}^1/_4$). The structural factor is

$$S_{g_1g_2g_3} = 1 + e^{-\pi i(g_2+g_3)} + e^{-\pi i(g_2+g_3)} + e^{-\pi i(g_1+g_2)} + e^{-{}^1/_2\pi i(g_1+g_2+g_3)} + e^{-{}^1/_2\pi i(g_1+3g_2+3g_3)} + e^{-{}^1/_2\pi i(3g_1+g_2+3g_3)} + e^{-{}^1/_2\pi i(3g_1+3g_2+g_3)} \tag{6.3}$$

Hence, it follows that

$$S_{100} = 0, \quad S_{110} = 0, \quad S_{111} = 6, \quad S_{200} = 0, \quad S_{211} = 0,$$
$$S_{220} = 8, \; S_{221} = 0.$$

Since the crystal lattices of germanium and silicon may be regarded as two face-centred cubic lattices displaced with respect to one another, their first Brillouin zone has the shape of a polyhedron with fourteen faces depicted in Fig. 4.5*a*[6].

## 4.7. Tight Binding Approximation

4.7.1 In the preceding section we considered the periodic potential $V$ (**r**) acting on the electron as a small perturbation of its free motion. The electron motion is strongly perturbed, only if certain conditions of interference (5.10) are met. This treatment is justified from the quantitative viewpoint, only if the electron kinetic energy is large as compared with the spatial variations of its potential energy $V$ (**r**). This situation is brought about, for example, when a crystal is irradiated by electrons of at least several hundred eV. On the other hand, it follows from the quantum mechanics virial theorem that the average kinetic energy of an electron in an atom, a molecule or a crystal should be of the order of the

[6] Other examples of calculations of structural factors and of Brillouin zones of complex lattices are contained in [4.6] and [4.7].

variations of its potential energy, and for this reason the weak binding approximation is inapplicable to the electrons in a crystal. There is an alternative approach that assumes that the electron's state in an isolated atom changes little, when the atoms join to form a crystal. This tight binding approximation of electrons is obviously more justified for the electrons of the deeper lying atomic energy levels, i.e., for those whose interaction with atoms in other sites is relatively weak. Of course, neither the tight binding nor the weak binding approximation of electrons describes the state of the electrons in the conduction band correctly from the quantitative point of view. For this reason neither of those approximations can be employed for quantitative calculations of the energy spectrum and of the wave functions of the conduction electrons in specific crystals. The essential point is, however, that they provide a good illustration of the general conclusions about the motion of the electron in a periodic field. In certain cases these illustrations provide new qualitative conclusions about the electron state in a periodic field.

**4.7.2** Let the electron wave function in an isolated atom $\psi_0(r)$ in the $s$-state satisfy the Schrödinger equation

$$-\frac{\hbar^2}{2m}\nabla^2\psi_0+\mathcal{U}(r)\,\psi_0=\varepsilon_0\psi_0, \tag{7.1}$$

where $\mathcal{U}(r)$ is the spherically symmetrical field of an isolated ion (for a metal) or of a neutral atom (for a semiconductor in which the conduction electron is a "surplus" one), $\varepsilon_0$ is the energy of the electron state in the ion (atom) being considered. Let

$$\int \psi_0^*\psi_0\,d\tau=1, \tag{7.1a}$$

i.e., the wave function $\psi_0$ is normalized to unity.

In the vicinity of the $n$-th lattice site the electron of an isolated atom is described by the wave function $\psi_0(|\mathbf{r}-\mathbf{a}_n|)$. In an ideal lattice all the $N=G^3$ sites of the principal region are absolutely equivalent, therefore the state of the electron with the energy $\varepsilon_0$ is $N$-fold degenerate (spins are not taken into account). The effect of the electron's interaction with all other atoms is to split the energy level $\varepsilon_0$ into a band, so that the degeneracy will be (at least partially) lifted.

As is established in the theory of perturbed degenerate states [4.1, § 50] and as seems obvious from general considerations, the wave function of the zeroth approximation should be built from all the degenerate wave functions in the form of the linear expression

$$\psi(\mathbf{r})=\sum_{\mathbf{n}} C_{\mathbf{n}}\psi_0(|\mathbf{r}-\mathbf{a}_n|). \tag{7.2}$$

The constant coefficients $C_{\mathbf{n}}$ are determined in compliance with the well-known rule [4.1, § 50]. To find the coefficients $C_{\mathbf{n}}$ we shall make use of a simple method making (7.2) satisfy the general requirements concerning the form of the Bloch function (3.1). It may easily be seen that to this end it suffices to make $C_{\mathbf{n}} = \exp(i\mathbf{k}\cdot\mathbf{a}_{\mathbf{n}})$, i.e.,

$$\psi(\mathbf{r}) = \sum_{\mathbf{n}} e^{i\mathbf{k}\cdot\mathbf{a}_{\mathbf{n}}} \psi_0(|\mathbf{r} - \mathbf{a}_{\mathbf{n}}|). \tag{7.2a}$$

Indeed, write expression (7.2a) down in the form

$$\psi(\mathbf{r}) = e^{i\mathbf{k}\cdot\mathbf{r}} \sum_{\mathbf{n}} e^{i\mathbf{k}\cdot(\mathbf{a}_{\mathbf{n}} - \mathbf{r})} \psi_0(|\mathbf{r} - \mathbf{a}_{\mathbf{n}}|). \tag{7.2b}$$

Prove that the multiplier in front of $e^{i\mathbf{k}\cdot\mathbf{r}}$ has the periodicity of the lattice, i.e., that it can be regarded as a modulating multiplier $u_{\mathbf{k}}(\mathbf{r})$ in the Bloch wave function (3.1). Substitute into this multiplier the vector $\mathbf{r} + \mathbf{a}_{\mathrm{m}}$ for $\mathbf{r}$; we obtain

$$\psi(\mathbf{r}) = \sum_{\mathbf{n}} e^{i\mathbf{k}\cdot(\mathbf{a}_{\mathbf{n}} - \mathbf{r} - \mathbf{a}_{\mathrm{m}})} \psi_0(|\mathbf{r} + \mathbf{a}_{\mathrm{m}} - \mathbf{a}_{\mathbf{n}}|).$$

Replace the summation over $\mathbf{n}$ by the summation over $\mathbf{l}$ making $\mathbf{a}_{\mathbf{n}} - a_{\mathrm{m}} = \mathbf{a}_{\mathbf{l}}$; then we obtain

$$\psi(\mathbf{r}) = \sum_{\mathbf{l}} e^{i\mathbf{k}\cdot(\mathbf{a}_{\mathbf{l}} - \mathbf{r})} \psi_0(|\mathbf{r} - \mathbf{a}_{\mathbf{l}}|),$$

this obviously coincides with the initial expression (7.2b). Thus we have at least demonstrated that the choice of the electron wave function in the form of (7.2a) satisfies the conditions of translational symmetry. It is, moreover, apparent that because of an exponentially rapid decline of the wave function $\psi_0$, the electron wave function in a crystal (7.2a) in the vicinity of the $n$-th site behaves approximately like

$$\psi(\mathbf{r}) \approx \text{const.} \times \psi_0(|\mathbf{r} - \mathbf{a}_n|),$$

i.e., like the atomic function of the **n**-th site.

If $V(\mathbf{r})$ is a self-consistent periodic potential acting on the electron (obviously $V(\mathbf{r}) \neq \sum_{\mathbf{n}} \mathcal{U}(\mathbf{r} - \mathbf{a}_{\mathbf{n}})$), then the precise single-electron wave function satisfies the equation

$$\hat{\mathcal{H}}\psi = -\frac{\hbar^2}{2m}\nabla^2\psi + V(\mathbf{r})\psi = \varepsilon\psi, \tag{7.3}$$

where $\varepsilon$ is the energy eigenvalue of an electron moving in the crystal. Premultiplying both sides of equation (7.3) by $\psi^*$ and integrating inside the crystal's principal region we obtain

$$\varepsilon = \frac{\int \psi^* \hat{\mathcal{H}} \psi \, d\tau}{\int \psi^* \psi \, d\tau} = \frac{\int \varphi^* \left[ -\frac{\hbar^2}{2m}\nabla^2 + V(\mathbf{r}) \right] \psi \, d\tau}{\int \psi^* \psi \, d\tau}. \tag{7.4}$$

Calculate the electron energy in the assumption that the wave function $\psi(\mathbf{r})$ is written in the form of expression (7.2a). Denoting $\mathbf{r} - \mathbf{a_n} = \boldsymbol{\rho}_\mathbf{n}$, we obtain

$$\hat{\mathcal{H}}\psi = \left[-\frac{\hbar^2}{2m}\nabla^2 + V(\mathbf{r})\right]\sum_{\mathbf{n}} e^{i\mathbf{k}\cdot\mathbf{a_n}}\psi_0(\boldsymbol{\rho}_\mathbf{n})$$

$$= \varepsilon_0 \sum_{\mathbf{n}} e^{i\mathbf{k}\cdot\mathbf{a_n}}\psi_0(\boldsymbol{\rho}_\mathbf{n}) + \sum_{\mathbf{n}} e^{i\mathbf{k}\cdot\mathbf{a_n}}[V(\mathbf{r}) - \mathcal{U}(\boldsymbol{\rho}_\mathbf{n})]\,\psi_0(\boldsymbol{\rho}_\mathbf{n}),$$

where we made use of equation (7.1) to substitute $-\frac{\hbar^2}{2m}\nabla^2\psi_0(\boldsymbol{\rho}_\mathbf{n})$. Substituting the expression obtained into (7.4) and substituting $\psi^*$ according to (7.2a) we obtain

$$\varepsilon = \varepsilon_0 + \frac{\sum_{\mathbf{m}}\sum_{\mathbf{n}} e^{i\mathbf{k}\cdot(\mathbf{a_n}-\mathbf{a_m})}\int \psi_0^*(\boldsymbol{\rho}_\mathbf{m})[V(\mathbf{r}) - \mathcal{U}(\boldsymbol{\rho}_\mathbf{n})]\,\psi_0(\boldsymbol{\rho}_\mathbf{n})\,d\tau}{\sum_{\mathbf{m}}\sum_{\mathbf{n}} e^{i\mathbf{k}\cdot(\mathbf{a_n}-\mathbf{a_m})}\int \psi_0^*(\boldsymbol{\rho}_\mathbf{m})\,\psi_0(\boldsymbol{\rho}_\mathbf{n})\,d\tau}.$$

Since all the sites are equivalent, both the numerator and the denominator do not depend on the **m** and **n** separately but only on their difference, i.e., on the relative position of the sites. Therefore, we may make $\mathbf{m} = 0$ (i.e., $\mathbf{a_m} = 0$, $\boldsymbol{\rho}_\mathbf{m} = \mathbf{r}$) and replace the summation over **m** both in the numerator and the denominator and multiply by the number of sites $N$ contained in the principal region. Hence,

$$\varepsilon = \varepsilon_0 + \frac{\sum_{\mathbf{n}} e^{i\mathbf{k}\cdot\mathbf{a_n}}\int \psi_0^*(\mathbf{r})[V(\mathbf{r}) - \mathcal{U}(\boldsymbol{\rho}_\mathbf{n})]\,\psi_0(\boldsymbol{\rho}_\mathbf{n})\,d\tau}{\sum_{\mathbf{n}} e^{i\mathbf{k}\cdot\mathbf{a_n}}\int \psi_0^*(\mathbf{r})\,\psi_0(\boldsymbol{\rho}_\mathbf{n})\,d\tau}. \tag{7.5}$$

Assume the decline of the atomic wave functions $\psi_0$ to be so rapid that their overlapping even between the neighbouring sites can be neglected, i.e., that

$$\int \psi_0^*(\mathbf{r})\,\psi_0(\boldsymbol{\rho}_\mathbf{n})\,d\tau = \delta_{0\mathbf{n}} = \begin{cases} 0, & \mathbf{n} \neq 0, \\ 1, & \mathbf{n} = 0. \end{cases} \tag{7.6}$$

Denote the integral in the numerator of the fraction (7.5) for $\mathbf{n} = 0$

$$\int \psi_0^*(\mathbf{r})[V(\mathbf{r}) - \mathcal{U}(\mathbf{r})]\,\psi_0(\mathbf{r})\,d\tau$$

$$= \int |\psi_0(\mathbf{r})|^2 [V(\mathbf{r}) - \mathcal{U}(\mathbf{r})]\,d\tau = -C < 0. \tag{7.7}$$

The negative value of the integral (7.7) may to some extent be explained as follows. In Figure 4.7 the solid line depicts the potential of an isolated atom $\mathcal{U}$, and the dashed line depicts the self-consistent periodic potential $V$. If we were to make a rather natural assumption that atomic interaction reduces the potential barriers for the electron as shown in the figure, we would see that the expression in square brackets in the integral (7.7) is everywhere negative, and therefore the integral itself is negative.

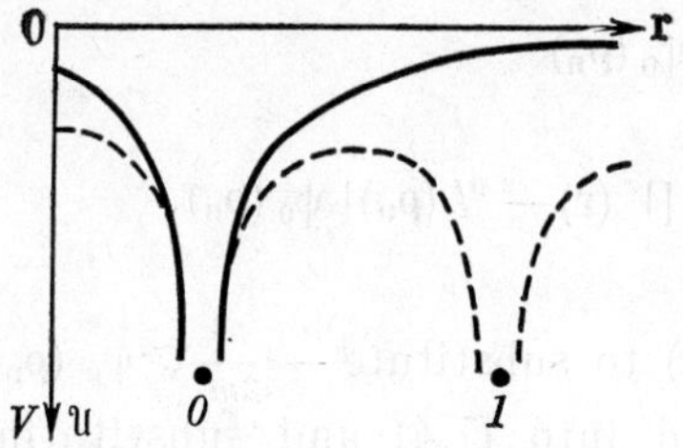

Fig. 4.7

Although we neglect the overlapping of the wave functions of the different sites themselves, it would be a consistent step to include the integrals in the numerator of the fraction (7.5) for $\mathbf{n} \neq 0$ for the neighbouring sites $\mathbf{n}_0$:

$$\int \psi_0^* (\mathbf{r}) [V(\mathbf{r}) - \mathcal{U}(\boldsymbol{\rho}_{\mathbf{n}_0})] \psi_0 (\boldsymbol{\rho}_{\mathbf{n}_0}) \, d\tau = -A_{\mathbf{n}_0}. \tag{7.8}$$

Indeed, although $\psi_0 (\boldsymbol{\rho}_{\mathbf{n}_0})$ is small close to the zeroth site, this is partly compensated for in this region by the large value of the difference $V(\mathbf{r}) - \mathcal{U}(\boldsymbol{\rho}_{\mathbf{n}_0})$. Certain general considerations that we shall not touch upon justify the assumption that in case (7.8) the integral is also negative. From relations (7.5)-(7.8) we obtain

$$\varepsilon = \varepsilon_0 - C - \sum_{\mathbf{n}_0} A_{\mathbf{n}_0} e^{i\mathbf{k}\cdot\mathbf{a}_{\mathbf{n}_0}}. \tag{7.9}$$

If the wave function $\psi_0 (\mathbf{r})$ belongs to the $s$-state, then $A_{\mathbf{n}_0}$ depends only on the distance from the zeroth site to the $\mathbf{n}_0$-th atom of the first coordination group, i.e., it is the same for all such atoms. In this case

$$\varepsilon = \varepsilon_0 - C - A \sum_{\mathbf{n}_0} e^{i\mathbf{k}\cdot\mathbf{a}_{\mathbf{n}_0}}. \tag{7.10}$$

In a simple cubic lattice every atom is surrounded by six nearest neighbours. Directing the $x$, $y$ and $z$ axes along the cube edges we obtain

$$\sum_{\mathbf{n}_0}^{1,6} e^{i\mathbf{k}\cdot\mathbf{a}_{\mathbf{n}_0}} = e^{ik_x a} + e^{-ik_x a} + e^{ik_y a} + e^{-ik_y a} + e^{ik_z a} + e^{-ik_z a}$$

$$= 2 [\cos ak_x + \cos ak_y + \cos ak_z],$$

where $a$ is the lattice parameter. Substituting the results into (7.10), we obtain

$$\varepsilon = \varepsilon_0 - C - 2A\,[\cos ak_x + \cos ak_y + \cos ak_z]. \tag{7.10a}$$

In the case of a body-centred cube every atom is surrounded by eight nearest neighbours, and instead of (7.10a) we obtain (Appendix 14)

$$\varepsilon = \varepsilon_0 - C - 8A \cos\frac{ak_x}{2}\cos\frac{ak_y}{2}\cos\frac{ak_z}{2}. \tag{7.10b}$$

The electron energy in a face-centred cubic lattice can be calculated in a similar way:

$$\varepsilon = \varepsilon_0 - C - 4A\left[\cos\frac{ak_x}{2}\cos\frac{ak_y}{2} + \cos\frac{ak_x}{2}\cos\frac{ak_z}{2} + \cos\frac{ak_y}{2}\cos\frac{ak_z}{2}\right]. \tag{7.10c}$$

It follows from expressions (7.10a)-(7.10c) that when a crystal is constituted from individual atoms, the electron energy of the isolated atom, $\varepsilon_0$, as the result of the interaction with the neighbouring atoms, is displaced by the amount $C$, and splits into an energy band inside which the electron energy is a periodic function of the wave vector $\mathbf{k}$. Every stationary energy level of an isolated atom experiences such a splitting; accordingly, the qualitative picture will be similar to that depicted in Fig. 4.2 for almost free electrons. The width of the allowed energy band, as we shall see below, is proportional to $A$, i.e., is determined to a great extent by the overlapping of the atomic wave functions $\psi_0(\mathbf{r})$ of the neighbouring atoms. For the outer valence electrons usually of interest to us this overlapping is great, so that the width of the energy band reaches several electron-volts, i.e., is of the order of, and even exceeds, the spacings between the energy levels of an isolated atom. Strictly speaking, this means that the method of strong bonding developed above is inapplicable in this case. For the electrons of the inner atomic shells this splitting is small; thus, for the $K$-electrons in the lattice of metallic sodium it is of the order of $2 \times 10^{-19}$ eV, so that the level remains practically a sharp one. As a result, the electron energy spectrum in a crystal assumes the form depicted in Fig. 4.8, where the allowed energy bands are shaded.

How can the electron motion through the lattice be imagined from the viewpoint of the tight binding approximation? In an isolated atom the electron spends an indefinitely long time on the stationary level $\varepsilon_0$. As such identical atoms are drawn together, and a lattice is formed, the possibility appears for the electron to go over from one lattice site to another by means of the quantum mechanics tunnel effect. Atomic interaction results in the widening of a sharp energy level into a band $\varepsilon_{ab}$ wide, which is related to the time an electron spends near a specified site $\tau$, by means of the indeterminacy relation $\varepsilon_{ab}\tau \approx \hbar$. For outer electrons $\varepsilon_{ab} \approx 10$ eV, and the corresponding times are $\tau \approx 10^{-15}$ s, but for the $K$-electrons of sodium, where $\varepsilon_{ab} \approx 2 \times 10^{-19}$ eV, an electron moves from one site to the neighbouring site on the average of once an hour. However, in the latter case the electron in a stationary state is distributed with an equal probability over all crystal lattice sites.

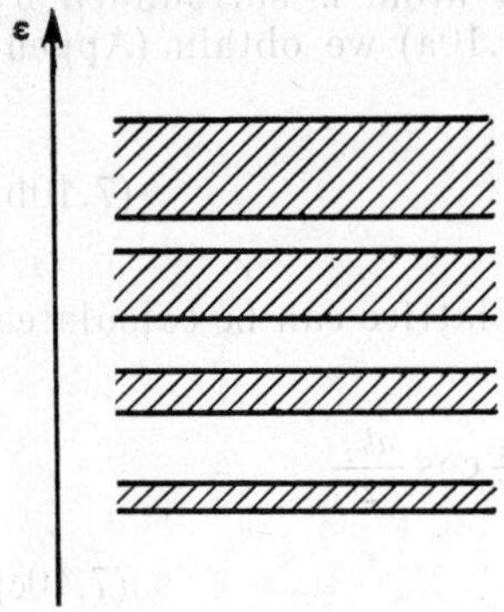


Fig. 4.8

**4.7.3** Let us illustrate the properties of the electron moving in an ideal periodic field (Section 4.3) by the example of tightly bound electrons in a simple cubic lattice. We shall measure the electron energy in expressions (7.10) from the level $\varepsilon_0 - C$, i.e., we shall make $\varepsilon_0 - C = 0$. It follows from (7.10a) that the electron energy $\varepsilon$ has a minimum equal to $\varepsilon_b = -6A$ at $k_x = k_y = k_z = 0$, i.e., in the centre of the Brillouin zone and a maximum equal to $\varepsilon_a = +6A$ at $k_x = k_y = k_z = \pi/a$, i.e., at the vertices of the zone's cube. Hence, in the case of a simple cubic lattice the bandwidth $\varepsilon_{ab} = \varepsilon_a - \varepsilon_b = 12A$.

For small $\mathbf{k}$, i.e., for $ak_x \ll 1$, $ak_y \ll 1$ and $ak_z \ll 1$, we obtain by expanding the cosines in series

$$\varepsilon = -2A\left[\left(1-\frac{(ak_x)^2}{2}\right)+\left(1-\frac{(ak_y)^2}{2}\right)+\left(1-\frac{(ak_z)^2}{2}\right)\right]$$
$$= -6A + Aa^2(k_x^2+k_y^2+k_z^2) = \varepsilon_b + Aa^2k^2. \tag{7.11}$$

Hence, for small $\mathbf{k}$ the electron energy, as in the case of weak bonding, is independent of the $\mathbf{k}$ direction and proportional to $k^2$. The scalar effective mass of the electron near the lower band edge is

$$m_n^*(b) = \frac{\hbar^2}{\left(\dfrac{\partial^2\varepsilon}{\partial k_x^2}\right)_b} = \frac{\hbar^2}{2Aa^2} = \frac{6\hbar^2}{a^2\varepsilon_{ab}}, \tag{7.12}$$

i.e., it is positive, and, if we approximately make $a = \text{const.}$ for different crystals, inversely proportional to the bandwidth $\varepsilon_{ab}$.

If the upper band edge we make $\frac{\pi}{a} - k_x = k'_x$, etc. and $ak'_x \ll 1$, etc., then

$$\varepsilon = -2A\,[\cos(\pi - ak'_x) + \cos(\pi - ak'_y) + \cos(\pi - ak'_z)]$$
$$= 2A\,[\cos ak'_x + \cos ak'_y + \cos ak'_z]$$
$$= 6A - Aa^2(k'^2_x + k'^2_y + k'^2_z) = \varepsilon_a - Aa^2k'^2. \qquad (7.11a)$$

Similarly the electron effective mass near the upper band edge is

$$m^*_n(a) = \frac{\hbar^2}{\left(\dfrac{\partial^2\varepsilon}{\partial k'^2_x}\right)_a} = -\frac{\hbar^2}{2Aa^2}, \qquad (7.12a)$$

i.e., it is negative. Consequently the effective mass of a hole

$$m^*_p(a) = -m^*_n(a) = \hbar^2/2Aa^2, \qquad (7.12b)$$

i.e., coincides with the effective mass of the electron near the lower band edge (7.12).

The shape of the constant-energy surfaces in the **k**-space near the zone centre and the cube vertices ($\pm k_x = \pm k_y = \pm k_z = \pi/a$) is obviously spherical. For intermediate energy values it is more

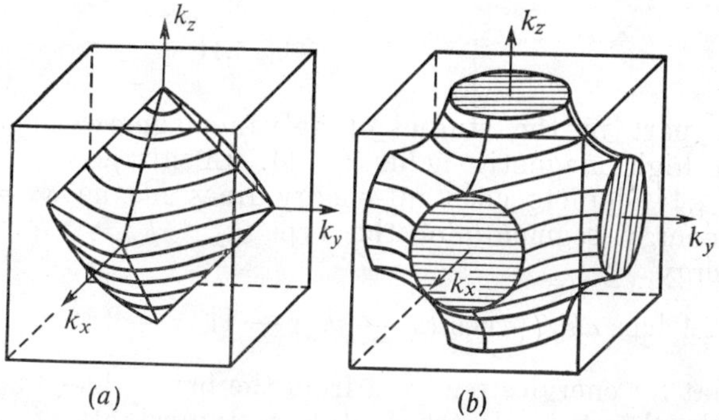


**Fig. 4.9**

intricate, as depicted in Fig. 4.9, where (*a*) corresponds to the energy $\varepsilon(\mathbf{k}) = -2A$ and (*b*) to the energy $\varepsilon(\mathbf{k}) = 0$ ($\varepsilon_0 - C = 0$, as before).

Since the energy $\varepsilon$ is periodic in the **k**-space with the periods $\mathbf{b}_i$, the shape of the $\varepsilon$ = const. surface is repeated in all the elementary cells (enlarged $2\pi$ times) of the reciprocal lattice space. For the case depicted in Fig. 4.9*b* the multiconnected constant-energy surface is of the shape shown in Fig. 4.10. In the case of the constant-energy surface $\varepsilon(\mathbf{k}) = -2A$ (Fig. 4.9*a*) the constant-energy surfaces of the neighbouring cells in the reciprocal lattice space have only common points of contact in the centre of reciprocal faces of the Brillouin zone. When the electron energy is less than $\varepsilon = -2A$, the closed constant-energy surfaces of the different cells of the reciprocal lattice have no common points of contact.

Such topology of the constant-energy surfaces for $\varepsilon$ (**k**) equal to the energy of the conduction Fermi-electrons in a metal plays an

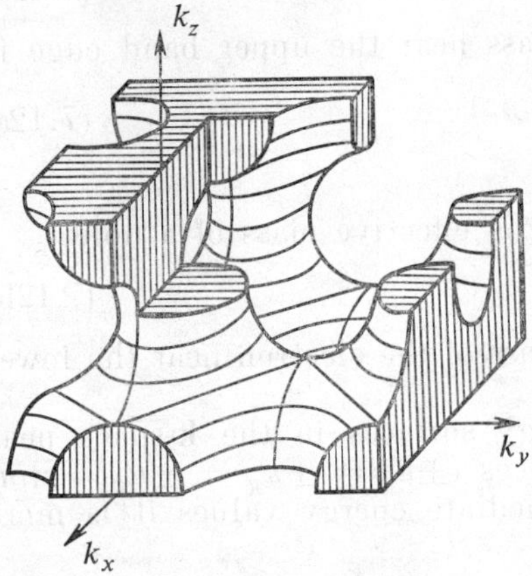


Fig. 4.10

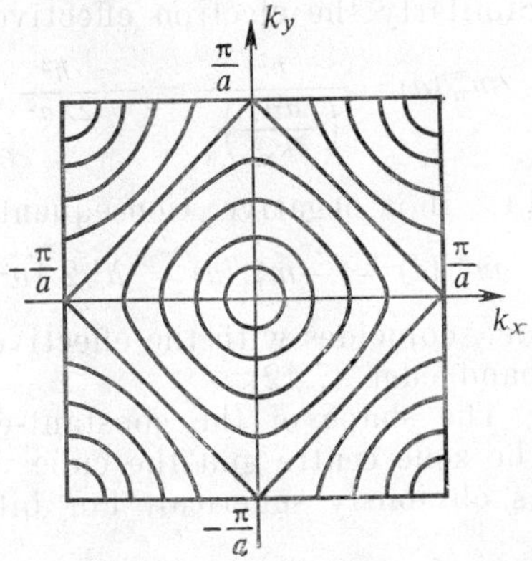


Fig. 4.11

important part in the studies of galvanomagnetic phenomena in metals in high magnetic fields (I. M. Lifshitz).

Figure 4.11 depicts constant-energy lines in the $k_z = 0$ plane. Constant-energy segments of the type $k_x + k_y = \pi/a$ correspond to the energy

$$\varepsilon = -2A\ [\cos ak_x + \cos(\pi - ak_x) + 1] = -2A.$$

We see that for energies removed from the band edges the behaviour of strongly bonded electrons differs appreciably from that of almost free electrons.

It follows from expressions (3.32) and (7.10a) that the mean quantum-mechanical velocity of the electron in the **k**-state is equal to

$$\mathbf{v} = \frac{2Aa}{\hbar} (\sin ak_x \mathbf{i}_0 + \sin ak_y \mathbf{j}_0 + \sin ak_z \mathbf{k}_0), \tag{7.13}$$

i.e., it depends not only on the magnitude of the wave vector **k**, but on its direction, as well. It will be seen from (7.13) that close to the lower and the upper band edges the velocity is zero. For the small **k**

$$\mathbf{v} = \frac{2Aa^2}{\hbar} (k_x \mathbf{i}_0 + k_y \mathbf{j}_0 + k_z \mathbf{k}_0) = \frac{2Aa^2\mathbf{k}}{\hbar} = \frac{\mathbf{p}}{m_n^*(b)}, \tag{7.13a}$$

where $\mathbf{p} = \hbar\mathbf{k}$ is the quasi-momentum, and the effective mass $m_n^*(b)$ is equal to (7.12).

Making $k_y = k_z = 0$ and $k_x = k$, we obtain for the motion along the $x$ axis

$$\varepsilon = -2A \cos ak, \tag{7.14}$$

$$v = \frac{2Aa}{\hbar} \sin ak. \tag{7.14a}$$

Figure 4.12 depicts the dependence of $\varepsilon$ vs $v$ and $k$ expressed by (7.14) and (7.14a). In accordance with (4.4)

$$\int_{-\pi/a}^{+\pi/a} v(k)\, dk = \frac{2Aa}{\hbar} \int_{-\pi/a}^{+\pi/a} \sin ak\, dk = 0,$$

i.e., the average velocity of the electron (its current) throughout the band is zero. Figure 4.12 shows especially clearly that $\varepsilon$ and $v$ have

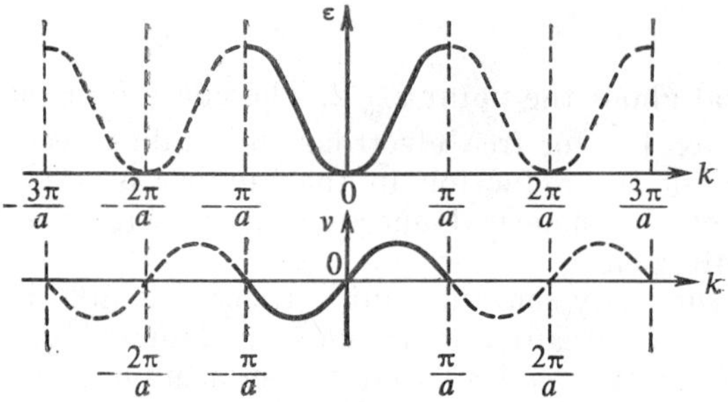


**Fig. 4.12**

a periodic dependence on the wave vector $k$, and that therefore they can be considered only inside the reduced band

$$-\frac{\pi}{a} \leqslant k \leqslant +\frac{\pi}{a}.$$

The density of states (per 1 cm³) according to expressions (3.26) and (7.10a) is equal to

$$g(\varepsilon) = \frac{1}{8\pi^3} \int \frac{d\sigma}{|\operatorname{grad}_{\mathbf{k}} \varepsilon|}$$

$$= \frac{1}{16\pi^3 Aa} \int \frac{d\sigma}{\sqrt{\sin^2 ak_x + \sin^2 ak_y + \sin^2 ak_z}}, \tag{7.15}$$

where the integration is performed over the surface

$$\varepsilon = -2A(\cos ak_x + \cos ak_y + \cos ak_z) = \text{const}. \tag{7.15a}$$

Making use of this relation we can reduce the integral (7.15) to a unidimensional one and compute it approximately. Curve *a* in Fig. 4.13 is the result of such a computation. Note that the curve

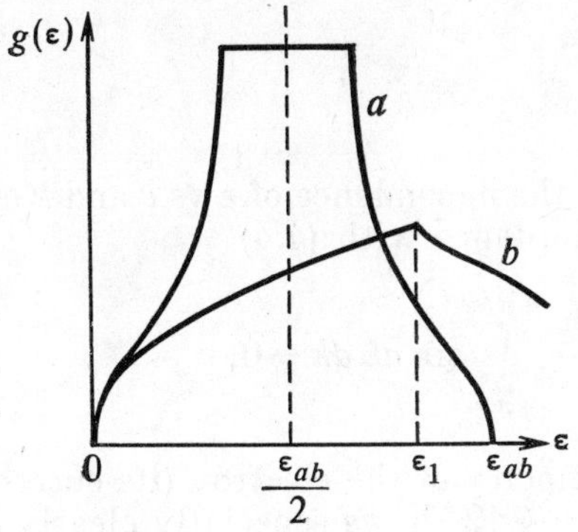


**Fig. 4.13**

is symmetrical about the point $\varepsilon_{ab}/2$. The curve *b* in the same figure depicts $g(\varepsilon) \propto \sqrt{\varepsilon}$ for free electrons, but takes account, starting from some $\varepsilon$, of the reduction in the number of states due to the intersection of the spherical energy surface with the faces of the cubic Brillouin zone.

**4.7.4** For the body-centred cubic lattice the first Brillouin zone has the shape of a regular dodecahedron depicted in Fig. 4.6*a*. It follows from relation (7.10b) that the minimum value of the electron energy[7] $\varepsilon_b = -8A$ is attained in the centre of the zone at $\mathbf{k} = 0$, and the maximum value $\varepsilon_a = +8A$ is attained at point $k_x = 2\pi/a$, $k_y = k_z = 0$ and at five other equivalent points in the centres of the square faces of the zone. Hence, the bandwidth is $\varepsilon_{ab} = \varepsilon_a - \varepsilon_b = 16A$, i.e., in this case, too, it is proportional to the overlap integral $A$. Constant energy $\varepsilon = 0$ corresponds to the plane $k_x = \pi/a$ and to five other equivalent planes. Figure 4.14 depicts constant energy lines formed by the intersection of the Brillouin zone with the following planes: (a) $k_z = 0$ and (b) $k_z = \pi/2a$.

In the same way as it has been done in the preceding section for a simple cubic lattice, we can obtain the expansion for the energy $\varepsilon$ near the lower and the upper band-edges, determine effective masses, average electron velocity $\mathbf{v}$ and the density of states $g(\varepsilon)$.

The shape of the first Brillouin zone for a face-centred cubic lattice is that of a truncated octahedron depicted in Fig. 4.5*a*. It follows

---

[7] We again make $\varepsilon_0 - C = 0$.

from equation (7.10c) that the minimum energy at $\mathbf{k} = 0$ is $\varepsilon_b = -12A$, and the maximum energy is $\varepsilon_a = 4A$ at $k_x = 2\pi/a$ and $k_y = k_z = 0$ and at five other equivalent points, so that the bandwidth is again $\varepsilon_{ab} = 16A$. In the centres of each of the hexagonal

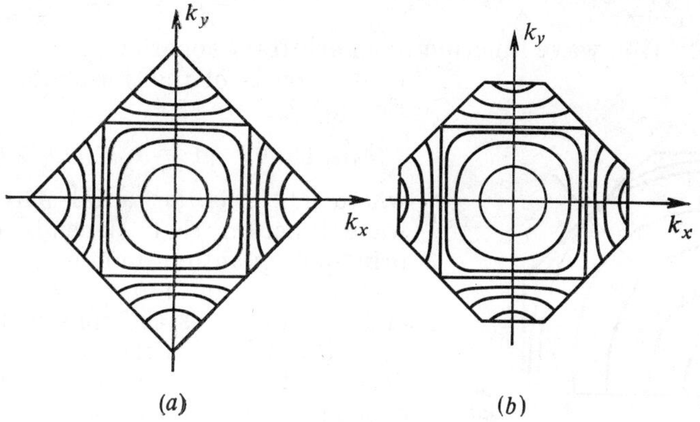


Fig. 4.14

faces ($\pm k_x = \pm k_y = \pm k_z = \pi/a$) the energy $\varepsilon = 0$. Figure 4.15 depicts the constant-energy surfaces for this case. Here (a) corresponds to the lower and (b) to the higher energy value $\varepsilon$. Figure 4.16

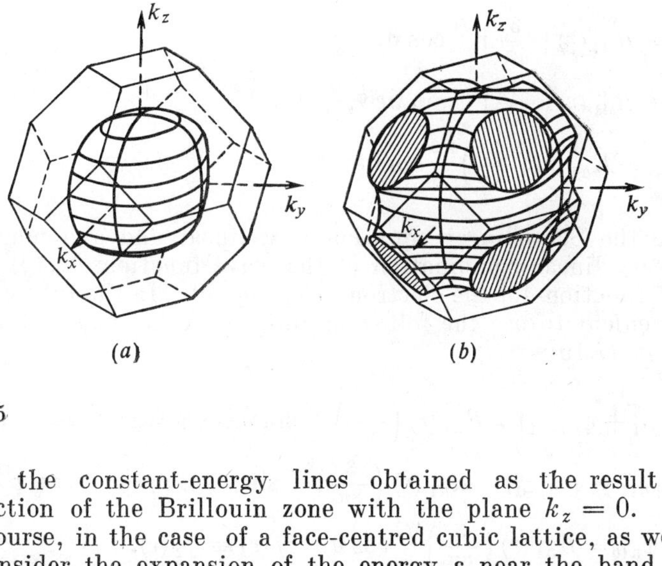


Fig. 4.15

depicts the constant-energy lines obtained as the result of the intersection of the Brillouin zone with the plane $k_z = 0$.

Of course, in the case of a face-centred cubic lattice, as well, we can consider the expansion of the energy $\varepsilon$ near the band edges, determine the effective masses, etc. The results obtained will be qualitatively similar to those for the simple and the body-centred cubic lattices.

**4.7.5** Consider the behaviour in a simple cubic lattice of an electron occupying a $p$-state in an isolated atom (ion) in the tight binding approximation.

The electron wave function in an arbitrary spherically symmetrical field $\mathcal{U}(r)$ is of the form [4.1, § 34, Eq. (34.7)]

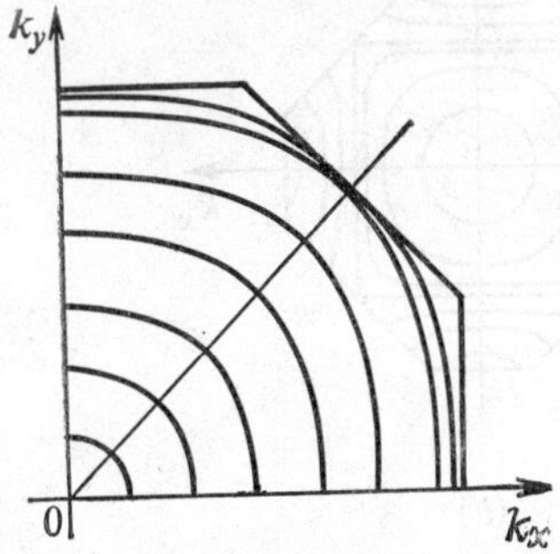


**Fig. 4.16**

$$\psi_{nlm}(r, \vartheta, \varphi) = R_{nl}(r) Y_{lm}(\vartheta, \varphi),$$

where $R_{nl}(r)$ is the radial part of the wave function that depends on the principal quantum number $n$ and the orbital quantum number $l \leqslant n-1$, and $Y_{lm}(\vartheta, \varphi)$ is the spherical function that describes the angular dependence of the wave function and itself depends on $l$ and the magnetic quantum number $m = 0, \pm 1, \pm 2, \ldots, \pm l$. The electron energy depends only on the quantum numbers $n$ and $l$ (in a Coulomb field—only on $n$); therefore there are three degenerate wave functions with $m = 0, +1, -1$ to correspond to the electron $p$-state with $l = 1$:

$$\psi_{n10} = R_{n1}(r)\left(\frac{3}{4\pi}\right)^{1/2} \cos\vartheta,$$

$$\psi_{n11} = R_{n1}(r)\left(\frac{3}{8\pi}\right)^{1/2} \sin\vartheta e^{i\varphi},$$

$$\psi_{n1,-1} = R_{n1}(r)\left(\frac{3}{8\pi}\right)^{1/2} \sin\vartheta e^{-i\varphi}. \tag{7.16}$$

Since the Schrödinger equation is a linear and a homogeneous one, every linear combination of the wave functions (7.16) is also a wave function for the electron in the $p$-state. In the future it will be convenient to use the following linear combinations of the wave functions (7.16):

$$\frac{1}{2}[\psi_{n11} + \psi_{n1,-1}] = R_{n1}(r)\left(\frac{3}{8\pi}\right)^{1/2} \sin\vartheta\cos\varphi = xf(r) \equiv \psi_x(\mathbf{r}),$$

$$\frac{1}{2i}[\psi_{n11} - \psi_{n1,-1}] = R_{n1}(r)\left(\frac{3}{8\pi}\right)^{1/2} \sin\vartheta\sin\varphi = yf(r) \equiv \psi_y(\mathbf{r}),$$

$$\frac{2}{\sqrt{2}}\psi_{n10} = R_{n1}(r)\left(\frac{3}{8\pi}\right)^{1/2} \cos\vartheta = zf(r) \equiv \psi_z(\mathbf{r}). \tag{7.16a}$$

Here $x = r \sin \vartheta \cos \varphi$, $y = r \sin \vartheta \sin \varphi$, $z = r \cos \vartheta$ are orthogonal coordinates of the electron, and $f(r)$ is a radially symmetric function that depends on the potential $\mathcal{U}(r)$.

The electron wave function in a crystal can be written in a form similar to (7.2a). However, now we shall have to take into account not only the translational degeneracy of the electron accounted for in (7.2a) by the multiplier $\exp i(\mathbf{k}\mathbf{a_n})$ and by the summation over all the equivalent lattice sites $\mathbf{n}$, but also the degeneracy (7.16). Accordingly, we make the electron wave function in a crystal equal to

$$\psi(\mathbf{r}) = \sum_{\mathbf{n}} e^{i\mathbf{k}\cdot\mathbf{a_n}} \{\alpha\psi_x(\mathbf{r}-\mathbf{a_n}) + \beta\psi_y(\mathbf{r}-\mathbf{a_n}) + \gamma\psi_z(\mathbf{r}-\mathbf{a_n})\}, \qquad (7.17)$$

where the coefficients $\alpha$, $\beta$ and $\gamma$ are to be determined from the general perturbation theory of degenerate states.

It can be demonstrated (Appendix 15) that in a simple cubic crystal the states $\psi_x$, $\psi_y$ and $\psi_z$ do not combine with one another, i.e., only one of the coefficients $\alpha$, $\beta$, or $\gamma$ can be nonzero. Hence, for a specified $\mathbf{k}$ there are three wave functions corresponding to the electron in a crystal, each with a different dependence of the energy $\varepsilon$ on $\mathbf{k}$.

The energy bands corresponding to these three wave functions overlap completely, so that the $p$-level degeneracy does not vanish in a simple cubic crystal.

Calculation (Appendix 15) yields the following result for the electron energy of $\alpha \neq 0$, $\beta = \gamma = 0$

$$\varepsilon_1 = \varepsilon_0 - C + 2A \cos ak_x - 2B(\cos ak_y + \cos ak_z). \qquad (7.18)$$

For $\beta \neq 0$, $\alpha = \gamma = 0$ the electron energy is

$$\varepsilon_2 = \varepsilon_0 - C + 2A \cos ak_y - 2B(\cos ak_x + \cos ak_z), \qquad (7.18a)$$

and for $\gamma \neq 0$, $\alpha = \beta = 0$ the electron energy is

$$\varepsilon_3 = \varepsilon_0 - C + 2A \cos ak_z - 2B(\cos ak_x + \cos ak_y). \qquad (7.18b)$$

Here $\varepsilon_0$ is the energy of a $p$-electron in an isolated atom,

$$C = -\int \psi_x^2(\mathbf{r})\,[V(\mathbf{r}) - \mathcal{U}(\mathbf{r})]\,d\tau, \qquad (7.19)$$

and

$$\int \psi_x(\mathbf{r})\,\psi_x(\mathbf{r}-\mathbf{a}_{\mathbf{n}_0})\,[V(\mathbf{r}) - \mathcal{U}(|\mathbf{r}-\mathbf{a}_{\mathbf{n}_0}|)]\,d\tau = \begin{cases} A \\ -B, \end{cases} \qquad (7.19a)$$

where the integral is equal to $A$ when its neighbouring atom $\mathbf{n}_0$ lies on the $x$ axis, and to $-B$ when the neighbouring atom $\mathbf{n}_0$ lies on the $y$ or the $z$ axis. Thanks to the cubic symmetry, the substitution $\psi_y$ or $\psi_z$ for $\psi_x$ does not change the value of the integrals (7.19) and (7.19a).

The meaning of the potential energies $V(\mathbf{r})$ and $\mathcal{U}(\mathbf{r})$ is the same as in expressions (7.7) and (7.8). The signs of the constants $A$ and $B$ cannot be determined quite uniquely. We shall consider two cases: I. $A > 0$ and $B > 0$ and II. $A > 0$ and $B < 0$. In the following we shall again measure the electron energy in all three states from the level $\varepsilon_0 - C$, we shall make $\varepsilon_0 - C = 0$.

**Case I:** $A > 0$, $B > 0$. For all three electron states (7.18), (7.18a), (7.18b) the energy minimum is equal to

$$\varepsilon_{1b} = \varepsilon_{2b} = \varepsilon_{3b} = -2A - 4B \tag{7.20}$$

In each of the three states two equivalent minima are located in the centres of two opposite square faces of the Brillouin zone. The maximum energy for all three states is also the same and equal to

$$\varepsilon_{1a} = \varepsilon_{2a} = \varepsilon_{3a} = +2A + 4B, \tag{7.20a}$$

so that the width of the three overlapping energy bands is equal to

$$\varepsilon_{ab} = \varepsilon_{1a} - \varepsilon_{1b} = 4A + 8B. \tag{7.20b}$$

Energy maxima for each of the three states are located at the mid-points of the four cube edges that are perpendicular to the faces bearing the corresponding minima. Hence, there are always 6 minima and 12 maxima in a Brillouin zone.

Expand energy $\varepsilon_1$ (7.18) in a series near the minimum $k_x = \pi/a$, $k_y = k_z = 0$ in small quantities $k'_x = \frac{\pi}{a} - k_x$, $k_y$ and $k_z$:

$$\begin{aligned} \varepsilon_1 &= 2A\cos(\pi - ak'_x) - 2B(\cos ak_y + \cos ak_z) \\ &= -2A\cos ak'_x - 2B(\cos ak_y + \cos ak_z) \\ &= -2A\left[1 - \frac{(ak'_x)^2}{2}\right] - 2B\left[\left(1 - \frac{(ak_y)^2}{2}\right) + \left(1 - \frac{(ak_z)^2}{2}\right)\right] \\ &= \varepsilon_{1b} + Aa^2k'^2_x + Ba^2(k_y^2 + k_z^2). \end{aligned} \tag{7.21}$$

We see that near this minimum the constant-energy surfaces are ellipsoids of revolution, the $x$ axis serving as their rotation axis. The components of the effective mass tensor are

$$\begin{aligned} m_{xx}^{-1} &= \frac{1}{\hbar^2}\left(\frac{\partial^2\varepsilon_1}{\partial k'^2_x}\right)_{\min} = \frac{2Aa^2}{\hbar^2} \equiv \frac{1}{m_1}, \\ m_{yy}^{-1} &= m_{zz}^{-1} = \frac{1}{\hbar^2}\left(\frac{\partial^2\varepsilon_1}{\partial k_y^2}\right)_{\min} = \frac{2Ba^2}{\hbar^2} \equiv \frac{1}{m_2}. \end{aligned} \tag{7.21a}$$

It follows from expressions (7.21) and (7.21a) that

$$\varepsilon_1 - \varepsilon_{1b} = \frac{\hbar^2 k'^2_x}{2m_1} + \frac{\hbar^2 k_y^2 + \hbar^2 k_z^2}{2m_2}. \tag{7.21b}$$

Similar expansions are possible near the five other energy minima. Hence, although each ellipsoid does not possess cubic symmetry by itself, the set of them satisfies such a symmetry. Similar expansions of the electron energy are also possible close to the maxima. In the $k_z = 0$ plane the electron energies are

$$\varepsilon_1 = 2A \cos ak_x - 2B \cos ak_y - 2B, \tag{7.22}$$

$$\varepsilon_2 = 2A \cos ak_y - 2B \cos ak_x - 2B, \tag{7.22a}$$

$$\varepsilon_3 = -2B (\cos ak_x + \cos ak_y) + 2A. \tag{7.22b}$$

Figure 4.17*a*, *b* and *c* depicts constant-energy lines corresponding to $\varepsilon_1$, $\varepsilon$ and $\varepsilon_3$. Letters *m* and *M* mark the energy minima and maxima,

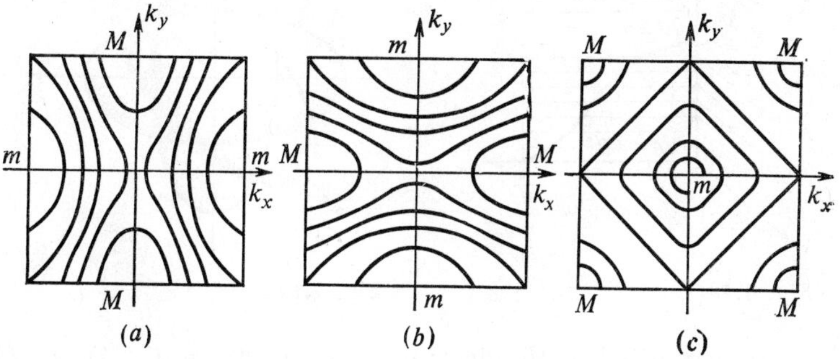


**Fig. 4.17**

respectively. For *a* and *b* a saddle point of the energy surface $\varepsilon (k_x, k_y)$ corresponds to the centre of the band ($k_x = k_y = 0$). Figure 4.18 depicts the dependence of the energy $\varepsilon_1 (k_x, k_y)$ along the axes $k_x$ and $k_y$.

**Case II:** $A > 0$ and $B < 0$. Denoting $B$ by $-B$ and making again $\varepsilon_0 - C = 0$, we obtain instead of equalities (7.18)

$$\varepsilon_1 = 2A \cos ak_x + 2B (\cos ak_y + \cos ak_z), \tag{7.23}$$

$$\varepsilon_2 = 2A \cos ak_y + 2B (\cos ak_x + \cos ak_z), \tag{7.23a}$$

$$\varepsilon_3 = 2A \cos ak_z + 2B (\cos ak_x + \cos ak_y). \tag{7.23b}$$

Hence it follows that the points corresponding to the energy minimum (7.20) in all three states are $\pm k_x = \pm k_y = \pm k_z = \pi/a$, i.e., the vertices of the Brillouin zone's cube. The energy maximum, the same for all three states, is attained in the zone centre at $k_x = k_y = k_z = 0$. Hence, the total energy bandwidth is the same as in the preceding case (7.20b). Expanding the electron energies

(7.23a), (7.23b) in series in $k_x$, $k_y$, $k_z$ near the common maximum we obtain

$$\varepsilon_1 = (2A + 4B) - Aa^2k_x^2 - Ba^2 (k_y^2 + k_z^2), \quad (7.24)$$

$$\varepsilon_2 = (2A + 4B) - Aa^2k_y^2 - Ba^2 (k_x^2 + k_z^2), \quad (7.24a)$$

$$\varepsilon_3 = (2A + 4B) - Aa^2k_z^2 - Ba^2 (k_x^2 + k_y^2). \quad (7.24b)$$

Hence, the constant-energy surfaces are identical ellipsoids with rotation axes directed along the $x$, $y$ and $z$ axes (Fig. 4.19). For

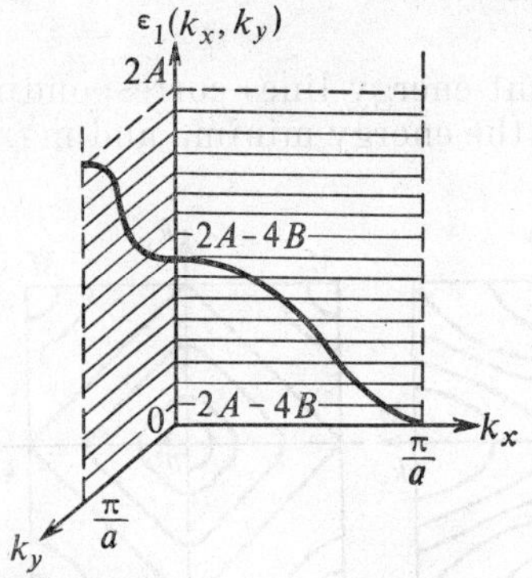


Fig. 4.18

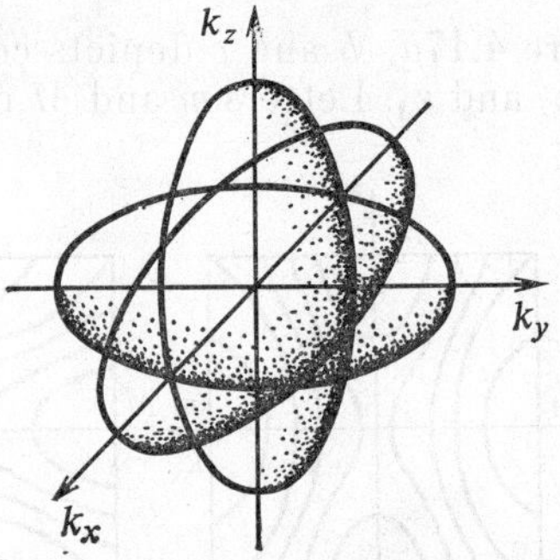


Fig. 4.19

a band almost filled with electrons the spectrum (7.24), (7.24a), (7.24b) with an opposite sign corresponds to holes. Making $k_y = k_z = 0$ we see that

$$\varepsilon_1 = \varepsilon_M - Aa^2k_x^2 \quad (7.25)$$

$$\varepsilon_2 = \varepsilon_3 = \varepsilon_M - Ba^2k_x^2. \quad (7.25a)$$

Figure 4.20 depicts the dependence of the energies $\varepsilon_1$, $\varepsilon_2$ and $\varepsilon_3$ on $k_x$ in the assumption that $A < B$, and that consequently the curve *2* corresponds to a doubly degenerate state ($\varepsilon_2 = \varepsilon_3$).

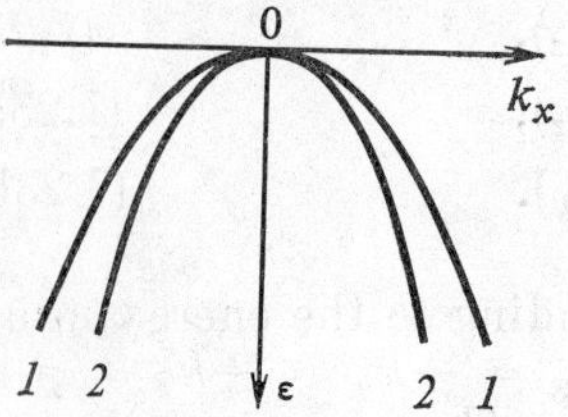


Fig. 4.20

In the $k_y$ direction, i.e., for $k_x = k_z = 0$, $\varepsilon_1$ and $\varepsilon_2$ are similar to (7.25a), and $\varepsilon_3$ is similar to (7.25). The state in the centre of the Brillouin zone (**k** = 0) is a six-fold degenerate one (with the spins taken into account).

In the absence of spin-orbit coupling such an energy spectrum would be observed for the holes of the valence band of germanium. The consideration of spin-orbit coupling results in a partial lifting of the degeneracy; a doubly degenerate state splits away and shifts to lower energy values, and the four-fold degenerate state branches into those of the light and heavy holes.

## 4.8 Structure of Energy Bands and Wave Function Symmetry in a Simple Cubic Lattice and in an Indium Antimonide Crystal

**4.8.1** The space group of the simple cubic lattice is symmorphic, therefore the irreducible representations of the wave vector group $G_{\mathbf{k}}$ for it can be classified with the aid of irreducible representations of $\Gamma$ $(R)$—a point group corresponding to the wave vector **k** (Section 2.9.4).

Since the conclusions about the vibration spectrum of a simple cubic lattice presented in Section 3.8 have been reached solely on the basis of symmetry considerations, they can be applied unaltered to the properties of the electron spectrum in a cubic crystal. The only difference is that in the case of vibrations we have to attribute to the centre of the Brillouin zone $\Gamma$ the irreducible representation $\Gamma_{15}$ (3.8.17), whereas in the case of the electron spectrum the choice of the electron state at point $\Gamma$ remains largely arbitrary. But should we presume that the electron state in the centre of the Brillouin zone is also determined by the irreducible representation $\Gamma_{15}$, all the conclusions obtained for the vibration spectrum in Section 3.8 would remain in force, we would only have to substitute the concept "energy band" for that of the "vibration branch".

Thus, for example the triple degenerate state $\Gamma_{15}$ splits up at point $\Delta$ (Fig. 3.13) into a nondegenerate state $\Delta_1$ and a doubly degenerate state $\Delta_5$ (3.8.19a). The interpretation of expressions (3.8.20), (3.8.21) and (3.8.22) for the case of the electron spectrum is similar. Other electron states in the centre $\Gamma$ and their splitting along the $\Delta$, $\Lambda$, $\Sigma$ axes may just as easily be considered with the aid of the compatibility relations displayed in Table 3.5[8].

**4.8.2** Compile Table 4.2 of transformations of the orthogonal coordinates $x$, $y$, $z$ operated on with the elements of the tetrahedral group $T_d$ (Section 2.3). The corresponding elements of the group $T_d$ are designated in the table by $R_i$ (. . .) ($i = 1, 2, 3, \ldots, 24$), the nature of the coordinate transformation is indicated in brackets. For instance, $R_5$ $(yzx)$ is a rotation about the $C_3$ axis (the space diagonal $Od$ in Fig. 2.8) through the angle $2\pi/3$ involving the transformations: $x \to y$, $y \to z$, $z \to x$; $R_2$ $(x\bar{y}\bar{z})$ is a rotation about the $x$ axis through the angle $\pi$ $(C_4^2)$ resulting in $x \to x$, $y \to -y$, $z \to -z$.

[8] For a more detailed group analysis of the electron spectra see the paper [4.8, p. 194] cited above (Secton 3.8.2).

Performing transformations $R_i$ with the function $f(x, y, z)$, we obtain, for example,

$$\hat{R}_2\hat{R}_5 f(x, y, z) = \hat{R}_2 f(y, z, x) = f(\overline{y}, \overline{z}, x).$$

According to Table 4.2 this is equivalent to the operation of the element $\hat{R}_7 f(x, y, z) = f(\overline{y}, \overline{z}, x)$, therefore, $R_2R_5 = R_7$; this can be checked directly by geometrical construction. In this way the whole multiplication table of the elements of group $T_d$ can be compiled. The operations corresponding to the elements $S_4$ are rotations through the angle $\pm\pi/2$ about the $x$, $y$, $z$ axes (Fig. 2.8b) followed by a reflection in the plane normal to the rotation axis. It may easily be seen that the transformation $S_4$ is equivalent to the transformation $JC_4$, if the rotations through the angle $\pi/2$ in the first and second cases are performed in the opposite directions; hence, $S_4 = JC_4$. The elements $\sigma$ are connected with a reflection in the planes $y = \pm z$, $z = \pm x$, $x = \pm y$; they can be represented as a rotation through the angle $\pi$ followed by an inversion [this follows directly from (2.3.2)]. For instance, $R_{19}$ $(xzy)$ involving the reflection in the plane $y = z$ can be represented as a rotation through the angle $\pi$ about the axis $y = -z$ $(x = 0)$ followed by an inversion (Fig. 4.21), therefore, $\sigma = JC_2$. The symmetry group of the tetrahedron consists of 24 elements and five classes: $E$, $3C_4^2$, $8C_3$, $6S_4 = 6JC_4$, $6\sigma = 6JC_2$ (the order of classes is as in Table 4.2, and not as in Table 2.7). The cubic group $\boldsymbol{O}$ isomorphic to $T_d$ is obtained, when the elements of the two last classes of $T_d$ are subjected to an inversion transformation; hence, the group $\boldsymbol{O}$ consists of the classes: $E$, $3C_4^2$, $8C_3$, $6C_4$, $6C_2$. To obtain the table of coordinate transformations of the group $\boldsymbol{O}$ we should perform an inversion in each of the twelve last elements

Table 4.2

| | |
|---|---|
| $E$ | $R_1(xyz)$ |
| $3C_4^2$ | $R_2(x\overline{y}\overline{z})$, $R_3(\overline{x}y\overline{z})$, $R_4(\overline{x}\overline{y}z)$ |
| $8C_3$ | $R_5(yzx)$, $R_6(\overline{y}z\overline{x})$, $R_7(\overline{y}\overline{z}x)$, $R_8(y\overline{z}\overline{x})$, $R_9(zxy)$, $R_{10}(\overline{z}\overline{x}y)$, $R_{11}(z\overline{x}\overline{y})$, $R_{12}(\overline{z}x\overline{y})$ |
| $6S_4 = 6JC_4$ | $R_{13}(\overline{x}z\overline{y})$, $R_{14}(\overline{x}\overline{z}y)$, $R_{15}(\overline{z}\overline{y}x)$, $R_{16}(z\overline{y}\overline{x})$, $R_{17}(y\overline{x}\overline{z})$, $R_{18}(\overline{y}x\overline{z})$ |
| $6\sigma = 6JC_2$ | $R_{19}(xzy)$, $R_{20}(x\overline{z}\overline{y})$, $R_{21}(zyx)$, $R_{22}(\overline{z}y\overline{x})$, $R_{23}(yxz)$, $R_{24}(\overline{yxz})$ |

$R_i$ ($i = 13, 14, \ldots, 24$) of Table 4.2 [with the result, for example, $R_{13}\ (\bar{x}z\bar{y}) \to R_{13}\ (x\bar{z}y)$].

The full cubic group $\boldsymbol{O}_h$ can be represented as a direct product of the $\boldsymbol{O}$ or the $\boldsymbol{T}_d$ group and of $\boldsymbol{C}_i = \{E, J\}$, i.e., $\boldsymbol{O}_h = \boldsymbol{O} \times \boldsymbol{C}_i = \boldsymbol{T}_d \times \boldsymbol{C}_i$. The group $\boldsymbol{O}_h$ is made up of 48 elements and 10 classes obtained as a result of multiplication of 5 classes $\boldsymbol{O}$ or $\boldsymbol{T}_d$ by the elements $E$ and $J$. In the first case ($\boldsymbol{O}_h = \boldsymbol{O} \times \boldsymbol{C}_i$) we obtain for $\boldsymbol{O}_h$ the following 10 classes: $E$, $3C_4^2$, $8C_3$, $6C_4$, $6C_2$, $J$, $3JC_4^2$, $8JC_3$, $6JC_4$, $6JC_2$; in the second case the multiplication of the $\boldsymbol{T}_d$ classes by $\boldsymbol{C}_i$ yields: $E$, $3C_4^2$, $8C_3$, $6S_4 = 6JC_4$, $6\sigma = 6JC_2$, $J$, $3JC_4^2$, $8JC_3$, $6J^2C_4 = 6C_4$, $6J^2C_2 = 6C_2$, the difference from the result of multiplication $\boldsymbol{O} \times \boldsymbol{C}_i$ being only in the order of classes.

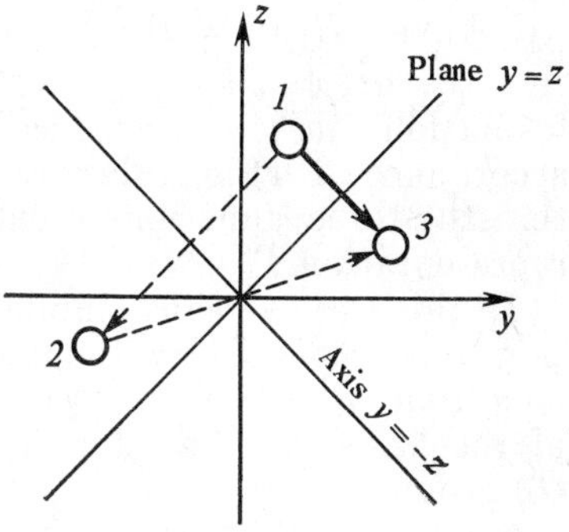


Fig. 4.21

To obtain a table of coordinate transformations for the $\boldsymbol{O}_h$ group, Table 4.2 should be supplemented by 24 transformations $R'_1$, $R'_2$, ..., $R'_{24}$ obtained from it as a result of inversion, so that, for example, $R'_{15} = JR_{15}\ (\bar{z}\bar{y}x) = R'_{15}\ (zy\bar{x})$ (note that $R'_{15}$ belongs to the class $6JJC_4 = 6C_4$).

With the aid of such a complete table for $\boldsymbol{O}_h$ we can find out how polynomials containing addends of the form $x^m y^n z^p$ are transformed. This presents the possibility to find such simplest polynomials (with the least $m$, $n$, $p$) that have the symmetry of the basis functions for the corresponding irreducible representations.

Thus, for example, the functions $\psi = xyz$ are transformed with the aid of the irreducible representation $\Gamma'_2$ (Table 3.2). Indeed, making use of Table 4.2 we can easily demonstrate that the result of the operation of the elements of class $3C_4^2$ is $xyz \to xyz$, i.e., $\psi \to \psi$ and that of the elements of class $6C_2$ is $xyz \to -xyz$, i.e., $\psi \to -\psi$; this is in accordance with the corresponding characters of the representation $\Gamma'_2$. It may easily be demonstrated that the representation $\Gamma'_2$ performs all the transformations performed by $\boldsymbol{O}_h$ on $\psi = xyz$.

In the same way it can be demonstrated that the three functions $\psi_1 = x$, $\psi_2 = y$, $\psi_3 = z$ are transformed with the aid of the irreducible representation $\Gamma_{15}$. Indeed for $C_4^2 = R_2\ (x\bar{y}\bar{z})$ we have

$$x' = 1 \cdot x + 0 \cdot y + 0 \cdot z,$$
$$y' = 0 \cdot x - 1 \cdot y + 0 \cdot z,$$
$$z' = 0 \cdot x + 0 \cdot y - 1 \cdot z.$$

The trace (character) of this transformation (just as for the other two $C_4^2$) is equal to $-1$, in accordance with Table 3.2. It may easily be checked that the functions $x$, $y$, $z$ yield correct values for all the characters of $\Gamma_{15}$.

Check in addition that the basis functions for the representation $\Gamma'_{25}$ are $\psi_1 = xy$, $\psi_2 = xz$, $\psi_3 = yz$. For instance, for $C_3 = R_{12}\,(\overline{zx}\overline{y})$ we have $\hat{R}_{12}\psi_1 = 0\cdot\psi_1 - 1\cdot\psi_2 + 0\cdot\psi_3$, $\hat{R}_{12}\psi_2 = 0\cdot\psi_1 + 0\cdot\psi_2 + 1\cdot\psi_3$, $\hat{R}_{12}\psi_3 = -1\cdot\psi_1 + 0\cdot\psi_2 + 0\cdot\psi_3$. The trace of this transformation (just as the trace of the other $C_3$) is equal to zero in accordance with the table of characters. It may easily be demonstrated that the three functions cited above yield all the characters of the representation $\Gamma'_{25}$.

In this way we can compile the table (see Table 4.3) of the lowest power polynomials (without the normalizing multiplier and for the condition $x^2 + y^2 + z^2 = \text{const.} = 1$) that determine the symmetry of the basis functions for all the irreducible representations of the $O_h$ group.[9]

4.8.2 In Section 1.2.5 we mentioned the fact that the indium antimonide compound (InSb) crystallizes in a diamond-like lattice which can be imagined as two face-centred cubic lattices (one made up of the In atoms, and the other made up of Sb atoms) displaced with respect to one another along a space diagonal of the cube to a distance of 1/4 of its length. InSb has no improper symmetry elements like germanium in which all the sites are occupied by identical atoms, and which for this reason is described by the symmorphic $T_d^2$ space group.

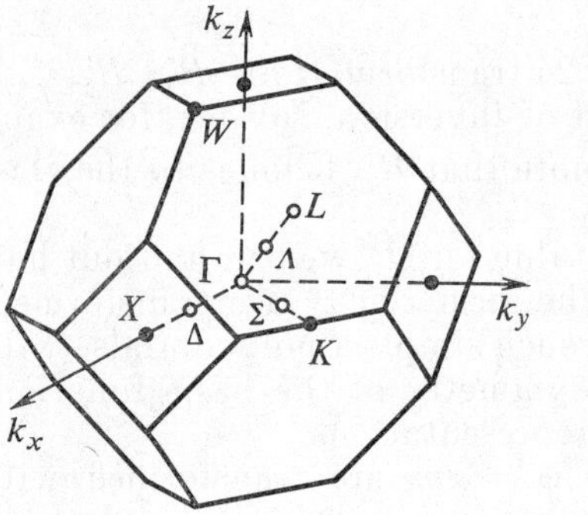


**Fig. 4.22**

For the face-centred cubic lattice the first Brillouin zone is represented by a polyhedron of fourteen faces described in Section 4.6.1 and depicted in Figs. 4.5 and 4.22.

The crystallographic group $\mathcal{F}$ for the InSb lattice is the point group of the tetrahedron $T_d$ described in Section 2.3. This group consists of 24 elements distributed over five classes (see Table 4.2), the $C_4$ axes coincide with the axes $k_x$, $k_y$ and $k_z$ passing through the centres of the six square faces of the polyhedron. The four $C_3$

9 The reader may ask how the basis functions are found in general. There is a very laborious method of projection operators (which we have not discussed) for determining basis functions. Nox and Gold [4.8, p. 41] note that "He [the reader] may have even gained the impression (from literature) that basis functions are obtained by luck, by educated guesses, or by black magic."

Table 4.3

| | |
|---|---|
| $\Gamma_1$ | 1; |
| $\Gamma_2$ | $x^4(y^2-z^2)+y^4(z^2-x^2)+z^4(x^2-y^2)$; |
| $\Gamma_{12}$ | $z^2-1/2(x^2+y^2)$, $x^2-y^2$; |
| $\Gamma'_{15}$ | $xy(x^2-y^2)$, $yz(y^2-z^2)$, $zx(z^2-x^2)$; |
| $\Gamma'_{25}$ | $xy$, $yz$, $zx$; |
| $\Gamma'_1$ | $xyz[x^4(y^2-z^2)+y^4(z^2-x^2)+z^4(x^2-y^2)]$; |
| $\Gamma'_2$ | $xyz$; |
| $\Gamma'_{12}$ | $xyz[z^2-1/2(x^2+y^2)]$, $xyz(x^2-y^2)$; |
| $\Gamma'_{15}$ | $x$, $y$, $z$; |
| $\Gamma_{25}$ | $z(x^2-y^2)$, $x(y^2-z^2)$, $y(z^2-x^2)$. |

axes pass through the centres of the eight hexagons (Fig. 4.22).

It follows from Section 2.9.4 that in case of symmorphic groups ($\alpha = 0$) the irreducible representation of the wave vector group $\Gamma(g)$ are described by the irreducible representations $\Gamma(R)$, where $R$ are the elements of the point group $\mathcal{F}_{\mathbf{k}}$. It follows from Table 2.7 of the $T_d$ group characters that only five electron states are possible at the centre of the Brillouin zone ($\mathbf{k} = 0$) of an InSb crystal: two nondegenerate ($\Gamma_1$, $\Gamma_2$), one doubly degenerate ($\Gamma_{12}$) and two triply degenerate ($\Gamma_{15}$, $\Gamma_{25}$) states. Of course, there can be numerous energy levels (bands) to correspond to each of those states, in the same way as there are many energy levels to correspond to an irreducible representation of an electron in an atom (with a specified orbital quantum number $l$).

Consider now a point group corresponding to point $\Delta$ lying on the $k_x$ axis (see Fig. 4.22); it consists of the elements $R_i$ (see Table 4.2) involving the transformation $x \to x$, i.e., $R_1 = E$, $R_2 = C_4^2$, $R_{19} = \sigma = JC_2$, $R_{20} = \sigma' = JC'_2$ (the planes $\sigma$ and $\sigma'$ pass through

Table 4.4

| | $E$ | $C_4^2$ | $JC_2$ | $JC_2'$ |
|---|---|---|---|---|
| $\Delta_1$ | 1 | 1 | 1 | 1 |
| $\Delta_2$ | 1 | 1 | $-1$ | $-1$ |
| $\Delta_3$ | 1 | $-1$ | 1 | $-1$ |
| $\Delta_4$ | 1 | $-1$ | $-1$ | 1 |
| $\Gamma_{15}$ | 3 | $-1$ | $-1$ | $-1$ |

the $k_x$ axis). Table 4.4 presents the characters of this fourth-order group consisting of four classes.

Consider point $X$ on the Brillouin zone surface (see Fig. 4.22). Since the wave vector $-k_X$ is equivalent to the vector $k_X$, to determine the wave vector $k_X$ the four elements of the group of point $\Delta$ should be supplemented by all the elements of the group correspond-

Table 4.5

| | $E$ | $C_4^2$ | $2C_4^2$ | $2JC_4$ | $2JC_2$ |
|---|---|---|---|---|---|
| $X_1$ | 1 | 1 | 1 | 1 | 1 |
| $X_2$ | 1 | 1 | 1 | $-1$ | $-1$ |
| $X_3$ | 1 | 1 | $-1$ | $-1$ | 1 |
| $X_4$ | 1 | 1 | $-1$ | 1 | $-1$ |
| $X_5$ | 2 | $-2$ | 0 | 0 | 0 |
| $\Gamma_{15}$ | 3 | $-1$ | $-1$ | 1 | $-1$ |

ing to the transformation $x \rightarrow -x$. It follows from Table 4.2 that such elements are: $R_3$, $R_4$, $R_{13}$, $R_{14}$, so that we obtain an eighth-order group consisting of five classes: $R_1 = E$, $R_2 = C_4^2$, $R_3 + R_4 = 2C_4^2$, $R_{13} + R_{14} = 2JC_4 = 2S_4$, $R_{19} + R_{20} = 2JC_2 = 2\sigma$.

Table 4.5 presents the characters of the representations of the wave vector $k_X$ group. The table of characters of wave vector groups for the other symmetrical points ($\Lambda$, $L$, $\Sigma$, $K$, $W$) of the Brillouin zone can be constructed in the same way (see Fig. 4.22).

Suppose the electron state in the centre of the Brillouin zone is described by the triple degenerate irreducible representation $\Gamma_{15}$. In that case, applying the general rule (Section 2.6.4), we shall be able to find out what will happen to $\Gamma_{15}$ at points $\Delta$ and $X$. In the same way as we have done in Section 3.8, where we studied the splitting of the vibration spectrum, we obtain, making use of Tables 4.4 and 4.5:

$$\Gamma_{15} = \Delta_1 + \Delta_3 + \Delta_4 \quad (8.1)$$

$$\Gamma_{15} = X_3 + X_5. \quad (8.2)$$

It will be demonstrated below that the states $\Delta_3$ and $\Delta_4$ do not split up on the $\Delta$ axis because of the additional symmetry connected with the invariance of the Schrödinger equation with respect to the reversal of time ($t \rightarrow -t$) (see below Section 4.14.4), therefore the electron spectrum in InSb can be of the form depicted in Fig. 4.23.

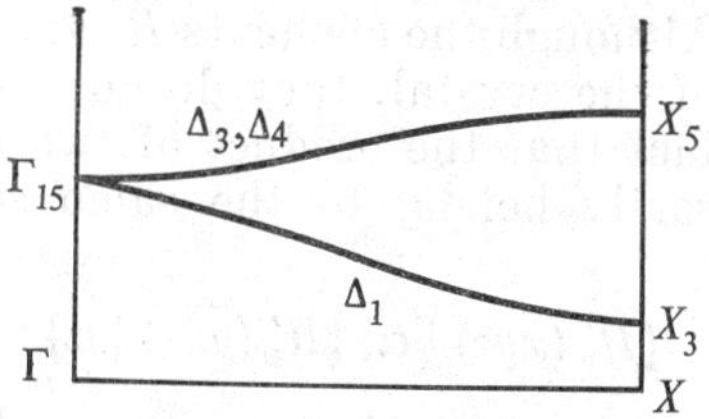


Fig. 4.23

Since we are able to determine not only the degree of degeneracy, but also the symmetry of the wave functions of the states at points $\Delta$ and $X$, the value of the information gained from group theory analysis for numerical calculations of the electron spectrum in InSb crystals becomes obvious.

## 4.9 Wave Vector Groups for a Germanium-Type Lattice

An essential peculiarity of the germanium lattice is the presence of symmetry elements containing the improper translation $\boldsymbol{\alpha} = \frac{a}{4}(1, 1, 1)$ (Section 2.5.2). The reason for this is that, in contrast to InSb, all the lattice sites are now occupied by atoms of one sort (Ge). The Brillouin zone in the case of germanium also has the shape of a polyhedron with fourteen faces depicted in Fig. 4.22.

The germanium lattice belongs to the crystal class $O_h$. The space group of germanium is made up of the elements $\{R_i \mid \mathbf{a}\}$ and $\{R_i' \mid \boldsymbol{\alpha} + \mathbf{a}\}$ ($i = 1, 2, \ldots, 24$); here $R_i$ are the elements of the

$\boldsymbol{T}_d$ point group (see Table 4.2) $R_i' = JR_i$ ($J$ is inversion), $\boldsymbol{\alpha} = \frac{a}{4}(1, 1, 1)$ is an improper translation and $\mathbf{a} \equiv \mathbf{a_n}$ is a lattice vector. Choose a lattice site $A$ in Fig. 1.12 as the origin of a right orthogonal coordinate system and direct the $x$, $y$, $z$ axes along the cube edges (the $z$ axis pointing downward); direct the basis vectors $\mathbf{a}_1$, $\mathbf{a}_2$, $\mathbf{a}_3$ from $A$ to the centres of adjoining faces, i.e., to the sites *2*, *4*, *3*. In that case we shall be able to write the basis vectors in terms of their orthogonal components as follows

$$\mathbf{a}_1 = \frac{a}{2}(1, 1, 0), \quad \mathbf{a}_2 = \frac{a}{2}(0, 1, 1), \quad \mathbf{a}_3 = \frac{a}{2}(1, 0, 1).$$

Although the elements $R_i$ and $\{R_i' \mid \boldsymbol{\alpha}\}$ are symmetry transformations of the crystal, they do not constitute groups. This follows from the fact that the product of two elements of such a set need not necessarily belong to the same set. For example:

$$\begin{aligned}&\{R_2'(\bar{x}yz) \mid \boldsymbol{\alpha}\}\{R_5'(\bar{y}\bar{z}\bar{x}) \mid \boldsymbol{\alpha}\}\\&= R_2'R_5' + \hat{R}_2'\frac{a}{4}(1, 1, 1) + \frac{a}{4}(1, 1, 1)\\&= R_7 + \frac{a}{4}(-1, 1, 1) + \frac{a}{4}(1, 1, 1) = R_7 + \frac{a}{2}(0, 1, 1),\end{aligned} \tag{9.1}$$

where $\frac{a}{2}(0, 1, 1) = \mathbf{a}_2$ is a lattice vector. (To find the product $R_2'R_5' = R_7$ and the operation $\hat{R}_2'\frac{a}{4}(1, 1, 1) = \frac{a}{4}(-1, 1, 1)$, we made use of the fact that $R_i' = JR_i$ and of the table of coordinate transformations (Table 4.2). It does indeed follow from (9.1) that the product $R_7 + \mathbf{a}_2$, since $\mathbf{a}_2$ is a lattice vector, does not belong to the set $R_i$, $\{R_i' \mid \alpha\}$, i.e., that the set does not constitute a group.

It follows from (2.9.28) and (2.9.35) that in the centre of the Brillouin zone ($\mathbf{k} = 0$) the irreducible representations of the wave vector group $\Gamma(g) = \Gamma(\{R \mid \boldsymbol{\alpha} + \mathbf{a}\})$ coincide with the irreducible representations of the crystallographic group $\Gamma(R)$. It follows from Table 3.2 of characters of the $\boldsymbol{O}_h$ group that in the centre of the Brillouin zone of germanium (and of other crystals of the cubic group) only the following electron states are possible: 4 nondegenerate ($\Gamma_1$, $\Gamma_2$, $\Gamma_1'$, $\Gamma_2'$), 2 doubly degenerate ($\Gamma_{12}$, $\Gamma_{12}'$) and four triply degenerate ($\Gamma_{15}$, $\Gamma_{25}$, $\Gamma_{15}'$, $\Gamma_{25}'$). The symmetry of these 10 states is shown in Table 4.3. In this case, too, we gain valuable information about the possible electron states in the centre of the Brillouin zone from the application of group theory.

Since point $\Delta$ is an internal one for the Brillouin zone, the group of the wave vector $\mathbf{k}_\Delta$ can be, in accordance with (2.9.28), described with the aid of the irreducible representations

$$\Gamma(g) = \Gamma(R) \exp(i\mathbf{k}_\Delta \cdot \boldsymbol{\alpha})$$

(the multiplier $\exp i\mathbf{k}_\Delta \cdot \boldsymbol{\alpha}$ common to all the irreducible representations can be dropped). Here $R$ are the elements of the point group corresponding to the transformation $x \to x$; they can be chosen with the aid of Table 4.2: $R_1 = E$, $R_2 = C_4^2$, $R_{19} = JC_2$, $R_{20} = JC_2$, $R_3' = JC_4^2$, $R_4' = JC_4^2$, $R_{13}' = C_4$, $R_{14}' = C_4$. This eight-order group is isomorphic to the groups $\boldsymbol{C}_{4v}$, $\boldsymbol{D}_4$ and $\boldsymbol{D}_{2d}$. Table 3.3 shows the characters of this group consisting of five classes. To obtain the table of characters of the wave vector group at point $\Delta$ we need only, in accordance with (2.9.28), multiply the third and fourth columns of Table 3.3 (related by an improper translation) by the factor $\exp(i\mathbf{k}_\Delta \cdot \boldsymbol{\alpha})$.

Let the irreducible representation $\Gamma_{15}$ correspond to the electron state in the centre of the Brillouin zone, i.e., let the electron be in the triply degenerate $p$-state. What will happen to it, if it moves to point $\Delta$? Since splitting of the $\Gamma_{15}$ state should take place for an arbitrary infinitesimal $\mathbf{k}_\Delta$, we should make $\mathbf{k}_\Delta \to 0$, i.e., make use of Table 3.3. By making

$$\Gamma_{15} = a_1\Delta_1 + a_2\Delta_2 + a_1'\Delta_1' + a_2'\Delta_2' + a_5\Delta_5 \tag{9.2}$$

we obtain in accordance with the general rule (2.6.34)

$$a_1 = a_5 = 1, \tag{9.3}$$

other coefficients in (9.2) being equal to zero. Hence

$$\Gamma_{15} = \Delta_1 + \Delta_5, \tag{9.4}$$

i.e., the state $\Gamma_{15}$ splits up at point $\Delta$ into a nondegenerate state $\Delta_1$ and a doubly degenerate state $\Delta_5$.

Consider point $X$ at the intersection of the $x$ axis with the Brillouin zone surface (see Fig. 4.22), i.e., consider the group of the wave vector $\mathbf{k}_X = \frac{2\pi}{a}(1, 0, 0)$, where $a$ is the edge of the direct lattice cube. Since the wave vector $\mathbf{k}_X$ group is nonsymmorphic, the results obtained in Section 2.9.4 are not applicable in this case.

The wave vector $\mathbf{k}_X$ group consists of the elements: $\{R_i \mid \mathbf{a}_m\}$, $\{R_i' \mid \boldsymbol{\alpha} + \mathbf{a}_m\}$, where $i = 1, 2, 3, 4, 13, 14, 19, 20$ (Table 4.2) and $R_i' = JR_i$ ($J$ is inversion). This is a point group $D_{4h}$ symmetry.

If $\mathscr{E}_n(\mathbf{k})$ is a $d_n$-fold degenerate energy level at point $X$ with the Bloch functions $\psi_{n\mathbf{k}_X j}(\mathbf{r})$ $(j = 1, 2, \ldots, d_n)$, then in accordance with (2.9.12) and (2.6.1a)

$$\hat{P}'_g \psi_{n\mathbf{k}_X j} = \{R_i \mid \mathbf{a}_m\} \psi_{n\mathbf{k}_X j} = e^{i\mathbf{k}_X \cdot \mathbf{a}_m} \hat{R}_i \psi_{n\mathbf{k}_X j}$$

$$= e^{i\mathbf{k}_X \cdot \mathbf{a}_m} \sum_{s=1}^{dn} \Gamma(R_i)_{sj} \psi_{n\mathbf{k}_X s}, \qquad (9.5)$$

$$\hat{P}'_{g'} \psi_{n\mathbf{k}_X j} = \{R'_i \mid \alpha + \mathbf{a}_m\} \psi_{n\mathbf{k}_X j} = e^{i\mathbf{k}_X \cdot \mathbf{a}_m} \{R'_i \mid \alpha\} \psi_{n\mathbf{k}_X j}$$

$$= e^{i\mathbf{k}_X \cdot \mathbf{a}_m} \sum_{s=1}^{dn} \Gamma(R'_i)_{sj} \psi_{n\mathbf{k}_X s}. \qquad (9.5a)$$

Note that the sixteen operations $R_i$ and $\{R'_i \mid \alpha\}$ ($i = 1, 2, 3, 4, 13, 14, 19, 20$) do not constitute a group, as was already remarked above.

Making use of the expressions for the basis vectors $\mathbf{a}_i$ ($i = 1, 2, 3$) in terms of their orthogonal components presented above write the lattice vector $\mathbf{a}_m = m_1\mathbf{a}_1 + m_2\mathbf{a}_2 + m_3\mathbf{a}_3$ in terms of its orthogonal components in the form $\mathbf{a}_m = \frac{a}{2}(m_1 + m_3, m_1 + m_2, m_2 + m_3)$. The exponential multiplier on the right-hand side of (9.5) and (9.5a) is equal to

$$e^{i\mathbf{k}_X \cdot \mathbf{a}_m} = e^{i\pi(m_1+m_3)} = \begin{cases} 1, & \text{if } m_1 + m_3 \text{ is even} \\ -1, & \text{if } m_1 + m_3 \text{ is odd.} \end{cases} \qquad (9.6)$$

Denoting $\mathbf{a}_g$ as an arbitrary translation for which $e^{i\mathbf{k}_X \cdot \mathbf{a}_g} = 1$, and $\mathbf{a}_n$ as an arbitrary translation for which $e^{i\mathbf{k}_X \cdot \mathbf{a}_n} = -1$, we obtain from (9.5) and (9.5a) that the corresponding irreducible representation for the elements of the group of the wave vector $\mathbf{k}_X$:

$$\left.\begin{aligned} A_i &\equiv \{R_i \mid \mathbf{a}_g\} \\ B_i &\equiv \{R_i \mid \mathbf{a}_n\} \\ C_i &\equiv \{R'_i \mid \alpha + \mathbf{a}_g\} \\ D_i &\equiv \{R'_i \mid \alpha + \mathbf{a}_n\} \end{aligned}\right\} \text{ is } \left\{\begin{aligned} &\tilde{\Gamma}(R_i) \\ -&\tilde{\Gamma}(R_i) \\ &\tilde{\Gamma}(R'_i) \\ -&\tilde{\Gamma}(R'_i) \end{aligned}\right. \qquad (9.7)$$

The 32 matrices $\tilde{\Gamma}(R_i)$, $-\tilde{\Gamma}(R_i)$, $\tilde{\Gamma}(R'_i)$, $-\tilde{\Gamma}(R'_i)$ ($i = 1, 2, 3, 4, 13, 14, 19, 20$)[10] constitute an irreducible representation of dimension $d_n$ of the thirty-second order group consisting of the elements $A_i$, $B_i$, $C_i$, $D_i$. It can easily be demonstrated that the product of any two elements (9.7) is an element of the same set. For instance,

$$A_2 D_3 = \{R_2 \mid \mathbf{a}_g\}\{R'_3 \mid \alpha + \mathbf{a}_n\}$$
$$= \{R_2 R'_3 \mid R_2\alpha + R_2\mathbf{a}_n + \mathbf{a}_g\} = \{R'_4 \mid \alpha + \mathbf{a}_n\} = D_4,$$

---

[10] Since we employ the definition of $\hat{P}'_R$ (2.6.1a), the irreducible representations form transposes to the matrices in (9.5) and (9.5a) [see (2.6.5c)].

where we made use of Table 4.2[11]

We can draw up the multiplication table for this group and making use of it find the group's classes. It can be demonstrated that the group consists of the following 14 classes:

$$\mathbb{C}_1 = A_1, \quad \mathbb{C}_2 = A_3 + B_3 + A_4 + B_4, \quad \mathbb{C}_3 = A_2,$$
$$\mathbb{C}_4 = C_{19} + D_{20},$$
$$\mathbb{C}_5 = C_{13} + D_{13} + C_{14} + D_{41}, \quad \mathbb{C}_6 = C_1 + D_1,$$
$$\mathbb{C}_7 = C_3 + D_3 + C_4 + D_4,$$
$$\mathbb{C}_8 = C_2 + D_2, \quad \mathbb{C}_9 = A_{19} + A_{20},$$
$$\mathbb{C}_{10} = A_{13} + B_{13} + A_{14} + B_{14},$$
$$\mathbb{C}_{11} = B_{19} + B_{20}, \quad \mathbb{C}_{12} = C_{20} + D_{19},$$
$$\mathbb{C}_{13} = B_2, \quad \mathbb{C}_{14} = B_1. \quad (9.8)$$

The table of characters will contain 14 irreducible representations. To find the table of characters (Table 4.6) we can apply the general considerations discussed in Section 2.6.3.

Table 4.6

| | $\mathbb{C}_1$ | $\mathbb{C}_2$ | $\mathbb{C}_3$ | $\mathbb{C}_4$ | $\mathbb{C}_5$ | $\mathbb{C}_6$ | $\mathbb{C}_7$ | $\mathbb{C}_8$ | $\mathbb{C}_9$ | $\mathbb{C}_{10}$ | $\mathbb{C}_{11}$ | $\mathbb{C}_{12}$ | $\mathbb{C}_{13}$ | $\mathbb{C}_{14}$ |
|---|---|---|---|---|---|---|---|---|---|---|---|---|---|---|
| $M_1$ | 1 | 1 | 1 | 1 | 1 | 1 | 1 | 1 | 1 | 1 | 1 | 1 | 1 | 1 |
| $M_2$ | 1 | 1 | 1 | −1 | −1 | 1 | 1 | 1 | −1 | −1 | −1 | −1 | 1 | 1 |
| $M_3$ | 1 | −1 | 1 | −1 | 1 | 1 | −1 | 1 | −1 | 1 | −1 | −1 | 1 | 1 |
| $M_4$ | 1 | −1 | 1 | 1 | −1 | 1 | −1 | 1 | 1 | −1 | 1 | 1 | 1 | 1 |
| $M_5$ | 2 | 0 | −1 | 0 | 0 | 2 | 0 | −2 | 0 | 0 | 0 | 0 | −2 | 2 |
| $M_6$ | 1 | 1 | 1 | 1 | 1 | −1 | −1 | −1 | −1 | −1 | −1 | 1 | 1 | 1 |
| $M_7$ | 1 | 1 | 1 | −1 | −1 | −1 | −1 | −1 | 1 | 1 | 1 | −1 | 1 | 1 |
| $M_8$ | 1 | −1 | 1 | −1 | 1 | −1 | 1 | −1 | 1 | −1 | 1 | −1 | 1 | 1 |
| $M_9$ | 1 | −1 | 1 | 1 | −1 | −1 | 1 | −1 | −1 | 1 | −1 | 1 | 1 | 1 |
| $M_{10}$ | 2 | 0 | −2 | 0 | 0 | −2 | 0 | 2 | 0 | 0 | 0 | 0 | −2 | 2 |
| $X_1$ | 2 | 0 | 2 | 0 | 0 | 0 | 0 | 0 | 2 | 0 | −2 | 0 | −2 | −2 |
| $X_2$ | 2 | 0 | 2 | 0 | 0 | 0 | 0 | 0 | −2 | 0 | 2 | 0 | −2 | −2 |
| $X_3$ | 2 | 0 | −2 | 2 | 0 | 0 | 0 | 0 | 0 | 0 | 0 | −2 | 2 | −2 |
| $X_4$ | 2 | 0 | −2 | −2 | 0 | 0 | 0 | 0 | 0 | 0 | 0 | 2 | 2 | −2 |

However, not all of the 14 irreducible representations can be employed to interpret the electron spectrum in germanium at point $X$.

[11] Note that all the $R_i$ and $R_i'$ in the set (9.7) transform $x \to \pm x$, therefore the operation with them on the lattice vectors $\mathbf{a}_n$ and $\mathbf{a}_g$ does not change the parity of the latter (so that $\mathbf{R}_2\mathbf{a}_n = \mathbf{a}_n$).

The characters of those representations must satisfy the conditions (9.7) and this, as we shall now see, is realized only in the case of four irreducible representations $X_1$, $X_2$, $X_3$ and $X_4$. For instance, the characters of the irreducible representations for the elements $A_i$ and $B_i$ are equal in magnitude, but opposite in sign (9.7), i.e., $\chi(A_i) = -\chi(B_i)$. If $A_i$ and $B_i$ belong to the same class, then simultaneously $\chi(A_i) = \chi(B_i)$, so that $\chi(A_i) = \chi(B_i) = 0$.

The situation is similar in the case of the irreducible representations of the elements $C_i$ and $D_i$. When we scrutinize the classes $\mathbb{C}_k$ (9.8), we are bound to see that for the "regular" irreducible representations the characters of the classes $\mathbb{C}_2$, $\mathbb{C}_5$, $\mathbb{C}_6$, $\mathbb{C}_7$, $\mathbb{C}_8$, $\mathbb{C}_{10}$ must be equal to zero, and this is the case only for the representations $X_1$, $X_2$, $X_3$ and $X_4$. Regular irreducible representations can be selected also on the basis of the characters of the class $\mathbb{C}_{14} = B_1$, since

$$\chi(B_1) = \chi[\{E\,|a_n\}] = -\sum_{s=1}^{dn} \tilde{\Gamma}(E)_{ss} = -\sum_{s=1}^{dn} \delta_{ss} = -d_n < 0.$$

It will be seen from Table 4.6 that in the case of $\mathbb{C}_{14}$ only four irreducible representations $X_1$, $X_2$, $X_3$ and $X_4$ meet this condition. Hence in the germanium-type crystals (diamond-like crystals with atoms of a single sort) at point $X$ there are only doubly degenerate states (see characters of $\mathbb{C}_1 = A_1$ for $X_i$).

The diamond-like lattice of indium antimonide (InSb) studied in the preceding paragraph is described by a symmorphic space group containing no nontrivial translations. The wave vector group at point $x$ consists, as we have seen, of eight elements $\{R_i \mid \mathbf{a}_m\}$, where $i = 1, 2, 3, 4, 13, 14, 19, 20$. In the case of InSb, in contrast to germanium, as may be seen from Table 4.5, both the nondegenerate and the doubly degenerate states are possible at point $X$.

## 4.10 Spin-Orbit Coupling and Double Groups

**4.10.1** Up to now we have studied the behaviour of the electron in a crystal without taking account of its spin.

We know, however, that in atoms the interaction of the electron magnetic moment with its orbital motion results in the splitting and the displacement of atomic energy levels. Thus, for example, the interaction of the intrinsic magnetic moment of the valence electron with its orbital motion results in a sodium atom in the splitting of the $D$-line equal to 0.002 eV (the sodium doublet). The energy level splitting is greater for heavier atoms. For rubidium atoms it is equal to 0.03 eV, and for quick-silver it is 0.23 eV.

The spin-orbit displacement of the conduction electron spectra in semiconductors is also the greater the greater their atomic number $Z$. Thus, in InSb ($Z_{\mathrm{In}} = 49$, $Z_{\mathrm{Sb}} = 51$) it is greater than in Ge

($Z_{Ge} = 32$), its magnitude in silicon ($Z_{Si} = 14$) being less than in germanium.

From Pauli's theory of the electron spin it is known that the proper vector of the spin operator is

$$\hat{\mathbf{S}} = \frac{\hbar}{2}\hat{\boldsymbol{\sigma}}, \tag{10.1}$$

where the two-row Pauli spin matrices $\{\hat{\sigma}_1, \hat{\sigma}_2, \hat{\sigma}_3\} = \boldsymbol{\sigma}$[12] in the representation in which the spin is directed along the $x_3$ axis are equal to

$$\hat{\sigma}_1 = \begin{pmatrix} 0 & 1 \\ 1 & 0 \end{pmatrix}, \quad \hat{\sigma}_2 = \begin{pmatrix} 0 & -i \\ i & 0 \end{pmatrix}, \quad \hat{\sigma}_3 = \begin{pmatrix} 1 & 0 \\ 0 & -1 \end{pmatrix}. \tag{10.2}$$

We see that the value of the spin's projection on the $x_3$ axis is equal to $\pm\hbar/2 = s\hbar$, where the spin coordinate $s = \pm 1/2$. It can easily be demonstrated that the Pauli matrices (10.2) satisfy the following relations:

$$\hat{\sigma}_k^2 = \mathbf{I}, \quad \hat{\sigma}_k\hat{\sigma}_l = -\hat{\sigma}_l\hat{\sigma}_k = i\hat{\sigma}_m, \tag{10.2a}$$

where $\mathbf{I}$ is a unit matrix of rank two and the indices $k$, $l$, $m$ assume the values 1, 2, 3 in cyclic order.

It follows from experiment, as well as from Dirac's rigorous relativistic theory, that a magnetic moment

$$\hat{\boldsymbol{\mu}} = -\mu_B\hat{\boldsymbol{\sigma}}, \tag{10.3}$$

where

$$\mu_B = e\hbar/2mc \tag{10.3a}$$

is the *Bohr magneton*, is connected with the electron spin.

Consider the interaction of the electron magnetic moment $\mu_B$ with its orbital motion.

When an electron is moving in an electric field of intensity $\mathbf{E}$ in the nonrelativistic approximation, a magnetic field [4.9].

$$\mathbf{H} = \mathbf{E} \times \frac{\mathbf{v}}{c} \tag{10.4}$$

acts on it in the frame connected with the electron. Here $\mathbf{v}$ is the electron velocity and $c$ is the velocity of light. If the electron is moving in a self-consistent periodic potential of a crystal $V(\mathbf{r})$, then

$$\mathbf{E} = \frac{1}{e}\nabla V(\mathbf{r}), \tag{10.5}$$

where $e$ is the electron charge.

---

[12] As usual, the indices 1, 2, 3 correspond to the $x$, $y$, $z$ axes.

The dipole energy of the moment $\mu$ in the magnetic field $\mathbf{H}$ is equal to $-\mu \cdot \mathbf{H}$ [4.10, Sec. 4.15], therefore the operator of the spin-orbit coupling, according to (10.3)-(10.5) is equal to

$$\hat{\mu} \cdot \hat{\mathbf{H}} = \mu_B \hat{\sigma} \cdot \left( \frac{1}{e} \nabla V \times \frac{\hat{\mathbf{p}}}{mc} \right) = -\frac{i\hbar^2}{2m^2c^2} \sigma \cdot (\nabla V \times \nabla),$$

where we took into account that the velocity operator $\hat{\mathbf{v}} = (1/m)\,\hat{\mathbf{p}} = -(i\hbar/m)\,\nabla$. A consistent relativistic deduction from Dirac's equation yields a twice smaller value of spin-orbit coupling, therefore the correct Hamiltonian for the spin-orbit coupling is

$$\hat{\mathscr{H}}_{so} = -\frac{i\hbar^2}{4m^2c^2} \hat{\sigma} \cdot (\nabla V \times \nabla). \tag{10.6}$$

Hence, the Hamiltonian for an electron in a crystal that includes spin-orbit coupling is

$$\hat{\mathscr{H}} (\hat{\sigma}, \mathbf{r}) \equiv \hat{\mathscr{H}} (\mathbf{r}) + \hat{\mathscr{H}}_{so}, \tag{10.7}$$

where

$$\hat{\mathscr{H}} (\mathbf{r}) = -\frac{\hbar^2}{2m} \nabla^2 + V (\mathbf{r}) \tag{10.7a}$$

and $\hat{\mathscr{H}}_{so}$ is equal to (10.6). Introduce the vector

$$\hat{\mathbf{P}} = \frac{\hbar}{2m^2c^2} (\nabla V \times \nabla), \tag{10.8}$$

then the Hamiltonian (10.7) can be written in the form

$$\hat{\mathscr{H}} (\hat{\sigma}, \mathbf{r}) = \hat{\mathscr{H}} (\mathbf{r}) - \frac{i\hbar}{2} \hat{\sigma} \cdot \hat{\mathbf{P}} = \hat{\mathscr{H}} (\mathbf{r}) - i\hat{\mathbf{S}} \cdot \hat{\mathbf{P}}. \tag{10.9}$$

The equation for the eigenvalues of the energy $\mathscr{E}$ is of the form

$$\hat{\mathscr{H}} (\hat{\sigma}, \mathbf{r}) \, \Psi (s, \mathbf{r}) = \mathscr{E} \Psi (s, \mathbf{r}), \tag{10.10}$$

where $s$ is the spin coordinate that assumes only the values $s = \pm 1/2$, the corresponding projections of the spin on the $x_3$ axis being equal to $\pm\hbar/2$. The wave function is

$$\Psi (s, \mathbf{r}) = \psi_1(\mathbf{r}) \, \nu_1 (s) + \psi_2 (\mathbf{r}) \, \nu_2 (s), \tag{10.11}$$

where $\nu_1 (s)$ and $\nu_2 (s)$ are the Pauli spin functions (spinors) equal to the probability amplitudes for the spin to point in the $+x_3$ or $-x_3$ direction. Therefore

$$\nu_1 (+1/2) = 1, \quad \nu_1 (-1/2) = 0, \quad \nu_2 (+1/2) = 0,$$
$$\nu_2 (-1/2) = 1. \tag{10.12}$$

The spin functions can be written down with the aid of single-column matrices

$$\nu_1(s) = \begin{pmatrix} 1 \\ 0 \end{pmatrix}, \quad \nu_2(s) = \begin{pmatrix} 0 \\ 1 \end{pmatrix} \tag{10.12a}$$

this being equivalent to (10.12). Accordingly, the full wave function (10.11) can be written in the form

$$\Psi(s, \mathbf{r}) = \begin{pmatrix} \psi_1(\mathbf{r}) \\ \psi_2(\mathbf{r}) \end{pmatrix}. \tag{10.11a}$$

The coordinate parts $\psi_1(\mathbf{r})$ and $\psi_2(\mathbf{r})$ of the wave function $\Psi(s, \mathbf{r})$ (10.11) are generally different if spin-orbit coupling is taken into account.

Making use of (10.2) and (10.12a) and of the matrices multiplication rule (A.3.15), we obtain

$$\begin{aligned} &\sigma_1\nu_1 = \nu_2, \quad \sigma_1\nu_2 = \nu_1, \quad \sigma_2\nu_1 = i\nu_2, \quad \sigma_2\nu_2 = -i\nu_1, \\ &\sigma_3\nu_1 = \nu_1, \quad \sigma_3\nu_2 = -\nu_2, \\ &\sigma^2\nu_1 = (\sigma_1^2 + \sigma_2^2 + \sigma_3^2)\,\nu_1 = 3\nu_1, \quad \sigma^2\nu_2 = 3\nu_2. \end{aligned} \tag{10.13}$$

From (10.12) the spin functions will be seen to be orthonormal:

$$\sum_{s=\pm 1/2} \nu_i(s)\,\nu_k(s) = \delta_{ik}. \tag{10.14}$$

Normalizing the full wave function (10.11) to unity, we obtain, making use of (10.14),

$$\begin{aligned} \sum_{s=-\frac{1}{2}}^{+1/2} \int \Psi^*(s, \mathbf{r})\,\Psi(s, \mathbf{r})\,d\mathbf{r} &= \int [\psi_1^*(\mathbf{r})\,\psi_1(\mathbf{r}) + \psi_2^*(\mathbf{r})\,\psi_2(\mathbf{r})]\,d\mathbf{r} \\ &= \int [|\psi_1(\mathbf{r})|^2 + |\psi_2(\mathbf{r})|^2]\,d\mathbf{r} = 1. \end{aligned} \tag{10.15}$$

The Hamiltonian $\hat{\mathscr{H}}(\hat{\boldsymbol{\sigma}}, \mathbf{r}) = \hat{\mathscr{H}}(\mathbf{r}) - i\hat{\mathbf{S}}\hat{\mathbf{P}}$ (10.9) is invariant under the transformations of the crystal's space group $\mathbf{G}$. Indeed, $\hat{\mathscr{H}}(\mathbf{r})$ (10.7a) and the scalar product of the axial vectors $\hat{\mathbf{S}}\hat{\mathbf{P}}$ do not change under such transformations (both vectors $\hat{\mathbf{S}}$ and $\hat{\mathbf{P}}$ change sign under improper rotations).

Hence, to be able to operate with the elements of the crystal's space group $g = \{R \mid \boldsymbol{\alpha} + \mathbf{a}\} \in \mathbf{G}$ on both sides of equation (10.10), we shall have to find out how $g$ acts on the wave function $\Psi(s, \mathbf{r})$, and to this end it is necessary to study the behavior of the spin functions (spinors) $\nu_1(s)$ and $\nu_2(s)$ in the process of coordinate frame rotation.

**4.10.2** Consider the rotation of the coordinate frame about the $x_3$ axis through an infinitesimal angle $\delta\alpha$. Denote the corresponding rotation operator by $R_{x_3}(\delta\alpha)$. Then

$$R_{x_3}(\delta\alpha)\, f(x_1, x_2, x_3) = f(x_1 + x_2\delta\alpha,\ x_2 - x_1\delta\alpha,\ x_3)$$
$$= f(x_1, x_2, x_3) + \frac{\partial f}{\partial x_1} x_2\delta\alpha - \frac{\partial f}{\partial x_2} x_1\delta\alpha$$
$$= \left[1 - \delta\alpha\left(x_1\frac{\partial}{\partial x_2} - x_2\frac{\partial}{\partial x_1}\right)\right] f(x_1, x_2, x_3). \tag{10.16}$$

Introduce the momentum operator

$$\hat{\mathbf{M}} = \mathbf{r}\times\hat{\mathbf{p}} = \mathbf{r}\times\frac{\hbar}{i}\nabla. \tag{10.17}$$

Then

$$x_1\frac{\partial}{\partial x_2} - x_2\frac{\partial}{\partial x_1} = (\mathbf{r}\times\nabla)|_3 = \frac{i}{\hbar} M_3 \tag{10.18}$$

and

$$R_{x_3}(\delta\alpha) = 1 - \frac{i}{\hbar}\hat{M}_3\delta\alpha. \tag{10.19}$$

The rotation through a finite angle $\alpha$ about the $x_3$ axis may be imagined as consisting of $n$ consecutive rotations through infinitesimal angles $\delta\alpha = \alpha/n$ for $n \to \infty$. It follows from (10.19) that

$$R_{x_3}(\alpha) = \lim_{n\to\infty}\left(1 - \frac{i}{\hbar}\hat{M}_3\frac{\alpha}{n}\right)^n = e^{-i\alpha\hat{M}_3/\hbar}, \tag{10.20}$$

if we make use of the well-known expression for the limit [4.11, p. 79].

The operator in the exponent is known to have the meaning of an infinite series obtained as a result of its expansion, i.e.,

$$R_{x_3}(\alpha) = e^{-i\alpha\hat{M}_3/\hbar} = 1 - \frac{i\alpha\hat{M}_3}{\hbar} + \frac{1}{2!}\left(\frac{i\alpha\hat{M}_3}{\hbar}\right)^2 - \ldots \tag{10.21}$$

The relation between the rotation operator $\hat{R}_{x_3}(\alpha)$ and the momentum operator $\hat{M}_3$ (10.20) was obtained for orbital motion (10.17). It can be demonstrated that it is valid in a more general sense, because it is connected with the isotropy of space. We shall assume that this relation exists also in the case of the intrinsic momentum, or spin, i.e., that

$$R_{x_i}(\alpha) = e^{-i\alpha\hat{S}_i/\hbar} = e^{\alpha\hat{\sigma}_i/2i}, \tag{10.22}$$

where $\hat{S}_i = (\hbar/2)\,\hat{\sigma}_i$ is the spin component along the $x_i$ axis.

Applying (10.22) to the spin function $v_1(s)$ and making use of (10.13), we obtain

$$R_{x_3}(\alpha)\, v_1(s) = e^{\alpha\hat{\sigma}_3/2i} v_1(s)$$

$$= \left[1 + \frac{\alpha\hat{\sigma}_3}{2i} + \frac{1}{2}\left(\frac{\alpha\hat{\sigma}_3}{2i}\right)^2 + \ldots\right] v_1(s) = e^{\alpha/2i} v_1(s). \tag{10.23}$$

The common rotation of the "primed" frame $(x_1', x_2', x_3')$ with respect to the "unprimed" frame $(x_1, x_2, x_3)$ can be expressed in terms of the three Euler angles $\alpha$, $\beta$, $\gamma$ (Fig. 4.24). Let both the "primed" and

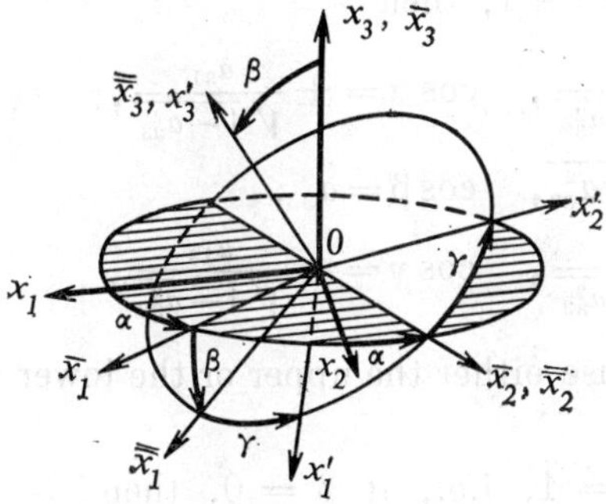


**Fig. 4.24**

"unprimed" frames coincide initially. Perform the following three rotations of the former: (1) rotation through a positive angle $\alpha$ about the common $x_3$ axis $(x_1x_2x_3 \to \bar{x}_1\bar{x}_2x_3)$; (2) rotation through a positive angle $\beta$ about the $\bar{x}_2$ axis $(\bar{x}_1\bar{x}_2x_3 \to \bar{\bar{x}}_1\bar{\bar{x}}_2\bar{\bar{x}}_3)$ and (3) rotation through a positive angle $\gamma$ about $\bar{\bar{x}}_3$ $(\bar{\bar{x}}_1\bar{\bar{x}}_2\bar{\bar{x}}_3 \to x_1'x_2'x_3')$.[13]

It is obvious that $0 \leqslant \alpha \leqslant 2\pi$, $0 \leqslant \beta \leqslant \pi$, $0 \leqslant \gamma \leqslant 2\pi$. Such a rotation $(x_1x_2x_3 \to x_1'x_2'x_3')$ transforms the orthogonal components of the position vector $\mathbf{r} = \{x_1, x_2, x_3\}$ as follows:

$$x_i' = \sum_{k=1}^{3} a_{ik} x_k. \tag{10.24}$$

---

[13] Note that different authors use different definitions for the Euler angles, and this obviously affects the form of the matrix $(a_{ik})$ (10.24a).

It can be demonstrated [4.12, Sec. 15-1] that the matrix $(a_{ik})$ is of the following form

$$[a_{ik}] = \begin{pmatrix} \cos\alpha\cos\beta\cos\gamma - \sin\alpha\sin\gamma & \sin\alpha\cos\beta\cos\gamma + \cos\alpha\sin\gamma & -\sin\beta\cos\gamma \\ -\cos\alpha\cos\beta\sin\gamma - \sin\alpha\cos\gamma & -\sin\alpha\cos\beta\sin\gamma + \cos\alpha\cos\gamma & \sin\beta\sin\gamma \\ \cos\alpha\sin\beta & \sin\alpha\sin\beta & \cos\beta \end{pmatrix}. \quad (10.24a)$$

The matrix elements $a_{ik}$ can be readily determined, if the angles $\alpha$, $\beta$, $\gamma$ are specified. The solution of the reciprocal problem when the matrix $(a_{ik})$ is specified and it is required to find the angles $\alpha$, $\beta$, $\gamma$ also presents no difficulties.

If $a_{33} = \cos\beta \neq \pm 1$, then

$$\sin\alpha = \pm\frac{a_{32}}{\sqrt{1-a_{33}^2}}, \quad \cos\alpha = \pm\frac{a_{31}}{\sqrt{1-a_{33}^2}},$$
$$\sin\beta = \pm\sqrt{1-a_{33}^2}, \quad \cos\beta = a_{33},$$
$$\sin\gamma = \pm\frac{a_{23}}{\sqrt{1-a_{33}^2}}, \quad \cos\gamma = \mp\frac{a_{13}}{\sqrt{1-a_{33}^2}}, \qquad (10.25)$$

where we should use either the upper or the lower sign for all expressions at a time.

If $a_{33} = \cos\beta = 1$, i.e., if $\beta = 0$, then

$$\cos(\alpha+\gamma) = a_{11} = a_{22}, \quad \sin(\alpha+\gamma) = a_{12} = -a_{21}. \qquad (10.25a)$$

If $a_{33} = \cos\beta = -1$, i.e., if $\beta = \pi$, then

$$\cos(\alpha-\gamma) = -a_{11} = a_{22}, \quad \sin(\alpha-\gamma) = -a_{12} = -a_{21}. \qquad (10.25b)$$

In the case of $a_{33} = \pm 1$, we can find $\beta$ and $\alpha \pm \gamma$, and this proves to be enough to determine the spin matrix $D^l$ (10.31).

Prove that

$$R_{\bar{\bar{x}}_3}(\gamma)\, R_{\bar{x}_2}(\beta)\, R_{x_3}(\alpha) = R_{x_3}(\alpha)\, R_{x_2}(\beta)\, R_{x_3}(\gamma). \qquad (10.26)$$

The left-hand side contains rotations through the Euler angles $\alpha$, $\beta$ and $\gamma$ as defined above, i.e., about the axes $x_3$, $\bar{x}_2$ and $\bar{\bar{x}}_3$. The right-hand side contains rotations through the angles $\gamma$, $\beta$, and $\alpha$ (in the reverse order) about static axes $x_3$, $x_2$, and $x_3$.

It will easily be seen that

$$R_{x_2}(\beta) = R^{-1}_{x_3}(\alpha)\, R_{\bar{x}_2}(\beta)\, R_{x_3}(\alpha). \qquad (10.27)$$

Indeed, on the right-hand side we rotate the mobile frame in turn about the $x_3$ axis through the angle $\alpha$, next about the new $\bar{x}_2$ axis through the angle $\beta$ and finally return the $\bar{x}_2$ axis to its initial position where it coincides with the axis $x_2$. This is obviously equivalent to the rotation through the angle $\beta$ about the $x_2$ axis, i.e., equality (10.27) has been proved. Premultiplying equality (10.27) by $R_{x_3}(\alpha)$, we obtain

$$R_{x_3}(\alpha)\, R_{x_2}(\beta) = R_{\bar{x}_2}(\beta)\, R_{x_3}(\alpha). \tag{10.27a}$$

It will easily be seen that

$$R_{x_3}(\gamma) = R_{x_3}^{-1}(\alpha)\, R_{\bar{x}_2}^{-1}(\beta)\, R_{\bar{\bar{x}}_3}(\gamma)\, R_{\bar{x}_2}(\beta)\, R_{x_3}(\alpha). \tag{10.28}$$

Indeed, the product of the second, third, and fourth multipliers on the right-hand side is equal to $R_{x_3}(\gamma)$ [the proof is similar to that of (10.27)]; all the rotations remaining on the right-hand side are performed about the $x_3$ axis, whence directly follows (10.28). Multiply now the left-hand and the right-hand sides of equalities (10.28) and (10.27a) and obtain relation (10.26), which we were required to prove.

It follows from (10.26) and (10.22) that the operator of the most general rotation $R_l(\alpha, \beta, \gamma)$ acting on the spin functions is equal to

$$R_l(\alpha, \beta, \gamma) = R_{x_3}(\alpha)\, R_{x_2}(\beta)\, R_{x_3}(\gamma) = e^{\alpha\hat{\sigma}_3/2i} e^{\beta\hat{\sigma}_2/2i} e^{\gamma\hat{\sigma}_3/2i}. \tag{10.29}$$

Apply the operator $R_l(\alpha, \beta, \gamma)$ to the spin functions $\nu_1(s)$ and $\nu_2(s)$. Making use of (10.13) and proceeding in the same way as when we obtained (10.23), we get after some calculations

$$R_l(\alpha, \beta, \gamma)\, \nu_1(s) = \cos\frac{\beta}{2}\, e^{-\frac{i}{2}(\alpha+\gamma)}\, \nu_1(s) + \sin\frac{\beta}{2}\, e^{\frac{i}{2}(\alpha-\gamma)}\, \nu_2(s),$$

$$R_l(\alpha, \beta, \gamma)\, \nu_2(s) = -\sin\frac{\beta}{2}\, e^{-\frac{i}{2}(\alpha-\gamma)}\, \nu_1(s) + \cos\frac{\beta}{2}\, e^{\frac{i}{2}(\alpha+\gamma)}\, \nu_2(s). \tag{10.30}$$

If $\hat{R}_l(\alpha, \beta, \gamma)$ is regarded as an element of the rotations group acting on the basis spin functions $\nu_1(s)$ and $\nu_2(s)$, then the two-row matrix on the right-hand side of (10.30) corresponds to the definition of the operator $\hat{P}'_R$ (A.6.1a). In this case the irreducible representation corresponding to the element $\hat{R}_l(\alpha, \beta, \gamma)$ will be the matrix $D^l$, a transpose of the matrix (10.30), i.e.,

$$D^l = \begin{pmatrix} \cos\frac{\beta}{2} e^{-\frac{i}{2}(\alpha+\gamma)} & -\sin\frac{\beta}{2} e^{-\frac{i}{2}(\alpha-\gamma)} \\ \sin\frac{\beta}{2} e^{\frac{i}{2}(\alpha-\gamma)} & \cos\frac{\beta}{2} e^{\frac{i}{2}(\alpha+\gamma)} \end{pmatrix} \tag{10.31}$$

with the character equal to

$$\mathrm{Tr}\,\{D^l\} = \chi_l = 2\cos\frac{\beta}{2}\cos\frac{\alpha+\gamma}{2}. \tag{10.31a}$$

Consider the transformation (10.30) for the case of the rotation of the frame about an arbitrary axis through the angle $2\pi$. Such a transformation of any coordinate-dependent function is equivalent to the unit element $E$. Make in (10.31), for example, $\alpha = 2\pi$, $\beta = \gamma = 0$; then

$$D^l = \begin{pmatrix} e^{-i\pi} & 0 \\ 0 & e^{i\pi} \end{pmatrix} = \begin{pmatrix} -1 & 0 \\ 0 & -1 \end{pmatrix} = -\begin{pmatrix} 1 & 0 \\ 0 & 1 \end{pmatrix}, \tag{10.32}$$

i.e.,

$$R_1(2\pi)\,\nu_1(s) = -\nu_1(s), \quad R_1(2\pi)\,\nu_2(s) = -\nu_2(s). \tag{10.33}$$

The same result will be obtained if we make $\beta = 2\pi$ and $\alpha = \gamma = 0$.[14] Thus, the operator $\hat{R}_1(2\pi)$ acting on the space coordinate dependent functions behaves as a unit element; at the same time it acts on the spin functions $\nu_1(s)$ and $\nu_2(s)$ and changes their signs (10.33). We shall denote

$$R_1(2\pi) \equiv \bar{E}. \tag{10.34}$$

Obviously

$$R_1(4\pi) = \overline{EE} = \bar{E}^2 = E,$$

where $E$ is a unit element with respect to the full wave function (10.11) that includes both the space functions $\psi_1(\mathbf{r})$, $\psi_2(\mathbf{r})$ and the spin functions $\nu_1(s)$, $\nu_2(s)$.

Let $\hat{R}_l$ be proper rotations, the corresponding matrices being $D^l$ (10.31); the corresponding matrices for the elements $E\bar{R}_l = \bar{R}_l$ will in that case be $-D^l$.

Of course, we may as well make the matrices $-D^l$ correspond to the element $R_l$ and the matrices $D^l$—to the element $\bar{E}R_l = \bar{R}_l$, since the only essential point is that a rotation through the angle $2\pi$ changes the sign of the matrix $D^l$. Therefore, it would be more con-

---

[14] Of course, the same will be true for a rotation through the angle $2\pi$ about an axis arbitrarily directed in space.

sistent to put the $\pm$ signs in front of the matrix on the right-hand side of (10.31).

Hence, as far as the spin is concerned, the operations $R_l$ do not coincide with $\overline{R}_l$. At the same time, if the Hamiltonian $\hat{\mathcal{H}}(\hat{\sigma}, \mathbf{r})$ is invariant under the rotations $R_l$, it will also be invariant under the operations $\overline{R}_l = \overline{E}R_l$. Hence, the space group of the Schrödinger-Pauli equation (10.10) contains twice as many elements, since in addition to $\{R_l \mid \boldsymbol{\alpha} + \boldsymbol{a}\}$ it contains also the elements $\{\overline{R}_l \mid \boldsymbol{\alpha} + \mathbf{a}\}$. These groups that account for transformational properties of the spin functions in the process of rotations became known as "*double*" *groups.* A double group contains twice as many elements as a simple one, however, the number of classes in it need not necessarily be twice as many. It can be demonstrated [4.8, p. 249] that the elements $C_2$ and $\overline{E}C_2$ belong to the same class, if there is an axis $C_2'$ perpendicular to $C_2$.

In matrix form equations (10.30) can be written down as follows

$$\boldsymbol{R}_l(\alpha, \beta, \gamma)\begin{pmatrix} \nu_1(s) \\ \nu_2(s) \end{pmatrix} = \tilde{\boldsymbol{D}}^l \begin{pmatrix} \nu_1(s) \\ \nu_2(s) \end{pmatrix}, \tag{10.35}$$

where $\tilde{\boldsymbol{D}}^l$ is the transpose of $\boldsymbol{D}_l$ (10.31).

Choosing, as was stated above, the matrix $-\boldsymbol{D}^l$ to correspond to the element $\overline{R}_l(\alpha, \beta, \gamma)$, we obtain

$$\overline{\boldsymbol{R}}^l(\alpha, \beta, \gamma)\begin{pmatrix} \nu_1(s) \\ \nu_2(s) \end{pmatrix} = -\tilde{\boldsymbol{D}}^l \begin{pmatrix} \nu_1(s) \\ \nu_2(s) \end{pmatrix}. \tag{10.35a}$$

Up to now we have been considering only the proper rotations $R_l(\alpha, \beta, \gamma)$. Demonstrate that if the point symmetry group contains improper rotations, the above considerations on the subject of double groups remain in force.

If the symmetry group contains improper rotations $JR_l$ in addition to all the proper rotation elements $R_l$, it can be represented as a direct product of the group of proper rotations $\{R_l\}$ and the group $\boldsymbol{C}_i = \{E, J\}$. We demonstrated in Section 2.6.5 that the irreducible representations of the direct product of two groups are equal to the direct product of their irreducible representations. Since the irreducible representations of the $\boldsymbol{C}_i$ group are equal to $+1$ and to $-1$ (see Table 2.3), and since the choice of sign of the matrix $\boldsymbol{D}^l$ for a simple group is, as we have seen, arbitrary, the above definition of the double groups remains valid.

If, on the other hand, the inversion operation $J$ is not part of the group (so that the group cannot be represented as a direct product of the proper rotation group $\{R_l\}$ and the group $\boldsymbol{C}_i = \{E, J\}$) but if the group includes improper rotations $JR_l$, then the distribution of the elements over classes is determined solely by the proper rotations and coincides with the distribution in isomorphic groups containing the elements $R_l$ instead of the elements $JR_l$. For this reason the representations of such isomorphic groups coincide.

## 4.11 Double Groups in InSb and Ge Crystals

**4.11.1** As already mentioned in Section 4.8.2, the electron states in an InSb crystal can be classified with the aid of irreducible representations $\Gamma(R)$ of the corresponding point groups. As we have seen, the following elements of Table 4.2 correspond to the wave vector group of point $\Delta$ (see Fig. 4.22): $R_1(xyz) = E$, $R_2(x\bar{y}\bar{z}) = C_4^2$, $R_{19}(xzy) = JC_2$ and $R_{20}(x\bar{z}\bar{y}) = JC_2'$. The following matrices $\mathbf{a} \equiv (a_{ik})$ of transformation (10.24) correspond to those $R_i$

$$\mathbf{a}^1 = \begin{pmatrix} 1 & 0 & 0 \\ 0 & 1 & 0 \\ 0 & 0 & 1 \end{pmatrix}, \quad \mathbf{a}^2 = \begin{pmatrix} 1 & 0 & 0 \\ 0 & -1 & 0 \\ 0 & 0 & -1 \end{pmatrix},$$

$$\mathbf{a}^{19} = \begin{pmatrix} 1 & 0 & 0 \\ 0 & 0 & 1 \\ 0 & 1 & 0 \end{pmatrix}, \quad \mathbf{a}^{20} = \begin{pmatrix} 1 & 0 & 0 \\ 0 & 0 & -1 \\ 0 & -1 & 0 \end{pmatrix}. \tag{11.1}$$

The determinants of the matrices of those transformations are equal to

$$|\mathbf{a}^1| = |\mathbf{a}^2| = 1, \quad |\mathbf{a}^{19}| = |\mathbf{a}^{20}| = -1, \tag{11.2}$$

whence it follows that proper rotations correspond to transformations $R_1$ and $R_2$ and improper rotations correspond to $R_{19}$ and $R_{20}$. The matrices of the transformations $JR_{19}$ and $JR_{20}$ (where $J$ is an inversion) are equal to

$$\hat{\mathbf{a}}_i^{19} = \begin{pmatrix} -1 & 0 & 0 \\ 0 & 0 & -1 \\ 0 & -1 & 0 \end{pmatrix}, \quad \hat{\mathbf{a}}_i^{20} = \begin{pmatrix} -1 & 0 & 0 \\ 0 & 0 & 1 \\ 0 & 1 & 0 \end{pmatrix}, \tag{11.3}$$

and the corresponding proper rotations are ($|\hat{\mathbf{a}}_i^{19}| = |\hat{\mathbf{a}}_i^{20}| = 1$). We can see directly that the group $R_1, R_2, R_{19}, R_{20}$ is isomorphic to the group $R_1, R_2, JR_{19}, JR_{20}$.

If $\mathscr{E}_n(\mathbf{k}_\Delta)$ is a $d_n$-fold degenerate energy level at point $\mathbf{k}_\Delta$ with spin-dependent eigenfunctions $\Psi_{n\mathbf{k}_\Delta j}(s, \mathbf{r})$ $(j = 1, 2, \ldots, d_n)$ (10.11), then the latter are basis functions of the irreducible representations of a double group of dimension $d_n$.

The correspondence of the matrices $D^l$ of different signs (10.35) and (10.35a) to the elements $R_l$ and $\bar{R}_l = \bar{E}R_l$ means that the corresponding matrices for the elements of the double wave vector $\mathbf{k}_\Delta$ group, $\{R_l \mid \mathbf{a_n}\}$ and $\{\bar{R}_l \mid \mathbf{a}_n\}$ $(l = 1, 2, 19, 20)$, are the matrices of the irreducible representation of rank $d_n$ $\Gamma(R_l)$ and $\Gamma(\overline{R}_l) = -\Gamma(R_l)$. The spin functions $v_1(s)$ and $v_2(s)$ are themselves basis functions of a two-dimensional double group, the corresponding matrices for whose elements $R_l$ and $\overline{R}_l$ are the matrices $D^l$ and $-D^l$ of rank two (10.31). Accordingly, we can use those matrices to draw up a multiplication table for our double group.

To determine the Euler angles $\alpha$, $\beta$, $\gamma$ corresponding to a proper rotation of the coordinate frame, we shall make use of the matrices $\mathbf{a}^1$, $\mathbf{a}^2$, $\hat{\mathbf{a}}_i^{19}$, and $\hat{\mathbf{a}}_i^{20}$ corresponding to the isomorphic group $R_1$, $R_2$, $JR_{19}$, $JR_{20}$. From (10.25a) and (10.25b), we obtain

$$\mathbf{a}^1: \alpha = \beta = \gamma = 0,$$
$$\mathbf{a}^2: \beta = \pi, \quad \gamma - \alpha = \pi,$$
$$\hat{\mathbf{a}}_i^{19}: \beta = \pi/2, \quad \alpha = \gamma = 3\pi/2,$$
$$\hat{\mathbf{a}}_i^{20}: \alpha = \beta = \gamma = \pi/2.$$

Making use of those values of $\alpha$, $\beta$, $\gamma$, we obtain from (10.31) the following two-dimensional representation of a double group:

$$D^1 = \pm\begin{pmatrix}1 & 0\\ 0 & 1\end{pmatrix}, \quad D^2 = \pm\begin{pmatrix}0 & -i\\ -i & 0\end{pmatrix},$$
$$D^{19} = \pm\frac{1}{\sqrt{2}}\begin{pmatrix}i & -1\\ 1 & -i\end{pmatrix}, \quad D^{20} = \pm\frac{1}{\sqrt{2}}\begin{pmatrix}-i & -1\\ 1 & i\end{pmatrix}, \tag{11.4}$$

where the upper sign is conventionally accepted for $R_i$, and the lower sign is conventionally accepted for $-\overline{R}_i$.[15] The eight matrices (11.4) represent a group of the eighth order. We can draw up a multiplication table for this group. For example,

$$\overline{R}_{19}R_2 = -\frac{1}{\sqrt{2}}\begin{pmatrix}i & -1\\ 1 & -i\end{pmatrix}\begin{pmatrix}0 & -i\\ -i & 0\end{pmatrix} = \frac{1}{\sqrt{2}}\begin{pmatrix}-i & -1\\ 1 & i\end{pmatrix} = R_{20},$$

and so on for all the elements. The result will be Table 4.7, in which the first left column corresponds to the first multiplicand, and the first upper row corresponds to the second multiplicand. This table serves for direct determination of inverse elements; thus, for example, $\overline{R}_{19}R_{19} = R_1 \equiv E$, therefore, $R_{19}^{-1} = \overline{R}_{19}$ and $\overline{R}_{19}^{-1} = R_{19}$. The group is not Abelian: $R_2R_{19} = R_{20}$, and $R_{19}R_2 = \overline{R}_{20}$, so that $R_2R_{19} \neq R_{19}R_2$.

[15] The conventionality is due to the fact that the addition of $2\pi$ to an arbitrary Euler angle, which is always possible when the angle is being defined, changes the sign of the matrix $D^l$.

Table 4.7

| | $R_1$ | $\bar{R}_1$ | $R_2$ | $\bar{R}_2$ | $R_{19}$ | $\bar{R}_{19}$ | $R_{20}$ | $\bar{R}_{20}$ |
|---|---|---|---|---|---|---|---|---|
| $R_1$ | $R_1$ | $\bar{R}_1$ | $R_2$ | $\bar{R}_2$ | $R_{19}$ | $\bar{R}_{19}$ | $R_{20}$ | $\bar{R}_{20}$ |
| $\bar{R}_1$ | $\bar{R}_1$ | $R_1$ | $\bar{R}_2$ | $R_2$ | $\bar{R}_{19}$ | $R_{19}$ | $\bar{R}_{20}$ | $R_{20}$ |
| $R_2$ | $R_2$ | $\bar{R}_2$ | $\bar{R}_1$ | $R_1$ | $R_{20}$ | $\bar{R}_{20}$ | $\bar{R}_{19}$ | $R_{19}$ |
| $\bar{R}_2$ | $\bar{R}_2$ | $R_2$ | $R_1$ | $\bar{R}_1$ | $\bar{R}_{20}$ | $R_{20}$ | $R_{19}$ | $\bar{R}_{19}$ |
| $R_{19}$ | $R_{19}$ | $\bar{R}_{19}$ | $\bar{R}_{20}$ | $R_{20}$ | $\bar{R}_1$ | $R_1$ | $R_2$ | $\bar{R}_2$ |
| $\bar{R}_{19}$ | $\bar{R}_{19}$ | $R_{19}$ | $R_{20}$ | $\bar{R}_{20}$ | $R_1$ | $\bar{R}_1$ | $\bar{R}_2$ | $R_2$ |
| $R_{20}$ | $R_{20}$ | $\bar{R}_{20}$ | $R_{19}$ | $\bar{R}_{19}$ | $\bar{R}_2$ | $R_2$ | $\bar{R}_1$ | $R_1$ |
| $\bar{R}_{20}$ | $\bar{R}_{20}$ | $R_{20}$ | $\bar{R}_{19}$ | $R_{19}$ | $R_2$ | $\bar{R}_2$ | $R_1$ | $\bar{R}_1$ |

Making use of Table 4.7 and of the definition of conjugate elements (2.2.3), we can demonstrate that the double group considered here consists of five classes

$$\mathbb{C}_1 = R_1,\ \mathbb{C}_2 = \bar{R}_1,\ \mathbb{C}_3 = R_2 + \bar{R}_2,\ \ \mathbb{C}_4 = R_{19} + \bar{R}_{19},$$
$$\mathbb{C}_5 = R_{20} + \bar{R}_{20}. \tag{11.5}$$

Since the number of irreducible representations is equal to the number of classes (2.6.30), and the sum of squares of the dimensions of the irreducible representations is equal to the group order (2.6.22), in our case

$$1^2 + 1^2 + 1^2 + 1^2 + 2^2 = 8,$$

i.e., the dimensions of the irreducible representations $\Delta_1$, $\Delta_2$, $\Delta_3$, $\Delta_4$ and $\Delta_5$ are equal to 1, 1, 1, 1, and 2, respectively; the same numbers must also stand in the first column $\mathbb{C}_1 = R_1$ of the table of characters.

Profitting by the property of orthonormality of the rows and of the columns draw up Table 4.8 of the characters of our double group. This table has been obtained by the usual method. However, for our double group the condition has to be satisfied $\chi(R_i) = -\chi(\bar{R}_i)$, therefore, if $R_i$ and $\overline{R}_i$ are part of the same class, its character must be zero. According to (11.5), the classes $\mathbb{C}_3$, $\mathbb{C}_4$ and $\mathbb{C}_5$ meet this condition, therefore, their characters must be zeros. It will be seen from Table 4.8 that this is realized only for the representation $\Delta_5$. InSb may be said to have only one (*spinor*) irreducible representation $\Delta_5$ at point $\Delta$.

Table 4.8

| | $\mathbb{C}_1$ | $\mathbb{C}_2$ | $\mathbb{C}_3$ | $\mathbb{C}_4$ | $\mathbb{C}_5$ |
|---|---|---|---|---|---|
| $\Delta_1$ | 1 | 1 | 1 | 1 | 1 |
| $\Delta_2$ | 1 | 1 | 1 | −1 | −1 |
| $\Delta_3$ | 1 | 1 | −1 | 1 | −1 |
| $\Delta_4$ | 1 | 1 | −1 | −1 | 1 |
| $\Delta_5$ | 2 | −2 | 0 | 0 | 0 |

Hence, we arrive at the conclusion that in an InSb crystal the electron, with its spin taken account of, at point $\Delta$ can be only in a doubly degenerate state $\Delta_5$.

The double groups in InSb at points $\Gamma$ and $X$ can be considered in the same way.

Tables 4.9 and 4.10 present spinor irreducible representations of a double group in InSb in the centre of the Brillouin zone $\Gamma$ and at point $X$. It will be seen from the tables that, when the spin is taken into account, the electron at point $\Gamma$ can be in two doubly degenerate states ($\Gamma_6$, $\Gamma_7$) and in one four-fold degenerate state ($\Gamma_8$), and at point $X$ in two doubly degenerate states ($X_6$, $X_7$).

The double groups at symmetry points of the Brillouin zone of Ge crystals, whose space group is not symmorphic, can be considered in a similar way.

Tables 4.11, 4.12, 4.13 present spinor irreducible representations of double groups at points $\Gamma$, $\Delta$, and $X$ for Ge crystals. It would be interesting to compare these tables with the appropriate Tables 3.2, 3.3 and 4.6 for simple groups in Ge at the same points of its Brillouin zone. For instance, when the spin is neglected, the electron can exist at point $X$ in four doubly degenerate states $X_1$, $X_2$, $X_3$, $X_4$ (Table

Table 4.9

| | $R_1$ | $\overline{R}_1$ | $R_2$-$R_4$, $\overline{R}_2$-$\overline{R}_4$ | $R_5$-$R_{12}$ | $\overline{R}_5$-$\overline{R}_{12}$ | $R_{13}$-$R_{18}$ | $\overline{R}_{13}$-$\overline{R}_{18}$ | $R_{19}$-$R_{24}$, $\overline{R}_{19}$-$\overline{R}_{24}$ |
|---|---|---|---|---|---|---|---|---|
| $\Gamma_6$ | 2 | $-2$ | 0 | 1 | $-1$ | $\sqrt{2}$ | $-\sqrt{2}$ | 0 |
| $\Gamma_7$ | 2 | $-2$ | 0 | 1 | $-1$ | $-\sqrt{2}$ | $\sqrt{2}$ | 0 |
| $\Gamma_8$ | 4 | $-4$ | 0 | $-1$ | 1 | 0 | 0 | 0 |
| $D\times\Gamma_{15}$ | 6 | $-6$ | 0 | 0 | 0 | $-\sqrt{2}$ | $\sqrt{2}$ | 0 |

Table 4.10

| | $R_1$ | $\overline{R}_1$ | $R_2$, $\overline{R}_2$ | $R_3$, $R_4$, $\overline{R}_3$, $\overline{R}_4$ | $R_{13}$, $R_{14}$ | $\overline{R}_{13}$, $\overline{R}_{14}$ | $R_{19}$, $R_{20}$, $\overline{R}_{19}$, $\overline{R}_{20}$ |
|---|---|---|---|---|---|---|---|
| $X_6$ | 2 | $-2$ | 0 | 0 | $\sqrt{2}$ | $-\sqrt{2}$ | 0 |
| $X_7$ | 2 | $-2$ | 0 | 0 | $-\sqrt{2}$ | $\sqrt{2}$ | 0 |

Table 4.11

| | $R_1$ | $\overline{R}_1$ | $R_2$-$R_4$, $\overline{R}_2$-$\overline{R}_4$ | $R_5$-$R_{12}$ | $\overline{R}_5$-$\overline{R}_{12}$ | $R_{13}$-$R_{18}$ | $\overline{R}_{13}$-$\overline{R}_{18}$ | $R_{19}$-$R_{24}$, $\overline{R}_{19}$-$\overline{R}_{24}$ |
|---|---|---|---|---|---|---|---|---|
| $\Gamma_6^+$ | 2 | $-2$ | 0 | 1 | $-1$ | $\sqrt{2}$ | $-\sqrt{2}$ | 0 |
| $\Gamma_6^-$ | 2 | $-2$ | 0 | 1 | $-1$ | $-\sqrt{2}$ | $\sqrt{2}$ | 0 |
| $\Gamma_7^+$ | 2 | $-2$ | 0 | 1 | $-1$ | $-\sqrt{2}$ | $\sqrt{2}$ | 0 |
| $\Gamma_7^-$ | 2 | $-2$ | 0 | 1 | $-1$ | $\sqrt{2}$ | $\sqrt{2}$ | 0 |
| $\Gamma_8^+$ | 4 | $-4$ | 0 | $-1$ | 1 | 0 | 0 | 0 |
| $\Gamma_8^-$ | 4 | $-4$ | 0 | $-1$ | 1 | 0 | 0 | 0 |

4.6); when spin is taken into account, it can exist at the same point only in one four-fold degenerate state $X_5$ (Table 4.13).

All the tables of characters for the spinor irreducible representations meet the condition specified above: $\chi(R_i) = -\chi(\overline{R}_i)$; therefore, if $R_i$ and $\overline{R}_i$ are part of the same class, $\chi(R_i) = \chi(\overline{R}_i) = 0$.

Table 4.11 (cont.)

| | $\{R'_1 \vert \alpha\}$ | $\{\bar{R}'_1 \vert \alpha\}$ | $\{R'_2$-$R'_4$, $\bar{R}'_2$-$\bar{R}'_4 \vert \alpha\}$ | $\{R'_5$-$R'_{12} \vert \alpha\}$ | $\{\bar{R}'_5$-$\bar{R}'_{12} \vert \alpha\}$ | $\{R'_{13}$-$R'_{18} \vert \alpha\}$ | $\{\bar{R}'_{13}$-$\bar{R}'_{18} \vert \alpha\}$ | $\{R'_{19}$-$R'_{24}$, $\bar{R}'_{19}$-$\bar{R}'_{24} \vert \alpha\}$ |
|---|---|---|---|---|---|---|---|---|
| $\Gamma_6^+$ | 2 | −2 | 0 | 1 | −1 | $\sqrt{2}$ | $-\sqrt{2}$ | 0 |
| $\Gamma_6^-$ | −2 | 2 | 0 | −1 | 1 | $\sqrt{2}$ | $-\sqrt{2}$ | 0 |
| $\Gamma_7^+$ | 2 | −2 | 0 | 1 | −1 | $-\sqrt{2}$ | $\sqrt{2}$ | 0 |
| $\Gamma_7^-$ | −2 | 2 | 0 | −1 | 1 | $-\sqrt{2}$ | $\sqrt{2}$ | 0 |
| $\Gamma_8^+$ | 4 | −4 | 0 | −1 | 1 | 0 | 0 | 0 |
| $\Gamma_8^-$ | −4 | 4 | 0 | 1 | −1 | 0 | 0 | 0 |

Table 4.12

| | $R_1$ | $\bar{R}_1$ | $R_2, \bar{R}_2$ | $R_{19}, R_{20}, \bar{R}_{19}, \bar{R}_{20}$ | $\{R'_3, \bar{R}'_3, R'_4, \bar{R}'_4 \vert \alpha\}$ | $\{R'_{13}, R'_{14} \vert \alpha\}$ | $\{\bar{R}'_{13}, \bar{R}'_{14} \vert \alpha\}$ |
|---|---|---|---|---|---|---|---|
| $\Delta_6$ | 2 | −2 | 0 | 0 | 0 | $\sqrt{2}\, e^{ik_\Delta \alpha}$ | $-\sqrt{2}\, e^{ik_\Delta \alpha}$ |
| $\Delta_7$ | 2 | −2 | 0 | 0 | 0 | $-\sqrt{2}\, e^{ik_\Delta \alpha}$ | $\sqrt{2}\, e^{ik_\Delta \alpha}$ |

Table 4.13

| | $R_1$ | $\bar{R}_1$ | $R_2, \bar{R}_2$ | $R_3, R_4, \bar{R}_3, \bar{R}_4$ | $R_{13}, R_{14}$ | $\bar{R}_{13}, \bar{R}_{14}$ | $R_{19}, R_{20}, \bar{R}_{19}, \bar{R}_{20}$ | $\{R'_1 \vert \alpha\}$ | $\{\bar{R}'_1 \vert \alpha\}$ |
|---|---|---|---|---|---|---|---|---|---|
| $X_5$ | 4 | −4 | 0 | 0 | 0 | 0 | 0 | −4 | 4 |

| | $\{R'_2, \bar{R}'_2 \vert \alpha\}$ | $\{R'_3, R'_4, \bar{R}'_3, \bar{R}'_4 \vert \alpha\}$ | $\{R'_{13}, R'_{14} \vert \alpha\}$ | $\{\bar{R}'_{13}, \bar{R}'_{14} \vert \alpha\}$ | $\{R'_{19}, \bar{R}'_{19} \vert \alpha\}$ | $\{R'_{20}, \bar{R}'_{20} \vert \alpha\}$ |
|---|---|---|---|---|---|---|
| $X_5$ | 0 | 0 | 0 | 0 | 0 | 0 |

The compatibility conditions, for instance, for the transition from point $\Gamma$ to line $\Delta$, can be determined from the tables of characters. For example, for InSb

$$\Gamma_8 = 2\Delta_5,$$

which follows unambiguously from the fact that the dimension of $\Gamma_8$ is 4, and that of $\Delta_5$ is 2. Putting for Ge

$$\Gamma_7^+ = a_1\Delta_6 + a_2\Delta_7,$$

we obtain from Tables 4.11 and 4.12 ($\mathbf{k}_\Delta \to 0$)

$$a_1 = \frac{1}{16}[2 \times 2 + (-2)(-2) + 2(\sqrt{2})(-\sqrt{2}) + 2(-\sqrt{2})(\sqrt{2})] = 0,$$

$$a_2 = \frac{1}{16}[2 \times 2 + (-2)(-2) + 2(-\sqrt{2})(-\sqrt{2}) + 2(\sqrt{2})(\sqrt{2})] = 1,$$

i. e.,

$$\Gamma_7^+ = \Delta_7.$$

The other compatibility relations presented in Table 4.14 for InSb (*a*) and for Ge (*b*) can be considered in the same way.[16]

Table 4.14

| $\Gamma_6$ | $\Gamma_7$ | $\Gamma_8$ | $X_6$ | $X_7$ |
|---|---|---|---|---|
| $\Delta_5$ | $\Delta_5$ | $\Delta_5\Delta_5$ | $\Delta_5$ | $\Delta_5$ |

(*a*)

| $\Gamma_6^+$ | $\Gamma_6^-$ | $\Gamma_7^+$ | $\Gamma_7^-$ | $\Gamma_8^+$ | $\Gamma_8^-$ | $X_5$ |
|---|---|---|---|---|---|---|
| $\Delta_6$ | $\Delta_6$ | $\Delta_7$ | $\Delta_7$ | $\Delta_6\Delta_7$ | $\Delta_6\Delta_7$ | $\Delta_6\Delta_7$ |

(*b*)

## 4.12 Spin-Orbit Splitting in InSb and in Ge Crystals

**4.12.1** The preceding paragraph was dedicated to the construction of double groups and to the determination of the corresponding spinor irreducible representations at some symmetrical points of the Brillouin zone in InSb and in Ge crystals. Let us now consider representations in the Brillouin zone centre Γ for InSb and for Ge, where the products of the spin functions and the Bloch functions are the basis functions. Such representations, as we shall see, may be both irreducible and reducible. In the latter case this will make it possible to determine the nature of spin-orbit splitting at point Γ in InSb and in Ge.

In the state $\mathscr{E}_n(k)$ is $d_n$-fold degenerate (spin is not taken into account), i.e., if there are $d_n$ Bloch functions $\psi_{n\mathbf{k}j}(\mathbf{r})$ ($j = 1, 2, \ldots, d_n$) corresponding to it, then the inclusion of the spin will make

[16] Double groups in InSb and in Ge are treated in more detail in [4.13], [4.8, p. 312], and [4.14].

it $2d_n$-degenerate, and the zeroth approximation wave functions will assume the form of $2d_n$ products

$$\psi_{n\mathbf{k}j}(\mathbf{r})\, \nu_i(s). \tag{12.1}$$

The Bloch wave functions $\psi_{n\mathbf{k}j}(\mathbf{r})$ satisfy the Schrödinger equation

$$\hat{\mathscr{H}}(\mathbf{r})\, \psi_{n\mathbf{k}j}(\mathbf{r}) = \mathscr{E}_n(\mathbf{k})\, \psi_{n\mathbf{k}j}(\mathbf{r}), \tag{12.2}$$

where the Hamiltonian

$$\hat{\mathscr{H}}(\mathbf{r}) = -\frac{\hbar^2}{2m}\nabla^2 + V(\mathbf{r})$$

is independent of the spin operators $\hat{\sigma}_i$, and we are, therefore, in a position to multiply the wave functions of equation (12.2) by the spin function $\nu_i(s)$, so that

$$\hat{\mathscr{H}}(\mathbf{r})\, [\nu_i(s)\, \psi_{n\mathbf{k}j}(\mathbf{r})] = \mathscr{E}_n(\mathbf{k})\, [\nu_i(s)\, \psi_{n\mathbf{k}j}(\mathbf{r})]. \tag{12.2a}$$

Let the element of the wave vector space group $G_\mathbf{k}$ be $g = \{R_l \mid \boldsymbol{\alpha}_l + \mathbf{a}_\mathbf{n}\}$, where $R_l$ is a proper or an improper rotation, $\boldsymbol{\alpha}_l$ is an improper translation corresponding to the element $R_l$, and $\mathbf{a}_\mathbf{n}$ is a lattice vector. Act with the operator $\hat{P}'_g$ (2.6.1a) on both sides of equation (12.2a). Since $\hat{\mathscr{H}}(r)$ is invariant under the operations of $\hat{P}'_g$, $\hat{P}'_g\, [\nu_i(s)\, \psi_{n\mathbf{k}j}(\mathbf{r})]$ will also be an eigenfunction of equation (12.2a) corresponding to the original value of $\mathscr{E}_n(k)$. Therefore, (2.7.15)

$$\hat{P}'_g\, [\nu_i(s)\, \psi_{n\mathbf{k}j}(\mathbf{r})] = \{R_l \mid \alpha_l + \mathbf{a}_\mathbf{n}\}\, [\nu_i(s)\, \psi_{n\mathbf{k}j}(\mathbf{r})]$$
$$= \sum_{r=1}^{2} \sum_{s=1}^{d_n} D^l_{ri}(\alpha, \beta, \gamma)\, \Gamma(g)_{sj}\, \psi_{n\mathbf{k}j}(\mathbf{r}), \tag{12.3}$$

where $\alpha$, $\beta$, $\gamma$ are Euler angles corresponding to the rotation $R_l$ (if the rotation is an improper one, it should be replaced by a proper rotation, as suggested at the end of Section 4.10.2), the spin matrix $\boldsymbol{D}^l$ is determined by expression (10.31), and $\Gamma(g)$ is an irreducible representation of the group $G_\mathbf{k}$ of dimension $d_n$ corresponding to the element $g$ of the space group of the wave vector $\mathbf{k}$.

It will be seen from the definition (A.3.49) that

$$D^l_{ri}(\alpha, \beta, \gamma)\, \Gamma(g)_{sj} = (\boldsymbol{D}^l \times \Gamma)_{rs,\, ij} \tag{12.4}$$

is a matrix element of the direct product of a two-row matrix $\boldsymbol{D}^l$ and the $d_n$-row matrix $\Gamma(g)$.

It follows from (12.3) and from the definition of the operator $\hat{P}'_g$ (2.6.1a) and (2.6.5c) that the part of the matrix representation of the wave vector group $g \in G_\mathbf{k}$ built on the basis functions (12.1) is played by the transpose of the matrix (12.4). The character of this representation is

$$\chi[D^l \times \Gamma(g)] = \chi(D^l) \cdot \chi[\Gamma(g)] = 2\cos\frac{\beta}{2}\cos\frac{\alpha+\gamma}{2}\chi[\Gamma(g)], \quad (12.5)$$

where we made use of the fact that the trace of the direct product of matrices is equal to the product of their traces (A.3.52) and of expression (10.31a). Since two signs can be attributed to the $D^l$ matrix (10.31), the elements $R_l$ and $\bar{R}_l$ of the double group can be made to correspond to the different signs of the characters.

Consider the centre Γ of the Brillouin zone in an InSb crystal. Since at point Γ the wave vector $\mathbf{k} = 0$, the electronic state without spin at this point can be classified with the aid of the irreducible representations of the point group $T_d$ (see Table 2.7).

To find the characters of the direct product (12.5), we should find the characters of the matrix $D^l$ corresponding to the different elements $R_l$ of the $T_d$ group. Making use of Table 4.2 and of equations (10.25a), (10.25b), find the corresponding Euler angles (they will prove to be equal for all the elements of one class of $T_d$); next find the characters of the matrix $D^l$ from equation (10.31a).

Taking into account both signs in (10.31a) and the fact that $\chi(\bar{R}_l) = -\chi(R_l)$, we can find the characters of the direct product (12.5) for all the classes of the double $T_d$ group.

If the representation of the double groups $[\widetilde{D^l \times \Gamma}(g)]$ is reducible, it will have important physical consequences. Let it be for the centre of the Brillouin zone in indium antimonide $\Gamma(g) = \Gamma\{R \mid \mathbf{a}_n\} = \Gamma(R) = \Gamma_{15}$, i.e., let the irreducible representation for the simple group be of the $p$-type and of dimension 3. In this case the direct product $D \times \Gamma_{15}$ is a $6 \times 6$ matrix. It will be seen, however, from Table 4.9 that the maximum dimension of the spinor irreducible representation of the double group at point Γ in InSb is 4, and for this reason the representation $D \times \Gamma_{15}$ is a reducible one. Making use of equation (12.5), find the characters of the direct product $D \times \Gamma_{15}$ for all the classes of the double group $T_d$; they are presented in the last row of Table 4.9, whence we see immediately that the characters of $D \times \Gamma_{15}$ are equal to the sum of the characters of $\Gamma_7$ and $\Gamma_8$, i.e.,

$$D \times \Gamma_{15} = \Gamma_7 + \Gamma_8. \quad (12.6)$$

Hence, the six-fold degenerate (with spin) state in the centre of the Brillouin zone of InSb consists of the double degenerate irreducible spinor representation $\Gamma_7$ and of the four-fold degenerate irreducible representation $\Gamma_8$. The spin-orbit coupling lifts this accidental degeneracy, and the level splits up in two, so that $\mathscr{E}(\Gamma_8) - \mathscr{E}(\Gamma_7) = \Delta$, where Δ is the spin-orbit splitting. The result (12.6) can be interpreted from the viewpoint of tight binding of electrons. The atomic electron in the $p$-state (orbital quantum number $l = 1$) is characterized by the quantum number $j$ which determines its total (with the spin) momentum equal to $j\hbar$. The possible values of $j$ for $l = 1$ are 3/2 and 1/2; the degeneracy of those levels equal to $2j+1$ will, accordingly, be equal to 4 and 2, respectively, similar to (12.6). Spin-orbit coupling causes the splitting of those levels in the atom. Studies of the atomic spectrum indicate that the four-fold degenerate term $P_{3/2}$ lies above the doubly degenerate term $P_{1/2}$. If a similar situation persists in the crystal, then $\mathscr{E}(\Gamma_8) - \mathscr{E}(\Gamma_7) = \Delta > 0$.

Calculating all the direct products $\boldsymbol{D}^l \times \Gamma_i(R_l)$, where $\Gamma_i$ are irreducible representations of the $\boldsymbol{T}_d$ group, and decomposing them in the same way as (12.6) into spinor irreducible representations $\Gamma_6, \Gamma_7, \Gamma_8$, we obtain Table 4.15. We presume the state of the electron in the valence and the conduction bands of InSb for $k = 0$ to be described (with no account taken of spin) by the irreducible representations $\Gamma_{15}$ and $\Gamma_1$.

Table 4.15

| | | |
|---|---|---|
| $\boldsymbol{D}\times\Gamma_1=\Gamma_6$, | $\boldsymbol{D}\times\Gamma_2=\Gamma_7$, | $\boldsymbol{D}\times\Gamma_{12}=\Gamma_8$, |
| $\boldsymbol{D}\times\Gamma_{15}=\Gamma_7+\Gamma_8$, | $\boldsymbol{D}\times\Gamma_{25}=\Gamma_6+\Gamma_8$ | |

Figure 4.25 depicts the pattern of electron energy in the centre of the Brillouin zone and along the $\Delta$ axis with spin-orbit coupling ($b$) and without it ($a$); here we made use of (8.1) and of Table 4.14.

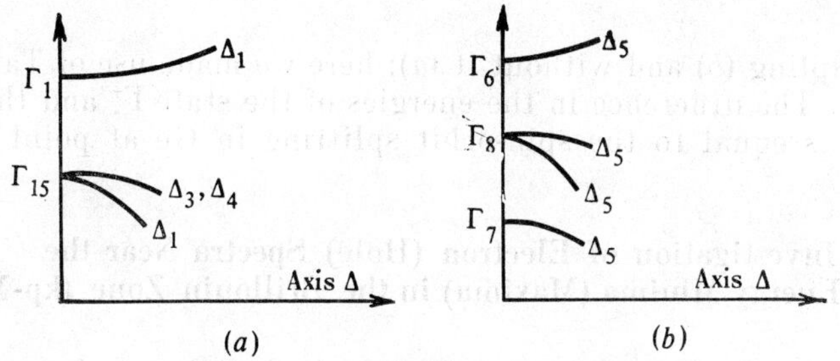


Fig. 4.25

Figure 4.25$b$ clearly shows the magnitude of the spin-orbit splitting.

In the same way as we did with Table 4.15, we can build Table 4.16 of decompositions of the direct product $\boldsymbol{D} \times \Gamma_i$, where $\Gamma_i$ are irreducible representations of the simple group in the centre of the

Brillouin zone of germanium; the spinor irreducible representations of the double group for Ge are presented in Table 4.11. We presume the state of the electron in the valence and the conduction bands of Ge for $\mathbf{k} = 0$ to be described (with no account taken of spin) by the irreducible representations $\Gamma'_{25}$ and $\Gamma'_2$.

Table 4.16

| | | | |
|---|---|---|---|
| $D\times\Gamma_1=\Gamma_6^+$, | $D\times\Gamma'_1=\Gamma_6^-$, | $D\times\Gamma_2=\Gamma_7^+$, | $D\times\Gamma'_2=\Gamma_7^-$, |
| $D\times\Gamma_{12}=\Gamma_8^+$, | $D\times\Gamma'_{12}=\Gamma_8^-$, | $D\times\Gamma_{15}=\Gamma_6^-+\Gamma_8^-$, | |
| $D\times\Gamma'_{15}=\Gamma_6^+\times\Gamma_8^+$, | $D\times\Gamma_{25}=\Gamma_7^-+\Gamma_8^-$, | $D\times\Gamma'_{25}=\Gamma_7^++\Gamma_8^+$ | |

Figure 4.26 is a schematic representation of the energies of the electron states at point $\mathbf{k} = 0$ and along the $\Delta$ axis in Ge with spin-

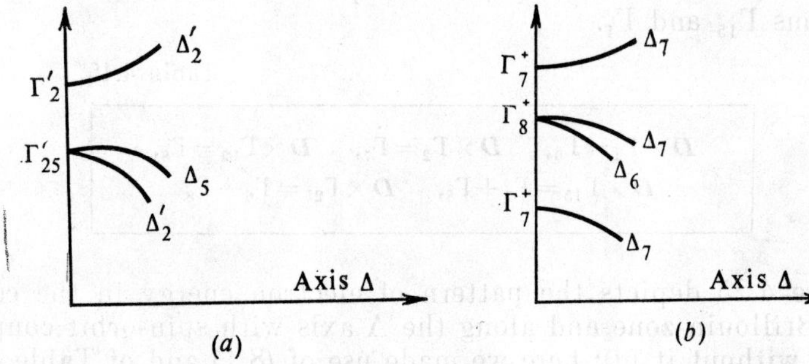


Fig. 4.26

orbit coupling ($b$) and without it ($a$); here we made use of Tables 3.5 and 4.14. The difference in the energies of the state $\Gamma_8^+$ and the lower state $\Gamma_7^+$ is equal to the spin-orbit splitting in Ge at point $\mathbf{k} = 0$.

## 4.13 Investigation of Electron (Hole) Spectra Near the Energy Minima (Maxima) in the Brillouin Zone (kp-Method)

**4.13.1** In Section 4.5 we considered the almost free electron approximation with the periodic lattice potential acting as a perturbation, and in Section 4.7 the tight binding approximation for which the unperturbed motion of the electrons is described by their states in noninteracting atoms. Both these approximations are obviously unsatisfactory for real crystals, therefore only qualitative results stemming from the lattice periodicity in space (the form of the Bloch function, the existence of allowed and forbidden energy bands, the periodic dependence of energy on the wave vector, etc.) were of essential importance.

In Sections 4.4.8-4.4.12 we applied group theory to study the electron spectrum at symmetrical points of the Brillouin zone. We were able to find the degree of degeneracy and the symmetry of wave functions at such points.

The problem of determining the electron spectrum $\varepsilon_n(\mathbf{k})$ throughout the Brillouin zone requires for its solution the integration in the Schrödinger equation (3.7). This problem is a very difficult one, even if computers are employed, largely because of the difficulty in specifying a self-consistent periodic potential $V(\mathbf{r})$.

However, in the case of the semiconductors, in which the number of charge carriers (electrons and holes) is usually small, only their energy spectrum in the vicinity of the energy minima and maxima in the Brillouin zone should be determined; indeed, if the number of electrons (holes) is small, at temperatures not too high the free charge carriers will assemble in the band near the energy minima (maxima). To determine the spectrum of free carriers near the energy extrema in the Brillouin zone, we can apply the perturbation theory. We shall at first consider this problem without taking into account the spin-orbit coupling. Equation (3.8) for the modulating function $u_{n\mathbf{k}}(\mathbf{r})$ is of the form

$$-\frac{\hbar^2}{2m}\nabla^2 u_{n\mathbf{k}} + V(\mathbf{r})\, u_{n\mathbf{k}} + \frac{\hbar}{m}(\mathbf{k}\hat{\mathbf{p}})\, u_{n\mathbf{k}} + \frac{\hbar^2 k^2}{2m} u_{n\mathbf{k}} = \varepsilon_n(\mathbf{k})\, u_{n\mathbf{k}}. \tag{13.1}$$

Here $n$ is the number of the allowed energy band, therefore $\varepsilon_n(\mathbf{k})$ is the electron energy in the $n$-th band at point $\mathbf{k}$; $\hat{\mathbf{p}} = -i\hbar\nabla$ is the electron's momentum operator, $m$ is the free electron mass. At point $\mathbf{k} = 0$ equation (13.1) assumes the form

$$-\frac{\hbar^2}{2m}\nabla^2 u_{n0} + V(\mathbf{r})\, u_{n0} = \varepsilon_n(0)\, u_{n0}. \tag{13.1a}$$

For small values of $\mathbf{k}$ the terms

$$\frac{\hbar^2 k^2}{2m} + \frac{\hbar}{m}(\mathbf{k}\hat{\mathbf{p}}) \tag{13.2}$$

in equation (13.1) can be regarded as a perturbation.

Let the energy extremum in the $l$-th band be located at point $\mathbf{k} = 0$, i.e., in the centre of the Brillouin zone, and let the state of the electron (hole) at this point with the energy $\varepsilon_l(0) \equiv \varepsilon_l^0$ be non-degenerate. Calculate from equation (13.1) the correction to the energy $\varepsilon_l(\mathbf{k}) - \varepsilon_l^0$ for small values of $\mathbf{k}$. If $\varepsilon_l^0$ corresponds to an energy extremum, the correction to the energy linear in $\mathbf{k}$ will be zero. The correction quadratic in **k** consists of corrections to the energy originating from $\frac{\hbar^2k^2}{2m}$ (the first approximation of the perturbation theory) and from $\left(\frac{\hbar}{m}\right)(\mathbf{k}\hat{\mathbf{p}})$ (the second approximation of the perturbation theory). The unperturbed wave functions $u_{l0}(\mathbf{r}) \equiv u_l$ satisfy equation (13.1a). Applying the general equations of the perturbation theory for a nondegenerate state [4.15, § 38], we obtain

$$\varepsilon_l(\mathbf{k}) - \varepsilon_l^0 = \left\langle u_l \left| \frac{\hbar^2k^2}{2m} \right| u_l \right\rangle + \frac{\hbar^2}{m^2} \sum_n{}' \frac{\langle u_l \,|\, \mathbf{k}\hat{\mathbf{p}} \,|\, u_n\rangle \langle u_n \,|\, \mathbf{k}\hat{\mathbf{p}} \,|\, u_l\rangle}{\varepsilon_l^0 - \varepsilon_n^0}$$

$$= \frac{\hbar^2k^2}{2m} + \frac{\hbar^2}{m^2} \sum_{\alpha\beta} k_\alpha k_\beta \sum_n{}' \frac{\langle u_l \,|\, \hat{p}_\alpha \,|\, u_n\rangle \langle u_n \,|\, \hat{p}_\beta \,|\, u_l\rangle}{\varepsilon_l^0 - \varepsilon_n^0}, \tag{13.3}$$

where the prime after the sum means that in the process of summation the term with $n=l$ should be dropped and $\alpha$ and $\beta$ run over the values $x, y, z$, so that, for example, $\hat{p}_\alpha = \hat{p}_x = -i\hbar\partial/\partial x$ for $\alpha = x$. The matrix element

$$\left\langle u_l \left| \frac{\hbar^2k^2}{2m} \right| u_l \right\rangle = \frac{\hbar^2k^2}{2m} \int u_l^* u_l \, d\tau = \frac{\hbar^2k^2}{2m} = \frac{\hbar^2}{2m} \sum_{\alpha\beta} k_\alpha k_\beta \delta_{\alpha\beta}, \tag{13.3a}$$

since the functions $u_l$ are presumed to be orthonormal. Making

$$\varepsilon_l(k) - \varepsilon_l^0 = \frac{\hbar^2}{2} \sum_{\alpha\beta} m_{\alpha\beta}^{-1} k_\alpha k_\beta, \tag{13.4}$$

where $m_{\alpha\beta}^{-1}$ is the inverse effective mass tensor, we obtain from (13.3) and (13.4)

$$m_{\alpha\beta}^{-1} = \frac{\delta_{\alpha\beta}}{m} + \frac{2}{m^2} \sum_n{}' \frac{\langle u_l \,|\, \hat{p}_\alpha \,|\, u_n\rangle \langle u_n \,|\, \hat{p}_\beta \,|\, u_l\rangle}{\varepsilon_l^0 - \varepsilon_n^0}, \tag{13.5}$$

where $\delta_{\alpha\beta}$ is the Kronecker delta. It would be interesting to compare (13.5) with expression (3.19) for the same quantity $m_{\alpha\beta}^{-1}$.

In a cubic crystal we obtain for the scalar effective mass

$$\frac{1}{m^*} = \frac{1}{m} + \frac{2}{m^2} \sum_n{}' \frac{|\langle u_l \,|\, \hat{p}_\alpha \,|\, u_n\rangle|^2}{\varepsilon_l^0 - \varepsilon_n^0}. \tag{13.6}$$

Since $\varepsilon_n^0$ may be both larger and smaller than $\varepsilon_l^0$ the effective mass $m^*$ may be both smaller and larger than the free electron mass $m$.

Equation (13.6) is applicable also when the energy extremum $\varepsilon_l(\mathbf{k})$ is located not in the Brillouin zone centre but at point $\mathbf{k}_0$; it should only be kept in mind that in this case the wave functions $u_l$ display the symmetry of the wave vector group $G_{\mathbf{k}_0}$; this is the reason why, for example, the function $u_l$ for the conduction electrons of germanium, which have their $\mathbf{k}_0$ lying on the [111] axis, displays not a cubic, but an axial symmetry, with the result that in a cubic crystal of $n$-Ge the effective mass becomes a tensor.

**4.13.2** Let now the state of the electron (of a hole) in the $l$-th band at point $\mathbf{k} = 0$, where the energy has an extremum, be a degenerate one. We are again interested in corrections to the energy quadratic in $\mathbf{k}$. It follows from the theory of perturbations of degenerate states [4.15, § 39] that in our case the corrections to the energy $\varepsilon^{(2)}$ originating from $\frac{\hbar}{m}(\mathbf{k}\hat{\mathbf{p}})$ in the second approximation are equal to the roots of the secular equation

$$\left|\frac{\hbar^2}{m^2}\sum_{ns}\frac{\langle l, r' \mid \mathbf{k}\hat{\mathbf{p}} \mid n, s\rangle \langle n, s \mid \mathbf{k}\hat{\mathbf{p}} \mid l, r\rangle}{\varepsilon_l^0-\varepsilon_n^0} - \varepsilon^{(2)}\delta_{rr'}\right| = 0. \tag{13.7}$$

Here $|\,l, r\rangle$ and $|\,l, r'\rangle$ are the unperturbed $f$-fold degenerate wave functions ($r, r' = 1, 2, \ldots, f$) satisfying equation (13.1a) for the energy eigenvalues $\varepsilon_l(0) \equiv \varepsilon_l^0$; $|n, s\rangle$ are wave functions for the energy level $\varepsilon_n^0$. The summation is performed over all $n \neq l$ and over $s$. The order of the determinant of the secular equation (13.7) is equal to the degree of degeneracy of the level $\varepsilon_l^0$.

The sum in the equation can be represented in the form

$$\frac{\hbar^2}{m^2}\sum_{\alpha\beta}k_\alpha k_\beta\sum_{ns}\frac{\langle l, r' \mid \hat{p}_\alpha \mid n, s\rangle \langle n, s \mid \hat{p}_\beta \mid l, r\rangle}{\varepsilon_l^0-\varepsilon_n^0} \equiv \mathcal{H}_{r'r}, \tag{13.7a}$$

where $\alpha$ and $\beta$ assume the values $x, y, z$. Write now the secular equation in the form

$$|\,\mathcal{H}_{r'r} - \varepsilon^{(2)}\delta_{r'r}\,| = 0. \tag{13.7b}$$

The corrections to the energy $\varepsilon_l^0$ of the first order in the term $\hbar^2k^2/2m$ are determined from the equation [4.15, § 39]

$$\left|\left\langle l, r'\left|\frac{\hbar^2k^2}{2m}\right|l, r\right\rangle - \varepsilon^{(1)}\delta_{r'r}\right| = 0 \tag{13.8}$$

or

$$\left|\left(\frac{\hbar^2k^2}{2m} - \varepsilon^{(1)}\right)\delta_{r'r}\right| = 0. \tag{13.8a}$$

This yields $f$ coinciding roots

$$\varepsilon^{(1)} = \hbar^2k^2/2m. \tag{13.9}$$

It follows from (13.7) and (13.9) that

$$\varepsilon_l(\mathbf{k}) = \varepsilon_l^0 + \varepsilon^{(1)} + \varepsilon_i^{(2)} = \varepsilon_l^0 + \frac{\hbar^2k^2}{2m} + \varepsilon_i^{(2)}, \tag{13.10}$$

where $\varepsilon_i^{(2)}$ is one of the $f$ roots of equation (13.7).

**4.13.3** Consider the structure of the secular equation (13.7b) for the case of the valence band of silicon or germanium. Theoretical and experimental considerations make very probable the assumption that the state of the hole in the centre of the zone ($\mathbf{k} = 0$) corresponds to the irreducible representation $\Gamma'_{25}$ of the $\boldsymbol{O}_h$ group. The basis functions of this irreducible representation of dimension 3 are (see Table 4.3): $X^+ = xy$, $Y^+ = yz$, $Z^+ = zx$. It can easily be demonstrated that the basis functions under the symmetry transformations of the group $\boldsymbol{O}_h$ behave in a way similar to that of simpler basis functions $X = x$, $Y = y$, $Z = z$ of the irreducible three-dimensional degenerate atomic $p$-state of the electron; therefore in the following we shall make use of the latter simpler functions with an $x$, $y$, $z$ symmetry.

It can be demonstrated that the sum $\sum\limits_{n,\,s}$ (13.7a) is transformed under the operation of the elements of the crystal's space group like the matrix element

$$\langle l,\ r' \,|\, \hat{p}_\alpha \hat{p}_\beta \,|\, l,\ r\rangle \equiv Q_{\alpha\beta}(r',\ r), \tag{13.11}$$

where the operators $\hat{p}_x$, $\hat{p}_y$, $\hat{p}_z$ are transformed like the coordinates $x$, $y$, $z$ (indeed, if under a given transformation $x \to -y$, then $\hat{p}_x = -i\hbar\partial/\partial x \to i\hbar\partial/\partial y = -\hat{p}_y$).

Consider the constant $Q_{\alpha\beta}(r',\ r)$ for the states $r' = r = X$ and demonstrate that for $\alpha \neq \beta$ it is equal to zero. Make, for example, $p_\alpha = p_x$, $p_\beta = p_y$, then (13.11) is transformed under the operation of the elements of the $\boldsymbol{O}_h$ group like the product $[xxyx]$; when the transformation $R_2\ (x\bar{y}\bar{z})$ (Table 4.2) is applied, we obtain

$$\hat{R}_2\,[xxyx] = -[xxyx],$$

whence it follows that $[xxyx] = 0$, i.e., that the corresponding constant of $Q_{xy}(X,\ X)$ is zero. If for the same states $r' = r = X$ we make $p_\alpha = p_\beta = p_x$, then (13.11) is transformed like the product $[xxxx] = [x^4]$; it will easily be seen that there can be no such transformation $R_i$ or $JR_i$ that would transform $[x^4]$ into $-[x^4]$, and because of that the corresponding constant of $Q_{xx}(X,\ X)$ is nonzero, we shall denote it by

$$L = \frac{\hbar^2}{m^2} \sum_{ns}{}' \frac{\langle l,\ X \,|\, \hat{p}_x \,|\, n,\ s\rangle \langle n,\ s \,|\, \hat{p}_x \,|\, l,\ X\rangle}{\varepsilon_l^0 - \varepsilon_n^0}$$

$$= \frac{\hbar^2}{m^2} \sum_{ns}{}' \frac{|\langle l,\ X \,|\, \hat{p}_x \,|\, n, s\rangle|^2}{\varepsilon_l^0 - \varepsilon_n^0}. \tag{13.12}$$

Let the states be as before $r' = r = X$, and let $p_\alpha = p_\beta = p_y$, then acting with $R_{13}\ (x\bar{z}\bar{y})$ on $[xyyx]$, we obtain

$$\hat{R}_{13}\,[xy^2x] = [xz^2x],$$

and this makes it possible to introduce a constant

$$M=\frac{\hbar^2}{m^2}\sum_{ns}{}'\frac{|\langle l,\ X\,|\,\hat{p}_y\,|\,n,\ s\rangle|^2}{\varepsilon_l^0-\varepsilon_n^0}=\frac{\hbar^2}{m^2}\sum_{ns}{}'\frac{|\langle l,\ X\,|\,\hat{p}_z\,|\,n,\ s\rangle|^2}{\varepsilon_l^0-\varepsilon_n^0}. \quad (13.13)$$

We can now write down the (1,1)-th term of the determinant (13.7b) in the form

$$\mathcal{H}_{xx}-\varepsilon^{(2)}=Lk_x^2+M\,(k_y^2+k_z^2)-\varepsilon^{(2)}.$$

Precisely in the same way it can be demonstrated that the (1,2)-th and the (2,1)-th terms of the determinant (13.7a) are equal to

$$\mathcal{H}_{xy}=\mathcal{H}_{yx}=Nk_xk_y,$$

where

$$N=\frac{\hbar^2}{m^2}\sum_{ns}{}'\frac{\langle l,\ X\,|\,\hat{p}_x\,|\,n,\ s\rangle\langle n,\ s\,|\,\hat{p}_y\,|\,l,\ Y\rangle}{\varepsilon_l^0-\varepsilon_n^0}. \quad (13.14)$$

It follows from Table 4.2 that $\hat{R}_5\,[x^4]=[y^4]$, $\hat{R}_5\,[xy^2x]=[yz^2y]$ and $\hat{R}_{17}\,[xy^2x]=[yx^2y]$; therefore the (2,2)-th term of the determinant (13.7b) is equal to

$$\mathcal{H}_{yy}-\varepsilon^{(2)}=Lk_y^2+M\,(k_x^2+k_z^2)-\varepsilon^{(2)}.$$

As a result the secular equation (13.7b) assumes the form

$$\begin{vmatrix} Lk_x^2+M\,(k_y^2+k_z^2)-\varepsilon^{(2)} & Nk_xk_y & Nk_xk_z \\ Nk_xk_y & Lk_y^2+M\,(k_x^2+k_z^2)-\varepsilon^{(2)} & Nk_yk_z \\ Nk_xk_z & Nk_yk_z & Lk_z^2+M\,(k_x^2+k_y^2)-\varepsilon^{(2)} \end{vmatrix}=0, \quad (13.15)$$

where the material constants $L$, $M$, $N$ are defined via (13.12), (13.13) and (13.14).

Secular equation (13.5) is of the third degree with respect of $\varepsilon^{(2)}$. We shall not solve it in the general form but shall restrict ourselves to the case of a specified direction of $\mathbf{k}$. Let $k_x\neq 0$, $k_y=k_z=0$; then (13.5) assumes the form

$$\begin{vmatrix} Lk_x^2-\varepsilon^{(2)} & 0 & 0 \\ 0 & Mk_x^2-\varepsilon^{(2)} & 0 \\ 0 & 0 & Mk_x^2-\varepsilon^{(2)} \end{vmatrix}=0, \quad (13.15a)$$

whence $[Lk_x^2-\varepsilon^{(2)}]\,[Mk_x^2-\varepsilon^{(2)}]^2=0$.

Hence, we have three roots of equation (13.15a):

$$\varepsilon_1^{(2)}=Lk_x^2,\quad \varepsilon_2^{(2)}=\varepsilon_3^{(2)}=Mk_x^2, \quad (13.15b)$$

which perfectly correspond to the case of tight binding (4.7.25):

$$\varepsilon_1 = \varepsilon_M - Aa^2k_x^2, \quad \varepsilon_2 = \varepsilon_3 = \varepsilon_M - Ba^2k_x^2.$$

Of course, the general solution of the secular equation (13.15) does not coincide with the simple expression (4.7.24); this is an indication of the fact that the $\mathbf{k}\hat{\mathbf{p}}$-method in the quadratic approximation goes beyond the limits of the simple tight-binding model that takes into account only the interaction between the neighbouring electrons.

**4.13.4** Equation (13.15) determines the spectrum of holes in germanium or silicon with no account taken of the spin-orbit coupling which, as we have learned in Section 4.4.12, results in the splitting of energy levels.

When spin is taken into account, there correspond to the state $\Gamma'_{25}$ in the centre of the valence band not the three functions transformed like $X$, $Y$ and $Z$ but six functions of the form: $X\nu_1$, $Y\nu_1$, $Z\nu_1$, $X\nu_2$, $Y\nu_2$, $Z\nu_2$, where $\nu_1(s)$ and $\nu_2(s)$ are spin functions (10.12). The secular equation (13.7) turns now into a determinant of the sixth order.

The zeroth approximation wave functions can obviously be chosen in the form of arbitrary linear combinations of the six functions $X\nu_1, Y\nu_1, \ldots, Z\nu_2$ mentioned above. It can be demonstrated that if such linear combinations are chosen in the form of

$$Y_{3/2}^{3/2} = -\frac{1}{\sqrt{2}}(x+iy)\nu_1, \quad Y_{3/2}^{1/2} = -\frac{1}{\sqrt{6}}[(x+iy)\nu_2 - 2z\nu_1],$$

$$Y_{3/2}^{-1/2} = \frac{1}{\sqrt{6}}[(x-iy)\nu_1 + 2z\nu_2], \quad Y_{3/2}^{-3/2} = \frac{1}{\sqrt{2}}(x-iy)\nu_2 \tag{13.16}$$

and

$$Y_{1/2}^{1/2} = -\frac{1}{\sqrt{3}}[(x+iy)\nu_2 + z\nu_1], \quad Y_{1/2}^{-1/2} = \frac{1}{\sqrt{3}}[(x-iy)\nu_1 - z\nu_2], \tag{13.16a}$$

they diagonalize the Hamiltonian of spin-orbit coupling $\hat{\mathscr{H}}_{so}$ (10.6). As a result, the secular equation with its sixth-order determinant splits up into two secular equations with determinants of the fourth and the second orders (see below).

The functions $Y_j^m$ for a unit sphere, i.e., for the condition $x^2 + y^2 + z^2 = 1$, are spherical functions serving as eigenfunctions of the operator of the projection of the total momentum

$$\hat{J}_z = \hat{M}_z + \hat{S}_z, \tag{13.17}$$

where $\hat{\mathbf{M}}$ and $\hat{\mathbf{S}}$ are determined by expressions (10.17) and (10.1); $m\hbar$ is the projection of the total momentum on the $z$ axis with the maximum value equal to $j\hbar$.

Acting, for instance, with the operator $\hat{J}_z$ (13.17) on the function $Y_{3/2}^{-1/2}$, we obtain

$$\hat{J}_z Y_{3/2}^{-1/2} = \frac{1}{\sqrt{6}}\left[\frac{\hbar}{i}\left(x\frac{\partial}{\partial y} - y\frac{\partial}{\partial x}\right)\right.$$

$$\left. + \frac{\hbar}{2}\hat{\sigma}_z\right][(x - iy)\,\nu_1 + 2z\nu_2]$$

$$= -\frac{\hbar}{2}\frac{1}{\sqrt{6}}[(x-iy)\,\nu_1 + 2z\nu_2] = -\frac{\hbar}{2}Y_{3/2}^{-1/2},$$

where we made use of the fact that $\hat{\sigma}_z\nu_1 = \nu_1$, and $\hat{\sigma}_z\nu_2 = -\nu_2$ (see (10.13)). We see that the eigenvalue of the operator $\hat{J}_z$ corresponding to the function $Y_j^m = Y_{3/2}^{-1/2}$, is indeed, equal to $m\hbar = -\hbar/2$, i.e., $m = -1/2$. It can be demonstrated that the wave functions (13.16a) on a unit sphere are mutually orthogonal and normalized (to $4\pi/3$).

To understand the structure of "regular" wave functions of the zeroth approximation (13.16a), we could also reason as follows: The representation of the binary group $\Gamma_{15}$ is equal to the direct product $\boldsymbol{D}^{1/2} \times \boldsymbol{D}^1$, where $\boldsymbol{D}^1$ is an irreducible representation of $\Gamma_{15}$ and $\boldsymbol{D}^{1/2}$ is an irreducible spinor representation (10.31). It can be demonstrated that

$$\boldsymbol{D}^{1/2} \times \boldsymbol{D}^1 = \boldsymbol{D}^{3/2} + \boldsymbol{D}^{1/2},$$

where $\boldsymbol{D}^{3/2}$, an irreducible representation of dimension 4 corresponding to the total momentum, is equal to $3\hbar/2$, and $\boldsymbol{D}^{1/2}$ is an irreducible representation of dimension 2 with a momentum $\hbar/2$. These representations are irreducible in the cubic group, as well, and they reduce to $\Gamma_8$ and $\Gamma_7$, respectively, in accordance with the decomposition (12.6): $\boldsymbol{D}^{1/2} \times \Gamma_{15} = \Gamma_8 + \Gamma_7$. Therefore, the basis functions of the four-fold degenerate state $\Gamma_8$ are the functions (13.16)—the eigenfunctions of the total momentum operator $J_z$ with the maximum eigenvalue $3\hbar/2$, and the basis functions of the doubly degenerate state $\Gamma_7$ are the functions (13.16a) corresponding to a momentum of $\hbar/2$.

It can be demonstrated that the nondiagonal matrix elements of the spin-orbit coupling operator $\hat{\mathcal{H}}_{so}$ (10.6) constructed on the functions (13.16) and (13.16a) are equal to zero. The correction due to spin-orbit coupling to the diagonal matrix elements $\mathcal{H}_{r'r}$ (13.7a) over the functions (13.16) is equal to $\Delta/3$, and for the functions

(13.16a) to $-2\Delta/3$, where $\Delta$ is the spin-orbit splitting of the bands $\Gamma_8$ and $\Gamma_7$ for $\mathbf{k} = 0$.[17]

If splitting of the bands $\mathscr{E}(\Gamma_8) - \mathscr{E}(\Gamma_7) = \Delta$ due to spin-orbit coupling is large as compared with the kinetic energy of the charge carriers, the secular equations for the bands $\Gamma_8$ and $\Gamma_7$ can be considered separately.

Denote by $\mathscr{H}'_{ik}$ the matrix elements of the secular equation for the determination of the perturbation energy in the second approximation of spin-orbit coupling, when the functions (13.16) have been chosen to serve as the "regular" wave functions of the zeroth approximation; choose the values $1 \to 3/2$, $2 \to 1/2$, $3 \to -1/2$, $4 \to -3/2$ to correspond to the indices $i, k = 1, 2, 3, 4$ of $\mathscr{H}'_{ik}$, then we obtain from (13.16)

$$\mathscr{H}'_{11} = \frac{1}{2}\langle (x+iy)\, v_1 \,|\, \hat{\mathscr{H}} \,|(x+iy)\, v_1\rangle$$

$$= \frac{1}{2}(\mathscr{H}_{xx} + \mathscr{H}_{yy} + i\mathscr{H}_{xy} - i\mathscr{H}_{yx}),$$

where $\mathscr{H}'_{rr}$ $(r', r = x, y, z)$ are the matrix elements (13.7a). Making use of notation (13.12), (13.13) and taking into account that $\mathscr{H}_{xy} = \mathscr{H}_{yx}$, we obtain

$$\mathscr{H}'_{11} = \frac{1}{2}(\mathscr{H}_{xx} + \mathscr{H}_{yy}) = \frac{1}{2}(L+M)(k_x^2 + k_y^2) - Mk_z^2$$

$$= Ak^2 + \frac{1}{2}B(k^2 - 3k_z^2) \equiv F, \qquad (13.18)$$

where

$$A = \frac{L+2M}{3} \quad \text{and} \quad B = \frac{L-M}{3}. \qquad (13.18a)$$

It will easily be seen that

$$\mathscr{H}'_{44} = \mathscr{H}'_{11} = F.$$

Next

$$\mathscr{H}'_{22} = \frac{1}{6}\langle (x+iy)\, v_2 - 2zv_1 \,|\, \hat{\mathscr{H}} \,|\, (x+iy)\, v_2 - 2zv_1\rangle$$

$$= \frac{1}{6}(\mathscr{H}_{xx} + \mathscr{H}_{yy} + i\mathscr{H}_{xy} - i\mathscr{H}_{yx} + 4\mathscr{H}_{zz}),$$

---

[17] Profiting by expression (10.6) for $\hat{\mathscr{H}}_{s0}$, by the explicit form of the functions (13.16), (13.16a), and by considerations of symmetry we can easily show that the displacement of the $\mathbf{k} = 0$ energy level of the $\Gamma_8$ band is proportional to $2\langle \mathbf{r} \mid \nabla N\rangle$ and that of the $\Gamma_7$ band is proportional to $(-4\langle \mathbf{r} \mid \nabla N\rangle)$, so that the splitting is $\Delta \propto 6\langle \mathbf{r} \mid \nabla V\rangle$.

where we made use of the orthogonality of spin functions. Substituting herein the values of the matrix elements (13.15), we obtain after some transformations

$$\mathcal{H}'_{22} = Ak^2 - \frac{1}{2} B (k^2 - 3k_z^2) \equiv G. \tag{13.19}$$

It can easily be demonstrated that

$$\mathcal{H}'_{33} = \mathcal{H}'_{22} = G.$$

Next

$$\mathcal{H}'_{12} = \frac{1}{2\sqrt{3}} \langle (x+iy)\, \nu_1 | \hat{\mathcal{H}} | (x+iy)\, \nu_2 - 2z\nu_1 \rangle$$

$$= -\frac{1}{\sqrt{3}} (\mathcal{H}_{xz} - i\mathcal{H}_{yz}) = -Dk_z (k_x - ik_y) \equiv H, \tag{13.20}$$

where $D = N/\sqrt{3}$,

$$\mathcal{H}'_{13} = -\frac{\sqrt{3}}{2} [B (k_x^2 - k_y^2) + iDk_x k_y] \equiv J. \tag{13.21}$$

It will easily be seen that

$$\mathcal{H}'_{14} = \mathcal{H}'_{23} = 0, \quad \mathcal{H}'_{24} = J, \quad \mathcal{H}'_{34} = -H.$$

Other elements of the matrix $\mathcal{H}'_{ik}$ can be easily obtained, if use is made of its Hermite properties

$$\mathcal{H}'_{ik} = \mathcal{H}'^{*}_{ki}.$$

The secular equation for the energy $\varepsilon \equiv \varepsilon^{(2)}$ is of the form

$$\begin{vmatrix} F-\varepsilon & H & J & 0 \\ H^* & G-\varepsilon & 0 & J \\ J^* & 0 & G-\varepsilon & -H \\ 0 & J^* & -H^* & F-\varepsilon \end{vmatrix} = 0, \tag{13.22}$$

where the quantities $F$, $G$, $H$, $J$ are equal to (13.18)-(13.21).

Decomposing (13.22) in the elements of the first row, we obtain after some algebraic operations

$$[(F - \varepsilon)(G - \varepsilon) - |H|^2 - |J|^2]^2 = 0,$$

or

$$(\varepsilon - F)(\varepsilon - G) - |H|^2 - |J|^2 = 0.$$

Solving this quadratic equation, we obtain

$$\varepsilon_{1,2} = \frac{F+G}{2} \pm \sqrt{\left(\frac{F+G}{2}\right)^2 - FG + |H|^2 + |J|^2}.$$

Substituting the values of $F$, $G$, $H$ and $J$, we obtain after lengthy but simple algebraic operations

$$\varepsilon_{1,2}=Ak^2\pm\sqrt{B^2k^4+C^2(k_x^2k_y^2+k_x^2k_z^2+k_y^2k_z^2)}\,, \tag{13.23}$$

where $A$ and $B$ are equal to (13.18a) and

$$C^2=D^2-3B^2=\frac{(N+L-M)(N-L+M)}{3} \tag{13.23a}$$

if use is made of the values of $B$ and $D$ (13.20).

Since the original equation for $\varepsilon$ is biquadratic, each of the roots (13.23) is doubly degenerate; this is the result of invariance with respect to the reversal of time.

To determine the energy spectrum in the doubly degenerate band $\Gamma_7$, which split off as the result of spin-orbit coupling, we should solve a secular equation of rank $2\times 2$ whose matrix elements are calculated on the wave functions (13.16a). Establishing the correspondence of the indices $i, k=1, 2$ of $\mathscr{H}'_{ik}$ to the values: $1\to 1/2$, $2\to -1/2$, we obtain from (13.16a)

$$\begin{aligned}\mathscr{H}'_{11}&=\frac{1}{3}\langle(x+iy)\nu_2+z\nu_1\,|\,\hat{\mathscr{H}}\,|\,(x+iy)\nu_2+z\nu_1\rangle\\&=\frac{1}{3}[\mathscr{H}_{xx}+\mathscr{H}_{yy}-i\mathscr{H}_{yx}+i\mathscr{H}_{xy}+\mathscr{H}_{zz}]\\&=\frac{L+2M}{3}(k_x^2+k_y^2+k_z^2)=Ak^2,\end{aligned} \tag{13.24}$$

where we made use of notation (13.12), (13.13) and (13.18a).

It can easily be demonstrated that

$$\mathscr{H}'_{22}=\mathscr{H}'_{11},\quad \mathscr{H}'_{12}=\mathscr{H}'_{21}=0. \tag{13.24a}$$

Making use of the values of $\mathscr{H}'_{ik}$ obtained above, write the secular equation

$$\begin{vmatrix}Ak^2-\Delta-\varepsilon & 0\\ 0 & Ak^2-\Delta-\varepsilon\end{vmatrix}=0. \tag{13.25}$$

Here $\Delta=\mathscr{E}(\Gamma_8)-\mathscr{E}(\Gamma_7)$ is the spin-orbit splitting of the levels $\Gamma_8$ and $\Gamma_7$ at point $\mathbf{k}=0$.

From (13.25) we obtain for the doubly degenerate root

$$\varepsilon_3=-\Delta+Ak^2, \tag{13.26}$$

i.e., a simple parabolic dispersion law operates in the $\Gamma_7$ band (constant-energy surfaces are spheres). The constant-energy surfaces corresponding to the dispersion law of band $\Gamma_8$ are *corrugated surfaces* shaped in the [100] (i.e., in the $k_xk_y$ plane) as depicted in Fig. 4.27.

**4.13.5** Cyclotron resonance, transport phenomena, and light absorption experiments in indium antimonide led to the disclosure of the following peculiarities of its band structure. The minimum of the conduction band and the maximum of the valence band are located at point $\mathbf{k} = 0$. The effective mass of the electrons at the bottom of the conduction band is very small and equal to $0.013m$, i.e., of the order of 0.01 of the free electron mass. Two hole branches, of light and heavy holes, are degenerate at point $\mathbf{k} = 0$. The effective mass of the heavy holes is of the order of $0.18m$, i.e., more than 10 times that of the electron mass and, probably, that of the light holes. The third hole branch (also of light holes) is split off, owing to spin-orbit coupling, from the upper edge of the valence band by the amount $\Delta \approx 0.9$ eV.

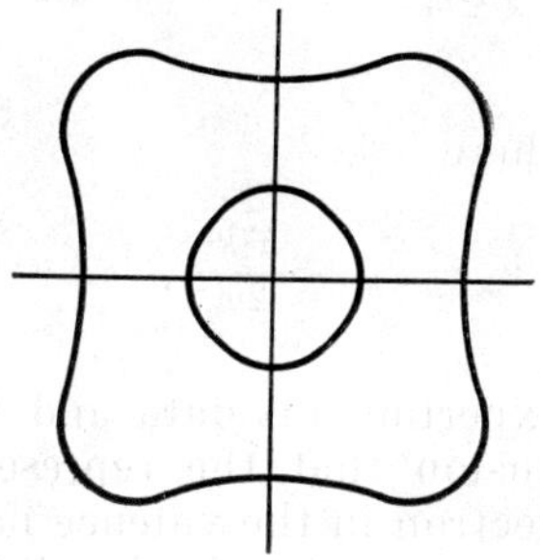

Fig. 4.27

Such a structure of the valence band in indium antimonide follows from the theory that makes use of the symmetry of the InSb crystal but takes no account of corrections to the energy linear in $\mathbf{k}$. The dispersion law $\varepsilon$ ($\mathbf{k}$) displays spherical symmetry for all the branches, but for light particles at high values of $\mathbf{k}$ it deviates appreciably from the simple parabola ($\varepsilon$ is not proportional to $k^2$).

The forbidden bandwidth $\varepsilon_G$ at room temperature is 0.17 eV and is strongly temperature-dependent, so that at 0 K, $\varepsilon_G = 0.23$ eV. Because of the narrowness of the forbidden band the electron state has to be considered simultaneously in the valence and in the conduction bands.

Such an "interaction" of the electron states in the valence and the conduction bands together with the inclusion of spin-orbit coupling provided a semiquantitative explanation of most of the features of its band structure (E. O. Cane, 1956).

The spin-orbit coupling operator (10.6) operating on a Bloch function yields two addends

$$\hat{\mathscr{H}}_{so}\,[u_{\mathbf{k}}(\mathbf{r})\,e^{i\mathbf{k}\cdot\mathbf{r}}] =$$

$$= e^{i\mathbf{k}\cdot\mathbf{r}}\frac{\hbar}{4m^2c^2}\,\hat{\boldsymbol{\sigma}}\cdot(\nabla V \times \hat{\mathbf{p}})\,u_{\mathbf{k}}(\mathbf{r}) + e^{i\mathbf{k}\cdot\mathbf{r}}\frac{\hbar^2}{4m^2c^2}\,\hat{\boldsymbol{\sigma}}\cdot(\nabla V \times \mathbf{k})\,u_{\mathbf{k}}(\mathbf{r}). \tag{13.27}$$

The order of magnitude of the first addend is $p/\hbar k$ times that of the second; here $p$ is the momentum of the electron in the atom (spin-orbit coupling is mainly due to the electrons of the lattice atoms), and $\hbar k$ is the quasi-momentum of the conduction electron; the ratio $p/\hbar k$ is quite great.

The first addend in (13.27) determines the spin-orbit splitting of the levels, and the second is responsible for the energy dependence of the effective mass.

Retaining only the first addend, we obtain instead of (13.1) (omitting the band number $n$)

$$\left\{\frac{\hat{p}}{2m}+V(\mathbf{r})+\frac{\hbar}{m}\mathbf{k}\cdot\hat{\mathbf{p}}+\frac{\hbar^2}{4m^2c^2}\hat{\sigma}\cdot(\nabla V\times\hat{\mathbf{p}})\right\}u_{\mathbf{k}}(\mathbf{r})=\varepsilon'_{\mathbf{k}}u_{\mathbf{k}}(\mathbf{r}), \tag{13.28}$$

where

$$\varepsilon'_{\mathbf{k}}=\varepsilon_{\mathbf{k}}-\frac{\hbar^2k^2}{2m}. \tag{13.28a}$$

Experimental data and theoretical considerations point to the conclusion that the representation corresponding to the state of the electron in the valence band at point $\mathbf{k}=0$ is the $p$-type irreducible representation $\Gamma_{15}$, and that corresponding to the electron state at point $\mathbf{k}=0$ of the conduction band is the $s$-type irreducible unit representation $\Gamma_1$.

When the conduction and the valence bands are considered simultaneously, it is expedient to choose the eight basis functions in the zeroth approximation as follows:

$$iS\nu_2,\quad \frac{1}{\sqrt{2}}(x-iy)\nu_1,\quad z\nu_2,\quad \frac{1}{\sqrt{2}}(x+iy)\nu_1, \tag{13.29}$$

$$iS\nu_1,\quad -\frac{1}{\sqrt{2}}(x+iy)\nu_2,\quad z\nu_1,\quad \frac{1}{\sqrt{2}}(x-iy)\nu_2. \tag{13.29a}$$

Here $S$ is a spherically symmetric function of the $\Gamma_1$ representation. The functions written one under the other in (13.29) and (13.29a) correspond, as we shall presently see, to degenerate electron states.

The structure of the basis functions (13.29), (13.29a) is founded on considerations similar to those discussed in connection with the choice of functions (13.16), (13.16a). The basis functions (13.29), (13.29a) provide for the simplest form of the $8\times 8$ matrix of the secular equation (13.7).

Choose an electron wave vector $\mathbf{k}$ in the direction of the $z$ axis ($k_z=k$). Then the $8\times 8$ matrix of the secular equation corresponding to the operator in the braces of equation (13.28) built on the wave functions (13.29), (13.29a) will be of the form

$$\begin{pmatrix}\mathscr{H} & 0\\ 0 & \mathscr{H}\end{pmatrix},$$

where

$$\mathcal{H} = \begin{vmatrix} \varepsilon_s & 0 & kP & 0 \\ 0 & \varepsilon_p - \Delta/3 & \sqrt{2}\,\Delta/3 & 0 \\ kP & \sqrt{2}\,\Delta/3 & \varepsilon_p & 0 \\ 0 & 0 & 0 & \varepsilon_p + \Delta/3 \end{vmatrix}. \tag{13.30}$$

Here the positive constant

$$\Delta = \frac{3\hbar i}{4m^2c^2} \left\langle x \left| \frac{\partial V}{\partial x} \hat{p}_y - \frac{\partial V}{\partial y} \hat{p}_x \right| y \right\rangle \tag{13.30a}$$

is equal to the spin-orbit splitting in the valence band.

The real quantity

$$P = -i \left( \frac{\hbar}{m} \right) \langle S \mid \hat{p}_z \mid z \rangle \tag{13.30b}$$

characterizes the "interaction" of the valence and the conduction bands, $\varepsilon_s$ and $\varepsilon_p$ are energies corresponding to the edges ($\mathbf{k} = 0$) of the conduction and the valence bands (without spin-orbit splitting).

The $8 \times 8$ matrix (13.30) is quasi-diagonal. Its left upper block $\mathcal{H}$ constructed on the basis functions (13.29) coincides with its right lower block constructed on the functions (13.29a).

To obtain the matrix (13.30) use was made of considerations of crystal symmetry ($T_d$) similar to those used in the determination of the structure of matrix (13.15). In addition, use was made of the orthonormality of the spin functions $\nu_1(s)$ and $\nu_2(s)$.

For example, the (1,3)-th term of the matrix (13.30) is equal to

$$\left\langle iS\nu_2 \left| \frac{\hat{p}^2}{2m} + V + \frac{\hbar}{m}(k\hat{p}_z) + \frac{\hbar}{4m^2c^2} \left[ \left( \frac{\partial V}{\partial y} \hat{p}_z - \frac{\partial V}{\partial z} \hat{p}_y \right) \hat{\sigma}_1 \right. \right. \right.$$
$$\left. \left. \left. + \left( \frac{\partial V}{\partial z} \hat{p}_x - \frac{\partial V}{\partial x} \hat{p}_z \right) \hat{\sigma}_2 + \left( \frac{\partial V}{\partial x} \hat{p}_y - \frac{\partial V}{\partial y} \hat{p}_x \right) \hat{\sigma}_3 \right] \right| z\nu_2 \right\rangle.$$

The matrix element of the first two addends of the operator written above is zero from considerations of symmetry: $\left\langle S \left| \frac{\hat{p}^2}{2m} + V \right| z \right\rangle = 0$; the matrix element of the first two addends in the square brackets is zero because of (10.13) and of the orthogonality of the spin functions; the matrix element of the last addend in the square brackets is zero from symmetry considerations $\left( \left\langle S \left| \frac{\partial V}{\partial x} \hat{p}_y - \frac{\partial V}{\partial y} \hat{p}_x \right| z \right\rangle = 0 \right)$. Hence, only the matrix element of the third addend remains, so that the (1,3)-th term is $-ik \left( \frac{\hbar}{m} \right) \langle S \mid \hat{p}_z \mid z \rangle = kP$, if use is made of the value of $P$ (13.30b).

The matrix (13.30) was obtained under the special assumption that electron wave vector **k** was directed along the $z$ axis. In the case of an arbitrary direction of **k** it is specified by the Euler angles (see Fig. 4.24): $\beta$, the polar angle, $\alpha$, the azimuthal angle, $\gamma = 0$.

The coordinate basis functions $x$, $y$, $z$, according to (10.24) and (10.24a), are transformed as follows:

$$\begin{pmatrix} x' \\ y' \\ z' \end{pmatrix} = \begin{pmatrix} \cos\alpha\cos\beta & \sin\alpha\cos\beta & -\sin\beta \\ -\sin\alpha & \cos\alpha & 0 \\ \cos\alpha\sin\beta & \sin\alpha\sin\beta & \cos\beta \end{pmatrix} \begin{pmatrix} x \\ y \\ z \end{pmatrix}. \tag{13.31}$$

The spin functions $\nu_1(s)$, $\nu_2(s)$, according to (10.30), are transformed as follows:

$$\begin{pmatrix} \nu_1' \\ \nu_2' \end{pmatrix} = \begin{pmatrix} \cos\frac{\beta}{2} e^{-i\alpha/2} & \sin\frac{\beta}{2} e^{i\alpha/2} \\ -\sin\frac{\beta}{2} e^{-i\alpha/2} & \cos\frac{\beta}{2} e^{i\alpha/2} \end{pmatrix} \begin{pmatrix} \nu_1 \\ \nu_2 \end{pmatrix}. \tag{13.31a}$$

We pointed out in Section 2.6 that there is a correspondence between a linear transformation of the basis functions and a similarity transformation of the appropriate matrices. On the other hand, it is demonstrated in Appendix 3, Section 4, that the matrix equations are invariant under similarity transformations; this means that the secular equation retains the same roots under a similarity transformation. For this reason we are entitled to use for the secular equation the particular form of the matrix (13.30).

Hence, corrections to the energy $\varepsilon'$ in the second approximation of perturbation theory can be found from the secular equation

$$\begin{vmatrix} \varepsilon_s - \varepsilon' & 0 & kP & 0 \\ 0 & \varepsilon_p - \frac{\Delta}{3} - \varepsilon' & \frac{\sqrt{2}}{3}\Delta & 0 \\ kP & \frac{\sqrt{2}}{3}\Delta & \varepsilon_p - \varepsilon' & 0 \\ 0 & 0 & 0 & \varepsilon_p + \frac{\Delta}{3} - \varepsilon' \end{vmatrix} = 0. \tag{13.32}$$

Since the structure of the matrix (13.30) consists of two identical blocks $\mathcal{H}$, the roots of equation (13.32) are doubly degenerate. Decomposing the determinant in the elements of the fourth row (column), we obtain

$$\left(\varepsilon_p + \frac{\Delta}{3} - \varepsilon'\right)\left[(\varepsilon_s - \varepsilon')\left(\varepsilon_p - \frac{\Delta}{3} - \varepsilon'\right)(\varepsilon_p - \varepsilon') - k^2P^2\left(\varepsilon_p - \frac{\Delta}{3} - \varepsilon'\right)(\varepsilon_s - \varepsilon')\right)\frac{2\Delta^2}{9}\right] = 0, \tag{13.33}$$

where the expression in square brackets is a third-order determinant obtained after the rejection of the fourth row and the fourth column. If the energy is measured from the upper edge of the valence band, i.e., if we write

$$\varepsilon_p + \frac{\Delta}{3} = 0, \quad \text{i.e.,} \quad \varepsilon_p = -\frac{\Delta}{3}, \tag{13.34}$$

and

$$\varepsilon_s = \varepsilon_G = \text{forbidden bandwidth},$$

then we obtain instead of (13.33)

$$\varepsilon' = 0, \tag{13.35}$$

$$\varepsilon' (\varepsilon' - \varepsilon_G)(\varepsilon' + \Delta) - k^2 P^2 \left(\varepsilon' + \frac{2\Delta}{3}\right) = 0. \tag{13.35a}$$

Making use of (13.28a), we obtain for small values of $k$ the following four solutions (up to $k^2$):[18]

$$\varepsilon_{v1} = \frac{\hbar^2 k^2}{2m}, \quad \varepsilon_{v2} = \frac{\hbar^2 k^2}{2m} - \frac{2P^2 k^2}{3\varepsilon_G},$$

$$\varepsilon_{v3} = -\Delta + \frac{\hbar^2 k^2}{2m} - \frac{P^2 k^2}{3(\varepsilon_G + \Delta)},$$

$$\varepsilon_c = \varepsilon_G + \frac{\hbar^2 k^2}{2m} + \frac{P^2 k^2}{3}\left(\frac{2}{\varepsilon_G} + \frac{1}{\varepsilon_G + \Delta}\right). \tag{13.36}$$

We can identify $\varepsilon_{v1}$ and $\varepsilon_{v2}$ with the energies in the heavy and light hole bands—$\varepsilon_{v3}$ with that of the light hole band split off as the result of spin-orbit coupling and $\varepsilon_c$ with that of the electron branch in the conduction band.

It should be pointed out at once that the expression for $\varepsilon_{v1}$ cannot be correct, because it does not even yield a correct sign (the hole energy is negative); this is proof that the two-band approximation that takes account only of the valence and the conduction bands is inadequate to interpret the heavy hole branch; in this case the upper and lower bands should also be taken into account.

Since the effective mass of the electron in the conduction band is almost 100 times less than that of the free electron $m$, the addend proportional to $P^2$ in the expression for $\varepsilon_c$ is about 100 times greater than the addend $\hbar^2 k^2/2m$, i.e., the dimensionless ratio $P^2 m/\hbar^2 \varepsilon_G$ is very great ($\sim 100$).

Knowing the values of $\varepsilon_G$ and $\Delta$, we can find the constant $P^2$ from comparison with experiment.

---

[18] In the zeroth approximation we neglect in (13.35a) the addend proportional to $k^2 P^2$ and then look for corrections of the order of $k^2$.

Making use of the fact that $P^2m/\hbar^2\varepsilon_G \gg 1$, we obtain from (13.36)

$$\varepsilon_{v2} = -\frac{2P^2k^2}{3\varepsilon_G}, \; \varepsilon_{v3} = -\Delta - \frac{P^2k^2}{3(\varepsilon_G+\Delta)},$$

$$\varepsilon_c = \varepsilon_G + \frac{P^2k^2}{3}\left(\frac{2}{\varepsilon_G} + \frac{1}{\varepsilon_G+\Delta}\right). \tag{13.36a}$$

It will be seen from these expressions that the order of magnitude of the effective mass of light holes is $\hbar^2\varepsilon_G/P^2$.

If $\Delta \gg kP$ and $\varepsilon_G$, equation (13.35a) can also be simplified. Dividing both its sides by $\Delta$ and neglecting in the zeroth approximation the quantities $\varepsilon_G/\Delta$ and $k^2P^2/\Delta$, we obtain in the zeroth approximation three roots $\varepsilon'_{01} = \varepsilon'_{02} = 0$, $\varepsilon'_{03} = -\Delta$. Introducing a small correction $\xi$ and making $\varepsilon'_1 = \varepsilon_2 = \xi$, $\varepsilon'_3 = -\Delta + \xi$, we obtain from (13.35a)

$$\varepsilon_{v2} = \frac{\hbar^2k^2}{2m} + \frac{\varepsilon_G - (\varepsilon_G^2 + 8P^2k^2/3)^{1/2}}{2},$$

$$\varepsilon_{v3} = -\Delta + \frac{\hbar^2k^2}{2m} - \frac{P^2k^2}{3\Delta},$$

$$\varepsilon_c = \frac{\hbar^2k^2}{2m} + \frac{\varepsilon_G + (\varepsilon_G^2 + 8P^2k^2/3)^{1/2}}{2}. \tag{13.36b}$$

The above expressions mean that the energy dispersion law for the electrons of $\varepsilon_c$ and for the light holes of $\varepsilon_{v2}$ is not parabolic.

Expanding the roots in (13.36b) up to $k^2$ for small values of the wave vector $\mathbf{k}$, we obtain (13.36) (for $\Delta \gg \varepsilon_G$). The terms linear in $k$ appear in the expression for the energy in the valence band of InSb in the case of the four-fold degenerate spinor irreducible representation $\Gamma_8$ (Table 4.9). Such linear terms appear in the second approximation of perturbation theory when the term $\mathbf{k}\cdot\hat{\mathbf{p}}$ and the spin-orbit coupling proportional to $\nabla V \times \hat{\mathbf{p}}$ (13.28) are taken into account. Note that such terms linear in $k$ do not appear in the conduction band of InSb and in the valence band of germanium (in the latter case due to the existence of inversional symmetry).

In InSb the linear terms are the cause of proportional to $k$ splitting of the heavy and light hole branches exhibiting spin degeneracy.

For the small $\mathbf{k}$ the terms linear in $k$ and not the terms proportional to $k^2$ or to higher powers of $k$ play the dominant part; for the large $\mathbf{k}$ the dominant part is played by terms proportional to $k^2$. The result is that the energy maxima in the valence band do not coincide with the point $\mathbf{k} = 0$, but are displaced with respect to it in the [111] directions. Experiments prove that the energy maxima lie only 0.015 eV above the upper edge of the valence band for $\mathbf{k} = 0$ and are located quite close to this point.

Similarly, it can be demonstrated that terms proportional to $k^3$ appear in the expression for the energy in the conduction band of indium antimonide; they too are responsible for the lifting of spin degeneracy in the conduction band.

## 4.14 Symmetry Involving Time Reversal

**4.14.1** It was demonstrated in Section 2.9.3 that in an equation complex conjugate to the time-dependent Schrödinger equation with a real Hamiltonian $\hat{\mathcal{H}}$ the evolution in time in the direction $-t$ of the state with the wave function $\psi^*$ is identical to that of the state $\psi$ in the $t$ direction. For a stationary state $\psi$ and $\psi^*$ describe degenerate states corresponding to the same energy $\mathcal{E}$.

In this section we intend to carry out more detailed studies within the framework of quantum mechanics of the symmetry involving time reversal. This problem is rather intricate, and for this reason several points will be presented without proof (in such cases the phrase "it can be demonstrated" will be used). The necessary proofs and supplements can be found in [2.3, Sec. 18].

**4.14.2** The time reversal operation $\hat{\mathcal{K}}$ may be said to turn a wave function $\psi$ $(\mathbf{r}, t)$ into a new function

$$\hat{\mathcal{K}}\psi\,(\mathbf{r},\ t) = \psi^*\,(\mathbf{r},\ -t) \tag{14.1}$$

that satisfies the same Schrödinger equation (if its Hamiltonian is real).

It is obvious that $\hat{\mathcal{K}}^2\psi = \hat{\mathcal{K}}\hat{\mathcal{K}}\psi = \hat{\mathcal{K}}\psi^* = \psi$, i.e., that

$$\hat{\mathcal{K}}^2 = 1. \tag{14.2}$$

For a stationary state

$$(\hat{\mathcal{H}} - \mathcal{E})\,\psi = 0. \tag{14.3}$$

For a real Hamiltonian $\hat{\mathcal{H}}$, $\psi$ and $\hat{\mathcal{K}}\psi = \psi^*$ are eigenfunctions of equation (14.3) corresponding to the same energy $\mathcal{E}$ eigenvalue; this may cause an additional degeneracy of a state with the energy $\mathcal{E}$. If the Hamiltonian $\hat{\mathcal{H}}$ is invariant under the transformation $\hat{g}$ of the group $\boldsymbol{G}$ $(g \in \boldsymbol{G})$, then $\psi$ and $\hat{g}\psi$ will also be eigenfunctions of equation (14.3) corresponding to the same energy $\mathcal{E}$, and this, too, may be the cause of the degeneracy of a state with the energy $\mathcal{E}$. However, we are unable to apply the theory developed in the preceding paragraphs to the case of symmetry that involves time reversal, because the time reversal operator $\hat{\mathcal{K}}$ is not a linear one. Indeed,

$$\hat{\mathcal{K}}(c_1\psi_1 + c_2\psi_2) = c_1^*\hat{\mathcal{K}}\psi_1 + c_2^*\hat{\mathcal{K}}\psi_2, \tag{14.4}$$

whereas for a linear operator $\hat{g}$ involving a coordinate transformation

$$\hat{g}\,(c_1\psi_1 + c_2\psi_2) = c_1\hat{g}\psi_1 + c_2\hat{g}\psi_2. \tag{14.4a}$$

It can easily be demonstrated that the operator $\hat{\mathscr{K}}$ commutes with all the $\hat{g}$ elements of the group $\boldsymbol{G}$. We have

$$\hat{g}\psi_i = \sum_j D_{ji}\psi_j, \tag{14.5}$$

where $D_{ji}\,(g)$ is the irreducible representation for the element $\hat{g}$. Then

$$\hat{\mathscr{K}}\hat{g}\psi_i = \hat{\mathscr{K}}\sum_j D_{ji}\psi_j = \sum_j D^*_{ji}\psi^*_j,$$

and

$$\hat{g}\hat{\mathscr{K}}\psi_i = \hat{g}\psi^*_i = \sum_j D^*_{ji}\psi^*_j,$$

whence

$$\hat{\mathscr{K}}\hat{g} = \hat{g}\hat{\mathscr{K}}. \tag{14.6}$$

The eigenfunctions $\psi$ and $\hat{\mathscr{K}}\psi$ satisfying equation (14.3) for the same energy eigenvalue $\mathscr{E}$ can be linearly independent; in this case there will be two independent sets of orthonormal eigenfunctions $\psi_i$ and $\hat{\mathscr{K}}\psi_i$ corresponding to one value of $\mathscr{E}$. If, on the other hand, $\psi_i$ and $\hat{\mathscr{K}}\psi_i$ are linearly interrelated, then

$$\hat{\mathscr{K}}\psi_i = \sum_j T_{ji}\psi_j, \tag{14.8}$$

where $\boldsymbol{T}$ is a unitary matrix providing for the orthonormality of the functions $\hat{\mathscr{K}}\psi_i$ in case the functions $\psi_i$ are orthonormal. It can be demonstrated that in the case (14.7) the representations $\boldsymbol{D}$ and $\boldsymbol{D}^*$ (14.5) are equivalent, i.e., that

$$\boldsymbol{D}^* = \boldsymbol{T}^{-1}\boldsymbol{D}\boldsymbol{T}. \tag{14.8}$$

In case the functions $\psi_i$ and $\hat{\mathscr{K}}\psi_i$ are linearly independent, i.e., not related by expression (14.7), their corresponding representations $\boldsymbol{D}$ and $\boldsymbol{D}^*$ may be either equivalent or nonequivalent [hence, (14.8) follows from (14.7), but the linear dependence (14.7) does not generally follow from the equivalence of the representations (14.8)].

Hence, three cases are possible:

(a) $\psi$ and $\hat{\mathscr{K}}\psi$ are linearly dependent: the representations $\boldsymbol{D}$ and $\boldsymbol{D}^*$ are equivalent, i.e., $\chi\,(g) = \chi^*\,(g)$;

(b) $\psi$ and $\hat{\mathscr{K}}\psi$ are linearly independent; $\boldsymbol{D}$ and $\boldsymbol{D}^*$ are nonequivalent, i.e., $\chi\,(g) \neq \chi^*\,(g)$;

(c) $\psi$ and $\hat{\mathscr{K}}\psi$ are linearly independent; $\boldsymbol{D}$ and $\boldsymbol{D}^*$ are equivalent, i.e., $\chi\,(g) = \chi^*\,(g)$. (14.9)

Here $\chi(g)$ is the character of the representation $D(g)$, and $\chi^*(g)$—that of the representation $D^*$. Since the linearly independent wave functions $\psi$ and $\hat{\mathcal{K}}\psi$ correspond to the same value of the energy $\mathcal{E}$, the invariance with respect to time reversal in cases (b) and (c) causes additional degeneracy. Therefore, in practice it is important to distinguish between the cases (14.9).

Prove that in cases (a) and (c), when the representations $D$ and $D^*$ are equivalent, i.e., related by expression (14.8), the requirement $T=\tilde{T}$ is a necessary and sufficient condition of the reality of $D$ (since $T$ is unitary, it follows that $\tilde{T}=T^{*-1}$). If, on the other hand, $T=-\tilde{T}$, the representation $D$ is essentially a complex one, i.e., there is no similarity transformation that can reduce it to a real form.

Taking the complex conjugate of (14.8), we obtain

$$D=T^{*-1}D^*T^*=T^{*-1}T^{-1}DTT^*=(TT^*)^{-1}D(TT^*)$$

or

$$(TT^*)D=D(TT^*),$$

i.e., the matrix $TT^*$ commutes with all the matrices $D(g)$. Then (in compliance with Schur's first lemma [4.16, Sec. 14.9-2]) it is a multiple of the unit matrix $I$:

$$TT^*=cI,\quad T=cT^{*-1}=c\tilde{T},$$

since $T$ is a unitary matrix. Hence, $\tilde{T}=cT$, and, therefore, $T=c\tilde{T}=c^2T$, whence $c=\pm 1$. Hence, two cases are possible: the first $T=\tilde{T}$ and the second $T=-\tilde{T}$. If $D(g)$ is real, it follows from (14.8) that $TD=DT$, and in accordance with Schur's first lemma $T=bI$ and $T=\tilde{T}$, i.e., the first case is realized. It can be proved that the condition $T=\tilde{T}$ is not only a necessary but also a sufficient one for $D(g)$ to be real. Thus we have proved the above statements. Profitting by the result prove that in case (a) of (14.9), when there is a linear dependence between $\psi$ and $\hat{\mathcal{K}}\psi$ (14.7), and when accordingly $T$ in (14.8) coincides with $T$ in (14.7), the representation $D$ is real.

It follows from (14.7) that

$$\hat{\mathcal{K}}^2\psi_i=\sum_j T^*_{ji}\hat{\mathcal{K}}\psi_j=\sum_{jl} T^*_{ji}T_{lj}\psi_l=\sum_l (TT^*)_{li}\psi_l.$$

Since $\hat{\mathcal{K}}^2=1$ (14.2), it follows that $(TT^*)_{li}=\delta_{li}$ and therefore $TT^*=I$ or $T=\tilde{T}$ (here again use is made of the fact that $T$ is a unitary matrix). Hence, in case (a) the representation $D$ is always real. And vice versa, if the representation $D$ is real, i.e., case (a) is realized, then the symmetry with respect to time reversal does not cause an additional degeneracy.

If, on the other hand, the representation $\boldsymbol{D}$ in (14.5) that transforms the functions is a complex one, then time reversal causes an additional double degeneracy, no matter whether the representations $\boldsymbol{D}$ and $\boldsymbol{D}^*$ are equivalent [case (c)] or not [case (b)].

**4.14.3** Up to now we have taken no account of the electron spin. The term in the Hamiltonian $\hat{\mathscr{H}}$ that accounts for the spin-orbit coupling is of the form (10.6)

$$\mathscr{H}_{s0} = -\frac{i\hbar^2}{4m^2c^2}\hat{\boldsymbol{\sigma}}\cdot(\nabla V \times \nabla), \tag{14.10}$$

where $\hat{\sigma} = \{\hat{\sigma}_1, \hat{\sigma}_2, \hat{\sigma}_3\}$ are Pauli spin matrices (10.2). In the case of time reversal, i.e., when complex conjugation takes place, (14.10) assumes the form

$$\hat{\mathscr{H}}^*_{s0} = \frac{i\hbar^2}{4m^2c^2}\hat{\boldsymbol{\sigma}}^*\cdot(\nabla V \times \nabla) = -\frac{i\hbar^2}{4m^2c^2}(-\hat{\boldsymbol{\sigma}}^*\cdot(\nabla V \times \nabla)). \tag{14.10a}$$

Here, in accordance with (10.2)

$$\hat{\sigma}^*_1 = \hat{\sigma}_1, \quad \hat{\sigma}^*_2 = -\hat{\sigma}_2, \quad \hat{\sigma}^*_3 = \hat{\sigma}_3. \tag{14.10b}$$

For the full Hamiltonian regarded as a functional of $\hat{\sigma}_i$

$$\hat{\mathscr{H}}^*(\hat{\sigma}_i) = -\hat{\mathscr{H}}(-\sigma^*_i). \tag{14.11}$$

The Schrödinger-Pauli equation for stationary states is of the form

$$[\hat{\mathscr{H}}(\hat{\sigma}_i) - \mathscr{E}]\,\Psi(\mathbf{r}, s) = 0, \tag{14.12}$$

where the wave function is [see (10.11) and (10.11a)]

$$\Psi(\mathbf{r}, s) = \sum_{i=1,2} \psi_i(\mathbf{r})\, \nu_i(s), \tag{14.13}$$

or

$$\Psi(\mathbf{r}, s) = \begin{pmatrix} \psi_1(\mathbf{r}) \\ \psi_2(\mathbf{r}) \end{pmatrix}. \tag{14.13a}$$

The operation of time reversal (of complex conjugation) (14.12) yields

$$[\hat{\mathscr{H}}^*(\hat{\sigma}_i) - \mathscr{E}]\,\Psi^*(\mathbf{r}, s) = 0. \tag{14.14}$$

Now we cannot say (as when the spin was not taken into account) that $\Psi(\mathbf{r}, s)$ and $\Psi^*(\mathbf{r}, s)$ are solutions of the same equation (14.12).

Subject equation (14.14) to a canonic transformation requiring that

$$S\hat{\mathscr{H}}^*(\hat{\sigma}_i)\, S^{-1} = S\hat{\mathscr{H}}(-\hat{\sigma}^*_i)\, S^{-1} = \hat{\mathscr{H}}(\hat{\sigma}_i), \tag{14.15}$$

where $S$ is the matrix of unitary transformation operating on the spin matrices $\hat{\sigma}_i$.

Obviously, to this end it should be

$$S\hat{\sigma}_1 S^{-1} = -\hat{\sigma}_1, \quad S\hat{\sigma}_2 S^{-1} = \hat{\sigma}_2, \quad S\hat{\sigma}_3 S^{-1} = -\hat{\sigma}_3, \tag{14.16}$$

in accordance with (14.10b).

Making use of (10.2a), we can easily check that those equations are satisfied if

$$S = S^{-1} = \hat{\sigma}_2. \tag{14.17}$$

The transformed wave function corresponding to the canonic transformation (14.15) is

$$S\Psi^* (\mathbf{r}, s) = \hat{\sigma}_2 \hat{\mathscr{K}}_0 \Psi (\mathbf{r}, s), \tag{14.18}$$

where $\hat{\mathscr{K}}_0$ is the complex conjugate operation.

Hence, the function $\hat{\sigma}_2 \hat{\mathscr{K}}_0 \Psi (\mathbf{r}, s)$ satisfies the same equation (14.12) as the function $\Psi (\mathbf{r}, s)$. In the case of the Schrödinger-Pauli equation (14.12) the operator corresponding to time reversal is

$$\hat{\mathscr{K}} = \hat{\sigma}_2 \hat{\mathscr{K}}_0. \tag{14.19}$$

Operating with it twice, we obtain

$$\hat{\mathscr{K}}^2 \Psi = \hat{\sigma}_2 \hat{\mathscr{K}}_0 \hat{\sigma}_2 \hat{\mathscr{K}}_0 \Psi = \hat{\sigma}_2 \hat{\mathscr{K}}_0 \hat{\sigma}_2 \Psi^* = \hat{\sigma}_2 \hat{\sigma}_2^* \Psi = -\Psi,$$

where we made use of (10.2a). Hence,

$$\hat{\mathscr{K}}^2 = -1 \tag{14.20}$$

in contrast to (14.2) for the spinless case.

In exactly the same way as it was done above, we can, making use of (14.20), prove that the necessary and sufficient condition for case (a) in (14.9) is $T = -\tilde{T}$. Hence, the representations corresponding to case (a) in (14.9) with the inclusion of spin are the essentially complex representations $\boldsymbol{D}(g)$.

In this case, too, as without spin, the invariance with respect to time reversal does not cause additional degeneracy. On the other hand, in cases, when the spinor wave functions are transformed with the aid of a double real representation [case (c) in (14.9)] or with the aid of a complex representation with complex characters [case (b) in (14.9)], the representations are doubled, i.e., an additional double degeneracy sets in.

**4.14.4** Frobenius and Schur have demonstrated that by making use of the properties of the $T$ matrix we can find out whether the representation will be a real or a complex one, knowing only its characters: if the sum of characters of the squares of the group's elements is equal to the number of elements $h$, then $T = \tilde{T}$, and the representation is real; if this sum is equal to $-h$, then $T = -\tilde{T}$, and the representations are complex and equivalent; and, finally, if it is zero, the representations are complex and nonequivalent. These results are displayed in Table 4.17, which takes into account the effect of time reversal on the degeneracy.

Table 4.17

| Frobenius-Schur criterion | Degeneracy | Relation between $D$ and $D^*$: In the absence of spin ($D$–ordinary representations) | Relation between $D$ and $D^*$: In the presence of spin ($D$–spinor representations) |
|---|---|---|---|
| (a) $\sum_g \chi(g^2) = \hat{\mathcal{K}}^2 h$ | No additional degeneracy | $D$ and $D^*$ can be made real and equal | $D$ and $D^*$ are equivalent, but essentially complex (i.e., cannot be made real) |
| (b) $\sum_g \chi(g^2) = 0$ | | $D$ and $D^*$ are not equivalent $\chi(g) \neq \chi^*(g)$ | |
| (c) $\sum_g \chi(g^2) = -\hat{\mathcal{K}}^2 h$ | Additional double degeneracy | $D$ and $D^*$ are equivalent, but essentially complex (i. e., cannot be made real) | $D$ and $D^*$ can be made real and equal |

*Note.* The value $\hat{\mathcal{K}}^2 = 1$ in the absence of spin (14.2) and $\hat{\mathcal{K}}^2 = -1$ in the presence of spin (14.20).

Let us show how to use this table with a simple example of the point group $C_3$ (Table 2.8). The Frobenius-Schur criterion for the irreducible representation $\Gamma_2$ yields

$$\sum_g \chi(g^2) = \chi(E^2) + \chi(C_3^2) + \chi(C_3^4) = \chi(E) + \chi(C_3^2) + \chi(C_3)$$

$$= 1 + \omega^2 + \omega = 1 + e^{-2\pi i/3} + e^{2\pi i/3} = 1 + 2\cos\frac{2\pi}{3} = 0.$$

It will be seen from the above table that for $\Gamma_2$ (as well as for $\Gamma_3$) there is an additional double degeneracy due to the symmetry with respect to time reversal. This means that the states $\Gamma_2$ and $\Gamma_3$ are united in one doubly degenerate state (this is the reason for their unification in Table 2.8).

4.14.5 Direct application of the Frobenius-Schur criterion (Table 4.17) to the investigation of the effect of time reversal on the electron energy levels in a crystal is impossible, since the summation in $\sum_g \chi(g^2)$ should embrace all the elements of the system's symmetry group, and the number of elements in the space group of a crystal is (practically) infinite.

Herring succeeded in transforming the Frobenius-Schur criterion to make it applicable to the energy bands in crystals expressing it in terms of characters of the elements of the wave vector group [4.8, p. 235.]. Herring's criterion that enables cases (a), (b) and (c) in Table 4.7 to be distinguished is of the form

$$\sum_{g_0} \chi(g_0^2) = \begin{cases} \hat{\mathscr{K}}^2 n \text{ case (a)}, \\ 0 \text{ case (b)}, \\ -\hat{\mathscr{K}}^2 n \text{ case (c)}. \end{cases} \tag{14.21}$$

Here $g_0$ is an element of the crystal's space group $G$ that does not contain trivial translations and transforms the wave vector $\mathbf{k}$ into $-\mathbf{k}$; accordingly, $g_0^2$ transforms $\mathbf{k} \to \mathbf{k}$, i.e., in one of the elements of the wave vector group $G_k$, which in the case of nonsymmorphic groups may, in general, contain a trivial translation, $n$ is the number of such elements. Note that $\chi(g_0^2)$ is the character of studied representation of the wave vector group $G_\mathbf{k}$ for $g_0^2$. In the end of Section 4.14.8 we have studied the energy spectrum of InSb. We have demonstrated that if the state corresponding to the centre of the Brillouin zone is the triply degenerate state $\Gamma_{15}$, then it splits up on the $\Delta$ axis as follows:

$$\Gamma_{15} = \Delta_1 + \Delta_3 + \Delta_4,$$

where $\Delta_1$, $\Delta_3$, and $\Delta_4$ are unidimensional irreducible representations of the group of the wave vector $\mathbf{k}_\Delta$. Next we noted that because of additional symmetry due to time reversal the states $\Delta_3$ and $\Delta_4$ fail to split up and form a doubly degenerate state.

Investigate this case applying the Herring criterion (14.21). The elements of the $T_d$ group, a point subgroup of the space group of the InSb crystal, are presented in Table 4.2. Four of them will easily be seen to transform $\mathbf{k}_\Delta \to -\mathbf{k}_\Delta$:

$$R_3(\bar{x}\bar{y}z) = C_4^2, \quad R_4(\bar{x}\bar{y}\bar{z}) = C_4'^2,$$
$$R_{13}(\bar{x}z\bar{y}) = JC_4, \quad R_{14}(\bar{x}\bar{z}y) = JC_4'.$$

Next

$$R_3^2 = R_1\,(xyz) = E, \quad R_4^2 = R_1\,(xyz) = E,$$
$$R_{13}^2 = R_2\,(x\bar{y}\bar{z}) = C_4^2, \quad R_{14}^2 = R_2\,(x\bar{y}\bar{z}) = C_4^2.$$

Turning to Table 4.4 of characters of the group of the wave vector $\mathbf{k}_\Delta$, we obtain for the representations $\Delta_3$ and $\Delta_4$

$$\sum_{g_0} \chi\,(g_0^2) = \chi\,(R_3^2) + \chi\,(R_4^2) + \chi\,(R_{13}^2) + \chi\,(R_{14}^2)$$
$$= \chi\,(E) + \chi\,(E) + \chi\,(C_4^2) + \chi\,(C_4^2) = 1 + 1 - 1 - 1 = 0,$$

i.e., in accordance with (14.21), case (b) is realized. It follows from (14.21) that both with or without the spin taken into account, the states $\Delta_3$ and $\Delta_4$ are doubly degenerate, i.e., they do not split up along the $\Delta$ axis.

If the crystal's space group contains inversion $J$ (Ge, Si), the sole element for the most general position of the wave vector $\mathbf{k}$ will be $g_0 = J$. In that case $\sum_{g_0} \chi\,(g_0^2) = \chi\,(J^2) = \chi\,(E) = 1$, and in (14.21) case (a) will be realized, this causing double degeneracy, when spin is taken into account. This point was noted, when expressions (13.23) and (13.26) were discussed.

## 4.15 Energy Band Structure of Some Semiconductors

Numerous experiments and theoretical investigations carried out recently helped to determine the energy band structure of several semiconductors: germanium, silicon, indium antimonide, etc. It was established that there was practically not a single case in which the energy spectrum of conduction electrons and holes would be of a simple parabolic type $\varepsilon = p^2/2m^*$, where $m^*$ is a scalar effective mass.

Maximum information about the energy band structure of semiconductors was gained in cyclotron resonance, in light absorption and in magnetoresistance experiments. Some of those experiments will be treated at length in the following chapters.

One of the most effective methods of calculating the energy band structure in semiconductors is the method of *orthogonalized plane waves* (K. Herring, 1940) which successfully combines the approximations of the almost free (see Section 4.5) and of the tightly bound (see Section 4.7) electrons. [19] Considerations stemming from group theory, i.e., symmetry properties discussed in the preceding sections, are of considerable aid in classifying the states and choosing wave

[19] A general notion about different methods of calculating the energy band structure in solids and on appropriate bibliography may be obtained from [4.17, Chap. II].

functions for conduction electrons (holes). Analytical methods of calculations supplemented by computer methods have been used.

Such calculations for germanium and silicon were carried out by Herman *et al.* (1953). In their calculations they had to limit themselves to the computation of the energy at some symmetrical points of the Brillouin zone, where they are greatly simplified. The values of the energy for intermediate points were obtained by interpolation.

The shape of the Brillouin zone for germanium and silicon is a polyhedron of fourteen faces as depicted in Fig. 4.22. The $x$, $y$, $z$ axes

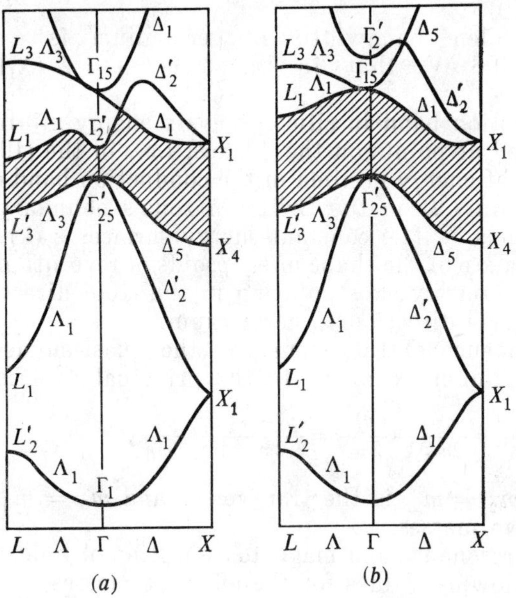


**Fig. 4.28**

pass through the centres of six hexagons and coincide with the crystallographic axes [100], [010], [001], etc. The axes directed to points $L$, the centres of the hexagons, coincide with the $[11\bar{1}]$, [111], etc. axes (the coordinates of point $L$ in Fig. 4.22 are: $\frac{2\pi}{a}\left[\frac{1}{2}\frac{1}{2}\frac{1}{2}\right]$, i.e., $k_x = k_y = k_z = \frac{\pi}{a}$, where $a$ is the edge of the direct lattice cube).

Figure 4.28 is a qualitative picture of the band structure of germanium ($a$) and silicon ($b$) (without spin-orbit splitting). The axes $\Gamma\Delta X$ and $\Gamma\Lambda L$ correspond to the directions [100] and [111]; the shaded part is the forbidden band.

Calculations and experimental results for Ge and Si indicate that the state of the holes in the centre of their valence bands ($\mathbf{k} = 0$) is the triply degenerate (without spin) state $\Gamma'_{25}$.

The state corresponding to the centre of the conduction band ($\mathbf{k} = 0$) in Ge is the nondegenerate state with the minimum energy $\Gamma'_2$, and in Si it is the triply degenerate state $\Gamma_{15}$.

Making use of the compatibility table (Table 3.5) we can easily find the type of splitting of the electron (hole) states resulting from the transition from the centre $\Gamma$ to the $\Delta$ or the $\Lambda$ axis (since the space groups of Ge and Si are nonsymmorphic, some care should be taken, when the surface of the Brillouin zone is reached; see the end of Section 4.9).

The forbidden bandwidth in germanium is $\varepsilon_G = \mathscr{E}(L_1) - \mathscr{E}(\Gamma'_{25}) = 0.8$ eV, that in silicon is $\varepsilon_G = \mathscr{E}(\Delta_1) - \mathscr{E}(\Gamma'_{25}) = 1.1$ eV.

The conduction band in silicon has six energy minima symmetrically arranged at points on the $\Delta$ axes, i.e., in the [100] directions. There are eight energy minima in the conduction band of germanium located at points $L$ on the Brillouin zone's boundary, i.e., in the [111] directions.[20] The constant-energy surfaces $\varepsilon(\mathbf{k}) = \text{const}$ near these minima are of the shape of ellipsoids of revolution (Section 3.3) with their symmetry axes pointing in the [100] direction in silicon and in the [111] direction in germanium.

The dependence of the energy on the quasi-momentum $\mathbf{p} = \hbar\mathbf{k}$ near the minimum expressed in the principal axes is of the form

$$(\mathbf{p}) = \frac{p_1^2}{2m_1} + \frac{p_2^2}{2m_2} + \frac{p_3^2}{2m_3} = \frac{p_1^2 + p_2^2}{2m_\perp} + \frac{p_3^2}{2m_\parallel}. \tag{15.1}$$

Here $m_1 = m_2 = m_\perp$ is the transverse, and $m_3 = m_\parallel$ the longitudinal effective masses.

Cyclotron resonance and magnetoabsorption of light in germanium yield the following values for the effective masses

$$m_\perp = 0.082m, \quad m_\parallel = 1.59m, \quad m_\parallel/m_\perp = 19, \tag{15.2}$$

where $m = 9.11 \times 10^{-28}$ g is the free electron mass.

For silicon

$$m_\perp = 0.19m, \quad m_\parallel = 0.92m, \quad m_\parallel/m_\perp = 4.8. \tag{15.3}$$

We have already mentioned that in the centre of the valence band ($\mathbf{k} = 0$) of Ge and Si the irreducible representation $\Gamma'_{25}$ is realized; to this representation the conclusions about the type of the energy spectrum arrived at in Section 4.13.3 are applicable.

---

[20] Since every minimum belongs to two Brillouin zones at a time (lies on their boundary), there are four minima to a zone.

The energy spectrum of holes as stipulated by (13.23) and (13.26) consists of three branches (bands):

$$\varepsilon_{v1} = -\frac{\hbar^2}{2m} [Ak^2 - \sqrt{B^2k^4 + C^2 (k_x^2k_y^2 + k_x^2k_z^2 + k_y^2k_z^2)}], \qquad (15.4)$$

$$\varepsilon_{v2} = -\frac{\hbar^2}{2m} [Ak^2 + \sqrt{B^2k^4 + C^2 (k_x^2k_y^2 + k_x^2k_z^2 + k_y^2k_z^2)}], \qquad (15.4a)$$

$$\varepsilon_{v3} = -\Delta - \frac{\hbar^2}{2m} Ak^2, \qquad (15.4b)$$

where instead of the constants $A$, $B$, $C$ that enter (13.23) we have introduced new dimensionless constants $A$, $B$, $C$ equal to the old ones divided by $(-\hbar^2/2m)$. Here $\varepsilon_{v1}$ is the heavy hole band, $\varepsilon_{v2}$ is the light hole band, and $\varepsilon_{v3}$ is the hole band split off from the upper edge of the valence band by an amount $\Delta$ because of the spin-orbit coupling. The simple parabolic law $\varepsilon_{v3} + \Delta \propto k^2$ and spherical constant-energy surfaces are characteristic only for the latter split-off band. In case (15.4) and (15.4a) the constant-energy surfaces have the shape of *corrugated* surfaces, the cross section of which with the (100) plane is depicted in Fig. 4.27.[21]

Introducing in the **k**-space spherical coordinates instead of the rectangular,[22] we obtain instead of (15.4)

$$\varepsilon_{v1} = -\frac{\hbar^2k^2}{2m} [A - \sqrt{B^2 + C^2 + \sin^2\theta (\sin^2\theta \sin^2\varphi \cos^2\varphi + \cos^2\theta)}]. \qquad (15.5)$$

Averaging isotropically over the angles, we obtain approximately

$$\varepsilon_{v1} \approx -\frac{\hbar^2k^2}{2m} [A - \sqrt{B^2 + C^2/5}] \qquad (15.5a)$$

and a similar expression with a plus sign for $\varepsilon_{v2}$. Expression (15.5a) and a similar one for $\varepsilon_{v2}$ enable the scalar effective masses to be introduced

$$m_{v1} = \frac{m}{A - \sqrt{B^2 + C^2/5}} \quad \text{and} \quad m_{v2} = \frac{m}{A + \sqrt{B^2 + C^2/5}} \qquad (15.5b)$$

for the heavy and light holes.

The study of cyclotron resonance in germanium yielded the following values for the constants: $A = 13.1 \pm 0.4$, $B = 8.3 \pm 0.6$, $C = 12.5 \pm 0.5$ and $\Delta = 0.3$ eV; the corresponding constants for silicon are: $A = 4.0 \pm 0.1$, $B = 1.1 \pm 0.4$, $C = 4.1 \pm 0.4$ and $\Delta = 0.04$ eV.

---

21 The values of the constants $A$, $B$, $C$ are for Ge.

22 The polar axis is directed along ellipsoid's symmetry axis, i.e., along $k_z$.

Making use of these numerical values, we obtain from (15.5a) for the effective masses of the heavy and light holes in

$$\text{Ge: } m_{v1} = 0.33m, \quad m_{v2} = 0.04m, \quad m_{v1}/m_{v2} = 8.0$$

and

$$\text{Si: } m_{v1} = 0.56m, \quad m_{v2} = 0.16m, \quad m_{v1}/m_{v2} = 3.5. \tag{15.6}$$

In Section 4.13.4 we studied the energy band structure of indium antimonide—a narrow band gap semiconductor. In that case there were two interesting properties characteristic of the energy bands: very small effective masses of charge carriers and deviation of the dispersion law from the simple parabolic law. There we also presented the numerical values of parameters characterizing the energy band structure in InSb.

Similar properties are also characteristic of several other compounds of elements of groups III and V of the Periodic Table: of indium arsenide (InAs), indium phosphide (InP), gallium arsenide (GaAs), aluminium antimonide (AlSb), aluminium phosphide (AlP). The notation for compounds of this type is $A^{III}B^{V}$.

At the same time some other semiconducting $A^{III}B^{V}$ compounds, for instance, gallium phosphide (GaP), have bands with the energy extremum for conduction electrons and holes located at different **k**-points of the Brillouin zone.[23]

[23] We shall not present here the numerical values of the parameters characterizing the energy band structure of the $A^{III}B^{V}$ compounds and discuss their important technical applications. For a detailed study of those subjects the reader is referred to [4.18].

# 5. Localized Electron States in Crystals

## 5.1 Wannier Functions. Electron Motion in the Field of an Impurity Atom

5.1.1 Studying the behaviour of conduction electrons in semiconductors, we frequently encounter the situation, when, in addition to the periodic potential of an ideal crystal, a field acts on the electron that can create localized (bound) electron states in the lattice. This takes place when a conduction electron moves in the field of an alien atom or an ion implanted in an ideal crystal lattice (*impurity atom* or *ion*).

The existence of a free surface of a crystal is also equivalent to the existence of some additional field near the surface. The Coulomb field of a hole acting on the electron and sometimes causing the formation of a bound electron-hole state, termed *exciton*, is also an example of such an additional field. Finally, the additional field due to the polarization of an ionic crystal caused by the action on the lattice of the electron itself is responsible for the so-called *polaron state*.

In some cases to describe localized electron states in the lattice it is expedient to use as the zeroth approximation not the Bloch functions

$$\psi_{n\mathbf{k}}(\mathbf{r}) = u_{n\mathbf{k}}(\mathbf{r})\, e^{i\mathbf{k}\cdot\mathbf{r}}, \tag{1.1}$$

where $n$ is the number of the allowed band, $\mathbf{k}$ is the wave vector, but the so-called *Wannier functions* (1.3) defined below. The Bloch functions (1.1) are periodic in the $\mathbf{k}$-space with the periods of the reciprocal lattice $\mathbf{b}_i$ ($i = 1, 2, 3$); this means that $\psi_{n\mathbf{k}}(\mathbf{r})$ can be expanded in the $\mathbf{k}$-space in a Fourier series

$$\psi_{n\mathbf{k}}(\mathbf{r}) = \frac{1}{\sqrt{N}} \sum_{l} \varphi_n(\mathbf{a}_l, \mathbf{r})\, e^{i\mathbf{k}\cdot\mathbf{a}_l}. \tag{1.2}$$

Indeed, the sum on the right-hand side does not change, when $\mathbf{k} + \mathbf{b}_i$ is substituted for $\mathbf{k}$:

$$\begin{aligned}\exp[i(\mathbf{k} + \mathbf{b}_i)\cdot\mathbf{a}_l] &= \exp[i\mathbf{k}\cdot\mathbf{a}_l]\exp[i\mathbf{b}_i\cdot\mathbf{a}_l] \\ &= \exp[i\mathbf{k}\cdot\mathbf{a}_l]\exp[i(2\pi \times \text{integer})] \\ &= \exp[i\mathbf{k}\cdot\mathbf{a}_l].\end{aligned}$$

The expansion coefficients $\varphi_n(\mathbf{a}_l, \mathbf{r})$ where $\mathbf{a}_l$ is the lattice vector of the **l**-th site are termed Wannier functions; they play the part of the "site representation" of the electron in an ideal lattice.

Multiplying both sides of equality (1.2) by $\exp[-i\mathbf{k}\cdot\mathbf{a}_m]$ and adding up over **k**, we obtain making use of (A.6.8)

$$\varphi_n(\mathbf{a}_m, \mathbf{r}) = \frac{1}{\sqrt{N}} \sum_{\mathbf{k}} e^{-i\mathbf{k}\cdot\mathbf{a}_m} \psi_{n\mathbf{k}}(\mathbf{r}) = \frac{1}{\sqrt{N}} \sum_{\mathbf{k}} u_{n\mathbf{k}} e^{i\mathbf{k}\cdot(\mathbf{r}-\mathbf{a}_m)}. \quad (1.3)$$

Since $u_{n\mathbf{k}}(\mathbf{r}) = u_{n\mathbf{k}}(\mathbf{r} - \mathbf{a}_m)$, it follows that $\varphi_n(\mathbf{a}_m, \mathbf{r}) = \varphi_n(\mathbf{r} - \mathbf{a}_m)$.

Demonstrate that each of the $N$ Wannier functions $\varphi_n(\mathbf{r} - \mathbf{a}_m)$ ($\mathbf{m} = 1, 2, \ldots, N$) is localized near its site **m**. To simplify the proof, consider a simple cubic lattice and approximate the Bloch function by a plane wave $\psi_{n\mathbf{k}}(\mathbf{r}) = \frac{1}{\sqrt{V}} e^{i\mathbf{k}\cdot\mathbf{r}}$. In this case (1.3) turns into

$$\varphi_n(\mathbf{r} - \mathbf{a}_m) = C \sum_{k_x} e^{ik_x\xi} \sum_{k_y} e^{ik_y\eta} \sum_{k_z} e^{ik_z\zeta}.$$

Here $C$ is the normalizing constant of the wave function $\varphi_n$; $\xi$, $\eta$, $\zeta$ are rectangular components of the position vector $\boldsymbol{\rho} = \mathbf{r} - \mathbf{a}_m$ and $k_i = 2\pi g_i/Ga$ $(-G/2 \leqslant g_i \leqslant G/2)$,[1] as stipulated by (4.3.3) for a simple cubic lattice with the edge equal to $a$.

Adding up the sum over $k_x$ (over $g$) as a geometrical progression, we obtain

$$\sum_{k_x} e^{ik_x\xi} = 1 + \sum_{g=1}^{G/2} e^{i\frac{2\pi\xi}{Ga}g} + \sum_{g=-1}^{-G/2} e^{i\frac{2\pi\xi}{Ga}g}$$

$$= \frac{e^{i\frac{\pi\xi}{a}} e^{i\frac{2\pi\xi}{Ga}} - e^{-i\frac{\pi\xi}{a}}}{e^{i\frac{2\pi\xi}{Ga}} - 1}. \quad (1.3a)$$

For a principal crystal region of sufficiently great dimensions it can be assumed that $\xi/Ga \ll 1$, therefore, $\exp i\frac{2\pi}{Ga}\xi \approx 1 + i\frac{2\pi}{Ga}\xi$. Then (1.3a) will be equal (up to a constant) to

$$\sum_{k_x} e^{ik_x\xi} \approx \frac{\sin\frac{\pi\xi}{a}}{\frac{\pi\xi}{a}}.$$

[1] In this case a large even number was chosen for $G$.

This expression has a maximum equal to unity for $\xi = 0$; as $\xi$ is increased it decreases, rapidly oscillating in the process. Hence, the Wannier function $\varphi_n(\mathbf{r} - \mathbf{a_m})$ has its maximum at point $\mathbf{r} = \mathbf{a_m}$, rapidly decreasing as $\mathbf{r} - \mathbf{a_m}$ is increased.

A similar proof can be carried out in a more general case, as well, for a crystal of arbitrary structure and for a Bloch function of a general form (it should only be assumed that $u_\mathbf{k}(\mathbf{r})$ depends weakly on $\mathbf{k}$). In this sense expression (1.2) is similar to the tight binding approximation (4.7.2a), but it is a precise expression. Expression (1.2) has an essential advantage over (4.7.2a) in that the Wannier functions (1.3) are mutually orthogonal both with respect to the band number $n$ and to the site number $\mathbf{m}$. Indeed,

$$\int_V \varphi_n^*(\mathbf{r} - \mathbf{a_m})\, \varphi_{n'}(\mathbf{r} - \mathbf{a_{m'}})\, d^3r$$

$$= \frac{1}{N} \sum_{\mathbf{kk'}} e^{i(\mathbf{k} \cdot \mathbf{a_m} - \mathbf{k'} \cdot \mathbf{a_{m'}})} \int \psi_{n\mathbf{k}}^*(\mathbf{r})\, \psi_{n'\mathbf{k'}}(\mathbf{r})\, d^3r$$

$$= \frac{1}{N} \sum_{\mathbf{kk'}} e^{i\mathbf{k} \cdot (\mathbf{a_m} - \mathbf{a_{m'}})} e^{i(\mathbf{k} - \mathbf{k'}) \cdot \mathbf{a_{m'}}} \delta_{nn'} \delta_{\mathbf{kk'}}$$

$$= \frac{\delta_{nn'}}{N} \sum_{\mathbf{k}} e^{i\mathbf{k} \cdot (\mathbf{a_m} - \mathbf{a_{m'}})} = \delta_{nn'} \delta_{\mathbf{mm'}}, \qquad (1.4)$$

where we made use of (4.3.4) and (A.6.8).

**5.1.2** Let in addition to the periodic potential $V(\mathbf{r})$ a field $\mathcal{U}(\mathbf{r})$ act on the conduction electron. In this case the Schrödinger equation will be of the form

$$-\frac{\hbar^2}{2m} \nabla^2 \Phi_i(\mathbf{r}) + [V(\mathbf{r}) + \mathcal{U}(\mathbf{r})]\, \Phi_i(\mathbf{r}) = \mathcal{E}_i \Phi_i(\mathbf{r}), \qquad (1.5)$$

where $\Phi_i(\mathbf{r})$ and $\mathcal{E}_i$ are the $i$-th eigenfunction and the corresponding energy eigenvalue.

Expand the eigenfunction $\Phi_i(\mathbf{r})$ in a closed system of orthonormal Wannier functions (1.3)

$$\Phi_i(\mathbf{r}) = \sum_{n'\mathbf{l'}} f_{n'}^i(\mathbf{a_{l'}})\, \varphi_{n'}(\mathbf{r} - \mathbf{a_{l'}}). \qquad (1.6)$$

Here $n'$ is the number of the band, $\mathbf{a_{l'}}$ is the vector of the site $\mathbf{l'}$, and $f_n^i$ are the expansion coefficients to be determined from (1.5). Substituting (1.6) into (1.5), premultiplying by $\varphi_n^*(\mathbf{r} - \mathbf{a_l})$ and integrating with respect to $d^3r = dx\,dy\,dz$, we obtain

$$\sum_{n'\mathbf{l'}} f_{n'}^i(\mathbf{a_{l'}}) \int \varphi_n^*(\mathbf{r} - \mathbf{a_l}) \left[ -\frac{\hbar^2}{2m} \nabla^2 + V(\mathbf{r}) + \mathcal{U}(\mathbf{r}) - \mathcal{E}_i \right] \times \varphi_{n'}(\mathbf{r} - \mathbf{a_{l'}})\, d^3r = 0. \qquad (1.7)$$

If the field $\mathcal{U}(\mathbf{r})$ varies smoothly and sufficiently slowly at a distance comparable to the lattice constant, so that we may make $\mathcal{U}(\mathbf{r}) \approx \approx \mathcal{U}(\mathbf{a}_{\mathbf{l}'})$, then the integral from the last two addends in the square brackets will be equal to

$$\int \varphi_n^*(\mathbf{r}-\mathbf{a}_\mathbf{l})[\mathcal{U}(\mathbf{r})-\mathscr{E}_i]\varphi_{n'}(\mathbf{r}-\mathbf{a}_{\mathbf{l}'})\,d^3r$$
$$=[\mathcal{U}(\mathbf{a}_{\mathbf{l}'})-\mathscr{E}_i)]\int \varphi_n^*(\mathbf{r}-\mathbf{a}_\mathbf{l})\varphi_{n'}(\mathbf{r}-\mathbf{a}_{\mathbf{l}'})\,d^3r$$
$$=[\mathcal{U}(\mathbf{a}_{\mathbf{l}'})-\mathscr{E}_i]\,\delta_{nn'}\delta_{\mathbf{l}\mathbf{l}'}, \tag{1.8}$$

where we made use of the orthonormality of Wannier functions (1.4). Substituting (1.8) into (1.7), we obtain

$$\sum_{n'\mathbf{l}'}[\mathcal{U}(\mathbf{a}_{\mathbf{l}'})-\mathscr{E}_i]\,f_{n'}^i(\mathbf{a}_{\mathbf{l}'})\,\delta_{nn'}\delta_{\mathbf{l}\mathbf{l}'}=[\mathcal{U}(\mathbf{a}_\mathbf{l})-\mathscr{E}_i]\,f_n^i(\mathbf{a}_\mathbf{l}). \tag{1.9}$$

The first two addends in the square brackets of (1.7) yield, when the Wannier functions are transformed into Bloch functions (1.3),

$$\sum_{n'\mathbf{l}'}f_{n'}^i(\mathbf{a}_{\mathbf{l}'})\frac{1}{N}\int\Big\{\sum_\mathbf{k}e^{i\mathbf{k}\cdot\mathbf{a}_\mathbf{l}}\psi_{n\mathbf{k}}^*(\mathbf{r})\Big[-\frac{\hbar^2}{2m}\nabla^2+V(\mathbf{r})\Big]$$
$$\times\sum_{\mathbf{k}'}e^{-i\mathbf{k}'\cdot\mathbf{a}_{\mathbf{l}'}}\psi_{n'\mathbf{k}'}(\mathbf{r})\Big\}\,d^3r$$
$$=\frac{1}{N}\sum_{n'\mathbf{l}'}f_{n'}^i(\mathbf{a}_{\mathbf{l}'})\sum_\mathbf{k}\sum_{\mathbf{k}'}e^{i(\mathbf{k}\cdot\mathbf{a}_\mathbf{l}-\mathbf{k}'\cdot\mathbf{a}_{\mathbf{l}'})}\varepsilon_{n'}(\mathbf{k}')$$
$$\times\int\psi_{n\mathbf{k}}^*\psi_{n'\mathbf{k}'}\,d^3r$$
$$=\frac{1}{N}\sum_{n'\mathbf{l}'}\sum_{\mathbf{k}\mathbf{k}'}f_{n'}^i(\mathbf{a}_{\mathbf{l}'})\,e^{i(\mathbf{k}\cdot\mathbf{a}_\mathbf{l}-\mathbf{k}'\cdot\mathbf{a}_{\mathbf{l}'})}\varepsilon_{n'}(\mathbf{k}')\,\delta_{nn'}\delta_{\mathbf{k}\mathbf{k}'}$$
$$=\frac{1}{N}\sum_{\mathbf{l}'}\sum_\mathbf{k}f_n^i(\mathbf{a}_{\mathbf{l}'})\,e^{i\mathbf{k}\cdot(\mathbf{a}_\mathbf{l}-\mathbf{a}_{\mathbf{l}'})}\varepsilon_n(\mathbf{k}), \tag{1.10}$$

where we made use of orthonormality of the Bloch functions and denoted the eigenvalue of the unperturbed Hamiltonian $-(\hbar^2/2m)\nabla^2 + + V(\mathbf{r})$ by $\varepsilon_{n'}(\mathbf{k}')$.

If we make $\mathbf{a}_\mathbf{l}-\mathbf{a}_{\mathbf{l}'}=\mathbf{a}_\mathbf{m}$, then the right-hand side of (1.10) will be equal to

$$\frac{1}{N}\sum_\mathbf{m}\sum_\mathbf{k}f_n^i(\mathbf{a}_\mathbf{l}-\mathbf{a}_\mathbf{m})\,e^{i\mathbf{k}\cdot\mathbf{a}_\mathbf{m}}\varepsilon_n(\mathbf{k}). \tag{1.11}$$

Expand the function $f_n^i(\mathbf{a}_l - \mathbf{r})$ in a Taylor series in $\mathbf{r}$ near point $\mathbf{a}_l$,

$$f_n^i(\mathbf{a}_l - \mathbf{r}) = f_n^i(\mathbf{a}_l) - (\mathbf{r}\cdot\nabla) f_n^i(\mathbf{a}_l) + \frac{1}{2}(\mathbf{r}\cdot\nabla)^2 f_n^i(\mathbf{a}_l) - \dots$$
$$= \left[1 - (\mathbf{r}\cdot\nabla) + \frac{1}{2}(\mathbf{r}\cdot\nabla)^2 - \dots\right] f_n^i(\mathbf{a}_l). \tag{1.12}$$

Here $\nabla f_n^i(\mathbf{a}_l) = [\nabla f_n^i(\mathbf{r})]_{\mathbf{r}=\mathbf{a}_l}$, and $\nabla\nabla f_n^i(\mathbf{a}_l)$ (whose typical term is of the form $[\partial^2 f_n^i(\mathbf{r})/\partial x\,\partial y]_{\mathbf{r}=\mathbf{a}_l}$) have similar meaning.

The expression in square brackets in (1.12) may be regarded as an operator $\exp(-\mathbf{r}\cdot\nabla)$.

Substituting $\mathbf{r} = \mathbf{a}_\mathbf{m}$ into (1.12), we obtain

$$f_n^i(\mathbf{a}_l - \mathbf{a}_\mathbf{m}) = e^{-\mathbf{a}_\mathbf{m}\cdot\nabla} f_n^i(\mathbf{a}_l). \tag{1.13}$$

It was noted in (4.3.10) that

$$\varepsilon_n(\mathbf{k}) = \sum_{\mathbf{m}'} c_{\mathbf{m}'} e^{-i\mathbf{k}\cdot\mathbf{a}_{\mathbf{m}'}} \tag{1.14}$$

(since in (4.3.10) $\mathbf{l} = \{l_1, l_2, l_3\}$ assume both positive and negative values, one can always change the sign before the exponential). Substituting (1.13) and (1.14) into (1.11), we obtain

$$\frac{1}{N}\sum_{\mathbf{m}}\sum_{\mathbf{k}}\sum_{\mathbf{m}'} e^{-i\mathbf{a}_\mathbf{m}\cdot\nabla} f_n^i(\mathbf{a}_l)\, c_{\mathbf{m}'} e^{i\mathbf{k}\cdot(\mathbf{a}_\mathbf{m} - \mathbf{a}_{\mathbf{m}'})}$$
$$= \sum_{\mathbf{m}}\sum_{\mathbf{m}'} e^{-i\mathbf{a}_\mathbf{m}\cdot\nabla} f_n^i(\mathbf{a}_l)\, c_{\mathbf{m}'}\delta_{\mathbf{m}\mathbf{m}'} = \left[\sum_{\mathbf{m}} c_\mathbf{m} e^{-\mathbf{a}_\mathbf{m}\cdot\nabla}\right] f_n^i(\mathbf{a}_l), \tag{1.15}$$

where we made use of (A.6.8). Comparing the latter expression with (1.14), we see that (1.11) is equal to

$$\varepsilon_n(-i\nabla) f_n^i(\mathbf{a}_l). \tag{1.16}$$

This expression is similar to (4.3.12).

Combining (1.16) with (1.9), we obtain

$$\varepsilon_n(-i\nabla) f_n^i(\mathbf{a}_l) + \mathcal{U}(\mathbf{a}_l) f_n^i(\mathbf{a}_l) = \mathscr{E}_i f_n^i(\mathbf{a}_l). \tag{1.17}$$

We obtain the following differential equation for the "smooth" function $f_n^i(\mathbf{r})$, which assumes the values $f_n^i(\mathbf{a}_l)$ at points $\mathbf{r} = \mathbf{a}_l$:

$$\varepsilon_n(-i\nabla) f_n^i(\mathbf{r}) + \mathcal{U}(\mathbf{r}) f_n^i(\mathbf{r}) = \mathscr{E}_i f_n^i(\mathbf{r}). \tag{1.18}$$

This is the equation that determines the expansion coefficients (1.6) sought. The meaning of the operator $\varepsilon_n(-i\nabla)$ is the same as in equation (4.3.13).

If the energy $\varepsilon_n(\mathbf{k})$ has a minimum (a maximum) at point $\mathbf{k} = 0$ and is spherically symmetrical, then in the quadratic approximation

$\varepsilon_n(-i\nabla) = -(\hbar^2/2m^*)\nabla^2$ [see (4.3.22a)], and equation (1.18) for a spherically symmetrical field $\mathcal{U}(\mathbf{r}) = \mathcal{U}(\mathbf{r})$ assumes the form

$$-\frac{\hbar^2}{2m^*}\nabla^2 f_n^i(\mathbf{r}) + \mathcal{U}(\mathbf{r}) f_n^i(\mathbf{r}) = \mathscr{E}_i f_n^i(\mathbf{r}), \tag{1.18a}$$

i.e., turns into the wave equation for a particle with the effective mass $m^*$ in a spherically symmetrical field $\mathcal{U}(r)$. A more general form has been considered in Section 4.3.3.

If the electron is attracted to an impurity centre in the field $\mathcal{U}(r)$ (a donor), it can be trapped on a localized level in the forbidden band below the conduction band. If the electron is repulsed by the impurity centre in the field $\mathcal{U}(r)$ (a negative acceptor ion), and if therefore a hole is attracted by this centre, the latter can be trapped on a localized state lying in the forbidden band above the valence band.

5.1.3 In the deduction of equation (1.18) it was assumed that the additional field $\mathcal{U}(\mathbf{r})$ changes slowly. Consider a method that can be used when $\mathcal{U}(\mathbf{r})$ changes so swiftly that it can be assumed to be nonzero only within the bounds of a single crystal cell. In this case it is more expedient to expand the perturbed electron wave function $\Phi_i(\mathbf{r})$ in Bloch functions $\psi_k(\mathbf{r}) = u_k(\mathbf{r}) \exp(i\mathbf{k}\cdot\mathbf{r})$ (we use Bloch functions of a single energy band)

$$\Phi_i(\mathbf{r}) = \sum_{\mathbf{k}'} f_i(\mathbf{k}')\psi_{\mathbf{k}'}(\mathbf{r}), \tag{1.19}$$

where the summation over $\mathbf{k}'$ embraces all its quasi-discrete values inside the first Brillouin zone. Substituting expression (1.19) into equation (1.5), premultiplying by $\psi_{\mathbf{k}}^*(\mathbf{r})$ and integrating with respect to $\mathbf{r}$, we obtain

$$\int \psi_{\mathbf{k}}^*(\mathbf{r})[\hat{\mathscr{H}}_0 + \mathcal{U}(\mathbf{r})] \sum_{\mathbf{k}'} f_i(\mathbf{k}')\psi_{\mathbf{k}'}(\mathbf{r})\,d\tau = \mathscr{E}_i \sum_{\mathbf{k}'} f_i(\mathbf{k}') \int \psi_{\mathbf{k}}^*(\mathbf{r})\psi_{\mathbf{k}'}(\mathbf{r})\,d\tau, \tag{1.20}$$

where $\hat{\mathscr{H}}_0 = -\frac{\hbar^2}{2m}\nabla^2 + V(\mathbf{r})$ is the unperturbed Hamiltonian.

Since $\hat{\mathscr{H}}_0\psi_{\mathbf{k}'} = \varepsilon(\mathbf{k}')\psi_{\mathbf{k}'}$ and, moreover, the Bloch functions are orthonormal, (1.20) is equal to

$$\varepsilon(\mathbf{k}) f_i(\mathbf{k}) + \sum_{\mathbf{k}'} f_i(\mathbf{k}') \int \psi_{\mathbf{k}}^*(\mathbf{r})\mathcal{U}(\mathbf{r})\psi_{\mathbf{k}'}(\mathbf{r})\,d\tau = \mathscr{E}_i f_i(\mathbf{k}). \tag{1.21}$$

Substitute in the integral of this equality Wannier functions (1.2) for the Bloch functions, then

$$\int \psi_{\mathbf{k}}^{*}(\mathbf{r})\, \mathcal{U}(\mathbf{r})\, \psi_{\mathbf{k}}(\mathbf{r})\, d\tau = \frac{1}{N} \sum_{\mathbf{nn}'} e^{\mathbf{k}(\mathbf{a}'_{\mathbf{n}}-\mathbf{a}_{\mathbf{n}})} \int \varphi^{*}(\mathbf{r}-\mathbf{a}_{\mathbf{n}})\, \mathcal{U}(\mathbf{r})\, \varphi(\mathbf{r}-\mathbf{a}'_{\mathbf{n}})\, d\tau. \qquad (1.22)$$

If the perturbation $\mathcal{U}(\mathbf{r})$ is nonzero only inside one crystal cell, for instance, the zeroth[2], only one term with $\mathbf{n} = \mathbf{n}' = 0$ need be taken into account in the double sum, and we obtain instead of (1.21)

$$\varepsilon(\mathbf{k})\, f_i(\mathbf{k}) + \frac{\mathcal{U}_0}{N} \sum_{\mathbf{k}'} f_i(\mathbf{k}') = \mathcal{E}_i f_i(\mathbf{k}), \qquad (1.23)$$

where $\mathcal{U}_0$ is the average value of the perturbation potential $\mathcal{U}(\mathbf{r})$ over the volume of the zeroth cell taken out of the integral in (1.22). Denoting the constant $\left(\frac{\mathcal{U}_0}{N}\right) \sum f_i(\mathbf{k}') = A$, we obtain from expression (1.23)

$$f_i(\mathbf{k}) = \frac{A}{\mathcal{E}_i - \varepsilon(\mathbf{k})}. \qquad (1.24)$$

Substituting this result into equation (1.23) and canceling out the unknown constant $A$ on both sides of the equality, we obtain

$$\sum_{\mathbf{k}'} \frac{1}{\mathcal{E}_i - \varepsilon(\mathbf{k}')} = \frac{N}{\mathcal{U}_0}, \qquad (1.25)$$

an equation for determining the energy eigenvalues $\mathcal{E}_i$ for a specified perturbation $\mathcal{U}_0$. The left-hand side of (1.25) is equal to the sum of fractions of the type $[\mathcal{E}_i - \varepsilon(\mathbf{k}_1)]^{-1}$, where $\mathbf{k}_1$ is one of the $N$ possible values of the wave vector $\mathbf{k}$ in the first Brillouin zone. If the perturbation is zero, the right-hand side of (1.25) is equal to infinity, and, therefore, the electron energy eigenvalues $\mathcal{E}_i$ in an ideal crystal coincide with one of the $N$ values of $\varepsilon(\mathbf{k})$. It can be demonstrated that there are $N$ eigenvalues $\mathcal{E}_i$ ($i = 1, 2, \ldots, N$) to correspond to every positive or negative value of $\mathcal{U}_0$ as well. As long as $|\mathcal{U}_0|$ remains less than a certain value, the eigenvalues $\mathcal{E}_i$ almost coincide with the unperturbed eigenvalues $\varepsilon(\mathbf{k})$. If on the other hand, for example, $\mathcal{U}_0 < 0$ and exceeds in magnitude the above mentioned value, then $\mathcal{E}_i$ corresponding to the lower edge of the energy band $\varepsilon(0)$ experiences a strong perturbation—the level splits off from the band edge. For $\mathcal{U}_0 > 0$, a similar situation exists

[2] For an infinite crystal, the result will obviously be independent of the cell in which the perturbation $\mathcal{U}(\mathbf{r})$ is concentrated.

in the case of the upper band edge. Figure 5.1 depicts the dependence of the eigenvalues $\mathcal{E}_i$ on the magnitude of the perturbation potential $\mathcal{U}_0$. The results obtained may be interpreted as follows. The electrons close to the lower (upper) band edge can be regarded as free

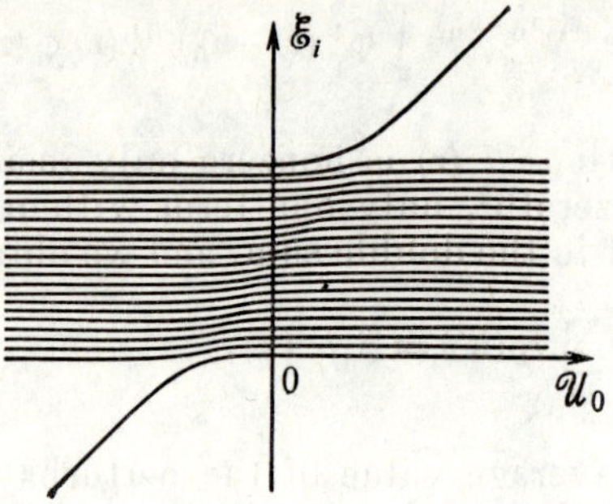


Fig. 5.1

electrons with a positive (a negative) mass $m^*$. It is an established fact that if the depth of a three-dimensional potential trough $|\mathcal{U}_0|$ is below a certain value, the particle has no bound states inside it.[3] In this case the perturbation $\mathcal{U}_0$ only scatters the particle, and little influences its energy spectrum.

## 5.2. Localized Electron States in a Nonideal Lattice

**5.2.1** In certain cases the electric properties of semiconductors and primarily the number of charge carriers in them depend substantially on various crystal lattice defects. Such defects include: (1) foreign (impurity) atoms that can either be part of the lattice substituting an equal number of the intrinsic atoms or occupy the interstitials; (2) intrinsic atoms of the lattice moved to the interstitials; (3) empty (vacant) lattice sites; (4) dislocations; (5) crystal surface, etc. Each such imperfection in the ideal lattice sets up an additional field that acts on the electron and can be treated with the aid of the methods discussed in the preceding section.

Consider, by way of an example, an arsenic atom with five valence electrons placed in one of the sites of the silicon lattice. Four valence electrons of the arsenic atom will, just as the four electrons of a silicon atom, take part in four directional lattice bonds. The fifth valence electron having weaker bonds with the lattice sites will move in the field of the lattice and of the singly charged arsenic ion. Since the Coulomb potential is a slowly-changing one, the pro-

---

[3] For a spherical potential trough of diameter $a$ this critical value is equal to $\pi^2\hbar^2/2m^*a^2$ [5.1, p. 87].

blem can be solved in the effective mass approximation discussed in Section 5.1.1.[4] The fifth electron of the arsenic atom which does not take part in the valence bonds, but provides for its electric neutrality, can exist either in a bound state near the ion or in a free state in the conduction band of silicon. We can try to account roughly for the effect of the crystal on the Coulomb field of an impurity ion assuming it to be immersed in a medium with the dielectric constant $\varepsilon_0$. In this case the additional field in (1.18a) will be equal to

$$\mathcal{U}(\mathbf{r}) = -e^2/\varepsilon_0 r, \tag{2.1}$$

For a simple band the electron energy is $\varepsilon = \hbar^2 k^2/2m^*$, and equation (1.18a) reduces to that for a hydrogen atom [5.2, Chap. 7] with the electron mass $m^*$ and the charge $e^* = \frac{e}{\sqrt{\varepsilon_0}}$. If the electron exists in a bound state near an impurity ion, its energy is

$$\mathcal{E}_i = \mathcal{E}_n = -\frac{m^*(e^*)^4}{2\hbar^2 n^2} = \frac{me^4}{2\hbar^2 n^2}\frac{m^*}{m}\frac{1}{\varepsilon_0^2} = -\frac{13.5}{n^2}\left(\frac{m^*}{m}\right)\frac{1}{\varepsilon_0^2}\,\text{eV}, \tag{2.2}$$

where $n$ is the principal quantum number and $m$ is the electron mass in a vacuum. Since it is assumed in (1.18a) that the energy $\mathcal{E}_0 = 0$ corresponds to the lower edge of the conduction band, then the negative energy levels of an electron in a bound state (2.2) are located in the forbidden band. If the impurity ion occupies the $\mathbf{l}$-th lattice site, we have for the fundamental quantum state

$$f_0(\mathbf{r}) = \frac{1}{\sqrt{\pi a_B^3}}\, e^{-\frac{|\mathbf{r}-\mathbf{a}_\mathbf{l}|}{a_B}}, \tag{2.3}$$

where the Bohr radius is

$$a_B = \frac{\hbar^2}{m^*(e^*)^2} = \frac{\hbar^2}{me^2}\left(\frac{m}{m^*}\right)\varepsilon_0 = 0.53\left(\frac{m}{m^*}\right)\varepsilon_0 \text{Å}.$$

It follows from (2.3) that

$$f_0(\mathbf{a}_\mathbf{n}) = \frac{1}{\sqrt{\pi a_B^3}}\, e^{-\frac{|\mathbf{a}_\mathbf{n}-\mathbf{a}_\mathbf{l}|}{a_B}}. \tag{2.3a}$$

Substituting expression (2.3a) into (1.6) we see that the electron wave function $\Phi_0(\mathbf{r})$ decays exponentially as the distance from the $\mathbf{l}$-th site in which the impurity ion is located is increased. An excited electron (by heat, light, etc.) can move from the bound state

[4] Strictly speaking, the additional field $\mathcal{U}(\mathbf{r})$ which in this case acts on the fifth electron is the difference between the field of the arsenic ion and the field of a silicon atom, however the latter is small in the region of the most probable location of the electron.

near an impurity ion to a free state—to the conduction band, where it turns into a charge carrier. Impurity centres that in an excited state can supply free charges to the conduction band are termed *donors.*

Now imagine an atom of Group III (boron, aluminium, indium or gallium) placed in one of the vacant sites of the silicon or germanium lattice. The atoms of Group III have three valence electrons in the *s*- and *p*-states, and to form a four-valent bond in the Si or Ge lattice they have to borrow one electron from the "pool" of valence electrons of the intrinsic atoms of the crystal. Thus, as a result of thermal or light excitation, an electron from the valence band can join a neutral impurity atom to form a negative ion.

Impurity centres that in an excited state can trap an electron from the valence band creating a positive hole in it are termed *acceptors.*

Hence, as a result of excitation donors turn into positive ions creating free electrons in the conduction band, and acceptors turn into negative ions creating positively charged holes in the valence band.

We can easily assess the energies of the ground state ($n = 1$) of donors and acceptors in germanium and silicon to be of the order of 0.01 eV. Such small values of $|\mathscr{E}_0|$ are due to the high dielectric constants of Ge ($\varepsilon_0 \approx 16$) and of Si ($\varepsilon_0 \approx 12$). Such levels which became known as *shallow* levels are actually observed in experiments with germanium and silicon, when they are doped by elements of Group III and Group V of the Periodic Table.

In the following chapter we shall consider in detail the statistical distribution of the electrons among the donors, the acceptors, the valence and the conduction bands. As we shall see, this distribution, and consequently, the number of charge carriers, is very sensitive to the crystal's temperature, this being a peculiar property of semiconductors.

5.2.2 When considering the states of the impurity electron in silicon, we should take into account the tensor nature of the effective mass and the existence of six equivalent energy minima in the conduction band on the [100] axes (Section 4.15). The latter circumstance is the cause of a six-fold degeneracy of the electron's wave function which is partially removed in the crystal field.

Consider, as we have done in the preceding section, the 1*s*-state of the electron in an impurity atom.

We have already noted that the Schrödinger equation in the effective mass approximation (4.3.23) (4.3.24) is actually satisfied by the smooth part $F(\mathbf{r})$ of the electron's wave function. Because of the existence of six equivalent energy minima in the conduction band the six-fold degenerate electron wave function is equal to

$$\psi^{(i)}(\mathbf{r}) = F^{(i)}(\mathbf{r})\,\psi_{\mathbf{k}_i}(\mathbf{r}), \quad i = 1, 2, \ldots, 6. \tag{2.4}$$

Here $\psi_{\mathbf{k}_i}(\mathbf{r})$ is the Bloch function of the conduction electron at the point of energy minimum $\mathbf{k}_i$; we assume the functions $\psi_{\mathbf{k}_i}(\mathbf{r})$ to be nondegenerate for the specific $i$. The functions $F^{(i)}(\mathbf{r})$ satisfy the equation with effective masses (4.3.23)

$$\left[-\frac{\hbar^2}{2m_{\parallel}}\frac{\partial^2}{\partial x_i^2}-\frac{\hbar^2}{2m_{\perp}}\left(\frac{\partial^2}{\partial y_i^2}+\frac{\partial^2}{\partial z_i^2}\right)-\frac{e^2}{\varepsilon_0 r}\right]F^{(i)}(\mathbf{r})=\mathscr{E}F^{(i)}(\mathbf{r}). \quad (2.5)$$

Here $m_{\parallel} = 0.92\ m$ is the longitudinal, and $m_{\perp} = 0.19\ m$ is the transverse effective electron masses in silicon ($m$ is the free electron mass) (Section 4.15). The $x_i$ axis is parallel to $\mathbf{k}_i$. The six degenerate wave functions (2.4) form the basis of a representation (in general, a reducible one) of the symmetry group of our system, namely, of the tetrahedron group $T_d$. To find the nature of the splitting of the six-fold degenerate level corresponding to (2.4), we should find the characters of the reducible representation and decompose it into irreducible representations of the $T_d$ group.

To find the characters of a reducible representation corresponding to the basis (2.4), we should investigate the mutual transformations of the functions $\psi_{\mathbf{k}_i}(\mathbf{r})$ under symmetry transformations of the $T_d$ group. The mutual transformations of the functions $\psi_{\mathbf{k}_i}(\mathbf{r})$ are obviously similar to those of the points of the minima $\mathbf{k}_i$. Since the points $\mathbf{k}_i$ are located on the axes $\pm x$, $\pm y$, $\pm z$ at equal distances from the origin, the functions $\psi_{\mathbf{k}_i}(\mathbf{r})$ may be said to be transformed like the six functions: $x$, $-x$, $y$, $-y$, $z$, $-z$ (if all the six coordinates are equal in magnitude).

The group $T_d$ has five classes: $E$, $8C_3$, $3C_4^2$, $6\sigma$, $6S_4$ (see Table 2.7).

A unit transformation $E$ results obviously in $x \to x$, $-x \to -x$, $y \to y$, . . .; therefore, the trace of the corresponding transformation matrix, i.e., its character, is equal to 6.

For the transformation $C_3$ (the axis [111]) we have $x \to y \to z \to x$, $-x \to -y \to -z \to -x$, i.e., there are zeros along the main diagonal of the matrix. Therefore, the corresponding character is zero. For the transformation $C_4^2$ (the axis [100]) we obtain $x \to x$, $-x \to -x$, $y \to -y \to y$, $z \to -z \to z$. We see that there are two unities on the main diagonal of the matrix; therefore, its character is two.

Precisely in the same way we can demonstrate that the characters of the classes $\sigma$ and $S_4$ are 2 and 0.

Making use of Table 2.7 and of the values of the characters obtained, we can draw up Table 5.1.

Making use of the general procedure of decomposition of a representation into irreducible representations (Section 2.6.4) or simply comparing the characters of the $1s$ representation with those of the irreducible representations of the $T_d$ group, we obtain

$$1s = A_1 + E + F_2, \quad (2.6)$$

Table 5.1

| | $E$ | $8C_3$ | $3C_4^2$ | $6\sigma$ | $6S_4$ |
|---|---|---|---|---|---|
| $A_1$ | 1 | 1 | 1 | 1 | 1 |
| $A_2$ | 1 | 1 | 1 | $-1$ | $-1$ |
| $E$ | 2 | $-1$ | 2 | 0 | 0 |
| $F_1$ | 3 | 0 | $-1$ | $-1$ | 1 |
| $F_2$ | 3 | 0 | $-1$ | 1 | $-1$ |
| $1s$ | 6 | 0 | 2 | 2 | 0 |

whence it follows that the six-fold degenerate state of a donor $1s$ splits up in the crystal field into nondegenerate, doubly degenerate and triply degenerate levels.

**5.2.3** We have considered the donor and acceptor states of electrons and holes in slowly varying impurity ion fields in which the effective mass method, i.e., equation (1.9), is applicable. As we have seen in Section 5.1.3, if the localized perturbation $\mathcal{U}_0$ reaches a certain value, it also causes the splitting off of an energy level from the bottom or the top of an energy band creating bound electron states of the donor or the acceptor type.

Since impurity ions (especially the negative ones) cannot be regarded as point defects, in some cases the field set up by them is more of the localized perturbation type than of a Coulomb potential type. In this case equation (2.2) is inapplicable, and, as a rule, deeper impurity levels are created.

For example, gold in silicon creates a donor level 0.35 eV above the valence band and an acceptor level 0.54 eV below the conduction band.[5]

After an electron joins a positive ion a neutral atom is formed whose field is to a good approximation described by a screening potential of the form

$$\mathcal{U}(r) = -A\frac{e^{-\alpha r}}{r}, \tag{2.7}$$

[5] The forbidden bandwidth in silicon at 300 K is equal to 1.12 eV.

where $1/\alpha$ is the effective radius of action of the field. If $1/\alpha$ is of the order of magnitude of the lattice parameter, the perturbation (2.7) can be regarded as localized. Since, if $1/\alpha$ are not too small, other bound states may exist in the potential field (2.7), such an atom can trap an additional electron, and for this reason in certain cases the impurity levels may be of the multiple charge type.

It is known from experiment that in germanium the monovalent Cu and Au can trap 1, 2 and even 3 electrons, and the bivalent Zn and Cd can trap 1 or 2 electrons.

Note that a qualitative theory of deep impurity levels has, as yet, not been developed.

**5.2.4** Possibly, the simplest type of localized perturbation of the periodic field of an ideal crystal is its surface. Bound electron states on the free surface of a crystal were first predicted and considered theoretically by I. E. Tamm (1932) and became known as *Tamm surface states.* Their existence was confirmed in experiments on contact phenomena in semiconductors.

We shall consider the surface states on the simplest unidimensional model of a finite crystal lattice (Fig. 5.2) that approximately describes

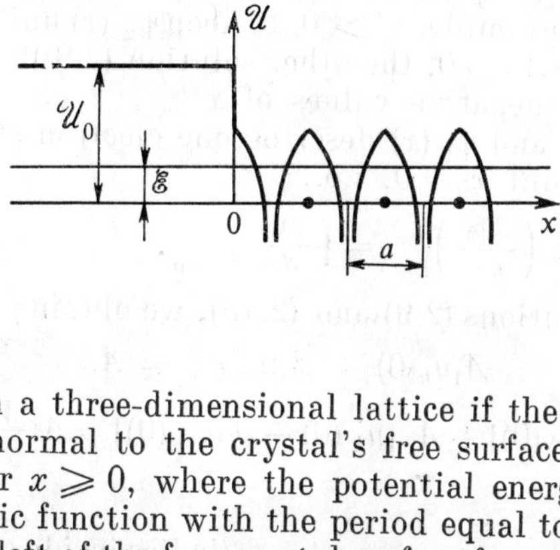


**Fig. 5.2**

the situation in a three-dimensional lattice if the only direction of interest is the normal to the crystal's free surface. Let the crystal fill the space for $x \geqslant 0$, where the potential energy of an electron $\mathcal{U}(x)$ is a periodic function with the period equal to the lattice parameter $a$. To the left of the free crystal surface for $x \leqslant 0$ the potential energy $\mathcal{U}(x)$ of an electron in a vacuum is constant equal to $\mathcal{U}_0$.

The Schrödinger equation for the unidimensional motion of the electron in the field $\mathcal{U}(x)$ is of the form

$$-\frac{\hbar^2}{2m}\frac{d^2\psi}{dx^2}+\mathcal{U}(x)\,\psi=\mathcal{E}\psi. \tag{2.8}$$

Consider the case $\mathcal{E}<\mathcal{U}_0$. Then the solution of (2.8) for $x \leqslant 0$ will be of the form

$$\psi_1(x)=A\exp\left(\frac{\sqrt{2m(\mathcal{U}_0-\mathcal{E})}}{\hbar}\,x\right). \tag{2.9}$$

In this equation we have dropped the addend proportional to $\exp\left(-\frac{\sqrt{2m(\mathcal{U}_0-\mathcal{E})}}{\hbar}x\right)$, because it tends to infinity as $x \to -\infty$. For $x \geqslant 0$ $\mathcal{U}(x)$ is a periodic function, and, consequently, the wave function is in the form of the Bloch function $u_k(x)\exp(ikx)$. The most general solution for a linear homogeneous differential equation of the second order can be made up of two linearly independent solutions, for instance,

$$\psi_2(x) = A_1 u_k(x) e^{ikx} + A_2 u_{-k}(x) e^{-ikx}. \tag{2.10}$$

The real values of the wave number (vector) $k$ correspond to the allowed values of the electron energy $\mathcal{E}(k)$ in an infinite periodic lattice. Indeed, if $k$ is complex, i.e., if $k = k' + ik''$, then

$$\psi_2(x) = A_1 u_k(x) e^{ik'x} e^{-k''x} + A_2 u_{-k}(x) e^{-ik'x} e^{k''x}. \tag{2.11}$$

It is quite obvious that, depending on the sign of $k''$, either the first or the second addend will tend to infinity as $x$ tends to $+\infty$ or to $-\infty$. For the wave function $\psi_2$ to be finite, the electron in an infinite lattice must not exist in a state with a complex $k$.

The situation is quite different in a semifinite lattice (see Fig. 5.2). If, for example, $k'' > 0$, to keep $\psi_2(x)$ finite for $x \to +\infty$, it suffices to make $A_2 = 0$, the other solution (2.9) finite for $x \to -\infty$ operating for the negative values of $x$.

To make $\psi_1(x)$ and $\psi_2(x)$ describe one electron state they should be matched at point $x = 0$, i.e.,

$$\psi_1(0) = \psi_2(0), \quad \left(\frac{d\psi_1}{dx}\right)_{x=0} = \left(\frac{d\psi_2}{dx}\right)_{x=0}. \tag{2.12}$$

Making use of solutions (2.9) and (2.10), we obtain

$$A_1 u_k(0) + A_2 u_{-k}(0) = A,$$
$$A_1[u'_k(0) + iku_k(0)] + A_2[u'_k(0) - iku_{-k}(0)] = A\frac{\sqrt{2m(U_0-\mathcal{E})}}{\hbar}. \tag{2.13}$$

If $k$ is real, and if, consequently, $\mathcal{E}(k)$ lies inside one of the allowed energy bands of an infinite crystal, then with $A_1 \neq 0$ and $A_2 \neq 0$ $\psi_2$ will be finite for $x \to \infty$. In this case two linear equations (2.13) for three unknowns $A_1$, $A_2$ and $A$ always have a solution. This means that all the electron states allowed in the infinite crystal can also be realized in a semifinite crystal.

If, on the other hand, $k$ is complex, and the corresponding energy $\mathcal{E}(k)$ lies inside the forbidden band of the infinite crystal, then to make $\psi_2$ stay finite for $x \to \infty$ it suffices to make (for $k'' > 0$) $A_2 = 0$. In this case (2.13) turns into a homogeneous linear system with two unknowns $A_1$ and $A$. For this system to have nonzero solutions its determinant must be zero, and this leads to some characteristic equation for the energy $\mathscr{E}$. This energy can be found as follows: in (2.13) make $A_2 = 0$, substitute $A = A_1 u_k(0)$ from the first equation into the second and cancel out $A_1$ in the latter to obtain

$$\mathscr{E} = \mathscr{U}_0 - \frac{\hbar^2}{2m}\left[\frac{u'_k(0)}{u_k(0)} + ik\right]^2, \tag{2.14}$$

where $k = k' + ik''$ $(k'' > 0)$.

The value of the energy (2.14) depends on the form of the periodic potential $\mathscr{U}(x)$ for $x \geqslant 0$. I. E. Tamm has demonstrated that if the periodic potential $\mathscr{U}(x)$ is approximated by rectangular barriers (the Kronig-Penny model), then under certain conditions there will be one surface level in each forbidden band. The wave function of the electron occupying this state decays exponentially both in vacuum (2.9) and inside the crystal (2.11), where $A_2 = 0$. Hence, the electron occupying a surface level remains localized near the surface. Since it is able to move along the surface, an additional surface conductivity appears.

In a three-dimensional crystal a surface level turns into a surface band, the number of states in which is of the order of $10^{15}$, i.e., of the order of the number of atoms per 1 $cm^2$ of the surface. Although for a specific crystal specimen the number of surface states may be comparable to the number of local impurity levels, their role is usually not great, because charging of the surface results in appreciable variations in its potential preventing the occupation of the surface levels by other electrons.

## 5.3 Excitons

**5.3.1** The band theory based on the single-electron approximation takes into account all the possible quantum states of the electron in a crystal. The action of all the atomic nuclei and of all the other electrons is in this approximation assumed to be reduced to the action of a certain external self-consistent three-dimensional periodic field. In a dielectric or in an intrinsic semiconductor the valence band is completely filled with electrons, so that the minimum excitation energy of an electron involves its transition from the filled valence band to the free conduction band. Such a transition results in the generation of charge carriers: of an electron in the conduction band and of a hole in the valence band.

Experiment shows, however, that the absorption of light in dielectrics at frequencies corresponding to the excitation of electrons is not always accompanied by the appearance of charge carriers (by photoconductivity). As was first demonstrated by Ya. I. Frenkel (1931), this may be the result of special currentless electron excitations of the crystal characterized by a quasi-momentum and an energy of translational motion. Ya. I. Frenkel termed such excitations *excitons*. Of course, strictly speaking, the exciton excitations transgress the precincts of the single-electron approximation, however, in a sense they can be incorporated into the band picture, if the additional interaction of the electron in the conduction band and the hole in the valence band is taken into account (Mott, Wannier).

In the following we shall be interested only in the bound states of the electron and the hole interacting with one another. If the dimensions of such an exciton are great compared with the lattice parameter, then the electron-hole interaction can to a good approximation be regarded as Coulomb interaction of two point charges attenuated $\varepsilon_0$ times, where $\varepsilon_0$ is the static dielectric constant of the crystal.

For the valence crystals of germanium and silicon we can, to an approximation adequate for our purpose, make $\varepsilon_0 = n^2$, where $n$ is the refraction index.

Consider the Schrödinger equation for an electron and for a hole moving in the crystal's periodic field and interacting in compliance with the Coulomb law.

Let $\mathbf{r}_n$ and $\mathbf{r}_p$ be position vectors of the electron and the hole, and let $\mathbf{k}_n$ and $\mathbf{k}_p$ be their respective wave vectors ($\mathbf{k}_p = -\mathbf{k}'_n$, where $\mathbf{k}'_n$ is the wave vector of the electron corresponding to the hole in the valence band). Denote by $\varepsilon_c(\mathbf{k}_n)$ and $\varepsilon_v(\mathbf{k}_p)$ the energies of the electron in the conduction band and of the hole in the valence band.

Generalizing the Wannier equation (4.3.13) for a system comprising an electron and a hole interacting in compliance with Coulomb's law, we have

$$\left[\varepsilon_c(-i\nabla_n) - \varepsilon_v(i\nabla_p) - \frac{e^2}{\varepsilon_0 |\mathbf{r}_n - \mathbf{r}_p|}\right] \psi(\mathbf{r}_n, \mathbf{r}_p) = \mathscr{E}\psi(\mathbf{r}_n, \mathbf{r}_p), \tag{3.1}$$

where $\nabla_n$ and $\nabla_p$ are operators in the coordinates of the electron and the hole, $\mathscr{E}$ is the energy of the complete system; the positive kinetic energy $\varepsilon_v$ is measured in the direction opposite to that in which the positive kinetic energy $\varepsilon_c$ is measured.

If both the energy minimum $\varepsilon_c(\mathbf{k}_n)$ and the energy maximum $\varepsilon_v(\mathbf{k}_p)$ are located in the centre of the Brillouin zone, and if the dispersion laws for the electron and the hole energies are characterized by scalar effective masses $m_n$ and $m_p$, then we have near the extrema

$$\varepsilon_c(\mathbf{k}_n) = \varepsilon_c(0) + \frac{\hbar^2 k_n^2}{2m_n}, \quad \varepsilon_v(\mathbf{k}_p) = \varepsilon_v(0) - \frac{\hbar^2 k_p^2}{2m_p}.$$

Substituting these expansions into (3.1), we obtain[6]

$$\left[-\frac{\hbar^2}{2m_n}\nabla_n^2-\frac{\hbar^2}{2m_p}\nabla_p^2-\frac{e^2}{\varepsilon_0\,|\,\mathbf{r}_n-\mathbf{r}_p\,|}\right]\psi(\mathbf{r}_n,\mathbf{r}_p)=(\mathscr{E}-\varepsilon_G)\,\psi(\mathbf{r}_n,\mathbf{r}_p), \quad (3.1a)$$

where the forbidden bandwidth is $\varepsilon_G = \varepsilon_c(0) - \varepsilon_v(0)$.

Introduce the position vectors of the centre of gravity of the electron-hole system $\mathbf{R}$ and of the relative positions of the electron and the hole $\mathbf{r}$:

$$\mathbf{R}=\frac{m_n\mathbf{r}_n+m_p\mathbf{r}_p}{m_n+m_p},\qquad \mathbf{r}=\mathbf{r}_n-\mathbf{r}_p. \quad (3.2)$$

Equation (3.1a) in the new coordinates assumes the form (see Appendix 16)

$$\left[-\frac{\hbar^2}{2M}\nabla_R^2-\frac{\hbar^2}{2\mu}\nabla_r^2-\frac{e^2}{\varepsilon_0 r}\right]\psi(\mathbf{R},\,\mathbf{r})=(\mathscr{E}-\varepsilon_G)\,\psi(\mathbf{R},\,\mathbf{r}). \quad (3.3)$$

Here $M = m_n + m_p$ is the exciton mass; $\mu$ is the reduced mass of the electron and the hole: $\frac{1}{\mu}=\frac{1}{m_n}+\frac{1}{m_p}$, $\nabla_R^2$ and $\nabla_r^2$ are Laplace operators in the variables $\mathbf{R}$ and $\mathbf{r}$.

Equation (3.3) is solved by separating the variables

$$\psi(\mathbf{R},\,\mathbf{r}) = \chi(\mathbf{R})\,\varphi(\mathbf{r}). \quad (3.4)$$

Substituting (3.4) into (3.3) and dividing both sides of the equation by $\chi(\mathbf{R})\,\varphi(\mathbf{r})$, we obtain

$$-\frac{\hbar^2}{2M}\frac{1}{\chi(\mathbf{R})}\nabla_R^2\chi(\mathbf{R})-\frac{1}{\varphi(\mathbf{r})}\left[\frac{\hbar^2}{2\mu}\nabla_r^2\varphi(\mathbf{r})+\frac{e^2}{\varepsilon_0 r}\varphi(\mathbf{r})\right]=\mathscr{E}-\varepsilon_G.$$

Since the first addend on the left-hand side depends solely on $\mathbf{R}$, and the second solely on $\mathbf{r}$, and since their sum equal to $\mathscr{E} - \varepsilon_G$ is a constant, both the first and the second addends should be constants, too, so that

$$-\frac{\hbar^2}{2M}\nabla_R^2\chi(\mathbf{R})=W\chi(\mathbf{R}), \quad (3.5)$$

$$-\frac{\hbar^2}{2\mu}\nabla_r^2\varphi(\mathbf{r})-\frac{e^2}{\varepsilon_0 r}\varphi(\mathbf{r})=\varepsilon\varphi(\mathbf{r}), \quad (3.5a)$$

where $W+\varepsilon=\mathscr{E}-\varepsilon_G$.

Equation (3.5) describes the free motion of a particle (exciton) of mass $M = m_n + m_p$ and with the energy

$$W = \hbar^2K^2/2M, \quad (3.6)$$

where $\mathbf{K}$ is the wave vector of a plane wave corresponding to the exciton's motion.

---

[6] As was already noted in Section 4.3, equation (3.1a) contains not the full wave function (3.1) but its smoothly varying part.

Equation (3.5a) describes the relative motion of the electron and the hole, which may be imagined as the motion of an electron of mass $\mu$ about a static hole. The energy values corresponding to the bound states of the electron and the hole are discrete negative values $\varepsilon = \varepsilon_i$. Those hydrogen-like terms are described by expression (2.2) in which the reduced mass $\mu$ should be substituted for the effective mass $m^*$. The exciton radius, similar to the Bohr radius of the hydrogen atom, is equal to $\frac{\hbar^2}{me^2}\left(\frac{m}{\mu}\right)\varepsilon_0 = 0.53\,\frac{m}{\mu}\,\varepsilon_0$ Å, where $m$ is the electron mass in vacuum. We see that excitons of large radii are formed in crystals with great dielectric constants $\varepsilon_0$.

**5.3.2** Excitons of small radii can be described in terms of the tight binding approximation (Ya. I. Frenkel, 1931). We shall use the Hartree-Fock method. Choose the electron wave functions in isolated lattice sites as the initial single-electron functions. Consider for the sake of simplicity a simple cubic lattice made up of identical atoms with one electron each, and let the ground unexcited state of the $l$-th electron in the isolated $n$-th site be described by the Schrödinger equation

$$\mathcal{H}_n \psi_n (\mathbf{r}_l) = \varepsilon_0 \psi_n (\mathbf{r}_l). \tag{3.7}$$

Here

$$\mathcal{H}_n = -\frac{\hbar^2}{2m} \nabla_l^2 + \mathcal{U}_n (\mathbf{r}_l), \tag{3.7a}$$

where $\nabla_l^2$ is the Laplace operator for the $l$-th electron, $\mathcal{U}_n(\mathbf{r}_l)$ is its potential energy in the $n$-th isolated lattice site and $\varepsilon_0$ is the energy of its ground state.

For the sake of simplicity, neglect the overlapping of the wave functions even of the neighbouring sites. Then we have by analogy with (4.2.7)

$$\int \psi_n^* (l)\, \psi_{n'} (l)\, d\tau_l = \delta_{nn'}, \tag{3.8}$$

where $l$ stands instead of $\mathbf{r}_l$ and $d\tau_l = dx_l dy_l dz_l$. Assume that spins of all the electrons are parallel; obviously, this is of no importance, if we are not interested in the magnetic properties of the system, the problems of the covalent bond, etc.

If the spin-orbit coupling is not taken into account, and if for this reason the wave function can be represented as the product of the coordinate part and the spin part, the former will in our case have to be antisymmetric because of the symmetry of the spin part. The correct antisymmetric wave function for the whole crystal constructed on the single-electron functions $\psi_n (l)$ takes the form of a determinant (4.2.8b)

$$\Phi(1, 2, \ldots, N) = \frac{1}{\sqrt{N!}} \begin{vmatrix} \psi_1(1) & \psi_1(2) & \ldots & \psi_1(N) \\ \cdot & \cdot & \cdot & \cdot \\ \psi_N(1) & \psi_N(2) & \ldots & \psi_N(N) \end{vmatrix}, \tag{3.9}$$

if there are $N$ sites in the crystal's principal region.

Making use of (3.8) we can easily demonstrate that, similar to (4.2.8a),

$$\int \Phi^*(1, 2, \ldots, N)\,\Phi(1, 2, \ldots, N)\,d\tau_1\,d\tau_2 \ldots d\tau_N$$

$$= \int \Phi^*\Phi\,d\tau = 1$$

$$(d\tau \equiv d\tau_1\,d\tau_2 \ldots d\tau_N). \tag{3.9a}$$

The Hamiltonian for the system of all the $N$ electrons

$$\hat{\mathscr{H}} = \sum_{n=1}^{N} \hat{\mathscr{H}}_n + V(1, 2, \ldots, N), \tag{3.10}$$

where $\sum_{n=1}^{N} \hat{\mathscr{H}}_n = \hat{\mathscr{H}}_0$ is the unperturbed Hamiltonian (for example, it may be assumed that $\hat{\mathscr{H}}_n = \frac{-\hbar^2}{2m}\nabla_n^2 + \mathscr{U}_n(r_n))$, and $V(1, 2, \ldots, N)$ includes the interaction of all the electrons and the energy of every $n$-th electron in the field of all the other $m$-th lattice sites $(m \neq n)$. In the zeroth approximation the system's energy is

$$\mathscr{E}_0 = \int \Phi^*\hat{\mathscr{H}}_0\Phi\,d\tau = N\varepsilon_0. \tag{3.11}$$

This is a direct corollary of expressions (3.9), (3.7) and (3.8)

The correction to the energy of the ground state in the first approximation of the perturbation theory is equal to

$$\mathscr{E}_1 = \int \Phi^* V \Phi\,d\tau. \tag{3.12}$$

Denote the wave function of the excited state of the $l$-th electron near the $n$-th site by $\psi_n'$; then

$$\hat{\mathscr{H}}_n\psi_n' = \varepsilon_1\psi_n', \tag{3.13}$$

where $\varepsilon_1$ is the energy of the excited state of the electron in the isolated site.

Assume that the excited wave functions of the specific site do not overlap any wave functions of the neighbouring sites. Moreover, for the given $n$

$$\int \psi_n^* \psi_n' \, d\tau = 0, \tag{3.14}$$

because the excited state $\psi_n'$ is orthogonal to the ground state $\psi_n$.

The electron wave function with the correct antisymmetric properties in a crystal in which the $n$-th site has been excited is of the form

$$\Phi_n(1, 2, \ldots, N) = \frac{1}{\sqrt{N!}} \begin{vmatrix} \psi_1(1) & \ldots & \psi_1(N) \\ \cdot & \cdot\ \cdot\ \cdot & \cdot \\ \psi_n'(1) & \ldots & \psi_n'(N) \\ \psi_N(1) & \ldots & \psi_N(N) \end{vmatrix}. \tag{3.15}$$

Making use of (3.8), where for $n \neq n'$ the wave functions may as well belong to the excited states, and of (3.14), we can easily demonstrate that

$$\int \Phi_m^* \Phi_n \, d\tau_1 \ldots d\tau_N = \delta_{mn}. \tag{3.16}$$

Expression (3.15), in contrast to (3.9), is not, as yet, a correct wave function of the zeroth approximation for an excited crystal. Indeed, because of the translational symmetry the excitation can be localized on any site, the crystal's energy remaining unchanged. Hence, the state is an $N$-fold degenerate one, and the correct wave function of the zeroth approximation should be constructed in the form of a linear superposition

$$\Phi'(1, 2, \ldots, N) = \sum_{n=1}^{N} C_n \Phi_n(1, 2, \ldots, N), \tag{3.17}$$

where $C_n$ are constants to be determined in accordance with the general rules of the perturbation theory of degenerate states (compare with Section 4.7.2).

Calculate the effect of the unperturbed operator $\hat{\mathscr{H}}_0 = \sum_{l=1}^{N} \hat{\mathscr{H}}_l(l)$ on the function $\Phi_n$ (3.15):

$$\hat{\mathscr{H}}_0 \Phi_n = [\hat{\mathscr{H}}_1(1) + \hat{\mathscr{H}}_2(2) + \ldots + \hat{\mathscr{H}}_l(l) + \ldots + \hat{\mathscr{H}}_N(N)] \times \frac{1}{\sqrt{N!}} \sum_p (-1)^{[p]} \hat{P} \{\psi_1(p_1) \ldots \psi_n'(p_n) \ldots \psi_N(p_N)\}, \tag{3.18}$$

where $p_1, p_2, \ldots, p_n, \ldots, p_N$ are numbers $1, 2, 3, \ldots, N$ arranged in a definite order. The determinant (3.15) has been written here in the form of a sum in which the summation embraces all possible permutations of the numbers $p_1, p_2, \ldots, p_n, \ldots, p_N$, $[p]$ denoting the number of disorders in such a permutation [5.3, §1].

We should differentiate between two cases of the operation of one of the operators $\hat{\mathscr{H}}_l(l)$ $(l = 1, 2, 3, \ldots, N)$ on the product of the single-electron functions $\{\psi_1(p_1) \ldots \psi_n'(p_n) \ldots \psi_N(p_N)\}$: (1) $p_n = l$; in that case the effect of the operator results, according to (3.13), in multiplication of $\{\ \}$ by $\varepsilon_1$; (2) $p_n \neq l$; in that case, according to (3.7), the effect of the operator is to multiply $\{\ \}$ by $\varepsilon_0$. Since for each $l$ the second case can be realized in $N - 1$ ways, it follows that

$$\hat{\mathscr{H}}_0\Phi_n = [(N-1)\,\varepsilon_0 + \varepsilon_1]\,\Phi_n. \tag{3.18a}$$

The energy of the crystal in this approximation is

$$\begin{aligned}\int \Phi^{*\prime}\hat{\mathscr{H}}_0\Phi'\,d\tau &= \sum_m \sum_n C_m^* C_n \int \Phi_m^*\hat{\mathscr{H}}_0\Phi_n\,d\tau \\ &= [(N-1)\,\varepsilon_0 + \varepsilon_1] \sum_m \sum_n C_m^* C_n \delta_{mn} \\ &= [(N-1)\,\varepsilon_0 + \varepsilon_1] \sum_n |C_n|^2 = (N-1)\,\varepsilon_0 + \varepsilon_1,\end{aligned} \tag{3.19}$$

if the wave function $\Phi'$ is normalized to unity, i.e., if we make

$$\sum_n |C_n|^2 = 1. \tag{3.19a}$$

The corresponding Schrödinger equation for the Hamiltonian (3.10) is

$$[\hat{\mathscr{H}}_0 + V(1, 2, \ldots, N)]\,\psi = W\psi. \tag{3.20}$$

Substitute herein instead of $\psi$ the approximate wave function $\Phi'$ (3.17). With reference to (3.18a) we obtain

$$\sum_n C_n \{[(N-1)\,\varepsilon_0 + \varepsilon_1]\,\Phi_n + V\Phi_n\} = W \sum_n C_n\Phi_n.$$

Premultiplying both sides of the equality by $\Phi_m^*$, integrating with respect to the coordinates of all electrons and taking account of (3.16), we obtain

$$\sum_{n=1}^{N} C_n V_{mn} - \mathscr{E}' C_m = 0, \tag{3.21}$$

where

$$V_{mn} = \int \Phi_m^* V \Phi_n\,d\tau, \tag{3.21a}$$

$$\mathscr{E}' = W - [(N-1)\,\varepsilon_0 + \varepsilon_1]. \tag{3.21b}$$

The linear homogeneous algebraic system of $N$ equations for $N$ unknown coefficients $C_n$ (3.21) in an expanded form is as follows

$$\begin{aligned}
&(V_{11}-\mathscr{E}')\,C_1+V_{12}C_2+\ldots+V_{1N}C_N=0,\\
&V_{21}C_1+(V_{22}-\mathscr{E}')\,C_2+\ldots+V_{2N}C_N=0,\\
&\ldots\ldots\ldots\ldots\ldots\ldots\ldots\ldots\ldots\ldots\\
&V_{N1}C_1+V_{N2}C_2+\ldots+(V_{NN}-\mathscr{E}')\,C_N=0.
\end{aligned}\tag{3.21c}$$

For this system to have nonzero solutions its determinant must be equated to zero

$$\begin{vmatrix}
(V_{11}-\mathscr{E}') & V_{12} & \ldots & V_{1N}\\
V_{21} & (V_{22}-\mathscr{E}') & \ldots & V_{2N}\\
\ldots & \ldots & \ldots & \ldots\\
V_{N1} & V_{N2} & \ldots & (V_{NN}-\mathscr{E}')
\end{vmatrix}=0. \tag{3.22}$$

This characteristic equation of the $N$-th power in $\mathscr{E}'$ determines in the first approximation the spectrum of the crystal's exciton excitation. We shall find the exciton spectrum making use of the following not rigorous, but obvious, method. In the same way as in the case of the tight binding single-electron approximation (Section 4.7), we can make the coefficients $C_n$, which take account of the translational degeneracy, equal to

$$C_n=C_0e^{i\mathbf{k}\cdot\mathbf{a}_n}, \tag{3.23}$$

where $\mathbf{k}$ is the wave vector of the quasi-particle—the exciton.

In this case it follows from equations (3.21)

$$\mathscr{E}'=\sum_n\frac{C_n}{C_m}V_{mn}=\sum_n e^{i\mathbf{k}(\mathbf{a}_n-\mathbf{a}_m)}V_{mn}. \tag{3.24}$$

Since the matrix elements $V_{mn}$, like the multipliers in front of them, depend only on the difference $(\mathbf{a}_n-\mathbf{a}_m)$ we can without loss of generality make $\mathbf{a}_m=0$, so that

$$\mathscr{E}'=\sum_n e^{i\mathbf{k}\cdot\mathbf{a}_n}V_{0n}. \tag{3.24a}$$

Taking into account only the six nearest neighbours in a simple cubic lattice ($n=1, 2, \ldots, 6$) and assuming for them $V_{0n}=-V_0$ to be the same, we obtain, similar to (4.7.10a),

$$\mathscr{E}'=-2V_0\,(\cos ak_x+\cos ak_y+\cos ak_z). \tag{3.24b}$$

Hence, in the tight binding approximation the exciton excitation in a simple cubic crystal is determined by an energy band (3.24b) similar to (4.7.10a).

Profitting by an analogy with the dependence (4.7.12) and (4.7.13), we can find the effective mass and the velocity of the exciton

$$m^*_{\text{exc}} = \hbar^2/2V_0a^2, \tag{3.25}$$

$$\mathbf{v}_{\text{exc}} = \frac{2V_0a}{\hbar}[\sin ak_x\,\mathbf{i}_0 + \sin ak_y\,\mathbf{j}_0 + \sin ak_z\mathbf{k}_0].$$

The exciton wave function is

$$\Phi'(1, 2, \ldots, N) = \frac{1}{\sqrt{N}} \sum_{n=1}^{N} e^{i\mathbf{k}\cdot\mathbf{a}n}\Phi_n \quad (1, 2, \ldots, N), \tag{3.26}$$

where we make $C_0 = 1/\sqrt{N}$ comply with the normalizing condition (3.19a). This expression reminds us of the single-electron wave function in the tight binding approximation (4.7.2a).

The value of $V_0$ is of the same order as that of $A$, which enters the expression (4.7.12); therefore, the effective mass of the exciton is, generally, of the same order as the effective masses of the electron and the hole.

## 5.4 Polarons

**5.4.1** In ionic crystals special states of conduction electrons can occur. These were first studied by S. I. Pekar (1946) and he suggested the term *polaron* for them.

The polarons appear as the result of polarization of the ionic lattice by the conduction electron. This polarization of the crystal causes a drop in the electron energy, i.e., results in the appearance of a potential trough at the place of the electron's location. The state of the conduction electron localized in this potential trough is described by a decaying wave function. Thus, a self-consistent state comes into being: the electron's localization causes the polarization of the crystal, and the latter in turn supports the localization of the electron. Evidently, this autolocalized state can move freely throughout the crystal.

If the dimensions of the electron $\psi$-cloud in the polaron state are great in comparison with the lattice parameter (polarons of large radius), the ionic crystal can be described as a continuous dielectric medium.

We should keep in mind that it is not the total crystal polarization that takes part in the formation of the polaron state, but only its inertial part. Indeed, the polarization of the ionic electron shells that follows the motion of the conduction electrons without inertia is part of the self-consistent potential acting on the electron [5.4]. The potential trough of the polaron state is due only to the inertial part of polarizability—to the displacement of heavy ions.

5.4.2 Consider the so-called *tight binding polarons* (see [5.4] and also [5.5]). The self-consistent state of the conduction electron in the polaron state is determined from the following Schrödinger equation:

$$-\frac{\hbar^2}{2m}\nabla^2\psi(\mathbf{r})+[V(\mathbf{r})+\mathcal{U}(\mathbf{r})]\,\psi(\mathbf{r})=\mathcal{E}\psi(\mathbf{r}). \tag{4.1}$$

Here $V(\mathbf{r})$ is the periodic potential of a crystal, $\mathcal{U}(\mathbf{r})$ is the electron energy in the field of a crystal polarized by the electron. For polarons of large radius the periodic potential $V(\mathbf{r})$ can be dropped, and simultaneously the effective electron mass $m^*$ can be substituted for the free electron mass $m$. Then

$$-\frac{\hbar^2}{2m^*}\nabla^2\psi+\mathcal{U}(\mathbf{r})\,\psi=\mathcal{E}\psi. \tag{4.2}$$

Equation (4.2) can be obtained by varying the functional

$$\mathcal{E}[\psi]=\frac{\hbar^2}{2m^*}\int(\nabla\psi)^2\,d\tau+\int\mathcal{U}\psi^2\,d\tau \tag{4.3}$$

with an additional normalizing condition[7]

$$\int\psi^2\,d\tau=1. \tag{4.4}$$

Indeed, subtract expression (4.4) multiplied by the indeterminate Lagrange multiplier $\lambda$ from (4.3). Varying the expression obtained in $\psi$ and equating it to zero we obtain

$$\frac{\hbar^2}{2m^*}\int 2\,(\nabla\psi\,\nabla\delta\psi)\,d\tau+\int\mathcal{U}2\psi\,\delta\psi\,d\tau-\int\lambda 2\psi\delta\psi\,d\tau=0,$$

where we made use of the fact that $\delta\nabla\psi=\nabla\delta\psi$.

In accordance with Green's theorem

$$\int(\nabla\psi\nabla\delta\psi)\,d\tau=-\int\delta\psi\nabla^2\psi\,d\tau+\oint\delta\psi\nabla\psi\,d\mathbf{S}.$$

The last surface integral vanishes as the surface $S\to\infty$, because $\psi$ decays rapidly in infinity.

Hence

$$\int\left[-\frac{\hbar^2}{2m^*}\nabla^2\psi+\mathcal{U}\psi-\lambda\psi\right]\delta\psi\,d\tau=0.$$

Since $\delta\psi$ is arbitrary, and $\lambda$ is identified with $\mathcal{E}$, this coincides with (4.2).

In some cases it is more convenient to use the variational principle, i.e., the functional $\mathcal{E}[\psi]$ (4.3), to solve the Schrödinger equa-

[7] We consider only the ground state of the polaron for which the wave function $\psi$ is real; therefore, $|\psi|^2=\psi^2$.

tion (4.2). This can be done when we know the general form of the wave function from physical considerations. Specifying the wave function in the form $\psi(\mathbf{r}; a_1, a_2, \ldots)$, where $a_1, a_2, \ldots$ are as yet indeterminate parameters, we can with the aid of (4.3) calculate the energy $\mathscr{E}$ as a function of $a_1, a_2, \ldots$. Finding the minimum of $\mathscr{E}(a_1, a_2, \ldots)$ as a function of the parameters $a_1, a_2, \ldots$, we shall determine the best energy eigenvalue corresponding to the chosen form of the wave function. This method even in the case of a very rough approximation of the wave function often yields fairly accurate eigenvalues.

The averaged energy of the electron interaction with the polarized crystal that enters (4.3)

$$\mathscr{U}_{av} = \int \mathscr{U}\psi^2 \, d\tau \tag{4.5}$$

in the case of a dielectric continuum can be determined as follows.

The electric induction vector resulting from the distributed charge of the electron $e\,|\psi(\mathbf{r}')|^2\,d\tau'$ is equal to

$$\mathbf{D}(\mathbf{r}) = e \int |\psi(\mathbf{r}')|^2 \frac{\mathbf{r}-\mathbf{r}'}{|\mathbf{r}-\mathbf{r}'|^3} \, d\tau' \tag{4.6}$$

and, of course, in a homogeneous dielectric is independent of the dielectric constant. The energy of a dipole with the moment $\mathbf{p}$ in the field $\mathbf{E}$, as is well known, is equal to $-(\mathbf{pE})$. The energy of the dipole $\mathbf{P}(\mathbf{r})\,d\tau$, where $\mathbf{P}$ is the crystal's polarization vector, in the electric field of the charge $e\,|\psi(\mathbf{r}')|^2\,d\tau'$ is equal to

$$-\mathbf{P}(\mathbf{r})\, d\tau e\, |\psi(\mathbf{r}')|^2 \, d\tau' \cdot \frac{\mathbf{r}-\mathbf{r}'}{|\mathbf{r}-\mathbf{r}'|^3}. \tag{4.7}$$

Integrating (4.7) with respect to all elements of the polarized crystal $d\tau$ and all elements of the distributed charge $d\tau'$, we obtain

$$\mathscr{U}_{av} = -e \int\int \mathbf{P}(\mathbf{r}) \cdot |\psi(\mathbf{r}')|^2 \frac{\mathbf{r}-\mathbf{r}'}{|\mathbf{r}-\mathbf{r}'|^3} \, d\tau \, d\tau'$$

$$= -\int \mathbf{D}(\mathbf{r}) \cdot \mathbf{P}(\mathbf{r}) \, d\tau. \tag{4.8}$$

The appearance of $\mathbf{D}$ in this expression instead of the field intensity $\mathbf{E}$ is explained by the fact that (4.7) is the energy of the dipole $\mathbf{P}\,d\tau$ in the field of the charge $e\,|\psi|^2\,d\tau'$ in vacuum and does not depend on the field of all other free or fixed charges in the crystal.

Substituting expression (4.8) into (4.3), we obtain

$$\mathscr{E}[\psi] = \frac{\hbar^2}{2m^*} \int (\nabla\psi)^2 \, d\tau - \int (\mathbf{D} \cdot \mathbf{P}) \, d\tau. \tag{4.9}$$

We must now take into account that $\mathbf{P}$ is not the total polarization $\mathbf{P}_0$, but only its inertial part.

If $\varepsilon$ is the total dielectric constant of the ionic crystal, then

$$\mathbf{P}_0 = \frac{\varepsilon - 1}{4\pi\varepsilon}\mathbf{D}. \tag{4.10}$$

The polarization $\mathbf{P}_e$, due to the deformation of the ionic electron shells in their equilibrium positions, which follows the motion of conduction electrons without any inertial lag is equal to

$$\mathbf{P}_e = \frac{n^2 - 1}{4\pi n^2}\mathbf{D}, \tag{4.11}$$

where $n$ is the long-wave refractive index.

Hence

$$\mathbf{P} = \mathbf{P}_0 - \mathbf{P}_e = \frac{1}{4\pi\varepsilon^*}\mathbf{D}, \tag{4.12}$$

where

$$\frac{1}{\varepsilon^*} = \frac{1}{n^2} - \frac{1}{\varepsilon}. \tag{4.12a}$$

The quantity $\varepsilon^*$ plays the part of the effective dielectric constant in the theory of polarons.

Note that in accordance with the variational principle the minimum of the functional $\mathscr{E}[\psi]$ (4.3) must be sought by varying $\psi$ in a constant field $\mathscr{U}$, i.e., for $P$ = const. Next $\mathbf{P}$ is found from equation (4.12) in terms of the corresponding $\mathbf{D}[\psi]$ (4.6). It can easily be demonstrated that the same result will be obtained if $\mathbf{P}$ in expression (4.9) is substituted according to (4.12), but a factor 1/2 is included in the integral. We shall obtain instead of (4.9) the functional

$$J[\psi] = \frac{\hbar^2}{2m^*}\int (\nabla\psi)^2\, d\tau - \frac{1}{8\pi\varepsilon^*}\int \mathbf{D}^2[\psi]\, d\tau. \tag{4.13}$$

Hence, and from the normalizing condition (4.4), we can find the wave function of the ground state of the polaron $\psi_0$ and the corresponding energy eigenvalue $\mathscr{E}_0 \equiv \mathscr{E}[\psi_0]$. Pekar uses the expression

$$\psi = A\,(1 + \alpha r + \beta r^2)\, e^{\alpha r} \tag{4.14}$$

as a test function of the ground $1s$-state of the polaron. Here $\alpha$ and $\beta$ are variational parameters and $A$ is the normalizing constant, which can be expressed in terms of the former.

Substituting expression (4.14) into (4.13), calculating the integrals and minimizing the expression obtained in $\alpha$ and $\beta$, we obtain

$$\psi_0 = 0.12\alpha_0^{3/2}\,(1 + \alpha_0 r + 0.45\alpha_0^2 r^2)\, e^{-\alpha_0 r}, \tag{4.15}$$

where

$$\alpha_0 \equiv \frac{1}{r_0} = 0.66 \frac{m^* e^2}{\hbar^2 \varepsilon^*} \tag{4.15a}$$

is the inverse radius of the polaron.[8]

From equations (4.9), (4.12), and (4.6) we obtain for the energy eigenvalue of the polaron's ground state the expressions

$$\mathscr{E}_0 \equiv \mathscr{E}[\psi_0] = -0.164 \frac{m^* e^4}{\hbar^2 \varepsilon^{*2}}, \tag{4.16}$$

which can be compared with the ground term of the hydrogen atom $-0.5\ (me^4/\hbar^2)$, where $m$ is the free electron mass. Usually, $m^* \leqslant m$, and since $\varepsilon^* > 1$, $\mathscr{E}_0$ is substantially less than the ionization energy of the hydrogen atom (13.5 eV).

Making use of expression (4.15) for $\psi_0$ and of (4.6) for $\mathbf{D}[\psi]$, we can demonstrate that [5.4]

$$\mathscr{U}_{av}[\psi_0] = \mathscr{U}^0_{av} = (4/3)\,\mathscr{E}_0, \tag{4.17}$$

$$\mathscr{T}[\psi_0] = (1/3)\mathscr{E}_0. \tag{4.18}$$

It follows from (4.3) and (4.13) that

$$\mathscr{E}[\psi] = \mathscr{T} + \mathscr{U}_{av}, \tag{4.19}$$

$$\mathscr{T}[\psi] = \mathscr{T} + (1/2)\,\mathscr{U}_{av}, \tag{4.20}$$

where $\mathscr{T} \equiv \left(\frac{1}{2m^*}\right) \int (\nabla\psi)^2\, d\tau$ is the electron kinetic energy.

On the other hand, the energy of an inertially polarized crystal is equal to

$$\mathscr{U}_p = \int \frac{\mathbf{E}\cdot\mathbf{D}}{8\pi}\, d\tau = \frac{1}{8\pi\varepsilon^*} \int \mathbf{D}^2\, d\tau, \tag{4.21}$$

and this yields for the ground state

$$\mathscr{U}_p[\psi_0] = \mathscr{U}^0_p = -(2/3)\,\mathscr{E}_0. \tag{4.21a}$$

Thermal dissociation of a polaron is theresult of a thermal fluctuation of the polarization trough, and, therefore, the activation energy is equal to the sum of $-\mathscr{E}_0$ and $-\mathscr{U}^0_p$, i.e.,

$$W_t = -\mathscr{E}_0 - \mathscr{U}^0_p = -\mathscr{E}_0 + (2/3)\,\mathscr{E}_0 = -(1/3)\mathscr{E}_0 = -\mathscr{T}[\psi_0]. \tag{4.22}$$

The red photodissociation threshold of the polaron is determined by the energy required for the electron's transition from the ground

[8] In the following the index zero will be used to mark the quantities relating to the ground state of the polaron.

state $-\mathscr{E}_0$ to the lower edge of the conduction band; the ionic configuration remains unchanged (the Frank-Condon principle), i.e.,

$$W_{ph} = -\mathscr{E}_0 = 3W_t. \tag{4.22a}$$

It should be stressed that the relations deduced above

$$|W_t| \div |\mathscr{U}_p| \div |W_{ph}| \div |\mathscr{U}_{av}| = 1 \div 2 \div 3 \div 4$$

hold only for tight binding polarons.

The ground state of the tight binding polaron can be determined with the aid of the following simple method. The minimum kinetic energy of an electron of mass $m^*$ in a spherical cavity of radius $r$ with impenetrable walls is equal to[9]

$$\mathscr{T} = \frac{\pi^2\hbar^2}{2m^*r^2}. \tag{4.23}$$

The potential energy of the interaction of this electron with an inertially polarizable continuum is of the order of

$$\mathscr{U}_{av} \approx -\frac{e^2}{\varepsilon^* r}. \tag{4.24}$$

The minimum of the total energy $\mathscr{E} = \mathscr{T} + \mathscr{U}_{av}$ as a function of $r$ is

$$\mathscr{E}_0 = -\frac{1}{\pi^2}\frac{m^*e^4}{2\varepsilon^{*2}\hbar^2} = -0.051\frac{m^*e^4}{\varepsilon^{*2}\hbar^2} \tag{4.25}$$

for

$$r_0 = \pi^2\frac{\varepsilon^*\hbar^2}{m^*e^2}. \tag{4.26}$$

Comparing expressions (4.25) and (4.26) with expressions (4.16) and (4.15a), we see that the approximate estimate yields practically the same equations as the variational method (the only difference is in the numerical factors).

A necessary condition for the formation of tight binding polarons is that the electron's motion in the polaron trough be fast as compared with the motion of the heavy ions (adiabatic approximation), i.e.,

$$\frac{r_0}{v} \ll \frac{1}{\omega}, \tag{4.27}$$

where $v$ is the electron velocity and $\omega$ is the maximum vibration frequency of the ions of the optical branch. Here the de Broglie wavelength ($\lambda\!\!\!^- = \lambda/2\pi$) for the electron in the ground state is equal to

$$\lambda\!\!\!^- \approx r_0 \ll \frac{v}{\omega} \equiv l, \tag{4.28}$$

---

[9] See [5.1, p. 87].

where $l$ is the distance traveled by the electron in the time of one ionic vibration. The electron kinetic energy is

$$\mathcal{T}_0 = (1/2)\, m^* v^2. \tag{4.29}$$

Using (4.29), (4.23) (for $r = r_0$), and (4.26) to exclude $v$ and $r_0$ from the inequality (4.27), we obtain

$$\frac{1}{\pi^3} \frac{m^* e^4}{\varepsilon^{*2}\hbar^2} \frac{1}{\hbar\omega} = \frac{2}{\pi} \frac{|\mathcal{E}_0|}{\hbar\omega} \gg 1, \tag{4.30}$$

which coincides with the usual condition of applicability of the adiabatic approximation (the energy of the fast subsystem greatly exceeds that of the slow subsystem).

Introducing the dimensionless parameter

$$\alpha^2 = \frac{\pi^2 |\mathcal{E}_0|}{\hbar\omega} = \frac{m^* e^4}{2\varepsilon^{*2}\hbar^3\omega}, \tag{4.31}$$

we obtain for the energy of the ground state of tight binding polarons

$$\mathcal{E}_0 = -\frac{1}{\pi^2}\alpha^2\hbar\omega \approx -\frac{1}{10}\alpha^2\hbar\omega, \tag{4.32}$$

if

$$\alpha \gg 1. \tag{4.33}$$

Hence, in this case the energy is proportional to $\alpha^2$.

5.4.3 Consider now the other extreme case—weak binding polarons (G. Frelich, G. Peltzer, S. Sinau, 1950), when the parameter

$$\alpha \ll 1. \tag{4.34}$$

Determine for this case the effect of the interaction of the electron with the ionic crystal polarized by it on the electron's energy spectrum.

Consider long optical waves in a binary ionic crystal in the continuous approximation (Section 3.9.2). It follows from (3.9.37) that the electric field $\mathbf{E}$ in a crystal is generated only by longitudinal vibrations $\mathbf{w}_l$. This appears quite natural, since only longitudinal vibrations are accompanied by the variations of the crystal's volume resulting in the liberation of fixed electric charges in it.

Substituting $\mathbf{E}$ from (3.9.37) into (3.9.32), we obtain for the polarization vector

$$\mathbf{P} = \omega_0 \sqrt{\frac{\varepsilon_0 - \varepsilon_\infty}{4\pi}}\,\mathbf{w} - \frac{\varepsilon_\infty - 1}{4\pi\varepsilon_\infty}\,\omega_0 \sqrt{4\pi(\varepsilon_0 - \varepsilon_\infty)}\,\mathbf{w}_l$$
$$= \sqrt{\frac{N_0 m_r \omega_l^2}{4\pi\varepsilon^*}}\,(\mathbf{u}_+ - \mathbf{u}_-). \tag{4.35}$$

As we shall see below, only the longitudinal vibrations interact with the electron; therefore, actually in (4.35) $\mathbf{w}_l = \mathbf{w}$. Here $N_0$

is the number of crystal cells per unit volume, $m_r = m_+m_-/(m_+ + m_-)$ is the reduced mass of the ions, $\omega_l = (\varepsilon_0/\varepsilon_\infty)^{1/2}\,\omega_0$ is the maximum frequency of longitudinal optical vibrations, $\frac{1}{\varepsilon^*} = \frac{1}{\varepsilon_\infty} - \frac{1}{\varepsilon_0}$, where $\varepsilon_\infty$ and $\varepsilon_0$ are the high-frequency and the static dielectric constants, $\mathbf{w} = \sqrt{N_0 m_r}\,\mathbf{s} = \sqrt{N_0 m_r}\,(\mathbf{u}_+ - \mathbf{u}_-)$, where $\mathbf{s} = \mathbf{u}_+ - \mathbf{u}_-$ is the displacement of the positive ion with respect to the negative ion. Making $k = 1$ for the positive ion and $k = 2$ for the negative ion, we have from (3.6.4) in the continuous approximation ($\mathbf{a}_n = \mathbf{r}$)

$$\mathbf{u}^k(\mathbf{r}) = \frac{1}{\sqrt{N_0 m_k}} \sum_{\mathbf{q},\, j} \mathbf{e}_{jk}(\mathbf{q})\, a_j(\mathbf{q}, t)\, e^{i\mathbf{q}\cdot\mathbf{r}}. \tag{4.36}$$

For long optical waves ($\mathbf{q}\to 0$) the polarization vectors $\mathbf{e}_{jk}$ are real, and the ions vibrate about the static centre of gravity of the cell, i.e.,

$$\sqrt{m_1}\,\mathbf{e}_{j1} + \sqrt{m_2}\,\mathbf{e}_{j2} = 0, \tag{4.37}$$

as stipulated by (3.5.22).

On the other hand, the normalizing condition (3.6.2) is of the form

$$\mathbf{e}_{j1}^2 + \mathbf{e}_{j2}^2 = 1. \tag{4.38}$$

It follows from (4.37) and (4.38) that

$$\mathbf{e}_{j1} = \sqrt{\frac{m_2}{m_1 + m_2}}\,\mathbf{i}_j, \quad \mathbf{e}_{j2} = -\sqrt{\frac{m_1}{m_1 + m_2}}\,\mathbf{i}_j, \tag{4.39}$$

where $\mathbf{i}_j$ is a unit vector parallel to $\mathbf{e}_{j1}$.

It follows from (4.35), (4.36) and (4.39) that

$$\mathbf{P} = \sqrt{\frac{\omega_l^2}{4\pi\varepsilon^*}} \sum_{\mathbf{q},\, j} \mathbf{i}_j \left(a_{j\mathbf{q}} e^{i\mathbf{q}\cdot\mathbf{r}} + a_{j\mathbf{q}}^* e^{-j\mathbf{q}\cdot\mathbf{r}}\right), \tag{4.40}$$

where we substituted the summation over $\mathbf{q}$ in the term $k = 2$ for the summation over $-\mathbf{q}$ and employed (3.6.5).

The Poisson equation for a scalar potential $\Phi$ related to the crystal's vibrations is of the form

$$\nabla^2\Phi = -4\pi\rho = 4\pi\,\mathrm{div}\,\mathbf{P}$$

$$= i\sqrt{\frac{4\pi\omega_l^2}{\varepsilon^*}} \sum_{\mathbf{q},\, j} (\mathbf{i}_j\cdot\mathbf{q}) \left(a_{j\mathbf{q}} e^{i\mathbf{q}\cdot\mathbf{r}} - a_{j\mathbf{q}}^* e^{-i\mathbf{q}\cdot\mathbf{r}}\right), \tag{4.41}$$

where the fixed charge $\rho = -\mathrm{div}\,\mathbf{P}$ [5.6, Chap. 2]. We see from this expression that the electric field is generated only by longitudinal vibrations for which $\mathbf{i}_j \parallel \mathbf{q}$, and, therefore, $\mathbf{i}_j \cdot \mathbf{q} = q$ (in the following we omit the index $j$ for longitudinal vibrations).

The solution of equation (4.41) is

$$\Phi = -i \sqrt{\frac{4\pi\omega_l^2}{\varepsilon^*}} \sum_{\mathbf{q}} \frac{1}{q} (a_q e^{i\mathbf{q}\cdot\mathbf{r}} - a_q^* e^{-i\mathbf{q}\cdot\mathbf{r}}). \tag{4.42}$$

This can be checked by direct substitution of (4.42) into (4.41). Since the expression in the parentheses is purely imaginary, the potential $\Phi$ is, as it should be, real.

To determine the probability of an electron transition involving the absorption or the emission of a phonon, we should calculate the appropriate transition matrix element from the perturbation energy $(-e\Phi)$. To this end it suffices to describe the electron state by a plane wave

$$\psi_{\mathbf{k}}(\mathbf{r}) = \frac{1}{\sqrt{V}} e^{i\mathbf{k}\cdot\mathbf{r}}, \tag{4.43}$$

where $\mathbf{k}$ is the electron's wave vector and $V = 1$ cm$^3$ is the volume of the crystal's principal region.

In this matrix element $\langle N'_{\mathbf{q}}, \mathbf{k}' \mid -e\Phi \mid N_{\mathbf{q}}, \mathbf{k}\rangle$ we can perform separately the integration with respect to the electron's coordinates and obtain for the first exponent in the parentheses of (4.42)

$$\frac{1}{V} \int_V e^{i(\mathbf{k}+\mathbf{q}-\mathbf{k}')\cdot\mathbf{r}}\, d\mathbf{r} = \delta_{\mathbf{k}',\, \mathbf{k}+\mathbf{q}},$$

this being the expression of the electron's wave vector conservation law in the process of phonon absorption:

$$\mathbf{k}' = \mathbf{k} + \mathbf{q}. \tag{4.44}$$

For the second exponent in the parentheses involving the emission of a phonon we obtain

$$\mathbf{k}' = \mathbf{k} - \mathbf{q}. \tag{4.44a}$$

The matrix elements from $a_q$ and $a_q^*$ involving the absorption and the emission of a phonon are, according to (3.10.25), (3.10.26) equal to

$$\langle N'_q | a_q | N_q \rangle = \sqrt{\frac{\hbar N_q}{2\omega_l}}, \quad \text{if } N'_{\mathbf{q}} = N_{\mathbf{q}} - 1 \tag{4.45}$$

and

$$\langle N'_q | a_q^* | N_q \rangle = \sqrt{\frac{\hbar (N_q + 1)}{2\omega_l}}, \quad \text{if } N'_{\mathbf{q}} = N_{\mathbf{q}} + 1. \tag{4.45a}$$

Hence, we obtain for the matrix element of interest

$$\langle N'_q,\ \mathbf{k}'\,|-e\Phi|\,N_{\mathbf{q}},\ \mathbf{k}\rangle=\langle N_{\mathbf{q}}\mp 1,\ \mathbf{k}\pm\mathbf{q}\,|-e\Phi|\,N_{\mathbf{q}},\ \mathbf{k}\rangle$$
$$=\pm i\sqrt{\frac{4\pi e^2\hbar\omega_l}{\varepsilon^*}}\,\frac{1}{q}\cdot\left\{\begin{matrix}\sqrt{N_q}\\ \sqrt{N_q+1}\end{matrix}\right., \tag{4.46}$$

where the upper signs (and the row) correspond to the absorption, and the lower to the emission of a phonon.

The interaction of the electron with the crystal polarized by it changes its energy. In the case of weak interaction ($\alpha \ll 1$) this change can be determined with the aid of the quantum-mechanics perturbation theory. The change in the energy eigenvalue of the $n$-th state in the first and the second approximations with respect to the perturbation energy $\mathscr{H}'$ is equal to [5.7, § 38]

$$\mathscr{E}_n-\mathscr{E}_n^{(0)}=\langle n\,|\,\mathscr{H}'\,|\,n\rangle+\sum_{m\neq n}\frac{|\langle m|\mathscr{H}'|n\rangle|^2}{\mathscr{E}_n^{(0)}-\mathscr{E}_m^{(0)}}. \tag{4.47}$$

Here $\mathscr{E}_n^{(0)}$ and $\mathscr{E}_m^{(0)}$ are the unperturbed energies of the $n$-th and $m$-th states.

Consider the change in the energy of a free electron $\varepsilon^{(0)} = \hbar^2k^2/2m^*$ caused by its interaction with a polarized crystal. If the crystal's absolute temperature is zero ($N_{\mathbf{q}} = 0$), the electron can only emit phonons ($N'_{\mathbf{q}} = N_{\mathbf{q}} + 1 = 1$; $\mathbf{k}' = \mathbf{k} - \mathbf{q}$). Since the matrix element (4.46) for $N'_{\mathbf{q}} = N_{\mathbf{q}}$ is zero ($\langle n \mid \mathscr{H}' \mid n\rangle = 0$), the electron energy experiences a change only in the second approximation of the perturbation theory. It follows from (4.47) and (4.46) that

$$\varepsilon(\mathbf{k})-\varepsilon^{(0)}=\sum_{\mathbf{q}}\frac{|\langle 1,\ \mathbf{k}-\mathbf{q}\,|-e\Phi\,|0,\ \mathbf{k}\rangle|^2}{\dfrac{\hbar^2k^2}{2m^*}-\dfrac{\hbar^2(\mathbf{k}-\mathbf{q})^2}{2m^*}-\hbar\omega_l}$$
$$=-\frac{2\pi e^2\hbar\omega_l}{\varepsilon^*}\int\frac{d\tau_q}{(2\pi)^3}\,\frac{1}{q^2}\,\frac{1}{\hbar\omega_l-\dfrac{\hbar^2}{2m^*}(2\mathbf{kq}-\mathbf{q}^2)}. \tag{4.48}$$

where we have replaced summation by integration. A photon $\hbar\omega_l$ and a wave vector $\mathbf{k}' = \mathbf{k} - \mathbf{q}$ correspond to the intermediate state $m$; we would like to remind the reader that the energy conservation law, generally, does not hold for the transition from the ground state $n$ to the intermediate state $m$.

Introduce in the integral polar coordinates with the axes directed along $\mathbf{k}$ ($d\tau_{\mathbf{q}} = 2\pi q^2\, dq \sin\vartheta\, d\vartheta$; $\mathbf{kq} = kq\cos\vartheta$) and expand the integrand in a series in the powers of $k$ up to $k^2$, inclusive. The integration of the first addend (independent of $k$) with respect to $\vartheta$ yields the factor 2, of the second addend (proportional to $k$) the factor zero, and of the third addend (proportional to $k^2$) the factor 2/3. As a result, the integrals of the first and of the second addends with respect to $q$ from 0 to $\infty$ reduce to integrals of rational fractions; they can be easily calculated or taken from the tables [5.8, p. 934].

Accordingly, we obtain from (4.48)

$$\varepsilon(k)=\frac{\hbar^2k^2}{2m^*}-\alpha\left(\hbar\omega_l+\frac{\hbar k^2}{2m^*}\right)=-\alpha\hbar\omega_l+\frac{\hbar^2k^2}{2m_{\text{pol}}}, \tag{4.49}$$

where the effective mass of the polaron is

$$m_{\text{pol}}=\frac{m^*}{1-\frac{\alpha}{6}}\approx m^*\left(1+\frac{\alpha}{6}\right). \tag{4.49a}$$

Here $\alpha$ is the parameter introduced in (4.31).

We see that for a weak binding polaron the energy level of the ground state is reduced by the amount $\alpha\hbar\omega_l$, its effective mass $m_{\text{pol}}$ increasing as the ratio $\frac{m_{\text{pol}}}{m^*}\approx 1+\frac{\alpha}{6}$.

Typical values of $\alpha$ for semiconductors with partially ionic bonds are less than unity. For the $A^{III}B^{V}$ compounds $\alpha$ lies within the limits of 0.015 (InSb) to 0.080 (InP), for the $A^{II}B^{VI}$ compounds from 0.39 (CdTe) to 0.65 (CdS). In calculations of $\alpha$ for those compounds the effective masses of the electrons $m^*$ were assumed to be small. For alkali-haloid crystals $\alpha > 1$; for instance, in LiI $\alpha = 2.4$ and in RbBr $\alpha = 6.6$. In calculations the effective electron mass was assumed to be equal to that of the free electron ($m = 0.9 \times \times 10^{-27}$ g). Since in both the case of tight and weak binding we made use of the continuous approximation, the above considerations apply only to polarons of large radius, i.e., when the polaron's radius exceeds (to be precise, greatly exceeds) the lattice parameter.

We do not consider here the theory of polarons of small radius when it is equal to or less than the lattice parameter [5.9].

# 6. Electric, Thermal, and Magnetic Properties of Solids

## 6.1 Metals, Dielectrics, and Semiconductors

In Chapter 4 we discussed the behaviour of an isolated electron in the periodic field of an ideal crystal. It was demonstrated that the electron in a stationary state possesses a nondecaying average velocity (4.3.28), i.e., that it moves freely throughout the crystal. At first glance, all materials should be excellent conductors (metals) with the conduction electrons equal in number to the total number of electrons in the body. Actually, even in metals the number of conduction electrons is much less—of the order of the number of atoms, and in dielectrics it is zero (at absolute zero).

To understand such a situation consider the collection of all the $NsZ$ electrons of a solid, where $N$ is the number of crystal cells, $s$ is the number of atoms per crystal cell, and $Z$ is the number of atomic electrons (the atomic number).[1] We shall take account of the interaction of electrons only through the self-consistent field, but we shall comply with the Pauli exclusion principle, which stipulates that in one quantum state (characterized by the wave vector **k**) there can be not more than two electrons and only with the oppositely directed spins. At absolute zero, when the system is in its lowest energy state, the electrons in the solid must occupy the lowest $NsZ/2$ states discussed in detail in Chapter 4. Here two cases are possible: either the highest level occupied by the electrons coincides with the upper edge of one of the allowed energy bands (Fig. 6.1*a*) or it finds itself inside such a band (Fig. 6.1*b*). In the latter case, if even a weak electric field is applied to the body, the electrons occupying the levels close to the band edge $\varepsilon_0$ would be accelerated and go over to other higher quantum states continuously adjoining $\varepsilon_0$, unoccupied by other electrons. This will create a difference in the numbers of electrons moving in the direction of the field and opposite to it—an electric current will appear; the body will behave like a metal. In the former case, when the electrons completely fill the upper energy band, the electric field is unable to bring about a redistribution of electrons, at least as long as it is not very high (less than $10^6$ V/cm) and does not cause

[1] For the sake of simplicity, we consider a simple (monatomic) substance.

electron transitions to higher allowed energy bands (dielectric breakdown). In this case there can be no electric current, and the body behaves like a dielectric (an insulator). The electrons of the deeper completely filled bands in metals behave in a similar fashion. Such a simple qualitative explanation of the difference between a metal and a dielectric was a great triumph for the energy band theory.

Every separate energy band has $2N$ quantum states; therefore, for a simple crystal (with $s = 1$) the number of bands filled with electrons is equal to $NZ : 2N = Z/2$. Hence, if $Z$ is odd, the number of the filled bands is not an integer, i.e., the elements with an odd atomic number that crystallize in a simple lattice are metals. The reciprocal conclusion that the elements with an even $Z$ (for $s = 1$) behave like dielectrics is wrong, since there is always the possibility that the energy bands may overlap (Section 4.5), the body behaving in this case like a metal. A more complex lattice structure (lower symmetry) and especially the presence of atoms of different kinds facilitate the separation of the energy bands, i.e., the formation of a dielectric. Solid hydrogen is a dielectric, although at first glance it ought to behave like an alkali metal. Such behaviour of hydrogen and of some other elements is due to the fact that in the process of crystallization the molecules are bonded together in the lattice by van der Waals forces ($s \neq 1$). From the energy band viewpoint we could obviously expect in this case the energy band to be completely filled with electrons. Hence, at absolute zero all solids behave either like metals or like dielectrics.

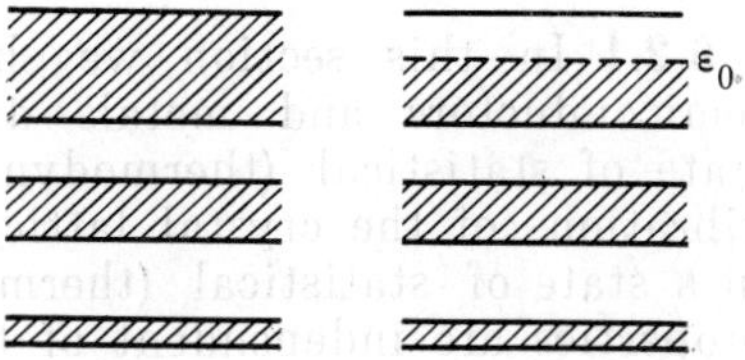

(a) (b)

Fig. 6.1

In the case of a dielectric in which the electrons completely fill the top band (the valence band) separated from the next allowed energy band (the conduction band) by a forbidden band of width $\varepsilon_G$, the electrons, as the temperature is raised, will start going over from the valence band to the conduction band. If $\varepsilon_G$ is of the order of 1 eV or less, pure crystals will at room temperature and above exhibit appreciable conductivity resulting from the motion of the electrons in the conduction band and of the holes in the valence band.

Such pure materials with a narrow forbidden band became known as *intrinsic semiconductors*. In practice, we have to deal more with *impurity semiconductors*, in which donors act as the suppliers of electrons for the conduction band and acceptors as the suppliers of holes for the valence band (Section 5.2.1).

By adding different impurity atoms in different quantities to the pure material, it is possible to obtain semiconductors with a great variety of electrical properties. Such processes of doping a semiconductor with different impurities were studied most extensively for germanium and silicon which gained widespread use in recent years.

## 6.2 Statistical Equilibrium of Free Electrons in Semiconductors and Metals

6.2.1 In this section we shall consider certain properties of semiconductors and metals due to free electrons (holes) in the state of statistical (thermodynamic) equilibrium with the thermal vibrations of the crystal lattice. An important feature of systems in a state of statistical (thermodynamic) equilibrium is that their properties are independent of the mechanism that establishes this equilibrium. Accordingly, in our case there is no need to consider the specific mechanism of interaction of free electrons and holes with the lattice vibrations and the processes of thermal excitation and recombination of electrons (holes). As we shall see in the following sections, those mechanisms are essential in the treatment of kinetic phenomena, i.e., of the processes of electric and of thermal conductivities, of galvano- and thermomagnetic phenomena, etc.

In the single-electron approximation the interaction between the electrons of a crystal is accounted for only by the self-consistent field in which every electron moves independently of the other electrons. From the statistical point of view the single-electron approximation corresponds to the ideal gas model. In the state of statistical equilibrium the ideal electron gas is described by the Fermi-Dirac statistics. In statistical equilibrium the average number of electrons in a definite quantum state characterized by three quantum numbers[2] $k_1$, $k_2$, $k_3$ with the energy $\varepsilon_{\mathbf{k}}$ at temperature $T$ is equal to [6.1, Chap. IX, § 2]

$$f_0(\varepsilon_{\mathbf{k}}) = \frac{1}{\exp\dfrac{\varepsilon_{\mathbf{k}}-\zeta}{k_0T}+1}. \tag{2.1}$$

Here $\zeta$ is the chemical potential per electron, $k_0$ is the Boltzmann constant. The function $f_0(\varepsilon_{\mathbf{k}})$ is called the *Fermi-Dirac distribution function*.

The total number of electrons in the volume $V$ is equal to

$$N = \sum_{\mathbf{k}} f_0(\varepsilon_{\mathbf{k}}) = \sum_{k} \frac{1}{\exp\dfrac{\varepsilon_{\mathbf{k}}-\zeta}{k_0T}+1}, \tag{2.2}$$

[2] $k_1$, $k_2$, $k_3$ are three discrete or quasi-discrete quantum numbers characterizing the orbital motion of the electron; very often we shall take them to mean the three components of the electron's wave vector $\mathbf{k}$ ($k_1 = k_x$, $k_2 = k_y$, $k_3 = k_z$). The quantum numbers of an atomic electron corresponding to them will be the principal quantum number $n$, the azimuthal quantum number $l$ and the magnetic quantum number $m$.

where the summation is performed over all the electron states in the volume $V$ with account taken of the spins, i.e., of the fact that every orbital quantum state (in accordance with the Pauli exclusion principle) can be occupied by two electrons with opposite spins.

Equation (2.2) defines the chemical potential $\zeta$ as a function of the electron concentration $n = N/V$ and of temperature $T$. If the number of electron quantum states (without spin) per $cm^3$ in the energy interval $d\varepsilon$ is $g(\varepsilon)\, d\varepsilon$, then the number of electrons with the energy in the range $\varepsilon$, $\varepsilon + d\varepsilon$ in a unit of volume is equal to

$$n(\varepsilon)\, d\varepsilon = 2 f_0(\varepsilon)\, g(\varepsilon)\, d\varepsilon. \tag{2.3}$$

If the state of the electrons in a crystal is characterized by the wave vector $\mathbf{k}$ ($k_1 = k_x$, $k_2 = k_y$, $k_3 = k_z$), then in certain cases it is more convenient to use the number of quantum states of an electron the end of whose wave vector $\mathbf{k}$ lies inside the element $d\tau_k$ of the $\mathbf{k}$-space instead of the function $g(\varepsilon)$. This number of quantum states per 1 $cm^3$, according to (3.5.27), is equal to $d\tau_k/(2\pi)^3$.[3] The number of electrons whose wave vector components lie inside the intervals $k_x$, $k_x + dk_x$; $k_y$, $k_y + dk_y$; $k_z$, $k_z + dk_z$ is equal to

$$n(\mathbf{k})\, dk_x\, dk_y\, dk_z = 2 f_0(\mathbf{k}) \frac{dk_x\, dk_y\, dk_z}{(2\pi)^3}. \tag{2.3a}$$

Here $f_0(\mathbf{k})$ is the distribution function (2.1) in which the energy $\varepsilon_{\mathbf{k}} = \varepsilon(\mathbf{k})$ is expressed in terms of the wave vector $\mathbf{k}$ (in the approximation of a scalar effective mass $\varepsilon(\mathbf{k}) = \hbar^2 k^2/2m^* = p^2/2m^*$).

We have seen that $g(\varepsilon)$ for electrons in a periodic field is determined by expression (4.3.25). If the electron energy is $\varepsilon = \frac{1}{2m^*} p^2$, then

$$g(\varepsilon) = \frac{\sqrt{2}}{2\pi^2} \frac{m^{*3/2}}{\hbar^3} \sqrt{\varepsilon}. \tag{2.4}$$

Hence, in this case the electron concentration is

$$n = 2 \int_0^\infty f_0(\varepsilon)\, g(\varepsilon)\, d\varepsilon = \frac{\sqrt{2}}{\pi^2} \frac{m^{*3/2}}{\hbar^3} \int_0^\infty \frac{\varepsilon^{1/2}\, d\varepsilon}{\exp \frac{\varepsilon - \zeta}{k_0 T} + 1}$$

$$= \frac{\sqrt{2}}{\pi^2} \frac{(m^* k_0 T)^{3/2}}{\hbar^3} \int_0^\infty \frac{x^{1/2}\, dx}{e^{x-z} + 1}, \tag{2.5}$$

where $x = \varepsilon/k_0 T$ and $z = \zeta/k_0 T$. This expression defines $\zeta$ as a function of $n$, $T$ and $m^*$.

---

[3] In the quasi-momentum space $\mathbf{p} = \hbar \mathbf{k}$. The corresponding number of quantum states differs by the factor $\hbar^{-3}$.

The integral on the right-hand side of (2.5) cannot be expressed in terms of elementary functions of $z$. The integrals of the type

$$\mathcal{F}_n(z)=\int_0^\infty \frac{x^n\,dx}{e^{x-z}+1} \tag{2.6}$$

are frequently used in the electron theory of crystals [in equation (2.5) $n=1/2$]. For $n=-1/2$, $1/2$, $3/2$ integrals (2.6) were tabulated by McDougall and Stoner [6.2]; for $n=5/2$, $7/2$, $9/2$, $11/2$ they were tabulated by Beer *et al.* [6.3], and for $n=1$, $2$, $3$, $4$ they were tabulated by Rhodes [6.4].

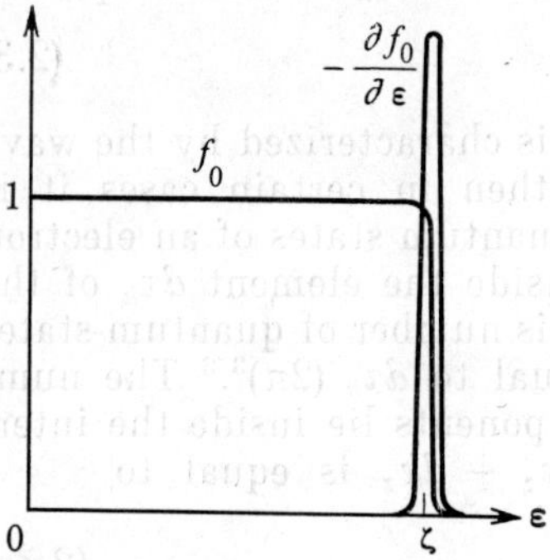

Fig. 6.2

Consider two important limiting cases in which the integral in expression (2.5) can be easily evaluated.

6.2.2 Consider the case of a strongly degenerate electron gas when $z \gg 1$, i.e., when the chemical potential $\zeta \gg k_0T$. As we shall see below, this case is realized in the metals in which the concentration of conduction electrons $n \approx 10^{22}$ cm$^{-3}$. In this case obviously $f_0(\varepsilon) \approx 1$ for $\varepsilon \ll \zeta$, and $f_0(\varepsilon) \approx 0$ for $\varepsilon \gg \zeta$; a sharp drop in $f_0(\varepsilon)$ takes place near the point $\varepsilon=\zeta$ in an interval of the order of $k_0T$.

The functions $f_0(\varepsilon)$ and $\left(-\frac{\partial f_0}{\partial \varepsilon}\right)$ are shown in Fig. 6.2. The lower the temperature, the closer is $\left(-\frac{\partial f_0}{\partial \varepsilon}\right)$ to the delta function. Indeed, $\left(-\frac{\partial f_0}{\partial \varepsilon}\right)$ is nonzero only near the point $\varepsilon=\zeta$, and

$$\int_0^\infty \left(-\frac{\partial f_0}{\partial \varepsilon}\right) d\varepsilon = -\int_0^\infty df_0 = f_0(0)-f_0(\infty)=1,$$

since $f_0(0)=1$ and $f_0(\infty)=0$.

Hence, $\left(-\frac{\partial f_0}{\partial \varepsilon}\right)=\delta(\varepsilon-\zeta_0)$, where $\zeta_0$ is the value of $\zeta$ for $T=0$.

Calculate the integral

$$I=\int_0^\infty \chi(\varepsilon)\, f_0(\varepsilon)\, d\varepsilon=\int_0^\infty f_0(\varepsilon)\, d\varphi(\varepsilon)$$

$$=f_0(\varepsilon)\,\varphi(\varepsilon)\Big|_0^\infty-\int_0^\infty \varphi(\varepsilon)\,\frac{\partial f_0}{\partial \varepsilon}\, d\varepsilon$$

$$= -\varphi(0) + \int_0^\infty \varphi(\varepsilon)\left(-\frac{\partial f_0}{\partial \varepsilon}\right) d\varepsilon, \tag{2.7}$$

where $\chi(\varepsilon)$ is an arbitrary function for the case of strong degeneracy. Here the first stage involves the transformation $\chi(\varepsilon)\,d\varepsilon = d\varphi(\varepsilon)$ typical of integration by parts. As we shall see below, for all practical cases $\varphi(0) = 0$; therefore,

$$I = \int_0^\infty \varphi(\varepsilon)\left(-\frac{\partial f_0}{\partial \varepsilon}\right) d\varepsilon. \tag{2.7a}$$

Make use of this equation to calculate expression (2.5) in the approximation $\left(-\frac{\partial f_0}{\partial \varepsilon}\right) \approx \delta(\varepsilon - \zeta_0)$. Since in (2.5) $\chi(\varepsilon) = 2g(\varepsilon)$, it follows that $\varphi(\varepsilon) = \frac{\sqrt{2}}{\pi^2}\frac{m^{*3/2}}{\hbar^3}\frac{2}{3}\varepsilon^{2/2}$. Hence,

$$n = \frac{\sqrt{2}}{\pi^2}\frac{m^{*3/2}}{\hbar^3}\frac{2}{3}\int_0^\infty \varepsilon^{3/2}\delta(\varepsilon - \zeta_0)\,d\varepsilon = \frac{(2m^*)^{3/2}}{2\pi^2\hbar^3}\zeta_0^{3/2}, \tag{2.8}$$

whence the chemical potential

$$\zeta_0 = \frac{\hbar^2}{2m^*}(3\pi^2 n)^{2/3} = \frac{(2\pi\hbar)^2}{2m^*}\left(\frac{3n}{8\pi}\right)^{2/3}, \tag{2.8a}$$

this precisely coinciding with the maximum kinetic energy $\varepsilon_0$ of the electrons of an ideal Fermi gas at absolute zero (Appendix 4).

In the next approximation (Appendix 17) the chemical potential is

$$\zeta = \zeta_0\left[1 - \frac{\pi^2}{12}\left(\frac{k_0 T}{\zeta_0}\right)^2\right]. \tag{2.9}$$

Making use of the zeroth approximation for the chemical potential, write the condition for strong degeneracy in explicit form:

$$\frac{\zeta_0}{k_0 T} = \frac{\hbar^2 (3\pi^2 n)^{2/3}}{2m^* k_0 T} \gg 1. \tag{2.10}$$

As can be seen from the distribution function $f_0(\varepsilon)$, the condition of degeneracy (2.10) is actually of an exponential type, i.e., $\exp(\zeta_0/k_0 T) \gg 1$; therefore, if $\zeta_0/k_0 T \approx 5 \div 7$, the degeneracy can be regarded as strong.

We see from (2.10) that high concentration $n$, small effective mass $m^*$, and low temperature $T$ facilitate degeneracy. For a typical metal with $n \approx 10^{22}\ \text{cm}^{-3}$, $m^* \approx 10^{-27}$ g at room temperature $\zeta_0/k_0 T \approx 10^2$, i.e., the degeneracy is very strong. The free electrons of a metal will easily be seen to remain in a state of strong degeneracy up to its melting point.

**6.2.3** We have considered the case of a positive chemical potential $\zeta$ satisfying the inequality $\exp(\zeta/k_0T) \gg 1$ (the degeneracy condition). Consider now an opposite case—that of a negative chemical potential satisfying the inequality $\exp(-\zeta/k_0T) \gg 1$. In this case the distribution function

$$f_0(\varepsilon) = \frac{1}{\exp[(\varepsilon-\zeta)/k_0T]+1} \approx \exp(\zeta/k_0T)\exp(-\varepsilon/k_0T) = A\exp(-\varepsilon/k_0T), \tag{2.11}$$

i.e., it reduces to the Maxwell-Boltzmann distribution with the normalizing constant $A \equiv \exp(\zeta/k_0T)$. The electron concentration in the conduction band, in accordance with (2.5), is

$$n = \frac{\sqrt{2}}{\pi^2}\frac{m^{*3/2}}{\hbar^3}\int_0^\infty A\exp(-\varepsilon/k_0T)\,\varepsilon^{1/2}\,d\varepsilon = \frac{(2\pi m^* k_0T)^{3/2}}{4\pi^3\hbar^3}A. \tag{2.12}$$

When we calculated the integral, we introduced the variable $x=\varepsilon^{1/2}$ and made use of equation (A.7.2). It follows from equation (2.12) that

$$A = \exp(\zeta/k_0T) = \frac{4\pi^3\hbar^3 n}{(2\pi m^* k_0T)^{3/2}}, \tag{2.12a}$$

$$\zeta = k_0T\ln\left[\frac{4\pi^3\hbar^3 n}{(2\pi m^* k_0T)^{3/2}}\right]. \tag{2.12b}$$

In contrast to the degenerate case (2.9), the chemical potential $\zeta$ in classical statistics will be seen from (2.12b) to depend fairly strongly on temperature.

Now the distribution function (2.11) can be represented in the form

$$f_0(\varepsilon) = \frac{4\pi^3\hbar^3 n}{(2\pi m^* k_0T)^{3/2}}\exp(-\varepsilon/k_0T). \tag{2.11a}$$

The applicability criterion for the classical statistics is

$$\exp\left(-\frac{\zeta}{k_0T}\right) = \frac{1}{A} = \frac{(2\pi m^* k_0T)^{3/2}}{4\pi^3\hbar^3 n} \gg 1, \tag{2.13}$$

which agrees with (2.10).

We see that the factors facilitating the application of classical statistics are low concentration $n$, high temperature $T$, and large effective mass $m^*$. For $n \approx 10^{17}$ cm$^{-3}$, $m^* \approx 10^{-27}$ g at room temperature $\frac{1}{A} \approx 300$, i.e., the criterion (2.13) is met with a margin.

The maximum concentration corresponding to $\frac{1}{A} \approx 1$ is $n \approx$ $\approx 10^{19}$ cm$^{-3}$.

**6.2.4** Normally in semiconductors the electron concentration in the conduction band is itself a function of temperature. The reason is thermal excitation of electrons on the impurity levels and in the valence band. The equations obtained above remain, of course, valid but of little use, because the explicit dependence $n(T)$ is unknown to us. Moreover, the excitation of electrons in the valence band and their transition to the conduction band leaves positively charged holes in the valence band, which together with electrons take part in transport phenomena. For this reason there should be another approach to the problem. Figure 6.3 shows the energy level diagram of an impurity semiconductor with the forbidden bandwidth $\varepsilon_G$ and with donor and acceptor levels below the bottom edge of the conduction band, $\varepsilon_D$ and $\varepsilon_A$, respectively. We shall measure all the energies from the zero level coinciding with the bottom edge of the conduction band. The probability that a quantum state with the energy $\varepsilon$ is unoccupied by an electron, i.e., is a hole by definition, is

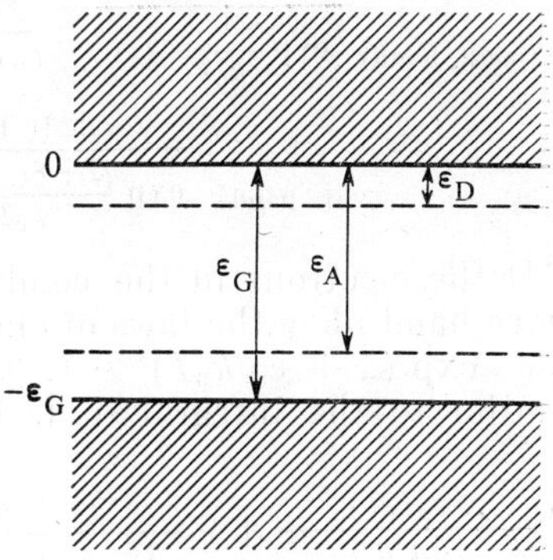


Fig. 6.3

$$f_0'(\varepsilon) = 1 - f_0(\varepsilon) = 1 - \frac{1}{\exp\frac{\varepsilon-\zeta}{k_0T}+1} = \frac{1}{\exp\frac{\zeta-\varepsilon}{k_0T}+1}, \tag{2.14}$$

where $f_0'(\varepsilon)$ is the Fermi-Dirac distribution function for holes.

If all the energies are measured from the bottom edge of the conduction band, the electron's energy in the conduction band is $\varepsilon =$ $= \hbar^2k^2/2m_n$, on the donor level $\varepsilon = -\varepsilon_D$, on the acceptor level $\varepsilon = -\varepsilon_A$, and in the valence band $\varepsilon = -\varepsilon_G - \varepsilon'$, where $\varepsilon' = \hbar^2k'^2/2m_p$ is the "kinetic" energy of the hole and $\varepsilon_G$ is the forbidden bandwidth.

Introducing the "chemical potential for holes" $\zeta' = -\varepsilon_G - \zeta$, we can write the hole distribution function (2.14) in the form

$$f_0'(\varepsilon) = \frac{1}{\exp\frac{\varepsilon'-\zeta'}{k_0T}+1}, \tag{2.14a}$$

by analogy with the electron distribution function (2.1).

The condition of electric neutrality for a semiconductor, from which its chemical potential $\zeta$ is determined, can be expressed in the following form: (the number of electrons in the conduction band) + (the number of electrons on acceptor levels) = (the number of holes in the valence band) + (the number of holes on the donor levels), i.e.,

$$\int\limits_{\text{(cond. band)}} \frac{2g(\varepsilon)\,d\varepsilon}{\exp\dfrac{\varepsilon-\zeta}{k_0T}+1}+\sum_{(A)}\frac{1}{\exp\dfrac{-\varepsilon_A-\zeta}{k_0T}+1}$$

$$=\int\limits_{\text{(val. band)}} \frac{2g(\varepsilon')\,d\varepsilon'}{\exp\dfrac{\zeta+\varepsilon_G+\varepsilon'}{k_0T}+1}+\sum_{(D)}\frac{1}{\exp\dfrac{\zeta+\varepsilon_D}{k_0T}+1}. \qquad (2.15)$$

If the electrons in the conduction band and the holes in the valence band obey the laws of classical statistics, $A\equiv\exp(\zeta/k_0T)\ll 1$ and $\exp(\zeta+\varepsilon_G/k_0T)\gg 1$, then the integrals in (2.15) will be similar to the integral in (2.12), so that

$$\frac{(2\pi m_n k_0T)^{3/2}}{4\pi^3\hbar^3}A+\frac{n_A}{\dfrac{1}{A}\exp\left(-\dfrac{\varepsilon_A}{k_0T}\right)+1}$$

$$=\frac{(2\pi m_p k_0T)^{3/2}}{4\pi^3\hbar^3}\frac{1}{A}\exp\left(-\frac{\varepsilon_G}{k_0T}\right)+\frac{n_D}{A\exp\left(+\dfrac{\varepsilon_D}{k_0T}\right)+1}. \qquad (2.15a)$$

Here $n_A$ and $n_D$ are acceptor and donor concentrations. It will be seen from (2.15a) that the electron concentration $n$ in the conduction band and the hole concentration $p$ in the valence band are

$$n=\frac{(2\pi m_n k_0T)^{3/2}}{4\pi^3\hbar^3}\exp\frac{\zeta}{k_0T}, \qquad (2.15b)$$

$$p=\frac{(2\pi m_p k_0T)^{3/2}}{4\pi^3\hbar^3}\exp\left(-\frac{\varepsilon_G+\zeta}{k_0T}\right). \qquad (2.15c)$$

The electric neutrality equation (2.15a) determines the quantity $A$ or the chemical potential $\zeta = k_0T\ln A$. Generally, this is an algebraic equation of the fourth degree in $A$. There are effective numerical and graphical methods for its solution. We shall limit ourselves to several specific but important cases in which the solution can be obtained in the form of simple equations.

1) Intrinsic semiconductor ($n_A = n_D = 0$).

We obtain from (2.15a)

$$A=\left(\frac{m_p}{m_n}\right)^{3/4}\exp\left(-\frac{\varepsilon_G}{2k_0T}\right), \qquad (2.16)$$

whence

$$\zeta = -\frac{\varepsilon_G}{2} + \frac{3}{4} k_0 T \ln\left(\frac{m_p}{m_n}\right). \tag{2.16a}$$

Since the factor $\frac{3}{4} \ln\left(\frac{m_p}{m_n}\right) \sim 1$, the second term in (2.16a) is of the order of $k_0T$, and therefore to within this quantity the chemical potential in an intrinsic semiconductor (dielectric) coincides with the middle of the forbidden band. If $m_p = m_n$, the above statement is accurate, and $\zeta$ is independent of temperature (as far as we ignore the temperature dependence of the forbidden bandwidth $\varepsilon_G$). The concentrations of the electrons in the conduction band and of the holes in the valence band are identical and

$$n_i = \frac{\left(2\pi \sqrt{m_n m_p}\, k_0 T\right)^{3/2}}{4\pi^3 \hbar^3} \exp\left(-\frac{\varepsilon_G}{2k_0T}\right) = n_T^0 \exp\left(-\frac{\varepsilon_G}{2k_0T}\right). \tag{2.16b}$$

The pre-exponential factor $n_T^0$ will easily be seen to be of the order of magnitude of the concentration of one particle per volume $\lambda^3$, where $\lambda$ is the de Broglie wavelength of an electron (hole) moving with thermal velocity. For $m_n = m_p = m = 0.9 \times 10^{-27}$ g and $T = 294$ K, $n_T^0 = 2.44 \times 10^{19}$ cm$^{-3}$. The concentration $n_T^0$ rises infinitely as $T \to \infty$ and is not limited by the number of electrons in the valence band because the finite width of the valence band is never taken into account when deriving (2.15a). The normalization condition in the case of an intrinsic semiconductor is limited to the statement that the number of electrons in the conduction band is equal to the number of vacant places (holes) in the valence band.

2) Impurity (extrinsic) semiconductor (for example, of the donor (or $n$-) type; $n_A = 0$) with a wide forbidden band ($\varepsilon_G \gg \varepsilon_D$).

From (2.15a) we have

$$\frac{(2\pi m_n k_0 T)^{3/2}}{4\pi^3\hbar^3} A\,[A \exp(\varepsilon_D/k_0T) + 1] = n_D, \tag{2.17}$$

where we neglected the term proportional to $\exp(-\varepsilon_G/k_0T)$ ($\varepsilon_G \gg \gg \varepsilon_D$!).

Consider the two following cases.

a) $A \exp \varepsilon_D/k_0T \gg 1$.

Since $A \ll 1$, this can only be when $\varepsilon_D \gg k_0T$. Neglecting unity in (2.17), we obtain

$$A = \frac{(2\pi\hbar)^{3/2} n_D^{1/2}}{\sqrt{2}\,(2\pi m_n k_0 T)^{3/4}} \exp(-\varepsilon_D/2k_0T) \tag{2.18}$$

and

$$\zeta = -\frac{\varepsilon_D}{2} + \frac{1}{2} k_0 T \ln \frac{(2\pi\hbar)^3 n_D}{2\,(2\pi m_n k_0 T)^{3/2}}. \tag{2.18a}$$

For a donor concentration $n_D = 10^{17}$ cm$^{-3}$, $m_n = 10^{-27}$ g and temperature $T = 300$ K, the quantity under the logarithm in (2.18a) is of the order of $10^{-2}$, i.e., the second term is of the order of $k_0T$. This means that the position of the chemical potential is approximately midway between the bottom edge of the conduction band and the donor level.

The concentration of conduction electrons is

$$n = \frac{\sqrt{2}\,(2\pi m_n k_0 T)^{3/4}\, n_D^{1/2}}{(2\pi\hbar)^{3/2}}\, e^{-\frac{\varepsilon_D}{2k_0T}}. \tag{2.18b}$$

Note that $n \propto \sqrt{n_D}$.

b) $A \exp(\varepsilon_D/k_0T) \ll 1$.

Since $A \ll 1$, this case is always realized, if $\varepsilon_D \lesssim k_0T$. Neglecting $A \exp(\varepsilon_D/k_0T)$ in the parentheses of (2.17), we obtain

$$A = \frac{4\pi^3\hbar^3 n_D}{(2\pi m_n k_0 T)^{3/2}}, \tag{2.19}$$

$$\zeta = k_0T \ln \frac{4\pi^3\hbar^3 n_D}{(2\pi m_n k_0 T)^{3/2}}. \tag{2.19a}$$

The concentration of conduction electrons is

$$n = n_D. \tag{2.19b}$$

The meaning of this result is obvious: if $k_0T \gtrsim \varepsilon_D$, almost all donors will be ionized, and $n \approx n_D$. In this case the chemical potential $\zeta$ will easily be seen to be negative and located below the donor level, i.e., $|\zeta| > \varepsilon_D$.

If $\varepsilon_D \gg k_0T$, we can always choose such small $n_D$ that condition (b) is met. For instance, if $\varepsilon_D = 10k_0T$, then for all donors to be ionized at room temperature it should be $n_D \ll 10^{15}$ cm$^{-3}$. This is also obvious, since the small donor concentration hampers the transition of the electrons on them from the valence band.

A semiconductor of the acceptor type, $n_D = 0$, can be considered in a way similar to case (2).

If in the statistical equilibrium of electrons the conduction band, the valence band and impurities of one type (for instance, donors) taken into account simultaneously, then (2.15a) turns into an algebraic equation of the third degree in $A$ and can be solved numerically or graphically.

Multiplying (2.15b) by (2.15c) and comparing the result with (2.16b), we obtain a useful equation

$$np = \frac{(2\pi \sqrt{m_n m_p}\, k_0 T)^3}{(4\pi^3\hbar^3)^2} \exp\left(-\frac{\varepsilon_G}{k_0T}\right) = n_i^2, \tag{2.20}$$

i.e., the product of the electron concentration and the hole concentration in an impurity semiconductor is equal to the square of the electron (or hole) concentration in the same semiconductor free from impurities (in the intrinsic semiconductor).

**6.2.5** In order to apply (2.15) to semiconductors with a complex energy band structure, such as, for example, germanium or silicon, we should find the density of states $g(\varepsilon)$.

In $n$-germanium and $n$-silicon the energy spectrum of the electrons consists of a set of equivalent minima symmetrically arranged in the Brillouin zone. Near each of these minima the energy is

$$\varepsilon(\mathbf{k}) = \frac{\hbar^2}{2}\left(\frac{k_x^2}{m_1} + \frac{k_y^2}{m_2} + \frac{k_z^2}{m_3}\right), \tag{2.21}$$

where $\mathbf{k}$ is the electron's wave vector measured from the minimum point; the $x$, $y$ and $z$ axes are directed along the principal axes of the energy ellipsoid $\varepsilon(\mathbf{k}) = \text{const.}$ (for germanium and for silicon the constant-energy surfaces are ellipsoids of revolution, so that $m_1 = m_2$).

Introduce in the $\mathbf{k}$-space the transformation

$$k_x = k'_x \sqrt{m_1}, \quad k_y = k'_y \sqrt{m_2}, \quad k_z = k'_z \sqrt{m_3}. \tag{2.21a}$$

Then the energy will be

$$\varepsilon = \hbar^2 k'^2/2. \tag{2.21b}$$

The number of states per cubic centimetre in the volume element $dk_x\, dk_y\, dk_z$ is

$$\frac{dk_x\, dk_y\, dk_z}{8\pi^3} = (m_1 m_2 m_3)^{1/2} \frac{dk'_x\, dk'_y\, dk'_z}{8\pi^3}.$$

Taking into account (2.21b), we obtain by analogy with (4.3.27)

$$g(\varepsilon) = \frac{\sqrt{2}}{2\pi^2} N_c \frac{(m_1 m_2 m_3)^{1/2}}{\hbar^3} \sqrt{\varepsilon}, \tag{2.22}$$

where $N_c$ is the number of equivalent minima. For germanium $N_c = 4$, for silicon $N_c = 6$.

We see the quantity $N_c^{2/3}(m_1 m_2 m_3)^{1/3}$ to be substituted for $m^*$ in (4.3.27).

For (2.22) to be of a form similar to (4.3.27), we should introduce the "effective mass for the density of states"

$$m_{\text{eff}} = N_c^{2/3}(m_1 m_2 m_3)^{1/3}. \tag{2.22a}$$

Hence, when calculating the chemical potential from (2.12b) we should substitute $m_{\text{eff}}$ for $m^*$.

In the case of holes in the valence band of germanium and silicon the energy of each of the "corrugated" surfaces can be reduced to the form (4.15.5)

$$\varepsilon = \frac{\hbar^2 k^2}{2m} \Phi(\vartheta, \varphi), \tag{2.23}$$

where $m$ is the electron mass and $\Phi(\vartheta, \varphi)$ is a certain function of the polar angles $\vartheta$ and $\varphi$. When determining $g(\varepsilon)$, we shall actually be interested in the transition from the variables $k$, $\vartheta$, $\varphi$ to $\varepsilon$, $\vartheta$, $\varphi$, i.e., from $k$ to $\varepsilon$, with $\vartheta$ and $\varphi$ remaining constant. The number of states per cubic centimeter in the volume $k^2 dk \sin\vartheta\, d\vartheta\, d\varphi = k^2\, dk\, d\Omega$ is

$$\frac{k^2\, dk\, d\Omega}{8\pi^3} = \frac{\sqrt{2}}{2\pi^2} \frac{m^{3/2}}{\hbar^3} \sqrt{\varepsilon}\, d\varepsilon \frac{1}{4\pi} \Phi^{-3/2}(\vartheta, \varphi)\, d\Omega,$$

where we made use of the relation (2.23) between $k$ and $\varepsilon$ for constant $\vartheta$ and $\varphi$. Hence,

$$g(\varepsilon) = \frac{\sqrt{2}}{2\pi^2} \frac{m^{3/2}}{\hbar^3} \sqrt{\varepsilon} \sum_n \frac{1}{4\pi} \int \Phi_n^{-3/2}(\vartheta, \varphi)\, d\Omega, \qquad (2.24)$$

where the integration is performed with respect to the full solid angle and the summation over the various constant-energy surfaces (heavy and light holes).

For holes the "effective mass for the density of states" is

$$m_{\text{eff}} = m \left[ \sum_n \frac{1}{4\pi} \int \Phi_n^{-3/2}(\vartheta, \varphi)\, d\Omega \right]^{2/3}. \qquad (2.24a)$$

Table 6.1

| Charge carriers | Ge (4 minima) | Si (6 minima) |
|---|---|---|
| Electrons | 0.412 | 1.129 |
| Holes (1) | 0.208 | 0.390 |
| Holes (2) | 0.0084 | 0.068 |
| All the holes | 0.216 | 0.458 |
| Mean geometrical value $\frac{m_{\text{eff}}^{3/2}}{m^{3/2}}$ for holes and electrons | 0.299 | 0.719 |

Table 6.1 presents the ratios $(m_{\text{eff}}/m)^{3/2}$ for electrons and holes in germanium and in silicon. The product of $m^{3/2}$ by the number taken from the table yields $m_{\text{eff}}^{2/3}$, which directly enters the expression for the density of states.

6.2.6 According to (2.16b) or (2.18b), the decrease in temperature down to absolute zero ($T = 0$) is accompanied by a decrease in the electron concentration to zero, and, consequently, the semiconductor's resistance rises infinitely. However, the resistance of most semiconductors at $T \to 0$ remains finite.

This is because the wave functions of localized electrons overlap, if the density of the impurity centres is high enough. This overlapping is accompanied by some widening of the energy levels of such electrons, and an *impurity band* is formed, enabling the electrons on impurity levels to wander about the crystal. Such a mechanism became known as *impurity conductivity*. The mobility of the electrons in the impurity band is usually much less than that of the electrons in the conduction band. The impurity conductivity depends little on temperature and rapidly decreases as impurity concentration is reduced. However, it is observed also with comparatively small concentrations of impurity centres, for example, in germanium down to concentrations of the order of $10^{15}$ cm$^{-3}$.

At higher impurity concentrations the impurity band merges with the conduction band and is occupied by electrons at all temperatures, the electron concentration in it being independent of temperature. Such materials are termed *semimetals*. The overlapping of the wave functions of impurity electrons starts the earlier the greater their Bohr radius, i.e., the greater the dielectric constant $\varepsilon$ and the less the electron effective mass $m^*$. This is why materials with high $\varepsilon$ and small $m^*$ become semimetals at comparatively low impurity concentrations. For example, in $n$-InSb already at $n_D \approx 10^{15}$ cm$^{-3}$ all impurities remain ionized at any temperature, a similar effect in $n$-Ge being observed only at impurity concentrations exceeding $5 \times 10^{18}$ cm$^{-3}$. In extremely pure specimens, for instance, in $n$- and $p$-germanium with the impurity concentration of $10^{13}$ cm$^{-3}$, there is no impurity band conductivity, and their resistance increases with a decrease in temperature in accordance with (2.18b).

## 6.3. Heat Capacity of Free Electrons in Metals and in Semiconductors

**6.3.1** In discussing the heat capacity of crystal lattices (Sec. 3.11) we have already noted that at high temperatures the heat capacity of metals obeys the Dulong and Petit law, and that this fact is in apparent contradiction with the assertion that, according to classical statistics, the free electrons must take part in the heat capacity of a body. This contradiction is resolved, if we accept that at temperatures at which the lattice vibrations obey the laws of classical statistics the free electrons in a metal are degenerate and obey the Fermi-Dirac statistics.

Consider the heat capacity of free electrons in a metal and their contribution to the total heat capacity of the crystal.

The energy of free electrons in a metal per cubic centimetre is

$$\mathscr{E} = 2\int_0^\infty \varepsilon f_0(\varepsilon)\, g(\varepsilon)\, d\varepsilon. \tag{3.1}$$

Calculate $\mathscr{E}$ up to $(k_0T/\zeta_0)^2$. To this end we should make use of equation (2.7a). In our case $\varphi(\varepsilon) = \frac{1}{5\pi^2}\left[\frac{(2m^*)^{3/2}}{\hbar^3}\right]\varepsilon^{5/2}$. Applying the general method expounded in Appendix 17, we obtain

$$\mathscr{E} = \frac{(2m^*)^{3/2}}{5\pi^2\hbar^3}\zeta^{5/2}\left[1+\frac{5\pi^2}{8}\left(\frac{k_0T}{\zeta}\right)^2\right]. \tag{3.2}$$

According to (2.9), $\zeta$ itself is temperature-dependent, so that up to the desired accuracy

$$\zeta^{5/2} = \zeta_0^{5/2}\left[1-\frac{\pi^2}{12}\left(\frac{k_0T}{\zeta_0}\right)^2\right]^{5/2} \approx \zeta_0^{5/2}\left[1-\frac{5\pi^2}{24}\left(\frac{k_0T}{\zeta_0}\right)^2\right], \tag{3.3}$$

we have used only the first two terms of the binomial expansion.

Substituting (3.3) into (3.2) (we can put in the square brackets of (3.2) $\zeta = \zeta_0$), and neglecting the terms of the order of $(k_0T/\zeta_0)^4$, we obtain

$$\mathscr{E} = \frac{(2m^*)^{3/2}}{5\pi^2\hbar^3}\zeta_0^{5/2}\left[1+\frac{5\pi^2}{12}\left(\frac{k_0T}{\zeta_0}\right)^2\right]. \tag{3.4}$$

The electron heat capacity per cubic centimetre at constant volume is

$$c_V = \frac{\partial\mathscr{E}}{\partial T} = \frac{\sqrt{2}}{3}\frac{m^{*3/2}}{\hbar^3}\zeta_0^{1/2}k_0^2T, \tag{3.5}$$

i.e., it is proportional to $T$.

At absolute zero the electron energy is

$$\mathscr{E}_0 = \frac{(2m^*)^{3/2}}{5\pi^2\hbar^3}\zeta_0^{5/2} = \frac{3^{5/3}\pi^{4/3}\hbar^2}{10m^*}n^{5/3}, \tag{3.4a}$$

which coincides with (A.4.5). The heat capacity at absolute zero is $c_V^{(0)} = \left(\frac{\partial\mathscr{E}}{\partial T}\right)_0 = 0$, in accordance with the Nernst heat theorem.

Expression (3.5) has an obvious meaning: only the electrons inside a narrow energy band $k_0T$ wide around the energy $\zeta_0$ take part in heat excitation, their number being of the order of $g(\zeta_0)\,k_0T$; every such electron contributes about $k_0$ to the heat capacity, and because of this their combined heat capacity is $\sim g(\zeta_0)\,k_0^2T$; this coincides in the order of magnitude with (3.5).

Assess the ratio of the electron heat capacity (3.5) to the lattice heat (per cubic centimeter). Since the electron concentration $n$ in a metal is of the order of the number of atoms per cubic centimeter the lattice heat capacity per cubic centimeter at high temperatures is of the order of $nk_0$, and the sought ratio is

$$\frac{m^{*3/2}\zeta_0^{1/2}k_0T}{\hbar^3 n} \approx \frac{k_0T}{\zeta_0}, \tag{3.6}$$

where we have substituted $n$ from (2.8). Since $\zeta_0$ is of the order of several electron volts, and $k_0T$ at room temperatures $\approx 0.03$ eV, the ratio (3.6) is of the order of a fraction of a percent. This explains why the free electrons take almost no part in the heat capacity of metals in a wide temperature range.

In the low temperature range the lattice heat capacity decreases at $T^3$ (3.11.17), i.e., more rapidly than in (3.5), and because of that below a definite temperature $T_0$ the free electron heat capacity exceeds that of the lattice. The temperature $T_0$ can be found by equating the electron heat capacity (3.5) to that of the lattice (3.11.17) (per cubic centimeter of the crystal).

**6.3.2** The heat capacity of a nondegenerate electron gas, whose concentration $n$ is independent of temperature, can be calculated if the Maxwell-Boltzmann distribution (2.11a) is substituted for $f_0(\varepsilon)$ in equation (3.1). Simple calculations yield the well-known result

$$c_V = 3/2\, nk_0. \tag{3.7}$$

A more interesting problem is the calculation of the free electron (hole) heat capacity in the case of a nondegenerate semiconductor when their concentration is temperature-dependent. It would appear at first glance that if the electron activation energy is of the order of $k_0T$, the electron heat capacity will be comparable to that of the lattice. Consider an intrinsic semiconductor with the forbidden bandwidth $\varepsilon_G$. The total energy of electrons and holes per cubic centimetre is

$$\mathscr{E} = 2\int_0^\infty \varepsilon f_0(\varepsilon)\, g(\varepsilon)\, d\varepsilon + 2\int_0^\infty (\varepsilon' + \varepsilon_G)\, f_0'(\varepsilon')\, g(\varepsilon')\, d\varepsilon', \tag{3.8}$$

where the first integral relates to electrons, and the second to holes. Substituting classical distribution functions for $f_0(\varepsilon)$ and $f_0'(\varepsilon')$ and expression (2.4) for $g(\varepsilon)$ and $g(\varepsilon')$, we obtain

$$\mathscr{E} = \frac{\sqrt{2}}{\pi^2}\frac{m_n^{3/2}}{\hbar^3} e^{\zeta/k_0T}\int_0^\infty \exp\left(-\frac{\varepsilon}{k_0T}\right)\varepsilon^{3/2}\, d\varepsilon$$
$$+\frac{\sqrt{2}}{\pi^2}\frac{m_p^{3/2}}{\hbar^3}\exp\left(-\frac{\zeta}{k_0T}\right)\exp\left(-\frac{\varepsilon_G}{k_0T}\right)\int_0^\infty$$
$$\times \exp\left(-\frac{\varepsilon'}{k_0T}\right)(\varepsilon' + \varepsilon_G)^{3/2}\, d\varepsilon'. \tag{3.8a}$$

Introducing into the integral a new variable $x = \sqrt{\varepsilon} = \sqrt{\varepsilon'}$ and applying (A.7.3), we obtain, making use of the value of the chemical potential (2.16),

$$\mathscr{E}=\frac{(2\pi\sqrt{m_n m_p}\,k_0T)^{3/2}}{4\pi^3\hbar^3}\exp\left(-\frac{\varepsilon_G}{2k_0T}\right)\times(3k_0T+\varepsilon_G)=n\,(3k_0T+\varepsilon_G), \qquad (3.8b)$$

where $n=p$ is the electron or hole concentration.

The electron and hole heat capacity turns out to be equal to

$$c_V=\frac{\partial\mathscr{E}}{\partial T}=k_0n\left[\frac{15}{2}+3\left(\frac{\varepsilon_G}{k_0T}\right)+\frac{1}{2}\left(\frac{\varepsilon_G}{k_0T}\right)^2\right], \qquad (3.9)$$

where $n=n^0{}_T\exp(-\varepsilon_G/2k_0T)$ [compare (2.16b)].

Formula (3.9) holds for $\varepsilon_G \geqslant k_0T$ (otherwise the degeneracy of free electrons and holes should be taken into account). It can easily be demonstrated that the maximum $c_V$ is attained for $\varepsilon_G\approx k_0T$; in this case its order of magnitude is $c_V\approx k_0n^0{}_T$.

The order of magnitude of lattice heat capacity (per cubic centimetre) at temperatures exceeding the Debye temperature is $c_V^{(L)}\approx\approx k_0n_0$, where $n_0$ is the number of atoms per cubic centimetre. Hence, $\frac{c_V}{c_V^{(L)}}\approx\frac{n_T^0}{n_0}\ll 1$.

Similar considerations operate in the case of impurity semiconductors as well. Hence, the electron and hole heat capacity in semiconductors is very small compared with the lattice heat capacity.

## 6.4. Magnetic Properties of Materials. Paramagnetism of Gases and of Conduction Electrons in Metals and Semiconductors

6.4.1 The subject of the magnetic properties of materials (magnetics) is very broad, and we shall deal only with certain fundamental concepts.[4]

If a body is placed in an external magnetic field, every element of its volume $dV$ acquires a magnetic moment

$$d\mathbf{M}=\mathbf{M}(\mathbf{H})\,dV. \qquad (4.1)$$

The magnetization, or the magnetic moment **M** per unit volume, depends on the magnitude and the direction of the magnetic field intensity **H**. In isotropic bodies **M** is parallel to **H**, its magnitude being independent of the field direction. For isotropic bodies, to the consideration of which we intend to limit ourselves, the quantity

$$\left(\frac{dM}{dH}\right)_{H=0}=\chi \qquad (4.1a)$$

[4] For a more detailed study the reader is referred to [6.5].

is termed the *magnetic susceptibility* of the material. It depends on the material, the temperature, and the pressure, but is independent of the magnetic field **H**.

If only the linear term of the expansion of **M** (**H**) into a series in **H** is retained, then

$$\mathbf{M} = \chi \mathbf{H}. \tag{4.1b}$$

It is easily seen that $\chi$ is a dimensionless quantity.

We distinguish between:

1) *Paramagnetic materials* for which $\chi > 0$, i.e., **M** is parallel to **H**. The oxygen gas is a paramagnetic, for it $\chi \propto 1/T$ (Curie's law); at room temperature and at atmospheric pressure $\chi_{O_2} = 0.14 \times 10^{-6}$.

Alkali metals are also paramagnetics but for them $\chi$ is practically independent of temperature; for sodium $\chi_{Na} = 0.58 \times 10^{-6}$.

2) *Diamagnetic materials* for which $\chi < 0$, i.e., **M** is antiparallel to the field **H**. The temperature dependence of $\chi$ of diamagnetics is a weak one (through the material's density). Inert gases are diamagnetics; for helium in normal conditions $\chi_{He} = -0.02 \times 10^{-6}$.

3) *Ferromagnetic materials* (iron, cobalt, nickel and some alloys). The properties of ferromagnetics are very complex and will not be discussed in this book. Above a definite temperature, termed the *Curie point*, equal to 1040 K for Fe, to 1404 K for Co and to 631 K for nickel, every ferromagnetic behaves like a paramagnetic. Below the Curie point the ferromagnetics exhibit some interesting properties: in pure single crystals even a weak magnetic field causes in certain crystallographic directions a very great magnetization **M** whose magnitude depends not only on the field intensity **H** but also on the prehistory of the specimen (*hysteresis*). The ferromagnetics retain a magnetic moment after the external magnetic field has been reduced to zero.

We shall derive some simple thermodynamic relations for the magnetics. Consider two bodies whose dimensions are small in comparison with the distance $R$ between them. If the magnetic moment of the first body is $\mathbf{M}'$ and it is parallel to $R$, then the first body sets up a magnetic field [6.6, Chap. 2]:

$$H = 2M'/R^3 \tag{4.2}$$

at the point of location of the second body (the field intensities of electric and magnetic fields are expressed by similar relations). This field induces in the second body a magnetic moment $M$ which is also parallel to $R$. The force of mutual attraction of the bodies is

$$\mathcal{K} = 6MM'/R^4. \tag{4.2a}$$

When $R$ is increased by $dR$ the work performed on the system is

$$dA = \mathcal{K}\, dR = -M\, dH, \tag{4.2b}$$

where we made use of (4.2a) and (4.2). Actually the expression for $dA$ is of a more general nature and is not limited to the model being considered. For a more general approach see [6.1, p. 92].

Consider a magnetic of fixed volume equal to 1 cm³. It follows from the first law of thermodynamics that the variation of the system's internal energy is

$$d\mathscr{E} = dQ + dA, \tag{4.3}$$

where $dQ$ is the heat delivered to the system by a reversible process, and $dA$ is the work performed on it. Making $dQ = T\,dS$ (where $dS$ is the variation of the system's entropy) and considering the work $dA$ (4.2b) performed in the course of the variation of the magnetic field, we obtain

$$d\mathscr{E} = T\,dS - M\,dH. \tag{4.3a}$$

The system's free energy is

$$\mathscr{F} = \mathscr{E} - TS. \tag{4.4}$$

Differentiating (4.4) and making use of (4.3a), we obtain

$$d\mathscr{F} = -S\,dT - M\,dH, \tag{4.4a}$$

whence

$$\left(\frac{\partial \mathscr{F}}{\partial T}\right)_H = -S, \quad \left(\frac{\partial \mathscr{F}}{\partial H}\right)_T = -M. \tag{4.4b}$$

Hence, and from (4.1a), we obtain

$$\chi = -\left(\frac{\partial^2 \mathscr{F}}{\partial H^2}\right)_{T;\,H=0}. \tag{4.5}$$

Thus, to find the magnetic susceptibility $\chi$, we have to find the expressions either for the magnetic moment $\mathbf{M}$ (4.1a), or for the free energy (4.4).

**6.4.2** Consider the theory of magnetization of a substance consisting of freely rotating particles (molecules) with permanent magnetic moments $\boldsymbol{\mu}_A$ (Langevin's theory of the paramagnetic gas). The magnetic field $\mathbf{H}$ forces the magnetic moments $\boldsymbol{\mu}_A$ to arrange themselves in the direction of the field, while thermal motion disrupts this arrangement. From the classical point of view the moment $\boldsymbol{\mu}_A$ can make an arbitrary angle with the field $\mathbf{H}$. The potential energy of $\boldsymbol{\mu}_A$ in the field $\mathbf{H}$ is equal to $-(\boldsymbol{\mu}_A\,\mathbf{H}) = \mu_A H \cos\vartheta$, where $\vartheta$ is the angle between $\boldsymbol{\mu}_A$ and $\mathbf{H}$. The probability that $\boldsymbol{\mu}_A$ makes an angle with $\mathbf{H}$ lying within the interval $\vartheta$, $\vartheta + d\vartheta$ is

$$dw = A \exp\frac{\mu_A H \cos\vartheta}{k_0 T} \sin\vartheta\, d\vartheta, \tag{4.6}$$

where $2\pi \sin\vartheta\, d\vartheta = d\Omega$ is the solid angle corresponding to the inter val $d\vartheta$, and $A$ is a constant determined from the normalization condition for the probability:

$$\int dw = A \int_0^{\pi} \exp \frac{\mu_A H \cos\vartheta}{k_0 T} \sin\vartheta\, d\vartheta = A \int_{-1}^{+1} e^{\alpha x}\, dx = 1, \tag{4.6a}$$

where $\alpha = \mu_A H/k_0 T$ and $x = \cos\vartheta$.

The mean value of the projection of the magnetic moment $\mu_A$ on the direction of the magnetic field is

$$\langle\mu_A\rangle = \int \mu_A \cos\vartheta\, dw = \mu_A \frac{\int_{-1}^{+1} x e^{\alpha x}\, dx}{\int_{-1}^{+1} e^{\alpha x}\, dx}. \tag{4.7}$$

The calculation of the integral in the denominator is quite simple, and the integral of the numerator is the derivative of the integral in the denominator with respect to $\alpha$. Hence,

$$\langle\mu_A\rangle = \mu_A L\left(\frac{\mu_A H}{k_0 T}\right), \tag{4.7a}$$

where

$$L(\alpha) = \frac{e^{\alpha} + e^{-\alpha}}{e^{\alpha} - e^{-\alpha}} - \frac{1}{\alpha} = \coth\alpha - \frac{1}{\alpha} \tag{4.7b}$$

is the Langevin function.

In weak fields, when $\mu_A H \ll k_0 T$, i.e., when $\alpha \ll 1$, we obtain, expanding the right-hand side of (4.7b) into a series,

$$L(\alpha) \approx \alpha/3. \tag{4.7c}$$

In this case the magnetization

$$M = n\langle\mu_A\rangle = \frac{\mu_A^2 n}{3k_0 T} H, \tag{4.8}$$

whence the magnetic susceptibility

$$\chi = M/H = \mu_A^2 n/3k_0 T, \tag{4.8a}$$

where $n$ is the concentration of magnetic particles.

We see that in accordance with Curie's law, $\chi \propto \frac{1}{T}$.

In the quantum mechanics theory of atomic and ionic paramagnetism we should take into account two important circumstances: the discrete nature of the space quantization of the electron's momentum and its spin.

It is demonstrated in quantum mechanics [6.7, Sec. 3.10] that an electron in a stationary state in an atom, having a definite angular momentum projection $\mathcal{L}_z = \hbar m$ ($m$ is the quantum magnetic number), has a magnetic moment $M_z = \mu_B m$, where

$$\mu_B = \frac{e\hbar}{2mc} = 0.927 \times 10^{-20} \text{ erg/Oe} \tag{4.9}$$

is the *Bohr magneton*, the elementary (minimum) magnetic moment existing in quantum mechanics.

Hence, for the orbital motion of the electron

$$\frac{M_z}{L_z} = \frac{e}{2mc}. \tag{4.9a}$$

The classical theory yields an identical value of the ratio of these quantities [see equation (5.2) in the following section].

The theory and experiments demonstrate that the free electron has a magnetic moment equal to the Bohr magneton $\mu_B$ and an angular momentum, the projections of which on a specified direction are $s_z = \pm\hbar/2$. These properties of the electrons became known as the *spin*.

The ratio $\mu_B / |s_z|$ for the electron spin is "anomalous", being twice that of (4.9a).

Taking the vector sum[5] of the magnetic moment and of the angular momenta of orbital motion and of spin, we can easily see that, because of the "anomalous" ratio $\mu_B/|s_z|$ for the spin, the direction of the resultant magnetic moment will not coincide with the direction of the resultant angular momentum.

This may be the cause of the anomalous Zeemann effect. It can be demonstrated [6.8, § 69] that the precession of the resultant magnetic moment about the resultant angular momentum creates an effective magnetic moment in the direction of the latter,

$$\mu_j = \mu_B j g, \tag{4.10}$$

where $j$ is the quantum number of the total angular momentum equal to $\hbar\sqrt{j(j+1)}$, and

$$g = 1 + \frac{j(j+1) + s(s+1) - l(l+1)}{2j(j+1)} \tag{4.10a}$$

is the *Lande splitting factor*. Here $l$ is the quantum number of the orbital angular momentum equal to $\hbar\sqrt{l(l+1)}$, and $s$ is the spin quantum number that assumes two values: $+1/2$ and $-1/2$.

The resultant angular momentum of the electron assumes in an external magnetic field $\mathbf{H}$ $2j + 1$ discrete orientations making angles

[5] The so-called *vector model of atoms* is introduced in quantum mechanics on the basis of group theory.

with the magnetic field: $\cos(\widehat{\mathbf{H}, j}) = m/j$, where $m$ is the magnetic quantum number that assumes the values $j, j-1, \ldots, 0, \ldots, j+1, -j$. The energy of the magnetic moment $\mu_j$ in a magnetic field is

$$u_\mu = -\mu_j H \cos(\widehat{\mathbf{H}, j}) = -\mu_B H\, gm. \tag{4.11}$$

The average value of the magnetic moment in the direction of the field is

$$\langle\mu\rangle = \frac{\sum\limits_{m=-j}^{+j} \mu_j \cos(\widehat{\mathbf{H}, j}) \exp(-u_\mu/k_0 T)}{\sum\limits_{m=-j}^{+j} \exp(-u_\mu/k_0 T)} = \mu_B g \frac{\sum\limits_{m=-j}^{+j} m e^{\alpha m}}{\sum\limits_{m=-j}^{+j} e^{\alpha m}}, \tag{4.12}$$

where $\alpha = \mu_B g H / k_0 T$.

Consider for the sake of simplicity the case of weak magnetic fields, when $\alpha \ll 1$, so that we can make $\exp(\alpha m) \approx 1 + \alpha m$. In this case it follows easily from (4.12) that

$$\langle\mu\rangle = \frac{\mu_B^2 g^2 j(j+1)}{3k_0 T} H, \tag{4.13}$$

so that the magnetic susceptibility is, by analogy with (4.8a),

$$\chi = n\langle\mu\rangle/H = \mu_{\text{eff}}^2 n/3k_0 T, \tag{4.13a}$$

where the effective magnetic moment $\mu_{\text{eff}} = \mu_B g \sqrt{j(j+1)}$. We see that in the quantum case, too, $\chi$ is inversely proportional to $T$.

**6.4.3** Consider the paramagnetism of free electrons (the electron gas) that is caused by the electron's magnetic moment (spin). In this case the orbital angular momentum is zero ($l = 0$) and, consequently, $j = s = 1/2$.

Hence, the Lande factor for free electrons is

$$g = 2 \tag{4.14}$$

as stipulated by (4.10a).

The paramagnetic susceptibility should, according to (4.13a), be

$$\chi = \mu_B^2 n / k_0 T, \tag{4.14a}$$

and this should result in a substantial and a strongly temperature-dependent paramagnetism of free electrons. However, as mentioned above, experiments prove that the paramagnetic susceptibility of metals is small and almost independent of temperature. The explanation for this circumstance was suggested by W. Pauli (1927) who considered the conduction electrons as a strongly degenerate Fermi gas. This work laid the foundation for the quantum theory of metals. It has been established in quantum mechanics that in the presence of an external magnetic field **H** the magnetic moment of the electron can point either in the direction of the field, its energy being $\varepsilon - \mu_B H$, or against the field, its energy being $\varepsilon + \mu_B H$, where $\varepsilon$ is the electron's energy in the absence of the field.

The total magnetic moment per cubic centimetre of the substance due to conduction electrons with moments pointing in the direction of the field is obviously

$$M_+ = \mu_B \int f_0(\varepsilon - \mu_B H)\, g(\varepsilon)\, d\varepsilon,$$

since $g(\varepsilon)\, d\varepsilon$ is the number of states in the energy interval $d\varepsilon$, having spins pointing in one direction.

Similarly, the total magnetic moment due to the electrons with spins pointing against the field is

$$M_- = \mu_B \int f_0(\varepsilon + \mu_B H)\, g(\varepsilon)\, d\varepsilon.$$

Hence, the resultant magnetic moment is

$$M = M_+ - M_- = \mu_B \int \{f_0(\varepsilon - \mu_B H) - f_0(\varepsilon + \mu_B H)\}\, g(\varepsilon)\, d\varepsilon. \quad (4.15)$$

For weak magnetic fields, $f_0(\varepsilon \mp \mu_B H)$ can be expanded into a series in $\mu_B H$, only the first term being retained; then

$$M = H\mu_B^2 2 \int \left(-\frac{\partial f_0(\varepsilon)}{\partial \varepsilon}\right) g(\varepsilon)\, d\varepsilon. \quad (4.15a)$$

It will be easily seen that $\frac{\partial f_0}{\partial \varepsilon} = -\frac{\partial f_0}{\partial \zeta}$, therefore,

$$\chi = \frac{M}{H} = \mu_B^2 \frac{\partial}{\partial \zeta}\left(2 \int f_0 g\, d\varepsilon\right) = \mu_B^2 \frac{\partial n}{\partial \zeta}, \quad (4.15b)$$

where $n$ is the concentration of free charges.

In the case of strongly degenerate conduction electrons of a metal the integral in (4.15a) can be calculated with the aid of (A.17.5), where in our case

$$\psi(\eta) = \frac{\sqrt{2}}{\pi^2} \frac{m^{*3/2}}{\hbar^3} (\zeta + k_0 T \eta)^{1/2}.$$

Hence

$$M = H\mu_B^2 \left[\psi(0) + \frac{\pi^2}{6} \psi''(0)\right]$$

$$= H\mu_B^2 \frac{\sqrt{2}}{\pi^2} \frac{m^{*3/2}}{\hbar^3} \zeta^{1/2} \left[1 - \frac{\pi^2}{24}\left(\frac{k_0 T}{\zeta}\right)^2\right].$$

Substituting $\zeta$ from (2.9), we obtain for $(k_0 T/\zeta_0)^2$

$$M = H\mu_B^2 g(\zeta_0) \left[1 - \frac{\pi^2}{12}\left(\frac{k_0 T}{\zeta_0}\right)^2\right]. \quad (4.16)$$

Making use of expression (2.8) for $\zeta_0$, we obtain

$$\chi = \frac{3^{1/3}}{\pi^{4/3}} \frac{\mu_B^2 m^* n^{1/3}}{\hbar^2} \left[ 1 - \frac{\pi^2}{12} \left( \frac{k_0 T}{\zeta_0} \right)^2 \right]. \tag{4.16a}$$

We see that the principal part of paramagnetic susceptibility of the conduction electrons in a metal is independent of temperature, which is in agreement with experiments; the small temperature correction to $\chi$ at room temperature is of the order of $(k_0 T/\zeta_0)^2 \approx 10^{-4}$. Comparing (4.16) with (4.8a) or (4.14a), we see that the thermal energy of molecules $k_0 T$ in Langevin's equation in the case of strongly degenerate Fermi gas is replaced by quantized energy $\hbar^2/m^* d^2$, where $d = n^{-1/3}$ is the average spacing between the particles.

For nondegenerate conduction electrons of a semiconductor

$$\frac{\partial n}{\partial \zeta} = \frac{\partial}{\partial \zeta} \exp \frac{\zeta - \varepsilon}{k_0 T} = \frac{n}{k_0 T}, \tag{4.17}$$

therefore (4.14a) follows from (4.15b).

**6.4.4** A method the use of which for solid-state research attracted considerable interest in recent years was the study of *paramagnetic resonance* of conduction electrons and electrons of impurity centres (E. K. Zavoisky, 1946).

According to (4.11) the energy level of an atomic electron splits up in a magnetic field into equidistant sublevels with a spacing of $g\mu_B H$ where $g$ is the Lande factor. The splitting of the electron level of an impurity centre or of a conduction electron level in a crystal can also be presumed to be equal to $g\mu_B H$, where the factor $g$, termed now *spectroscopic splitting factor*, generally accounts both for the electron's orbital angular momentum and for its interaction with the lattice. Accordingly, $g$ for the conduction electrons in a crystal differs from 2 and can be anisotropic, i.e., have different values for different orientations of the magnetic field with respect to the crystal.

The energy level of a conduction electron splits up into two sublevels, because of the electron's spin ($s = \pm 1/2$). However, in the case of a multiplet splitting of the electron's energy level in the magnetic field selection rules permit only transitions between adjacent sublevels ($\Delta m = \pm 1$). For this reason the resonance absorption of high-frequency radio waves takes place at the frequency $\omega$ satisfying the condition

$$\hbar\omega = g\mu_B H. \tag{4.18}$$

Hence, by determining the position of the resonance peaks, we can find the factor $g$, and draw certain conclusions about the electron's state in the crystal. The measurement of the width and shape of the resonance peak helps in study of the interaction of the impurity electron with the magnetic moments of the atoms in neighbouring sites, of the interaction of the electron with lattice vibrations, etc.

A similar paramagnetic resonance picture is observed in the atomic nuclei, in particular when they are in the crystal lattice.

Consider the semiphenomenological theory of paramagnetic resonance (Bloch, 1946).

Let **M** be the magnetization vector, i.e., the magnetic moment of a unit volume of the material, and let **L** be the resultant angular momentum of the electrons in a unit of volume, then

$$\frac{d\mathbf{L}}{dt} = \mathbf{M} \times \mathbf{H}. \tag{4.19}$$

This can be obtained by adding up the equations of motion of the individual particles in a unit volume.

Since from the most general conclusions arrived at in quantum mechanics

$$\mathbf{M} = \gamma \mathbf{L}, \tag{4.20}$$

where

$$\gamma = eg/2mc, \tag{4.20a}$$

it follows that

$$\frac{d\mathbf{M}}{dt} = \gamma \mathbf{M} \times \mathbf{H}. \tag{4.21}$$

Let a strong permanent magnetic field be applied in the direction of the $z$ axis (so that $H_z \equiv H_0$) and a small alternating high-frequency magnetic field in the direction of the $x$ axis: $H_x = H_1 \exp(i\omega t)$ $(H_1 \ll H_0)$. We shall seek the solution of (4.21) in the form of

$$M_x = M_{1x} e^{i\omega t}, \quad M_y = M_{1y} e^{i\omega t}, \quad M_z = M_{0z} + M_{1z} e^{i\omega t}, \tag{4.22}$$

$M_{1x}$, $M_{1y}$, $M_{1z}$ being of the order of $H_1$ and $M_{0z}$ of the order of $H_0$.

Substituting (4.22) into (4.21), we obtain

$$\begin{aligned} i\omega M_x &= \gamma M_y H_z, \\ i\omega M_y &= \gamma (M_z H_x - M_x H_z) \approx \gamma (M_{0z} H_x - M_x H_z), \\ \frac{d}{dt} M_z &= i\omega M_{1z} e^{i\omega t} = -\gamma M_y H_x \approx 0, \end{aligned} \tag{4.23}$$

if we neglect the quantities of the second order of smallness in $H_1$. Hence, the last equation can be ignored.

Excluding $M_y$ from the first two equations in (4.23), we obtain

$$\chi_x = \frac{M_x}{H_x} = \frac{\chi_0}{1 - (\omega/\omega_0)^2}, \tag{4.24}$$

where

$$\chi_0 = \frac{M_{0z}}{H_z} \quad \text{and} \quad \omega_0 = \gamma H_0. \tag{4.24a}$$

Since we took no account of the "forces of friction", i.e., the processes of relaxation of the precessional motion of **M**, the amplitude of the "alternating susceptibility" $\chi_x$ tends to infinity at the resonance frequency $\omega \approx \omega_0$.

## 6.5. Diamagnetism of Atoms and of Conduction Electrons. Magnetic Properties of Semiconductors

6.5.1 In the preceding section we pointed out that there are substances (diamagnetics) for which the susceptibility $\chi < 0$, i.e., the induced magnetic moment is directed against the field **H**. We shall demonstrate in this section that quantum systems of moving charges (atomic and molecular electrons, conduction electrons) always possess diamagnetic properties that in some cases are concealed by a stronger paramagnetism.

Consider at first an isolated atom with $Z$ electrons in an external magnetic field **H**. Place the origin of a rectangular coordinate system in the atom's nucleus and let the $z$ axis point in the direction of the magnetic field **H**. In accordance with the well-known Larmor theorem [6.9, Sec. 7.7], the effect of the magnetic field on the electrons to the first approximation consists of a uniform rotation of the electron system as a whole about the $z$ axis (about the field **H**) with a constant angular velocity

$$\omega_L = eH/2mc, \tag{5.1}$$

termed the *Larmor frequency*. Here $e$ and $m$ are the charge and the mass of the electron, and $c$ is the velocity of light. If we look in the $+z$ direction (i.e., in the direction of the field **H**), we will see the electrons revolve clockwise, and, accordingly, the magnetic moment $\mu_0$ associated with the respective current will be directed against the magnetic field **H**. There is a universal relation between $\mu_0$ and the electron's angular momentum $l$ [6.9, Sec. 7.6]

$$\mu_0 = \frac{e}{2mc}\, l. \tag{5.2}$$

Since

$$l = m\omega_L \sum_{i=1}^{Z} \overline{(x_i^2 + y_i^2)}, \tag{5.3}$$

where $\overline{x_i^2 + y_i^2}$ is the sum of squares of the coordinates of the $i$th electron averaged over the atomic volume, it follows that

$$\mu_0 = \frac{e^2 H}{4mc^2} \sum_{i=1}^{Z} \overline{(x_i^2 + y_i^2)}. \tag{5.4}$$

The diamagnetic susceptibility of a gas of such atoms is

$$\chi_{\text{diam}} = \frac{n\mu_0}{H} = \frac{e^2 n}{4mc^2} \sum_{i=1}^{Z} \overline{(x_i^2 + y_i^2)}, \tag{5.5}$$

where $n$ is the number of atoms per cubic centimeter.

The quantity $\sum_{i=1}^{Z} \overline{(x_i^2 + y_i^2)}$ can be calculated with the aid of quantum mechanics; its order of magnitude is $Za^2$, where $a$ are the atomic linear dimensions.

If the atom has a constant magnetic moment equal to the Bohr magneton (4.9), then the paramagnetic susceptibility of a gas of such atoms is $\chi_{\text{par}} = \mu_B^2 n / 3k_0 T$. The order of magnitude of the ratio of the diamagnetic and the paramagnetic susceptibilities is

$$\frac{\chi_{\text{diam}}}{\chi_{\text{par}}} \approx \frac{e^2 n Z a^2}{mc^2} \div \frac{\mu_B^2 n}{k_0 T} = Z k_0 T \div \frac{e^2}{a}, \tag{5.6}$$

where the Bohr radius $\hbar^2/me^2$ has been substituted for the factor $a$.

Since $e^2/a$ is an energy of the order of the atomic energy, the ratio (5.6) at room temperature is of the order of $10^{-2}$ if the atomic number $Z$ is not too high. This is why diamagnetic properties are exhibited only by the materials whose atoms have no constant magnetic moment.

**6.5.2** At first glance the diamagnetism of free electrons (conduction electrons) can be calculated in a similar fashion. A free electron with a velocity $v$ acted upon by a magnetic field will describe in the plane perpendicular to the magnetic field a circular orbit of a radius

$$r = mcv_\perp / eH = v_\perp / \omega_c, \tag{5.7}$$

where $v_\perp$ is the electron's velocity component in the plane perpendicular to the magnetic field, and

$$\omega_c = eH/mc \tag{5.7a}$$

is the so-called *cyclotron frequency* equal to $2\omega_L$, where $\omega_L$ is the Larmor frequency (5.1). According to (5.2), the magnetic moment corresponding to this circular motion is

$$\mu = \frac{erv_\perp}{2c} = \frac{\frac{mv_\perp^2}{2}}{H}, \tag{5.8}$$

where we excluded $r$ by making use of equation (5.7).[6] For an electron gas described by classical statistics $m\overline{v_{\perp}^2}/2 = k_0T$; therefore, the diamagnetic susceptibility is

$$\chi = n\,\frac{k_0T}{H^2}. \tag{5.9}$$

This expression cannot be true, because it is independent of the electron's charge and is inversely proportional to $H^2$. It can be demonstrated (Bohr, 1911) that we obtained this erroneous result, because we failed to consider the surface (boundary) of the body inside which the electrons move. The electrons cannot describe complete circular orbits close to the surface, and this results in the appearance of a surface current in the specimen, which fully compensates the magnetic moments of the circular orbits inside the body.

Demonstrate in a general form that the diamagnetic susceptibility of an ideal electron gas, described by classical statistics, is zero. Let the magnetic field be [6.6, Sec. 4.5]

$$\mathbf{H} = \operatorname{curl}\mathbf{A}, \tag{5.10}$$

that is,

$$H_x = \frac{\partial A_z}{\partial y} - \frac{\partial A_y}{\partial z}, \quad H_y = \frac{\partial A_x}{\partial z} - \frac{\partial A_z}{\partial x}, \quad H_z = \frac{\partial A_y}{\partial x} - \frac{\partial A_x}{\partial y}, \tag{5.10a}$$

where $\mathbf{A}(x, y, z)$ is the *vector potential* defined up to a gradient transformation.

Making use of equations (5.10a), we can easily demonstrate that for a magnetic field directed along the $z$ axis ($H_x = H_y = 0$, $H_z = H$) the vector potential can be chosen in the following form:

$$A_x = 0, \quad A_y = xH, \quad A_z = 0. \tag{5.11}$$

The Hamilton function for the electron (with the charge of $-e$) in a magnetic field is equal to [6.10, § 16]

$$\begin{aligned}\mathcal{H} &= \frac{1}{2m}\left(\mathbf{p} + \frac{e}{c}\mathbf{A}\right)^2 + \mathcal{U}(\mathbf{r}) \\ &= \frac{1}{2m}\left[\left(p_x + \frac{e}{c}A_x\right)^2 + \left(p_y + \frac{e}{c}A_y\right)^2 \right.\\ &\quad \left. + \left(p_z + \frac{e}{c}A_z\right)^2\right] + \mathcal{U}(x, y, z),\end{aligned} \tag{5.12}$$

where $\mathcal{U}(\mathbf{r})$ is the electron's potential energy.

[6] We exclude $r$ and not $v_{\perp}$ since we assume that the velocity $v$ is specified.

The free energy of an ideal gas described by classical statistics is equal to [6.1, §3]

$$\mathcal{F} = -k_0 T \ln Z. \tag{5.13}$$

Here the statistical integral

$$Z = \left[ \iiint_{-\infty}^{+\infty} dp_x\, dp_y\, dp_z \iiint_{(V)} dx\, dy\, dz\, e^{-\mathcal{H}/k_0 T} \right]^N, \tag{5.13a}$$

where $\mathcal{H}$ is the Hamilton function for a single particle, and $N$ is the number of particles.

The Hamilton function $\mathcal{H}$ for electrons in a magnetic field is identical to (5.12). The statistical integral is equal to the product of $N$ identical six-fold integrals, since all the $N$ electrons are independent and identical.

Substitute new variables

$$\pi_x = p_x + \frac{e}{c} A_x; \quad \pi_y = p_y + \frac{e}{c} A_y; \quad \pi_z = p_z + \frac{e}{c} A_z \tag{5.14}$$

for the integration variables $p_x$, $p_y$, $p_z$. Then $dp_x dp_y dp_z = d\pi_x d\pi_y d\pi_z$; the integration limits remain as before, and, since a definite integral is independent of the integration variables, $Z$ and for this reason the free energy $\mathcal{F}$ are also independent of the magnetic field $H$. In this case, according to (4.5), the magnetic susceptibility $\chi$ is zero.

In exactly the same way it can be demonstrated that the diamagnetic susceptibility of an electron gas described by the Fermi statistics is also zero. Hence, a quasi-classical consideration of the electron gas shows that it does not exhibit diamagnetism.

We may ask why is there a contradiction between this result and the nonzero diamagnetism of atoms (5.5). Actually, we obtained (5.5) on the basis of an essentially quantum mechanics assumption about the stationary state of the atomic electrons that do not obey the laws of classical statistics.

**6.5.3** L.D. Landau (1930) demonstrated that a quantum mechanics treatment of the motion of free electrons in a magnetic field yields for their diamagnetic susceptibility a value equal to 1/3 of their paramagnetic susceptibility. This fact does not contradict the preceding result, because it is due to the quantization of the motion of the free electron in a magnetic field.

Let a uniform magnetic field pointing in the direction of the $z$ axis act on the conduction electron and on the periodic potential $V(\mathbf{r})$. The vector potential of the magnetic field can be chosen in the form (5.11). It follows from (5.12) that the Schrödinger equation should be of the form[7]

$$\frac{1}{2m}\left[\left(\frac{\hbar}{i}\frac{\partial}{\partial x}\right)^2+\left(\frac{\hbar}{i}\frac{\partial}{\partial y}+\frac{e}{c}Hx\right)^2+\left(\frac{\hbar}{i}\frac{\partial}{\partial z}\right)^2\right]\psi(\mathbf{r})+V(\mathbf{r})\,\psi(\mathbf{r})=\mathscr{E}\psi(\mathbf{r}). \tag{5.15}$$

The crystal's periodic potential $V(\mathbf{r})$ can be dropped if the effective mass $m^*$ is substituted for the free electron mass $m$. Then, instead of (5.15), we obtain

$$-\frac{\hbar^2}{2m^*}\left[\frac{\partial^2\psi}{\partial x^2}+\left(\frac{\partial}{\partial y}+\frac{ieHx}{\hbar c}\right)^2\psi+\frac{\partial^2\psi}{\partial z^2}\right]=\mathscr{E}\psi. \tag{5.16}$$

Since this equation does not contain $y$ and $z$ in an explicit form, we shall look for a solution in the form of

$$\psi(x,\,y,\,z)=\varphi(x)\,e^{i(k_y y+k_z z)}. \tag{5.17}$$

Substituting (5.17) into (5.16) and canceling out the exponential factor, we obtain after a simple transformation

$$-\frac{\hbar^2}{2m^*}\frac{d^2\varphi}{dx^2}+\frac{1}{2}m^*\omega_0^2(x-x_0)^2\varphi=\mathscr{E}_1\varphi, \tag{5.18}$$

where

$$\omega_0=\frac{eH}{m^*c},\qquad x_0=-\frac{\hbar c}{eH}k_y\quad\text{and}\quad\mathscr{E}_1=\mathscr{E}-\frac{\hbar^2k_z^2}{2m^*}. \tag{5.18a}$$

We see that the electron's motion along the $x$ axis is described by the Schrödinger equation for a linear harmonic oscillator with the mass $m^*$, the natural frequency $\omega_0=2\omega_L^*$ ($\omega_L^*$ is the Larmor frequency of a particle of mass $m^*$), oscillating about its equilibrium position $x_0$. By analogy with (3.10.4), the energy eigenvalues $\mathscr{E}_1$ for the oscillator are

$$\mathscr{E}_1=\mathscr{E}-\frac{\hbar^2k_z^2}{2m^*}=\left(N+\frac{1}{2}\right)\hbar\omega_0=(2N+1)\,\mu^*H, \tag{5.19}$$

where $\mu^*=\left(\frac{m}{m^*}\right)\mu_B$, and the eigenfunctions of (5.18) are of the form

$$\varphi(x)=\frac{\exp\left[-\frac{1}{2}\left(\frac{x-x_0}{\lambda}\right)^2\right]}{\sqrt{\lambda}}H_N\left(\frac{x-x_0}{\lambda}\right), \tag{5.19a}$$

---

[7] The Schrödinger equation is obtained in the well-known way: $\hat{\mathscr{H}}\psi=\mathscr{E}\psi$, where $\hat{\mathscr{H}}$ is the Hamiltonian obtained from the Hamilton function by substituting $\frac{\hbar}{i}\frac{\partial}{\partial x}$ for $p_x$, etc.

where $\lambda \equiv \left(\frac{\hbar}{m^*\omega_0}\right)^{1/2} = \left(\frac{\hbar c}{eH}\right)^{1/2}$ is the so-called magnetic length,[8] and $H_N$ is the Hermite polynomial of the $N$-th order.

The form of (5.17) apparently suggests the physical equivalence of the directions $y$ and $z$ and not of $x$ and $y$, as is actually the case (the magnetic field is directed along the $z$ axis). This is due to the fact that the operators corresponding to the "centre of the circle" of the moving electron (one of such operators is $x_0 = -ep_y/eH$) do not commute with one another.[9] We can obtain a more symmetrical, although a less convenient for applications, expression for the wave function $\psi(\mathbf{r})$ if we solve the problem in cylindrical coordinates.

The eigenvalues of (5.16)

$$\mathscr{E}(N, k_z) = (2N+1)\,\mu^* H + \frac{\hbar^2 k_z^2}{2m^*} \tag{5.19b}$$

are degenerate in the quantum number $k_y$.

To calculate the number of the electron's quantum states $g(\mathscr{E})\,d\mathscr{E}$ in the energy interval $\mathscr{E}$, $\mathscr{E} + d\mathscr{E}$, let the wave function (5.17) satisfy cyclic conditions along the $y$ and $z$ axes, i.e., let the same wave functions correspond to the coordinates $y$, $z$ and $y \pm L_2$, $z \pm L_3$, respectively. The necessary condition for this will easily be seen to be

$$k_y = \frac{2\pi}{L_2} n_2, \quad k_z = \frac{2\pi}{L_3} n_3, \tag{5.20}$$

where $n_2$ and $n_3$ are arbitrary integers. Indeed, $\exp ik_y(y \pm L_2) = \exp(ik_y y)\exp(\pm i2\pi n_2) = \exp(ik_y y)$.

Let $L_1$ be the $x$-dimension of the body. We shall not apply the cyclic conditions to the wave function along the $x$ axis (this would be inconvenient, since expression (5.17) is not periodic in $x$), but we shall assume that the solution of (5.17) exists only in the region

$$0 < |x_0| < L_1, \tag{5.21}$$

so that $|x_0|^{\max} = L_1$.

The number of quantum states for a specified value of $\mathscr{E}(N, k_z)$ is determined by the degeneracy in $k_y$ and is equal to $\left(\frac{L_2}{2\pi}\right) k_y^{\max}$.

---

[8] If $H$ is expressed in oersteds, $\lambda = \frac{10^{-3}}{4\sqrt{H}}$ cm.

[9] For a more detailed exposition of the subject see [6.11, §112].

It follows from (5.18a) and (5.21) that $k_y^{\max} = \left(\frac{eH}{\hbar c}\right) | x_0 |^{\max} = \frac{eH}{\hbar c} L_1$, so that the degeneracy we are interested in is equal to

$$\frac{eH}{2\pi\hbar c} L_1 L_2. \tag{5.22}$$

Determine now the number of quantum states whose energy $Z(\mathscr{E})$ is below $\mathscr{E}$. The number of quantum states with energies from $\mathscr{E}_1 = (2N + 1)\mu^* H$ to $\mathscr{E}$ is equal to

$$\frac{2 | k_z | L_3}{2\pi} = \frac{2 (2m^*)^{1/2} L_3}{2\pi\hbar} \sqrt{\mathscr{E} - (2N+1)\mu^* H}, \tag{5.22a}$$

where the factor 2 accounts for both signs of $k_z$ when the latter is determined from (5.19). The number of quantum states with energies below $\mathscr{E}$ for all possible values of $N$ is

$$Z(\mathscr{E}) = \frac{2 (2m^*)^{1/2} L_3}{2\pi\hbar} \sum_N \sqrt{\mathscr{E} - (2N + 1)\mu^* H}. \tag{5.22b}$$

Here the summation over $N$ embraces all the nonnegative values of the expression under the radical sign (including zero). Since the degeneracy of each state (5.22b) is (5.22), the total number of quantum states with an energy lower than $\mathscr{E}$ is

$$Z(\mathscr{E}) = \frac{2 (2m^*)^{1/2} eHV}{(2\pi\hbar)^2 c} \sum_N [\mathscr{E} - (2N+1)\mu^* H]^{1/2}, \tag{5.23}$$

where $V = L_1 L_2 L_3$ is the body's volume.

The number of states per $V = 1\ \text{cm}^3$ (with no account of spin) is

$$g(\mathscr{E}) = \frac{dZ(\mathscr{E})}{d\mathscr{E}} = \frac{(2m^*)^{1/2} eH}{(2\pi\hbar)^2 c} \sum_N [\mathscr{E} - (2N+1)\mu^* H]^{-1/2}. \tag{5.23a}$$

The free energy per cubic centimetre is [6.1, equations VIII.1.37 and IX.3.3]

$$\mathscr{F} = n\zeta - 2k_0 T \int^{\infty} \ln\left(1 + e^{(\zeta - \mathscr{E})/k_0 T}\right) g(\mathscr{E})\, d\mathscr{E}$$

$$= n\zeta - 2k_0 T \int^{\infty} \ln\left(1 + e^{(\zeta - \mathscr{E})/k_0 T}\right) \frac{dZ(\mathscr{E})}{d\mathscr{E}}\, d\mathscr{E}. \tag{5.24}$$

Here $n$ is the free electron concentration, and the factor 2 takes account of the spin. The lower integration limit for each term in (5.23a) is $(2N + 1)\mu^* H$.

Integrating by parts, we obtain

$$\mathcal{F} = n\zeta - 2\int^{\infty} f_0(\mathscr{E})\, Z(\mathscr{E})\, d\mathscr{E}, \tag{5.24a}$$

where $f_0(\mathscr{E})$ is the Fermi distribution function.

Introduce a new integration variable and the notation

$$\varepsilon = \frac{\mathscr{E}}{2\mu^* H}, \quad \varepsilon_0 = \frac{\zeta}{2\mu^* H}, \quad \Theta = \frac{k_0 T}{2\mu^* H}. \tag{5.25}$$

Integrating (5.24a) again by parts, we obtain in the new notation

$$\mathcal{F} = n\zeta + A\int^{\infty} \Phi(\varepsilon)\frac{d}{d\varepsilon}\left(\frac{1}{e^{\varepsilon-\varepsilon_0/\Theta}+1}\right) d\varepsilon, \tag{5.25a}$$

where

$$A = \frac{16 m^{3/2} (\mu^* H)^{5/2}}{3\pi^2 \hbar^3}, \qquad \Phi(\varepsilon) = \sum_N (\varepsilon - N - 1/2)^{3/2}, \tag{5.25b}$$

and the lower integration limit of each term in $\Phi(\varepsilon)$ is determined from the condition of reality of the corresponding radical.

Making use of the Poisson equation, we obtain (Appendix 18)

$$\Phi(\varepsilon) = \frac{2}{5}\varepsilon^{5/2} - \frac{1}{16}\varepsilon^{1/2} = \frac{3}{8\pi^2}\sum_{l=1}^{\infty}\frac{(-1)^l}{l^{5/2}}$$

$$\times [\sin(2\pi l\varepsilon)\cdot S(\sqrt{4l\varepsilon}) + \cos(2\pi l\varepsilon)\cdot C(\sqrt{4l\varepsilon})], \tag{5.26}$$

where $S(u)$ and $C(u)$ are the Fresnel integrals which are given in an explicit form in (A.18.8a) and (A.18.8b).

In the case of a strong degeneracy of the electron gas the derivative

$$\frac{\partial}{\partial\varepsilon}\left(\frac{1}{\exp\frac{\varepsilon-\varepsilon_0}{\Theta}+1}\right) = -\delta(\varepsilon-\varepsilon_0) \tag{5.27}$$

behaves like a delta function with an effective width of the maximum of the order of $\Theta$.

The terms in the sum (5.26) with a specified $l$ oscillate with a period of $\varepsilon = 1/l$, i.e., with the maximum period for $l = 1$; therefore, for $\Theta \gg 1$, i.e., for

$$k_0 T \gg 2\mu^* H, \tag{5.28}$$

the sum over $l$ in (5.26) in the integrand of (5.25a) vanishes, and we obtain

$$\mathcal{F} = n\zeta_0 - A\left(\frac{2}{5}\varepsilon_0^{5/2} - \frac{1}{16}\varepsilon_0^{1/2}\right), \tag{5.29}$$

where $\zeta_0$ is the Fermi energy at absolute zero.

We may ask whether or not the dependence of the free energy $\mathscr{F}$ on the magnetic field, stemming from the effect of the latter on the chemical potential $\zeta$, should be taken into account. If the chemical potential $\zeta = \zeta^0 + \Delta\zeta$, where $\zeta^0$ is the chemical potential in the absence of a magnetic field, then

$$\mathscr{F}(\zeta) = \mathscr{F}(\zeta^0 + \Delta\zeta) = \mathscr{F}(\zeta^0) + \left(\frac{\partial \mathscr{F}}{\partial \zeta}\right)_0 \Delta\zeta + \frac{1}{2}\left(\frac{\partial^2 \mathscr{F}}{\partial \zeta^2}\right)_0 (\Delta\zeta)^2. \quad (5.30)$$

In statistical equilibrium $(\partial\mathscr{F}/\partial\zeta)_0 = 0$; therefore, the variation of the free energy is

$$\mathscr{F}(\zeta) - \mathscr{F}(\zeta^0) = \frac{1}{2}\left(\frac{\partial^2 \mathscr{F}}{\partial \zeta^2}\right)_0 (\Delta\zeta)^2. \quad (5.30a)$$

Since in an isotropic body the variation of the chemical potential $\Delta\zeta$ is independent of the direction of the magnetic field, it follows that $\Delta\zeta \propto H^2$, and, therefore, in the lowest approximation

$$\mathscr{F}(\zeta) - \mathscr{F}(\zeta^0) \propto H^4. \quad (5.31)$$

Hence, in an approximation that takes into account in the free energy $\mathscr{F}$ the terms of the order of $H^2$, $\zeta$ can be assumed to be independent of the magnetic field.

Taking into account the values of $A$ (5.25b) and $\varepsilon_0$ (5.25), we see that the second term in (5.29) is also independent of the magnetic field.

For the diamagnetic susceptibility $\chi$, we obtain from (5.29)

$$\chi = -\frac{\partial^2 \mathscr{F}}{\partial H^2} = -\frac{1}{3}\frac{\sqrt{2}}{\pi^2}\frac{m^{*3/2}\mu^{*2}\zeta_0^{1/2}}{\hbar^3}. \quad (5.32)$$

If we substitute into (5.32) $\zeta_0$ from (2.8a), we shall see immediately that the magnitude of the diamagnetic susceptibility (5.32) is $\frac{1}{3}\left(\frac{m}{m^*}\right)^2$ times the temperature-independent paramagnetic susceptibility (4.16a). In the case opposite to (5.28), when $\Theta \ll 1$, we have

$$k_0 T \ll 2\mu^* H. \quad (5.33)$$

In this case the period of oscillations in (5.26) (at least for $l = 1$) exceeds the effective width $\Theta$ of the $\delta$-function's maximum in the integral (5.25a), and because of that oscillating terms of the type

$$\sin\left(\frac{\pi\zeta_0}{\mu^* H}\right) S\left(\sqrt{\frac{2\zeta_0}{\mu^* H}}\right) + \cos\left(\frac{\pi\zeta_0}{\mu^* H}\right) C\left(\sqrt{\frac{2\zeta_0}{\mu^* H}}\right) \quad (5.34)$$

appear in the free energy $\mathscr{F}$.

Oscillating terms similar to (5.34) will obviously also appear in the magnetic susceptibility $\chi = -\partial^2\mathscr{F}/\partial H^2$. The oscillation periods of the trigonometric functions and of the Fresnel integrals in

(5.34) are determined by the condition that $1/H$ changes by $2\mu^*/\zeta_0$, or by a quantity of the order of $\mu^*/2\zeta_0$.

Such oscillations of magnetic susceptibility were first observed for bismuth and were termed the *de Haas-van Alphen effect* after the scientists who discovered this phenomenon in 1930.

Consider now the diamagnetic susceptibility of nondegenerate conduction electrons in a semiconductor. Expression (5.25a) for the free energy remains, of course, valid in this case too, but it can be simplified: the Boltzmann distribution $\exp\left(\frac{\varepsilon_0-\varepsilon}{\Theta}\right)$ can be substituted for the Fermi distribution function. Hence,

$$\mathscr{F}=n\zeta-\frac{Ae^{\varepsilon_0/\Theta}}{\Theta}\sum_{N=0}^{\infty}\int_{N+1/2}^{\infty}\left(\varepsilon-N-\frac{1}{2}\right)^{3/2}e^{-\varepsilon/\Theta}\,d\varepsilon, \tag{5.35}$$

where the lower integration limit is determined from the condition that the expression in the parentheses in (5.35) should be positive.

Introduce a new integration variable

$$x=\frac{\varepsilon-N-\frac{1}{2}}{\Theta}; \tag{5.36}$$

then

$$\mathscr{F}=n\zeta-A\Theta^{3/2}\exp\frac{\zeta}{k_0T}\exp\left(-\frac{1}{2\Theta}\right)\sum_{N=0}^{\infty}e^{-N/\Theta}\int_0^{\infty}x^{3/2}e^{-x}\,dx. \tag{5.37}$$

The integral, according to (A.7.12), is

$$\int_0^{\infty}x^{3/2}e^{-x}\,dx=\frac{3\sqrt{\pi}}{4}. \tag{5.38}$$

The infinite decreasing geometric progression

$$\sum_{N=0}^{\infty}e^{-N/\Theta}=(1-e^{-1/\Theta})^{-1}. \tag{5.39}$$

Hence

$$\mathscr{F}=n\zeta-\frac{3\sqrt{\pi}}{4}A\Theta^{5/2}\exp\frac{\zeta}{k_0T}\left[\frac{2ze^{-z}}{1-e^{-2z}}\right], \tag{5.40}$$

where

$$z=1/2\Theta=\mu^*H/k_0T. \tag{5.40a}$$

Since the factor $A\Theta^{5/2}$ is independent of the magnetic field $H$, the dependence of the free energy $\mathscr{F}$ on $H$, if the weak dependence of $\zeta$

on the magnetic field is ignored, is determined by the square brackets in (5.40).

The limiting case (5.33), i.e., $z \gg 1$, is of little interest in the case of nondegenerate semiconductors[10]. For this reason we shall consider the other limiting case (5.28) when $z \ll 1$.

Expanding the expression in the square brackets in (5.40) into a series in $z$ up to the terms of the order of $z^2$, we obtain

$$\left[\frac{2ze^{-z}}{1-e^{-2z}}\right]=\frac{2z}{e^z-e^{-z}}=\frac{z}{\sinh z}\approx\frac{z}{z+\frac{z^3}{3!}+\ldots}\approx 1-\frac{1}{6}z^2. \quad (5.41)$$

Substituting this result into (5.40), we obtain to within $H^2$

$$\mathscr{F}=n\zeta-\frac{3\sqrt{\pi}}{4}A\Theta^{5/2}e^{\zeta/k_0T}\left[1-\frac{1}{6}\left(\frac{\mu^*H}{k_0T}\right)^2\right]. \quad (5.42)$$

Making use of (5.25), (5.25b) and (2.12a), we obtain for the diamagnetic susceptibility

$$\chi=-\frac{\partial^2\mathscr{F}}{\partial H^2}=-\frac{1}{3}\frac{\mu^{*2}n}{k_0T}=-\frac{1}{3}\left(\frac{m}{m^*}\right)^2\frac{\mu_B^2 n}{k_0T}, \quad (5.43)$$

[cf. (4.14a)]. Hence, the ratio of the diamagnetic and paramagnetic susceptibilities of the free charge carriers is the same both for strongly degenerate electrons of metals (5.30), and for nondegenerate electron gas of semiconductors (5.43). It can be demonstrated that this remains true for any electron gas degeneracy.

6.5.4 The magnetic susceptibility of an atomic semiconductor is

$$\chi=\chi_A+\chi_L+\chi_S+\chi_T, \quad (5.44)$$

i.e., is the sum of the diamagnetic susceptibility of the basis lattice $\chi_A$, the paramagnetic and diamagnetic susceptibilities of the charge carriers $\chi_L$, the susceptibility of impurity centres $\chi_S$ and, as was demonstrated in recent studies, the susceptibility of thermal crystal imperfections $\chi_T$ (dislocations, surface levels, faces, etc.) [6.12, p. 174].

The holes, like the electrons, have an intrinsic magnetic moment equal to $\mu_B$ (and, consequently, exhibit paramagnetic properties). This follows from the fact that when an electron is removed from the valence band, its magnetic moment changes by $\mu_B$.

Hence, the nondegenerate charge carriers in a semiconductor of mixed conductivity will have a magnetic susceptibility

$$\chi_L=\frac{\mu_B^2 n}{k_0T}\left[1-\frac{1}{3}\left(\frac{m}{m_n}\right)^2\right]+\frac{\mu_B^2 p}{k_0T}\left[1-\frac{1}{3}\left(\frac{m}{m_p}\right)^2\right], \quad (5.45)$$

where $n$ and $p$ are electron and hole concentrations, and $m_n$ and $m_p$ are their effective masses. If the dispersion law for the electron's

[10] In this case $\chi \propto \exp(-z)$, i.e., it is exponentially small.

(hole's) energy is expressed by an effective mass tensor, then the ratios $m^2/m_n^2$ and $m^2/m_p^2$ in (5.45) are replaced by the ratios of $m^2$ to some simple combination of the effective mass tensor components.

Very little can be said about the values of $\chi_A$, $\chi_S$ and $\chi_T$. In any case, at present they cannot be theoretically calculated with adequate accuracy. This is the reason why the studies of the magnetic susceptibility of semiconductors yield relatively little information about the properties of the charge carriers in them.

## 6.6. Cyclotron (Diamagnetic) Resonance

6.6.1 The most complete and reliable information about the energy spectra of charge carriers (electrons and holes) in crystals was obtained in the study of *cyclotron resonance*. Specifically, the structure of the energy bands of germanium and silicon, described in Section 4.15, was to a large extent verified by the studies of cyclotron resonance in those materials. We shall formulate the physical principles underlying the cyclotron resonance phenomenon and give a simplified theoretical treatment.

A quasi-free electron (hole) with a scalar effective mass $m^*$ moves in a magnetic field $H$ like a free electron describing a circular orbit in the plane perpendicular to the magnetic field. Its radius is (5.7)

$$r = v_{\perp}/\omega_c, \tag{6.1}$$

where

$$\omega_c = eH/m^*c \tag{6.1a}$$

is the *cyclotron frequency*, and $v_{\perp}$ is the electron velocity component perpendicular to the magnetic field.

The independence of $\omega_c$ from $v$ (nonrelativistic approximation) was used in the design of the accelerator of charge of particle termed *cyclotron*. We can obviously try to realize the principle of cyclotron acceleration in relation to the quasi-free charge carriers in a crystal. To this end we must apply, in addition to the constant magnetic field **H**, a high-frequency electromagnetic field oscillating in a plane perpendicular to **H**. If the frequency $\omega$ of this field coincides with $\omega_c$, the electron will be accelerated by the voltage of the high-frequency field on both semicircles of its orbit. As a result, its velocity (energy) will rise and it will move in an unwinding spiral like a charged particle in a cyclotron. This effect, which became known as the *cyclotron* (*diamagnetic*) *resonance* (Dorfman and Dingle, 1951), can be discovered from the appearance of a maximum in the absorption spectrum of high-frequency radiation. If we determine that $\omega \approx \omega_c$ for such a resonance, we will be able to determine the effective mass $m^*$ in the most straightforward way from (6.1a).

In addition to its interaction with the permanent magnetic and the high-frequency fields, the conduction electron also interacts with the lattice vibrations (phonons) and with static lattice imperfections (impurity centres, etc.). All these factors combine to establish a definite mean free time of the electron $\tau$. The interaction with the lattice vibrations and the impurities can be approximated by a friction force $\mathbf{F}_R$ applied to the electron. If the electron's mean free time $\tau$ is identified with the relaxation time of its velocity $\mathbf{v}$ due to the effect of the friction force $\mathbf{F}_R$, i.e., if it is assumed that after the external forces have ceased to act (at $t = 0$) the electron's velocity decreases in accordance with the law

$$\mathbf{v} = \mathbf{v}_0 e^{-t/\tau}, \tag{6.2}$$

where $\mathbf{v}_0$ is the velocity at $t = 0$, then the friction force will be

$$\mathbf{F}_R = m^* \frac{d\mathbf{v}}{dt} = -\frac{m^*\mathbf{v}}{\tau}, \tag{6.2a}$$

i.e., is proportional to the magnitude of the velocity $\mathbf{v}$ and acts in the direction opposite to it.

In the following chapter we shall present a more rigorous derivation of the relaxation time from the kinetic equation. In the process $\tau$ will be seen in general to depend on $v$. In this chapter we shall assume, for the sake of simplicity, that $\tau$ does not depend on $v$, this having no effect on the qualitative picture to be obtained here.

The necessary condition for the cyclotron resonance to be observed is

$$\tau \gg 1/\omega_c, \tag{6.3}$$

since only in this case a circular orbit, on which synchronous acceleration of the particle can take place, can be established.

Assess the values of some quantities involved in the cyclotron resonance. Make the frequency of the alternating electromagnetic field equal to 24 000 MHz. At resonance $\omega_c/2\pi = 24\,000$ MHz, whence $\omega_c = 1.5 \times 10^{11}$ rad/s. For a real ratio $m^*/m \approx 0.3$ the magnetic field, according to (6.1a), should be $H = 2 \times 10^3$ Oe. At a temperature of $T = 4$ K the mean thermal velocity of the electron is $v_T = \left(\frac{8k_0T}{\pi m^*}\right)^{1/2} \approx 2.4 \times 10^6$ cm/s, and the radius of its orbit (for $v_\perp \approx v_T$), according to (6.1), is $r \approx 5 \times 10^{-5}$ cm.

**6.6.2** Consider the elementary theory of the cyclotron resonance and assume that the charge $e$ has a scalar effective mass $m^*$ and moves in compliance with the laws of classical mechanics. If a constant magnetic field $\mathbf{H}$, a high-frequency electric field $\mathbf{E}$ (the effect of the magnetic field of high-frequency radiation can be ignored) and a friction force $\mathbf{F}_R$ (6.2a) act on the electron, its equations of motion will be of the form

$$m^* \frac{d\mathbf{v}}{dt} = -\frac{m^*\mathbf{v}}{\tau} + e\left(\mathbf{E} + \frac{1}{c}\mathbf{v}\times\mathbf{H}\right). \qquad (6.3a)$$

If $\mathbf{H}$ is directed along the $z$ axis, $\mathbf{E} = \mathbf{E}_0 \exp(-i\omega t)$ along the $x$ axis, and $\mathbf{v}$ is proportional to $\exp(-i\omega t)$, we obtain in projections on the $x$ and the $y$ axes

$$m^*\left(-i\omega + \frac{1}{\tau}\right) v_x = eE_x + \frac{e}{c} v_y H, \qquad (6.4)$$

$$m^*\left(-i\omega + \frac{1}{\tau}\right) v_y = -\frac{e}{c} v_x H. \qquad (6.4a)$$

According to Ohm's law in the differential form, the current density is

$$j = \sigma E_x = nev_x, \qquad (6.5)$$

where $\sigma$ is the conductivity, and $n$ is the concentration of charge carriers. Excluding $v_y$ from (6.4) and (6.4a), we obtain for the complex conductivity

$$\sigma = \frac{nev_x}{E_x} = \sigma_0 \frac{1 - i\omega\tau}{1 + (\omega_c^2 - \omega^2)\tau^2 - 2i\omega\tau}, \qquad (6.6)$$

where

$$\sigma_0 = ne^2\tau/m^* = ne\mu \qquad (6.6a)$$

is the conductivity for $\omega = \omega_c = 0$, i.e., in a constant electric field $E_x = E_0$ and in the absence of a magnetic field ($H = 0$). The quantity

$$\mu = e\tau/m^* = v_x/E_0, \qquad (6.6b)$$

equal to the mean velocity of the electron in a constant electric field of unit intensity, is called the *mobility*.

The absorption of high-frequency electromagnetic radiation in a specimen is proportional to the real part of the conductivity $\sigma$ (6.6), i.e.,

$$\sigma_{Re} = \sigma_0 \frac{1 + \nu^2 + \nu_c^2}{(1 + \nu_c^2 - \nu^2)^2 + 4\nu^2}, \qquad (6.7)$$

where $\nu = \omega\tau$ is proportional to the frequency of the electromagnetic field, and $\nu_c = \omega_c\tau = \mu H/c$ is proportional to the cyclotron frequency. Figure 6.4 depicts the graphs of the $\sigma_{Re}/\sigma_0$ vs $\nu_c/\nu = \omega_c/\omega$ dependence for various values of $\nu = \omega\tau$. For $\omega\tau = 2$, when condition (6.3) necessary for the observation of the cyclotron resonance is

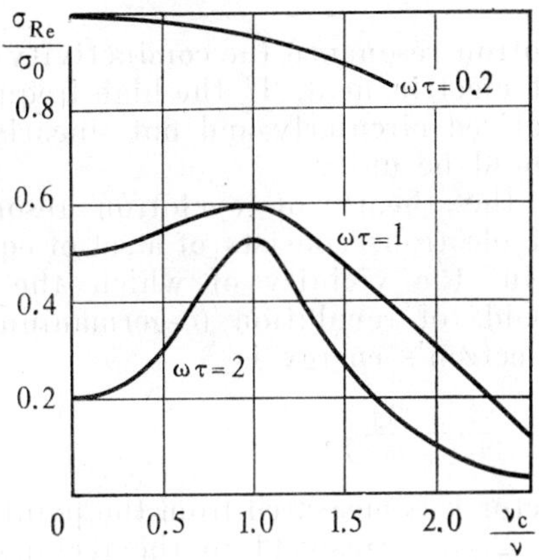


Fig. 6.4

to a certain extent satisfied, the graph exhibits a sharp maximum at $\omega = \omega_c$.

Consider the following cases.

1. $\nu_c \gg \nu$, $\nu_c \gg 1$. This case is realized in a low-frequency alternating electric field and in strong magnetic fields. It is seen from (6.7) that

$$\sigma_{Re} \approx \sigma_0/\nu_c^2 \propto 1/H^2, \tag{6.8}$$

i.e., the absorption decreases in proportion to $H^2$.

2. $\nu_c \gg \nu$, $\nu_c \ll 1$. This is the case of a low-frequency alternating electric field and a weak magnetic field (or of low $\tau$ due to high temperatures). In this case

$$\sigma_{Re} \approx \sigma_0 (1 - \nu_c^2). \tag{6.9}$$

If we introduce the specific resistance $\rho = 1/\sigma$, so that $\Delta\rho/\rho = -\Delta\sigma/\sigma$, we shall obtain from (6.9)

$$\Delta\rho/\rho_0 = \nu_c^2 = (\mu H/c)^2. \tag{6.9a}$$

This result, as we shall see in Chap. 8, coincides up to a numerical factor with that obtained from the kinetic equation.

3. $\nu \gg 1$, $\nu_c = 0$. The specimen is irradiated with a high-frequency radiation (for instance, infrared radiation) in the absence of a magnetic field. In this case the absorption coefficient is proportional to

$$\sigma_{Re} \approx \sigma_0/\nu^2. \tag{6.10}$$

4. $\nu = \nu_c \gg 1$. This is the condition for the cyclotron resonance. We have from (6.7)

$$\sigma_{\mathrm{Re}} \approx \frac{1}{2}\sigma_0. \tag{6.11}$$

Hence, at the cyclotron resonance the conductivity is two times less than in a constant electric field. If the high-frequency electromagnetic field is polarized circularly and not linearly, as in our case ($E_y=0$), $\sigma_{\mathrm{Re}}/\sigma_0$ would be unity.

6.6.3 Consider the theory of cyclotron resonance when the energy spectrum of electrons consists of a set of equivalent minima in the **k**-space, in the vicinity of which the constant-energy surfaces are ellipsoids of revolution ($n$-germanium and $n$-silicon). In this case the electron's energy is

$$\varepsilon(\mathbf{k}) = \frac{\hbar^2}{2}\left(\frac{k_1^2}{m_1} + \frac{k_2^2}{m_2} + \frac{k_3^2}{m_3}\right), \tag{6.12}$$

where the wave vector **k** is measured from the point of minimum $\mathbf{k}_0$, and the indices 1, 2, 3 correspond to the rectangular coordinates coinciding with the major axes of the energy ellipsoid. In the case of $n$-germanium and $n$-silicon the constant-energy surfaces (near the minima) are the ellipsoids of revolution, so that $m_1 = m_2 = m_\perp$ and $m_3 = m_\parallel$, where $m_\perp$ and $m_\parallel$ are, respectively, the transverse and the longitudinal masses of the effective mass tensor. Making use of (6.12), we obtain for the velocity components

$$v_i = \frac{1}{\hbar}\frac{\partial \varepsilon}{\partial k_i} = \frac{\hbar k_i}{m_i} = \frac{p_i}{m_i} \quad (i = 1, 2, 3), \tag{6.13}$$

where the quasi-momentum $\mathbf{p} = \hbar\mathbf{k}$.

Since the high-frequency field and the friction force $\mathbf{F}_R$ are small perturbations, it is sufficient to consider the quasi-classical motion equations for the electron in a magnetic field **H**:

$$\frac{d(\hbar\mathbf{k})}{dt} = -\frac{e}{c}\mathbf{v}\times\mathbf{H}. \tag{6.14}$$

Making use of (6.13), we obtain

$$m_1\frac{dv_1}{dt} = -\frac{e}{c}v_2H_3 + \frac{e}{c}v_3H_2,$$

$$m_2\frac{dv_2}{dt} = -\frac{e}{c}v_3H_1 + \frac{e}{c}v_1H_3,$$

$$m_3\frac{dv_3}{dt} = -\frac{e}{c}v_1H_2 + \frac{e}{c}v_2H_1, \tag{6.14a}$$

where $H_1 = H\alpha_1$, $H_2 = H\alpha_2$, $H_3 = H\alpha_3$ are the projections of the magnetic field on the coordinate axes, and $\alpha_1$, $\alpha_2$, $\alpha_3$ are the respective direction cosines.

Assuming that the electron velocity is $\mathbf{v} \propto \exp(-i\omega t)$, i.e., making

$$v_1 = v_{10}e^{-i\omega t}, \quad v_2 = v_{20}e^{-i\omega t}, \quad v_3 = v_{30}e^{-i\omega t}, \tag{6.15}$$

we obtain, substituting (6.15) into (6.14a),

$$\begin{aligned} -icm_1\omega v_{10} + eH_3v_{20} - eH_2v_{30} &= 0, \\ eH_3v_{10} + icm_2\omega v_{20} - eH_1v_{30} &= 0, \\ eH_2v_{10} - eH_1v_{20} - icm_3\omega v_{30} &= 0. \end{aligned} \tag{6.16}$$

This system of algebraic homogeneous linear equations in the unknown amplitudes $v_{10}$, $v_{20}$, $v_{30}$ has a nonzero solution only if the determinant of the system (6.16) is equal to zero [6.13, p. 258, Theorem 3]. Equating the determinant of (6.16) to zero, we obtain the third-degree characteristic equation for $\omega$, whose roots are

$$\omega = 0 \tag{6.17}$$

and

$$\omega^2 = \omega_1^2\alpha_1^2 + \omega_2^2\alpha_2^2 + \omega_3^2\alpha_3^2 \equiv \omega_c^2, \tag{6.17a}$$

where

$$\omega_1 = \frac{eH}{c\sqrt{m_2m_3}}, \quad \omega_2 = \frac{eH}{c\sqrt{m_1m_3}}, \quad \omega_3 = \frac{eH}{c\sqrt{m_1m_2}}. \tag{6.17b}$$

The frequency $\omega = 0$ (6.17) corresponds to the electron's motion along the magnetic field; the frequency $\omega = \omega_c$ (6.17a) is the cyclotron frequency corresponding to the electron's motion in a plane perpendicular to the magnetic field. If the magnetic field coincides, for example, with the first axis, then $\alpha_1 = 1$, $\alpha_2 = \alpha_3 = 0$, and $\omega_c = \omega_1$. Determining the cyclotron frequency (from the condition of maximum absorption in the resonator) for different orientations of the magnetic field with respect to the crystal, i.e., for different $\alpha_1$, $\alpha_2$, and $\alpha_3$, we can determine $\omega_1$, $\omega_2$ and $\omega_3$, i.e., the components of the effective mass tensor $m_1$, $m_2$, and $m_3$.

As was pointed out earlier, constant-energy surfaces for *n*-germanium and *n*-silicon are ellipsoids of revolution, so that $m_1 = m_2 = m_\perp$ and $m_3 = m_\parallel$. If the magnetic field makes an angle $\theta$ with the third axis, then $\alpha_3^2 = \cos^2\theta$, $\alpha_1^2 + \alpha_2^2 = 1 - \alpha_3^2 = \sin^2\theta$, and it follows from (6.17a) that

$$\omega_c^2 = \left(\frac{eH}{c}\right)^2 \left[\frac{\cos^2\theta}{m_\perp^2} + \frac{\sin^2\theta}{m_\perp m_\parallel}\right]. \tag{6.18}$$

Measuring $\omega_c$ for different $\theta$, we can find $m_\perp$ and $m_\parallel$. Since there are several equivalent energy minima in the **k**-space, the experimental absorption curves have numerous peaks corresponding to different minima. We can introduce the *cyclotron effective mass* $m_c^*$, making

$$\omega_c = eH/m_c^* c. \tag{6.19}$$

Then in the case of (6.18)

$$\left(\frac{1}{m_c^*}\right)^2 = \frac{\cos^2\theta}{m_\perp^2} + \frac{\sin^2\theta}{m_\perp m_\parallel}, \tag{6.20}$$

i.e., $m_c^*$ depends not only on $m_\perp$ and $m_\parallel$, but also on the angle $\theta$, which defines the direction of the magnetic field.

6.6.4 Derive the general expression for the cyclotron effective mass $m_c^*$, valid for an arbitrary dispersion law for the electrons (holes), that would enable the energy $\varepsilon$ vs $\mathbf{k}$ dependence for such particles to be determined from the cyclotron resonance experiments. Our equations, in particular, will be valid for *p*-germanium and *p*-silicon for which $\varepsilon(\mathbf{k})$ takes the form (4.15.4) making, strictly speaking, the introduction of the effective mass tensor impossible.

Consider the motion of an electron or a hole with an arbitrary dispersion law $\varepsilon(\mathbf{k})$ in a uniform magnetic field $\mathbf{H}$. The quasi-classical equations of motions are in this case of the form

$$\frac{d}{dt}(\hbar\mathbf{k}) = \frac{e}{c}\,\mathbf{v}\times\mathbf{H}, \tag{6.21}$$

$$\mathbf{v} = \frac{1}{\hbar}\nabla_{\mathbf{k}}\varepsilon(\mathbf{k}), \tag{6.21a}$$

where $e$ is now the algebraic value of the charge and $\nabla_{\mathbf{k}} \equiv \mathrm{grad}_{\mathbf{k}}$.

If $\mathbf{H}$ points in the direction of the $z$ axis, then

$$\hbar\frac{dk_z}{dt} = \frac{e}{c}\,\mathbf{v}\times\mathbf{H}_z = 0,$$

and therefore

$$k_z = \text{const}. \tag{6.22}$$

On the other hand, from (6.21) and (6.21a) we have

$$\frac{d\varepsilon(\mathbf{k})}{dt} = \nabla_{\mathbf{k}}\varepsilon\,\frac{d\mathbf{k}}{dt} = \mathbf{v}\,\frac{e}{c}\,\mathbf{v}\times\mathbf{H} = 0,$$

whence

$$\varepsilon(\mathbf{k}) = \text{const}. \tag{6.23}$$

This means that the motion of the charge in the $\mathbf{k}$-space follows a curve resulting from the intersection of a constant-energy surface (6.23) and a plane (6.22).

Form the projection of both sides of equation (6.21) in the **k**-space on the plane $k_z$ = const.:

$$\hbar \frac{dl}{dt} = \frac{e}{c} v_\perp H, \tag{6.24}$$

where $dl$ is an element of the "trajectory" of motion in the **k**-space, and

$$v_\perp = \sqrt{v_x^2 + v_y^2} \tag{6.24a}$$

is the projection of the velocity on a plane normal to the direction of the magnetic field (i.e., to the $z$ axis).

If the trajectory in the **k**-space is closed, the period of revolution of the charge is

$$T_c = \oint dt = \frac{\hbar c}{eH} \oint \frac{dl}{v_\perp}, \tag{6.25}$$

if $dt$ is substituted from (6.24).

On the other hand, the cross-sectional area bounded by the closed trajectory is

$$S = \int\int dk_x dk_y. \tag{6.26}$$

Perform in the double integral the transformation from the differentials $dk_x$, $dk_y$ to $dl$ and $dk_\perp$; $dk_\perp$ is an element of the normal to $dl$ in the plane $k_z$ = const. From the general equation (6.21a) we have

$$v_\perp = \frac{1}{\hbar} \frac{\partial \varepsilon}{\partial k_\perp}, \tag{6.27}$$

whence

$$dk_\perp = \frac{1}{\hbar} \frac{d\varepsilon}{v_\perp}. \tag{6.27a}$$

Hence

$$S = \int\int dl\, dk_\perp = \frac{1}{\hbar} \int d\varepsilon \oint \frac{dl}{v_\perp} \tag{6.28}$$

or

$$\hbar \frac{\partial S}{\partial \varepsilon} = \oint \frac{dl}{v_\perp}. \tag{6.28a}$$

Substituting this into (6.25), we obtain

$$T_c = \frac{\hbar^2 c}{eH} \frac{\partial S}{\partial \varepsilon}. \tag{6.29}$$

Since the cyclotron frequency is $\omega_c = 2\pi/T_c$, the cyclotron effective mass (6.19) is

$$m_c^* = \frac{\hbar^2}{2\pi} \frac{\partial S}{\partial \varepsilon}. \tag{6.30}$$

Measuring $m_c^*$ at different orientations of the magnetic field with respect to the crystal, we can, in principle, reconstruct the shape of the surface $\varepsilon(\mathbf{k}) = \text{const}$. If the dispersion law $\varepsilon(\mathbf{k})$ is specified in an analytical form but contains some unknown parameters, as, for example, in cases (6.12) or (4.15.4), then (6.30) can be used for the experimental determination of such parameters. By way of an example, consider a simple case in which constant-energy surfaces are spheres, so that

$$\varepsilon = \hbar^2 k^2/2m^*, \tag{6.31}$$

where $m^*$ is the scalar effective mass. In this case

$$S = \pi k_\perp^2 = \pi (k^2 - k_z^2) = \pi \left( \frac{2m^*\varepsilon}{\hbar^2} - k_z^2 \right). \tag{6.31a}$$

Substituting this into (6.30), we obtain

$$m_c^* = m^*, \tag{6.32}$$

which agrees with (6.1a). Expression (6.20) can be obtained from (6.30) just as easily. In the case of a complex constant-energy surface $\varepsilon(\mathbf{k}) = \text{const}$. (other than a sphere or an ellipsoid) the derivative $\partial S/\partial \varepsilon$ generally depends on $k_z$, so that $m_c^*(k_z)$. Since the crystal contains electrons with different $k_z$'s, $m_c^*$ corresponding to the maximum of the cyclotron absorption curve will be $m_c^*(\bar{k}_z)$, where $\bar{k}_z \approx \bar{k}_x \approx \bar{k}_y$, their order of magnitude following from the condition $\varepsilon \approx k_0 T$.

6.6.5 At temperatures low enough to satisfy inequality (5.33) only magnetic oscillators with small quantum numbers $N$ will be excited, and for this reason the quasi-classical description of the particle's motion (6.21) cannot be correct. Making, as before, $\omega_c = 1.5 \times 10^{11}$ rad/s and taking into account that $2\mu^* H = \hbar\omega_c$, we obtain for (5.33)

$$T \leqslant 1 \text{ K}. \tag{6.33}$$

Hence, at liquid helium temperatures at which the cyclotron resonance experiments are usually conducted we find ourselves close to the limits of applicability of the quasi-classical theory.

From a quantum mechanics point of view, the cyclotron resonance should be regarded as electron (hole) transitions from one quantum orbit to another, caused by the high-frequency field. It can easily be shown that in this case only transitions accompanied by the variation of the quantum number of the magnetic oscillator $\Delta N = \pm 1$ are possible. From this point of view it is natural to apply the term *diamagnetic resonance* to this phenomenon.

In the quantum mechanics theory the cyclotron resonance frequency $\omega_c^q$ is determined by the difference of adjacent terms $\mathscr{E}(N + 1) - \mathscr{E}(N) \approx \hbar\omega_c^q$. In the case, when the constant-energy surfaces are spheres, it follows from (5.19) that $\omega_c^q = 2\mu^* H/\hbar = eH/m^*c = \omega_c$, i.e., is independent of the quantum number $N$ coinciding with the classical cyclotron frequency determined from equation (6.29). Both frequencies coincide also in the case when the constant-energy surfaces are ellipsoids. Hence, in these cases equation (6.29) is valid in the entire temperature range.

For more complex constant-energy surfaces, as for instance in $p$-Ge, the difference $\mathscr{E}(N+1) - \mathscr{E}(N)$ depends on the level's number $N$, and $\omega_c^q$ coincides with $\omega_c$ (6.29) only for large values of the quantum numbers $N$.

The quantum mechanics theory of the diamagnetic (cyclotron) resonance is discussed in [6.14].

## 6.7. Metal-Semiconductor Contact. Rectification

**6.7.1** Consider an impurity semiconductor of the electron type the dissociation energy of whose donors is $\varepsilon_D \ll k_0 T$, so that they are practically all ionized. If the semiconductor is brought in contact with a metal, the potential[11] of its surface in contact with the metal

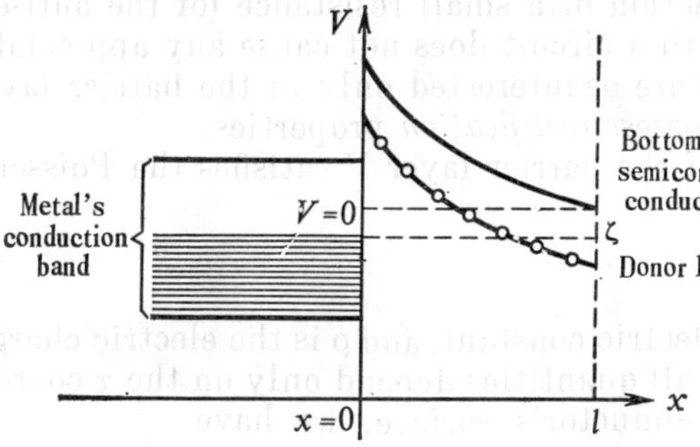


Fig. 6.5

will, as the result of a redistribution of the semiconductor's electrons, coincide with the potential of the metal surface. Since the semiconductor and the metal freely exchange electrons and are in a state of statistical equilibrium, their chemical potentials coincide. The *work function* of an electron in a metal is known to be equal to the difference between the electron's potential energy in a vacuum and its chemical potential inside the body. If the work function of the metal $w_m$ exceeds that of the semiconductor $w_s$, the potential in the contact layer of the semiconductor will increase, and the energy bands will be deflected, as shown in Fig. 6.5. The potential on the surface of the

[11] In the following we, for the sake of brevity, use the term potential for the electron's potential energy $V = -e\varphi$, where $\varphi$ is the electrostatic potential.

semiconductor (metal) will be $V_0 = w_m - w_s > 0$, if the potential energy of the free electrons inside the semiconductor is assumed to be zero, i.e., if all the energies are measured from the bottom edge of the semiconductor's conduction band. In the case being considered the free electron concentration in the bulk of the semiconductor will be higher than in the contact layer, which we shall term *barrier* (exhaustion) *layer*. If $w_m < w_s$, the potential will be $V_0 < 0$, i.e., it will be below the bottom edge of the semiconductor's conduction band, so that the energy bands close to the semiconductor's surface will be deflected in the direction opposite to that shown in Fig. 6.5. The electron concentration in the bulk of the semiconductor will be lower than in the contact layer, and we shall use the term *anti-exhaustion layer* to describe it.

In the first case ($w_m > w_s$), depicted in Fig. 6.5, the electrostatic potential at the surface is obviously $\varphi_0 = -\frac{V_0}{e} < 0$; in the second case ($w_m < w_s$) — $\varphi_0 > 0$. It is also obvious that for a hole-type semiconductor the first case results in the formation of an anti-exhaustion layer, and the second case results in the formation of a barrier layer.

Since the connection of a small resistance (of the anti-exhaustion layer) in series with a circuit does not cause any appreciable effects, we shall in the future be interested only in the barrier layers which, as we shall see, possess *rectification* properties.

The potential of the barrier layer $V$ satisfies the Poisson equation

$$\nabla^2 V = \frac{4\pi}{\varepsilon} e\rho, \tag{7.1}$$

where $\varepsilon$ is the dielectric constant, and $\rho$ is the electric charge density. If we assume that all quantities depend only on the $x$ coordinate normal to the semiconductor's surface, we have

$$\frac{d^2 V}{dx^2} = \frac{4\pi e^2}{\varepsilon} [n_D - n(x)], \tag{7.2}$$

where $n_D$ and $n(x)$ are the concentrations of the positively charged (ionized) donors and of the free electrons, respectively. It follows from the consideration of electrical neutrality that inside the semiconductor $n(x) = n = n_D$. The electron concentration in the barrier layer, in accordance with the Boltzmann law, is

$$n(x) = n \exp\left(-\frac{V(x)}{k_0 T}\right). \tag{7.3}$$

Introducing the dimensionless potential

$$\Phi(x) = \frac{V(x)}{k_0 T} \tag{7.4}$$

and the characteristic *Debye length* (*radius*)

$$l_D=\sqrt{\frac{\varepsilon k_0T}{4\pi e^2n}}, \tag{7.4a}$$

write (7.2) in the form

$$\frac{d^2\Phi}{dx^2}=\frac{1}{l_D^2}[1-e^{-\Phi(x)}]. \tag{7.4b}$$

The last equation contains the single parameter $l_D$; therefore, from considerations of dimensionality, it follows that the thickness of the barrier layer should be of the order of $l_D$. The Debye length $l_D$ plays the obvious part of the *screening radius of the potential* (of the charge).

If, for example, $\Phi(x) \ll 1$, i.e., if $V(x) \ll k_0T$ (this case is rarely encountered in practice), then expanding the exponent in (7.4b) into a series, we obtain

$$\frac{d^2V}{dx^2}=\frac{1}{l_D^2}V. \tag{7.5}$$

Taking into account that $V(0) = V_0$ and $V(\infty) = 0$, we obtain

$$V(x) = V_0e^{-x/l_D}, \tag{7.6}$$

i.e., $l_D$ is the length on which the potential (the charge) decreases $e$ times.

For $T = 300$ K, $n = 10^{16}$ cm$^{-3}$ and $\varepsilon = 16$ (Ge) the Debye length is $l_D=4\times 10^{-6}$ cm, i.e., many times greater than the lattice parameter. For normal metals $l_D$ is of the order of the lattice parameter, so that, strictly speaking, this concept cannot be introduced in this case.

**6.7.2** The electric current density in the presence of an electric field $E_x$ and a concentration gradient of free electrons $\frac{dn(x)}{dx}$ is

$$j=\sigma E_x+(-e)\left(-D\frac{dn(x)}{dx}\right), \tag{7.7}$$

where $D$ is the diffusion coefficient. The specific conductivity is

$$\sigma = en(x)\mu, \tag{7.7a}$$

where $\mu$ is the electron mobility (6.6b), which we shall consider in greater detail in Chapter 9.

The term for current $\sigma E_x$ in expression (7.7) is *field-induced* and for current $(-e)\left(-D\frac{dn}{dx}\right)$ is *diffusion current*.

The total current in conditions of statistical equilibrium is always zero. This means, for example, that in the barrier layer the field-induced current is compensated by the diffusion current.[12] From (7.7)

[12] This case is considered in greater detail in Section 6.7.3.

and (7.7a) we obtain in this case

$$-en(x)\,\mu\frac{d\varphi}{dx}+eD\frac{dn(x)}{dx}=0, \tag{7.8}$$

where $-\frac{d\varphi}{dx}$ has been substituted for $E_x$. Integrating (7.8), we obtain

$$n(x) = ne^{\mu\varphi/D}, \tag{7.9}$$

since $n(x) = n$ at $\varphi = 0$. On the other hand, in accordance with the Boltzmann distribution for a nondegenerate electron gas,

$$n(x)=n_0\exp\left(-\frac{V}{k_0T}\right)=n\exp\frac{e\varphi}{k_0T}. \tag{7.10}$$

Comparing (7.9) with (7.10), we arrive at the *Einstein relation* connecting the mobility and the diffusion coefficient

$$\mu=\frac{e}{k_0T}D. \tag{7.11}$$

6.7.3 Consider now the phenomena which take place when an electric current flows through the barrier layer. The electrons moving in the semiconductor collide with the phonons (lattice vibrations), impurity centers and with other imperfections of the crystal lattice. Such collisions result in drastic changes in the direction of motion of the electrons. We can introduce the concept of the *mean free path l* for a conduction electron; it is understood to be the mean path traveled between successive collisions.[13] Obviously, $l = v\tau$, where $v$ is the mean velocity and $\tau$ is the mean free time of the electron (see Section 6.6.1).

When considering the flow of current through the barrier layer, we should distinguish between two cases: the first, when the electron's mean free path greatly exceeds the barrier layer width, and second, when the reverse is true. Because of the great resistance of the barrier layer depleted of the charge carriers, practically all the potential difference $U$ falls across it. In the first case (Bathe's diode rectification theory; 1942), when the electron's mean free path is large, the calculation of the electron flux falling on the surface ($x = 0$) of the semiconductor out of its inner region is quite simple. The number of electrons in the semiconductor with speeds lying in the intervals $v_x$, $v_x + dv_x$, $v_y$, $v_y + dv_y$ and $v_z$, $v_z + dv_z$, passing through a square centimeter of its surface normal to the $x$ axis per one second, is

$$dn=n\left(\frac{m^*}{2\pi k_0T}\right)^{3/2}\exp\left(-\frac{m^*(v_x^2+v_y^2+v_z^2)}{2k_0T}\right)v_x dv_x dv_y dv_z, \tag{7.12}$$

[13] The concept of the mean free path will be rigorously defined in the next chapter from the kinetic equation.

where $n$ is the electron concentration in the semiconductor's bulk [6.1, p. 63].

To find the electron flux reaching the semiconductor's surface (at $x = 0$) we must integrate (7.12) with respect to $v_y$ and $v_z$ from $-\infty$ to $+\infty$ and with respect to $v_x$ from the value determined by the equation

$$m^* v_x^2/2 = V_0 - eU = -e(\varphi_0 + U) > 0 \tag{7.13}$$

to $v_x = \infty$.

Here $U$ is the variation of $\varphi_0$ at $x = 0$ due to the voltage across the barrier layer. Indeed, if $|v_x|$ is less than the value obtained from (7.13), then an electron with such a velocity will be unable to negotiate the potential barrier of the barrier layer and reach the semiconductor's surface.

Hence, the electron flux from the semiconductor to the metal is

$$Q = n\left(\frac{m^*}{2\pi k_0 T}\right)^{3/2} \int\limits_{\sqrt{2(V_0 - eU)/m^*}}^{\infty} v_x \exp\left(-\frac{m^* v_x^2}{2k_0 T}\right) dv_x$$

$$\times \iint\limits_{-\infty}^{\infty} \exp\left(-\frac{m^*(v_y^2 + v_z^2)}{2k_0 T}\right) dv_y dv_z$$

$$= \frac{1}{4} n v_T \exp\left(-\frac{V_0}{k_0 T}\right) \exp\frac{eU}{k_0 T} = Q_0 \exp\frac{eU}{k_0 T}, \tag{7.14}$$

where $v_T = \left(\frac{8k_0 T}{\pi m^*}\right)^{1/2}$ is the mean thermal velocity of the electrons, $n v_T$ is the electron flux density in the bulk of the semiconductors, and $Q_0 = \left(\frac{1}{4}\right) n v \exp\left(-\frac{V_0}{k_0 T}\right)$ is the density of the electron flux crossing the metal-semiconductor boundary at $x = 0$ for $U = 0$.

The integrals with respect to $v_y$ and $v_z$ can be calculated by equation (2.6.1); the calculation of the integral with respect to $v_x$ can be performed quite easily if we introduce a new integration variable $t = v_x^2$ with the aid of the transformation $v_x dv_x = 1/2\, dv_x^2$. Since in equilibrium the resultant electron flux crossing the boundary at $x = 0$ is zero, and since the potential $U$ applied to the barrier layer does not affect the state of the electrons in the metal, the electron flux from the metal to the semiconductor is always equal to $Q_0$.

If $U > 0$, i.e., the metal serves as the positive electrode and the semiconductor as the negative, the height of the barrier $V_0$ is decreased, so that $Q > Q_0$. The resultant electron flux flows from the semiconductor to the metal, the direction of the electric current being opposite. If we agree to regard this current as positive, then its value will be

$$j = e(Q - Q_0) = j_0\left(\exp\frac{eU}{k_0T} - 1\right), \qquad (7.15)$$

where $j_0 = eQ_0$.

This direction of the applied voltage is termed *forward*, the same term that is used for the current. For $U < 0$, when the metal serves as the negative electrode, the potential barrier at the semiconductor-metal boundary rises, and the resultant electron flux flows from the metal to the semiconductor. In this case the electric current is determined by the same expression (7.15), which now yields for it a negative value.

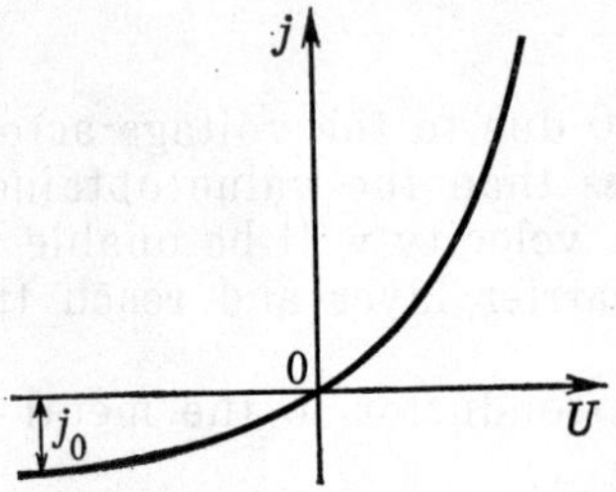

Fig. 6.6

In this case we speak of the *reverse direction* of the voltage and use the term *reverse*, or *back*, current.

Figure 6.6 is a schematic representation of the volt ampere *characteristic of a diode rectifier*, which satisfies (7.15).

For $U < 0$, as $|U|$ is increased, the reverse current tends to saturation $(-j_0)$. For positive $U \gg \frac{k_0T}{e}$ the current rises exponentially however, only as long as $U < |\varphi_0|$, i.e., as long as the barrier layer still exists. For $U$ close to $|\varphi_0|$ the resistance of the barrier drops so drastically that a substantial part of the applied voltage drops in the semiconductor's bulk. Expression (7.15), which takes no account of this effect, is no longer applicable for such voltages.

Consider now the so-called *diffusion theory of rectification* (Davydov, Schottky, Pekar, 1939) applicable in the case when the electron's mean free path $l$ is much less than the barrier layer width $l_D$. In this case the expression for the current is (7.7). In addition, we must assume that on the metal-semiconductor boundary

$$\varphi(0) = \varphi_0 + U. \qquad (7.16)$$

From (7.7) we have

$$\frac{dn(x)}{dx} - \frac{e}{k_0T}\frac{d\varphi}{dx}n(x) - \frac{j}{\mu k_0T} = 0, \qquad (7.17)$$

where we made use of the Einstein relation (7.11).

The distribution of the electrostatic potential $\varphi(x)$, is, of course, determined by the concentration distribution $n(x)$. However, formally we can consider (7.17) as a first-order linear differential equation for the unknown function $n(x)$, assuming $\frac{d\varphi}{dx}$ to be a specified function of $x$. The solution of (7.17) is of the form [6.15, § 1, i. 6]

$$n(x)=n\exp\frac{e\varphi(x)}{k_0T}-\frac{j}{\mu k_0T}\int_x^\infty \exp\frac{e}{k_0T}[\varphi(x)-\varphi(\xi)]\,d\xi, \tag{7.18}$$

where the single integration constant has been chosen so as to satisfy the boundary condition $n(\infty)=n$ ($n$ is the constant concentration in the semiconductor's bulk). The solution (7.18) can be checked by direct substitution into (7.17). When differentiating the integral in (7.18) with respect to $x$ we should keep in mind that both the integrand and the lower integration limit depend on $x$ [6.15, § 8, i. 83].

Apply now the solution (7.18) to the semiconductor's surface at $x=0$. Then

$$n_0\equiv n(0)=n\exp\frac{e\varphi_0}{k_0T}\exp\frac{eU}{k_0T}$$
$$-\frac{j}{\mu k_0T}\int_0^\infty \exp\frac{e}{k_0T}[\varphi(0)-\varphi(\xi)]\,d\xi, \tag{7.19}$$

where we made use of (7.16).

Since in the case being considered each of the electron fluxes—from the metal to the semiconductor and the reverse one—are much greater than the resultant flux $(1/e)\,j$, the electron concentration at the boundary $x=0$ remains almost equal to its equilibrium value, i.e.,

$$n_0=n\exp\frac{e\varphi_0}{k_0T}. \tag{7.20}$$

Solving (7.19) with respect to $j$ and making use of (7.20), we obtain

$$j=j_0\left[\exp\frac{eU}{k_0T}-1\right], \tag{7.21}$$

where

$$j_0=\frac{n_0\mu k_0T}{\int_0^\infty \exp\frac{e}{k_0T}[\varphi(0)-\varphi(\xi)]\,d\xi}. \tag{7.21a}$$

As we shall see below, the integral that determines $j_0$ depends little on $U$, and because of that the relation between $j$ and $U$ in the diffusion theory of rectification (7.21) takes the same form as in the diode theory (7.15). Inside the barrier layer the electrostatic potential $\varphi(x)$ is negative, being maximum in magnitude at $x=0$; therefore, the exponent in the integral (7.21a) is negative and rapidly grows in magnitude with the increase in $\xi$. This makes it possible to calculate the integral by expanding φ (ξ) into a series in ξ and retaining only the linear terms:

$$\varphi(\xi)=\varphi(0)+\left(\frac{d\varphi}{d\xi}\right)_0\xi=\varphi(0)+\left(\frac{d\varphi}{dx}\right)_0\xi. \tag{7.22}$$

Then

$$\int_0^\infty \exp\left\{\frac{e}{k_0T}[\varphi(0)-\varphi(\xi)]\right\}d\xi$$

$$\approx\int_0^\infty \exp\left[-\frac{e}{k_0T}\left(\frac{d\varphi}{dx}\right)_0\xi\right]d\xi=\frac{k_0T}{e}\frac{1}{\left(\frac{d\varphi}{dx}\right)_0}. \tag{7.23}$$

Substituting (7.23) into (7.21a), we obtain

$$j_0=en_0\mu\left(\frac{d\varphi}{dx}\right)_0, \tag{7.24}$$

the field-induced current at the semiconductor's boundary. The weak dependence of $\left(\frac{d\varphi}{dx}\right)_0$ on $U$ in (7.21) can be neglected as compared with the exponential term. In this case (7.21) is similar to (7.15).

Hence, in the case of the diffusion rectification theory, too, the volt-ampere characteristic retains the shape depicted in Fig. 6.6; however, in this case the current $j_0$ for a barrier of equal height $\varphi_0$ is much less.

The formation of a barrier layer at the semiconductor's surface is not always due to the difference in work functions of the metal and the semiconductor. In a variety of cases the barrier layer is formed because electrons localized on surface levels (Section 5.2.3), charge the semiconductor's surface. In this case the surface barrier $V_0$ is independent of the nature of the metal contact (J. Bardeen, 1947).

In numerous cases important for technical applications the barrier layer is of chemical origin (oxide films, shellac, etc.). The rectification theory for this case (N. F. Mott) is similar to the one presented above.

## 6.8. Properties of p-n Junctions

6.8.1 In the preceding section we studied the electrical properties of the metal-semiconductor contact. It would seem appropriate now to consider the contact between two semiconductors, for example, of different conductivity types—the electron and the hole type. The contact between an $n$- and a $p$-type semiconductor can be realized in its most unmitigated form if two adjacent regions, one of the electron-type conductivity and the other of the hole-type conductivity, are created inside the same crystal. The boundary between the $p$ and $n$ regions inside a semiconductor is termed a $p$-$n$ *junction*.

Most semiconductor radio devices employ $p$-$n$ junctions in germanium or silicon. The technology of manufacturing such devices, which we shall not stop to deal with, has reached a high degree of perfection.

Consider an ideal $p$-$n$ junction in Ge: the acceptor concentration, for instance, of indium atoms to the left of the plane $x = 0$ is constant and equal to $n_A$; to the right of $x = 0$ there is a constant donor concentration (for instance, of antimony atoms) equal to $n_D$. At temperatures not too low all impurities are ionized, so that in the bulk of the left $p$-region the hole concentration is $p_p \approx n_A$, and in the bulk of the right $n$-region the concentration of conduction electrons is $n_n \approx n_D$. The mobile holes diffuse from the left $p$-region to the right $n$-region; vice versa, the conduction electrons will cross the plane $x = 0$ from the right to the left. As a result a diffuse layer of a negative charge is formed to the left of $x = 0$, and a layer of a positive charge is formed to the right of $x = 0$. The double layer thus formed results in a potential jump that prevents any further hole flux from the left to the right and electron flux from the right to the left.

In addition to impurity conductivity, germanium exhibits some intrinsic conductivity. Let $n_p$ be the equilibrium concentration of electrons in the bulk of the $p$-region and $p_n$ the equilibrium concentration of holes in the bulk of the $n$-region. At any point of the $p$-$n$ junction being considered here the electron and hole concentrations satisfy the relation (2.20):

$$np = n_i^2, \tag{8.1}$$

where $n_i$ is the intrinsic carrier concentration. It follows from (8.1) that

$$\log p - \log n_i = \log n_i - \log n. \tag{8.1a}$$

Figure 6.7 is a representation on a logarithmic scale of the electron and hole concentrations in a $p$-$n$ junction. We see the curves $\log p\ (x)$ and $\log n\ (x)$ to be symmetrical about the straight line $\log n_i$, this is a direct result of (8.1a). The planes $x_p$ and $x_n$ define the boundaries of the double layer or of rapid variation of the potential in the $p$-$n$ junction. In all practical cases $n_p \ll p_p$ and $p_n \ll n_n$.[14] If, for example,[15] $n_A = p_p = 10^{16}$ and $n_D = n_n = 10^{14}$, then it follows from (8.1) that

$$n_p = \frac{n_i^2}{p_p} \approx 10^{10}, \quad p_n = \frac{n_i^2}{n_n} \approx 10^{12},$$

[14] Therefore we make $p_p = n_A$ and $n_n = n_D$.

[15] Here and below the concentrations are given in $cm^{-3}$.

since in germanium at room temperature $n_i \approx 10^{13}$ cm$^{-3}$. As in the case of the barrier layer, we can write for the p-n junction the Poisson equation (7.2) that can be solved in certain limiting cases. Again denoting the electrostatic potential by $\varphi(x)$ and putting in the bulk

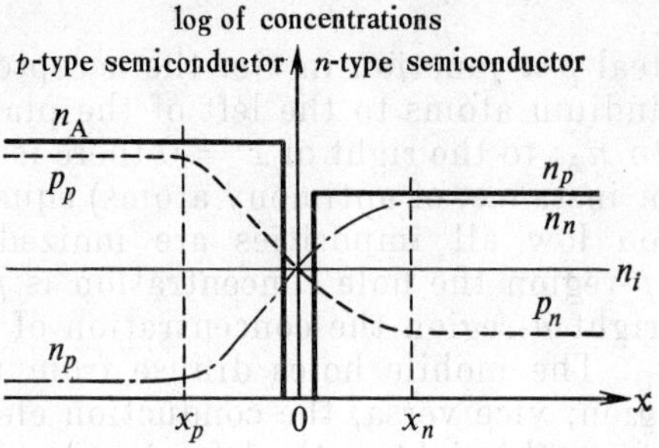


**Fig. 6.7**

of the n-region $\varphi(+\infty) = 0$, we obtain from the Boltzmann distribution

$$p(x) = p_n \exp\left(-\frac{e\varphi(x)}{k_0 T}\right), \tag{8.2}$$

$$n(x) = n_n \exp\left(+\frac{e\varphi(x)}{k_0 T}\right). \tag{8.2a}$$

If $\varphi(-\infty) \equiv \varphi_0$, the potential jump across the p-n junction is

$$\varphi_0 = -\frac{k_0 T}{e} \log\frac{p_p}{p_n} = \frac{k_0 T}{e} \log\frac{n_p}{n_n}. \tag{8.3}$$

At room temperature and for the concentrations cited above

$$\varphi_0 = -0.026 \cdot \log 10^4 = -0.24 \text{ V}. \tag{8.3a}$$

Since we have been considering a p-n junction in the state of statistical equilibrium, the level of the chemical potential $\zeta$ in both p- and n-regions should be identical. Figure 6.8 depicts the electron's potential energy $V(x) = -e\varphi(x)$ in the p-n junction. The dimensions of the region of substantial variation of the potential $(x_n - x_p)$ are of the order of the Debye length $l_D$.

6.8.2 Consider now the current through the p-n junction and demonstrate that the latter exhibits rectifying properties.[16] Retaining the potential of the n-region at the zero level (for example, earthing it), apply a positive electrostatic potential $U$ to the p-region. Since

[16] For the sake of simplicity, we consider the case when $n_A = n_D$, and, accordingly, $x_n = |x_p|$.

the resistance of the $p$-$n$ junction itself is relatively high, we shall assume the entire potential difference $U$ to be applied to the region $(x_n - x_p)$. A positive potential $U$ reduces the jump of the negative potential $\varphi_0$ (8.3) and causes an additional hole flux to flow from the $p$-region, which on entering the $n$-region gradually recombines with the electrons, so that the hole concentration drops to its equilibrium value $p_n$. A similar situation is observed for electrons that move in

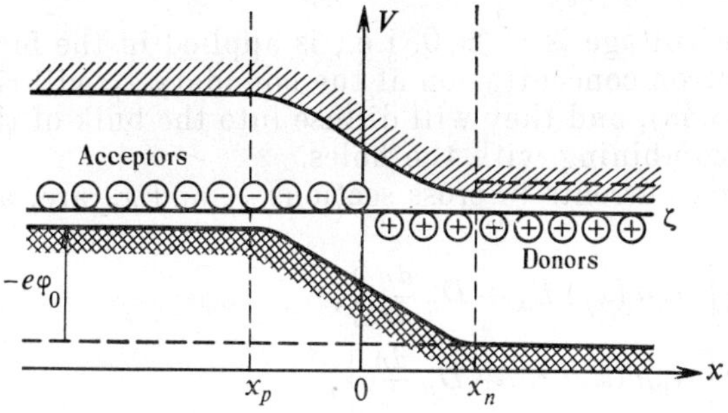


Fig. 6.8

the $n$-region in the direction opposite to the holes and enter the $p$-region. As a result, a purely hole current in the $p$-region changes to a purely electron current in the $n$-region. In the stationary case the sum of both currents $j_p + j_n$ should be constant at all points of the $p$-$n$ junction. The penetration of minority charge carriers into a semiconductor (of holes into an $n$-type semiconductor and of electrons into a $p$-type semiconductor) is termed *carrier injection.* Most semiconductor devices operate on the principle of carrier injection.

In the case under consideration, when the holes in the $p$-region and the electrons in the $n$-regions move in opposite directions, the resistance of the $p$-$n$ junction falls just as it did when electrons were flowing from the semiconductor to the metal (Sec. 6.7). Hence, the current will grow more rapidly than stipulated by the linear Ohm's low, i. e., this will be the forward direction $U > 0$.

We shall consider the forward current in conditions of a low recombination rate of minority carriers, when they penetrate deep into the semiconductor (W. Shockley). To be more precise, the transitional region in which the current $j_p$ is replaced by the current $j_n$ is much wider than the $p$-$n$ junction "width" $x_n - x_p$. This means that the current components at points $x_n$ and $x_p$ are $j_n(x_p) = j_n(x_n)$ and $j_p(x_p) = j_p(x_n)$. For $n_n = p_p$, $j_n = j_p$. At $U = 0$ in the state of equilibrium

$$n(x_p) = n_p = n_n \exp\frac{e\varphi_0}{k_0T} \quad (\varphi_0 < 0), \tag{8.4}$$

since the electron density obeys the Boltzmann law (8.2a). At $U > 0$ the electron flux from the $n$- to the $p$-region exceeds the electron flux flowing in the opposite direction, but the difference between the fluxes remains small in comparison to each of them, and because of that the Boltzmann law is still approximately valid for the carrier concentration distribution:

$$n(x_p) = n_n \exp \frac{e\varphi_0 + eU}{k_0 T} = n_p \exp \frac{eU}{k_0 T}. \qquad (8.4a)$$

Hence, if the voltage is $U > 0$, i.e., is applied in the forward direction, the electron concentration at the boundary $x_p$ will rise in accordance with (8.4a), and they will diffuse into the bulk of the $p$-region, gradually recombining with the holes.

The electron currents in cross sections $x_p$ and $x_n$ are, according to (7.7),

$$j_n(x_p) = e\left[\mu_n n(x_p) E_x + D_n \frac{dn}{dx}\right], \qquad (8.5)$$

$$j_n(x_n) = e\left[\mu_n n(x_n) E_x + D_n \frac{dn}{dx}\right], \qquad (8.5a)$$

where all the quantities in the right-hand sides of (8.5) and (8.5a) refer to the respective cross sections $x_p$ and $x_n$.

In the $n$-region, where the electron concentration is high, even a weak electric field $E_x$ will cause a large conductivity current, so that the diffusion current can be neglected, and $j_n(x_n) \approx e\mu_n n_n E_x$. However, if $\exp\left[\frac{e(\varphi_0+V)}{k_0 T}\right] \ll 1$, then $\frac{n(x_p)}{n_n} \ll 1$, and for this reason the conductivity current in $j_n(x_p)$ is small as compared with $e\mu_n n_n E_x = j_n(x_n)$. Hence, $j_n(x_p) \approx$ $\approx j_n(x_n)$, and the current $j_n(x_p)$ is practically a diffusion current.

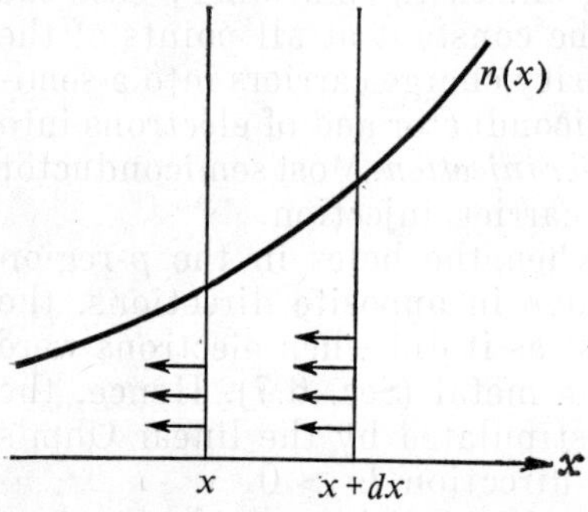


Fig. 6.9

Calculate the diffusion electron current in the cross section $x_p$. Consider first the electron balance in the layer $dx$ shown in Fig. 6.9. The electron flux entering the layer $dx$ through the cross section $x+dx$ is $D_n \frac{dn(x+dx)}{dx}$, where $D_n$ is the diffusion coefficient, and the electron flux flowing out of the cross section $x$ is $D_n \frac{dn(x)}{dx}$. Therefore in the stationary case

$$D_n \frac{dn(x+dx)}{dx} - D_n \frac{dn(x)}{dx} = (\gamma np - g)\,dx, \qquad (*)$$

where $g$ is the number of electrons (holes) generated in cubic centimeter per second as the result of thermal excitation, and $\gamma np$ is the number of electrons annihilated in cubic centimeter per second as the result of their recombination with holes ($\gamma$ is the volume recombination coefficient); $g$ is the same everywhere in the $p$-$n$ junction; in the bulk of the $p$-region the generation and recombination fully compensate each other. Therefore,

$$g = \gamma n_p p_p.$$

At $x \approx x_p$ the hole concentration is $p \approx p_p$. Since

$$\frac{dn(x+dx)}{dx} = \frac{dn(x)}{dx} + \frac{d^2n(x)}{dx^2}dx,$$

it follows from (*) that

$$D_n \frac{d^2n(x)}{dx^2} = \gamma p_p [n(x) - n_p]$$

or

$$\frac{d^2n(x)}{dx^2} = \frac{1}{\mathscr{L}_n^2}[n(x) - n_p], \tag{8.6}$$

where

$$\mathscr{L}_n = \sqrt{\frac{D_n}{\gamma p_p}} \tag{8.6a}$$

is the *diffusion length for the electrons* in the $p$-region.

In the transient case in the absence of a current the electron annihilation rate $-\left(\frac{\partial n}{\partial t}\right)$ is proportional to the deviation of the electron concentration $n$ from its equilibrium value[17]:

$$-\frac{\partial n}{\partial t} = \frac{n - n_p}{\tau_n}, \tag{8.7}$$

where $\tau_n$ is the *lifetime of the electrons* in the $p$-region.

On the other hand,

$$-\frac{\partial n}{\partial t} = \gamma n p_n - g = \gamma p_p (n - n_p). \tag{8.8}$$

Comparing (8.7) with (8.8), we obtain

$$\tau_n = 1/\gamma p_p, \tag{8.9}$$

but then

$$\mathscr{L}_n = \sqrt{D_n \tau_n}. \tag{8.10}$$

[17] At any rate, for small deviations of concentration from its equilibrium value $n_p$.

Now we can formulate the condition of low recombination or deep injection, used by us:

$$\mathcal{L}_n \gg | x_p |. \tag{8.11}$$

The solution of (8.6) that satisfies the boundary condition $n(-\infty) = n_p$ is of the form

$$n(x) = n_p + C \exp \frac{x}{\mathcal{L}_n}, \tag{8.12}$$

where $C$ is the integration constant.

The electron flux through the cross section $x_p$ from right to left is

$$D_n \frac{dn(x)}{dx}\bigg|_{x=x_p} = \frac{D_n}{\mathcal{L}_n} C e^{x/\mathcal{L}_n} = \frac{D_n}{\mathcal{L}_n} [n(x_p) - n_p], \tag{8.13}$$

therefore, the current is

$$j_n(x_p) = \frac{eD_n}{\mathcal{L}_n} [n(x_p) - n_p]. \tag{8.14}$$

In statistical equilibrium $n(x_p) = n_p$, and the current $j_n = 0$. A rise in the electrostatic potential at point $x_p$ by the amount $U$ reduces the potential energy of an electron by $-eU$ and increases the electron concentration in accordance with (8.4a). It follows from (8.14) and (8.4a) that

$$j_n(x_p) \approx \frac{eD_n n_p}{\mathcal{L}_n} \left[\exp \frac{eU}{k_0 T} - 1\right]. \tag{8.15}$$

From similar considerations we obtain

$$j_p(x_n) \approx \frac{eD_p p_n}{\mathcal{L}_p} \left[\exp \frac{eU}{k_0 T} - 1\right]. \tag{8.16}$$

As we have pointed out above, $j_p(x_n) \approx j_p(x_p)$, therefore, the total forward current is

$$j = j_n(x_p) + j_p(x_p) = j_0 [e^{eU/k_0 T} - 1], \tag{8.17}$$

where

$$j_0 \equiv \frac{eD_n n_p}{\mathcal{L}_n} + \frac{eD_p p_n}{\mathcal{L}_p}. \tag{8.17a}$$

For $U < 0$ (8.17) determines the reverse current. It will be seen from (8.17) that the volt-ampere characteristic of a *p-n* junction is of the same type as in the case of barrier layer rectification (see Fig. 6.6).

The assumptions made in the *p-n* junction rectification theory are well founded in the case of germanium, the proof of this being the agreement of the theory with the experiment depicted in Fig. 6.10.

We are not in a position to deal here with numerous semiconductor devices whose operation is based on the *p-n* junctions: transistors, tunnel diodes, *p-n* junction photocells, phototransistors, semicon-

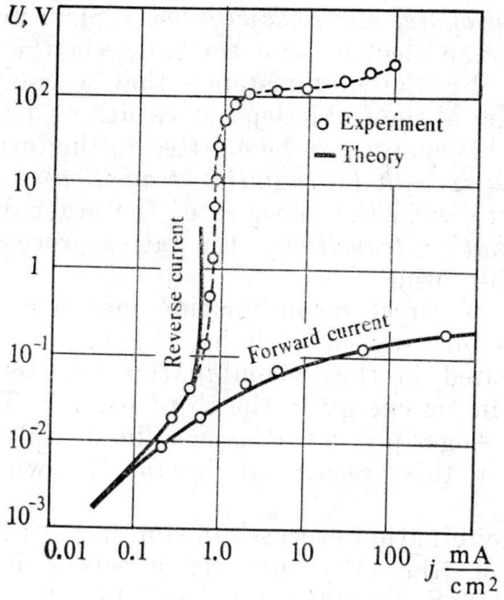


Fig. 6.10

ductor lasers, etc. This extensive field of semiconductor electronics has grown into a separate discipline, and the study of it requires specialized literature.[18]

## 6.9. Generation and Recombination of Charge Carriers. Quasi-Fermi Levels

6.9.1 When we considered the properties of a semiconductor in the state of statistical equilibrium (see Sec. 6.2), we pointed out that they are independent of the mechanism of the interaction of charge carriers with one another and with the lattice. However, the interaction mechanism becomes essential when nonequilibrium properties, specifically the free charge transport phenomena, begin to be considered.

In intrinsic semiconductors free from impurity centres the free electrons and holes appear as the result of electron transitions from the valence band to the conduction band. Such *generation of electron-hole pairs* can be caused by thermal excitation of electrons in the

[18] A sound treatment of the subject can be found in [6.16].

valence band or by the absorption by them of electromagnetic radiation supplied by an external source, or by the absorption of the energy of conduction electrons accelerated in a strong electric field, etc. The annihilation of such electron-hole pairs takes place as the result of the *recombination of free electrons with holes*. Such processes of *direct recombination* of the electron and the hole via the forbidden band are complicated by the circumstance that a comparatively high energy of the order of the forbidden bandwidth $\varepsilon_G$ has to be liberated in the process. This energy can be emitted in the form of a photon of the frequency $\omega_s \approx \varepsilon_G/\hbar$ (a *radiative transition*) or in the form of numerous phonons with the energies of the order of $\hbar\omega_q \approx k_0T \ll \ll \varepsilon_G$ (a *nonradiative transition*), the latter process being far less probable than the former.

The processes of direct recombination may also involve a third free particle (an electron or a hole). In this triple collision the energy $\varepsilon_G$ liberated in the recombination process is transmitted in the form of kinetic energy to the third particle. This phenomenon is similar to the Auger process (P. Auger) in an isolated atom [6.17, p. 299], and for this reason it became known as the Auger recombination.

Finally, a recombination process can simultaneously involve a photon and a phonon (the latter usually affects only the momentum of a free particle). Such indirect transitions will be considered in more detail when we deal with the absorption of light by semiconductors (Sec. 7.3). The variation of the number of free electrons (holes) per cubic centimetre per second as a result of direct recombination (without a third particle being involved) is obviously proportional to the rate of electron-hole collisions, and the latter in turn is proportional to their concentrations $n$ and $p$. Hence,

$$-\left(\frac{dn}{dt}\right)_r = -\left(\frac{dp}{dt}\right)_r = \gamma np - g. \tag{9.1}$$

Here $\gamma$ is the volume recombination coefficient with the dimensionality $cm^3\ s^{-1}$, $g$ is the electron-hole pair generation rate per cubic centimeter as a result of the thermal excitation of electrons in the valence band.[19]

In the state of statistical equilibrium of the semiconductor $n = n_0$, $p = p_0$ and $\gamma n_0 p_0 - g = 0$, i.e., the thermal generation rate is

$$g = \gamma n_0 p_0, \tag{9.2}$$

[19] The right-hand side of (9.1) could be supplemented by the term describing the generation of electron-hole pairs, for example, as a result of absorption of light.

this, of course, being true in the nonequilibrium case as well. Hence, (9.1) takes the form

$$-\left(\frac{dn}{dt}\right)_r = \gamma\,(np - n_0 p_0). \tag{9.3}$$

In Sec. 5.2 we mentioned shallow and deep impurity levels in germanium and in silicon. Deep impurity levels can substantially facilitate the process of recombination of a free electron with a hole, since the transition of an electron (a hole) to a deep impurity level results in the liberation of lower energy than in the case of the direct band-to-band recombination. Such *indirect* (*step*) recombination can, for example, be imagined as follows: a deep donor level traps a free electron and then the neutral donor traps a hole; the result is the recombination of an electron with a hole. It should be pointed out that the probability of a multiphonon electron (hole) transition to an impurity level is much greater than that of multiphonon radiation in the course of direct recombination (this is due to the deformation of the lattice near an impurity centre).

An equation for the recombination rate similar to (9.3) can be written for indirect recombination. If the semiconductor contains, for example, donors, the number of free electrons trapped by donors per unit time is proportional to the electron concentration and to the concentration of vacant (positively charged) donors; the recombination rate of electrons trapped by donors with holes is proportional to the concentration of occupied (neutral) donors and to the hole concentration.

Indirect recombination can take place via several trapping levels on impurity centres as well as with the participation of a third particle (Auger processes).

All such cases, which we are unable to discuss here, have been extensively studied both experimentally and theoretically (e.g., see [6.16, Chap. 6] and [6.18, Chaps. 5 and 6]).

**6.9.2** If the electron and hole concentrations are not equilibrium concentrations, in the majority of cases we can consider the electrons and the holes to be in statistical equilibrium. This is due to the fact that the characteristic time between collisions for charge carriers in each band is of the order of $10^{-12}$ s, while the average time for recombination processes varies in the wide range from $10^{-3}$ to $10^{-11}$ s. Thus, the electrons (holes) inside the band reach equilibrium much quicker than they recombine.

If a semiconductor is in statistical equilibrium, the distribution functions for electrons (2.1) and for holes (2.14) depend on a single parameter, the chemical potential $\zeta$.

When electrons in the conduction band and holes in the valence band are separately in statistical equilibrium, we can introduce different chemical potentials: $\zeta_n$ for the electrons and $\zeta_p$ for the holes (these potentials are called the *quasi-Fermi levels*). Then the expressions for the nonequilibrium concentrations of electrons and holes in the nondegenerate case will be

$$n=\frac{(2\pi m_n k_0 T)^{3/2}}{4\pi^3\hbar^3}\exp\frac{\zeta_n}{k_0 T}, \tag{9.4}$$

$$p=\frac{(2\pi m_p k_0 T)^{3/2}}{4\pi^3\hbar^3}\exp\left(-\frac{\varepsilon_G+\zeta_p}{k_0 T}\right). \tag{9.4a}$$

In equilibrium the concentrations are obviously $n = n_0$, $p = p_0$, and the quasi-Fermi levels coincide with the equilibrium chemical potential, i.e., $\zeta_n = \zeta_p = \zeta$. Vice versa, the greater the difference between the concentrations and their equilibrium values the more $\zeta_n$ and $\zeta_p$ deviate from $\zeta$. Indeed, it follows from (9.4), (9.4a), (2.15b), and (2.15c) that

$$\zeta_n-\zeta=k_0T\ln\frac{n}{n_0}, \qquad \zeta-\zeta_p=k_0T\ln\frac{p}{p_0}. \tag{9.5}$$

Consider now the situation when an electric field is applied to a semiconductor with nonequilibrium carrier concentrations, so that electron and hole currents flow in it.

If the field $\mathbf{E} = -\text{grad}\,\Phi$ applied to the conductor at a specified point $\mathbf{r}$ is much less than the crystal field, i.e., much less than several hundred thousand volts per centimeter, then the energy band structure of the semiconductor will remain intact, with the band edges acquiring in the electric field $\mathbf{E}$ an equal slope leaving the forbidden bandwidth unaltered, so that grad $\varepsilon_G = 0$.

The conduction electron's energy at point $\mathbf{r}$ in the electric field $\mathbf{E} = -\text{grad}\,\Phi(\mathbf{r})$, where $\Phi(\mathbf{r})$ is the electrostatic potential, is $\varepsilon_{\mathbf{k}} - e\Phi(\mathbf{r})$, (where $\varepsilon_{\mathbf{k}}$ is the "kinetic" energy of the electron, which depends on the wave vector $\mathbf{k}$). Similarly, the hole's energy is $\varepsilon_{\mathbf{k}'} + e\Phi(\mathbf{r})$, where the hole's kinetic energy $\varepsilon_{\mathbf{k}'} < 0$. For a standard band

$$\varepsilon_k = (\hbar^2/2m_n)\,k^2, \quad \text{and} \quad \varepsilon_{k'} = -(\hbar^2/2m_p)\,k'^2. \tag{9.6}$$

If we substitute now total energies that include potential energies in the field $\Phi(\mathbf{r})$ for the kinetic energies in (2.1) and (2.14), expressions (9.4) and (9.4a) will assume the form

$$n(\mathbf{r})=\frac{(2\pi m_n k_0 T)^{3/2}}{4\pi^3\hbar^3}\exp\frac{\zeta_n(\mathbf{r})+e\Phi(\mathbf{r})}{k_0 T}, \tag{9.7}$$

$$p(\mathbf{r})=\frac{(2\pi m_p k_0 T)^{3/2}}{4\pi^3\hbar^3}\exp\left(-\frac{\varepsilon_G+\zeta_p(\mathbf{r})+e\Phi(\mathbf{r})}{k_0 T}\right). \tag{9.7a}$$

It is obvious that if in the general case the electron $n(\mathbf{r})$ and hole $p(\mathbf{r})$ concentrations are the functions of the coordinates, the quasi-Fermi levels are also the functions of the coordinates.

In the three-dimensional case the combined conductivity and diffusion currents of electrons and holes are, according to (7.7),

$$\mathbf{j}_n = en\mu_n \mathbf{E} + k_0 T \mu_n \operatorname{grad} n, \tag{9.8}$$

$$\mathbf{j}_p = ep\mu_p \mathbf{E} - k_0 T \mu_p \operatorname{grad} p, \tag{9.8a}$$

where we made use of the Einstein relation (7.11), and $\mu_n$ and $\mu_p$ are the electron and hole mobilities. Find grad $n$ (**r**) and grad $p$ (**r**) from (9.7) and (9.7a); take into account that grad $\Phi = -\mathbf{E}$. Then from (9.8) and (9.8a), we obtain

$$\mathbf{j}_n = n\mu_n \operatorname{grad} \zeta_n, \tag{9.9}$$

$$\mathbf{j}_p = p\mu_p \operatorname{grad} \zeta_p. \tag{9.9a}$$

Hence, the gradient of the quasi-Fermi level accounts for both the current due to the gradient of electrostatic potential (conductivity current) and the current due to the concentration gradient (diffusion current).

# 7. Optical Phenomena in Semiconductors

## 7.1. Kramers-Kronig Dispersion Relations

**7.1.1** In this section we shall discuss certain problems of optics of isotropic media from the viewpoint of classical electrodynamics. The electric induction vector in an optically isotropic medium (for instance, in a cubic crystal) in a static or in a slowly varying electromagnetic field is

$$\mathbf{D} = \varepsilon \mathbf{E} = \mathbf{E} + 4\pi \mathbf{P}, \tag{1.1}$$

where **E** is the average electric field intensity in the medium, **P** is the polarization vector, and $\varepsilon$ is the dielectric permittivity (dielectric constant).

Relation (1.1) is based on the assumptions of the medium's isotropy, linear relationship between **D** and **E**, localization and synchronism in the sense that $\mathbf{D}(\mathbf{r}, t)$ is a function of $\mathbf{E}(\mathbf{r}, t)$ at the same point in space and at the same moment of time. If the relationship between $\mathbf{D}(\mathbf{r}, t)$ and $\mathbf{E}(\mathbf{r}, t)$ is not of a localized and synchronous nature, this is referred to as *dispersion in space and time*. We shall demonstrate how the relationship between **D** and **E** changes in a rapidly varying electromagnetic field, when the latter condition—that of synchronism—is violated. We shall further investigate whether or not the condition of localized interaction can remain valid in a rapidly varying electromagnetic field, i.e., whether or not the space periodicity can in this case greatly exceed the dimensions of the atomic structure of the lattice. The term rapidly varying electromagnetic field will be used in the sense that the field's frequency $\omega > \omega_P$, the polarization frequency of the medium ($1/\omega_P \approx \tau_P$ is the polarization time). The maximum frequency connected with the redistribution of atomic electrons is of the order of $\omega_P \approx v_a/a$, where $v_a$ is the velocity of motion of atomic electrons and $a$ is the atomic dimension. The maximum wavelength of the electromagnetic field corresponding to $\omega = \omega_P$ is $\lambda = \dfrac{c}{\omega_P} \approx \dfrac{c}{v_a} \cdot a \gg a$, since the velocity $v_a$ is at least two orders of magnitude lower than the velocity of light $c$. Hence, in the case of rapidly varying electromagnetic fields with a frequency $\omega \gg \omega_P$, the wavelength $\lambda$ can still exceed the atomic dimensions. This makes it possible to apply macroscopic equations of electrodynamics, i.e., to use the localized description of phenomena.[1]

The considerations presented below refer both to dielectrics and to semiconductors and metals.

If we assume that the relationship between **D** and **E** is linear, as in (1.1), but the frequency of the electromagnetic field is so high that the value of **D** $(t)$ is determined not solely by the value of **E** $(t)$ at the same instant of time but also by preceding values **E** $(t')$ $(-\infty < t' \leqslant t)$, we obtain

$$\mathbf{D}(t) = \mathbf{E}(t) + \int_{-\infty}^{t} f(t-t')\,\mathbf{E}(t')\,dt'. \tag{1.2}$$

Here $f(t)$ is a function determined only by the properties of the medium. As we shall see below, it is convenient to separate the term **E** $(t)$ on the right-hand side of (1.2). The fact that **D** $(t)$ is determined by the values of **E** $(t')$ at preceding instants of time agrees with the *causality principle.*

The integral on the right-hand side of (1.2) can be written in the form

$$\int_{-\infty}^{\infty} f(t-t')\,\mathbf{E}(t')\,dt', \tag{1.2a}$$

if we make an additional assumption that $f(t-t') = 0$ for $t' > t$.

Let us go over to the Fourier representation of the functions **D** $(t)$, **E** $(t)$, and $f(t)$:

$$\mathbf{D}(t) = \frac{1}{2\pi}\int_{-\infty}^{\infty} \mathbf{D}(\omega)\,e^{-i\omega t}\,d\omega, \tag{1.3}$$

$$\mathbf{E}(t) = \frac{1}{2\pi}\int_{-\infty}^{\infty} \mathbf{E}(\omega)\,e^{-i\omega t}\,d\omega, \tag{1.3a}$$

$$f(t) = \frac{1}{2\pi}\int_{-\infty}^{\infty} f(\omega)\,e^{-i\omega t}\,d\omega. \tag{1.3b}$$

Here it is presumed that for $t > 0$ the function $f(t) = 0$. The inverse transformations are of the form

$$\mathbf{D}(\omega) = \int_{-\infty}^{\infty} \mathbf{D}(t)\,e^{i\omega t}\,dt, \tag{1.4}$$

those for **E** $(\omega)$ and $f(\omega)$ being similar.

[1] It should be noted that close to the exciton absorption fringe and when $\lambda \gg a$ we have to take into account dispersion in space.

Substituting (1.3), (1.3a), (1.3b) into (1.2), we obtain

$$\frac{1}{2\pi}\int_{-\infty}^{\infty} \mathbf{D}(\omega)\, e^{-i\omega t}\, d\omega = \frac{1}{2\pi}\int_{-\infty}^{\infty} \mathbf{E}(\omega)\, e^{-i\omega t}\, d\omega$$

$$+\frac{1}{(2\pi)^2}\int_{-\infty}^{\infty} dt' \int_{-\infty}^{\infty} d\omega \int_{-\infty}^{\infty} d\omega' f(\omega)\, \mathbf{E}(\omega')\, e^{-i\omega(t-t')} e^{-i\omega' t'}.$$

Since

$$\frac{1}{2\pi}\int_{-\infty}^{\infty} dt'\, e^{i(\omega-\omega')t'} = \delta(\omega-\omega'),$$

it follows that

$$\int_{-\infty}^{\infty} \{\mathbf{D}(\omega) - [1+f(\omega)]\,\mathbf{E}(\omega)\}\, e^{-i\omega t}\, d\omega = 0.$$

Since the integral must be zero for all values of $t$, it follows that

$$\mathbf{D}(\omega) = \tilde{\varepsilon}(\omega)\,\mathbf{E}(\omega), \tag{1.5}$$

where

$$\tilde{\varepsilon}(\omega) = 1 + f(\omega) = 1 + \int_{0}^{\infty} f(t)\, e^{i\omega t}\, dt. \tag{1.5a}$$

We see from (1.5) that in the case of a rapidly varying electromagnetic field the connection between the Fourier components of the electric induction $\mathbf{D}$, of the field intensity $\mathbf{E}$ and the dielectric permittivity $\tilde{\varepsilon}$ is the same as in the stationary case (1.1).

The term *frequency dispersion* of the dielectric permittivity $\tilde{\varepsilon}(\omega)$ means its frequency dependence.

Since $f(t)$ is a real function, it follows from (1.5a) that $\tilde{\varepsilon}(\omega)$ is a complex function, i.e.,

$$\tilde{\varepsilon}(\omega) = \varepsilon_1(\omega) + i\varepsilon_2(\omega), \tag{1.6}$$

where $\varepsilon_1(\omega)$ is its real and $\varepsilon_2(\omega)$ its imaginary parts.

It follows from (1.5a) that the change of sign of $\omega$ transforms $\tilde{\varepsilon}(\omega)$ into $\tilde{\varepsilon}^*(\omega)$, i.e., $\varepsilon_1(-\omega) + i\varepsilon_2(-\omega) = \varepsilon_1(\omega) - i\varepsilon_2(\omega)$, whence

$$\varepsilon_1(-\omega) = \varepsilon_1(\omega), \quad \varepsilon_2(-\omega) = -\varepsilon_2(\omega). \tag{1.6a}$$

Hence, the real part of the dielectric permittivity is an even function of $\omega$, and the imaginary part is an odd function of $\omega$. As we shall see below, the relationship of $\varepsilon_1(\omega)$ and $\varepsilon_2(\omega)$ to the optical constants of a medium—the refractive index and the absorption coefficient—is a simple one.

Important relations exist between $\varepsilon_1(\omega)$ and $\varepsilon_2(\omega)$ (see Appendix 19):

$$\varepsilon_1(\omega)-1=\frac{1}{\pi}\fint_{-\infty}^{\infty}\frac{\varepsilon_2(x)}{x-\omega}\,dx=\frac{2}{\pi}\fint_{0}^{\infty}\frac{x\varepsilon_2(x)}{x^2-\omega^2}\,dx, \tag{1.7}$$

$$\varepsilon_2(\omega)=-\frac{1}{\pi}\fint_{-\infty}^{\infty}\frac{\varepsilon_1(x)-1}{x-\omega}\,dx=-\frac{2\omega}{\pi}\fint_{0}^{\infty}\frac{\varepsilon_1(x)}{x^2-\omega^2}\,dx, \tag{1.7a}$$

termed *Kramers-Kronig dispersion relations* (1927). The integrals in (1.7) and (1.7a) should be interpreted in the sense of their principal value [7.1, i. 26].

The deviation of relations (1.7) and (1.7a) is based fundamentally on the assumption that $f(t-t')=0$ for $t'>t$, i.e., on the causality principle. Relations (1.7) and (1.7a) can be used to process experimental data, since, as we have already stated, $\varepsilon_1(\omega)$ and $\varepsilon_2(\omega)$ are directly related to the optical constants of a medium: its refractive index and absorption coefficient (see below).

It follows from the dispersional relations (1.7) that if we know (for instance, from the experiment) the $\varepsilon_2(\omega)$ dependence in the entire frequency range, we can find the dependence $\varepsilon_1(\omega)$, and vice versa. Of course, we never know the dependence $\varepsilon_2(\omega)$ in the entire frequency range. However, if we are interested in $\varepsilon_1(\omega)$ for the frequency $\omega$, we can neglect the contribution of $\varepsilon_2(x)$ for the values of $x$ greatly differing from $\omega$. The shape of the $\varepsilon_1(\omega)$ curve at the point $\omega$ is determined by the values of $\varepsilon_2(x)$ for $x$ close to $\omega$. This results in definite relations between the curves $\varepsilon_1(\omega)$ and $\varepsilon_2(\omega)$ in the vicinity of $\omega$. Velicky [7.2] has demonstrated that

peak of refractive index $\leftrightarrow$ absorption fringe, (1.8)

decline of refractive index $\leftrightarrow$ absorption peak. (1.9)

Such correspondence is observed in experiments.

**7.1.2** J. C. Maxwell has demonstrated that the dielectric constant of a medium is equal to the square of its refractive index. We showed in Section 7.1.1 that with dispersion the dielectric permittivity becomes complex-valued and we pointed out that its real and imaginary parts are related to the optical characteristic of the medium.

Introduce the complex-valued refractive index

$$\tilde{n}=n+ik. \tag{1.10}$$

To establish the physical meaning of its real part $n$ and its imaginary part $k$, assume that it is connected with the complex-valued dielectric permittivity $\tilde{\varepsilon}(\omega)$ by the Maxwell relation

$$\tilde{\varepsilon}(\omega) = \varepsilon_1(\omega) + i\varepsilon_2(\omega) = \tilde{n}^2 = n^2 - k^2 + i\cdot 2nk. \tag{1.11}$$

Hence

$$\varepsilon_1(\omega) = n^2 - k^2, \tag{1.12}$$

$$\varepsilon_2(\omega) = 2nk. \tag{1.12a}$$

Consider the propagation of a plane electromagnetic wave in the direction of the $x$ axis in a medium with a complex-valued dielectric permittivity $\tilde{\varepsilon}$. As we shall see, in this case the wave vector $\varkappa$ must also be complex-valued. For the electric field of the wave we have

$$\mathbf{E} = \mathbf{E}_0 \exp i(\varkappa x - \omega t). \tag{1.13}$$

Substituting this expression into the wave equation [7.3, Sec. 7.10, Eq. (7.83)]

$$\nabla^2\mathbf{E} = \frac{\tilde{\varepsilon}}{c^2}\ddot{\mathbf{E}}$$

with a complex-valued dielectric constant $\tilde{\varepsilon}$, we obtain $\tilde{\varkappa}^2 = (\omega/c)^2\tilde{\varepsilon} = (\omega/c)^2\,\tilde{n}^2$, or

$$\tilde{\varkappa} = \frac{\omega}{c}\tilde{n} = \frac{\omega}{c}(n + ik). \tag{1.13a}$$

Substituting this value of $\tilde{\varkappa}$ into (1.13), we obtain

$$E = E_0 \exp\left(-\frac{\omega k}{c}x\right)\exp\left[i\omega\left(\frac{x}{c_1} - t\right)\right], \tag{1.13b}$$

where $c_1 = c/n$ is the velocity of light in the medium. We see the real part $n$ of the complex-valued refractive index to be the usual refractive index. The imaginary part $k$ is termed *extinction coefficient* and determines the absorption in the medium (the amplitude of the electric field decreases $e \approx 2\cdot 718$ times for $x = c/\omega k)^2$.

When light propagates in the $x$ direction in an absorbing medium, the decrease in its intensity is $dI(x) = -\alpha I\,dx$, where $\alpha$ is the absorption coefficient, whence

$$I(x) = I_0 e^{-\alpha x}. \tag{1.14}$$

Here $I_0 = I(0)$. Since $I \propto E^2$, it follows from (1.14) and (1.13b) that

$$\alpha = 2\omega k/c. \tag{1.15}$$

---

[2] We hope the reader will not be confused by the use here of the conventional notation $n$ and $k$ for the refractive index and the extinction coefficient, despite the fact that in other sections of the book it is used for the free carrier concentration and for the electron's wave vector, respectively.

It follows from (1.12) and (1.12a) that for absorption $\varepsilon_1(\omega) \approx n^2$, and it is determined by the imaginary part of the dielectric permittivity $\varepsilon_2(\omega)$.

A method often used for experimental determination of the optical constants $n$ and $k$ is the measurement of the reflection of light from the surface of a body. For normal incidence the reflection coefficient is [7.3, Sec. 7.1]

$$R = \left| \frac{\tilde{n}-1}{\tilde{n}+1} \right| = \frac{(n-1)^2+k^2}{(n+1)^2+k^2}. \tag{1.16}$$

For low absorption

$$R = \frac{(n-1)^2}{(n+1)^2} \tag{1.16a}$$

and, consequently, by measuring the reflection coefficient we can find the refractive index $n$.

## 7.2. Interband Absorption of Light Involving Direct Transitions

**7.2.1** Information about the behaviour of electrons in atoms and molecules was largely obtained as the result of spectroscopic studies.

Until recently the interpretation of the results of spectroscopic studies carried out for solids, specifically for semiconductors, encountered great difficulties, because the laws governing electron motion in solids are more diversified and complicated than in atoms.

In most semiconductors (germanium, silicon, indium antimonide, gallium antimonide, lead sulphides, selenides and tellurides, etc.) the optical absorption exhibits the following features: as long as the frequency of light is $\omega < \varepsilon_G/\hbar$, where $\varepsilon_G$ is the forbidden bandwidth, the absorption is very small, but when the photon's energy reaches the value $\hbar\omega \geqslant \varepsilon_G$, the absorption increases rapidly by several orders of magnitude because the photon transmits its energy to the electrons of the valence band, ejecting them to the conduction band (*interband absorption*).

The study of the interband absorption spectrum near the threshold of its rapid increase (near the *absorption fringe*[3]) can obviously provide information on the structure of the electron's energy spectrum near the top edge of the valence band and the bottom edge of the conduction band; this information is of essential importance for the determination of the semiconductor's electrical properties.

[3] The wavelength corresponding to the absorption fringe for $\varepsilon_G = 1$ eV is $\lambda_G = 2\pi c/\omega_G = 2\pi c\hbar/\varepsilon_G = 10^{-4}$ cm $= 1$ μm.

The wave vector of the photon, corresponding to the absorption fringe, is usually much less than that of the electron (hole) generated in the result of absorption. Indeed, the photon's wave vector is $\varkappa = 2\pi\lambda_G = 10^5$ cm, and the electron's wave vector is $k = \frac{1}{\hbar}\sqrt{2m(\hbar\omega - \varepsilon_G)}$, its order of magnitude for $\hbar\omega - \varepsilon_G \approx k_0 T$ and $T \sim 100°C$ being $k \approx 10^7$ cm.

Hence, in the case of a photon absorption by an electron it can be assumed that $\varkappa \ll k$. Since the absorption of a photon by an electron requires not only the energy conservation law, but also the wave vector (quasi-momentum) conservation law to be satisfied, the electron's wave vector **k** experiences almost no change, i.e., in the course of the transition from the valence to the conduction band the electron moves almost to the same point of the Brillouin zone (to the same point **k**). These are the so-called *direct transitions*.

It will be seen from Fig. 7.28*a* that the maximum of the valence band of germanium is located in the centre of the Brillouin zone ($\mathbf{k} = 0$), and the minimum of the conduction band—on the Brillouin zone's surface at point $L$ (see Fig. 7.22, $k_x = k_y = k_z = \pi/a$).

Experiment shows that in germanium *indirect transitions* of the valence electron from the centre of the Brillouin zone to the point $L$ on its surface can be observed in addition to the direct transitions. Such a great change in the electron's wave vector is possible only if the absorption of a photon is accompanied by the emission or the absorption of a phonon with a wave vector equal to $q \approx \pi/a$.

Experiment proves that the frequency dependence of the absorption coefficient in some semiconductors is affected by the Coulomb interaction of the electron and the hole generated in the act of photon absorption.

At frequencies lower than the frequency corresponding to the absorption fringe it is possible to observe the absorption of light by conduction electrons, involving electron transitions inside the conduction band. It should be kept in mind that a free electron in the absence of interaction with the imperfections of a regular lattice cannot completely absorb a photon. Indeed, the energy and the momentum conservation laws for this case are of the form[4]

$$\frac{p^2}{2m^*} + \hbar\omega = \frac{p'^2}{2m^*}, \qquad \mathbf{p} + \left(\overrightarrow{\frac{\hbar\omega}{c}}\right) = \mathbf{p}',$$

where $\mathbf{p} = m^*\mathbf{v}$ and $\mathbf{p}' = m^*\mathbf{v}'$ are the electron's momenta prior to and after the absorption of a photon of the frequency $\omega$, $m^*$ is the electron's effective mass, $c$ is the velocity of light, and $\left(\overrightarrow{\frac{\hbar\omega}{c}}\right)$ is the

[4] For simplicity, we consider here a nonrelativistic case and a standard band.

photon's momentum. Substituting $\mathbf{p}'$ from the second equation into the first equation, we obtain for the cosine of the angle between the directions of $\mathbf{p}$ and $\overrightarrow{\left(\frac{\hbar\omega}{c}\right)}$

$$\cos\vartheta = \frac{m^*c}{p}\left[1 - \frac{\hbar\omega}{2m^*c^2}\right] \approx \frac{m^*c}{p} = \frac{c}{v} \gg 1,$$

which cannot be satisfied for any angle $\vartheta$.

Therefore, for simple bands the absorption of light by free carriers can take place only if accompanied by the simultaneous emission or absorption of a photon or by transmission of the momentum to an impurity atom.

Important information on semiconductors can be gained from the study of the absorption of light in them in the presence of electric and especially magnetic fields.

**7.2.2** The electromagnetic field of a light wave in a vacuum can be described with the aid of the *vector potential* $\mathbf{A}(\mathbf{r}, t)$ and the *scalar potential* $\varphi(\mathbf{r}, t)$ [7.3, Secs. 7.4 and 7.6]. The electromagnetic potentials $\mathbf{A}(\mathbf{r}, t)$ and $\varphi(\mathbf{r}, t)$ are defined to within a gradient transformation, and this makes it possible to put the scalar potential $\varphi(\mathbf{r}, t) \equiv 0$; then the electric field $\mathbf{E}(\mathbf{r}, t)$ and the magnetic field $\mathbf{H}(\mathbf{r}, t)$ will be

$$\mathbf{E} = -\frac{1}{c}\frac{\partial \mathbf{A}}{\partial t}, \tag{2.1}$$

$$\mathbf{H} = \operatorname{curl} \mathbf{A}. \tag{2.2}$$

We can also enforce the Lorentz condition

$$\operatorname{div} \mathbf{A} = 0 \tag{2.3}$$

for the vector potential $\mathbf{A}(\mathbf{r}, t)$. In this case $\mathbf{A}(\mathbf{r}, t)$ will satisfy the wave equation.

The Hamiltonian for the electron with the charge $-e$ in a periodic crystal field $V(\mathbf{r})$ takes the form (6.5.12)

$$\hat{\mathscr{H}} = \frac{1}{2m}\left[\hat{\mathbf{p}} + \frac{e}{c}\mathbf{A}(\mathbf{r},\ t)\right]^2 + V(\mathbf{r}), \tag{2.4}$$

where $\hat{\mathbf{p}} = -i\hbar\nabla$. Next

$$\left[\hat{\mathbf{p}} + \frac{e}{c}\mathbf{A}\right]^2 = \hat{\mathbf{p}}^2 + \frac{e}{c}\hat{\mathbf{p}}\mathbf{A} + \frac{e}{c}\mathbf{A}\hat{\mathbf{p}} + \frac{e^2}{c^2}\mathbf{A}^2 = \hat{\mathbf{p}}^2 + 2\frac{e}{c}\mathbf{A}\hat{\mathbf{p}} + \frac{e^2}{c^2}\mathbf{A}^2, \tag{2.5}$$

where we made use of the commutation relations [7.4, § 24] $\hat{\mathbf{p}}\mathbf{A} = \mathbf{A}\hat{\mathbf{p}} - i\hbar \operatorname{div} \mathbf{A}$ and condition (2.3). The order of magnitude of the ratio of the third term to the second on the right-hand side of (2.5) is

$$\frac{e}{c}\frac{|\mathbf{A}|}{p}=\frac{e|\mathbf{E}_0|}{\omega p}\approx\frac{e}{\omega p}\left(\frac{8\pi S}{c}\right)^{1/2}, \tag{2.6}$$

where we made use of (2.10); $S=\frac{c}{8\pi}|\mathbf{E}_0|^2$ is the density of the light flux (Poynting's vector). For a powerful light flux of $S = 1\ \mathrm{W/cm^2}$ (this is ten times the flux of solar radiation at the boundary of the atmosphere), for the frequency of light $\omega = 10^{14}\ \mathrm{s^{-1}}$ and at a normal temperature of $T \approx 300$ K (for which the average momentum $p \approx \hbar k \approx 10^{-20}\ \mathrm{g \cdot cm \cdot s^{-1}}$) the ratio (2.6) is of the order of $10^{-4}$, therefore the last term in (2.5) can be neglected.

Hence, the perturbation in the Hamiltonian (2.4) due to the effect of the light wave is

$$\mathcal{H}'=\frac{e}{mc}\mathbf{A}\hat{\mathbf{p}}=-\frac{ie\hbar}{mc}\mathbf{A}\nabla. \tag{2.7}$$

For a plane monochromatic wave the vector potential is

$$\mathbf{A}(\mathbf{r},\ t)=A_0\mathbf{e}\exp i(\boldsymbol{\varkappa}\cdot\mathbf{r}-\omega t), \tag{2.8}$$

where $A_0$ is the amplitude (generally, a complex-valued one), $\mathbf{e}$ is a unit vector, $\omega$ and $\boldsymbol{\varkappa}$ are the frequency and the wave vector of the electromagnetic wave.

The electric field is, according to (2.1),

$$\mathbf{E}=-\frac{1}{c}\frac{\partial\mathbf{A}}{\partial t}=\frac{i\omega}{c}\mathbf{A}, \tag{2.9}$$

so that the polarization vector $\mathbf{e}$ points in the direction of the electric field. The amplitude of the electric field is

$$|E_0|=\frac{\omega}{c}|A_0|. \tag{2.10}$$

It follows from (2.8) and (2.3) that

$$\operatorname{div}\mathbf{A}=iA_0\exp[i(\boldsymbol{\varkappa}\cdot\mathbf{r}-\omega t)](\mathbf{e}\cdot\boldsymbol{\varkappa})=0, \tag{2.11}$$

i.e., $\boldsymbol{\varkappa}$ is perpendicular either to $\mathbf{e}$ or $\mathbf{E}$ (the electromagnetic waves are transverse waves).

To express the modulus of the vector potential's amplitude $|A_0|$ in terms of photon energy per cubic centimeter $N\hbar\omega$ ($\hbar\omega$ is the photon energy, $N$ is the number of photons per cubic centimeter), find the real values of the fields[5] making the vector potential equal to

$$\begin{aligned}A(\mathbf{r},\ t)&=A_0\exp i(\boldsymbol{\varkappa}\cdot\mathbf{r}-\omega t)+A_0^*\exp[-i(\boldsymbol{\varkappa}\cdot\mathbf{r}-\omega t)]\\&=2|A_0|\cos(\boldsymbol{\varkappa}\cdot\mathbf{r}-\omega t),\end{aligned} \tag{2.12}$$

[5] It has been established that complex field values cannot be used directly in expressions quadratic in the energy density [7.3, p. 398].

where, without loss of generality, we assume $A_0 = A_0^* = |A_0|$. The real value of the electric field intensity is

$$E = -\frac{1}{c}\frac{\partial A}{\partial t} = 2\frac{\omega}{c}|A_0| \sin(\varkappa \cdot \mathbf{r} - \omega t).$$

For nonmagnetic optically isotropic media to which the cubic crystals (Ge, Si, InSb, etc.) belong the energy density of the electromagnetic wave averaged over time is

$$\frac{1}{8\pi}[n^2\overline{E^2} + \overline{H^2}] = \frac{1}{4\pi} n^2\overline{E^2} = \frac{1}{4\pi}\cdot 4\left(\frac{\omega}{c}\right)^2 |A_0|^2 \overline{\sin^2(\varkappa \cdot \mathbf{r} - \omega t)}$$
$$= \frac{1}{2\pi}\left(\frac{\omega}{c}\right)^2 |A_0|^2,$$

where $n$ is the refractive index [7.3, Sec. 7.2]. Here we made use of the fact that in a plane harmonic wave the average electric energy $n^2\overline{E^2}$ is equal to the average magnetic energy $\overline{H^2}$ and that $\overline{\sin^2(\varkappa \cdot \mathbf{r} - \omega t)} = 1/2$.

Equating the expression obtained to the photons' energy density $N\hbar\omega$, we obtain

$$|A_0| = \sqrt{2\pi N\hbar\omega/\varkappa} \tag{2.12a}$$

Here the wave number $\varkappa = \omega/v$, where the phase velocity $v = c/n$. For the perturbation energy (2.7) we obtain

$$\hat{\mathscr{H}}' = -\frac{i\hbar e}{mc} A_0 \exp[-i(\omega t - \varkappa \cdot \mathbf{r})(\mathbf{e}\cdot\nabla), \tag{2.13}$$

where $|A_0|$ is given by (2.12a).[6]

The electron's state in the conduction and the valence bands is described by the Bloch wave function

$$\psi_{n\mathbf{k}}(\mathbf{r}) = \exp\left[-\frac{i}{\hbar}\varepsilon_n(\mathbf{k})\, t\right] u_{n\mathbf{k}}(\mathbf{r}) \exp i\mathbf{k}\cdot\mathbf{r}, \tag{2.14}$$

where $n$ is the band number, $\mathbf{k}$ is the electron's reduced wave vector, $\varepsilon_n(\mathbf{k})$ is the electron's energy, $u_{nk}(\mathbf{r})$ is a periodic function with the period of the lattice.

---

[6] Strictly speaking, we should have used in (2.13) for $\mathbf{A}(\mathbf{r}, t)$ the real binomial (2.12), the second term of which, proportional to $\exp i\omega t$, takes into account photon emission. In normal waves, when the number of electrons excited in the act of interband absorption is small, this process can be neglected; in special cases (lasers) this cannot be done.

The matrix element of the perturbation energy (2.13) for the transition $(n_1, \mathbf{k}_1) \to (n_2, \mathbf{k}_2)$, connected with photon absorption, is

$$\langle n_2, \mathbf{k}_2|\hat{\mathscr{H}}'|n_1, \mathbf{k}_1\rangle = \int d^3r \psi^*_{n_2\mathbf{k}_2}\hat{\mathscr{H}}'\psi_{n_1\mathbf{k}_1}$$

$$= -\frac{i\hbar e}{mc} A_0 \int d^3r \{u^*_{n_2\mathbf{k}_2} \exp[-i(\mathbf{k}_2 - \boldsymbol{\varkappa})\cdot\mathbf{r}]\, \mathbf{e}\cdot\nabla u_{n_1\mathbf{k}_1} \exp i\mathbf{k}_1\cdot\mathbf{r}\}$$

$$\times \exp\left\{\frac{i}{\hbar}[\varepsilon_{n_2}(\mathbf{k}_2) - \varepsilon_{n_1}(\mathbf{k}_1) - \hbar\omega]\, t\right\}$$

$$= \mathscr{P}_{12} \exp\left\{\frac{i}{\hbar}[\varepsilon_{n_2}(\mathbf{k}_2) - \varepsilon_{n_1}(\mathbf{k}_1) - \hbar\omega]\, t\right\}, \qquad (2.15)$$

where $\mathscr{P}_{12}$ is the time-independent part of the matrix element.

The integrand in (2.15) is

$$[\ldots]\exp i(\ldots) \equiv [u^*_{n_2\mathbf{k}_2}\mathbf{e}\cdot\nabla u_{n_1\mathbf{k}_1} + i(\mathbf{e}\cdot\mathbf{k}_1)\, u^*_{n_2\mathbf{k}_2}u_{n_1\mathbf{k}_1}]$$
$$\times \exp\{i(\mathbf{k}_1 + \boldsymbol{\varkappa} - \mathbf{k}_2)\cdot\mathbf{r}\}. \qquad (2.16)$$

Make the radius vector equal to $\mathbf{r} = \mathbf{a_m} + \mathbf{r}'$, where $\mathbf{a_m}$ is the vector of the $\mathbf{m}$-th lattice site. Since the functions $u_{n\mathbf{k}}(\mathbf{r})$ are periodic, the integral (2.15) takes the form

$$\sum_{\mathbf{m}} \exp\{i(\mathbf{k}_1 + \boldsymbol{\varkappa} - \mathbf{k}_2)\cdot\mathbf{a_m}\} \int [\ldots]\exp\{i(\mathbf{k}_1 + \boldsymbol{\varkappa} - \mathbf{k}_2)\cdot\mathbf{r}'\}\, d^3r'. \qquad (2.17)$$

Here [...] denotes the expression in the square brackets of (2.16), and the integration with respect to $d^3r'$ is performed over the volume of the unit cell $\Omega_0$. It follows from (A.6.4) that

$$\sum_{\mathbf{m}} \exp[(i\mathbf{k}_1 + \boldsymbol{\varkappa} - \mathbf{k}_2)\cdot\mathbf{a_m}] = N\delta_{0,\,\mathbf{k}_1+\boldsymbol{\varkappa}-\mathbf{k}_2}$$

$$= \frac{V}{\Omega_0}\delta_{0,\,\mathbf{k}_1+\boldsymbol{\varkappa}-\mathbf{k}_2} = \frac{1}{\Omega_0}\delta_{0,\,\mathbf{k}_1+\boldsymbol{\varkappa}-\mathbf{k}_2}. \qquad (2.18)$$

Here $\delta_{0,\,\mathbf{k}_1+\varkappa-\mathbf{k}_2}$ is the Kronecker delta, $N$ is the number of unit cells of the volume $\Omega_0$ in the crystal's principal region $V = 1\,\mathrm{cm}^3$. In (A.6.4) we made the reciprocal lattice vector $\mathbf{b_g} = 0$, since both $\mathbf{k}_1$ and $\mathbf{k}_2$ lie in the first Brillouin zone, and $\boldsymbol{\varkappa} \ll \mathbf{k}_1$, $\boldsymbol{\varkappa} \ll \mathbf{k}_2$. Equation (2.18) expresses the wave vector (quasi-momentum) conservation law and, since $\boldsymbol{\varkappa}$ is small, reduces to

$$\boldsymbol{\varkappa} \approx 0, \quad \mathbf{k}_1 = \mathbf{k}_2 = \mathbf{k}. \qquad (2.19)$$

Condition (2.19) enables us to make the exponent in (2.17) equal to unity;[7] then the integral of the second term in [...] is zero since the

[7] This corresponds to the dipole approximation for optical transitions in an atom.

functions $u_{n_1,\mathbf{k}_1}$ and $u_{n_2,\mathbf{k}_2}$ are orthogonal.

Hence, the time-independent matrix element $\mathcal{P}_{12}$ in (2.15) is

$$\mathcal{P}_{\mathrm{vc}}=-\frac{i\hbar e}{mc}A_0\frac{1}{\Omega_0}\int\limits_{\Omega_0} d^3r u^*_{\mathrm{c}\mathbf{k}}(\mathbf{r})\,\mathbf{e}\cdot\nabla u_{\mathrm{v}\mathbf{k}}(\mathbf{r})=A_0\frac{e}{mc}(\mathbf{e}\cdot\mathbf{p}_{\mathrm{cv}}), \quad (2.20)$$

where we substituted the indices v (valence band) and c (conduction band) for the indices $n_1$ and $n_2$, made $\mathbf{k}_1=\mathbf{k}_2=\mathbf{k}$ and dropped the prime at the integration variable $\mathbf{r}'$. Here the matrix transition element for the momentum is

$$\mathbf{p}_{\mathrm{cv}}=i\hbar\frac{1}{\Omega_0}\int\limits_{\Omega_0} d^3r u^*_{\mathrm{c}\mathbf{k}}(\mathbf{r})\,\nabla u_{\mathrm{v}\mathbf{k}}(\mathbf{r}) \quad (2.20a)$$

(the factor $1/\Omega_0$ has been introduced for convenience).

The number of v $\to$ c transitions per unit time per unit volume, for which the energy ($\varepsilon_{\mathrm{c}}=\varepsilon_{\mathrm{v}}+\hbar\omega$) and the quasi-momentum (or the wave vector) ($\mathbf{k}_1=\mathbf{k}_2$) conservation laws are satisfied, is (see (A.20.10a))

$$\mathcal{W}_{\mathrm{vc}}=\frac{2\pi}{\hbar}\int|\mathcal{P}_{\mathrm{vc}}|^2\,\delta[\varepsilon_{\mathrm{c}}(\mathbf{k})-\varepsilon_{\mathrm{v}}(\mathbf{k})-\hbar\omega]\frac{2}{(2\pi)^3}d^3k. \quad (2.21)$$

Here $[2/(2\pi)^3]\,d^3k$ is the number of electron states (with the account of the spin) in the region $d^3k$; the integration is performed over $d^3k$ with regard to the energy conservation law. We assume that band 1 (the valence band) to be completely filled with electrons and band 2 (the conduction band) to be completely free; such an assumption is correct if $k_0T\ll\varepsilon_{\mathrm{G}}$. The number of photons absorbed in cubic centimetre per second is obviously equal to the number of transitions $\mathcal{W}_{\mathrm{vc}}$ (2.21). If a plane light wave propagates in a semiconductor, the density of the photon flux (the number of photons per cubic centimetre per second) is $Nv=Nc/n$, when $N$ is the number of photons in cubic centimetre, and the phase velocity is equal to the light velocity in a vacuum divided by the refractive index: $v=c/n$.

The absorption coefficient due to direct interband transitions is[8]

$$\alpha=\frac{\mathcal{W}_{\mathrm{vc}}}{Nv}=\frac{1}{\pi}\frac{e^2}{m^2cn\omega}\int d^3k|\mathbf{e}\cdot\mathbf{p}_{\mathrm{cv}}|^2\,\delta[\varepsilon_{\mathrm{c}}(\mathbf{k})-\varepsilon_{\mathrm{v}}(\mathbf{k})-\hbar\omega]. \quad (2.22)$$

We use the value of $|A_0|^2$ (2.12a) and make the photon's wave vector $\varkappa=\omega/v=\omega n/c$.

Let the minimum of the conduction band and the maximum of the valence band be located at the same point $\mathbf{k}_0$ in the Brillouin zone. Such an energy-band structure is peculiar to the semiconductors PbTe, PbSe and PbS. The points of the corresponding extrema are located along the ⟨111⟩ axes on the Brillouin zone boundaries (see Fig. 4.22, points on $L$).

[8] This definition obviously coincides with (1.14).

If the constant-energy surfaces for the electron and the hole in the vicinities of the corresponding extrema are ellipsoids with parallel axes (which is also true for the aforementioned semiconductors), then

$$\varepsilon_c(\mathbf{k}) = \varepsilon_c(\mathbf{k}_0) + \frac{\hbar^2}{2}\left[\frac{(k_1-k_{01})^2}{m_{c1}} + \frac{(k_2-k_{02})^2}{m_{c2}} + \frac{(k_3-k_{03})^2}{m_{c3}}\right], \quad (2.23)$$

$$\varepsilon_v(\mathbf{k}) = \varepsilon_v(\mathbf{k}_0) - \frac{\hbar^2}{2}\left[\frac{(k_1-k_{01})^2}{m_{v1}} + \frac{(k_2-k_{02})^2}{m_{v2}} + \frac{(k_3-k_{03})^2}{m_{v3}}\right]. \quad (2.23a)$$

Here the axes 1, 2, 3 point in the direction of the ellipsoids' principal axes, $m_{ci}^{-1}$ and $m_{vi}^{-1}$ are components of the inverse effective mass tensors for the electrons in the conduction band and for the holes in the valence band. The argument for the δ-function in (2.22) is

$$\varepsilon_c(\mathbf{k}) - \varepsilon_v(\mathbf{k}) - \hbar\omega = \frac{\hbar}{2}\left[\frac{(k_1-k_{01})^2}{\mu_1} + \frac{(k_2-k_{02})^2}{\mu_2} + \frac{(k_3-k_{03})^2}{\mu_3}\right] - (\hbar\omega - \varepsilon_0), \quad (2.24)$$

where

$$\varepsilon_0 = \varepsilon_c(\mathbf{k}_0) - \varepsilon_v(\mathbf{k}_0), \quad (2.24a)$$

and the reduced components of the inverse effective mass tensor are

$$\frac{1}{\mu_i} = \frac{1}{m_{ci}} + \frac{1}{m_{vi}}. \quad (2.24b)$$

For values of $\mathbf{k}$ close to $\mathbf{k}_0$ we have

$$\mathbf{e}\cdot\mathbf{p}_{cv}(\mathbf{k}) = \mathbf{e}\cdot\mathbf{p}_{cv}(\mathbf{k}_0) + \left[\frac{\partial}{\partial\mathbf{k}}\mathbf{e}\cdot\mathbf{p}_{cv}(\mathbf{k})\right]_{\mathbf{k}=\mathbf{k}_0}(\mathbf{k}-\mathbf{k}_0). \quad (2.25)$$

If $\mathbf{e}\cdot\mathbf{p}_{cv}(\mathbf{k}_0) \neq 0$, this transition is termed an *allowed transition.* Retaining only the first term in (2.25), we obtain from (2.22), (2.24) and (2.25)

$$\alpha_{al} = \frac{1}{\pi}\frac{e^2}{m^2cn\omega}\sum_{\mathbf{k}_0}|\mathbf{e}\cdot\mathbf{p}_{cv}(\mathbf{k}_0)|^2\int\delta\left\{\frac{\hbar^2}{2}\left(\frac{(k_1-k_{01})^2}{\mu_1} + \frac{(k_2-k_{02})^2}{\mu_2} + \frac{(k_3-k_{03})^2}{\mu_3}\right) - (\hbar\omega - \varepsilon_0)\right\}dk_1\,dk_2\,dk_3,$$

where $\sum_{\mathbf{k}_0}$ is the sum of values of $|\mathbf{e}\cdot\mathbf{p}_{cv}(\mathbf{k}_0)|^2$ for all points corresponding to the Brillouin zone extrema being considered.

Going over to the integration variables $k_i' = (k_i - k_{0i})/\sqrt{\mu_i}$, we obtain

$$\alpha_{al} = \frac{1}{\pi}\frac{e^2}{m^2cn\omega}\sum_{\mathbf{k}_0}|\mathbf{e}\cdot\mathbf{p}_{cv}(\mathbf{k}_0)|^2\sqrt{\mu_1\mu_2\mu_3} \times \int\delta\left\{\frac{1}{2}\hbar^2k'^2 - (\hbar\omega - \varepsilon_0)\right\}4\pi k'^2\,dk',$$

where we put $dk'_1\, dk'_2\, dk'_3 = 4\pi k'^2\, dk'$. Introducing the integration variable $x = \hbar^2 k'^2/2$ and making use of the δ-function, we obtain

$$\alpha_{\mathrm{al}} = \frac{4\sqrt{2}\, e^2}{m^2 c n \omega} \sum_{\mathbf{k}_0} |\mathbf{e}\cdot\mathbf{p}_{\mathrm{cv}}(\mathbf{k}_0)|^2 \frac{\sqrt{\mu_1\mu_2\mu_3}}{\hbar^3} (\hbar\omega - \varepsilon_0)^{1/2}. \quad (2.26)$$

It is obvious that for $\hbar\omega < \varepsilon_0$, i.e., below the absorption fringe, $\alpha_{\mathrm{al}}$ is zero.

In Section 7.4 we shall see that $\alpha_{\mathrm{al}}$ is determined by the real part of the conductivity σ which is isotropic in cubic crystals, and for this reason $\sum_{\mathbf{k}_0} |\mathbf{e}\cdot\mathbf{p}_{\mathrm{cv}}(\mathbf{k}_0)|^2$ should also be isotropic. This could be checked by direct calculation which takes into account the mutual orientation of the ⟨111⟩ axes coinciding with the directions of $\mathbf{k}_0$. The isotropic invariant will, however, be easily seen in this case to be

$$\sum_{\mathbf{k}_0} |\mathbf{e}\cdot\mathbf{p}_{\mathrm{cv}}(\mathbf{k}_0)|^2 = 4\,\frac{p^2_{\mathrm{cv}x} + p^2_{\mathrm{cv}y} + p^2_{\mathrm{cv}z}}{3}. \quad (2.26\mathrm{a})$$

In the case of a standard band at the point $\mathbf{k}_0 = 0$ (2.26) takes the form

$$\alpha_{\mathrm{al}} = \frac{2e^2}{m^2 c n \omega} |\mathbf{e}\cdot\mathbf{p}_{\mathrm{cv}}(0)|^2 \left(\frac{2\mu}{\hbar^2}\right)^{3/2} (\hbar\omega - \varepsilon_0)^{1/2}, \quad (2.26\mathrm{b})$$

where $\mu = \mu_1 = \mu_2 = \mu_3$, $1/\mu = 1/m_c + 1/m_v$, and $\varepsilon_0 = \varepsilon_c(0) - \varepsilon_v(0)$. To avoid complicating the discussion, consider for the forbidden transitions the case of a standard band with extrema at the point $\mathbf{k}_0 = 0$. Fundamental dependences of the absorption coefficient, specifically, its dependence on the frequency of light, obtained for this case will be similar to those obtained in the case of a non-standard band.

For the forbidden transitions $\mathbf{e}\cdot\mathbf{p}_{\mathrm{cv}}(0) = 0$, therefore, the expansion (2.25) starts from the second term, the square of whose modulus is

$$\left|\frac{\partial}{\partial\mathbf{k}}[\mathbf{e}\cdot\mathbf{p}_{\mathrm{cv}}(0)]\,\mathbf{k}\right|^2 = \left|\frac{\partial}{\partial\mathbf{k}}[\mathbf{e}\cdot\mathbf{p}_{\mathrm{cv}}(0)]\right|^2 k^2\cos^2\vartheta.$$

Here ϑ is the angle between the vectors $\frac{\partial}{\partial\mathbf{k}}[\mathbf{e}\cdot\mathbf{p}_{\mathrm{cv}}(0)]$ and $\mathbf{k}$. The argument for the δ-function in (2.22) in the case of a standard band is $(\hbar^2k^2/2\mu) - (\hbar\omega - \varepsilon_0)$.

Substituting the above expressions in (2.22) and putting in polar coordinates $d^3k = \sin\vartheta\, d\vartheta\, d\varphi k^2 dk$ (the polar axis coincides with the vector $\frac{\partial}{\partial\mathbf{k}}[\mathbf{e}\cdot\mathbf{p}_{\mathrm{cv}}(0)]$), we obtain

$$\alpha_{\mathrm{for}} = \frac{1}{\pi}\frac{e^2}{m^2 c n \omega}\left|\frac{\partial}{\partial\mathbf{k}}[\mathbf{e}\cdot\mathbf{p}_{\mathrm{cv}}(0)]\right|^2 \times \int \delta\left[\frac{\hbar^2k^2}{2\mu} - (\hbar\omega - \varepsilon_0)\right] k^2\cos^2\vartheta\,\sin\vartheta\, d\vartheta\, d\varphi\, k^2\, dk.$$

Integration with respect to the angles yields $4\pi/3$, and integration with respect to $k$ is performed exactly in the same way as in the preceding case. As the result, we obtain for the coefficient

$$\alpha_{\text{for}} = \frac{2e^2}{3m^2cn\omega}\left(\frac{2\mu}{\hbar^2}\right)^{5/2}\left|\frac{\partial}{\partial \mathbf{k}}[\mathbf{e}\cdot\mathbf{p}_{cv}(0)]\right|^2(\hbar\omega-\varepsilon_0)^{3/2}. \quad (2.27)$$

Comparing this expression with (2.26), we see that in the case of allowed transitions $\alpha_{\text{al}} \propto (\hbar\omega - \varepsilon_0)^{1/2}$, whereas in the case of forbidden transitions $\alpha_{\text{for}} \propto (\hbar\omega - \varepsilon_0)^{3/2}$; in the latter case this is due to the appearance of an additional multiplier $k^2$ in the integrand of (2.22).

It follows from (1.12a) and (1.15) that the imaginary part of the dielectric permittivity is $\varepsilon_2(\omega) = (nc/\omega)\alpha$. Substituting herein the values (2.26) or (2.27) instead of $\alpha$, we obtain the expression for $\varepsilon_2(\omega)$ for the allowed and forbidden interband transitions, which does not contain the refractive index.

Experiment proves, however, the dependence of the absorption coefficient on the frequency of light $\omega$ for the allowed and forbidden transitions is generally of a more complex character than stipulated by (2.26) or (2.27). As has been established, the light absorption coefficient can be substantially affected by the Coulomb interaction of the electron and the hole generated in the act of absorption of a photon, i.e., by the free and bound exciton states. The effect of the electron-hole interaction on light absorption in direct and indirect interband transitions was considered by R. Elliot [7.5].

**7.2.3** In Section 5.3.1 we presented an elementary theory of excitons of large radius.

Suppose the exciton was generated in the course of photon absorption by an electron in the valence band. Since the photon's wave vector $\varkappa$ is small, the wave vector conservation law in the case of exciton generation reduces to the requirement that the wave vectors of the electron $\mathbf{k}_n$ and the hole $\mathbf{k}_p$ be equal in magnitude and opposite in direction, so that the exciton's wave vector $\mathbf{K} = \mathbf{k}_n + \mathbf{k}_p \approx 0$, therefore, the kinetic energy of the exciton as a whole (5.36) $W = 0$.

We see the total exciton energy to be (5.3.5b)

$$\mathscr{E} = \varepsilon_0 + \varepsilon, \quad (2.28)$$

where we denoted the difference $\varepsilon_c(0) - \varepsilon_v(0) = \varepsilon_0$; in the case of direct transitions the latter may differ from $\varepsilon_G$.

The Schrödinger equation which describes the relative motion of an electron and a hole is of the form (5.3.5a)

$$-\frac{\hbar^2}{2\mu}\nabla_{\mathbf{r}}^2\varphi(\mathbf{r}) - \frac{e^2}{n^2 r}\varphi(\mathbf{r}) = e\varphi(\mathbf{r}). \quad (2.29)$$

Here $\mu$ is the reduced mass of the electron and hole, and the static dielectric constant $\varepsilon_0$ has been assumed to be equal to the square of the refractive index ($\varepsilon_0 = n^2$). This equation coincides with the equation for a hydrogen-like atom with the nuclear charge $e/n$ and the electron mass $\mu$. For bound states of the exciton

$$\mathscr{E} = \varepsilon_0 + \varepsilon = \varepsilon_0 - \frac{\varepsilon_{ex}}{\nu^2}, \tag{2.30}$$

where the principal quantum number $\nu = 1, 2, 3, \ldots$, and

$$\varepsilon_{ex} = \mu e^4/2\hbar^2 n^4. \tag{2.30a}$$

In addition to the bound states, we can consider the free states of an exciton, for which the corresponding values of energy $\varepsilon$ in (2.29) are positive. The solutions of (2.29) for $\varepsilon > 0$ are known from the theory of the hydrogen-like atom. In this case the relative state of motion of the generated electron and hole is described by the wave vector $|\mathbf{k}| = \frac{1}{\hbar}\sqrt{2\mu\varepsilon}$, where $\mu$ is the reduced mass. In the derivation of (2.26) and (2.27) we assumed that the electron and the hole move independently after the generation and each of them is described by a plane wave. Actually, when the correlation of their motion is taken into account, the absorption coefficient $\alpha$ increases in proportion to the increase in the probability density of the state, when the electron occupies the place of the hole (or vice versa, the hole occupies the place of the electron), as compared with the state described by a plane wave.

It can be shown [7.6, p. 566][9] that for the nonbound states the ratio of the squared modulus of the wave function $\varphi(\mathbf{r})$ at the zero point to the squared modulus of the plane wave that describes free motion is

$$\frac{|\varphi(0)|^2}{|\psi_{\mathbf{k}}|^2} = \frac{\pi\beta \exp(\pi\beta)}{\sinh \pi\beta}, \tag{2.31}$$

where

$$\beta = \left(\frac{2\mu}{\hbar^2}\right)^{1/2} \frac{\varepsilon_{ex}^{1/2}}{k} = \frac{1}{ka_B}, \tag{2.31a}$$

and the Bohr radius is

$$a_B = \hbar^2 n^2/\mu e^2. \tag{2.31b}$$

It appears natural (and is confirmed by Elliot's detailed analysis of the problem) that the weight factor $|\varphi(0)|^2/|(\psi_{\mathbf{k}})|^2$ (2.31) should be introduced in the integrands for $\alpha_{al}$ and $\alpha_{for}$ to account for the electron-hole Coulomb interaction.

---

[9] It should be kept in mind that (136.11) is written in atomic units ($a_B = 1$).

Hence, we obtain for a standard band

$$\alpha_{\rm al}^{\rm Coul}=\frac{1}{\pi}\,\frac{e^2}{m^2c\omega}\,|\mathbf{e}\cdot\mathbf{p}_{\rm cv}(0)|^2\int\delta\left[\frac{\hbar^2k^2}{2\mu}-(\hbar\omega-\varepsilon_0)\right]\times 4\pi k^2\,\frac{\dfrac{\pi}{a_{\rm B}k}\exp\dfrac{\pi}{a_{\rm B}k}}{\sinh\dfrac{\pi}{a_{\rm B}k}}\,dk.$$

Making use, as in the preceding case, of the δ-function to compute the integral, we obtain by analogy with (2.26a)

$$\alpha_{\rm al}^{\rm Coul}=\frac{2\pi e^2}{nm^2\omega c}\left(\frac{2\mu}{\hbar^2}\right)^{3/2}|\mathbf{e}\cdot\mathbf{p}_{\rm cv}(0)|^2\,\varepsilon_{\rm ex}^{1/2}\,\frac{e^z}{\sinh z},\qquad(2.32)$$

where

$$z=\pi\sqrt{\frac{\varepsilon_{\rm ex}}{\hbar\omega-\varepsilon_0}}.\qquad(2.32a)$$

Close to the absorption fringe $\hbar\omega-\varepsilon_0\ll\varepsilon_{\rm ex}$, i.e., $z\gg1$, and the effect of the Coulomb interaction is great. In the limiting case, when $\hbar\omega\to\varepsilon_0$, i.e., $z\to\infty$, the absorption coefficient does not tend to zero as does $\alpha_{\rm al}$ (2.26), but in accordance with (2.32) tends to the value

$$\alpha_{\rm al}^{\rm Coul}\big|_{\hbar\omega\to\varepsilon_0}=\frac{4\pi e^2}{nm^2c\omega}\left(\frac{2\mu}{\hbar^2}\right)^{3/2}|\mathbf{e}\cdot\mathbf{p}_{\rm cv}(0)|^2\,\varepsilon_{\rm ex}^{1/2}.\qquad(2.33)$$

Far from the absorption fringe, when $\hbar\omega-\varepsilon_0\gg\varepsilon_{\rm ex}$, the Coulomb interaction is less important. However, even in the case of relatively large kinetic energies of the electron and the hole, for instance, for $\hbar\omega-\varepsilon_0=\pi^2\varepsilon_{\rm ex}\approx10\varepsilon_{\rm ex}$, i.e., $z=1$, the Coulomb interaction between them makes $\alpha_{\rm al}$ 4.6 times as large:

$$\alpha_{\rm al}^{\rm Coul}/\alpha_{\rm al}=2ze^z/\sinh z=2e/\sinh 1=4.6.$$

Elliot has demonstrated that the weight factor in the integral for $\alpha_{\rm al}^{\rm Coul}$ connected with forbidden transitions to the state $\varphi(\mathbf{r})$ is equal to $|\partial\varphi/\partial\mathbf{r}|^2_{\mathbf{r}=0}$. Computation similar to the above yields

$$\alpha_{\rm for}^{\rm Coul}=\frac{2\pi e^2}{3nm^2c\omega}\left(\frac{2\mu}{\hbar^2}\right)^{5/2}\left|\frac{\partial}{\partial\mathbf{k}}[\mathbf{e}\cdot\mathbf{p}_{\rm cv}(0)]\right|^2\varepsilon_{\rm ex}^{3/2}\times\left(1+\frac{\pi^2}{z^2}\right)\frac{e^z}{\sinh z}.\qquad(2.34)$$

Far from the absorption fringe, when $\hbar\omega-\varepsilon_0\gg\varepsilon_{\rm ex}$, i.e., $z\ll1$ (2.34) reduces to (2.27). Close to the absorption fringe, when $\hbar\omega\to\varepsilon_0$, i.e., $z\to\infty$, we obtain from (2.34)

$$\alpha_{\rm for}^{\rm Coul}\big|_{\hbar\omega\to\varepsilon_0}=\frac{4\pi e^2}{3nm^2c\omega}\left(\frac{2\mu}{\hbar^2}\right)^{5/2}\left|\frac{\partial}{\partial\mathbf{k}}[\mathbf{e}\cdot\mathbf{p}_{\rm cv}(0)]\right|^2\varepsilon_{\rm ex}^{3/2}.\qquad(2.35)$$

Having absorbed a photon, the electron and hole can occupy not only states with a continuous spectrum, but also discrete levels of the bound states of the exciton. There will be separate spectral lines just below the absorption fringe to correspond to such transitions. In the paper cited above Elliot has shown that the relative intensities of these discrete absorption lines for allowed transitions are proportional to $|\varphi_{\nu lm}(0)|^2$, where $\varphi_{\nu lm}(\mathbf{r})$ is the wave function of the exciton's bound state, and $\nu$, $l$, $m$ are the principal, the orbital and the magnetic quantum numbers of this state.

It will be seen from the expression for $\varphi_{\nu lm}(\mathbf{r})$ [7.6, § 36, Eq. (36.14)] that for $\mathbf{r} = 0$ only the $s$-states with $l = m = 0$ differ from zero, and that in this case

$$|\varphi_{\nu 00}(0)|^2 = \frac{1}{\pi a_B^3}\frac{1}{\nu^3}, \tag{2.36}$$

where the principal quantum number is $\nu = 1, 2, 3, \ldots$, and the Bohr radius $a_B$ is given by (2.31b).

It can be shown that in this case the absorption coefficient is

$$\alpha_{al}^{ex} = \frac{8\pi}{cn}\frac{e^2 s^2}{\hbar a_B^3 \nu^3}\delta(\omega - \omega_\nu). \tag{2.37}$$

[7.7, p. 547]. Here $s = \sqrt{\varepsilon_G/2\mu}$, $\omega_\nu$ is the frequency corresponding to the discrete exciton term (2.30). The relative intensity of discrete absorption lines for forbidden transitions is proportional to $|\partial\varphi_{\nu lm}(\mathbf{r})/\partial\mathbf{r}|^2_{\mathbf{r}=0}$.

It can be shown that for a hydrogen-like atom this quantity is nonzero only for the $p$-states ($l = 1$). Representing the three orthonormalized functions of the $p$-state in the form $X = xf(r)$, $Y = yf(r)$, and $Z = zf(r)$, we obtain

$$\left|\frac{\partial\varphi_{\nu 1 m}}{\partial\mathbf{r}}\right|^2_{\mathbf{r}=0} = \left|\frac{\partial X(0)}{\partial x}\right|^2 + \left|\frac{\partial Y(0)}{\partial y}\right|^2 + \left|\frac{\partial Z(0)}{\partial z}\right|^2 = \frac{1}{3\pi a_B^5}\frac{\nu^2-1}{\nu^5}, \tag{2.38}$$

where the principal quantum number is $\nu = 2, 3, \ldots$.

The absorption coefficient for this case is [7.7, p. 548]

$$\alpha_{for}^{ex} = \frac{16\pi}{3cn}\frac{e^2\hbar\theta^2}{m^{*2}a_B^3\omega}\frac{\nu^2-1}{\nu^5}\delta(\omega - \omega_\nu). \tag{2.39}$$

Here $\theta$ is a dimensionless constant of the order of unity, and $\omega_\nu$ is used in the sense of (2.37).

**7.2.4** The calculation of the matrix element $\mathcal{P}_{vc}$ (2.20) presents considerable difficulties, since usually we are ignorant about the electron wave functions in the valence and conduction bands. However, in many cases we need only establish whether the integral

$$\mathbf{e}\cdot\mathbf{M}_{cv} = \mathbf{e}\cdot\int \psi_{c\mathbf{k}}^*(\mathbf{r})\,\hat{\mathbf{p}}\psi_{v\mathbf{k}}(\mathbf{r})\,d^3r, \tag{2.40}$$

to which the matrix element $\mathcal{P}_{vc} = \mathcal{P}_{12}$ (2.15) is proportional, has a nonzero value.

Since the transformation of the operator $\hat{\mathbf{p}} = \frac{\hbar}{i}\frac{\partial}{\partial \mathbf{r}}$ under symmetry operations is the same as that of $\mathbf{r}$, the selection rules (Section 2.10) can easily be applied to the integral (2.40). It can also be shown that integral (2.40) can be expressed in terms of the system's dipole moment. Indeed, the operator is

$$\hat{\mathbf{p}} = \frac{1}{m}\hat{\mathbf{v}} = \frac{1}{m}\frac{\partial \mathbf{r}}{\partial t} = \frac{1}{m}\frac{1}{i\hbar}(\mathbf{r}\hat{\mathcal{H}} - \hat{\mathcal{H}}\mathbf{r}), \quad (2.41)$$

where we used by the expression for the time derivative of an operator [7.4, § 31]; here $\hat{\mathcal{H}}$ is the system's Hamiltonian.

Substituting (2.41) into (2.40), we obtain

$$\mathbf{M}_{cv} = \frac{1}{im\hbar}\{(\hat{\mathbf{r}}\hat{\mathcal{H}})_{cv} - (\hat{\mathcal{H}}\hat{\mathbf{r}})_{cv}\} = \frac{1}{im\hbar}\sum_l (\mathbf{r}_{cl}\mathcal{H}_{lv} - \mathcal{H}_{cl}\mathbf{r}_{lv})$$

$$= \frac{1}{im\hbar}(\mathcal{E}_v\mathbf{r}_{cv} - \mathcal{E}_c\mathbf{r}_{cv}) = \frac{\mathcal{E}_c - \mathcal{E}_v}{im\hbar}\langle c|\mathbf{r}|v\rangle. \quad (2.42)$$

We made use of the fact that

$$\mathcal{H}_{lv} = \int \psi_l^*\hat{\mathcal{H}}\psi_v d^2r = \mathcal{E}_v \int \psi_l^*\psi_v d^3r = \mathcal{E}_v\delta_{lv} \quad (2.42a)$$

and similarly $\mathcal{H}_{cl} = \mathcal{E}_c\delta_{cl}$.

Hence,

$$\mathbf{e}\cdot\mathbf{M}_{cv} = \text{const}\cdot\int \psi_{c\mathbf{k}}^* r_e \psi_{v\mathbf{k}} d^3r, \quad (2.43)$$

where $r_e$ is the projection of the position vector on the direction of the electric field $\mathbf{e}$.

We see that $\mathcal{P}_{vc}$ is proportional to the matrix element of the dipole transition (2.43).

The selection rules, including the specific case of a dipole transition in a field with a symmetry of the $\boldsymbol{O}$ group, were considered in Section 2.10.

Consider the selection rules for a dipole transition in the centre of the Brillouin zone ($\mathbf{k} = 0$) for germanium and silicon. Let the transitions take place from the state $\Gamma'_{25}$ in the valence band (see Fig. 4.27). In a field of cubic symmetry of the $\boldsymbol{O}_h$ group $x$, $y$, $z$ belong to the irreducible representation $\Gamma_{15}$ (see Table 3.2). Constructing, in accordance with the general rule (Section 2.10), the direct product $\Gamma_{15} \times \times \Gamma'_{25}$ and expanding it into the irreducible representations of the $\boldsymbol{O}_h$ group, we see that only the following transitions are possible: $\Gamma'_{25} \leftrightarrow \leftrightarrow \Gamma_{15}$, $\Gamma'_{25} \leftrightarrow \Gamma_{25}$, $\Gamma'_{25} \leftrightarrow \Gamma'_{12}$, $\Gamma'_{25} \leftrightarrow \Gamma'_2$, other transitions from $\Gamma'_{25}$ are forbidden.

The situation is more complicated in bismuth telluride, where the selection rules depend on the polarization of the electromagnetic wave. The wave vector group for point $\mathbf{k} = 0$ in $Bi_2Te_3$ is the point group $D_{3d} = D_3 \times C_i$; its characters are given in Table 7.1. It can

Table 7.1

| $C_{3v} \times C_i$ | $E$ | $2C_3$ | $3J\sigma_v$ | $J$ | $2JC_3$ | $3\sigma_v$ |
|---|---|---|---|---|---|---|
| $D_3 \times C_i$ | $E$ | $2C_3$ | $3C_2'$ | $J$ | $2JC_3$ | $3JC_2'$ |
| $L_1$ | 1 | 1 | 1 | 1 | 1 | 1 |
| $L_2$ | 1 | 1 | −1 | 1 | 1 | −1 |
| $L_3$ | 2 | −1 | 0 | 2 | −1 | 0 |
| $L_1'$ | 1 | 1 | 1 | −1 | −1 | −1 |
| $L_2'$ | 1 | 1 | −1 | −1 | −1 | 1 |
| $L_3'$ | 2 | −1 | 0 | −2 | 1 | 0 |

easily be established that the coordinate $z$ (parallel to the crystal's $c$ axis) is transformed according to the transformation $L_1'$, and the coordinates $x$, $y$—according to the transformation $L_3'$. Let the electric field of the wave point be in the direction of $z$, then $L_1 \leftrightarrow L_1'$ is an allowed transition (indeed, the direct product $L_1' \times L_1$ contains the irreducible representation $L_1'$). If the electric field is perpendicular to $z$, the transition $L_1 \leftrightarrow L_1'$ will easily be seen to be forbidden, but the transition $L_1 \leftrightarrow L_3'$ to be allowed.

If we want to take into account the dependence of the wave function on the spin, we must consider binary groups and the corresponding tables of characters. The appropriate procedure is more complicated, but remains identical in principle [7.8, p. 93].

## 7.3. Indirect Interband Transitions

**7.3.1** The shape of the energy bands of some semiconductors (Ge, Si) is as depicted in Fig. 4.28; the deepest minimum of the conduction band is located at a point that does not coincide with the position of the maximum of energy in the valence band at point $O$ corresponding to $\mathbf{k} = 0$. In germanium, the minima $L$ in the conduction band are located on the Brillouin zone boundary in the $\langle 111 \rangle$ direction, and on silicon—on the $\Delta$ axis in the $\langle 100 \rangle$ direction. To be definite, we shall in the future consider germanium.

The transition of an electron from the valence band (near $O$) to the conduction band near $L$ as a result of absorption of a light quantum is impossible. Indeed, let $\mathbf{k}_1$ and $\mathbf{k}_2$ be the wave vectors of the electron in the initial and the final states, respectively, and $\varkappa$ be the wave vector of the photon. Then it follows from the wave vector (quasi-momentum) conservation law that $\mathbf{k}_2 - \mathbf{k}_1 - \varkappa \approx \mathbf{k}_2 - \mathbf{k}_1 \approx 0$, since $\varkappa \ll \mathbf{k}_2$, which is of the order of $\mathbf{k}_L$. If we take into account that $\mathbf{k}_1$ is small, we see that the wave vector conservation law cannot be satisfied. If, however, the electron simultaneously with the absorption of a photon will absorb (emit) a phonon with the wave vector $\mathbf{q} = \pm(\mathbf{k}_2 - \mathbf{k}_1)$, then the wave vector conservation law can be satisfied.

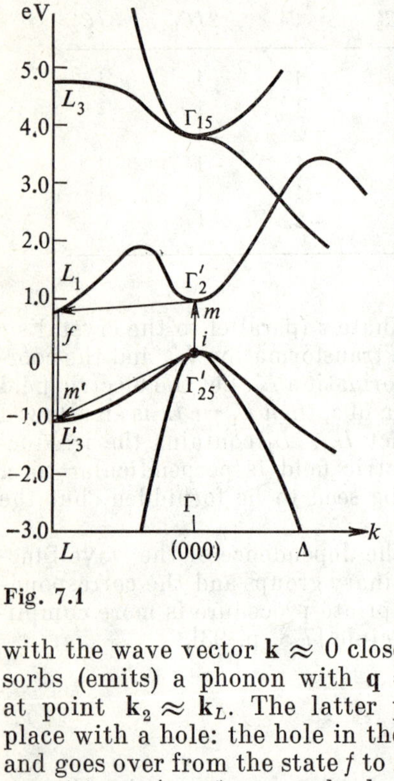

Fig. 7.1

Figure 7.1 depicts the energy bands in germanium and two electron transition channels *imf* and *im′f* from the initial state *i* in the valence band (close to $\mathbf{k} \approx 0$) to the final state *f* in the conduction band (close to point *L*). Along *imf* the electron goes over from the initial state *i* to the intermediate state *m* by absorbing a photon (the direct interband transition was discussed in Section 7.2) and then to the final state *f* by absorbing (emitting) a phonon with the wave vector $\mathbf{q} = \pm(\mathbf{k}_2 - \mathbf{k}_1) \approx \pm\mathbf{k}_L$. Along *im′f* the electron with the wave vector $\mathbf{k}_2 \approx \mathbf{k}_L$ from the valence band absorbs a photon and goes over to the conduction band in the vicinity of point *L*, and then another electron with the wave vector $\mathbf{k} \approx 0$ close to point *O* in the valence band absorbs (emits) a phonon with $\mathbf{q} = \pm(\mathbf{k}_2 - \mathbf{k}_1)$ and occupies a hole at point $\mathbf{k}_2 \approx \mathbf{k}_L$. The latter process can be imagined as taking place with a hole: the hole in the conduction band absorbs a phonon and goes over from the state *f* to *m′* (this is equivalent to the electron transition $m' \to f$); next the hole absorbs a phonon and goes over from the state *m′* to *i*, the two hole transitions being equivalent to the transition of an electron from *i* to *f*. Since the latter course *im′f* can be regarded both as an electron or a hole transition, it is not essential whether the state *m′* is free or occupied by an electron.

Such processes involving a simultaneous absorption of a photon and absorption (emission) of a phonon are dealt with in the second approximation of the quantum mechanics theory of transitions (see A.20.18). The total number of such transitions via an intermediate state $m$ per unit time per unit volume is

$$\mathscr{W}=\frac{2\pi}{\hbar}\sum_{if}\sum_{m}\frac{|M_{fm}^{\text{phon}}|^2\,|M_{mi}^{\text{phot}}|^2}{(\varepsilon_m-\varepsilon_i-\hbar\omega)^2}\,\delta(\varepsilon_f-\varepsilon_i-\hbar\omega\mp\hbar\omega_{\mathbf{q}}). \tag{3.1}$$

Here the subscripts $i$, $f$ and $m$ denote the initial, final, and intermediate states, $M_{mi}^{\text{phot}}$ is the matrix element of the electron-photon interaction for the $i\to m$ transition, $M_{mf}^{\text{phon}}$ is the matrix element of the electron-phonon interaction for the $m\to f$ transition, $\hbar\omega$ is the photon's energy and $\hbar\omega_{\mathbf{q}}$ is the energy of the phonon with the wave vector $\mathbf{q}$.

It is established in the theory of quantum transitions that the energy conservation law has to be satisfied only for the initial $i$ and the final $f$ states. The wave vector conservation law applies in the case of intermediate transitions as well.

We can to a good approximation use scalar effective masses $m_c$ and $m_v$ to describe the electrons and holes in germanium and in silicon in the centre of the Brillouin zone, i.e., near point $O$ ($\mathbf{k}=0$). Then we have for the holes near the energy maximum in the valence band

$$\varepsilon_i\equiv\varepsilon_v(\mathbf{k})=\varepsilon_v(0)-\frac{\hbar^2k^2}{2m_v}=\varepsilon_v(0)-\tilde{\varepsilon}_v, \tag{3.2}$$

and for the electrons near the energy minimum in the conduction band

$$\varepsilon_m\equiv\varepsilon_c(\mathbf{k})=\varepsilon_c(0)+\frac{\hbar^2k^2}{2m_c}=\varepsilon_c(0)+\tilde{\varepsilon}_c^0. \tag{3.3}$$

Here $\tilde{\varepsilon}_v$ and $\tilde{\varepsilon}_c^0$ are "kinetic" energies of the holes and the electrons, proportional to $k^2$.

The constant-energy surfaces for electrons in germanium in the vicinity of $L$, i.e., of the absolute minima, are, as we know from Section 4.15, ellipsoids of revolution with the transverse effective mass $m_t$ and the longitudinal effective mass $m_l$. The electron energy close to point $L$ is

$$\varepsilon_f\equiv\varepsilon_c(\mathbf{k}')=\varepsilon_c(\mathbf{k}_L)+\frac{\hbar^2}{2}\left(\frac{k_1^2+k_2^2}{m_t}+\frac{k_3^2}{m_l}\right)=\varepsilon_c(\mathbf{k}_L)+\tilde{\varepsilon}_c. \tag{3.4}$$

Here $\tilde{\varepsilon}_c$ is the "kinetic" energy of the electron close to point $L$, which in the major axes is a function of the squares of the rectangular components of the wave vector.

The summation over the initial state $i$, i.e., the integration with respect to $\mathbf{k}$ (or $\tilde{\varepsilon}_v$), is at the same time the summation over the intermediate state $m$, since the wave vector $\mathbf{k}$ is conserved during the $i\to m$ transition.

Making use of (3.2)-(3.4), we obtain for the energy differences in (3.1)

$$\varepsilon_m - \varepsilon_i - \hbar\omega = \tilde{\varepsilon}_c(\mathbf{k}) - \tilde{\varepsilon}_v(\mathbf{k}) - \hbar\omega = \varepsilon_0 + \tilde{\varepsilon}_c^0 + \tilde{\varepsilon}_v - \hbar\omega. \tag{3.5}$$

Here $\hbar\omega$ is the photon's energy and $\varepsilon_0 = \varepsilon_c(0) - \varepsilon_v(0)$ is the forbidden bandwidth at point $O$ (for the direct transition). Next

$$\begin{aligned}\varepsilon_f = \varepsilon_i - \hbar\omega \mp \hbar\omega_{\mathbf{q}} &= \varepsilon_c(\mathbf{k}) - \varepsilon_v(\mathbf{k}) - \hbar\omega \mp \hbar\omega_{\mathbf{q}} \\ &= \varepsilon_c(\mathbf{k}_L) + \tilde{\varepsilon}_c - \varepsilon_v(0) + \tilde{\varepsilon}_v - \hbar\omega \mp \hbar\omega_{\mathbf{q}} \\ &= \varepsilon_G + \tilde{\varepsilon}_c + \tilde{\varepsilon}_v - \hbar\omega \mp \hbar\omega_{\mathbf{q}}.\end{aligned} \tag{3.6}$$

Here $\mp\hbar\omega_{\mathbf{q}}$ is the phonon's energy corresponding to its absorption or emission, $\varepsilon_G = \varepsilon_c(\mathbf{k}_L) - \varepsilon_v(0)$ is the forbidden bandwidth in Ge; $\varepsilon_0 - \varepsilon_G$ is temperature-dependent and is equal to $\varepsilon_0 - \varepsilon_G \approx \approx (0.80 - 0.66)$ eV $= 0.14$ eV at 300 K.

For $M_{mi}^{\text{phot}}$ we have, according to (2.20),

$$M_{mi}^{\text{phot}} = A_0\left(\frac{e}{mc}\right)(\mathbf{e}\cdot\mathbf{p}_{cv}(\mathbf{k})) = \left(\frac{2\pi\hbar e^2 N}{m^2 c n \varkappa}\right)^{1/2}(\mathbf{e}\cdot\mathbf{p}_{cv}(\mathbf{k})), \tag{3.7}$$

where $\mathbf{p}_{cv}(\mathbf{k})$ is equal to (2.20a), and $|A_0|$ is equal to (2.12a). For allowed transitions $\mathbf{e}\cdot\mathbf{p}_{cv}(\mathbf{k}) \approx \mathbf{e}\cdot\mathbf{p}_{cv}(0)$ and is independent of $\mathbf{k}$.

When we study the interaction of the electron with the lattice vibrations in Chapter 8, we shall see that the square of the matrix element for phonon absorption or emission (8.3.10), (8.3.10a) contains the factor $N_{\mathbf{q}}$ and $N_{\mathbf{q}} + 1$, where $N_{\mathbf{q}}$ is Planck's function (3.11.10a):

$$N_{\mathbf{q}} = \frac{1}{\exp\dfrac{\hbar\omega_{\mathbf{q}}}{k_0 T} - 1}. \tag{3.8}$$

At temperatures above the Debye temperature ($k_0 T > \hbar\omega_{\mathbf{q}}$)

$$N_{\mathbf{q}} \approx N_{\mathbf{q}} + 1 \approx \frac{k_0 T}{\hbar\omega_{\mathbf{q}}}, \tag{3.8a}$$

at temperatures $T \to 0$

$$N_{\mathbf{q}} \to 0, \quad N_{\mathbf{q}} + 1 \to 1, \tag{3.8b}$$

i.e., the probability of phonon absorption tends to zero but the probability of phonon emission remains finite.

In the most general case we may write

$$|M_{fm}^{\text{phon}}|^2 = C_{\mathbf{q}}^{(j)}(N_{\mathbf{q}} + 1/2 \mp 1/2). \tag{3.9}$$

Here $j$ is the number of the vibration branch. The upper sign refers to the absorption and the lower—to the emission of a phonon. Expression (3.9) can be regarded as being independent of $\mathbf{q}$. Indeed, experimental data on the vibrational spectrum of germanium indicate that the energy $\hbar\omega_{\mathbf{q}}$ of longitudinal acoustic waves interacting with the electrons varies from zero (for $\mathbf{q} = 0$) to $2.75 \times 10^{-2}$ eV (for $\mathbf{q} \approx \mathbf{k}_L$), this being only 0.03 from the value of $\varepsilon_G = 0.7$ eV, and because of that we can regard the square of the matrix element (3.9) and the difference (3.6) as being independent of $\mathbf{q}$ and make, for example, $\mathbf{q} = \mathbf{k}_L$.[10]

Consider the absorption due to indirect allowed transitions involving the emission of phonons of the $j$-th vibration branch.

From the definition of the absorption coefficient of light (2.22) we obtain

$$\alpha_{\text{al}}^{\text{ind}} = 4\pi^2 \frac{e^2}{m^2 c n \omega} |\mathbf{e}\cdot\mathbf{p}_{cv}(0)|^2 C_{\mathbf{k}_L}^{(j)} (N_{\mathbf{k}_L j} + 1)$$
$$\times \frac{2}{(2\pi)^3} \int d^3k \frac{2}{(2\pi)^3} \int d^3k' \frac{1}{(\varepsilon_m - \varepsilon_i - \hbar\omega)^2}$$
$$\times \delta(\varepsilon_f - \varepsilon_i - \hbar\omega + \hbar\omega_{\mathbf{q}}) + \dots . \qquad (3.10)$$

Here the plus sign takes into account the similar terms connected with the interaction with other vibration branches $j$, as well as with phonon absorption processes. However, it should be kept in mind, as we intend to show at the end of this section, that there are definite selection rules for the matrix elements $M_{fm}^{\text{phon}}$, so that not all the phonons can take part in indirect transitions.

The summation over the initial state $i$ in (3.10) and the final state $f$ in (3.1) has been replaced by integration with respect to $\mathbf{k}$ and $\mathbf{k}'$, respectively:

$$\sum_i \to \frac{2}{(2\pi)^3} \int d^3k = \frac{2}{(2\pi)^3} \cdot 2\pi \frac{(2m_v)^{3/2}}{\hbar^3} \int \sqrt{\tilde{\varepsilon}_v}\, d\tilde{\varepsilon}_v, \qquad (3.10a)$$

$$\sum_f \to \frac{2}{(2\pi)^3} \int d^3k' = \frac{2}{(2\pi)^3} \cdot 2\pi \frac{(2m_{\text{eff}})^{3/2}}{\hbar^3} \int \sqrt{\tilde{\varepsilon}_c}\, d\tilde{\varepsilon}_c. \qquad (3.10b)$$

Here we made use of (6.2.19a) and (6.2.22), and $m_{\text{eff}} = N_c^{2/3} (m_t^2 m_l)^{1/3}$.

Absorption involving indirect transitions is much weaker (second approximation of the perturbation theory!) than the absorption involving direct transitions, and for this reason we should be interested in (3.10) only in the vicinity of the indirect absorption fringe: $\hbar\omega \approx \varepsilon_G$. Since $\varepsilon_0 > \varepsilon_G$, and since we are interested only in

[10] Actually, the situation is, so to say, some ten times more favourable, since $\varepsilon_G$ should be compared not with $\hbar\omega(\mathbf{k}_L) - \hbar\omega(0)$ but only with the spread in the values of $\mathbf{q}$ as $\mathbf{k}$ approaches point $L$.

the values of $\hbar\omega$ slightly exceeding $\varepsilon_G$, we can write for the denominator in (3.10)

$$(\varepsilon_m - \varepsilon_i - \hbar\omega)^2 = (\varepsilon_0 + \tilde{\varepsilon}_c^0 + \tilde{\varepsilon}_v - \hbar\omega)^2 \approx (\varepsilon_0 - \hbar\omega)^2.$$

As a result, (3.10) acquires the form

$$\alpha_{al}^{ind} = \frac{B_j^{em}}{\omega(\varepsilon_0 - \hbar\omega)^2} \times \iint \sqrt{\tilde{\varepsilon}_v \tilde{\varepsilon}_c}\, \delta[\tilde{\varepsilon}_c + \tilde{\varepsilon}_v + \varepsilon_G + \hbar\omega_{\mathbf{k}_L} - \hbar\omega]\, d\tilde{\varepsilon}_v\, d\tilde{\varepsilon}_c + \ldots, \quad (3.11)$$

where $B_j^{em}$ is a constant corresponding to the emission of a phonon of the $j$-th vibration branch. It is independent of $\omega$ but depends on the temperature. Making use of the $\delta$-function in (3.11) for the integration with respect to $\tilde{\varepsilon}_v$, we obtain for the double integral [7.9, equations (2.261) and (2.262)]

$$\int_0^b \sqrt{\tilde{\varepsilon}_c (b - \tilde{\varepsilon}_c)}\, d\tilde{\varepsilon}_c = \frac{\pi}{8} b^2, \quad (3.11a)$$

where

$$b = \hbar\omega - \varepsilon_G - \hbar\omega_{\mathbf{k}_L}. \quad (3.11b)$$

Taking into account not only the processes of phonon emission but of absorption as well, and summing over different vibration branches $j$, we obtain, making use of (3.11a) and (3.11b),

$$\alpha_{al}^{ind} = \frac{B^{em}}{\omega(\varepsilon_0 - \hbar\omega)^2} (\hbar\omega - \varepsilon_G - \hbar\omega_{\mathbf{k}_L})^2 + \frac{B^{ab}}{\omega(\varepsilon_0 - \hbar\omega)^2} (\hbar\omega - \varepsilon_G + \hbar\omega_{\mathbf{k}_L})^2. \quad (3.12)$$

For forbidden transitions $\mathbf{e}\cdot\mathbf{p}_{cv}(\mathbf{k}) = \frac{\partial}{\partial \mathbf{k}} [\mathbf{e}\cdot\mathbf{p}_{cv}(\mathbf{k})]_{\mathbf{k}=0}\mathbf{k}$, and because of that the square of the matrix element $M_{mi}^{phot}$ contains an additional factor $k^2$. The presence of this factor in the integrand in (3.10) results in the coefficient of light absorption $\alpha_{forb}^{ind}$ containing terms proportional to $(\hbar\omega - \varepsilon_G \pm \hbar\omega_{\mathbf{k}_L})^3$.

**7.3.2** Exciton states affect the light absorption coefficient $\alpha^{ind}$ not only in the case of direct but also in the case of indirect transitions. This point was considered by R. Elliott [7.5] cited above. The electron-hole interaction in the case of indirect transitions, same as in the case of direct transitions, is accounted for by the weight factor determining the probability of the location of the electron and the hole at the same point in space (Section 7.2.2). We shall limit ourselves to the presentation of the results obtained.

It follows from (3.12) that if the exciton states are not taken into account, the absorption coefficient $\alpha_{\mathrm{al}}^{\mathrm{ind}}$ for indirect transitions is made up of the sum of terms proportional to $(\hbar\omega - \varepsilon_{\mathrm{G}} \mp \hbar\omega_{\mathrm{q}})^2$. If we take into account the electron-hole interaction in the course of a transition followed by the formation of an unbound state, the result is the change in the exponent in the terms determining $\alpha_{\mathrm{al}}^{\mathrm{ind}}$ from 2 to 3/2. If in the course of an indirect transition a bound electron-hole (exciton) state is formed below the level $\varepsilon_{\mathrm{G}}$, then there will be absorption with a continuous spectrum, described by a sum of terms with an exponent 1/2. At the same time for direct transitions we shall in this case obtain a series of discrete lines. The explanation for the continuous absorption spectrum for indirect transitions is that a phonon takes part in the process, which makes possible the transitions to any point of the exciton band.

Finally, the exponent in the power dependence of $\alpha_{\mathrm{forb}}^{\mathrm{ind}}$ on the energy $\hbar\omega - \varepsilon_{\mathrm{G}} \mp \hbar\omega_{\mathbf{k}_L}$ in the case of forbidden indirect transitions changes from 3 to 5/2 above $\varepsilon_{\mathrm{G}}$ (nonbound states) and to 3/2 below $\varepsilon_{\mathrm{G}}$ (bound states).

**7.3.3** Consider the application of the group theory to the derivation of selection rules for indirect transitions in germanium in which the indirect transitions have been extensively studied in experiment. The conduction band of germanium is known to consist of eight minima (see Section 4.15) whose centres are located at points $L$ on the Brillouin zone boundaries (see Fig. 4.22).

Figure 4.28*a* schematically shows the energy bands in germanium and indicates the irreducible representations used to transform the electron wave function at symmetrical points of the Brillouin zone. Theoretical and experimental studies indicate that irreducible representations corresponding to the valence and the conduction bands at point $\mathbf{k} = 0$ are $\Gamma'_{25}$ and $\Gamma'_2$ of the $\boldsymbol{O}_{\mathrm{h}}$ group. The wave vector group at point $L$ (see Fig. 4.22) is a subgroup of $\boldsymbol{O}_h = \boldsymbol{T}_{\mathrm{d}} \times \boldsymbol{C}_{\mathrm{i}}$. The vector $\mathbf{k}_L$ is easily seen to transform into itself under the following elements of the $\boldsymbol{T}_{\mathrm{d}}$ group: $E$, $2C_3$, $3\sigma_{\mathrm{v}}$ which constitute the group $\boldsymbol{C}_{3\mathrm{v}}$. Since the end point of the vector $\mathbf{k}_L$ lies on the Brillouin zone boundary, inversion $J$ transforms it into an equivalent vector. Hence, the group of the wave vector $\mathbf{k}_L$ is the group $\boldsymbol{D}_{3\mathrm{d}} = \boldsymbol{C}_{3\mathrm{v}} \times \boldsymbol{C}_{\mathrm{i}}$ of order twelve consisting of six classes: $E$, $2C_3$, $3\sigma_{\mathrm{v}}$, $J$, $2JC_3$, $3J\sigma_{\mathrm{v}}$.

We have represented the group $\boldsymbol{D}_{3\mathrm{d}}$ of the wave vector $\mathbf{k}_L$ as a direct product: $\boldsymbol{D}_{3\mathrm{d}} = \boldsymbol{C}_{3\mathrm{v}} \times \boldsymbol{C}_{\mathrm{i}}$. This, as we shall see below, is convenient, since the elements of the group $\boldsymbol{T}_{\mathrm{d}} \times \boldsymbol{C}_{\mathrm{i}}$ and, consequently, the elements of its subgroup $\boldsymbol{C}_{3\mathrm{v}} \times \boldsymbol{C}_{\mathrm{i}}$ involving inversion $J$ are from Section 2.5.2 known to be symmetry operations of the germanium lattice only if they are accompanied by a nontrivial translation $\boldsymbol{\alpha} = \frac{a}{4}(1,1,1)$. Table 7.1 presents the characters of the $\boldsymbol{D}_{3\mathrm{d}}$

group with identical symmetry elements corresponding to the columns but expressed once in terms of $D_3 \times C_i$ and the second time in terms of $C_{3v} \times C_i$; indeed, $J\sigma_v = C'_2$, and as a result $\sigma_v = JC'_2$ (it should be remembered that $C'_2 \perp C_3$ and $\sigma_v$ passes through the $C_3$ axis).

It follows from theory and from experiment that the irreducible representations of the electron wave function in the valence band and in the conduction band at point $L$ are $L'_3$ and $L_1$, respectively (see Fig. 7.1).

We see from Fig. 7.1 that in germanium indirect transitions via the nearest bands through the channels, *imf* ($\Gamma'_{25} \to \Gamma'_2 \to L_1$) and *im'f* ($\Gamma'_{25} \to L_3 \to L_1$), are possible. We showed in Section 7.2.3 that the direct transition $\Gamma'_{25} \to \Gamma'_2$ involving the absorption of a photon is an allowed transition. It can just as easily be demonstrated that the direct transition $L'_3 \to L_1$ is an allowed transition. Hence, the only thing that remains to be done is to consider the selection rules for the transitions $\Gamma'_2 \to L_1$ and $\Gamma'_{25} \to L'_3$, involving the absorption (emission) of a phonon with the wave vector $\mathbf{q}_L$. To this end, first determine the irreducible representations that transform the normal vibrations of the germanium lattice at point $L$. We should proceed as suggested in Section 3.8. First, we should find the complete reducible representation corresponding to all the vibrational degrees of freedom at point $\mathbf{q}_L$. The number of atoms in a unit cell of germanium is $s = 2$, therefore, the dimension of this complete reducible representation is $3s = 3 \times 2 = 6$. To find the characters of this complete representation we shall make use of equations (3.8.13)-(3.8.15).

As was noted above, the elements of group $D_{3d} = C_{3v} \times C_i$ of the wave vector $\mathbf{q}_L$, involving inversion $J$, must be accompanied by a nontrivial translation $\boldsymbol{\alpha} = \frac{a}{4}(1, 1, 1)$. Since in the nontrivial translation $\boldsymbol{\alpha}$ there is no atom in a unit cell that retains its position ($n_{CJ} = 0$), it follows from (3.8.15) that

$$\chi^u (J \mid \boldsymbol{\alpha}) = \chi^u (J\sigma_v \mid \boldsymbol{\alpha}) = \chi^u (JC_3 \mid \boldsymbol{\alpha}) = 0.$$

On the other hand, for the transformations $E$ and $C_3$, $n_{C(\varphi)} = 2$, therefore, it follows from (3.8.13) that

$$\chi^u (E \mid 0) = (1 + 2 \cos 0°)\cdot 2 = 6,$$
$$\chi^u (C_3 \mid 0) = (1 + 2 \cos 120°)\cdot 2 = 0.$$

Finally, for the transformation $\sigma_v$ we obtain from (3.8.14) ($n_{S(0)} = 2$)

$$\chi^u (\sigma_v \mid 0) = \chi^u (S_{-}(0) \mid 0) = (-1 + 2 \cos 0°)\cdot 2 = 2.$$

Combining these results, write out the characters of the complete representation of the vibrations at point $L$ in Table 7.2. Expanding

Table 7.2

| $C_{3v} \times C_i$ | $E$ | $2C_3$ | $3J\sigma_v$ | $J$ | $2JC_3$ | $3\sigma_v$ |
|---|---|---|---|---|---|---|
| $\chi^u$ | 6 | 0 | 0 | 0 | 0 | 2 |

Table 7.3

| $D_{3d}$ | $E$ | $2C_3$ | $3C_2'$ | $J$ | $2JC_3$ | $3JC_2'$ |
|---|---|---|---|---|---|---|
| $\Gamma_2' \times L_1$ | 1 | 1 | −1 | −1 | −1 | 1 |

this reducible representation in the irreducible representations of Table 7.1, we obtain

$$\chi^u = L_1 + L_3 + L_2' + L_3'. \tag{3.13}$$

Experimental studies of neutron scattering in germanium have shown that the following types of phonons correspond to the normal vibrations (3.13):

$L_1$—longitudinal optical ($LO$) phonons,
$L_3$—transverse acoustic ($TA$) phonons, (3.14)
$L_2'$—longitudinal acoustic ($LA$) phonons,
$L_3'$—transverse optical ($TO$) phonons.

Let us turn now to the selection rules for the matrix element $M_{fm}^{\text{phon}}$ that enters (3.1). Considering the transition $\Gamma_2' \to L_1$ we note that the matrix element for this transition contains under the integral sign electron functions transformed by the irreducible representations $\Gamma_2'$ and $L_1$, oscillator wave functions of the crystal's normal vibrations and the operator $\hat{\mathcal{U}}$ of the electron-phonon interaction. The integration with respect to the oscillator functions, which can be performed independently of the integration with respect to the electron coordinates, results in selection rules (3.10.25), (3.10.26) according to which only transitions involving the emission or the absorption of a single phonon are possible.

It can be snown that the irreducible representations that transform the operator $\hat{\mathcal{U}}$ are the same as those that transform normal coordinates, i.e., representations (3.13). We shall approach the problem from a somewhat different point of view. In the case of an allowed transition involving the emission of a phonon the initial state of the system consists of an electron described by a wave function with a $\Gamma_2'$ symmetry, and the final state consists of an electron in the state $L_1$

and a phonon whose wave function must have the same symmetry $\Gamma'_2$. This means that the direct product of $\Gamma'_2 \times L_1$ by one of the irreducible representations (3.13) must contain the irreducible representation $\Gamma'_2$ or, to put it another way, the direct product $\Gamma'_2 \times L_1$ must contain one of the irreducible representations (3.13).

The characters of $\Gamma'_2$ corresponding to the classes of the group $D_{3d}$ can be obtained from Table 3.2. The characters of the direct product $\Gamma'_2 \times L_1$ are given in Table 7.3. Expanding $\Gamma'_2 \times L_1$ in irreducible representations of the $D_{3d}$ group, we obtain

$$\Gamma'_2 \times L_1 = L'_2, \tag{3.15}$$

as can easily be checked with the aid of Table 7.1.

Considering the second channel, we obtain in precisely the same way

$$\Gamma'_{25} \times L'_3 = L'_1 + L'_2 + 2L'_3. \tag{3.16}$$

Comparing (3.15), (3.16) with (3.13) and (3.14), we see that the following phonons can take part in indirect transitions:

$$\begin{aligned} &L'_2\text{—longitudinal acoustic } (LA) \text{ phonon,} \\ &2L'_3\text{—transverse optical } (TO) \text{ phonon.} \end{aligned} \tag{3.17}$$

The transition involving the $LA$-phonon can take place via the bands $\Gamma'_2$ or $L'_3$, whereas the transition involving the $TO$-phonon can take place only via the band $L'_3$. The factor 2 in front of $L'_3$ in (3.16) indicates that there are two linearly independent matrix elements determining the probability of this transition. At the same time transitions involving the emission or absorption of longitudinal optical and of transverse acoustic phonons are forbidden and cannot take part in the process considered here, no matter whether we consider transitions via the bands $\Gamma'_2$ and $L'_3$ or via other $\Gamma$ and $L$ bands. This follows from the fact that the product of the representations $\Gamma'_{25}$, $L_1$, and representation $\Gamma'_{15}$ corresponding to the absorption of a photon (Section 7.2.3) does not contain the representations $L_3$ or $L'_2$ that are connected with the absorption of phonons.

## 7.4. Absorption of Light in Semiconductors by Free Charge Carriers

At frequencies below the frequency corresponding to the interband absorption edge in semiconductors a weak intraband absorption by free carriers can be observed.

Consider a simple phenomenological theory of light absorption by free carriers.

We shall base our consideration on the Maxwell equations [7.3, Sec. 7.1] for a homogeneous, isotropic, conducting, and nonmagnetic ($\mu = 1$) medium:

$$\text{curl}\,\mathbf{H} = \frac{\varepsilon}{c}\dot{\mathbf{E}} + \frac{4\pi\sigma}{c}\mathbf{E}, \tag{4.1a}$$

$$\text{curl}\,\mathbf{E} = -\frac{1}{c}\dot{\mathbf{H}}, \tag{4.1b}$$

$$\text{div}\,\mathbf{E} = 0, \tag{4.1c}$$

$$\text{div}\,\mathbf{H} = 0. \tag{4.1d}$$

Here $\mathbf{E}$ and $\mathbf{H}$ are the intensities of the electric and magnetic fields, $\varepsilon$ is the dielectric constant that determines the induction vector $D = \varepsilon\mathbf{E}$, $\sigma$ is the conductivity that determines the current $\mathbf{j}$ in accordance with Ohm's law ($\mathbf{j} = \sigma\mathbf{E}$). There are no free charges in the right-hand side of (4.1c).

In order to exclude $\mathbf{H}$ from (4.1a) and (4.1b), take the curl of both sides of (4.1b) and the time derivative of both sides of (4.1a). Making use of the identity [7.10, p. 59] curl curl = grad div $-\nabla^2$ and (4.1c), we obtain

$$\nabla^2\mathbf{E} = \frac{\varepsilon}{c^2}\ddot{\mathbf{E}} + \frac{4\pi\sigma}{c^2}\dot{\mathbf{E}}. \tag{4.2}$$

For a nonconducting medium ($\sigma = 0$) this equation turns into the wave equation.

Equation (4.2) has a particular solution in the form of a plane wave with a complex wave vector $\tilde{\varkappa}$ (1.13). Substituting (1.13) into (4.2), we obtain

$$\tilde{\varkappa}^2 = \frac{\omega^2}{c^2}\left(\varepsilon + \frac{4\pi\sigma}{\omega}i\right) = \frac{\omega^2}{c^2}\tilde{\varepsilon}. \tag{4.3}$$

Hence, the *complex-valued dielectric permittivity*

$$\tilde{\varepsilon} = \tilde{n}^2 = (n + ik)^2 = \varepsilon + \frac{4\pi\sigma}{\omega}i, \tag{4.4}$$

where $\tilde{n} = n + ik$ is the *complex-value refractive index.*

For low $\omega$ the imaginary part of $\tilde{\varepsilon}$ is equal to $4\pi\sigma/\omega$, where $\sigma$ is the real static conductivity independent of the frequency $\omega$.

We shall show that at high frequencies $\omega$ the conductivity begins to be dependent on $\omega$ and becomes complex-valued. As we saw in Section 7.1.1, free carriers interacting only with an electromagnetic wave cannot absorb light. Express the effect of the interaction of free carriers with lattice imperfections (thermal vibrations, impurities, etc.) in terms of the relaxation time $\tau$, as it was done in Section 6.6. In (6.6.4) make the magnetic field $\mathbf{H} = 0$. Then

$$v_x = \frac{eE_x}{m^*} \frac{\tau}{1 - i\omega\tau}. \tag{4.5}$$

The current density along the $x$ axis is

$$j_x = \tilde{\sigma} E_x = N_0 e v_x = \frac{N_0 e^2 \tau}{m^*} \frac{E_x}{1 - i\omega\tau}, \tag{4.6}$$

where $N_0$ is the free carriers' concentration.

Hence,

$$\tilde{\sigma} = \sigma_0 \frac{1}{1 - i\omega\tau} = \sigma_0 \frac{2 + i\omega\tau}{1 + \omega^2\tau^2}, \tag{4.7}$$

where $\sigma_0 = N_0 e^2 \tau / m^*$ is the *dc* conductivity.

Substituting (4.7) into (4.4), we obtain

$$\tilde{\varepsilon} = \varepsilon_1 + i\varepsilon_2 = (n + ik)^2$$
$$= \varepsilon - \frac{4\pi\sigma_0}{\omega} \frac{\omega\tau}{1 + \omega^2\tau^2} + i \frac{4\pi\sigma_0}{\omega} \frac{1}{1 + \omega^2\tau^2}. \tag{4.8}$$

Hence, the real and imaginary parts of the dielectric permittivity are

$$\varepsilon_1 = n^2 - k^2 = \varepsilon - \frac{4\pi\sigma_0}{\omega} \frac{\omega\tau}{1 + \omega^2\tau^2}$$
$$= \varepsilon - \frac{4\pi}{\omega} \operatorname{Im} \tilde{\sigma} = \varepsilon \left(1 - \frac{\omega_p^2}{\omega_0^2 + \omega^2}\right), \tag{4.9}$$

$$\varepsilon_2 = 2nk = \frac{4\pi\sigma_0}{\omega} \frac{1}{1 + \omega^2\tau^2}$$
$$= \frac{4\pi}{\omega} \operatorname{Re} \tilde{\sigma} = \varepsilon \frac{\omega_0}{\omega} \frac{\omega_p^2}{\omega_0^2 + \omega^2}. \tag{4.9a}$$

Here $\omega_0 = 1/\tau$ is the frequency of collisions with lattice imperfections and $\omega_p = \sqrt{4\pi N_0 e^2 / \varepsilon m^*}$ is the *plasma frequency*.

It follows from (4.9a) and (1.14) that the absorption coefficient is

$$\alpha = \frac{2\omega k}{c} = \frac{4\pi}{nc} \operatorname{Re} \tilde{\sigma} = \frac{4\pi\sigma_0}{nc} \frac{1}{1 + \omega^2\tau^2}, \tag{4.9b}$$

i.e., $\alpha$ is proportional to $\operatorname{Re} \tilde{\sigma}$. This fact was already discussed in Section 6.6.

It follows from (4.9a), (4.9b) that absorption is proportional to $\sigma_0$, i.e., to the free carrier's concentration. In the frequency range $\omega^2\tau^2 \gg 1$ (but, of course, $\omega < \omega_G$ —the frequency of the interband absorption fringe) the coefficient of absorption $\alpha \propto \omega^{-2} \propto \lambda^2$, i.e., is proportional to the square of the wavelength. This has been substantiated by experiment on $n$-type germanium.

The classical theory of light absorption by free charge carriers developed above is valid only in the range $\hbar\omega < k_0 T$ (or $\hbar\omega < \varepsilon_F$ —the Fermi energy for degenerate semiconductors). A consistent quantum mechanics theory of light absorption by free carriers, based on the second approximation of the theory of perturbations, was developed by H. Fröhlich and formulated by H. Y. Fan et al. In the limiting case $\hbar\omega \gg k_0 T$ for a simple band and for scattering by phonons of the acoustic branch the quantum theory yields $\alpha \propto \omega^{-2} (\hbar\omega/k_0 T)^{1/2}$. A review of theoretical results and a comparison with experimental data can be found in a paper by Fan [7.11, Chap. 9].

## 7.5. Polaritons

**7.5.1** Consider some new elementary perturbations in a solid—a peculiar hybrid of phonons of the optical branch of crystal vibrations with photons (quanta of the electromagnetic field). Quasiparticles of such elementary perturbations became known as *polaritons.*

In Section 3.9 we applied a semiphenomenological method to study long-wave vibrations of the optical branch of an ionic cubic crystal containing two oppositely charged ions per unit cell. The classical equations of relative motion of two oppositely charged ions are of the form (3.9.31)

$$\frac{d^2\mathbf{w}}{dt^2} = -\omega_0^2\mathbf{w} + \omega_0 \sqrt{\frac{\varepsilon_0 - \varepsilon_\infty}{4\pi}}\,\mathbf{E}. \tag{5.1}$$

Here $\mathbf{w} = \sqrt{N_0 m_r}\,\mathbf{s}$ is the "normalized" ion displacement, where $\mathbf{s}$ is the displacement of the positive ion with respect to the negative ion, $N_0$ is the number of cells per unit volume of the crystal, $m_r$ is the reduced mass of the ions, $\omega_0^2 = \frac{\varkappa}{m_r} - \frac{4\pi N_0 e^*(\varepsilon_\infty + 2)}{9m_r}$ is the frequency of mechanical vibrations of the ions, where $\varkappa$ is the coefficient of the quasi-elastic force of ionic interaction, $e^*$ is the effective charge of the ions, $\varepsilon_0$ and $\varepsilon_\infty$ are the static and the high-frequency dielectric constants, $\mathbf{E}$ is the average electric field in the crystal.

According to (3.9.32), the polarization vector is

$$\mathbf{P} = \omega_0 \sqrt{\frac{\varepsilon_0 - \varepsilon_\infty}{4\pi}}\,\mathbf{w} + \frac{\varepsilon_\infty - 1}{4\pi}\,\mathbf{E}. \tag{5.2}$$

Having separated the displacement $\mathbf{w}$ into the transverse $\mathbf{w}_t$ and the longitudinal $\mathbf{w}_l$ ($\mathbf{w} = \mathbf{w}_t + \mathbf{w}_l$) (3.9.33), we have shown that the frequencies of the transverse and longitudinal vibrations are, respectively, (3.9.39) and (3.9.40)

$$\omega_t = \omega_0, \quad \omega_l = \omega_0 \sqrt{\frac{\varepsilon_0}{\varepsilon_\infty}}. \tag{5.3}$$

**7.5.2** To construct the theory of polaritons we should supplement the equation of mechanical vibrations (5.1) with the equations of the electromagnetic field [7.3, Sec. 7.1]

$$\operatorname{div} \mathbf{D} = \operatorname{div} (\mathbf{E} + 4\pi\mathbf{P}) = 0, \tag{5.4}$$

$$\operatorname{div} \mathbf{H} = 0, \tag{5.5}$$

$$\operatorname{curl} \mathbf{E} = -\frac{1}{c} \dot{\mathbf{H}}, \tag{5.6}$$

$$\operatorname{curl} \mathbf{H} = \frac{1}{c} \dot{\mathbf{D}} = \frac{1}{c} (\dot{\mathbf{E}} + 4\pi\dot{\mathbf{P}}). \tag{5.7}$$

We assume that the medium is nonmagnetic and the free charges and conductivity currents are absent.

In Section 3.9 we made use of only one of these equations (5.4), which takes no account of the time lag. Write

$$\left.\begin{array}{l} \mathbf{w} = \mathbf{w}_0 \\ \mathbf{P} = \mathbf{P}_0 \\ \mathbf{E} = \mathbf{E}_0 \\ \mathbf{H} = \mathbf{H}_0 \end{array}\right\} \times \exp i\,(\mathbf{k}\cdot\mathbf{r} - \omega t) \tag{5.8}$$

where $\mathbf{w}_0$, $\mathbf{P}_0$, $\mathbf{E}_0$ and $\mathbf{H}_0$ are generally complex-valued amplitudes.

Substituting (5.8) into (5.1), (5.4)-(5.7), we obtain

$$-\omega^2\mathbf{w} = -\omega_0^2\mathbf{w} + \omega_0 \sqrt{\frac{\varepsilon_0 - \varepsilon_\infty}{4\pi}}\,\mathbf{E}, \tag{5.9}$$

$$\mathbf{k}\cdot(\mathbf{E} + 4\pi\mathbf{P}) = 0, \tag{5.10}$$

$$\mathbf{k}\cdot\mathbf{H} = 0, \tag{5.11}$$

$$\mathbf{k}\times\mathbf{E} = \frac{\omega}{c}\mathbf{H}, \tag{5.12}$$

$$\mathbf{k}\times\mathbf{H} = -\frac{\omega}{c}(\mathbf{E} + 4\pi\mathbf{P}). \tag{5.13}$$

In contrast to the electrostatic case (Section 3.9), the identity $\mathbf{E} = 0$ is impossible. Indeed, if $\mathbf{E} = 0$, then it follows from (5.12) that $\mathbf{H} = 0$, but in that case it follows from (5.13) that $\mathbf{P} = 0$ and, finally, from (5.2) that $\mathbf{w} = 0$. Hence $\mathbf{E} = 0$ corresponds to the trivial case when $\mathbf{E} = \mathbf{H} = \mathbf{P} = \mathbf{w} = 0$.

It follows from (5.9) that

$$\mathbf{w}=\frac{\omega_0}{\omega_0^2-\omega^2}\sqrt{\frac{\varepsilon_0-\varepsilon_\infty}{4\pi}}\,\mathbf{E}. \tag{5.14}$$

Substituting this into (5.2), we obtain

$$\mathbf{P}=\left\{\frac{\omega_0^2}{\omega_0^2-\omega^2}\,\frac{\varepsilon_0-\varepsilon_\infty}{4\pi}+\frac{\varepsilon_\infty-1}{4\pi}\right\}\mathbf{E}. \tag{5.15}$$

Hence, and from (5.10), we obtain

$$(\mathbf{k}\cdot\mathbf{E})\left\{\frac{\omega_0^2}{\omega_0^2-\omega^2}(\varepsilon_0-\varepsilon_\infty)+\varepsilon_\infty\right\}=0. \tag{5.16}$$

Consider two cases. In the first the expression in braces in (5.16) is zero, i.e.,

$$\frac{\omega_0^2}{\omega_0^2-\omega^2}(\varepsilon_0-\varepsilon_\infty)+\varepsilon_\infty=0, \tag{5.17}$$

and it follows from (5.15) that $\mathbf{E}+4\pi\mathbf{P}=0$, but then by (5.13) $\mathbf{k}\times\mathbf{H}=0$, which together with (5.11) yields $\mathbf{H}=0$. Now it follows from (5.12) that $\mathbf{k}\times\mathbf{E}=0$, but as we have noted above, $\mathbf{E}\neq 0$, so that $\mathbf{E}\parallel\mathbf{k}$. Making use of (5.14) and (5.15), we obtain

$$\mathbf{w}\parallel\mathbf{P}\parallel\mathbf{E}\parallel\mathbf{k}, \tag{5.18}$$

i.e., all the vibrating vectors are longitudinal, and the frequency, as determined from (5.17), is

$$\omega^2=\left(\frac{\varepsilon_0}{\varepsilon_\infty}\right)\omega_0^2=\omega_1^2. \tag{5.19}$$

Hence, this solution coincides with the electrostatic solution for longitudinal vibrations (3.9.42).

In the second case in (5.16) $\mathbf{k}\cdot\mathbf{E}=0$, but since $\mathbf{E}\neq 0$, it follows that $\mathbf{E}\perp\mathbf{k}$. Now it follows from (5.12) that $\mathbf{k}$, $\mathbf{E}$, $\mathbf{H}$ constitute in the order as written here a right-handed orthogonal system of vectors, and, therefore,

$$\mathbf{k}\cdot\mathbf{E}=\frac{\omega}{c}H. \tag{5.20}$$

The vibrations are now transverse, since

$$\mathbf{H}\perp(\mathbf{E}\parallel\mathbf{P}\parallel\mathbf{w})\perp\mathbf{k}, \tag{5.21}$$

and (5.11) is satisfied automatically. Substituting (5.15) and (5.20) into (5.13), we obtain after canceling $E$

$$\frac{c^2k^2}{\omega^2}=\frac{\omega_0^2}{\omega_0^2-\omega^2}(\varepsilon_0-\varepsilon_\infty)+\varepsilon_\infty \tag{5.22}$$

(here we took into account that $\mathbf{k} \times \mathbf{H}$ is directed opposite to $\mathbf{E}$). Solving this quadratic equation, we obtain

$$\omega^2 = \frac{\varepsilon_0\omega_0^2 + c^2k^2 \pm \sqrt{(\varepsilon_0\omega_0^2 + c^2k^2)^2 - 4\varepsilon_\infty\omega_0^2c^2k^2}}{2\varepsilon_\infty}, \tag{5.23}$$

i.e., two dispersion branches.

Denote the branch corresponding to the plus sign in front of the radical by $\omega_1^2(k)$, and denote the branch corresponding to the minus sign by $\omega_2^2(k)$. Then we obtain from (5.23) for $k \to 0$

$$\omega_1(0) = \left(\frac{\varepsilon_0}{\varepsilon_\infty}\right)^{1/2} \omega_0 = \omega_l, \qquad \omega_2(k \to 0) = \frac{c}{\sqrt{\varepsilon_0}} k \tag{5.24}$$

and for $k \to \infty$

$$\omega_1(\infty) = \frac{c}{\sqrt{\varepsilon_\infty}} k, \qquad \omega_2(\infty) = \omega_0 \tag{5.24a}$$

When we calculate $\omega_2(\infty)$ we should take into account that the term $c^2k^2$ in front of the radical cancels with $(c^2k^2)^2$ under the radical sign, and for this reason the radical should be calculated to within the next approximation in $(c^2k^2)^{-1}$.

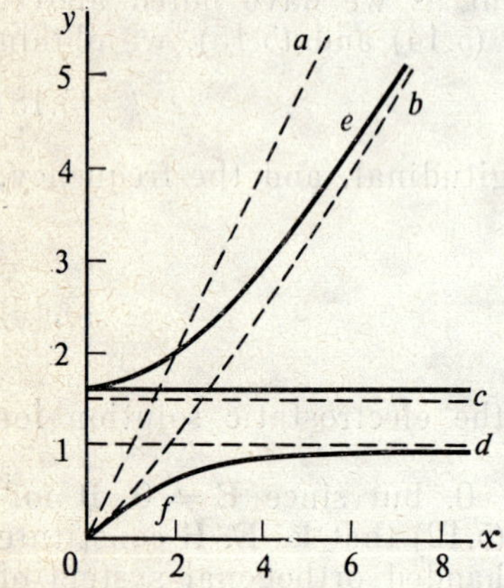


Fig. 7.2

Figure 7.2 shows the functions $y = y(x)$, where $y = \omega/\omega_0$ and $x = \dfrac{k}{\omega_0/c} = ck/\omega_0$. The dashed straight line $a$ whose equation is $y = x$ (or $\omega = ck$) corresponds to the propagation of light in a vacuum.

In the case of high frequencies ($k \to \infty$) light propagates in the medium according to the law $y = x/\sqrt{\varepsilon_\infty}$, i.e., $\omega = (c/\sqrt{\varepsilon_\infty})\, k$ [see (5.24a)], and in the case of low frequencies ($k \to 0$) according to the law $y = x\sqrt{\varepsilon_0}$, i.e., $\omega = (c/\sqrt{\varepsilon_0})\, k$ [see (5.24)]. The propagation of light in the medium is shown in Fig. 7.2 by the dashed line $b$. The horizontal dashed lines $c$ and $d$ depict longitudinal and transverse mechanical vibrations with frequencies $\omega_l = (\varepsilon_0/\varepsilon_\infty)^{1/2}\,\omega_0$ and $\omega_t = \omega_0$. The solid lines $c$, $e$, $f$ are dispersion curves of polaritons corresponding to the solutions (5.19) and (5.23).

The solid straight line $c$ corresponds to longitudinal polaritons (5.18), (5.19) of purely mechanical origin. The branches $e$ and $f$ describe transverse polaritons of the mixed electromagnetic-mechanical type; they originate from the hybridization of vibrations corresponding to the dashed straight lines *b*, *c* and *d*. K. Huang [7.12] has calculated the mechanical energy of the transverse polaritons in branches *e* and *f* as part of the total energy. He showed that for the same values of material constants for which the graphs in Fig. 7.2 have been calculated part of the mechanical energy in the lower branch *f* grows with the increase in $x = ck/\omega_0$ from 30% for $x = 0$ to 95% for $x = 6$; on the contrary, part of the mechanical energy of polaritons in the upper branch *e* drops with an increase in $x$ from 70% for $x = 0$ to 5% for $x = 6$.

The polariton branch *e* was studied in experiment.

There are other hybrids of quanta of the electromagnetic photon field with excitations of solids, for instance, with excitons, in addition to the elementary excitations discussed above. Such quasiparticles, also termed polaritons, have been observed in experiment. An essential contribution to this theory has been made by S. I. Pekar.

## 7.6. Faraday's Rotation Effect

**7.6.1** In 1846 the great British physicist Michael Faraday discovered the following phenomenon. When a transparent body is placed in a strong magnetic field, and a plane polarized ray of light is sent through it in the direction of the field, the polarization plane of the ray rotates through an angle proportional to the intensity of the magnetic field and to the path traveled by the ray in the magnetic field.

This phenomenon which Faraday, being convinced that light and electromagnetism were of common origin, sought to discover year after year became an important milestone on the road to electromagnetic theory of light. For semiconductors, the Faraday effect became in some cases a convenient tool for measuring the effective mass of charge carriers.

We shall consider the Faraday effect for the simplest case of free charge carriers with a scalar effective mass $m^*$.

First we shall have to generalize the theory of the cyclotron resonance discussed in Section 6.6.1. Consider this time not a plane polarized wave but two circularly polarized waves. In this case equation (6.6.3a) assumes the form

$$m^* \left( \dot{\mathbf{v}} + \frac{1}{\tau} \mathbf{v} \right) = -eE_0 \mathbf{e} \exp(-i\omega t) - \frac{e}{c} \mathbf{v} \times \mathbf{H}. \tag{6.1}$$

Here $\mathbf{v}$ is the velocity of the electron, $\tau$ is its relaxation time, $\mathbf{H}$ is the intensity of a constant magnetic field pointing in the direction of the $z$ axis, $E_0$ is the amplitude of the high-frequency electric field, $-e$ is the electron charge, and the polarization vector is $\mathbf{e} = \{1, \gamma, 0\}$[11], where $\gamma = \pm i = \cos\frac{\pi}{2} \pm i \sin\frac{\pi}{2} = \exp\left(\pm i\,\frac{\pi}{2}\right)$.

In (6.6.3a) $\gamma = 0$, and because of that the electric field had only the component $E = E_0 \exp(-i\omega t)$ along the $x$ axis. In the case of (6.1) the electric field, in addition to the component $E_0 \exp(-i\omega t)$ along the $x$ axis, has also the component $E_0 \exp -i\left(\omega t \mp \frac{\pi}{2}\right)$ along the $y$ axis. It is established in elementary optics that the summation of oscillations with equal amplitudes and with a phase difference of $\pm\pi/2$, one oscillation taking place along the $x$ axis and the other along the $y$ axis, results in a circularly polarized oscillation. If $\gamma = +i$, we speak of *left-hand polarization*[12], and if $\gamma = -i$, of *right-hand polarization*.

Assume that $\mathbf{v}$ is also proportional to $\exp(-i\omega t)$. Find the projections of (6.1) on the axes $x$ and $y$ ($\mathbf{H} \parallel z$):

$$\frac{m^*}{\tau}(1 - i\omega\tau)\, v_x = -eE - \frac{e}{c} H v_y,$$

$$\frac{m^*}{\tau}(1 - i\omega\tau)\, v_y = -eE\gamma + \frac{e}{c} H v_x. \tag{6.2}$$

Solving these equations for $v_x$, we obtain for the current

$$j_x = \tilde{\sigma} E = -eN_0 v_x = \sigma_0 \frac{1 - (i\omega + \gamma\omega_c)\,\tau}{(1 - i\omega\tau)^2 + \omega_c^2\tau^2}\, E, \tag{6.3}$$

where $N_0$ is the concentration of free electrons, $\sigma_0 = e^2 N_0 \tau / m^*$ is the dc conductivity and $\omega_c = eH/m^*c$ is the cyclotron frequency.

For $\gamma = 0$ (6.3) coincides with (6.6.6). Making in (6.3) $\gamma = \pm i$ and separating the real part from the imaginary one, we obtain for the complex conductivity

$$\tilde{\sigma} = \sigma_{\mathrm{Re}} + i\sigma_{\mathrm{Im}}, \tag{6.4}$$

where the real part is

$$\sigma_{\mathrm{Re}} = \sigma_0 \frac{1 + (\omega_c^2 \pm \omega^2)\,\tau^2}{[1 + (\omega_c^2 - \omega^2)\,\tau^2]^2 + 4\omega^2\tau^2}, \tag{6.4a}$$

and the imaginary part is

$$\sigma_{\mathrm{Im}} = \sigma_0 \frac{(\omega \pm \omega_c)(\omega_c^2 - \omega^2)\,\tau^3 - (\omega \pm \omega_c)\,\tau}{[1 + (\omega_c^2 - \omega^2)\,\tau^2]^2 + 4\omega^2\tau^2}. \tag{6.4b}$$

As we know from (4.9b), the absorption coefficient $\alpha$ is proportional to $\sigma_{\mathrm{Re}}$, i.e., to the real part of the conductivity. If $\omega_c\tau \gg 1$ (and

---

[11] Note that $\mathbf{e}$ is not a unit vector.

[12] In the case of left-hand polarization the electric vector in the wave rotates clockwise if we look in the direction of the wave.

just such are the conditions in which cyclotron resonance experiments are carried out), it follows from (6.4a) that at resonance, when $\omega \approx \omega_c$,

$$\sigma_{\text{Re}} \approx \begin{cases} \sigma_0 & \text{for } \gamma = +i, \qquad (6.5a) \\ \dfrac{\sigma_0}{4\omega_c^2\tau^2} \approx 0 & \text{for } \gamma = -i, \qquad (6.5b) \end{cases}$$

since the upper sign in (6.4a) corresponds to $\gamma = +i$ and the lower, to $\gamma = -i$. Hence, the absorption peak for electrons can be observed only in the case of a left-hand polarization. The holes rotate in the direction opposite to that of the electrons, and because of that for them an absorption peak can be observed only in the case of $\gamma = -i$, i.e., with a right-hand polarization.

Hence, if we know the direction of circular polarization of a high-frequency field, we can establish whether electrons or holes are responsible for the cyclotron resonance.

**7.6.2** It follows from (4.4) that the square of the complex-valued refractive index is

$$\tilde{n}^2 = (n+ik)^2 = \varepsilon + i\,\frac{4\pi\tilde{\sigma}}{\omega} = \varepsilon + i\,\frac{4\pi}{\omega}\,(\sigma_{\text{Re}} + i\sigma_{\text{Im}})$$

$$= \varepsilon - \frac{4\pi}{\omega}\,\sigma_{\text{Im}} + i\,\frac{4\pi}{\omega}\,\sigma_{\text{Re}}, \qquad (6.6)$$

whence

$$n^2 - k^2 = \varepsilon - \frac{4\pi}{\omega}\,\sigma_{\text{Im}}, \qquad (6.6a)$$

$$2nk = \frac{4\pi}{\omega}\,\sigma_{\text{Re}}. \qquad (6.6b)$$

If absorption is so small that it can be neglected in (6.6a), then

$$n = \sqrt{\varepsilon - \frac{4\pi}{\omega}\,\sigma_{\text{Im}}}. \qquad (6.7)$$

If we have two rays circularly polarized in opposite directions ($\gamma = \pm i$), there are two different values $\sigma_{\text{Im}}^{\pm}$ in (6.4b) to correspond to them, and this means that the refractive indices (6.7) of these rays are also different:

$$n^+ = \sqrt{\varepsilon - \frac{4\pi}{\omega}\,\sigma_{\text{Im}}^+}, \qquad (6.7a)$$

$$n^- = \sqrt{\varepsilon - \frac{4\pi}{\omega}\,\sigma_{\text{Im}}^-}. \qquad (6.7b)$$

A plane-polarized wave can be represented as a superposition of two waves with equal amplitudes circularly polarized in opposite directions. The refractive index for the right-hand wave can differ from the refractive index for the left-hand polarized wave, in this case their phase difference will change as the waves propagate in the medium.

Denote the phases of both circularly polarized waves by

$$\varphi_{\pm}=\omega\left(\frac{n\pm z}{c}-t\right). \tag{6.8}$$

The complex-valued component of the combined electric field of both waves along the $x$ axis is

$$E_x=E_0\left(e^{i\varphi_+}+e^{i\varphi_-}\right)=2E_0e^{i\frac{\varphi_++\varphi_-}{2}}\cos\frac{\varphi_+-\varphi_-}{2}, \tag{6.9}$$

which can be easily checked if exponential functions are substituted for the cosine on the right-hand side.

Since the phases of $E_y$ in the circularly polarized waves are $\varphi_++\frac{\pi}{2}$ and $\varphi_--\frac{\pi}{2}$, it follows that

$$E_y=E_0\left(e^{i\left(\varphi_++\frac{\pi}{2}\right)}+e^{i\left(\varphi_--\frac{\pi}{2}\right)}\right]=2E_0e^{i\frac{\varphi_++\varphi_-}{2}}\cos\left(\frac{\pi}{2}+\frac{\varphi_+-\varphi_-}{2}\right). \tag{6.9a}$$

If the angle between the polarization plane and the $x$ axis is $\theta$, then

$$\tan\theta=\frac{\operatorname{Re}\{E_y\}}{\operatorname{Re}\{E_x\}}=-\frac{\sin\frac{\varphi_+-\varphi_-}{2}}{\cos\frac{\varphi_+-\varphi_-}{2}}=-\tan\frac{\varphi_+-\varphi_-}{2}, \tag{6.10}$$

where Re $\{E_y\}$ and Re $\{E_x\}$ are the real parts of $E_y$ and $E_x$. In (6.10) use was made of the coincidence of the real parts of identical complex-valued multipliers on the right-hand sides of (6.9) and (6.9a).

The magnitude of the rotation angle of the polarization plane for a distance $z=d$ traveled by the ray in the magnetic field is, according to (6.10) and (6.8),

$$\theta=\frac{|\varphi_+-\varphi_-|}{2}=\frac{\omega d}{2c}(n^+-n^-). \tag{6.11}$$

It follows from (6.7a), (6.7b) that

$$(n^+)^2-(n^-)^2=\frac{4\pi}{\omega}(\sigma_{\mathrm{Im}}^- -\sigma_{\mathrm{Im}}^+)$$

or (6.12)

$$n^+-n^-=\frac{2\pi}{\omega\bar{n}}(\sigma_{\mathrm{Im}}^- -\sigma_{\mathrm{Im}}^+),$$

where $\bar{n}=\frac{1}{2}(n^++n^-)$.

The usual frequency range used to determine the effective mass $m^*$ with the aid of the Faraday effect is $\omega \gg \omega_c$, $1/\tau$. For this case we obtain from (6.4b)

$$\sigma_{\rm Im}^{-} - \sigma_{\rm Im}^{+} = \sigma_0 \frac{2\omega_c}{\omega^2\tau^2}, \tag{6.13}$$

where we must substitute the values $\sigma_0 = e^2 N_0 \tau/m^*$ and $\omega_c = eH/m^*c$.

From (6.11)-(6.13) we obtain for the polarization-plane rotation angle

$$\theta = \frac{2\pi e^3 N_0 H d}{c^2 \bar{n} \omega^2 m^{*2}}. \tag{6.14}$$

For $\bar{n}$ we can take the medium's refractive index in the absence of the magnetic field.

An important point to note is that (6.14) is independent of $\tau$, i.e., of the scattering mechanism. It is valid also in the case $\omega_c \tau \ll 1$, when the cyclotron resonance method is inapplicable. Specifically, this makes it possible to measure the effective mass $m^*$ at room temperature. Equation (6.14) can be generalized for the cases of complex band patterns and for different types of charge carriers; of course, in such cases it yields only average effective mass values.

The Faraday effect was observed and the results quantitatively compared with the results of other experiments with numerous semiconductors: Ge, $n$-InSb, AlSb, GaP, GaAs, InAs, etc.[13]

## 7.7. Theory of Interband Absorption of Light in a Quantizing Magnetic Field

**7.7.1** In Section 7.2 we discussed the absorption of light due to direct interband transitions. For direct allowed transitions the absorption coefficient $\alpha_{\rm al}$ is determined by (2.26).

Consider now the theory of interband transitions in a semiconductor with a standard band placed in a strong (quantizing) magnetic field.

In Section 6.5.3 we considered the motion of an electron with a scalar effective mass $m^*$ in a quantizing magnetic field (L. D. Landau, 1930) and showed that its wave function is

$$F_{Nk_yk_z}(x, y, z) = e^{i(k_y y + k_z z)} \varphi_N (x - x_0). \tag{7.1}$$

Here $\varphi_N (x - x_0)$ is the oscillator function for the equilibrium position $x_0 = -(\hbar c/eH) k_y$, $N$ is the oscillator's quantum number. For the sake of simplicity we assume that the linear dimensions of the crystal

[13] For a review of experimental results with a detailed bibliography see [7.13, p. 366].

in all three directions are equal to unity. The energy eigenvalues corresponding to the eigenfunction (7.1) are

$$\varepsilon_{Nk_z} = (2N+1)\,\mu^* H + \frac{\hbar^2 k_z^2}{2m^*}. \tag{7.2}$$

Here $\mu^* = e\hbar/2m^*c$ is the "effective" Bohr magneton. At the same time $2\mu^* H = \hbar\omega_c$, where $\omega_c = eH/m^*c$ is the cyclotron frequency. The energy eigenvalues (7.2) are degenerate with respect to the quantum number $k_y$.

J. M. Luttinger and W. Kohn (1955) have demonstrated that the wave function of an electron in the periodic field of a crystal placed in a quantizing magnetic field in the first approximation, i.e., with no account taken of the interaction with other bands, is

$$\Psi\,(x,\ y,\ z) = u_{n0}\,(\mathbf{r})\,F_{Nk_yk_z}\,(x,\ y,\ z), \tag{7.3}$$

where $u_{n0}\,(\mathbf{r})$ is the Bloch electron function $u_{n\mathbf{k}}\,(\mathbf{r})\,e^{i\mathbf{k}\cdot\mathbf{r}}$ at point $\mathbf{k}=0$ (the band is assumed to be a simple one with the energy minimum at point $\mathbf{k}=0$).

In the presence of the field of a light wave and an external magnetic field the Hamiltonian (2.4) is

$$\hat{\mathscr{H}} = \frac{1}{2m}\left[\hat{\mathbf{p}} + \frac{e}{c}\mathbf{A}^0 + \frac{e}{c}\mathbf{A}\,(\mathbf{r},\ t)\right]^2 + V\,(\mathbf{r}), \tag{7.4}$$

where the vector potential for the constant magnetic field directed along the $z$ axis is $\mathbf{A}^0 = \{0,\ Hx,\ 0\}$ [see (6.5.11)].

The part of perturbation in (7.4), as in (2.4), is played by the field of the electromagnetic wave $\mathbf{A}\,(\mathbf{r},\ t)$ (2.8). Proceeding in the same way as in (2.5), we obtain instead of (2.7) the perturbation

$$\hat{\mathscr{H}}' = \frac{e}{m}\,\mathbf{A}\,(\mathbf{r},\ t)\left[\hat{\mathbf{p}} + \frac{e}{c}\mathbf{A}^0\right]. \tag{7.5}$$

The matrix element of this perturbation between the wave functions (7.3) for the valence band and the conduction band is

$$\mathscr{P}_{\mathrm{vc}} = \langle \Psi_{\mathrm{c}} \mid \hat{\mathscr{H}}' \mid \Psi_{\mathrm{v}} \rangle = \int u^*_{\mathrm{c}0} F^*_{\mathrm{v}'} \hat{\mathscr{H}}' u_{\mathrm{v}0} F_{\mathrm{v}}\, d^3r. \tag{7.6}$$

Here $\Psi_c$ and $\Psi_v$ are wave functions (7.3) for the conduction and the valence bands, $\nu \equiv \{N,\ k_y,\ k_z\}$ are electron quantum numbers in the quantizing magnetic field.

The integrand in (7.6) contains the functions $u^*_{c0}\,(\mathbf{r})$ and $u_{v0}\,(\mathbf{r})$ rapidly varying inside the crystal cell $\Omega_0$ and slowly varying factors $F_{\nu'}\,(\mathbf{r})$ and $F_\nu\,(\mathbf{r})$. Replace the integral in (7.6) over the crystals' principal region $V = Z\Omega_0$ ($V = 1$ cm$^3$) with the sum of integrals over the volumes of the crystal cells $\Omega_0$:

$$\mathcal{P}_{vc} = \int_V u^*_{c0} F^*_{v'} \frac{e}{mc} \mathbf{A} \cdot \left( \hat{\mathbf{p}} + \frac{e}{c} \mathbf{A}^0 \right) u_{v0} F_v \, d^3r$$

$$= \frac{e}{mc} \int_V u^*_{c0} F^*_{v'} \Big[ (\mathbf{A} \cdot \hat{\mathbf{p}} u_{v0}) F_v + (\mathbf{A} \cdot \hat{\mathbf{p}} F_v) u_{v0}$$

$$+ \frac{e}{c} \mathbf{A} \cdot \mathbf{A}^0 u_{v0} F_v \Big] d^3r = \frac{e}{mc} \sum_z \Big\{ F^*_{v'} F_v$$

$$\times \int_{\Omega_0} u^*_{c0} (\mathbf{A} \cdot \hat{\mathbf{p}} u_{v0}) \, d^3r + (F^*_{v'} \mathbf{A} \cdot \hat{\mathbf{p}} F_v)$$

$$\times \int_{\Omega_0} u^*_{c0} u_{v0} \, d^3r + \frac{e}{c} \mathbf{A} \cdot \mathbf{A}^0 F^*_{v'} F_v \int_{\Omega_0} u^*_{c0} u_{v0} \, d^3r \Big\}. \tag{7.7}$$

Here we took the slowly varying functions out of the integrals over the unit cells $\Omega_0$.

Because of the orthogonality of the functions $u_{c0}$ and $u_{v0}$, the last two terms in the braces in (7.7) vanish, and

$$\mathcal{P}_{vc} = \sum_z F^*_{v'} F_v \Omega_0 \cdot \frac{1}{\Omega_0} \int_{\Omega_0} u^*_{c0} \frac{e}{mc} \mathbf{A} \cdot \hat{\mathbf{p}} u_{v0} \, d^3r$$

$$= \int_V F^*_{v'} F_v d^3r \cdot \frac{1}{\Omega_0} \int_{\Omega_0} u^*_{c0} \frac{e}{mc} \mathbf{A} \cdot \hat{\mathbf{p}} u_{v0} \, d^3r, \tag{7.8}$$

where we have replaced the summation over $z$ with the integration over the volume $V$.

Making $\mathbf{A}$ equal to (2.8), $p = -i\hbar\nabla$ and assuming that the photon's wave vector $\varkappa = 0$, as in Section 7.2, we obtain, using (7.1)

$$\mathcal{P}_{vc} = \frac{e}{mc} A_0 (\mathbf{e} \cdot \mathbf{p}_{cv}) \int e^{i(k_y - k'_y)y} \, dy \int e^{i(k_z - k'_z)z} \, dz$$

$$\times \int \varphi^*_{N'} (x - x'_0) \, \varphi_N (x - x_0) \, dx, \tag{7.9}$$

where $\mathbf{p}_{cv}$ is given by (2.20a).

The integral over the region $L_y = 1$ cm is

$$\int_0^1 e^{i(k_y - k'_y)y} \, dy = \begin{Bmatrix} 1, & \text{if} & k_y = k'_y \\ 0, & \text{if} & k_y \neq k'_y \end{Bmatrix} = \delta_{k_y k'_y},$$

i.e., is the Kronecker delta. A similar result can be obtained for the integral with respect to $dz$. Since the oscillator functions $\varphi_N (x - x_0)$ are orthonormalized, and $x'_0 = x_0$ (because $k_y = k'_y$), the last integral in (7.9) is equal to $\delta_{NN'}$. Accordingly, we obtain instead of (7.9)

$$\mathcal{P}_{vc} = \frac{e}{mc} A_0 (\mathbf{e} \cdot \mathbf{p}_{cv}) \delta_{k_y k_y'} \delta_{k_z k_z'} \delta_{NN'}. \tag{7.9a}$$

To calculate the rate of transitions per unit volume, caused by the perturbation $\hat{\mathcal{H}}'$ (7.5) we must, when calculating $\mathcal{W}_{vc}$, sum (integrate) both over the initial and the final states, i.e., over the quantum states $\nu = \{N, k_y, k_z\}$ and $\nu' \equiv \{N', k_y', k_z'\}$. Equation (7.9a) makes it possible to reduce this double summation to the summation over $N$, $k_y$, $k_z$.

Hence, we obtain by analogy with (2.21)

$$\mathcal{W}_{vc} = \frac{2\pi}{\hbar} \sum_{N,\, k_y,\, k_z} |\mathcal{P}_{vc}|^2 \delta(\varepsilon_c - \varepsilon_v - \hbar\omega)$$
$$= \frac{2\pi}{\hbar} \frac{e^2}{m^2c^2} |A_0|^2 |\mathbf{e} \cdot \mathbf{p}_{cv}(0)|^2 \sum_N \int \frac{2dk_y}{2\pi} \int \frac{2dk_z}{2\pi}$$
$$\times \delta\left[\hbar\omega - \varepsilon_G - (2N+1)\mu_r H - \frac{\hbar^2 k_z^2}{2m_r}\right], \tag{7.10}$$

where the inverse reduced effective mass is $1/m_r = 1/m_c + 1/m_v$, and the "reduced effective Bohr magneton" is $\mu_r = e\hbar/2m_r c$. In the argument of the $\delta$-function we put $N' = N$ (since $k_y' = k_y$) and $k_z' = k_z$.

Since $k_y = -\frac{eH}{\hbar c} x_0$, and $0 \leqslant x_0 \leqslant 1$, the integration with respect to $k_y$ yields

$$\int dk_y = \frac{eH}{\hbar c}. \tag{7.11}$$

Next

$$\int dk_z \delta\left[B - \frac{\hbar^2 k_z^2}{2m_r}\right] = \frac{1}{\hbar}\left(\frac{m_r}{2}\right)^{1/2} B^{-1/2}, \tag{7.12}$$

which can easily be checked if $\zeta = \hbar^2 k_z^2/2m_r$ is substituted for the integration variable $k_z$.

The light absorption coefficient $\alpha_H$ is equal to the transition rate $\mathcal{W}_{vc}$ (7.10) divided by the photon flux $Nv = Nc/n$.[14] Substituting

---

[14] Here $N$ is the number of photons per cubic centimeter.

into (7.10) (2.12a) for $A_0$ and making use of (7.11) and (7.12), we obtain

$$\alpha_H = \frac{2e^2}{m^2 \omega n c} \,|\, \mathbf{e} \cdot \mathbf{p}_{cv}(0) \,|^2 \left(\frac{2m_r}{\hbar^2}\right)^{3/2} \mu_r H \sum_N \times [\hbar\omega - \varepsilon_G - (2N+1)\,\mu_r H]^{-1/2}. \quad (7.13)$$

The sum over $N$ in (7.13) embraces all values of $N$ for which the values of the expression in the square brackets are nonnegative. The values of $\omega$, $N$ and $H$ for which this expression is zero determine the singular points of the absorption coefficient $\alpha_H$. These points correspond to the condition

$$\hbar\omega_{max} - \varepsilon_G = (2N+1)\,\mu_r H, \quad (7.14)$$

where $\omega_{max}$ is the frequency of light corresponding to maximum absorption. Of course, in actual fact the absorption coefficient $\alpha_H$ does not become infinite at such points, since there are several factors (for instance, the interaction of the electrons with the lattice vibrations) that "smear" the infinite peaks. Such finite peaks (oscillations) of $\alpha_H$ are observed in experiment.

It is seen from (7.14) that for a specified $N$ the frequency of light $\omega_{max}$ depends linearly on the magnetic field $H$. Figure 7.3 depicts the frequency of light $\omega_{max}$ vs magnetic field $H$ dependences for different $N$'s. Any vertical straight line $H$ = const. intersects the rays for different N's and thereby determines the values of $\omega_{max}$ corresponding to the absorption peaks for a specified $H$. The points of intersection of a horizontal straight line $\omega$ = const. determine for a given $\omega_{max}$ the values of the magnetic field $H$ corresponding to the absorption peaks.

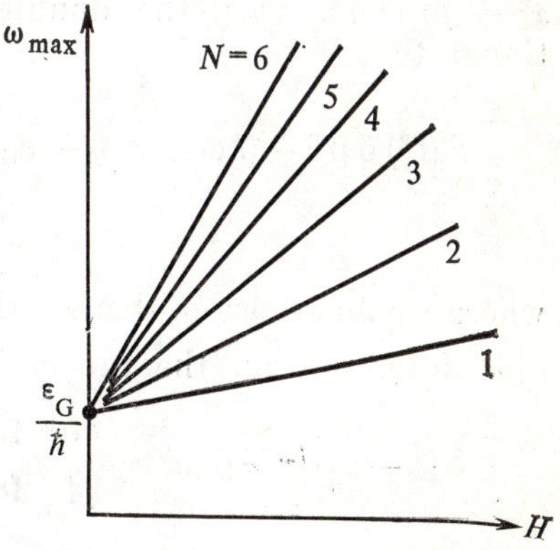


Fig. 7.3

The picture observed in experiment is frequently a more complicated one: the absorption curve contains not only peaks but steps as well, because, in addition to the direct allowed transitions considered above, there is a possibility of indirect transitions involving phonons and forbidden transitions that behave differently depending on whether $\mathbf{e} \parallel \mathbf{H}$ or $\mathbf{e} \perp \mathbf{H}$ [$\mathbf{e}$ is the polarization vector in (2.8)].

Consider briefly indirect transitions. In the case of indirect transitions we must sum in (3.1) separately over $N$ and over $N'$ and integrate with respect to $k_y$ and $k'_y$; $k_z$ and $k'_z$ [in this case equation

(7.9a) does not hold, since a phonon with its own wave vector $\mathbf{q}$ takes part in the interaction]. Double integration with respect to $k_y$ and $k'_y$ yields, according to (7.11), the factor $(eH/\hbar c)^2$. Omitting the constant factor, we obtain

$$\alpha_{\text{ind}}^H \propto \left(\frac{eH}{\hbar c}\right)^2 \sum_{NN'} \int\int dk_z\, dk'_z \delta \times \left[(2N'+1)\,\mu_c H + \frac{\hbar^2 k'^2_z}{2m_c} + \varepsilon_G + (2N+1)\,\mu_v H + \frac{\hbar^2 k_z^2}{2m_v} \mp \hbar\omega_q - \hbar\omega\right]. \quad (7.15)$$

In comparison with (7.10), an additional term $\mp\hbar\omega_q$, connected with the absorption (upper sign) or the emission (lower sign) of a phonon, appears in the argument of the δ-function.

Introduce the variables $(\hbar^2 k_z^2/2m_v) = x^2$, $(\hbar^2 k'^2_z/2m_c) = y^2$, and $x^2 + y^2 = r^2$. Then the double integral in (7.15) will be proportional to

$$\int_0^\infty d\,(r^2)\,\delta\,[r^2 - (\hbar\omega \pm \hbar\omega_q - \varepsilon_G) - (2N+1)\,\mu_v H - (2N'+1)\,\mu_c H], \quad (7.15\text{a})$$

where we made use of the fact that $dk_z dk'_z \propto dx\, dy \propto r\, dr = \frac{1}{2}\, d\,(r^2)$.

It follows from the properties of the δ-function that

$$\int_0^\infty \delta\,(x-a)\,dx = \theta\,(a) = \begin{cases} 0 & \text{for } a < 0, \\ 1 & \text{for } a > 0. \end{cases} \quad (7.16)$$

The step function θ is used in mathematics alongside the δ-function.

Eventually, we can write for (7.15)

$$\alpha_{\text{ind}}^H \propto \left(\frac{eH}{\hbar c}\right)^2 \sum_{NN'} \theta\,[\hbar\omega \pm \hbar\omega_q - \varepsilon_G - (2N+1)\,\mu_v H - (2N'+1)\,\mu_c H]. \quad (7.17)$$

Consider the dependence of $\alpha_{\text{ind}}^H$ on the photon's energy $\hbar\omega$ in the case of a constant magnetic field. Choose the term with $N = N' = 0$ in the sum (7.17). For small $\hbar\omega$, as long as the argument of the function $\theta\,(u)$ is negative, $\theta\,(u) = 0$. Suppose that, as $\hbar\omega$ increases, the argument of the function $\theta\,(u)$ turns zero at point *1* in Fig. 7.4. Then for greater values of $\hbar\omega$ the function $\theta\,(u) = 1$, and $\alpha_{\text{ind}}^H$ makes the jump depicted in the figure. At point *2* the argument of the function $\theta(u)$ for another term, for example, for $N = 1$, $N' = 0$, turns zero, and a second step appears on the graph of $\alpha^H_{\text{ind}}$. Since $\mu_v \neq \mu_c$, and the phonon energy $\hbar\omega_q$ is different vibration branches and can enter the argument of $\theta(u)$ either with a plus or a minus sign, the steps on the graph of $\alpha^H_{\text{ind}}$ form an irregular pattern as shown in Fig. 7.4. The direct forbidden transitions, which shall not be discussed here, introduce an additional complication into the picture of light absorption in semiconductors in a magnetic field.

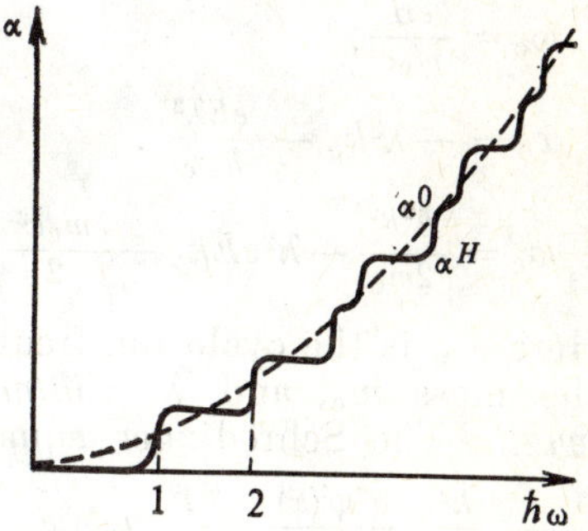

Fig. 7.4

7.7.2 New useful information can be gained from the study of light absorption in crossed electric and magnetic fields (A. G. Aronov, 1963).

Let the magnetic field **H** point in the direction of the $z$ axis and let the electric field **E** point in the direction of the $x$ axis. Choose the vector potential, as was done in Section 6.5.3 in the form $\mathbf{A} \equiv \{0, Hx, 0\}$.

The Schrödinger equation for an electron with the effective mass $m_c$ and the charge $-e$ in the conduction band is of the form

$$\hat{\mathscr{H}} F(\mathbf{r}) = \varepsilon_c F(\mathbf{r}), \tag{7.18}$$

where the Hamiltonian is

$$\hat{\mathscr{H}} = -\frac{\hbar^2}{2m_c}\frac{\partial^2}{\partial x^2} + \frac{1}{2m_c}\left(\frac{\hbar}{i}\frac{\partial}{\partial y} + \frac{e}{c}Hx\right)^2 - \frac{\hbar^2}{2m_c}\frac{\partial^2}{\partial z^2} + eEx. \tag{7.18a}$$

Here the potential energy of the electron in the electric field is $-(-e)Ex = eEx$.

Since the Hamiltonian $\hat{\mathscr{H}}$ does not contain $y$ and $z$ in an explicit form, we must search for the solution of (7.18), as we have done in Section 6.5.3, in the form

$$F(x, y, z) = \varphi(x)\, e^{i(k_y y + k_z z)}. \tag{7.19}$$

Operating with the Hamiltonian $\hat{\mathscr{H}}$ on (7.19) and canceling the exponential factor, we obtain

$$-\frac{\hbar^2}{2m_c}\frac{d^2\varphi(x)}{dx^2} + \left[\frac{1}{2m_c}\left(\frac{eH}{c}\right)^2 x^2 + \left(\frac{\hbar eH}{m_c c}k_y + eE\right)x + \frac{\hbar^2(k_y^2 + k_z^2)}{2m_c}\right]\varphi(x). \tag{7.20}$$

This expression can be reduced to the form

$$-\frac{\hbar^2}{2m_c}\frac{d^2\varphi(x)}{dx^2}+\frac{1}{2}m_c\omega_c^2(x-x_c)^2\varphi(x)+w_c\varphi(x), \tag{7.21}$$

where $\omega_c$, $x_c$ and $w_c$ are constants independent of $x$. Comparing the coefficients of $x^2$, $x^1$, $x^0 = 1$ in (7.20) and (7.21), we obtain

$$\omega_c=\frac{eH}{m_c c}, \tag{7.21a}$$

$$x_c=-\lambda^2 k_y-\frac{eE\lambda^2}{\hbar\omega_c}, \tag{7.21b}$$

$$w_c=\frac{\hbar^2 k_z^2}{2m_c}-\lambda^2 eEk_y-\frac{m_c c^2}{2}\left(\frac{E}{H}\right)^2. \tag{7.21c}$$

Here $\omega_c$ is the cyclotron frequency for the conduction electron with the mass $m_c$, and $\lambda = (\hbar/m_c\omega_c)^{1/2} = (\hbar c/eH)^{1/2}$ is the *magnetic length*. The Schrödinger equation for $\varphi(x)$ takes the form

$$-\frac{\hbar^2}{2m_c}\frac{d^2\varphi(x)}{dx^2}+\frac{1}{2}m_c\omega_c^2(x-x_c)^2\varphi(x)=(\varepsilon_c-w_c)\varphi(x) \tag{7.22}$$

which is similar to (6.5.18).

Hence, the energy eigenvalues are

$$\begin{aligned}\varepsilon_c&=\varepsilon(0)+(2N+1)\mu_c H+w_c\\&=\varepsilon_c(0)+(2N+1)\mu_c H+\frac{\hbar^2k_z^2}{2m_c}-\lambda^2 eEk_y-\frac{m_c c^2}{2}\left(\frac{E}{H}\right)^2,\end{aligned} \tag{7.23}$$

where $\varepsilon_c(0)$ is the energy level corresponding to the bottom edge of the conduction band, $\mu_c = e\hbar/2m_c c$ is the effective Bohr magneton of the electron, and $N = 0, 1, 2$ is the oscillator's quantum number. We see that in crossed electric and magnetic fields the energy eigenvalues $\varepsilon_c$ depend on $k_y$, i.e., the degeneracy with respect to $k_y$ that existed in the case of the magnetic field is removed.

The eigenfunctions of (7.22) are

$$\varphi(x)=\varphi_N(x-x_c)=\frac{\exp\left[-\frac{1}{2}\left(\frac{x-x_c}{\lambda}\right)^2\right]}{\sqrt{\pi}}H_N\left(\frac{x-x_c}{\lambda}\right), \tag{7.24}$$

i.e., are of the same form (6.5.19a) as in the case of the magnetic field acting alone [the value of $x_c$ (7.21b) is, of course, different]; $H_N$ are Hermite polynomials.

To calculate the light absorption coefficient $\alpha_{H/E}$ due to direct allowed transitions in crossed electric and magnetic fields, we should, in compliance with (7.9), additionally know the wave functions (7.19) and the energy eigenvalues (7.23) for the holes in the valence band.

Instead of (7.21a)-(7.21c); we obtain for holes with the charge $+e$ and the effective mass $m_v$

$$\omega_v = \frac{eH}{m_v c}, \tag{7.25a}$$

$$x_v = -\lambda^2 k'_y + \frac{eE\lambda^2}{\hbar\omega_v}, \tag{7.25b}$$

$$w_v = \frac{\hbar^2 k'^2_z}{2m_v} + \lambda^2 eEk'_y - \frac{m_v c^2}{2}\left(\frac{E}{H}\right)^2, \tag{7.25c}$$

where we have substituted $e$ for $-e$ and $m_v$ for $m_c$. For the energy eigenvalues of the hole we obtain instead of (7.23)

$$\begin{aligned}\varepsilon_v &= \varepsilon_v(0) - (2N'+1)\mu_v H - w_v \\ &= \varepsilon_v(0) - (2N'+1)\mu_v H - \frac{\hbar^2 k'^2_z}{2m_v} - \lambda^2 eEk'_y + \frac{m_v c^2}{2}\left(\frac{E}{H}\right)^2.\end{aligned} \tag{7.26}$$

Here $\mu_v = e\hbar/2m_v c$ is the effective Bohr magneton for the hole. The hole's energy is measured in the negative direction down from the top edge of the valence band $\varepsilon_v(0)$. We could instead consider the electron state in the valence band with a negative effective mass $-m_v$ and obtain the same expression (7.26) for $\varepsilon_v$. The matrix element $\mathcal{P}_{vc}$ (7.9) will now be

$$\mathcal{P}_{vc} = \frac{e}{mc} A_0 (\mathbf{e}\cdot\mathbf{p}_{cv}(0))\, \delta_{k_y k'_y} \delta_{k_z k'_z} \int_{-\infty}^{\infty} \varphi_N(x - x_c)\, \varphi_{N'}(x - x_v)\, dx. \tag{7.27}$$

We see that, as is the case of (7.9), $k_y = k'_y$, $k_z = k'_z$, but now $x_c \neq x_v$ (since $m_c \neq m_v$). Since the centres of gravity of the oscillator functions $x_c$ and $x_v$ no longer coincide, they will not be mutually orthogonal, i.e., the integral generally is not equal to zero for $N \neq N'$.

To calculate it, we can use the equation [7.9, p. 838, equation 7.377]

$$\int_{-\infty}^{\infty} e^{-x^2} H_m(x+y)\, H_n(x+z)\, dx = 2^n \sqrt{\pi}\, m!\, z^{n-m} \mathcal{L}_m^{n-m}(-2yz), \tag{7.28}$$

where $m \leqslant n$, and $\mathcal{L}_m^{n-m}$ are the generalized Laguerre polynomials [7.14, p. 100]. Transform the integral in (7.27) making $(x - x_c)/\lambda =$

$= \xi - a$ and choose $a$ such that the term in the exponent linear in $\xi$ vanishes. In this case the integral in (7.27) becomes

$$\exp\left[-\left(\frac{x_c - x_v}{2\lambda}\right)^2\right] \int_{-\infty}^{\infty} e^{-\xi^2} H_N\left(\xi - \frac{x_c - x_v}{2\lambda}\right) H_{N'}\left(\xi + \frac{x_c - x_v}{2\lambda}\right) d\xi$$

$$= \exp\left[-\left(\frac{x_c - x_v}{2\lambda}\right)^2\right] \sqrt{\pi}\, 2^N N'! \left(\frac{x_v - x_c}{2\lambda}\right)^{N-N'} \times \mathscr{L}_{N'}^{N-N'}\left[2\left(\frac{x_c - x_v}{2\lambda}\right)^2\right], \quad (7.29)$$

where $N' \leqslant N$ (when $N' > N$, we must interchange $N$ with $N'$ and $x_v$ with $x_c$); $\mathscr{L}_{N'}^{N-N'}(u)$ are polynomials of degree $N'$ in $u$, the first of them being equal to

$$\mathscr{L}_0^p(u) = 1, \quad \mathscr{L}_1^p(u) = 1 + p - u,$$

$$\mathscr{L}_2^p(u) = \frac{1}{2}\left[(1+p)(2+p) - 2(2+p)u + u^2\right].$$

Now we can write $\mathscr{P}_{vc}$ (7.27) in the form

$$\mathscr{P}_{vc} = \frac{e}{mc} A_0 (\mathbf{e} \cdot \mathbf{p}_{cv}(0))\, \delta_{k_y k'_y} \delta_{k_z k'_z} \sqrt{\pi}\, e^{-\gamma^2} 2^N N'! \times \gamma^{N-N'} \mathscr{L}_{N'}^{N-N'}(2\gamma^2), \quad (7.30)$$

where

$$\gamma = \frac{x_v - x_c}{2\lambda} = \frac{(m_c + m_v)\, c\lambda E}{2\hbar H}. \quad (7.30a)$$

The argument for the $\delta$-function entering $\mathscr{W}_{vc}$ (7.10) is

$$\varepsilon_c - \varepsilon_v - \hbar\omega = \varepsilon_G + (2N+1)\mu_c H + (2N'+1)\mu_v H + \frac{\hbar^2 k_z^2}{2m_r} - \frac{(m_c + m_v)\, c^2}{2}\left(\frac{E}{H}\right)^2 - \hbar\omega, \quad (7.31)$$

where we made use of the fact that $k_y = k'_y$ and $k_z = k'_z$. Next, proceeding in the same way as in Section 7.7.1, we obtain for the absorption coefficient due to direct allowed transitions in crossed electric and magnetic fields

$$\alpha_{H/E} = \frac{2e^2}{m^2 \omega n c} |\mathbf{e} \cdot \mathbf{p}_{cv}(0)|^2 \left(\frac{2m_r}{\hbar^2}\right)^{3/2} (\mu_r H)\, e^{-2\gamma^2} \times \sum_{NN'} \pi 2^{2N} [N'!]^2 \gamma^{2(N-N')} [\mathscr{L}_{N'}^{N-N'}(2\gamma^2)]^2$$

$$\times \left[\hbar\omega - \varepsilon_G - (2N+1)\mu_c H - (2N'+1)\mu_v H + \frac{(m_c + m_v)\, c^2}{2} \frac{E^2}{H^2}\right]^{-1/2} \quad (7.32)$$

where $N' \leqslant N$.

Note the principal features of light absorption in crossed electric and magnetic fields:

a) there are no selection rules for the quantum numbers of the magnetic oscillators $N$ and $N'$, i.e., in principle, transitions can take place between any Landau sublevels in the valence and conduction bands;

b) $\alpha_{H/E}$ decreases exponentially with the increase in the electric field. At strong fields $\alpha_{H/E} \propto E^{2(N'+N))}e^{-\beta E^2}$. This sets a limitation on the magnitude of the maximum electric field at which absorption can still be observed: $E_{max} \equiv \hbar H/(m_c + m_v)\, c\lambda$ $(\gamma^2 \sim 1)$;

c) the displacement of the absorption peak in the electric field for allowed transitions at $\mathbf{E} = 0$ $(N = N')$ is $\Delta\hbar\omega = (m_c + m_v) \times \times c^2E^2/2H^2$. Measuring this displacement, we can find the sum of the effective masses $m_c + m_v$. Since the relative effective mass $m_r = m_c m_v/(m_c + m_v)$ can be found from the measurements of $\alpha_H$ (7.14), the combined measurements of $\alpha_{H/E}$ and of $\alpha_H$ make it possible to determine $m_c$ and $m_v$ individually, another method of finding these masses is from the frequencies of transitions between the levels $N \neq N'$.

The effective masses $m_c$ and $m_v$ can also be found from the differences of frequencies of transitions that are forbidden at $\mathbf{E} = 0$ but allowed in a nonzero electric field.

## 7.8. Absorption of Light in Semiconductors in a Homogeneous Electric Field (Franz-Keldysh Effect)

7.8.1 In 1958 W. Franz and L. V. Keldysh independently published the results of their studies of light absorption in a semiconductor placed in a homogeneous electric field.

The problem can be solved in accordance with the general procedure discussed in Section 7.7. We shall consider a semiconductor with a simple band and assume that the electron in the conduction band is described by a scalar effective mass $m_c$ and that in the valence band by a negative effective mass $-m_v$.

For the function $F_{Nk_yk_z}$ in (7.3) we must now take a stationary wave function of a free electron in a homogeneous electric field $\mathbf{E}$.

It will be evident from arguments to be presented below that the quantum mechanics problem about the motion of a free electron in a homogeneous electric field can easier be solved not in the $x$-**representation** but in the $p$-representation. This simply means that in the Schrödinger equation we should consider as operators not the momentum components ($\hat{p}_x = -i\hbar\partial/\partial x$, etc.) but the coordinates ($\hat{x} = i\hbar\partial/\partial p_x$, etc.). In this case the stationary wave functions $F$ will depend not on $x$, $y$, $z$ but on $p_x$, $p_y$, $p_z$ [7.4, § 13]. If the electric field $\mathbf{E}$ points in the direction of the $x$ axis, the potential energy of an electron with the charge $-e$ will be $-(-e)\, Ex = eEx$. The

Schrödinger equation for an electron of mass $m_c$ in the conduction band in a field $\mathbf{E} \parallel x$ takes in the $p$-representation the form

$$\left[\frac{p_x^2+p_y^2+p_z^2}{2m_c}+eEi\hbar\frac{\partial}{\partial p_x}\right]F_c(p_x, p_y, p_z)=\varepsilon_c F_c(p_x, p_y, p_z). \quad (8.1)$$

Here $F_c(p_x, p_y, p_z)$ is the electron wave function in the $p$-representation and $\varepsilon_c$ is its energy eigenvalue.

It follows from (8.1) that in the $p$-representation the wave function satisfies a first-order differential equation, whereas in the $x$-representation it satisfies a second-order differential equation, which complicates the mathematics of the problem [7.6, §24].

Making

$$F_c(p_x, p_y, p_z) = \psi_c(p_x)\,\varphi_c(p_y)\,\chi_c(p_z), \quad (8.2)$$

substituting this expression into (8.1) and dividing both sides of the equation by $F_c(p_x, p_y, p_z)$, we obtain

$$\left(\frac{p_x^2}{2m_c}+\frac{1}{\psi_c(p_x)}eEi\hbar\frac{d\psi_c}{dp_x}\right)+\frac{p_y^2}{2m_c}+\frac{p_z^2}{2m_c}=\varepsilon_c. \quad (8.3)$$

The first term in the parentheses depends only on $p_x$, the second only on $p_y$, and the third only on $p_z$. Since their sum is constant and equal to $\varepsilon_c$, each term is also constant. To satisfy this condition make $p_y = p_{yc}$ and $p_z = p_{zc}$, where $p_{yc}$ and $p_{zc}$ are eigenvalues of the momentum components along the $y$ and the $z$ axes. Then we obtain, instead of (8.3),

$$\frac{d\psi_c}{dp_x}-\frac{i}{\hbar eE}\left(\frac{p_x^2}{2m_c}-\varepsilon_c+\frac{1}{2m_c}p_{\perp c}^2\right)\psi_c=0, \quad (8.4)$$

where $p_{\perp c}^2 = p_{yc}^2 + p_{zc}^2$.

The wave function (8.2) is

$$F_c(p_x, p_y, p_z) = \psi_c(p_x)\,\delta_{p_y p_{yc}}\,\delta_{p_z p_{zc}}, \quad (8.5)$$

where $\delta_{p_y p_{yc}}$ and $\delta_{p_z p_{zc}}$ are Kronecker's deltas (we assume that the crystal is finite, so that $p_y$ and $p_z$ assume quasi-discrete values).

The solution of the first order linear differential equation (8.4) is of the form [7.15, p. 23]

$$\psi_c(p_x)=C\exp\left\{\frac{i}{\hbar eE}\int_0^{p_x}\left(\frac{p_x'^2}{2m_c}-\varepsilon_c+\frac{1}{2m_c}p_{\perp c}^2\right)dp_x'\right\}$$

$$=C\exp\left\{\frac{1}{\hbar eE}\left[\frac{p_x^3}{6m_c}-\left(\varepsilon_c-\frac{1}{2m_c}p_{\perp c}^2\right)p_x\right]\right\}. \quad (8.6)$$

Find the factor $C$ from the condition that the function $\psi_c(p_x)$ is normalized to a $\delta$-function of energy:

$$\int_{-\infty}^{\infty} \psi_c^*(p_x, \varepsilon_c') \psi_c(p_x, \varepsilon_c)\, dp_x = \delta(\varepsilon_c - \varepsilon_c'). \tag{8.7}$$

Substituting (8.6) into the left-hand side of (8.7), we obtain

$$|C|^2 \int_{-\infty}^{\infty} \exp\left\{\frac{i}{\hbar eE}(\varepsilon_c - \varepsilon_c')\, p_x\right\} dp_x$$

$$= |C|^2 (\hbar eE) \int_{-\infty}^{\infty} \exp\{i(\varepsilon_c - \varepsilon_c')\,\xi\}\, d\xi = |C|^2 (2\pi\hbar eE)\, \delta(\varepsilon_c - \varepsilon_c').$$

Here we made use of the standard definition of the $\delta$-function [7.6, § 15, Eq. (15.7)]. Comparing this with (8.7), we see that

$$C = 1/\sqrt{2\pi\hbar eE}. \tag{8.8}$$

For an electron in the valence band we must solve an equation similar to (8.3), substituting in it $-m_v$ for $m_c$ and $\varepsilon_v + \varepsilon_G$ for $\varepsilon_c$ ($\varepsilon_G$ is the forbidden bandwidth); the aim of the latter operation is to reckon the energy in the valence and the conduction bands from a common level.

Then

$$F_v(p_x, p_y, p_z) = \psi_v(p_x)\, \delta_{p_y p_{yv}} \delta_{p_z p_{zv}}, \tag{8.9}$$

where

$$\psi_v(p_x) = C \exp\left\{-\frac{i}{\hbar eE}\left[\frac{p_x^3}{6m_v} + \left(\varepsilon_v + \varepsilon_G + \frac{1}{2m_v} p_{\perp v}^2\right) p_x\right]\right\}. \tag{8.10}$$

The value of the normalization constant $C$ in (8.10) is the same as in (8.8). Now calculate the matrix element $\langle F_c \mid F_v\rangle$ that enters $\mathcal{P}_{vc}$ (7.8). We have

$$\langle F_c \mid F_v\rangle = \sum_{p_y p_z} \delta_{p_y p_{yc}} \delta_{p_z p_{zc}} \delta_{p_y p_{yv}} \delta_{p_z p_{zv}}$$

$$\times |C|^2 \int_{-\infty}^{\infty} dp_x \exp\left\{-\frac{i}{\hbar eE}\left[\frac{p_x^3}{6\mu} + \left(\varepsilon_v - \varepsilon_c + \varepsilon_G + \frac{1}{2m_c} p_{\perp c}^2 + \frac{1}{2m_v} p_{\perp v}^2\right) p_x\right]\right\}. \tag{8.11}$$

Here $\mu$ is the reduced mass:

$$\frac{1}{\mu}=\frac{1}{m_c}+\frac{1}{m_v}. \tag{8.11a}$$

Summation over $p_y$ and $p_z$ yields $\delta_{p_{yc}p_{yv}}\delta_{p_{zc}p_{zv}}=\delta_{p_{\perp c}p_{\perp v}}$. Substitute a new integration variable for $p_x$ and make

$$p_x=(2\mu\hbar eE)^{1/3}u. \tag{8.12}$$

Then we shall obtain instead of (8.11)

$$\langle F_c \mid F_v\rangle=\frac{\delta_{p_{\perp c}p_{\perp v}}}{\hbar\omega_E}\frac{1}{2\pi}\int_{-\infty}^{\infty}\exp\left[-i\left(\frac{u^3}{3}+xu\right)\right]du, \tag{8.13}$$

where the frequency

$$\omega_E=(eE)^{2/3}/(2\mu\hbar)^{1/3}, \tag{8.13a}$$

and

$$x=\frac{1}{\hbar\omega_E}\left[\varepsilon_v+\varepsilon_c+\varepsilon_G+\frac{1}{2m_c}p_{\perp c}^2+\frac{1}{2m_v}p_{\perp v}^2\right]. \tag{8.13b}$$

Since

$$\int_{-\infty}^{\infty}\exp\left[-i\left(\frac{u^3}{3}+xu\right)\right]du=\int_0^{\infty}\left\{\exp i\left(\frac{u^3}{3}+xu\right)\right.$$

$$\left.+\exp\left[-i\left(\frac{u^3}{3}+xu\right)\right]\right\}du=2\int_0^{\infty}\cos\left(\frac{u^3}{3}+xu\right)du,$$

it follows that

$$\langle F_c \mid F_v\rangle=\frac{\delta_{p_{\perp v}p_{\perp c}}}{\hbar\omega_E}\operatorname{Ai}(x), \tag{8.14}$$

where the *Airy function* [7.16, p. 508, Equation (5)]

$$\operatorname{Ai}(x)=\frac{1}{\pi}\int_0^{\infty}\cos\left(\frac{u^3}{3}+xu\right)du. \tag{8.14a}$$

By analogy with (7.8),

$$\mathscr{P}_{vc}=\frac{e}{mc}A_0(\mathbf{e}\cdot\mathbf{p}_{cv}(0))\langle F_c \mid F_v\rangle$$

$$=\frac{e}{mc}A_0(\mathbf{e}\cdot\mathbf{p}_{cv}(0))\,\delta_{p_{\perp c}p_{\perp v}}\frac{1}{\hbar\omega_E}\operatorname{Ai}(x), \tag{8.15}$$

where we made use of (8.14). The presence of the function $\delta_{p_{\perp v}p_{\perp c}}$ reduces the calculation of $\mathscr{W}_{vc}$ (7.10) to the summation (integration) only over $p_{\perp c} = p_{\perp v} \equiv p_{\perp}$, $\varepsilon_v$ and $\varepsilon_c$. Hence, the coefficient of light absorption in the electric field is

$$\alpha_E = \frac{\mathscr{W}_{vc}}{Nc/n} = \frac{2\pi}{\hbar}\frac{n}{Nc}\sum_{\varepsilon_v \varepsilon_c,\, p_\perp} |\mathscr{P}_{vc}|^2 \delta(\varepsilon_c - \varepsilon_v - \hbar\omega)$$

$$= \frac{2\pi}{\hbar}\frac{n}{Nc}\frac{e^2}{m^2c^2} |A_0|^2 |\mathbf{e}\cdot\mathbf{p}_{cv}(0)|^2 \frac{1}{(\hbar\omega_E)^2}$$

$$\times \int d\varepsilon_v \int d\varepsilon_c 2 \int_0^\infty \frac{2\pi p_\perp\, dp_\perp}{(2\pi\hbar)^2} \mathrm{Ai}^2(x)\,\delta(\varepsilon_c - \varepsilon_v - \hbar\omega), \tag{8.16}$$

where the factor 2 in front of the last integral takes into account the electron's spin.

When integrating with respect to $\varepsilon_v$, in $x$ we make $\varepsilon_v = \varepsilon_c - \hbar\omega$ and obtain from (8.13b)

$$x = \frac{1}{\hbar\omega_E}\frac{1}{2\mu} p_\perp^2 = \frac{\omega_G - \omega}{\omega_E}. \tag{8.16a}$$

For an electric field of 3000 V/cm and $2\mu = 10^{-27}$ g the frequency $\omega_E = 3 \times 10^{12}$ s$^{-1}$. In (8.16) $\omega_G = \varepsilon_G/\hbar$ is the frequency corresponding to the forbidden bandwidth; for $\varepsilon_G \approx 1$ eV, $\omega_G \approx 10^{15}$ s$^{-1}$, i.e., $\omega_G$ is two or three orders of magnitude higher than $\omega_E$.

For a specimen of the length $L_x = 1$ cm in the $x$ direction the energy $\varepsilon_c$ for a fixed value of $p_\perp$ varies from $\varepsilon_c(0)$ to $\varepsilon_c(L_x) = \varepsilon_c(0) + eEL_x = \varepsilon_c(0) + eE$; therefore, the integration with respect to $\varepsilon_c$ yields the factor

$$\int_{\varepsilon_c(0)}^{\varepsilon_c(L_x)} d\varepsilon_c = eE. \tag{8.16b}$$

Introducing with the aid of (8.16a) the integration variable $x$ instead of $p_\perp$ and substituting (2.12a) for $A_0$, we obtain from (8.16)

$$\alpha_E = R(\hbar\omega_E)^{1/2}\pi \int_\beta^\infty \mathrm{Ai}^2(x)\,dx, \tag{8.17}$$

where

$$R = \frac{2e^2}{m^2cn\omega} |\mathbf{e}\cdot\mathbf{p}_{cv}(0)|^2 \left(\frac{2\mu}{\hbar^2}\right)^{3/2} \tag{8.17a}$$

and

$$\beta = \frac{\omega_G - \omega}{\omega_E}. \tag{8.17b}$$

The factor following $R$ in (8.17) takes into account the dependence of $\alpha_E$ on the electric field (via $\omega_E \propto E^{2/3}$).

7.8.2 Consider the expression for $\alpha_E$ (8.17) in the two limiting cases.

In the first case $\beta > 0$ $(\omega < \omega_G)$ and $\beta \gg 1$. In this case only large positive values of $x$ are essential in the integral (8.17), and for this reason we can use the asymptotic expansion [7.16, p. 511, Eq. (20)]

$$\mathrm{Ai}(x) \approx \frac{1}{2\sqrt{\pi}} x^{-1/4} \exp\left(-\frac{2}{3} x^{3/2}\right)\left[1 - \frac{5}{48}\frac{1}{x^{3/2}} + \dots\right]. \quad (8.18)$$

For $x \gg 1$ the expression in the square brackets can be equated to unity. Next substituting (8.18) into the integral of (8.17) and integrating by parts, we obtain

$$\int_{\beta}^{\infty} \frac{1}{4\pi} x^{-1/2} \exp\left(-\frac{4}{3} x^{3/2}\right) dx$$

$$= -\left[\frac{1}{8\pi} \exp\left(-\frac{4}{3} x^{3/2}\right) \cdot \frac{1}{x}\Big|_{\beta}^{\infty} + \int_{\beta}^{\infty} \exp\left(-\frac{4}{3} x^{3/2}\right) \frac{1}{x^2}\, dx\right].$$

Integrating the integral on the right-hand side by parts, we can easily show that it is $\sim \beta^{3/2}$ times smaller than the first term. Neglecting it, we obtain

$$\alpha_E = R(\hbar\omega_E)^{1/2} \frac{1}{8\beta} \exp\left(-\frac{4}{3}\beta^{3/2}\right). \quad (8.19)$$

We see that in this case

$$\alpha_E \propto \exp\left[-\gamma \frac{(\omega_G - \omega)^{3/2}}{E}\right] \quad (8.19a)$$

where $\gamma = \dfrac{4\sqrt{2\mu\hbar}}{3e}$ is a constant independent both of the frequency $\omega$ and of the electric field.

In the second case $\beta < 0$ $(\omega > \omega_G)$ and $|\beta| \gg 1$. The asymptotic expansions of the Airy functions for large negative values of $x = -\xi$ $(\xi \gg 1)$ are of the form [7.16, p. 508, Eq. (13)]

$$\mathrm{Ai}(-\xi) \approx \frac{1}{\sqrt{\pi}} \xi^{-1/4}\left[\sin\left(\frac{2}{3}\xi^{3/2} + \frac{\pi}{4}\right) - \frac{5}{48}\xi^{-3/2} \cos\left(\frac{2}{3}\xi^{3/2} + \frac{\pi}{4}\right)\right]. \quad (8.20)$$

Since for $\beta < 0$, $x$ in the integral of (8.17) takes on both the positive and negative values including zero, it is impossible to apply equations (8.18) and (8.20) directly.

Derive an expression that could be used in this case. The Airy function satisfies the differential equation [7.6, § 24, Eq. (24.3)]

$$\frac{d^2\mathrm{Ai}\,(x)}{dx^2} = x\mathrm{Ai}\,(x). \tag{8.21}$$

Multiplying both sides of the equation by $2d\ \mathrm{Ai}\ (x)/dx = 2\mathrm{Ai}'$, we obtain

$$[\mathrm{Ai}'^2]' = x\,[\mathrm{Ai}^2]'. \tag{8.22}$$

Integrating both sides of this equation from 0 to $\beta = -\xi$ and calculating the integral on the right-hand side by parts, we obtain

$$-\int_0^{-\xi} \mathrm{Ai}^2\,(x)\,dx = \xi\mathrm{Ai}^2\,(-\xi) + [\mathrm{Ai}'\,(-\xi)]^2. \tag{8.23}$$

If we make $\beta = -\xi$, we obtain for the integral in (8.17)

$$\int_{-\xi}^{\infty} \mathrm{Ai}^2\,(x)\,dx = \int_{-\xi}^{0} \mathrm{Ai}^2\,(x)\,dx + \int_0^{\infty} \mathrm{Ai}^2\,(x)\,dx = -\int_0^{-\xi} \mathrm{Ai}^2\,(x)\,dx$$
$$+\int_0^{\infty} \mathrm{Ai}^2\,(x)\,dx = \xi\mathrm{Ai}^2\,(-\xi) + [\mathrm{Ai}'\,(-\xi)]^2 + \int_0^{\infty} \mathrm{Ai}^2\,(x)\,dx. \tag{8.24}$$

For $\xi \gg 1$ make use of the asymptotic expansion (8.20). Then we obtain in the lowest approximation

$$\xi\mathrm{Ai}^2\,(-\xi) + [\mathrm{Ai}'\,(-\xi)]^2 = \frac{1}{\pi}\,\xi^{1/2}\sin^2\left(\frac{2}{3}\,\xi^{3/2} + \frac{\pi}{4}\right)$$
$$+\frac{1}{\pi}\,\xi^{1/2}\cos^2\left(\frac{2}{3}\,\xi^{3/2} + \frac{\pi}{4}\right) = \frac{1}{\pi}\,\xi^{1/2} = \frac{1}{\pi}\,(-\beta)^{1/2}$$
$$= \frac{1}{\pi}\left(\frac{\omega - \omega_G}{\omega_E}\right)^{1/2}. \tag{8.25}$$

Substituting this into (8.24), we obtain for (8.17)

$$\alpha_E = R\left[\,(\hbar\omega - \varepsilon_G)^{1/2} + (\hbar\omega_E)^{1/2}\,\pi\int_0^{\infty} \mathrm{Ai}^2\,(x)\,dx\,\right]. \tag{8.26}$$

For an electric field $E \to 0$ we obtain

$$\alpha_{E=0} = R\,(\hbar\omega - \varepsilon_G)^{1/2};$$

this coincides with (2.26b).

**7.8.3** The exponential form of the absorption coefficient $\alpha_E$ (8.19a) can be understood from physical considerations.

When a homogeneous electric field much weaker than the atomic lattice fields is applied to a semiconductor, the electron's energy bands become inclined as shown in Fig. 7.5, where the electron's energy $\varepsilon$ is along the vertical axis. An electron with a constant energy can move in the valence band between the points $A$ and $B$ bouncing off the band edges (we ignore the fact that the valence band is filled with electrons obeying the Pauli principle). However, since the electron possesses wave properties, it can tunnel through the forbidden band $BC$ and penetrate the conduction band $CD$. The term for this phenomenon is *internal field emission* or the *Zener effect*.

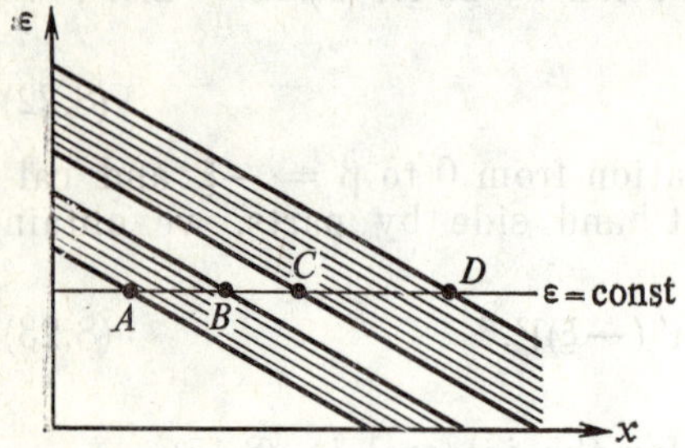


Fig. 7.5

The probability of tunneling decreases exponentially with the increase in the bandwidth $BC$; this increase is the result of an increase in the forbidden bandwidth $\varepsilon_G$ or of a decrease in the electric field $E$. Calculation yields for the tunneling probability

$$W(B \to C) \propto \exp\left(-c\frac{\varepsilon_G^{3/2}}{E}\right), \tag{8.27}$$

where $c$ is a constant dependent on the effective masses $m_c$ and $m_v$. Expression (8.27) is similar to (8.19a) for $\alpha_E$. If $\omega > \omega_G$, the newly generated electron and hole in the course of their motion are reflected by the barrier established by the electric field. The interference of the reflected and the incident waves results in the oscillations of the absorption coefficient.

The account of the Coulomb interaction of electrons and holes changes the quantitative pattern of light absorption but does not change the qualitative picture of the phenomenon (I.A. Merkulov, V.I. Perel, 1973). Thus, if $\omega < \omega_G$, the exponential nature of the dependence (8.19) remains unaltered, but the coefficient in front of $\beta^{3/2}$ in the exponent changes to a new one, and an important additional factor appears in front of the exponent. The former is due to a change in the shape of the barrier, and the latter is due to the change, resulting from the Coulomb interaction, in the probability of locating the electron and the hole at a specified point.

When the electron-hole Coulomb interaction results in a bound state, an exciton, a line spectrum appears in the absorption coefficient at a frequency close to the exciton generation threshold. In an electric field those lines exhibit the Stark effect similar to the Stark effect in the hydrogen atom; moreover, the width of the lines becomes finite and equal to the inverse exciton ionization probability for the given state (I. A. Merkulov, 1974; A. G. Aronov, A. S. Ioselevich, 1977).

The light absorption coefficient which takes account of the electron-hole interaction can also be calculated for the case when $\omega > \omega_G$ (A. G. Aronov, A. S. Ioselevich, 1977).

If we compare the results of Section 7.7.2 with those of the Franz-Keldysh theory, we may ask how the results can be matched when the magnetic field tends to zero. It will be seen from (7.31) that for $H \to 0$ the shift in the absorption fringe equal to $\frac{(m_c+m_v)c^2}{2}\left(\frac{E}{H}\right)^2 \to \infty$. Obviously, for $H \to 0$ the conditions of the applicability of the equations obtained in Section 7.8.2 no longer hold.

It can be shown that the theory developed in Section 7.7.2 is valid, provided that

$$\frac{(m_c+m_v)\,c^2}{2}\left(\frac{E}{H}\right)^2 \ll \varepsilon_G, \tag{8.28}$$

or

$$E/H \ll s/c, \tag{8.28a}$$

where the velocity $s = \sqrt{\frac{2\varepsilon_G}{m_c+m_v}}$. As long as inequality (8.28) is satisfied, the character of electron and hole motion remains finite. The Landau quantization remains in force. When this inequality is violated, the electron and hole move infinitely, there is no Landau quantization, and the Franz-Keldysh effect in a transverse magnetic field takes place (A. G. Aronov, G. E. Pikus, 1965).

# 8. Kinetic Equation and Relaxation Time of Conduction Electrons in Crystals

## 8.1. Transport Phenomena and Boltzmann Equation

**8.1.1** In the first five sections of Chapter 6 we considered the thermal and magnetic properties of semiconductors and metals due to free electrons and holes in the state of statistical (thermodynamical) equilibrium. As was already noted above, an essential distinction of systems in a state of statistical equilibrium is the independence of their properties of the interaction mechanism in the system.

Of great theoretical and practical interest, in addition to the study of such equilibrium states, is the study of nonequilibrium conduction electrons (holes) moving in a crystal under the effect of applied external fields: the electric, the magnetic, and the temperature field. Such processes involving the transport of electrons and holes are termed *transport phenomena* or *kinetic effects*.

In cases, when the quantities that describe the transport phenomenon—the current density, the heat flux, the electric field intensity, etc.—are independent of time, the process is termed *stationary*. For a stationary current to flow in the presence of an electric field that accelerates the electrons, the conduction electrons must collide (must be scattered) with some lattice inhomogeneities (atomic vibrations or crystal defects) and lose energy gained in the electric field. As we shall see below, in the majority of cases the electron collisions (scattering) can be regarded as elastic, and the electrical resistance will be determined by the average rate of variation of the component of the electron's momentum (velocity) in the direction of the electric field in the result of scattering.

An important property of the nonequilibrium processes is their strong dependence on the interaction mechanism in the system, in our case it is the conduction electron's interaction with the lattice vibrations and with crystal defects.

In Chapter 6 we learned that the electrons in a state of thermodynamical equilibrium in the classical case are described by the Boltzmann equilibrium distribution function $f_0(\mathscr{E}) = \exp\left(\frac{\zeta-\mathscr{E}}{k_0T}\right)$, where the total energy is $\mathscr{E} = \frac{mv^2}{2} + \mathscr{U}(x, y, z)$ [$\mathscr{U}(x, y, z)$ is the potential energy which we usually equated to zero]. Similarly, for the electrons in a nonequilibrium state we can introduce a nonequilibrium distribution function

$$f(v_x, v_y, v_z, x, y, z, t)\, dv_x\, dv_y\, dv_z\, dx\, dy\, dz = f(\mathbf{v}, \mathbf{r}, t)\, d^3v\, d^3\mathbf{r} \tag{1.1}$$

giving the number of electrons with speeds lying inside the intervals $v_x$, $v_x + dv_x$, etc.[1] at the instant $t$ in the volume $d^3r = dx\, dy\, dz$ at point $\mathbf{r}$.

It is possible to describe the electrons by simultaneously specifying their coordinates and velocities (conjugate momenta) only insofar as their motion obeys the laws of classical mechanics.

Knowing the function $f(\mathbf{v}, \mathbf{r}, t)$, we can calculate the current density at point $\mathbf{r}$ at the instant $t$. Figure 8.1 depicts an area of 1 cm² perpendicular to the plane of the figure and to the $x$ axis and a cylinder of the height $v_x\, dt$ constructed on this area. The number of $\mathbf{v}$-electrons inside the cylinder is $f(\mathbf{v}, \mathbf{r}, t)\, d^3v v_x\, dt$. All these electrons in the time $dt$ will travel a distance $v_x\, dt$ in the $x$ direction and, therefore, will cross the area. The total number of all the electrons crossing the area in the time $dt$ is

$$dt \iiint_{-\infty}^{+\infty} f(\mathbf{v}, \mathbf{r}, t)\, v_x\, dv_x\, dv_y\, dv_z,$$

if we take into account both the electrons crossing the area from left to right and from right to left. Since the charge of each electron is $-e$, the current density in the $x$ direction will be

$$j_x = -e \iiint_{-\infty}^{+\infty} f(\mathbf{v}, \mathbf{r}, t)\, v_x\, dv_x\, dv_y\, dv_z. \tag{1.2}$$

Write the nonequilibrium distribution function in the form

$$f(\mathbf{v}, \mathbf{r}, t) = f_0(\mathscr{E}) + f_1(\mathbf{v}, \mathbf{r}, t). \tag{1.3}$$

Since $f_0(\mathscr{E})$ is an even function of $v_x$ (depends on $v_x^2$), the integral (1.2) of $f_0(\mathscr{E})\, v_x$ with respect to $dv_x$ is zero, and (1.2) yields

$$j_x = -e \iiint_{-\infty}^{+\infty} f_1(\mathbf{v}, \mathbf{r}, t)\, v_x\, dv_x\, dv_y\, dv_z. \tag{1.2a}$$

The principal problem in the theory of kinetic phenomena is the determination of the nonequilibrium distribution function $f(\mathbf{v}, \mathbf{r}, t)$.

---

[1] We shall use the term $\mathbf{v}$-electrons for such electrons.

8.1.2 Deduce an equation for the function $f(\mathbf{v}, \mathbf{r}, t)$. Consider the variation of the number of v-electrons in the time $dt$ moving in the usual (geometrical) space. Figure 8.2 depicts an element of volume $d^3\mathbf{r} = dx\, dy\, dz$. Consider the variation of the number of v-electrons due to the arrival of electrons through the left face $dy\, dz$ and to the exit of electrons through the right face $dy\, dz$ (we

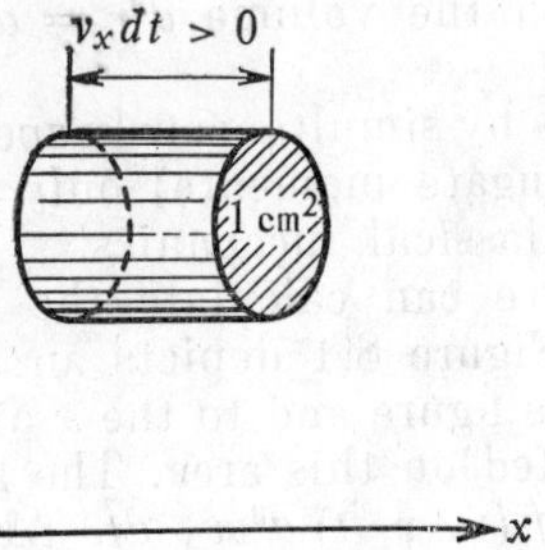


Fig. 8.1

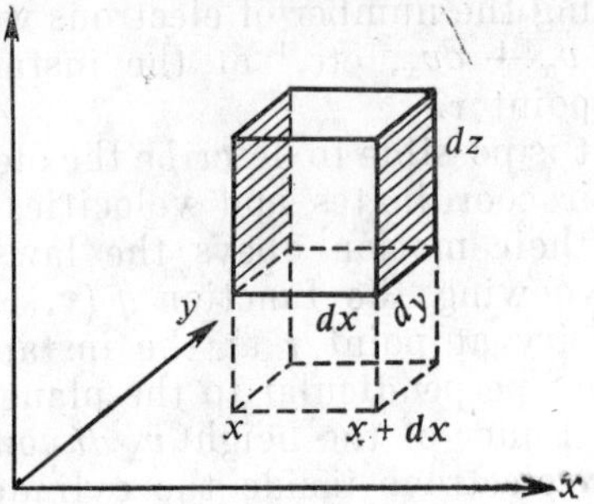


Fig. 8.2

assume that $v_x > 0$). The number of v-electrons entering through the left face in the time $dt$ is $f(\mathbf{v}, x, y, z, t) \times d^3v\, dy\, dz v_x\, dt$. The number of v-electrons leaving in the time $dt$ through the right face is $f(\mathbf{v}, x+dx, y, z, t)\, d^3v\, dy\, dz v_x\, dt$. The increase in the number of v-electrons in the volume $d^3\mathbf{r}$ in the result of this process is

$$
\begin{aligned}
f(\mathbf{v}, x, y, z, t)\, d^3v\, dy\, dz v_x\, dt - f(\mathbf{v}, x+dx, y, z, t)\, d^3v\, dy\, dz v_x\, dt \\
= -v_x \{f(\mathbf{v}, x+dx, y, z, t) - f(\mathbf{v}, x, y, z, t)\}\, dy\, dz\, d^3v\, dt \\
= -v_x \frac{\partial f}{\partial x}\, dx\, dy\, dz\, d^3v\, dt.
\end{aligned}
$$

The increase in the number of v-electrons in the volume $d^3\mathbf{r}$ due to the motion of the electrons through all six faces of the volume $d^3\mathbf{r}$ is

$$
\begin{aligned}
-\left(v_x \frac{\partial f}{\partial x} + v_y \frac{\partial f}{\partial y} + v_z \frac{\partial f}{\partial z}\right) d^3\mathbf{v}\, d^3\mathbf{r}\, dt = -(\mathbf{v} \cdot \nabla_{\mathbf{r}} f)\, d^3\mathbf{v}\, d^3\mathbf{r}\, dt \\
= -\mathbf{v} \frac{\partial f}{\partial \mathbf{r}}\, d^3\mathbf{v}\, d^3\mathbf{r}\, dt. \qquad (1.4)
\end{aligned}
$$

In a similar way we can consider the variation of the number of v-electrons in the volume $d^3v = dv_x\, dv_y\, dv_z$ due to their "motion" in the v-space with the "velocities" $\dot{v}_x$, $\dot{v}_y$, $\dot{v}_z$. The increase in the number of v-electrons moving in the v-space is

$$-(\dot{\mathbf{v}}\cdot\nabla_{\mathbf{v}}f)\,d^3\mathbf{v}\,d^3\mathbf{r}\,dt=-\dot{\mathbf{v}}\cdot\frac{\partial f}{\partial \mathbf{v}}\,d^3\mathbf{v}\,d^3\mathbf{r}\,dt$$

$$=-\frac{1}{m}(\mathbf{F}\cdot\nabla_{\mathbf{v}}f)\,d^3\mathbf{v}\,d^3\mathbf{r}\,dt, \qquad (1.4\text{a})$$

since the acceleration $\dot{\mathbf{v}}=\frac{d\mathbf{v}}{dt}=\frac{1}{m}\mathbf{F}(\mathbf{r},\ t)$, where $\mathbf{F}(\mathbf{r},\ t)$ is the force acting on the electron at point $\mathbf{r}$ at the instant $t$.

An additional variation in the number of v-electrons is the result of their collisions with the lattice vibrations (phonons) and with crystal defects. Every collision of a v-electron removes it from the volume $d^3\mathbf{v}$, because it results in a drastic change in its velocity. In the following we shall consider only the elastic collisions that result only in the change in the direction of the velocity, but not in its magnitude, so that if the velocity after the collision is $\mathbf{v}'$, it means that $v'=v$.

Let $W(\mathbf{v},\ \mathbf{v}')\,d^3v'dt$ be the probability for an electron with the velocity $\mathbf{v}$ to experience an elastic scattering in the time $dt$ and turn into a $\mathbf{v}'$-electron. In principle, the probability $W(\mathbf{v},\ \mathbf{v}')$ can be a function of $\mathbf{r}$ and $t$.

The total number of v-electrons vanishing in the time $dt$ as a result of collisions is

$$-\int_{\mathbf{v}'}[f(\mathbf{v},\mathbf{r},t)\,d^3v\,d^3rW(\mathbf{v},\mathbf{v}')\,dt]\,d^3v',$$

where the integration is performed over all values of $\mathbf{v}'$.

On the other hand, the number of v-electrons will increase as a result of transformations of various $\mathbf{v}'$-electrons into v-electrons caused by collisions (in the same volume $d^3r$). This increase in the number of v-electrons in the time $dt$ is

$$\int_{\mathbf{v}'}[f(\mathbf{v}',\mathbf{r},t)\,d^3rW(\mathbf{v}',\mathbf{v})\,d^3v\,dt]\,d^3v',$$

where the integration is performed over all values of $\mathbf{v}'$.

The eventual increase in the number of v-electrons in the time $dt$ caused by collisions is

$$d^3v\,d^3r\,dt\int_{v'}[f(\mathbf{v}',\mathbf{r},t)\,W(\mathbf{v}',\mathbf{v})-f(\mathbf{v},\mathbf{r},t)\,W(\mathbf{v},\mathbf{v}')]\,d^3v'. \qquad (1.4\text{b})$$

On the other hand, the increase in the number of v-electrons in the time $dt$ is

$$f(\mathbf{v},\mathbf{r},t+dt)\,d^3v\,d^3r-f(\mathbf{v},\mathbf{r},t)\,d^3v\,d^3r=\frac{\partial f}{\partial t}\,d^3v\,d^3r\,dt. \qquad (1.5)$$

Equating (1.5) with the sum of (1.4), (1.4a) and (1.4b), we obtain

$$\frac{\partial f}{\partial t} = -(\mathbf{v}\cdot\nabla_{\mathbf{r}} f) - \frac{1}{m}(\mathbf{F}\nabla_{\mathbf{v}} f)$$
$$+ \int \{f(\mathbf{v}', \mathbf{r}, t)\, W(\mathbf{v}', \mathbf{v}) - f(\mathbf{v}, \mathbf{r}, t)\, W(\mathbf{v}, \mathbf{v}')\}\, d^3v'. \tag{1.6}$$

The equation is called the *Boltzmann kinetic equation.*

The so-called *field* term

$$\left(\frac{\partial f}{\partial t}\right)_{\mathrm{f}} = -(\mathbf{v}\cdot\nabla_{\mathbf{r}} f) - \frac{1}{m}(\mathbf{F}\cdot\nabla_{\mathbf{v}} f) \tag{1.7}$$

determines the rate of variation of the distribution function $f$ as the result of continuous motion of the electrons in the **r**- and the **v**-spaces, and the *collision* term

$$\left(\frac{\partial f}{\partial t}\right)_{\mathrm{c}} = \int \{f(\mathbf{v}', \mathbf{r}, t)\, W(\mathbf{v}', \mathbf{v}) - f(\mathbf{v}, \mathbf{r}, t)\, W(\mathbf{v}, \mathbf{v}')\}\, d^3v' \tag{1.7a}$$

determines the rate of variation of $f$ as the result of collisions (scattering) of the electrons.

In the stationary case

$$\frac{\partial f}{\partial t} = \left(\frac{\partial f}{\partial t}\right)_{\mathrm{f}} + \left(\frac{\partial f}{\partial t}\right)_{\mathrm{c}} = 0 \tag{1.8}$$

or

$$\mathbf{v}\cdot\nabla_{\mathbf{r}} f + \frac{1}{m}\mathbf{F}\cdot\nabla_{\mathbf{v}} f = \int \{f(\mathbf{v}', \mathbf{r})\, W(\mathbf{v}', \mathbf{v})$$
$$- f(\mathbf{v}, \mathbf{r})\, W(\mathbf{v}, \mathbf{v}')\}\, d^3v'. \tag{1.8a}$$

Since the right-hand side of (1.8a) contains the integral of an unknown function $f(\mathbf{v}')$, the Boltzmann kinetic equation is an integro-differential equation. Of course, in order to solve it, we should know the force **F** and the transition probability $W(\mathbf{v}, \mathbf{v}')$.

8.1.3 In the case of equilibrium the distribution $f = f_0(\mathscr{E}) = \exp\left\{\frac{\zeta - \mathscr{E}}{k_0 T}\right\}$, where the total energy $\mathscr{E} = \varepsilon + \mathscr{U}$, i.e., is equal to the sum of the kinetic energy $\varepsilon = mv^2/2$ and the potential energy $\mathscr{U}(\mathbf{r})$. In this case the left-hand side of equation (1.8a) vanishes. Indeed,

$$\mathbf{v}\cdot\nabla_{\mathbf{r}} f_0(\varepsilon + \mathscr{U}) = \frac{1}{k_0 T} f_0 \mathbf{v}\cdot(-\nabla_{\mathbf{r}}\mathscr{U}) = \frac{1}{k_0 T} f_0(\mathbf{v}\cdot\mathbf{F}),$$

and

$$\frac{1}{m}\mathbf{F}\cdot\nabla_{\mathbf{v}} f_0(\varepsilon + \mathscr{U}) = -\frac{1}{m}\mathbf{F}\cdot\frac{m\mathbf{v}}{k_0 T} f_0 = -\frac{1}{k_0 T} f_0(\mathbf{v}\mathbf{F}).$$

Hence,

$$\left(\frac{\partial f}{\partial t}\right)_c = \int \{f_0(\mathscr{E}')\,W(\mathbf{v}',\mathbf{v}) - f_0(\mathscr{E})\,W(\mathbf{v},\mathbf{v}')\}\,d^3v'$$
$$= f_0(\mathscr{E}) \int \{W(\mathbf{v}',\mathbf{v}) - W(\mathbf{v},\mathbf{v}')\}\,d^3v' = 0, \quad (1.9)$$

since $\mathscr{E} = \mathscr{E}'$. The integral can be equal to zero for an arbitrary $\mathbf{v}$ only if

$$W(\mathbf{v}, \mathbf{v}') = W(\mathbf{v}', \mathbf{v}). \quad (1.10)$$

The last equation is a corollary of the general principle of detailed equilibrium that stipulates the equality of the probabilities of the direct and the reciprocal processes. In quantum mechanics equation (1.10) is a direct corollary of its laws.

If the only force acting on the electrons is the electric field $\mathbf{E}$ pointing in the direction of the $x$ axis, the correction to the distribution function, as will be demonstrated below, can be represented in the form[2]

$$f_1 = -\frac{\partial f_0}{\partial \varepsilon}\,\chi(\varepsilon)\,v_x, \quad (1.11)$$

where $\chi(\varepsilon)$ is a function of the energy $\varepsilon = \frac{m}{2}(v_x^2 + v_y^2 + v_z^2)$. Substituting (1.11) into (1.7a), assuming that the collisions are elastic ($\varepsilon' = \varepsilon$) and taking into account equality (1.10), we obtain

$$\left(\frac{\partial f}{\partial t}\right)_c = -f_1 \int W(\mathbf{v},\mathbf{v}')\left(1 - \frac{v_x'}{v_x}\right) d^3v'. \quad (1.12)$$

Since we assume that the electron scattering is of the elastic type, it follows that

$$W(\mathbf{v}, \mathbf{v}') = W_0(v, \theta)\,\delta(v - v'), \quad (1.12a)$$

where $\delta(v - v')$ accounts for the fact that $v = v'$, and $\theta$ is the angle between the directions of the velocities $\mathbf{v}$ and $\mathbf{v}'$.

Choosing $\mathbf{v}$ as the polar axis in the velocity space $\mathbf{v}'$, as depicted in Fig. 8.3, we obtain

$$d^3v' = v'^2 \sin\theta\, d\theta\, d\Phi = v'^2\, d\Omega, \quad (1.12b)$$

where $\theta$ and $\Phi$ are the polar and the azimuthal angles determining the direction of the vector $\mathbf{v}'$, and $d\Omega = \sin\theta\, d\theta\, d\Phi$ is the solid

---

[2] I.e., in the form of a product of some function of the energy $\varepsilon$ by $v_x$; the multiplier $\left(-\frac{\partial f_0}{\partial \varepsilon}\right)$ is written out separately for convenience in future calculations.

angle in the direction of $\mathbf{v}'$. The $x$ axis coincides in the direction with the electric field $\mathbf{E}$. We have

$$v'_x = v' \cos \alpha = v \cos \alpha, \quad v_x = v \cos \vartheta. \tag{1.12c}$$

It is established in spherical trigonometry that [8.1]

$$\cos \alpha = \cos \vartheta \cos \theta + \sin \vartheta \sin \theta \cos \Phi. \tag{1.12d}$$

Substituting (1.12a)-(1.12d) into (1.12), we obtain

$$\left(\frac{\partial f}{\partial t}\right)_c = -f_1 \int W(\theta)(1-\cos\theta)\,d\Omega, \tag{1.13}$$

where $W(\theta)\,d\Omega = v^2 W_0(v, \theta)\,d\Omega$ is the probability of elastic scattering of an electron with the velocity $v$ into the solid angle $d\Omega$ in the time of 1 s.

Since the dimensionality of the integral in (1.12) is (time)$^{-1}$, we can introduce the *relaxation time* $\tau$ making

$$\frac{1}{\tau} = \int W(\theta)(1-\cos\theta)\,d\Omega. \tag{1.13a}$$

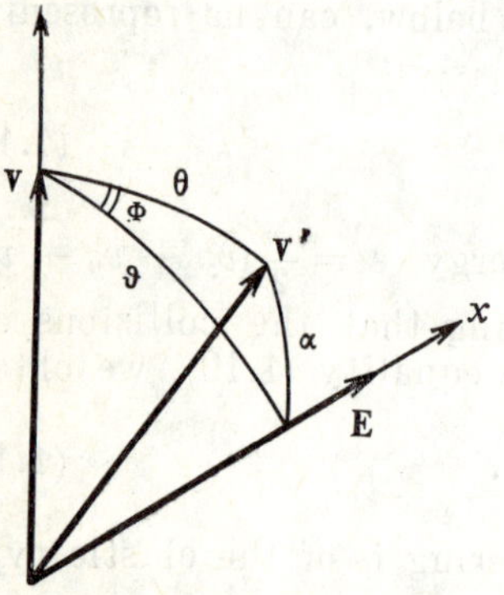

Fig. 8.3

The relaxation time is a function solely of the electron's velocity $v$ (or of its energy $\varepsilon$) and of the scattering machanism. It follows from (1.13) and (1.13a) that

$$\left(\frac{\partial f}{\partial t}\right)_c = -\frac{f_1}{\tau} = -\frac{f-f_0}{\tau}. \tag{1.14}$$

We can grasp the physical meaning of relaxation time if we consider the process of the establishment of statistical equilibrium in a homogeneous system in which, in the absence of forces acting on it, the velocity distribution at the initial instant $t = 0$ was not an equilibrium one. Making $\nabla_r f = 0$ and $\mathbf{F} = 0$, we have as a result of expressions (1.6), (1.7a), (1.14)

$$\left(\frac{\partial f}{\partial t}\right)_c = \frac{\partial f}{\partial t} = -\frac{f-f_0}{\tau}.$$

Integrating the last equality, we obtain

$$(f - f_0) = (f - f_0)_{t=0} e^{-t/\tau}. \tag{1.14a}$$

We see that $\tau$ is the time following the removal of the external field in which the difference $(f - f_0)$ decreases $e$ times. Since a system approaches equilibrium as the result of electron collisions (with lattice vibrations and crystal defects), several collisions being enough for the electrons to reach the equilibrium state, the order of magnitude of the relaxation time is that of the electron's free transit time. We define the electron's mean free path as

$$l = v\tau. \tag{1.15}$$

As we shall see below, in many cases of practical interest the correction to the equilibrium distribution function $f_1 \ll f_0$, and because of that we can with an accuracy up to the first order of magnitude in $f_1$ substitute for $f$ on the left-hand side of equa-

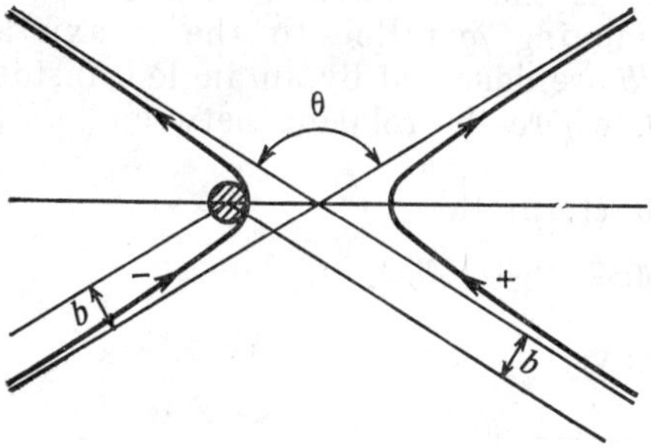


**Fig. 8.4**

tion (1.8a) the equilibrium function $f_0$. If the electric field acts alone, we have from (1.8a) and (1.14)

$$f_1 = \frac{e}{m} E \frac{\partial f_0}{\partial v_x} \tau = -eE \frac{f_0}{k_0 T} \tau v_x. \tag{1.16}$$

Hence, if we know the relaxation time $\tau$, we can use equation (1.2a to find the current density. At the same time we see that $f_1$, indeed is of the form (1.11), where $\chi(\varepsilon) = -eE\tau(v)$.

8.1.4 Calculate the relaxation time from equation (1.13a) for electrons (holes) colliding with (scattered by) impurity ions uniformly distributed throughout the semiconductor's volume. Consider the motion of a hole in the Coulomb field of a donor

$$\mathcal{U}(r) = \frac{e^2}{\varepsilon_0 r} \tag{1.17}$$

in accordance with the laws of classical mechanics. Here $e$ is the elementary charge, $\varepsilon_0$ is the dielectric constant.

It can be demonstrated [8.2, p. 331] that the hole moves along a hyperbola, and that its impact parameter is

$$b = \frac{e^2}{\varepsilon_0 m v^2} \cot \frac{\theta}{2}, \tag{1.18}$$

where $\theta$ is the hole scattering angle (Fig. 8.4). Expression (1.18) is also valid for the case of an electron scattered by a donor, when the potential energy of attraction is equal to (1.17) with a minus sign. Of course, in this case the motion takes place along a hyperbola only if the electron total energy $\frac{mv^2}{2} + \mathcal{U}(r) > 0$[3].

The differential effective cross section $\sigma(\theta)$ is by definition

$$\sigma(\theta)\, d\Omega = \frac{\text{number of electrons deflected by angle } \theta \text{ into solid angle } d\Omega}{\text{number of electrons striking } 1\text{ cm}^2 \text{ in the same time}}. \tag{1.19}$$

Its dimensionality is that of an area (cm$^2$).

All particles moving parallel to the $x$ axis and striking the ring of area $2\pi b\, db$ are deflected by an angle $\theta$ inside the solid angle $d\Omega = 2\pi \sin\theta\, d\theta$, where the relation between $|db|$ and $d\theta$ follows from (1.18).

It follows from (1.19) that

$$\sigma(\theta)\, 2\pi \sin\theta\, d\theta = 2\pi b\, |db|. \tag{1.19a}$$

From (1.18) we have

$$db = -\frac{e^2}{2\varepsilon_0 mv^2} \frac{1}{\sin^2 \frac{\theta}{2}}\, d\theta. \tag{1.18a}$$

From expressions (1.19a), (1.18) and (1.18a), we obtain the Rutherford equation

$$\sigma(\theta) = \left(\frac{e^2}{2\varepsilon_0 mv^2}\right)^2 \frac{1}{\sin^4 \frac{\theta}{2}} \tag{1.20}$$

used by him in his studies of the scattering of $\alpha$-particles by the nuclei of heavy elements.

Define the integral cross section as

$$\sigma = \int \sigma(\theta)\, d\Omega = 2\pi \int_0^{\pi} \sigma(\theta) \sin\theta\, d\theta, \tag{1.21}$$

i.e., as the total number of scattered particles per unit flux density of incident particles.

It can easily be demonstrated that the integral cross section $\sigma$ for a Coulomb field, when the differential cross section is of the form (1.20), is infinite. This is a specific property of the Coulomb potential (1.17) and is due to its slow decrease with distance. Establish the relationship between $\sigma(\theta)$ and the probability

---

[3] The right-hand branch of the hyperbola in Fig. 8.4 depicts the trajectory of a hole, and the left-hand branch depicts the trajectory of an electron.

$W(\mathbf{v}, \mathbf{v}')$, which in the case of elastic scattering depends on the angle $\theta$ between the velocities $\mathbf{v}$ and $\mathbf{v}'$ and on their magnitudes.

Let the volume $V$ contain a Coulomb centre and $N$ electrons moving in all directions with velocities $v$. The electron flux striking this centre in the time of 1 s will be $(N/V)\,v$ (we choose for the electrons moving in each direction an area of 1 cm$^2$ perpendicular to their direction of motion). The total number of electrons scattered in the time of 1 s by the angle from $\theta$ to $\theta + d\theta$ is $(N/V)\,v\sigma(\theta)\,d\Omega$.

On the other hand, this number is equal to $NW(\theta)\,d\Omega$. Hence,

$$W(\theta) = v\sigma(\theta)/V. \tag{1.22}$$

We can easily check that the dimensionality on the right-hand side is s$^{-1}$.

Substituting (1.22) into (1.13a) and making use of the equality $d\Omega' = 2\pi \sin\theta\, d\theta$, we obtain

$$\frac{1}{\tau} = \frac{2\pi v}{V} \int \sigma(\theta)(1-\cos\theta)\sin\theta d\theta. \tag{1.23}$$

If the volume contains $N_I$ independently scattering ions, then

$$\frac{1}{\tau} = 2\pi v n_I \int \sigma(\theta)(1-\cos\theta)\sin\theta\, d\theta, \tag{1.23a}$$

where $n_I = N_I/V$ is the concentration of the ions.

If expression (1.20) is substituted for $\sigma(\theta)$, the integral for the lower limit $\theta = 0$ diverges logarithmically, this (as in the case of the divergence of the integral cross section $\sigma$) is the result of a slow decrease in the Coulomb potential with distance. The divergence of expression (1.23a) can be eliminated if we find a way to take into account the field established by the other conduction electrons that screen the ionic fields and cut off the effect of the Coulomb potential. From this point of view it seems natural to limit the sphere of action of each scattering centre to one-half the average distance between the ions. Then the maximum impact parameter will be $b_{\max} = 1/2\, n_I^{-1/3}$, and the minimum scattering angle $\theta_{\min}$ will be found from equation (1.18):

$$\cot\frac{\theta_{\min}}{2} = \frac{\varepsilon_0 m v^2}{2 n_I^{1/3} e^2}. \tag{1.24}$$

It follows from expressions (1.23a) and (1.20) that

$$\frac{1}{\tau} = 2\pi\, v n_I \left(\frac{e^2}{2\varepsilon_0 m v^2}\right)^2 \int_{\theta_{\min}}^{\pi} \frac{(1-\cos\theta)\sin\theta\, d\theta}{\sin^4(\theta/2)}.$$

Substituting $2\sin^2(\theta/2)$ for $1 - \cos\theta$ and $2\sin(\theta/2)\cos\theta/2$ for $\sin\theta$ and making use of equality (1.24), we obtain

$$\frac{1}{\tau}=\frac{2\pi n_I e^4}{\varepsilon_0^2 m^2 v^3}\log\left\{1+\left[\frac{\varepsilon_0 m v^2}{2e^2 n_I^{1/3}}\right]^2\right\}. \tag{1.25}$$

This expression for the relaxation time $\tau$ due to the scattering of charge carriers by the impurity ions is often termed the *Conwell-Weisskopf equation* (1946).

Since the logarithm in (1.25) is a slowly varying function of $v$, the relaxation time is practically equal to

$$\tau \propto v^3 \propto \varepsilon^{3/2}. \tag{1.25a}$$

Making use of expressions (1.25) and (1.16), we can from equation (1.2a) calculate the current density and hence the conductivity.

## 8.2. Kinetic Equation for Electrons in a Crystal

**8.2.1** In the preceding paragraph the motion of the electrons was assumed to obey the laws of classical mechanics; accordingly, their state was described in the **r**- and **v**-spaces, and the equation was used

$$\frac{d\mathbf{v}}{dt}=\frac{1}{m}\mathbf{F}. \tag{2.1}$$

The last equation is also true in the quasi-classical approximation if the electron energy $\varepsilon = \hbar^2k^2/2m^* = m^*v^2/2$, where $m^*$ is the effective mass and $\mathbf{v} = \hbar\mathbf{k}/m^*$.

Hence, if the electron energy $\varepsilon(\mathbf{k})$ is in the above form, all the equations of the preceding section will be applicable when we substitute $m^*$ for $m$.

In the case of an arbitrary dispersion law $\varepsilon(\mathbf{k})$, the electron's velocity in the crystal (4.3.32), is

$$\mathbf{v}=\frac{1}{\hbar}\nabla_{\mathbf{k}}\varepsilon(\mathbf{k})=\frac{1}{\hbar}\frac{\partial\varepsilon(\mathbf{k})}{\partial\mathbf{k}} \tag{2.2}$$

and not $\hbar\mathbf{k}/m^*$, therefore the question arises, what form will the kinetic equation (1.8a) take in this case?

In the quasi-classical case a trajectory can be attributed to the electron along which it moves with the velocity (2.2). The electron's wave vector **k** which describes its quantum state, in the quasi-classical approximation satisfies equation (4.3.35):

$$\frac{d\mathbf{k}}{dt}=\frac{1}{\hbar}\mathbf{F}. \tag{2.3}$$

Considering the electrons in the **r**- and **k**-spaces, introduce the distribution function $f(\mathbf{k}, \mathbf{r}, t)$ such that

$$f(\mathbf{k}, \mathbf{r}, t)\frac{d^3k}{4\pi^3} \tag{2.4}$$

would be equal to the number of electrons per unit volume with wave vector components $k_x$, $k_x + dk_x$, etc. ($d^3k = dk_x\, dk_y\, dk_z$) at the instant $t$ at point **r**.

In the state of statistical equilibrium (6.2.1)

$$f = f_0(\varepsilon) = (e^{\frac{\varepsilon-\zeta}{k_0T}} + 1)^{-1}, \tag{2.5}$$

where $\varepsilon$ is a specified function of **k**.

Since now instead of equation (2.1) we have (2.3), the kinetic equation (1.8a), for the case of an electric **E** and a magnetic **H** field acting on the electron, will take the form

$$\mathbf{v}\cdot\nabla_{\mathbf{r}}f - \frac{e}{\hbar}\left(\mathbf{E} + \frac{1}{c}[\mathbf{v}\times\mathbf{H}]\right)\cdot\nabla_{\mathbf{k}}f = \left(\frac{\partial f}{\partial t}\right)_{\mathrm{c}}, \tag{2.6}$$

where **v** is determined by expression (2.2). The collision term $(\partial f/\partial t)_{\mathrm{c}}$ can be expressed in a form similar to that in the preceding section. Let $W(\mathbf{k}, \mathbf{k}')$ be the probability that an electron will go from state **k** to state **k**′ in the time of 1 s. Then we have by analogy with (1.7a)

$$\left(\frac{\partial f}{\partial t}\right)_{\mathrm{c}} = \sum_{\mathbf{k}'}\{W(\mathbf{k}', \mathbf{k})f(\mathbf{k}')[1 - f(\mathbf{k})] - W(\mathbf{k}, \mathbf{k}')f(\mathbf{k})[1 - f(\mathbf{k}')]\}, \tag{2.7}$$

where we took into account the Pauli principle, i.e., the probability of the transition $\mathbf{k}\to\mathbf{k}'$ was supposed to be proportional to $[1 - f(\mathbf{k}')]$ (the probability of the state **k**′ to be unoccupied).

It follows from the principle of detailed equilibrium that in conditions of statistical equilibrium the electron flux $\mathbf{k}\to\mathbf{k}'$ should be equal to the reverse flux, i.e.,

$$W(\mathbf{k}', \mathbf{k})f_0(\varepsilon')[1 - f_0(\varepsilon)] = W(\mathbf{k}, \mathbf{k}')f_0(\varepsilon)[1 - f_0(\varepsilon')].$$

Making use of the Fermi distribution function in explicit form, we obtain

$$W(\mathbf{k}', \mathbf{k})e^{\varepsilon/k_0T} = W(\mathbf{k}, \mathbf{k}')e^{\varepsilon'/k_0T}. \tag{2.7a}$$

In the case of elastic scattering $\varepsilon = \varepsilon'$, and

$$W(\mathbf{k}', \mathbf{k}) = W(\mathbf{k}, \mathbf{k}'). \tag{2.7b}$$

In this case

$$\left(\frac{\partial f}{\partial t}\right)_c = \sum_{\mathbf{k}'} \{W(\mathbf{k}', \mathbf{k}) f(\mathbf{k}') - W(\mathbf{k}, \mathbf{k}') f(\mathbf{k})\}$$

$$= \sum_{\mathbf{k}'} W(\mathbf{k}, \mathbf{k}') \{f(\mathbf{k}') - f(\mathbf{k})\}, \tag{2.8}$$

i.e., the collision term coincides with one that ignores the Pauli principle because the inclusion of the Pauli principle results in identical increases in the numbers of electrons going from state **k** to **k**′ and in the reverse direction from state **k**′ to **k**.

As we shall see in Section 9.2, the nonequilibrium distribution function can be represented in the form

$$f(\mathbf{k}) = f_0(\varepsilon) + f_1(\mathbf{k}) = f_0(\varepsilon) - \frac{\partial f_0}{\partial \varepsilon} \boldsymbol{\chi}(\varepsilon) \cdot \mathbf{k}, \tag{2.9}$$

where $\boldsymbol{\chi}(\varepsilon)$ is an unknown vector function of the electron energy[4] $\varepsilon$.

It will be demonstrated that the nonequilibrium additional term is of the form

$$f_1(\mathbf{k}) = -\frac{\partial f_0}{\partial \varepsilon} \boldsymbol{\chi}(\varepsilon) \cdot \mathbf{k} \tag{2.9a}$$

in the presence of both electric and magnetic fields and of a temperature gradient, provided the energy $\varepsilon \propto k^2$, and $\tau$ depends only on $\varepsilon$.

Since in the case of elastic scattering $\varepsilon' = \varepsilon$, it follows that

$$f(\mathbf{k}') - f(\mathbf{k}) = f_0(\varepsilon') - \frac{\partial f_0}{\partial \varepsilon'} \boldsymbol{\chi}(\varepsilon') \mathbf{k}' - f_0(\varepsilon) + \frac{\partial f_0}{\partial \varepsilon} \boldsymbol{\chi}(\varepsilon) \cdot \mathbf{k}$$

$$= \frac{\partial f_0}{\partial \varepsilon} \boldsymbol{\chi}(\varepsilon) \cdot \mathbf{k} \left[1 - \frac{\mathbf{k}'_\chi}{\mathbf{k}_\chi}\right] = f_1(\mathbf{k}) \frac{\Delta k_\chi}{k_\chi}, \tag{2.10}$$

where $k_\chi$ is the projection of **k** on the vector $\boldsymbol{\chi}$ and $\Delta k_\chi = k'_\chi - k_\chi$. From (2.8) and (2.10) we obtain

$$\left(\frac{\partial f}{\partial t}\right)_c = -f_1(\mathbf{k}) \sum_{\mathbf{k}'} W(\mathbf{k}, \mathbf{k}') \left[1 - \frac{k'_\chi}{k_\chi}\right]. \tag{2.11}$$

Introduce the relaxation time $\tau$, making

$$\frac{1}{\tau(\mathbf{k})} = \sum_{\mathbf{k}'} W(\mathbf{k}, \mathbf{k}') \left[1 - \frac{k'_\chi}{k_\chi}\right] = -\sum_{\mathbf{k}'} W(\mathbf{k}, \mathbf{k}') \frac{\Delta k_\chi}{k_\chi}. \tag{2.12}$$

---

[4] The multiplier $\partial f_0/\partial \varepsilon$, which depends only on the energy $\varepsilon$, has been introduced for convenience of calculations.

Then (2.11) can be written in the form

$$\left(\frac{\partial f}{\partial t}\right)_c = -\frac{f_1(\mathbf{k})}{\tau} = -\frac{f-f_0}{\tau}. \tag{2.11a}$$

If the electron's constant-energy surfaces $\varepsilon(\mathbf{k}) = \text{const.}$ are not spheres but ellipsoids, then, strictly speaking, we should take into account the anisotropy of electron scattering, and in this case the relaxation time $\tau$, provided it can be introduced, will no longer be a scalar but will turn into a rank 2 tensor [8.3], [8.4].

In cases of complex laws of dispersion of the energy $\varepsilon(\mathbf{k})$, for example, such as operate in the cases of holes in *p*-Ge and *p*-Si (Fig. 4.15*a* and *b*), the relaxation time cannot be rigorously introduced at all, and this makes the theory of kinetic phenomena extremely complicated.

**8.2.2** In the following chapter we shall use the kinetic equation (2.6) to consider various transport phenomena in the general case in the presence of electric and magnetic fields and of a temperature gradient. Here we shall, for the purpose of illustration, calculate the conductivity in the presence of only the electric field **E**.

If only the electric field **E** is present, we obtain from (2.6) and (2.11a) with an accuracy up to the first order of magnitude in $f_1 \propto E$, $f = f_0$

$$-\frac{e}{\hbar}\mathbf{E}\cdot\nabla_{\mathbf{k}} f_0 = -\frac{f_1(\mathbf{k})}{\tau(\mathbf{k})}. \tag{2.13}$$

Making use of (2.2), we obtain from (2.13)

$$f_1(\mathbf{k}) = e\tau(\mathbf{k})\frac{\partial f_0}{\partial \varepsilon}(\mathbf{v}\cdot\mathbf{E}). \tag{2.13a}$$

The electric current density is, by analogy with (1.2), equal to

$$\begin{aligned}\mathbf{j} &= -\frac{e}{4\pi^3}\int \mathbf{v} f(\mathbf{k})\, d^3k = -\frac{e}{4\pi^3}\int \mathbf{v} f_1(\mathbf{k})\, d^3k \\ &= -\frac{e^2}{4\pi^3}\int \tau(\mathbf{k})\frac{\partial f_0}{\partial \varepsilon}\mathbf{v}(\mathbf{v}\cdot\mathbf{E})\, d^3k,\end{aligned} \tag{2.14}$$

since the current corresponding to the equilibrium function $f_0$ is zero. Introducing the suffixes $i$ and $l$ to denote the components of **v** and **E** in a rectangular coordinate frame, we obtain

$$j_l = -\frac{e^2}{4\pi^3}\int \tau(\mathbf{k})\frac{\partial f_0}{\partial \varepsilon} v_l \sum_i v_i E_i\, d^3k = \sum_i \sigma_{li} E_i. \tag{2.14a}$$

The electric conductivity tensor

$$\sigma_{li} = -\frac{e^2}{4\pi^3}\int \tau(\mathbf{k})\frac{\partial f_0}{\partial \varepsilon} v_l v_i\, d^3k, \tag{2.14b}$$

where $v_l$ as a function of **k** is determined by (2.2).

It can easily be demonstrated that if the relaxation time $\tau$ depends only on the magnitude of $|\mathbf{k}|$, the conductivity tensor (2.14b) reduces to a scalar.

## 8.3. Scattering of Electrons by Acoustic Lattice Vibrations

**8.3.1** We have seen that the stationary state of the electron in a periodic crystal field is characterized by a time-independent velocity (2.2). Since the electron carries a charge $-e$, in the absence of any electric field there will be an undamped current corresponding to this velocity. The resistance of an ideal crystal may be said to be zero. A finite resistance of a crystal connected with a finite relaxation time $\tau$ or with a mean free path $l = v\tau$ is due to various deviations of the crystal field from strict periodicity. One of the major causes of violations of crystal field periodicity are thermal lattice vibrations discussed in Chapter 3.

Now we shall calculate the relaxation time connected with the scattering of conduction electrons by thermal vibrations in a simple monoatomic cubic lattice.

**8.3.2** Employ Bloch's hypothesis of deformable ions (1928) to determine the variation (perturbation) of the electron's potential energy in a periodic field $V(\mathbf{r})$ due to atomic vibrations.

Assume that every point of the crystal $\mathbf{r}$ is "displaced" in the course of atomic vibrations in accordance with equation (3.6.27) in which $\mathbf{r}$ is substituted for $\mathbf{a}_n$. In other words the "displacement" of any point of interatomic space is regarded as an interpolation between the values of displacements of the surrounding atoms. Such interpolation is the more legitimate (in the succeeding applications) the closer the vibration phases of the neighbouring atoms, i.e., the greater the wavelength of the acoustic wave.

In accordance with the aforesaid, we write

$$\mathbf{u} = \frac{1}{\sqrt{NM}} \sum_{qj}{}' \mathbf{e}_{qj} \{a_{\mathbf{q}j} e^{i\mathbf{q}\cdot\mathbf{r}} + a^*_{\mathbf{q}j} e^{-i\mathbf{q}\cdot\mathbf{r}}\}, \tag{3.1}$$

where we introduce the notation $\mathbf{e}_j(\mathbf{q}) \equiv e_{\mathbf{q}j}$ and $a_j(\mathbf{q}) \equiv a_{\mathbf{q}j}$. According to (3.6.2)

$$\mathbf{e}_{\mathbf{q}j}\mathbf{e}_{\mathbf{q}j'} = \delta_{jj'}. \tag{3.2}$$

The essence of the hypothesis of deformable ions is the assumption that the potential energy of the electron at point $\mathbf{r}$, which prior to the crystal's deformation was $V(\mathbf{r})$, after the deformation moves to point $\mathbf{r} + \mathbf{u}$, where the pre-deformation potential energy was

$V(\mathbf{r}+\mathbf{u})$. Hence, the variation of the electron's potential energy at point $\mathbf{r}+\mathbf{u}$ in the linear approximation in $\mathbf{u}$ is[5]

$$\mathcal{U} \equiv \Delta V = V(\mathbf{r}) - V(\mathbf{r}+\mathbf{u}) = -(\operatorname{grad} V \cdot \mathbf{u}). \tag{3.3}$$

We have expanded the function $V(\mathbf{r}+\mathbf{u})$ in a series in the displacement $\mathbf{u}$ ($u_x$, $u_y$, $u_z$) and have retained only the first term in the expansion. Expression (3.1) can be substituted for $\mathbf{u}$ into (3.3).

In order to determine the relaxation time from equation (2.12), we must know the transition probability $W(\mathbf{k}, \mathbf{k}')$, which we shall calculate with the aid of Dirac's theory of quantum transitions (see Appendix 20).

Consider a system consisting of an electron in a periodic field and of normal lattice vibrations. The system's wave function in the zeroth approximation, i.e., without the interaction (3.3) being taken into account, is equal to the product of Bloch's wave function of the electron in an ideal crystal (4.3.5) and the wave functions of the oscillators (3.10.4b), i.e.,

$$\Psi^{(0)}_{\mathbf{k},\,N_{\mathbf{q}j}} = \psi_{\mathbf{k}}(\mathbf{r}) \prod_{\mathbf{q}j} \psi_{N_{\mathbf{q}j}}(Q_{\mathbf{q}j}) = \frac{1}{\sqrt{N}} u_{\mathbf{k}}(\mathbf{r})\, e^{i\mathbf{k}\cdot\mathbf{r}} \prod \psi_{N_{\mathbf{q}j}}(Q_{\mathbf{q}j}), \tag{3.4}$$

where $N = G^3$ is the number of atoms in the principal region. Thanks to the presence of the factor $N^{-1/2}$ in Bloch's function,

$$\int |u_{\mathbf{k}}(\mathbf{r})|^2\, d^3r_0 = 1, \tag{3.4a}$$

where the integration is performed over the volume of the crystal's elementary cell (4.3.6).

Expand the perturbed wave function $\Psi(t)$ in the closed system[6] of unperturbed functions (3.4):

$$\Psi(t) = \sum_{\mathbf{k}'N'_{\mathbf{q}j}} a(\mathbf{k}', N'_{\mathbf{q}j}, t)\, \Psi^{(0)}_{\mathbf{k}'N'_{\mathbf{q}j}} \times \exp\left\{-\frac{i}{\hbar} + \left[\varepsilon_{k'} + \sum_{\mathbf{q}j} (N'_{\mathbf{q}j} + 1/2)\, \hbar\omega_{\mathbf{q}j}\right] t\right\}, \tag{3.5}$$

where $a(\mathbf{k}', N'_{\mathbf{q}j}, t)$ are unknown coefficients of the expansion[7], and

$$\varepsilon_{\mathbf{k}'} + \sum_{\mathbf{q}j} \left(N'_{\mathbf{q}j} + \frac{1}{2}\right) \hbar\omega_{\mathbf{q}j} = \mathscr{E}_{\mathbf{k}',\,N'_{\mathbf{q}j}} \tag{3.5a}$$

[5] This quantity coincides with the linear approximation in $\mathbf{u}$ for the variation of the electron's potential energy at point $\mathbf{r}$.

[6] Strictly speaking, a closed system of electron wave functions includes states belonging to other energy bands, as well; however, the "admixture" of such states in the processes being considered here is almost nonexistent.

[7] The exponential multipliers have been written out separately in (3.5) for the sake of convenience.

is the energy of the unperturbed system equal to the sum of the electron energy and the energies of all normal vibrations. If at the initial instant $t = 0$ the electron was in the state $\mathbf{k}$, and the normal vibrations were characterized by their quantum numbers $N'_{\mathbf{q}j}$, then

$$a(\mathbf{k}', N'_{\mathbf{q}j}, 0) = \begin{cases} 1 \text{ for } \mathbf{k}' = \mathbf{k} \text{ and } N'_{\mathbf{q}j} = N_{\mathbf{q}j}, \\ 0 \text{ in all other cases.} \end{cases} \quad (3.6)$$

From the general theory we obtain [8.5, Sec. 29]

$$i\hbar \frac{da(\mathbf{k}', N'_{\mathbf{q}j}, t)}{dt} = \langle \mathbf{k}', N'_{\mathbf{q}j} | \Delta V | \mathbf{k}, N_{\mathbf{q}j} \rangle \times \exp\left\{ \frac{i}{\hbar} \left[ \varepsilon_{\mathbf{k}'} - \varepsilon_{\mathbf{k}} + \sum_{\mathbf{q}j} (N'_{\mathbf{q}j} - N_{\mathbf{q}j}) \hbar\omega_{\mathbf{q}j} \right] t \right\}, \quad (3.7)$$

where the matrix element for the transition from state $\mathbf{k}$, $N_{\mathbf{q}j}$ to state $\mathbf{k}'$, $N'_{\mathbf{q}j}$ caused by the perturbation (3.3) is equal to

$$\langle \mathbf{k}', N'_{\mathbf{q}j} | \Delta V | \mathbf{k}, N_{\mathbf{q}j} \rangle = \int \Psi^{*(0)}_{\mathbf{k}', N'_{\mathbf{q}j}} \Delta V \Psi^{(0)}_{\mathbf{k}, N_{\mathbf{q}j}} \, d^3r \prod_{\mathbf{q}j} dQ_{\mathbf{q}j}. \quad (3.7a)$$

Here the integration is performed with respect to the three electron coordinates $d^3r$ over the crystal's principal region and with respect to $3N$ normal coordinates $\prod_{\mathbf{q}j} dQ_{\mathbf{q}j}$ of lattice vibrations. The exponent of expression (3.7) contains the difference between the energies of the final and the initial states of the system.

The calculation of the matrix element (3.7a) is rather laborious and is presented in Appendix 21. It can be demonstrated that the electron interacts only with one longitudinal vibration branch for which $\mathbf{e}_{\mathbf{q}j}$ is parallel to $\mathbf{q}$[8]. Calculation shows the matrix element (3.7a) to be nonzero, i.e., the transitions $\mathbf{k}, N_{\mathbf{q}j} \to \mathbf{k}', N'_{\mathbf{q}j}$ to be possible, only in two cases:

1) the electron's wave vector and its energy in the final state are equal to

$$\mathbf{k}' = \mathbf{k} + \mathbf{q} \text{ and } \varepsilon_{\mathbf{k}'} = \varepsilon_{\mathbf{k}} + \hbar\omega_{\mathbf{q}}, \quad (3.8)$$

the number of $\mathbf{q}$-phonons in the final state being

$$N'_{\mathbf{q}} = N_{\mathbf{q}} - 1; \quad (3.8a)$$

2) the electron's wave vector and its energy are equal to

$$\mathbf{k}' = \mathbf{k} - \mathbf{q}, \quad \varepsilon_{\mathbf{k}'} = \varepsilon_{\mathbf{k}} - \hbar\omega_{\mathbf{q}}; \quad (3.9)$$

and

$$N'_{\mathbf{q}} = N_{\mathbf{q}} + 1. \quad (3.9a)$$

---

[8] For the sake of simplicity, the polarization index $j$ will be omitted in the future.

The first process will naturally be interpreted as the absorption of a phonon by the electron, and the second as the emission of a phonon by the electron, with quasi-momentum and energy conservation laws being observed both in the acts of phonon emission and absorption. In the first case the matrix element (3.7a) is equal to[9]

$$-\frac{2i}{3}\frac{Cq}{\sqrt{N}}\sqrt{\frac{\hbar N_q}{2M\omega_q}}, \tag{3.10}$$

in the second case

$$+\frac{2i}{3}\frac{Cq}{\sqrt{N}}\sqrt{\frac{\hbar (N_q+1)}{2M\omega_q}}. \tag{3.10a}$$

Here

$$C=\frac{\hbar^2}{2m}\int |\mathrm{grad}\, u_k|^2\, d^3r_0 \tag{3.10b}$$

and the integration is performed over the volume of the crystal's elementary cell. The constant $C$ with the dimensionality of energy characterizes the intensity of the electron's interaction with lattice vibrations. In order to assess the order of magnitude of $C$, make $|\mathrm{grad}\, u_k| \approx u_k/a$, where $a$ is the lattice parameter and obtain for the order of magnitude of $C$

$$C\approx\left(\frac{\hbar^2}{2ma^2}\right)\int |u_k|^2\, d^3r_0=\left(\frac{\hbar^2}{2ma^2}\right),$$

where use was made of (3.4a). For $a\approx 10^{-8}$ cm $C\approx 5$ eV, i.e., is of the order of magnitude of bonding energy. According to the general theory (Appendix 20), the probability of transition per 1 s is

$$W(\mathbf{k},\mathbf{k}')=W(\mathbf{k},\mathbf{k}\pm\mathbf{q})$$
$$=\frac{2\pi}{\hbar}|\langle\mathbf{k}', N'_{\mathbf{q}}|\Delta V|\mathbf{k}, N_{\mathbf{q}}\rangle|^2\,\delta(\varepsilon_{\mathbf{k}'}-\varepsilon_{\mathbf{k}}\mp\hbar\omega_{\mathbf{q}}). \tag{3.11}$$

In the case of phonon absorption

$$W^+(\mathbf{k},\mathbf{q})=w(q)\,N_{\mathbf{q}}\delta(\varepsilon_{\mathbf{k}+\mathbf{q}}-\varepsilon_{\mathbf{k}}-\hbar\omega_{\mathbf{q}}), \tag{3.11a}$$

and in the case of emission

$$W^-(\mathbf{k},\mathbf{q})=w(q)\,(N_{\mathbf{q}}+1)\,\delta(\varepsilon_{\mathbf{k}-\mathbf{q}}-\varepsilon_{\mathbf{k}}+\hbar\omega_{\mathbf{q}}), \tag{3.11b}$$

where

$$w(q)=\frac{4\pi}{9N}\frac{C^2q^2}{M\omega_{\mathbf{q}}}. \tag{3.11c}$$

---

[9] The additional multiplier $1/\sqrt{M}$, as compared with (A.21.3), stems from expression (3.1) for $\mathbf{u}$.

In accordance with (1.22), the transition probability is inversely proportional to the crystal's volume (to the number of atoms in the principal region). The delta-function $\delta(\varepsilon_{\mathbf{k}\pm\mathbf{q}} - \varepsilon_{\mathbf{k}} \mp \hbar\omega_{\mathbf{q}})$ expresses the energy and the wave vector conservation laws.

In order to calculate the relaxation time $\tau$ due to the scattering of electrons by the lattice vibrations, we should substitute (3.11a) and (3.11b) into equation (2.12).

## 8.4. Relaxation Time of Conduction Electrons in an Atomic Semiconductor and in a Metal

**8.4.1** Calculate the relaxation time (mean free path) of the conduction electron interacting with the lattice vibrations in an atomic semiconductor. As we shall demonstrate presently, the electron scattering in this case is almost elastic, so that equation (2.12) can be used to calculate $\tau$.

If the electron energy is proportional to $k^2$, i.e., if $\varepsilon_{\mathbf{k}} = \hbar^2k^2/2m^*$, and if the frequency of longitudinal acoustic waves is $\omega_{\mathbf{q}l} = v_0 q$, where $v_0$ is the velocity of the longitudinal acoustic waves, then it follows from the conservation laws (3.8) and (3.9) that

$$\frac{\hbar^2(\mathbf{k}\pm\mathbf{q})^2}{2m^*} = \frac{\hbar^2k^2}{2m^*} \pm \hbar v_0 q, \tag{4.1}$$

where the upper sign refers to the case of phonon absorption and the lower, to the case of phonon emission. Hence,

$$q = \mp 2k\cos\vartheta \pm \frac{2m^*}{\hbar} v_0, \tag{4.1a}$$

where $\vartheta$ is the angle between the vectors **k** and **q**.

Assess the order of magnitude of the ratio of the second addend to the first on the right-hand side of expression (4.1a):

$$\frac{m^*v_0}{\hbar k} = \frac{m^*v_0}{p} \approx \frac{m^*v_0}{\sqrt{m^*k_0T}} = \sqrt{\frac{T_{cr}}{T}}, \tag{4.1b}$$

where the average thermal value of the quasi-momentum $p = \hbar k$ has been substituted for the quasi-momentum itself, and the critical temperature $T_{cr} = m^*v_0^2/k_0$. Making $m^* = 10^{-27}$ g, $v_0 = 2 \times 10^5$ cm/s, we obtain $T_{cr} \approx 1$ K; therefore, in the entire temperature range $T \gg 1$ K we can neglect the second addend and write

$$q = \mp 2k\cos\vartheta. \tag{4.1c}$$

Hence, it follows from the conservation laws that the electrons absorb and emit phonons with $q \approx k$.

It follows from expression (4.1) that

$$q_{min} = 0, \quad q_{max} = 2k. \tag{4.1d}$$

Since the average value of $k$ for electrons at room temperature is of the order of $10^7$ cm$^{-1}$, and the maximum value of the wave vector $\mathbf{q}$, according to Debye's theory (3.11.13), is equal to $q_0 = (6\pi^2/\Omega_0)^{1/3} \approx 10^8$ cm$^{-1}$, it follows that $q_{max} \ll q_0$. The electrons interact only with longwave phonons, and because of this, the linear dispersion law ($\omega = v_0 q$) is true.

Neglecting the second addend in (4.1a) means obviously neglecting the phonon's energy in (4.1), i.e., the circumstance that the scattering is not perfectly elastic. In the case of elastic scattering the $\delta$-functions in (3.11a) and (3.11b) take the form

$$\delta(\varepsilon_{\mathbf{k}\pm\mathbf{q}}-\varepsilon_{\mathbf{k}})=\delta\left(\pm\frac{\hbar^2kq}{m^*}\cos\vartheta+\frac{\hbar^2q^2}{2m^*}\right)=\frac{m^*}{\hbar^2kq}\delta\left(\frac{q}{2k}\pm\cos\vartheta\right), \tag{4.2}$$

where we made use of the property $\delta(a\pm bx)=\frac{1}{b}\delta\left(\frac{a}{b}\pm x\right)$ [8.5, p. 56].

We can now compute the relaxation time $\tau$ from equation (2.12). When an electron is scattered by lattice vibrations,

$$\mathbf{k}' = \mathbf{k} \pm \mathbf{q}, \tag{4.3}$$

where the upper sign corresponds to the absorption and the lower sign corresponds to the emission of a phonon. We have from (2.12) and (4.3)

$$\frac{1}{\tau}=-\sum_{\mathbf{q}} W^+(\mathbf{k},\mathbf{q})\frac{q_\chi}{k_\chi}+\sum_{\mathbf{q}} W^-(\mathbf{k},\mathbf{q})\frac{q_\chi}{k_\chi}, \tag{4.4}$$

where the first sum on the right-hand side takes into account the absorption and the second the emission of phonons. We have replaced the summation over $\mathbf{k}'$ in (2.12) with the summation over $\mathbf{q}$ for a specified $\mathbf{k}$.

Replace the summation over $\mathbf{q}$ by integration over the $\mathbf{q}$-space in spherical coordinates with the polar axis coinciding in direction with $\mathbf{k}$:

$$\sum_{\mathbf{q}}\to\frac{V}{(2\pi)^3}\int_{q_{min}}^{q_{max}} q^2\,dq\int_0^{\pi}\sin\vartheta\,d\vartheta\int_0^{2\pi} d\ . \tag{4.5}$$

Figure 8.5 depicts the vectors $\mathbf{k}$, $\mathbf{q}$ and $\chi$ and the angles between them It is an established fact that [8.1, Chap. I]

$$\cos\alpha = \cos\vartheta\cos\beta + \sin\vartheta\sin\beta\cos\varphi. \tag{4.6}$$

Since $q_\chi = q\cos\alpha$, and $k_\chi = k\cos\beta$, the inverse relaxation time is

$$\frac{1}{\tau} = -\frac{V}{(2\pi)^3}\frac{m^*}{\hbar^2 k^2}\int_{q_{\min}}^{q_{\max}} q^2\,dq \int_0^\pi \sin\vartheta\,d\vartheta \int_0^{2\pi} d\varphi$$
$$\times\left\{w(q)\,N_q\delta\left(\frac{q}{2k}+\cos\vartheta\right)\frac{\cos\alpha}{\cos\beta} - w(q)\,(N_q+1)\,\delta\right.$$
$$\left.\times\left(\frac{q}{2k}-\cos\vartheta\right)\frac{\cos\alpha}{\cos\beta}\right\}. \tag{4.7}$$

Only $\cos\alpha$ depends on $\varphi$; therefore,

$$\int_0^{2\pi} \cos\alpha\,d\varphi = 2\pi\cos\vartheta\cos\beta. \tag{4.8}$$

Integrating with respect to $\sin\vartheta\,d\vartheta = -d\cos\vartheta$ and making use of the $\delta$-functions in the first and second addends in the braces in (4.7), we obtain

$$\frac{1}{\tau} = \frac{V}{8\pi^2}\frac{m^*}{\hbar^2 k^3}\int_{q_{\min}}^{q_{\max}} w(q)\,(2N_q+1)\,q^3\,dq, \tag{4.9}$$

since $q_{\min}$ and $q_{\max}$ for the processes of phonon absorption and emission are identical.

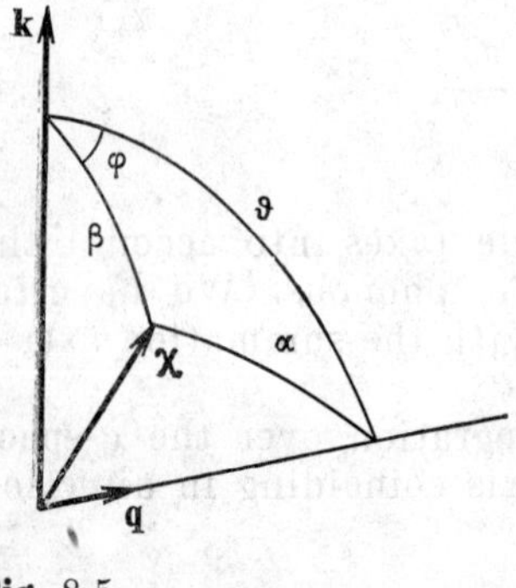


Fig. 8.5

In order to be able to use the probabilities (3.11) in (4.4), we must make some assumptions about the numbers occupying the phonons $N_q$. As we shall see, in most cases in the processes involving electron transport the phonon distribution deviates little from the distribution corresponding to thermodynamical equilibrium and described by Planck's equation (3.11.10a). We shall learn in the next chapter that in the presence of a temperature gradient, when considering electron kinetic phenomena, we have sometimes to take into account the deviation of the phonon distribution function from its equilibrium value ("phonon drag" effects).

Hence, we make

$$N_q = \langle N_q\rangle_{\text{eq}} = \frac{1}{\exp\frac{\hbar v_0 q}{k_0 T}-1}. \tag{4.10}$$

Taking into account (4.1d), we see that the power of the exponent is

$$\frac{\hbar v_0 q}{k_0 T} \approx \frac{\hbar v_0 k}{k_0 T} = \frac{v_0 p}{k_0 T} \approx \frac{v_0 \sqrt{m^* k_0 T}}{k_0 T} = \sqrt{\frac{T_{cr}}{T}} \ll 1. \tag{4.10a}$$

Therefore, (4.10) can be expanded into a series, and we obtain

$$\langle N_q \rangle_{eq} \approx \langle N_q + 1 \rangle_{eq} = \frac{k_0 T}{\hbar v_0 q}. \tag{4.10b}$$

Substituting this value of $N_q$ and $w(q)$ from (3.11c) into (4.9), we obtain after elementary integration with respect to $q$ in the limits (4.1d)

$$\tau = \frac{9\pi}{4} \frac{M v_0^2 \hbar^3}{\Omega_0 C^2 m^* k_0 T} \frac{1}{k} = \frac{9\pi}{4\sqrt{2}} \frac{M v_0^2 \hbar^4}{\Omega_0 C^2 m^{*3/2} k_0 T} \frac{1}{\sqrt{\varepsilon}} \tag{4.11}$$

or

$$\tau = \frac{\tau_{0k}}{k} = \frac{\tau_0}{\sqrt{\varepsilon}}, \tag{4.11a}$$

where

$$\tau_{0k} = \frac{9\pi}{4} \frac{M v_0^2 \hbar^3}{\Omega_0 C^2 m^* k_0 T} \quad \text{and} \quad \tau_0 = \frac{\hbar}{\sqrt{2m^*}} \tau_{0k}.$$

Note that the relaxation time of charge carriers in atomic semiconductors $\tau \propto T^{-1} \varepsilon^{-1/2}$. The mean free path

$$l = v\tau = \frac{9\pi}{4} \frac{M v_0^2 \hbar^4}{\Omega_0 C^2 m^{*2} k_0 T} \tag{4.12}$$

is independent of the electron energy.

**8.4.2** Consider the relaxation time of conduction electrons in a metal. It will be demonstrated in the following chapter that practically the only electrons taking part in transport phenomena in a metal are those with the energy close to the Fermi energy $\zeta_0$. The wave vector corresponding to such electrons is

$$k(\zeta_0) = \left(\frac{1}{\hbar}\right) \sqrt{2m^* \zeta_0} \approx 10^8 \text{ cm}^{-1}, \tag{4.13}$$

i.e., a vector an order of magnitude greater than $k$ for semiconductors at room temperature. Therefore relation (4.1b) for metals takes the form

$$\frac{m^* v_0}{\hbar k(\zeta_0)} = \sqrt{\frac{k_0 T_k}{\zeta_0}} \ll 1 \tag{4.13a}$$

for all temperatures. Hence, the electrons in a metal, same as in a semiconductor, are scattered elastically. The same expressions (3.11a) and (3.11b) hold for the transition probability, and because of this the relaxation time is calculated precisely in the same way as it was done above. However, when integrating with respect to $q$ in (4.9) the following should be kept in mind. As we have seen above, the maximum value of the phonon wave vector is, according to Debye (3.11.13), equal to $q_0 \approx 10^8$ cm$^{-1}$, i.e., it is of the same order as $k$ $(\zeta_0)$. Therefore the following two cases should be distinguished: (1) $k(\zeta_0) < \frac{q_0}{2}$ and (2) $k(\zeta_0) > \frac{q_0}{2}$. In the former which is realized in metals with low conduction electron concentrations (semimetals), the integration with respect to $q$ should be performed up to $2k(\zeta_0)$, so that the relaxation time will be equal to (4.11). In the latter which is usually realized in good metals, the integration with respect to $q$ should be performed up to $q_0$, this being the only difference from the procedure developed for semiconductors. Accordingly, $\tau$ for a metal can be obtained from expression (4.11) multiplied by $(2k)^4$ and divided by $q_0^4$.[10] Hence, for metals we have

$$\tau = \frac{1}{\pi^3} \frac{\Omega_0 M k_0 T_c \hbar}{m^* C^2} \left(\frac{T_c}{T}\right) k^3 = \frac{2\sqrt{2}}{\pi^3} \frac{\Omega_0 M m^{*1/2} k_0 T_c}{\hbar^2 C^2} \left(\frac{T_c}{T}\right) \varepsilon^{3/2} \tag{4.14}$$

and

$$l = v\tau = \frac{4}{\pi^3} \frac{\Omega_0 M k_0 T_c}{\hbar^2 C^2} \left(\frac{T_c}{T}\right) \varepsilon^2. \tag{4.14a}$$

Here $T_c = \hbar v_0 q_0 / k_0$ is the Debye temperature [we have introduced an error when we assumed that the velocity of longitudinal acoustic waves was equal to their average velocity (3.9.16)].

Assess the order of magnitude of the ratio of mean free paths of electrons in an atomic semiconductor (4.11a) and a metal (4.14a):

$$\frac{l_s}{l_m} \approx \left(\frac{\hbar^2}{m^* a^2}\right) : \varepsilon \approx \left(\frac{\hbar^2}{m^* a^2}\right) : \zeta_0, \tag{4.15}$$

i.e., it is of the order of unity ($a$ is the lattice parameter equal to $\Omega_0^{1/3}$). Assess the mean free path of electrons $l_s \approx l_m$. From (4.12) we have

$$l \approx \frac{M v_0^2}{C} \frac{\hbar^2/m^* a^2}{C} \frac{\hbar^2/m^* a^2}{k_0 T} a. \tag{4.16}$$

Since $Mv_0^2$ and $\hbar^2/m^*a^2$ are of the order of the bonding energy per atom, the first and second multipliers are of the order of 1 and the third multiplier is much greater than 1; therefore, $l \gg a$. In real atomic semiconductors $l$ can be tens of times and in the low temperature range thousands of times greater than the lattice parameter $a$. From the point of view of classical mechanics the electron's mean free path in a crystal should be of the order of mag-

[10] In (4.9) the integration both in the cases of a semiconductor and a metal is performed with respect to $q^3$.

nitude of the lattice parameter, since the interatomic distance in a solid is of the order of atomic dimensions. Great values of the mean free paths of electrons following from the theory and observed in experiment prove that the motion of electrons obeys the laws of quantum mechanics.

## 8.5. Theory of Deformation Potential in Cubic Crystals with a Simple Energy-Band Structure

8.5.1 A special method of calculating the relaxation time of conduction electrons in atomic crystals has been developed and has become known as the *theory of deformation potentials* (Bardeen and Shockley, 1950). The results obtained with the aid of this method in the case of a simple electron energy-band structure coincide with those obtained on the basis of the hypothesis of deformable ions (8.3) or of Nordheim's hypothesis of rigid ions [8.6]. However, the theory of deformation potentials boasts of several advantages, the most important of which is the simplicity of deduction that makes it possible to extend the theory to more complex cases, for example, to the case of a complex energy-band structure of charge carriers. We shall consider the theory of deformation potentials for a simple cubic monoatomic crystal with a simple band structure in which the energy of the conduction electron is

$$\varepsilon = \hbar^2 \mathbf{k}^2/2m^*. \qquad (5.1)$$

For a better understanding of the peculiarities of the method of deformation potentials consider the formation of energy bands in a crystal as isolated atoms are brought closer together. Imagine the atoms of some element arranged in a "lattice" corresponding to the type of lattice peculiar to the element but at great interatomic distances $a$ from one another.

As long as $a$ greatly exceeds the true (equilibrium) lattice parameter $a_0$, the atoms do not interact, and their energy spectra remain discrete. It will be seen from Fig. 8.6$a$ that for $a > 3a_0$ the $3s$- and the $3p$-levels of sodium are practically discrete. As $a$ is decreased, i.e., as the atoms are brought closer together, the atomic energy levels, because of interaction, widen into energy bands. As can be seen from Fig. 8.6$a$, at $a = 1.7a_0$ the widened $3p$ and $3s$ terms begin to overlap forming at $a = 1.7a_0$ the sodium metal's conduction band.

Another situation occurs, for example, in the case of the diamond crystal. The discrete $2s$- and $2p$-states of the isolated carbon atom widen as $a$ is decreased, and for a definite value of $a > a_0$ the $2s$- and the $2p$-bands overlap (Fig. 8.6.$b$). However calculation shows that as the carbon atoms are brought closer together, a hybridization of the $s$- and $p$-states sets in (due to the formation of four valence bonds with the nearest neighbors), and at $a = a_0$ the allowed bands of the mixed $s$-$p$ type are again separated by a forbidden band.

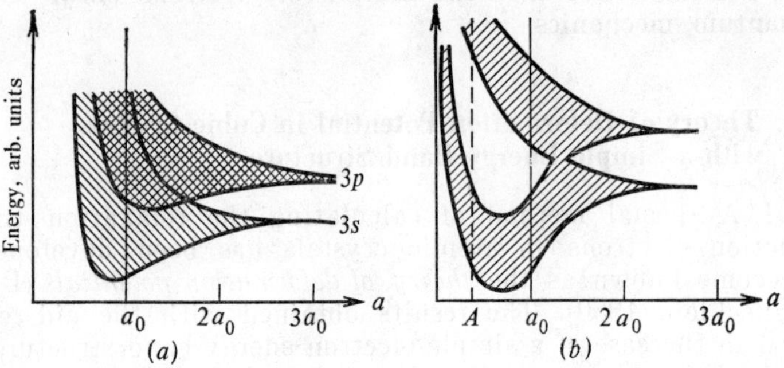


**Fig. 8.6**

It will be seen from Fig. 8.6*b* that in the case of the uniform compression of the crystal resulting in a decrease in the lattice parameter $a$ the bottom edge of the conduction band shifts upward and the top edge of the valence band shifts downward, thus increasing the

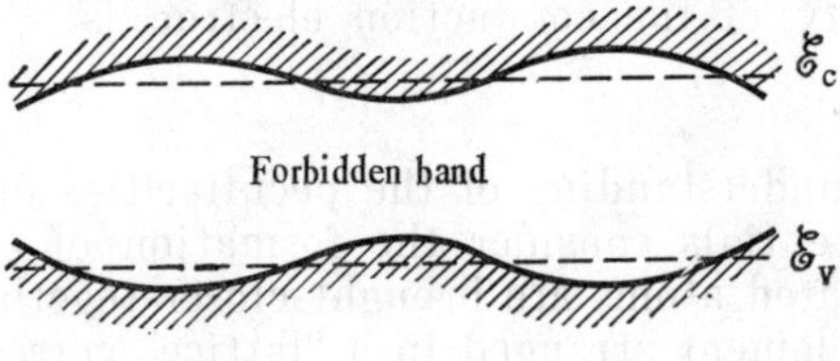


**Fig. 8.7**

width of the forbidden band. Should the lattice parameter $a_0$ correspond to point $A$, an opposite situation would set in. In this respect the effect of compression and extension can be substantially different from the effect of the external electric field that always displaces the edges of the conduction and valence bands in the same direction. Figure 8.7 depicts the pattern of wave-like variations of the forbidden bandwidth accompanying the passage of an acoustic compression wave.

In the approximation of a continuous medium the state of a deformed crystal is characterized by the components of the *deformation* (strain) *tensor*

$$\varepsilon_{ij} = \frac{1}{2}\left(\frac{\partial u_j}{\partial x_i} + \frac{\partial u_i}{\partial x_j}\right), \tag{5.2}$$

where $x_i$ ($x_1 \equiv x$, $x_2 \equiv y$, $x_3 \equiv z$) are rectangular coordinates of the point of the continuous medium, and $u_i$ ($i = 1, 2, 3$) are rectangular coordinates of the projection of the displacement $\mathbf{u}(x_1, x_2, x_3)$ of the point of the continuous medium $x_1, x_2, x_3$.

In the general case the position of the bottom edge of the conduction band $\mathscr{E}_c$ (of the upper edge of the valence band $\mathscr{E}_v$) can be regarded as a function of the components of the deformation tensor $\varepsilon_{ij}$. Expanding $\mathscr{E}_c$ into a series in $\varepsilon_{ij}$, we obtain

$$\mathscr{E}_c(\varepsilon_{ij}) = \mathscr{E}_c(0) + \sum_{i,j} a_{ij}\varepsilon_{ij} = \mathscr{E}_c(0) + a_{11}\varepsilon_{11} + a_{12}\varepsilon_{12} + \dots \tag{5.3}$$

The quantities $a_{ij}$ and $\varepsilon_{ij}$ are functions of the orientation of the coordinate axes $x_i$ with respect to crystallographic axes. Moreover, $a_{ij}$ obviously depend on the type of crystal. Place the origin of our rectangular coordinate frame at the vertex of the cube of the undeformed crystal cell and direct the coordinate axes along the cube's edges. The nondiagonal coefficients $a_{ij}$ ($i \neq j$) in the case of a crystal of cubic symmetry can easily be shown to vanish. Indeed, rotate the coordinate frame about the axis $x_3 \equiv z$ through the angle $\pi/2$. Then we have in the new (primed) frame $x_1' = x_2$, $x_2' = -x_1$, and, therefore, in accordance with (5.2), $\varepsilon_{12}' = -\varepsilon_{12}$. Such rotation of the coordinate frame does not change the coefficients $a_{ij}$, since the crystal's orientation remains the same for both positions of the frame, so that $a_{12}' = a_{12}$. Let the crystal's deformation be such that only the tensor's component $\varepsilon_{12}$ is nonzero. In this case the displacement of the conduction band's edge expressed in terms of both coordinate frames rotated with respect to each other will be

$$\mathscr{E}_c(\varepsilon_{12}) - \mathscr{E}_c(0) = a_{12}\varepsilon_{12} = a_{12}'\varepsilon_{12}' = -a_{12}\varepsilon_{12},$$

whence $a_{12} = 0$. In the same way it can be demonstrated that all the other nondiagonal coefficients $a_{ij}$ are zeros. Since in a cubic crystal the axes $x_1$, $x_2$, $x_3$ are equivalent, it follows that $a_{11} = a_{22} = a_{33} = \mathscr{E}_1$, and we obtain from (5.3)

$$\mathscr{E}_c(\varepsilon_{ij}) = \mathscr{E}_c(0) + \mathscr{E}_1 \Delta, \tag{5.4}$$

where

$$\Delta = \varepsilon_{11} + \varepsilon_{22} + \varepsilon_{33} = \frac{\partial u_1}{\partial x_1} + \frac{\partial u_2}{\partial x_2} + \frac{\partial u_3}{\partial x_3} = \operatorname{div} \mathbf{u} = \frac{\delta V}{V} \tag{5.4a}$$

is the relative variation of volume at the given point [8.7, Sec. 1].

Bardeen and Shockley's theory of deformation potentials contains the proof that the scattering of electrons by lattice vibrations can be described not in terms of Bloch's wave function, but in terms of a plane wave, provided the scattering potential (3.3) is replaced by the expression

$$\mathscr{U} = \mathscr{E}_1 \Delta = \mathscr{E}_1 \operatorname{div} \mathbf{u}(\mathbf{r}). \tag{5.5}$$

We shall discover the constant $\mathscr{E}_1$, which has the dimensionality of energy closely related to the interaction constant $C$ determined in (3.10b).[11]

8.5.2 Determine the relaxation time by making use of the scattering potential (5.5) and of a plane wave to describe the electron:

$$\psi_{\mathbf{k}}(\mathbf{r})=\frac{1}{\sqrt{V}}e^{i\mathbf{k}\cdot\mathbf{r}}, \tag{5.6}$$

where $V$ is the volume of the crystal's principal region.

The displacement $\mathbf{u}\cdot(\mathbf{r})$ of the crystal's point $\mathbf{r}$ in the continuous medium approximation is, according to (3.1), equal to

$$\mathbf{u}(\mathbf{r})=\frac{1}{\sqrt{NM}}\sum_{\mathbf{q},j}{}' \mathbf{e}_{\mathbf{q}j}\{a_{\mathbf{q}j}e^{i\mathbf{q}\cdot\mathbf{r}}+a^*_{\mathbf{q}j}e^{-i\mathbf{q}\cdot\mathbf{r}}\}. \tag{5.7}$$

Here $N$ is the number of atoms in the principal region, $\mathbf{e}_{\mathbf{q}j}$ are orthonormal polarization vectors, $a_{\mathbf{q}j}$ are complex normal coordinates with a harmonic time-dependence characterized by a frequency $\omega_{\mathbf{q}j}=v_{0j}q$, where $v_{0j}$ is the velocity of acoustic waves with the polarization $j$. Hence, the relative variation of volume is

$$\Delta=\operatorname{div}\mathbf{u}=\frac{i}{\sqrt{NM}}\sum_{\mathbf{q}}{}' (\mathbf{e}_{\mathbf{q}j}\cdot\mathbf{q})\{a_{\mathbf{q}j}e^{i\mathbf{q}\cdot\mathbf{r}}-a^*_{\mathbf{q}j}e^{-i\mathbf{q}\cdot\mathbf{r}}\}. \tag{5.8}$$

Since the scattering potential (5.5) is proportional to $\Delta$, it follows from (5.8) that the conduction electrons interact only with longitudinal acoustic waves for which $\mathbf{e}_{\mathbf{q}j}\parallel\mathbf{q}$, and $\mathbf{e}_{\mathbf{q}j}\cdot\mathbf{q}=q$. In the future we shall drop the polarization suffix $j$ of the longitudinal wave.

In order to determine the transition probability $W(\mathbf{k},\mathbf{k}')$ from equation (3.11), calculate the matrix element of the perturbation energy (5.5) on the electron wave functions in the plane wave approximation (5.6), i.e., in the effective mass approximation (5.1):

$$M_{\mathbf{k}\mathbf{k}'}=\langle\mathbf{k}'\,|\,\mathcal{U}\,|\,\mathbf{k}\rangle=\int_V \psi^*_{\mathbf{k}'}(\mathbf{r})\,\mathcal{U}(\mathbf{r})\,\psi_{\mathbf{k}}(\mathbf{r})\,d\tau. \tag{5.9}$$

Making use of (5.5) and (5.8), we obtain

$$M_{\mathbf{k}\mathbf{k}'}=\frac{i\mathscr{E}_1}{\sqrt{NM}}\sum_{\mathbf{q}j}{}' q\left\{a_{\mathbf{q}}\frac{1}{V}\int_V e^{i(\mathbf{k}+\mathbf{q}-\mathbf{k}')\cdot\mathbf{r}}\,d\tau-\right.$$

$$\left.-a^*_{\mathbf{q}}\frac{1}{V}\int_V e^{i(\mathbf{k}-\mathbf{q}-\mathbf{k}')\cdot\mathbf{r}}\,d\tau\right\}. \tag{5.10}$$

[11] M. F. Deigen and S. I. Pekar simultaneously and independently used expression (5.4) for the displacement of the edge of the conduction band in their calculations of the "condenson" electron state in a cubic atomic crystal. Titeica in 1935 used (5.5) for the electron scattering potential in metals in his calculations of galvanomagnetic phenomena in strong magnetic fields. The achievement of Bardeen and Shockley is that they laid the foundation for the application of the theory of deformation potentials within the framework of the effective mass method.

The first integral in braces is nonzero and equal to $V$ only if

$$\mathbf{k}+\mathbf{q}=\mathbf{k}', \tag{5.10a}$$

and the second only if

$$\mathbf{k}-\mathbf{q}=\mathbf{k}'. \tag{5.10b}$$

Case (5.10a) corresponds to the process in which the electron absorbs a phonon with the wave vector $\mathbf{q}$ and the energy $\hbar\omega_{\mathbf{q}}$, and case (5.10b) corresponds to the process involving phonon emission.

In both cases

$$|M_{\mathbf{kk}'}|^2=\frac{\mathscr{E}_1^2}{NM}q^2|a_q|^2, \tag{5.11}$$

since $|a_q|^2=|a_{\mathbf{q}}^*|^2$.

It follows from (3.6.27) that the energy of one normal vibration corresponding to a longitudinal acoustic wave is equal to $2\omega_q^2|a_q|^2=2v_0^2q^2|a_q|^2$, where $v_0$ is the velocity of longitudinal acoustic waves. In the state of statistical equilibrium (in the temperature range above the Debye temperature)

$$2v_0^2q^2|a_q|^2=k_0T, \tag{5.12}$$

whence

$$|a_q|^2=k_0T/2v_0^2q^2. \tag{5.12a}$$

It follows from (5.11) and (5.12a) that

$$|M_{\mathbf{kk}'}|^2=\mathscr{E}_1^2k_0T/2NMv_0^2. \tag{5.13}$$

It follows from (3.10) and (3.10a) that at temperatures above the Debye temperature

$$\langle\mathbf{k}', N'_{qj}|\Delta V|\mathbf{k}, N_{qj}\rangle=\frac{4}{9}\frac{C^2q^2}{N}\frac{\hbar}{2M\omega_q}\frac{k_0T}{\hbar\omega_q}=\frac{4}{9}\frac{C^2k_0T}{2NMv_0^2}. \tag{5.14}$$

Here we took into account that $N_q\approx N_q+1=k_0T/\hbar\omega_q$, and $\omega_q=v_0q$. If we assume that the squares of the moduli of the matrix elements (5.13) and (5.14) are equal, we obtain

$$\mathscr{E}_1=2/3C, \tag{5.15}$$

where the constant $C$ is determined by expression (3.10b).

The transition probability (3.11) that enables us to calculate the electron's relaxation time $\tau$ in the same way as was done in 8.4 follows directly from (5.14). The relaxation time obtained in this way will obviously coincide with expression (4.11) in which it remains only to substitute $3/2\ \mathscr{E}_1$ for $C$.

Of course, the above argument cannot be regarded as direct proof of relationship (5.15), because it follows from the comparison of ultimate results of calculations based on different physical assumptions. A direct proof of relationship (5.15) based on the quantum mechanics theory of perturbations for a deformed crystal is presented in Appendix 22.

## 8.6. Scattering of Conduction Electrons by Lattice Vibrations in Ionic Crystals

**8.6.1** In an ionic crystal (provided it is not a piezoelectric crystal) the conduction electron (hole) interacts with the optical vibrations much more intensely than with the acoustic vibrations. This is because in an ionic crystal optical vibrations are accompanied by the appearance in each crystal cell of a dipole electric moment that interacts strongly with a conduction electron (a hole).

When we studied weak binding polarons (5.4.3), we calculated the matrix elements of the electron's interaction with long-wave optical vibrations of an ionic crystal in the continuous medium approximation. We have seen the correction to the electron energy in the first approximation of the theory of perturbations to be zero. Calculate in the same approximation the electron scattering on optical vibrations of an ionic crystal. It follows from (5.4.46) and (3.11) that the probability of an electron transition $\mathbf{k} \to \mathbf{k}'$ involving the absorption or the emission of an optical phonon $\hbar\omega_l$ is equal to

$$W(\mathbf{k}, \mathbf{k}') = W(\mathbf{k}, \mathbf{k} \pm \mathbf{q}) =$$
$$= \frac{2\pi}{\hbar} \,|\, \langle N'_{\mathbf{q}}, \mathbf{k}' \,|\, -e\Phi \,|\, N_{\mathbf{q}}, \mathbf{k}\rangle \,|^2\, \delta(\varepsilon_{\mathbf{k}'} - \varepsilon_{\mathbf{k}} \mp \hbar\omega_l) =$$
$$= w(q) \begin{Bmatrix} N_{\mathbf{q}} \\ N_{\mathbf{q}}+1 \end{Bmatrix} \delta(\varepsilon_{\mathbf{k}\pm\mathbf{q}} - \varepsilon_{\mathbf{k}} \mp \hbar\omega_l), \qquad (6.1)$$

where

$$w(q) = \frac{4\pi^2 e^2 \omega_l}{\varepsilon^*} \frac{1}{q^2}. \qquad (6.1a)$$

Here the upper signs (and the upper line in the braces) correspond to the absorption, and the lower to the emission of a phonon. The energy conservation law is expressed by the $\delta$-function in (6.1). For a conduction electron (a hole) characterized by an effective mass $m^*$, the energy conservation law takes the form

$$\frac{\hbar^2 (\mathbf{k} \pm \mathbf{q})^2}{2m^*} = \frac{\hbar^2 k^2}{2m^*} \pm \hbar\omega_l,$$

or

$$\frac{\hbar^2 q^2}{2m^*} \pm \frac{2\hbar^2 kq}{2m^*} \cos\vartheta \mp \hbar\omega_l = 0. \qquad (6.2)$$

Here $\vartheta$ is the angle between the wave vectors $\mathbf{k}$ and $\mathbf{q}$. Solving the quadratic equation (6.2) for $q$, we obtain two roots

$$q_1 = -k\cos\vartheta \pm \sqrt{k^2\cos^2\vartheta + \varkappa^2}, \quad q_2 = k\cos\vartheta \pm \sqrt{k^2\cos^2\vartheta - \varkappa^2}, \tag{6.3}$$

where

$$\hbar^2\varkappa^2/2m^* = \hbar\omega_l. \tag{6.3a}$$

Consider now various temperature intervals.

**A. High temperatures:** $k_0T \gg \hbar\omega_l$ or $k \gg \varkappa$. In this case $\varkappa^2$ under the root in (6.3) can be neglected, and in phonon absorption and emission, we have

$$q_{\min} = 0, \quad q_{\max} = 2k. \tag{6.4}$$

Next, for $k_0T \gg \hbar\omega_l$

$$N_{\mathbf{q}} = \langle N_{\mathbf{q}}\rangle_{eq} = \frac{k_0T}{\hbar\omega_l} \gg 1. \tag{6.5}$$

Since for $k_0T \gg \hbar\omega_l$ the scattering is elastic, the relaxation time is determined by equation (4.9):

$$\frac{1}{\tau} = \frac{1}{8\pi^2}\frac{m^*}{\hbar^2k^3}\int_{q_{\min}}^{q_{\max}} w(q)(2N_q+1)\,q^3\,dq. \tag{6.6}$$

Substituting herein (6.1$a$), (6.4) and (6.5), we obtain for the relaxation time of conduction electrons (holes) in ionic crystals at high temperatures

$$\tau = \frac{\sqrt{2}}{2}\frac{\hbar^2\varepsilon^*}{e^2(m^*)^{1/2}k_0T}\,\varepsilon^{1/2}. \tag{6.7}$$

Hence, $\tau \propto \sqrt{\varepsilon}/T$. The mean free path of an electron having a thermal energy $\varepsilon = m^*v^2/2 = 3/2k_0T$ is

$$l = \tau v = \frac{3}{2}\varepsilon^*\frac{\hbar^2}{m^*e^2}. \tag{6.7a}$$

If the electron's effective mass $m^*$ is equal to the free electron mass $m$, then $\hbar^2/me^2$ is the radius of the Bohr orbit, and the mean free path $l$ is of the order of the lattice parameter. Expression (6.7) is inapplicable in this case, it can be applied only if the ratio $\varepsilon^*/m^*$ is greater.

**B. Low temperatures:** $k_0T \ll \hbar\omega_l$ or $k \ll \varkappa$. In this case practically the only processes taking place are the phonon absorption processes, and in accordance with (6.3), we have

$$q_{\min} = \sqrt{k^2+\varkappa^2} - k \ (\vartheta = 0), \quad q_{\max} = \sqrt{k^2+\varkappa^2} + k \ (\vartheta = \pi). \tag{6.8}$$

At low temperatures the scattering is inelastic, and, therefore, in general the relaxation time cannot be introduced with the aid of equation (4.9). However, as was demonstrated by B.I. Davydov and I.M. Shmushkevich in 1940, in the low-temperature case as well the relaxation time can be introduced, provided a correct calculation procedure is followed.

Qualitatively this can be explained as follows. At low temperatures, when $k_0T \ll \hbar\omega_l$, the absolute majority of the electrons are able only to absorb the phonons. Such absorption of a phonon results in the electron going over to the energy interval from $\hbar\omega_l$ to $2\hbar\omega_l$. Such an electron will immediately emit a phonon, because the ratio of the emission probability to the absorption probability is equal, according to (6.1), to $\frac{N_q+1}{N_q} \approx \exp\frac{\hbar\omega_l}{k_0T} \gg 1$. The variation of the electron energy in the result of such an absorption and an almost immediate emission of a phonon will be very small (only at the expense of the $\omega_l$ vs $q$ dependence), but the variation of its wave vector will be substantial. This makes it possible to regard the electron scattering in a definite sense as elastic and to introduce the relaxation time.

Calculate the relaxation time $\tau$ by using equation (4.4) and by taking into account only the first sum corresponding to the absorption of a phonon.

The $\delta$-function in $W^+_{\mathbf{k},\mathbf{k}'}$ is of the form

$$\delta\left(\frac{\hbar^2q^2}{2m^*}-\frac{\hbar^2\varkappa^2}{2m^*}+\frac{\hbar^2kq}{m^*}\cos\vartheta\right)=\frac{m^*}{\hbar^2kq}\,\delta\left(\frac{q^2-\varkappa^2}{2kq}+\cos\vartheta\right). \tag{6.9}$$

Performing operations similar to those that led to expression (4.9), we obtain

$$\frac{1}{\tau}=\frac{1}{8\pi^2}\,\frac{m^*}{\hbar^2k^3}\int\limits_{q_{\min}}^{q_{\max}} w(q)\,N_q\,\frac{q^2-\varkappa^2}{q}\,q^2\,dq. \tag{6.10}$$

Substituting herein (6.1a) and (6.8) and taking into account that at low temperatures

$$N_{\mathbf{q}}=\langle N_{\mathbf{q}}\rangle_{\mathrm{eq}}=\frac{1}{\exp\frac{\hbar\omega_l}{k_0T}-1}\approx\exp\left(-\frac{\hbar\omega_l}{k_0T}\right), \tag{6.11}$$

we obtain

$$\frac{1}{\tau}=\frac{e^2m^*\omega_l\exp\left(-\frac{\hbar\omega_l}{k_0T}\right)}{\hbar^2\varepsilon^*k^3}\left\{2k\sqrt{k^2+\varkappa^2}-\varkappa\log\frac{\sqrt{k^2+\varkappa^2}+k}{\sqrt{k^2+\varkappa^2}-k}\right\}. \tag{6.12}$$

Carrying $\varkappa^2$ out of the braces and expanding the remaining expression into a series in $k/\varkappa \ll 1$, we see that the first nonzero member is of the order of $(k/\varkappa)^3$, so that the expression in the braces is equal to

$$\frac{4}{3} k^3 \varkappa^2. \tag{6.12a}$$

From (6.12) and (6.12a) we obtain for the relaxation time of the conduction electron (hole) in an ionic crystal at low temperatures

$$\tau = \frac{3\sqrt{2}}{2} \frac{\hbar^2 \varepsilon^* \exp\frac{\hbar\omega_l}{k_0 T}}{e^2 (m^*)^{1/2} (\hbar\omega_l)^{1/2}}. \tag{6.13}$$

Hence, $\tau \propto \exp(\hbar\omega_l/k_0T)$ and is independent of the electron energy. The mean free path of an electron with the thermal energy $\varepsilon = (m^* v^2/2) = {}^3/_2 k_0 T$ is equal to

$$l = \tau v = \frac{3\sqrt{6}}{2} \varepsilon^* \left(\frac{k_0 T}{\hbar\omega_l}\right)^{1/2} \exp\frac{\hbar\omega_l}{k_0 T} \frac{\hbar^2}{m^* e^2}. \tag{6.13a}$$

Here $\hbar^2/m^*e$ is the radius of the "effective" Bohr orbit. Thanks to the great value of the multiplier $\exp(\hbar\omega_l/k_0T)$, $l$ is much larger than the lattice parameter, and expression (6.13) for the relaxation time $\tau$ is in general applicable in the low temperature range.

**8.6.2** In piezoelectric semiconductors (Section 3.1.3), for instance in zinc blende ZnS, the electron scattering by acoustic vibrations may be comparable to the scattering by optical vibrations discussed in Section 8.6.1. This is because in piezoelectrics long acoustic waves (that produce elastic stresses) are accompanied by the electric polarization of the crystal.

Meijer and Polder [8.8] were the first to consider piezoelectric scattering by acoustic vibrations. We shall present some of their results in a simplified form sufficient for the determination of the dependence of the electron's relaxation time $\tau$ on its energy $\varepsilon = \hbar^2 k^2/2m^*$ and on the temperature $T$.

Piezoelectric crystals are ionic lattices without an inversion (symmetry) centre. It follows from (3.6.4a) that the displacements of the positive and negative ions (labels $k = 1, 2$) are equal in the continuous medium approximation to

$$\mathbf{u}^k(\mathbf{r}) = \frac{1}{\sqrt{N m_k}} \sum_{\mathbf{q}, j}{}' \{\mathbf{e}_{jk}(\mathbf{q}) a_j(\mathbf{q}) e^{i\mathbf{q}\cdot\mathbf{r}} + \mathbf{e}^*_{jk}(\mathbf{q}) a^*_j(\mathbf{q}) e^{-i\mathbf{q}\cdot\mathbf{r}}\}, \tag{6.14}$$

where the summation over $\mathbf{q}$ is performed inside one-half of the Brillouin zone. The dipole moment appearing in a unit cell with the volume $\Omega_0$ as the result of vibration is equal to the geometrical sum of the ionic displacements multiplied by their effective charges $\pm e^*$.

As we shall see, only the long acoustic waves for which the vector **q** may be assumed to be the same for both ions in the cell will be of interest to us. The polarization vector **P** (**r**) equal to the dipole moment per unit volume can be written in the form

$$\mathbf{P}(\mathbf{r})=\frac{e^*}{\Omega_0}\frac{1}{\sqrt{N}}\sum_{\mathbf{q},j}{}'\left(\frac{\mathbf{e}_{j1}(\mathbf{q})}{\sqrt{m_1}}-\frac{\mathbf{e}_{j2}(\mathbf{q})}{\sqrt{m_2}}\right)a_j(\mathbf{q})\,e^{i\mathbf{q}\cdot\mathbf{r}}+\text{c.c.} \tag{6.15}$$

where c.c. denotes a quantity complex-conjugate of the first addend.

The polarization (6.15) results in a fixed charge distributed in space $\rho=-\operatorname{div}\mathbf{P}(\mathbf{r})$ [8.9, Sec. 2.2] which serves as a source of an electric field whose potential $\Phi$ satisfies the Poisson equation

$$\nabla^2\Phi=-4\pi\rho=4\pi\operatorname{div}\mathbf{P}(\mathbf{r})$$

$$=i\frac{4\pi e^*}{\Omega_0}\frac{1}{\sqrt{N}}\sum_{\mathbf{q},j}{}'\mathbf{q}\left(\frac{\mathbf{e}_{j1}}{\sqrt{m_1}}-\frac{\mathbf{e}_{j2}}{\sqrt{m_2}}\right)a_j(\mathbf{q})\,e^{i\mathbf{q}\cdot\mathbf{r}}+\text{c.c.} \tag{6.16}$$

Adopting a procedure precisely similar to that used for solving equation (5.4.41), we obtain

$$\Phi=-i\frac{4\pi e^*}{\Omega_0\sqrt{N}}\sum_{\mathbf{q},j}{}'(\mathbf{q}_1\cdot\mathbf{h}_j)\,a_j(\mathbf{q})\,e^{i\mathbf{q}\cdot\mathbf{r}}+\text{c.c.}, \tag{6.17}$$

where the unit vector $\mathbf{q}_1=\mathbf{q}/q$ and $\mathbf{h}_j(\mathbf{q})=\frac{1}{q}\left(\frac{\mathbf{e}_{j1}(\mathbf{q})}{\sqrt{m_1}}-\frac{\mathbf{e}_{j2}(\mathbf{q})}{\sqrt{m_2}}\right)$.

The difference between expressions (6.17) and (5.4.42) is that in the case of long-wave optical vibrations of an ionic crystal only the longitudinal wave interacts with the electron, whereas in the case of acoustic vibrations of a piezoelectric crystal generally all three vibration branches interact with the electron ($\mathbf{h}_j\neq 0$ for all $j$).

The perturbation energy corresponding to the potential (6.17) is $-e\Phi$.

As we intend to demonstrate now, the first sum in (6.17) that contains the multiplier $a_j(\mathbf{q})\exp(i\mathbf{q}\cdot\mathbf{r})$ describes the processes of absorption by the conduction electron of a phonon $\hbar\omega_{\mathbf{q}j}$, therefore according to (3.11), the phonon $\hbar\omega_{\mathbf{q}j}$ absorption probability is

$$W_j^+(\mathbf{k},\mathbf{k}')$$

$$=\frac{2\pi}{\hbar}\,|\langle\mathbf{k}',N'_{\mathbf{q}j}\,|-e\Phi\,|\,\mathbf{k},N_{\mathbf{q}j}\rangle|^2\,\delta(\varepsilon_{\mathbf{k}'}-\varepsilon_{\mathbf{k}}-\hbar\omega_{\mathbf{q}j})$$

$$=\frac{32\pi^3e^{*2}e^2}{\hbar\Omega_0^2N}\left|\langle\mathbf{k}',N'_{\mathbf{q}j}\,|\sum_{\mathbf{q}}{}'(\mathbf{q}_1\mathbf{h}_j)\,a_j(\mathbf{q})\,e^{i\mathbf{q}\cdot\mathbf{r}}\,|\,\mathbf{k},N_{\mathbf{q}j}\rangle\right|^2\times$$

$$\times\,\delta(\varepsilon_{\mathbf{k}'}-\varepsilon_{\mathbf{k}}-\hbar\omega_{\mathbf{q}j}). \tag{6.18}$$

Making use of plane waves for the electron wave functions, as was done in (5.4.43), we obtain

$$\mathbf{k}' = \mathbf{k} + \mathbf{q}. \tag{6.19}$$

This corresponds to the process of absorption by an electron of a phonon with the wave vector **q**. Accordingly, the matrix elements $a_j(\mathbf{q})$ will be nonzero only for $N'_{\mathbf{q}j} = N_{\mathbf{q}j} - 1$, in which case they will be equal to (5.4.45):

$$\langle N'_{\mathbf{q}j} \mid a_j(\mathbf{q}) \mid N_{\mathbf{q}j} \rangle = \sqrt{\frac{\hbar N_{\mathbf{q}j}}{2\omega_{\mathbf{q}j}}}. \tag{6.20}$$

The c.c. member in (6.17) that contains the multiplier $a_j^*(\mathbf{q}) \exp(-i\mathbf{q}\cdot\mathbf{r})$ describes the process of emission by an electron of a phonon for which

$$\mathbf{k}' = \mathbf{k} - \mathbf{q}, \tag{6.19a}$$

and

$$\langle N'_{\mathbf{q}j} \mid a_j^*(\mathbf{q}) \mid N_{\mathbf{q}j} \rangle = \sqrt{\frac{\hbar (N_{\mathbf{q}j}+1)}{2\omega_{\mathbf{q}j}}}, \tag{6.20a}$$

if $N'_{\mathbf{q}j} = N_{\mathbf{q}j} + 1$.

For long-wave acoustic phonons $\omega_{\mathbf{q}j} = v_{\mathbf{q}j} q$, where $v_{\mathbf{q}j}$ is the sound velocity that depends on the direction of the vector **q**. The electron-phonon interaction (at all $T \gg 1$ K) is an elastic interaction (see Section 8.4), the δ-function in (6.18) is equal to (4.2), and, in accordance with (4.10b),

$$\langle N_{\mathbf{q}j} \rangle_{eq} \approx \langle N_{\mathbf{q}j} + 1 \rangle_{eq} = k_0 T / \hbar v_{\mathbf{q}j} q. \tag{6.21}$$

In this approximation the probability of phonon emission $W_j^-$ is equal to the absorption probability, so that the total probability of electron-phonon interaction $W_j = W_j^+ + W_j^- = 2W_j^+$. From (4.2)-(4.7), (6.18), (6.20) and (6.21) we obtain for the relaxation time for which the $j$-th branch is responsible

$$\frac{1}{\tau_j} = -\sum_{\mathbf{q}} 2W_j^+(\mathbf{k}, \mathbf{k}+\mathbf{q}) \frac{q_\chi}{k_\chi} = -\frac{e^{*2} e^2 m^* k_0 T}{\hbar^3 \Omega_0 k^2}$$
$$\times \int_{q_{min}}^{q_{max}} dq \int_0^{\pi} \sin\vartheta \, d\vartheta \int_0^{2\pi} d\varphi \left\{ \left[ \frac{|\mathbf{q}_1 \mathbf{h}_j|^2}{v_{\mathbf{q}j}^2} \right] \frac{\cos\alpha}{\cos\beta} \delta\left( \frac{q}{2k} + \cos\vartheta \right) \right\}, \tag{6.22}$$

where we took into account that $V/N = \Omega_0$.

General considerations and concrete calculations performed by Meijer and Polder [8.8] demonstrate that the multiplier $\left[\frac{|\mathbf{q}_1 \cdot \mathbf{h}_j|^2}{v_{\mathbf{q}j}^2}\right]$ is independent of the magnitude of the vector $\mathbf{q}$ being a function only of its direction, i.e., of the angles $\vartheta$ and $\varphi$. To simplify expression (6.12), carry the value of $\overline{\left[\frac{|\mathbf{q}_1 \cdot \mathbf{h}_j|^2}{v_{\mathbf{q}j}^2}\right]}$ averaged over the directions in space out of the integral; then integration with respect to $\varphi$ will yield

$$\int_0^{2\pi} \frac{\cos\alpha}{\cos\beta}\, d\varphi = 2\pi \cos\vartheta,$$

and integration with respect to $\vartheta$, as in the case of (4.7), will yield the multiplier $q/2k$. Finally, we obtain from (6.22)

$$\frac{1}{\tau_j} = \frac{\pi e^{*2} e^2 m^* k_0 T}{\hbar^3 \Omega_0 k^3} \overline{\left[\frac{|\mathbf{q}_1 \cdot \mathbf{h}_j|^2}{v_{\mathbf{q}j}^2}\right]} \int_0^{2k} q\, dq = \frac{2\pi e^{*2} e^2 m^*}{\hbar^3 \Omega_0} \overline{\left[\frac{|\mathbf{q}_1 \cdot \mathbf{h}_j|^2}{v_{\mathbf{q}j}^2}\right]} \frac{k_0 T}{k}. \tag{6.23}$$

Until the multiplier $\left[\frac{|\mathbf{q}_1 \cdot \mathbf{h}_j|^2}{v_{\mathbf{q}j}^2}\right]$ has been calculated, expression (6.23) presents interest only because of its dependence on the temperature $T$ and on the electron energy $\varepsilon = \hbar^2 k^2/2m^*$. This dependence is obviously the same for all three acoustic branches. Since $1/\tau_{ac} = \sum_{j=1}^{3} 1/\tau_j$, where $\tau_{ac}$ is the relaxation time of interaction with all acoustic branches, it follows that

$$\frac{1}{\tau_{ac}} \propto \frac{T}{k} \propto \frac{T}{\sqrt{\varepsilon}}. \tag{6.24}$$

We see the dependence of the relaxation time $\tau_{ac}$ on the temperature $T$ and on the electron energy $\varepsilon$ in a piezoelectric to be the same as that for the electron's interaction with the optical vibrations in ionic crystals of the NaCl type in the high-temperature range (6.7). In the paper cited above Meijer and Polder attempted to calculate $\mathbf{q}_1\mathbf{h}_j$ and $v_{\mathbf{q}j}$ by employing the theory of elasticity for piezoelectrics. They succeeded in expressing the values of those quantities in terms of elastic constants and the piezoelectric modulus only for the principal crystallographic directions in the cubic crystal (100), (110) and (111), using those values for the approximate calculation of the relaxation time $\tau_{ac}$. Substituting the numerical values of elastic constants and of the piezoelectric modulus for ZnS, they obtained

$$\frac{1}{\tau_{ac}} = 5.1 \times 10^{10} \sqrt{\frac{T}{x}}, \tag{6.25}$$

where $x = \varepsilon / k_0 T$.

The inverse relaxation time of the electron's interaction with optical vibrations in the low-temperature range (6.13) for the same material is

$$\frac{1}{\tau_{op}} = 5.2 \times 10^{13} e^{-538/T}. \tag{6.26}$$

Table 8.1 shows the values of the two inverse relaxation times for an electron with $\varepsilon = 3k_0T/2$ at different temperatures. We witness $1/\tau_{ac}$ to reach the same order of magnitude as $1/\tau_{op}$ at a temperature of 125 K with the scattering by piezoacoustic vibrations becoming increasingly predominant as the temperature is decreased.

Table 8.1

| $T$ K | $\frac{1}{\tau_{ac}} \times 10^{11}$, s$^{-1}$ | $\frac{1}{\tau_{op}} \times 10^{11}$, s$^{-1}$ |
|---|---|---|
| 250 | 6.6 | 60 |
| 200 | 5.9 | 53 |
| 150 | 5.1 | 14 |
| 125 | 4.5 | 7.0 |
| 100 | 4.2 | 2.4 |
| 75 | 3.6 | 0.39 |

## 8.7. Scattering of Conduction Electrons by Charged and Neutral Impurity Atoms

**8.7.1** At the end of Section 8.1 we considered from the point of view of classical mechanics the problem of charge carrier scattering by the impurity ions. We took the Coulomb potential in a medium with the dielectric constant $\varepsilon_0$ for the scattering potential (1.17). To guarantee the convergence of the integral (1.23) that determines the relaxation time, we performed the integration with respect to the scattering angle $\theta$ from $\theta_{min}$ corresponding to the maximum impact parameter $b_{max} = \frac{1}{2} n_I^{-1/3}$, where $n_I$ is the impurity ion concentration. It was pointed out that such a limitation of the impact parameter to the value equal to one-half of the average distance between neighbouring impurity ions reflects the screening of ionic charges by carriers of opposite sign. Here we shall present a more consistent calculation of the effect of such screening.

Let $n$ be the average (homogeneous) concentration of free electrons in a crystal, and $n'(\mathbf{r})$ their concentration in the impurity ion's field. If $\varphi$ is the potential field established by the positive ion charge $+e$ placed in the origin and by the negative charge of excess electrons $-e(n'-n)$, then Poisson's equation will assume the form

$$\nabla^2\varphi = \frac{4\pi e}{\varepsilon_0}(n'-n), \tag{7.1}$$

where $\varepsilon_0$ is the dielectric constant and

$$\lim_{r\to 0}\varphi = e/\varepsilon_0 r, \tag{7.1a}$$

since a charge $+e$ is placed in the origin.

According to (6.2.5)

$$n = \frac{(2m^* k_0 T)^{3/2}}{2\pi^2\hbar^3}\mathcal{F}_{1/2}(z), \tag{7.2}$$

where $\mathcal{F}_{1/2}(z)$ is the Fermi integral (6.2.6) with the index 1/2, and $z = \zeta/k_0T$ ($\zeta$ is the chemical potential). For a smoothly varying field $\varphi$ (Section 5.1) the electron potential energy $(-e\varphi)$ adds up with its energy $\varepsilon = \hbar^2k^2/2m^*$ in its distribution function, so that

$$f_0 = \left[\exp\left(\frac{\varepsilon - e\varphi - \zeta}{k_0T}\right)+1\right]^{-1}. \tag{7.3}$$

The latter expression is equivalent to one in which $\zeta + e\varphi$ is substituted for $\zeta$, therefore

$$n' = \frac{(2m^* k_0 T)^{3/2}}{2\pi^2\hbar^3}\mathcal{F}_{1/2}(z+u), \tag{7.4}$$

where $u = e\varphi/k_0T$.

If $e\varphi \ll \zeta$ (which is not always true), (7.4) can be expanded into a series in the powers of $u$, so that only the first member is retained. Then

$$n' = n + \frac{(2m^* k_0 T)^{3/2}}{2\pi^2\hbar^3}\mathcal{F}'_{1/2}(z)\,u \tag{7.5}$$

where

$$\mathcal{F}'_{1/2}(z) \equiv \frac{\partial \mathcal{F}_{1/2}(z)}{\partial z}.$$

Substituting this into (7.1), we obtain

$$\nabla^2\varphi = q^2\varphi, \tag{7.6}$$

where

$$q^2 = \frac{4\sqrt{2}}{\pi}\,\frac{e^2 m^{*3/2}(k_0T)^{1/2}}{\varepsilon_0\hbar^3}\,\mathcal{F}'_{1/2}\left(\frac{\zeta}{k_0T}\right). \tag{7.6a}$$

The solution of equation (7.6) possessing the property of spherical symmetry and satisfying the condition (7.1a) is of the form

$$\varphi = \frac{e}{\varepsilon_0 r} e^{-qr}. \tag{7.6b}$$

We can check this directly by substituting (7.6b) into (7.6), taking into account that in the spherically symmetrical case

$$\nabla^2 \varphi = \frac{1}{r^2} \frac{d}{dr} \left( r^2 \frac{d\varphi}{dr} \right).$$

For $r \approx 1/q = r_0$ the potential $\varphi$, in accordance with (7.6b), decreases substantially ($e$ times) and for this reason $r_0$ is termed the *screening radius.*

Find $r_0$ for the cases of weak and strong degeneracies of free electrons.

**A. Nondegenerate electrons** (Section 6.2.3).

In this case

$$\mathscr{F}'_{1/2}(z) = \frac{d}{dz} \int_0^\infty e^{z-x} x^{1/2} \, dx = \frac{\sqrt{\pi}}{2} e^z, \tag{7.7}$$

$$e^z = e^{\zeta/k_0 T} = \frac{4\pi^3 n \hbar^3}{(2\pi m^* k_0 T)^{3/2}}. \tag{7.7a}$$

It follows from (7.6a) and (7.7) that the screening radius

$$r_0 = \frac{1}{q} = \sqrt{\frac{\varepsilon_0 k_0 T}{4\pi e^2 n}} \tag{7.8}$$

coincides with the Debye radius (6.7.4a).

Writing (7.8) in the form

$$\frac{r_0}{d} = \sqrt{\frac{k_0 T}{4\pi e^2/\varepsilon_0 d}}, \tag{7.8a}$$

where $d = n^{-1/3}$ is the average spacing between the electrons, we see the order of magnitude of $r_0/d$ to be equal to the square root of the ratio of the electron thermal energy to the Coulomb energy of interaction of an electron pair separated by a distance $d$.

**B. Degenerate electrons** (Section 6.2.2).

In this case

$$\mathscr{F}'_{1/2}(z) = z^{1/2} = \frac{\pi \hbar}{(2m^* k_0 T)^{1/2}} \left( \frac{3n}{\pi} \right)^{1/3}. \tag{7.9}$$

Substituting (7.9) into (7.6a), we obtain for the screening radius

$$r_0 = \frac{1}{q} = \left[ \frac{\varepsilon_0 \hbar^2}{4m^* e^2} \left( \frac{\pi}{3n} \right)^{1/3} \right]^{1/2}. \tag{7.10}$$

In this case the order of magnitude of the screening radius is found from the condition

$$\frac{\hbar^2}{2m^* r_0^2} \approx \frac{e^2}{\varepsilon_0 d}, \tag{7.11}$$

i.e., the energy of an electron in a region whose linear dimensions are $r_0$ is of the order of the Coulomb energy of interaction of electrons separated by a distance $d$.

**8.7.2** According to (1.23a), the relaxation time is

$$\frac{1}{\tau} = 2\pi v n_I \int \sigma(\theta)(1-\cos\theta)\sin\theta\, d\theta, \tag{7.12}$$

where $v$ is the electron's velocity, $\sigma(\theta)$ is the differential scattering cross section, and $n_I$ is the impurity ion concentration.

To determine $\sigma(\theta)$, we shall make use of Born's approximation [8.10, § 108] of the quantum mechanics theory of scattering. In Born's method the electron is regarded as free, and the scattering is considered to be a small perturbation. This means that Born's method is applicable only to sufficiently fast electrons. When the free electron $k$ is scattered by a potential of the type (7.6b), its wave vector changes by a value of the order of $q$; therefore, Born's method is applicable only when $q \ll k$. For electrons with energies $\hbar^2 k^2/2m^* \approx k_0 T$ and $q$ determined by (7.8) or (7.10), Born's method will easily be seen to be applicable only at high enough temperatures.

The scattering cross section of the potential (7.6b) can easily be demonstrated to be equal in Born's approximation to [8.11, Chap. 9, Problem 1]

$$\sigma(\theta) = \left[\frac{e^2/\varepsilon_0}{m^* v^2(1-\cos\theta) + \dfrac{\hbar^2 q^2}{2m^*}}\right]^2. \tag{7.13}$$

For $q = 0$ the potential (7.6b) reduces to a purely Coulomb potential, and the cross section (7.13), to the Rutherford equation (1.20). Substituting (7.13) into (7.12) and introducing a new integration variable $t = \cos\theta$ ($dt = -\sin\theta\, d\theta$), we obtain as the result of elementary integration

$$\tau = \frac{\varepsilon_0^2 m^{*2} v^3}{2\pi e^4 n_I \Phi(\eta)} = \frac{\sqrt{2m^*}\,\varepsilon_0^2 \varepsilon^{3/2}}{\pi e^4 n_I \Phi(\eta)}, \tag{7.14}$$

where the energy $\varepsilon = m^{*2} v^2/2$, and

$$\Phi(\eta) = \ln(1+\eta) - \frac{\eta}{1+\eta}, \tag{7.14a}$$

is a slowly varying function of the argument

$$\eta = 4m^{*2} v^2/\hbar^2 q^2 = 8m^* \varepsilon/\hbar^2 q^2. \tag{7.14b}$$

Since $\Phi(\eta)$ is a slowly varying function, we can consider $\tau \propto \varepsilon^{3/2}$ to be an adequately accurate approximation, as in the case of the Conwell-Weisskopf equation (1.25).

**8.7.3** The charge carrier scattering by neutral impurity atoms can be approximately considered as the scattering of slow electrons of mass $m^*$ by the hydrogen atom immersed in a medium with the dielectric constant $\varepsilon_0$ [8.12].

When the velocity of the scattered electrons is small, the scattering is spherically symmetrical and only the zero phase should be taken into account in the quantum mechanics equations of the theory of scattering. Only a numerical calculation of the cross section is possible if we want to calculate the effects of electron exchange and of polarization of the scattering atom. The result obtained for the relaxation time is

$$\frac{1}{\tau}=\frac{20\varepsilon_0\hbar^3}{m^{*2}e^2}n_0, \qquad (7.15)$$

where $n_0$ is the concentration of neutral impurity atoms.

Note that the relaxation time for the scattering by neutral impurities is independent of the energy. As will be demonstrated below, this is tantamount to the independence of the charge carrier mobility of the temperature (provided, of course, $n_0$ is temperature-independent). Scattering by neutral atoms can be demonstrated to be comparable in some cases to scattering by impurity ions.

**8.7.4** Consider the problem of calculating relaxation time $\tau$ in the case of the simultaneous action of several scattering mechanisms.

For the sake of definiteness, suppose that the probability per unit time of a conduction electron going from state $\mathbf{k}$ to state $\mathbf{k}'$ as the result of its interaction with the acoustic vibrations is $W_L(\mathbf{k}, \mathbf{k}')$, and the probability of its being scattered by impurity ions is $W_I(\mathbf{k}, \mathbf{k}')$. If both processes are alternative, the total scattering probability per unit time will be $W(\mathbf{k}, \mathbf{k}') = W_L(\mathbf{k}, \mathbf{k}') + W_I(\mathbf{k}, \mathbf{k}')$.

In the case of elastic collisions we can introduce the relaxation time by using equation (2.12):

$$\frac{1}{\tau}=-\sum_{k'} W(\mathbf{k}, \mathbf{k}')\frac{\Delta k_\chi}{k_\chi}$$

$$=-\sum_{k'} W_L(\mathbf{k}, \mathbf{k}')\frac{\Delta k_\chi}{k_\chi}-\sum_{k'} W_I(\mathbf{k}, \mathbf{k}')\frac{\Delta k_\chi}{k_\chi}=\frac{1}{\tau_L}+\frac{1}{\tau_I}. \qquad (7.16)$$

Hence, in the case of simultaneous action of several scattering mechanisms in order to calculate the inverse resultant (effective) relaxation time, we should add up the inverse relaxation times corresponding to the individual scattering mechanisms.

In the general case

$$\frac{1}{\tau}=\sum_i \frac{1}{\tau_i}, \tag{7.17}$$

where $\tau_i$ is the relaxation time due to the $i$-th scattering mechanism. In all the cases considered in this chapter the relaxation time can be expressed in the form

$$\tau = \tau_0 \varepsilon^r, \tag{7.18}$$

where the power $r$ assumes different values for different scattering mechanisms. For example, in the case of scattering by acoustic vibrations and by impurity ions

$$\tau_L=\frac{\tau_{0L}}{k_0T}\,\varepsilon^{-1/2}, \quad \tau_I=\tau_{0I}\varepsilon^{3/2}, \tag{7.18a}$$

i.e., $r$ is equal to $-1/2$ and $3/2$, respectively.

According to (7.16), the effective relaxation time due to the combined action of both mechanisms is

$$\tau=\frac{\tau_{0L}\tau_{0I}\varepsilon^{3/2}}{\tau_{0L}+\tau_{0I}\,(k_0T)\,\varepsilon^2} \tag{7.19}$$

and no longer exhibits the simple form (7.18).

Note that if some scattering mechanism becomes inactive, its relaxation time $\tau\to\infty$. Hence, if the scattering by impurity ions is negligible, $\tau_{0I}\to\infty$, and it follows from (7.19) that

$$\tau=\frac{\tau_{0L}}{k_0T}\,\varepsilon^{-1/2}=\tau_L. \tag{7.19a}$$

# 9. Kinetic Processes (Transport Phenomena) in Semiconductors

## 9.1. Introduction

9.1.1 The kinetic processes, or transport phenomena, performed in solids by electrons and holes are very interesting and important both from a theoretical and practical point of view.

As was already noted in Section 8.1, the kinetic processes, in contrast to the states of statistical equilibrium, are affected by the interaction mechanisms operating in the system. This makes the theory of kinetic processes more complicated and less reliable than the theory of equilibrium systems. However, the study of kinetic processes makes it possible to clarify the interaction mechanisms operating in a system and to investigate properties that cannot be established in studies of equilibrium systems.

On the other hand, the practical uses of semiconductors are also based on the utilization of their electron and hole currents conduction properties, i.e., on transport phenomena. In modern devices employing semiconducting materials wide use is made not only of electric fields, but also of magnetic fields and temperature and charge carrier concentration gradients; hence, a practical need arises for the study of all the various kinetic processes: of electric and heat conductivities, of thermoelectric and galvanomagnetic phenomena, etc.

Because of the complex nature of the conditions affecting the course of kinetic processes, their study splits up into a great variety of cases.

First, already the number of principal kinetic effects is great. To enumerate the more important: electric and heat conductivities, thermoelectric power, the Thomson and Peltier effects, the Hall effect, the magnetoresistive effect, the longitudinal and transverse Nernst effects.

Second, in some cases we come up against several types of charge carriers (electrons, holes, heavy and light holes).

Third, different mechanisms of charge carrier scattering have to be taken into account: atomic lattice vibrations, ionized and neutral impurities, ionic lattice vibrations, etc.

Fourth, we have to distinguish between the cases in which the conduction electrons are in a degenerate state and in which they are in a nondegenerate state.

Fifth, a factor of great importance is the type of dependence of the electron (hole) energy on the wave vector **k**; the constant energy surfaces may turn out to be spheres, ellipsoids, corrugated surfaces (Section 4.15).

If the problem of the nonequilibrium phonon distribution "dragging" conduction electrons in thermoelectric and thermomagnetic phenomena and the problem of "hot" electrons (in strong electric fields) are added to the list, the total number of different cases that can be considered will exceed several hundred. It is not only impossible to consider all theoretically feasible cases but it is also useless. The evolution of semiconductor physics in the last 20 years has established the fact of extreme complexity and of great variety of their internal structure, and, consequently, it is *a priori* useless to try to describe all the cases that are known at present or may be discovered in the future.

We shall adopt the following procedure in treating transport phenomena in semiconductors:

(1) nondegenerate semiconductors with a simple energy-band pattern;

(2) semiconductors with a simple energy-band pattern in the case of charge carrier degeneracy;

(3) semiconductors of the germanium type;

(4) semiconductors with a spherical nonparabolical energy band (the Kane model);

(5) the phonon drag effect;

(6) quantum mechanics theory of galvanomagnetic and thermomagnetic phenomena.

**9.1.2** For the current of conduction electrons in a semiconductor to be zero, they must be in a state of thermodynamical equilibrium. To this end, not only their temperature $T$ should be everywhere the same, but their chemical potential $\zeta^* = \zeta - e\varphi$ as well, where $\zeta$ is the so-called electrochemical potential dependent on the concentration and the temperature of the electrons at a specified point in the semiconductor, $\varphi$ $(x, y, z)$ is the electrostatic potential at the same point, and $(-e\varphi$ is the electron energy).[1]

The electron current appears when $T$ or $\frac{\zeta}{e} - \varphi = \frac{\zeta^*}{e}$ vary in the semiconductor's volume. In the stationary case the electron

[1] The equality $\zeta^* =$ const. in conditions of statistical equilibrium is a corollary of general postulates of physical statistics (see, for example, [9.1, p. 219]). Up to now we have employed the electrochemical potential $\zeta$ under the term of simply chemical potential (for instance, in Section 6.2), and this practice will be continued.

current in an isotropic semiconductor in the linear approximation consists obviously of members proportional to $\nabla T$ and $\nabla\left(\frac{\zeta}{e}-\varphi\right)$, i.e., the expression for the dc current density is

$$\mathbf{j}=\sigma\nabla\left(\frac{\zeta}{e}-\varphi\right)-\beta\nabla T. \tag{1.1}$$

In a homogeneous semiconductor in isothermic conditions $\zeta$ is the same throughout its volume, and, therefore, $\nabla\left(\frac{\zeta}{e}-\varphi\right)=-\nabla\varphi=\mathbf{E}$, i.e., coincides with the intensity of the electrostatic field $\mathbf{E}$. If a nonhomogeneous semiconductor is in a state of statistical equilibrium ($\nabla T=0$, $\mathbf{j}=0$), electric fields with intensity $\mathbf{E}=-\nabla(\zeta/e)$ are established in it. Relationship (1.1) is of a localized character, i.e., all the quantities in (1.1) are, generally, functions of the coordinates of the chosen point $\mathbf{r}$.

In addition to the electric current density $\mathbf{j}$, consider the density of the "kinetic" or thermal energy flux $\mathbf{q}=\mathbf{w}-\left(\varphi-\frac{\zeta}{e}\right)\mathbf{j}$, where we subtracted from the total energy flux $\mathbf{w}$ the part connected with the "potential" energy carried by the electron flux. In the linear approximation

$$\mathbf{q}=\mathbf{w}-\left(\varphi-\frac{\zeta}{e}\right)\mathbf{j}=\gamma\nabla\left(\frac{\zeta}{e}-\varphi\right)-\varkappa\nabla T. \tag{1.2}$$

In the case of an anisotropic semiconductor the scalar kinetic coefficients $\sigma$, $\beta$, $\gamma$, $\varkappa$ turn into rank 2 tensors, and (1.1) and (1.2) take the form

$$j_i=\sum_k \sigma_{ik}\nabla_k\left(\frac{\zeta}{e}-\varphi\right)-\sum_k \beta_{ik}\nabla_k T, \tag{1.1a}$$

$$q_i=\sum_k \gamma_{ik}\nabla_k\left(\frac{\zeta}{e}-\varphi\right)-\sum_k \varkappa_{ik}\nabla_k T. \tag{1.2a}$$

Here the suffixes $i$, $k=1, 2, 3$ or $i$, $k=x, y, z$. The coefficients in (1.1a) and (1.2a) satisfy the symmetry relations or Onsager's principle [9.1, Chap. X, § 6]

$$\sigma_{ik}=\sigma_{ki}, \tag{1.3}$$

$$\varkappa_{ik}=\varkappa_{ki}, \tag{1.4}$$

$$\gamma_{ik}=T\beta_{ki}. \tag{1.5}$$

If the semiconductor is placed in a magnetic field, the kinetic coefficients become functions of the magnetic field intensity $\mathbf{H}$. The relationships (1.3)-(1.5) are generalized to cover this case as follows:

$$\sigma_{ik}(\mathbf{H}) = \sigma_{ki}(-\mathbf{H}), \tag{1.3a}$$

$$\varkappa_{ik}(\mathbf{H}) = \varkappa_{ki}(-\mathbf{H}), \tag{1.4a}$$

$$\gamma_{ik}(\mathbf{H}) = T\beta_{ki}(-\mathbf{H}). \tag{1.5a}$$

The relationship (1.5a), which makes it possible to determine the energy flux in an electric field instead of calculating the electric field in the presence of a temperature gradient, is especially useful.

In the case of the charge carriers' motion obeying the laws of classical mechanics the kinetic coefficients can in principle be determined from the kinetic equation. This will form the main subject of this chapter.

Note that the Onsager relations (1.3a)-(1.5a) are also valid in the case of the quantum mechanics theory of transport phenomena.

## 9.2. Determination of Nonequilibrium Distribution Function for Conduction Electrons in the Case of a Spherically Symmetric Band

**9.2.1** In the case of a spherically symmetrical energy band the electron energy

$$\varepsilon = \varepsilon(k) \tag{2.1}$$

depends on the magnitude of the wave vector $k = |\mathbf{k}|$, i.e., on the constant-energy surface of the sphere.

In this case the electron's velocity is

$$\mathbf{v} = \frac{1}{\hbar}\frac{\partial \varepsilon(k)}{\partial \mathbf{k}} = \frac{1}{\hbar}\frac{\partial \varepsilon(k)}{\partial k}\frac{\mathbf{k}}{k} = u(k)\,\mathbf{k}, \tag{2.2}$$

where

$$u(k) = \frac{1}{\hbar k}\frac{\partial \varepsilon(k)}{\partial k}. \tag{2.2a}$$

Hence, in this case $\mathbf{v} \parallel \mathbf{k}$.

In accordance with (8.2.9a) make the nonequilibrium distribution function equal to

$$f(\mathbf{k}) = f_0(\varepsilon) + f_1(\mathbf{k}), \quad \text{with} \quad f_1(\mathbf{k}) = -\frac{\partial f_0}{\partial \varepsilon}\chi(\varepsilon)\,\mathbf{k}, \tag{2.3}$$

where the equilibrium distribution function is

$$f_0(\varepsilon) = \frac{1}{\exp\dfrac{\varepsilon - \zeta}{k_0 T} + 1}. \tag{2.3a}$$

The form of the member $f_1(\mathbf{k})$ in (2.3) will be dealt with below.

We shall determine the correction $f_1(\mathbf{k})$ from the kinetic equation (8.2.6) and (8.2.11a):

$$\mathbf{v}\nabla_r f-\frac{e}{\hbar}\left(\mathbf{E}+\frac{1}{c}\,\mathbf{v}\times\mathbf{H}\right)\nabla_k f=-\frac{f_1(\mathbf{k})}{\tau(\varepsilon)}, \tag{2.4}$$

i.e., in the relaxation time $\tau(\varepsilon)$ approximation.

If we are interested in the currents proportional to $\mathbf{E}$ and $\nabla T$ (linear transport phenomena), the correction $f_1(\mathbf{k})$ should also be determined in the linear approximation in $\mathbf{E}$ and $\nabla T$. To this end, we must substitute $f\approx f_0$ on the left-hand side of equation (2.4). Hence,

$$\nabla_r f\approx\nabla_r f_0=\frac{\partial f_0}{\partial\varepsilon}\left\{\frac{\zeta-\varepsilon}{T}\nabla T-\nabla\zeta\right\}, \tag{2.5}$$

where we assume that local thermodynamical equilibrium is in force, i.e., $T=T(\mathbf{r})$ and $\zeta=\zeta(\mathbf{r})$.

On the other hand,

$$\nabla_k f\approx\nabla_k f_0=\frac{\partial f_0}{\partial\varepsilon}\frac{\partial\varepsilon}{\partial\mathbf{k}}=\frac{\partial f_0}{\partial\varepsilon}\hbar\mathbf{v}. \tag{2.6}$$

We see the addend containing the magnetic field on the left-hand side of (2.4) vanishing in the approximation $f\approx f_0$, because $(\mathbf{v}\times\mathbf{H})\cdot\mathbf{v}\equiv 0$. Hence, in order to take into account the magnetic field in equation (2.4), we must calculate $\nabla_k f$ in the next approximation:

$$\nabla_k f=\nabla_k\left[f_0-\frac{\partial f_0}{\partial\varepsilon}\boldsymbol{\chi}(\varepsilon)\,\mathbf{k}\right]$$
$$=\frac{\partial f_0}{\partial\varepsilon}\hbar\mathbf{v}-(\boldsymbol{\chi}\cdot\mathbf{k})\,\nabla_k\left(\frac{\partial f_0}{\partial\varepsilon}\right)-\frac{\partial f_0}{\partial\varepsilon}\nabla_k(\boldsymbol{\chi}\cdot\mathbf{k}). \tag{2.6a}$$

Since

$$\nabla_k\left(\frac{\partial f_0}{\partial\varepsilon}\right)=\left(\frac{\partial^2 f_0}{\partial\varepsilon^2}\right)\nabla_k\varepsilon=\left(\frac{\partial^2 f_0}{\partial\varepsilon^2}\right)\hbar\mathbf{v}, \tag{2.6b}$$

it follows that not only the first addend in (2.6a) but also the second one, vanishes when multiplied by $\mathbf{v}\times\mathbf{H}$.

To determine the third addend on the right-hand side of (2.6a), calculate the $k_x$ axis component of $\nabla_k(\boldsymbol{\chi}\cdot\mathbf{k})$:

$$\frac{\partial}{\partial k_x}(\boldsymbol{\chi}\cdot\mathbf{k})=\frac{\partial}{\partial k_x}(k_x\chi_x+k_y\chi_y+k_z\chi_z)=\chi_x+\left(k_x\frac{\partial\chi_x}{\partial k_x}+k_y\frac{\partial\chi_y}{\partial k_x}+k_z\frac{\partial\chi_z}{\partial k_x}\right)$$
$$=\chi_x+\left(k_x\frac{\partial\chi_x}{\partial\varepsilon}\frac{\partial\varepsilon}{\partial k_x}+k_y\frac{\partial\chi_y}{\partial\varepsilon}\frac{\partial\varepsilon}{\partial k_x}+k_z\frac{\partial\chi_z}{\partial\varepsilon}\frac{\partial\varepsilon}{\partial k_x}\right)$$
$$=\chi_x+\left(k_x\frac{\partial\chi_x}{\partial\varepsilon}+k_y\frac{\partial\chi_y}{\partial\varepsilon}+k_z\frac{\partial\chi_z}{\partial\varepsilon}\right)\hbar v_x.$$

Hence

$$\nabla_{\mathbf{k}} (\boldsymbol{\chi} \cdot \mathbf{k}) = \boldsymbol{\chi} + \left( k_x \frac{\partial \chi_x}{\partial \varepsilon} + k_y \frac{\partial \chi_y}{\partial \varepsilon} + k_z \frac{\partial \chi_z}{\partial \varepsilon} \right) \hbar \mathbf{v}. \tag{2.6c}$$

Substituting (2.5), (2.6a)-(2.6c) into (2.4), we obtain

$$\frac{\partial f_0}{\partial \varepsilon} \left\{ \frac{\zeta - \varepsilon}{T} \nabla T - \nabla \zeta - e\mathbf{E} - \frac{e}{\hbar c} \mathbf{H} \times \boldsymbol{\chi} \right\} \mathbf{v} = \frac{1}{\tau(\varepsilon)} \left( \frac{\partial f_0}{\partial \varepsilon} \right) \frac{1}{u(k)} \boldsymbol{\chi} \cdot \mathbf{v}. \tag{2.7}$$

On the left-hand side we made use of the cyclic property of the triple product [9.2, p. 214, equation (4)]: $(\mathbf{v} \times \mathbf{H}) \cdot \boldsymbol{\chi} = (\mathbf{H} \times \boldsymbol{\chi}) \cdot \mathbf{v}$; and on the right-hand side we substituted (2.3) for $f_1(\mathbf{k})$ making $\mathbf{k} = \mathbf{v}/u(k)$. Expression (2.7) justifies the form of (2.3) chosen for $f_1(\mathbf{k})$. Since $\mathbf{v}$ is arbitrary, it follows from (2.7) that for the conduction electrons

$$\boldsymbol{\chi}_n(\varepsilon) = -\tau_n(\varepsilon)\, u_n(k) \left\{ \frac{\varepsilon - \zeta}{T} \nabla T + \nabla(\zeta - e\varphi) - \frac{e}{\hbar c} \mathbf{H} \times \boldsymbol{\chi}_n \right\}. \tag{2.8}$$

Here $\varphi(\mathbf{r})$ is the electrostatic potential of the applied electric field $\mathbf{E} = -\nabla\varphi$, the suffix $n$ here and below means that the corresponding quantities refer to the negative charge carriers—electrons.

In Section 6.2.4 we saw that the statistical behavior of holes is equivalent to that of electrons if the holes are described by an energy $\varepsilon'$ and a chemical potential $\zeta'$ equal, respectively, to

$$\varepsilon' = \frac{\hbar^2 k'^2}{2m_p}, \tag{2.9}$$

$$\zeta' = -\varepsilon_G - \zeta, \tag{2.9a}$$

where $\varepsilon_G$ is the forbidden bandwidth. Then we obtain for the holes instead of (2.8)

$$\boldsymbol{\chi}_p(\varepsilon') = -\tau_p(\varepsilon')\, u_p(k') \left\{ \frac{\varepsilon' + \varepsilon_G + \zeta}{T} \nabla T - \nabla(\zeta - e\varphi) + \frac{e}{\hbar c} \mathbf{H} \times \boldsymbol{\chi}_p \right\}, \tag{2.10}$$

where for the positive holes the sign of the charge has been changed as compared with (2.8). The suffix $p$ indicates that the corresponding quantities characterize positive holes.

For (2.8) and (2.10) to determine nonequilibrium distribution functions of electrons and holes, the relaxation times $\tau_n$ and $\tau_p$ must greatly exceed the average lifetime of the electron (hole) determined by recombination processes.

When the magnetic field $\mathbf{H} = 0$, $\chi_n$ and $\chi_p$ follow directly from (2.8) and (2.10); when $\mathbf{H} \neq 0$, the corresponding equations must be solved for $\chi_n$ and $\chi_p$. It can be demonstrated (Appendix 23) that in this case

$$\chi_n = -\tau_n u_n \times \frac{\left\{\frac{\varepsilon-\zeta}{T}\nabla T+\nabla(\zeta-e\varphi)\right\}+\frac{e\tau_n u_n}{\hbar c}\left[\mathbf{H}\left\{\frac{\varepsilon-\zeta}{T}\nabla T+\nabla(\zeta-e\varphi)\right\}\right]}{1+\left(\frac{e\tau_n u_n}{\hbar c}\mathbf{H}\right)^2} + \frac{\left(\frac{e\tau_n u_n}{\hbar c}\right)^2\left(\left\{\frac{\varepsilon-\zeta}{T}\nabla T+\nabla(\zeta-e\varphi)\right\}\mathbf{H}\right)\mathbf{H}}{1+\left(\frac{e\tau_n u_n}{\hbar c}\mathbf{H}\right)^2} \tag{2.11}$$

and

$$\chi_p = -\tau_p u_p \times \frac{\left\{\frac{\varepsilon'+\varepsilon_G+\zeta}{T}\nabla T-\nabla(\zeta-e\varphi)\right\}-\frac{e\tau_p u_p}{\hbar c}\left[\mathbf{H}\left\{\frac{\varepsilon'+\varepsilon_G+\zeta}{T}\nabla T-\nabla(\zeta-e\varphi)\right\}\right]}{1+\left(\frac{e\tau_p u_p}{\hbar c}\mathbf{H}\right)^2} + \frac{\left(\frac{e\tau_p u_p}{\hbar c}\right)^2\left(\left\{\frac{\varepsilon'+\varepsilon_G+\zeta}{T}\nabla T-\nabla(\zeta-e\varphi)\right\}\mathbf{H}\right)\mathbf{H}}{1+\left(\frac{e\tau_p u_p}{\hbar c}\mathbf{H}\right)^2}. \tag{2.12}$$

These complicated expressions, as we shall see below, can be substantially simplified in particular cases, when the magnetic field $\mathbf{H}$ is zero or weak (the precise definition of a weak magnetic field is given below), when the temperature gradient $\nabla T = 0$, etc.

In the case of a simple electron energy band

$$\varepsilon = \hbar^2 k^2/2m_n, \tag{2.13}$$

therefore, in accordance with (2.2a),

$$u_n = \frac{1}{\hbar k}\frac{\partial\varepsilon}{\partial k} = \frac{\hbar}{m_n}, \tag{2.13a}$$

a similar expression for $u_p$ being valid in case of holes.

**9.2.2** Calculate the electric current of nonequilibrium electrons. It follows from (7.2.14) that the current density is

$$\mathbf{j}_n = -\frac{e}{4\pi^3}\int \mathbf{v} f_1(\mathbf{k})\, d^3k = -\frac{e}{4\pi^3}\int\left[-\frac{\partial f_0}{\partial\varepsilon}u_n(k)\right](\chi_n\cdot\mathbf{k})\,\mathbf{k}\, d^3k, \tag{2.14}$$

where we made use of (2.2) and (2.3); here $\chi_n$ is equal to (2.11). The integral on the right-hand side of (2.14) is equal in the polar coordinates (the polar axis $z \parallel \chi_n$) to

$$\int \left[ -\frac{\partial f_0}{\partial \varepsilon} u_n(k) \right] \chi_n k \cos\vartheta \{\mathbf{i}_0 k \sin\vartheta \cos\varphi + \mathbf{j}_0 k \sin\vartheta \sin\varphi + \mathbf{k}_0 k \cos\vartheta\} k^2\, dk \sin\vartheta\, d\vartheta\, d\varphi.$$

Here $\mathbf{i}_0$, $\mathbf{j}_0$, $\mathbf{k}_0$ are unit vectors pointing in the direction of the axes $x$, $y$, $z$. Integration of the first and second addends in traces with respect to $\varphi$ from 0 to $2\pi$ yields zero, therefore, the integration with respect to $\varphi$ and $\vartheta$ yields

$$\int_0^{2\pi} d\varphi \int_0^{\pi} \cos^2\vartheta \sin\vartheta\, d\vartheta = \frac{4\pi}{3}.$$

Hence, we obtain for (2.14)

$$\mathbf{j} = -\frac{e}{3\pi^2} \int_0^{\infty} \left[ -\frac{\partial f_0}{\partial \varepsilon} u_n(k) \right] \chi_n(k) k^4\, dk. \tag{2.15}$$

For the hole current, we must substitute in this expression $e$ for $-e$, the equilibrium distribution function for holes, for $f_0$, and $\chi_p$ for $\chi_n$ (2.12).

The energy $\varepsilon$ and $u_n(k)$ for electrons in the case of a simple energy band are equal to (2.13) and to (2.13a), respectively. In this case

$$\chi_n = -\frac{\hbar e}{m_n} \chi_n^*, \tag{2.16}$$

where $\chi_n$ is equal to (2.11); then

$$\chi_n^* = \frac{\tau_n \mathbf{P}_n + \gamma_n \tau_n^2 \mathbf{H} \times \mathbf{P}_n + \gamma_n^2 \tau_n^3 (\mathbf{H} \cdot \mathbf{P}_n) \mathbf{H}}{1 + (\gamma_n \tau_n H)^2} \tag{2.17}$$

with

$$\mathbf{P}_n = \frac{\varepsilon - \zeta}{eT} \nabla T + \nabla \left( \frac{\zeta}{e} - \varphi \right) \tag{2.17a}$$

and

$$\gamma_n = e/m_n c. \tag{2.17b}$$

Calculate the electric current set up by the "field" $\chi_n$ equal to (2.16). It follows from (6.2.11a) that for a nondegenerate semiconductor

$$\frac{\partial f_0}{\partial \varepsilon} = \frac{4\pi^3 \hbar^3 n}{(2\pi m_n)^{3/2} (k_0 T)^{5/2}} e^{-\varepsilon/k_0 T}. \tag{2.18}$$

Substituting this value into the integral (2.15), introducing in it with the aid of (2.13) a new integration variable $x = \varepsilon/k_0T$ instead of $k$, and making use of (2.16), we obtain

$$\mathbf{j}_n = \frac{ne^2}{m_n} \frac{4}{3\sqrt{\pi}} \int_0^\infty \chi_n^* e^{-x} x^{3/2}\, dx. \tag{2.19}$$

The following averaging symbol can conveniently be introduced

$$\langle \chi_n^* \rangle = \frac{4}{3\sqrt{\pi}} \int_0^\infty \chi_n^* e^{-x} x^{3/2}\, dx. \tag{2.20}$$

For $\mathbf{g}$ = const. in accordance with (A.7.12), we obtain

$$\langle \mathbf{g} \rangle = \mathbf{g} \frac{4}{3\sqrt{\pi}} \int_0^\infty e^{-x} x^{3/2}\, dx = \mathbf{g} \frac{4}{3\sqrt{\pi}} \Gamma\left(\frac{5}{2}\right) = \mathbf{g}. \tag{2.20a}$$

Hence, the current density is

$$\mathbf{j}_n = \frac{ne^2}{m_n} \langle \chi_n^* \rangle. \tag{2.21}$$

The electric current carried by holes is calculated in a similar way. If we were to introduce for the holes a vector $\chi_p^*$ connected with $\chi_p$ (2.12) by the relationship

$$\chi_p = \frac{\hbar e}{m_p} \chi_p^*, \tag{2.22}$$

we would obtain

$$\chi_p^* = \frac{\tau_p \mathbf{P}_p - \gamma_p \tau_p^2 \mathbf{H} \times \mathbf{P}_p + \gamma_p^2 \tau_p^3 (\mathbf{H} \cdot \mathbf{P}_p) \mathbf{H}}{1 + (\gamma_p \tau_p \mathbf{H})^2}, \tag{2.23}$$

where

$$\mathbf{P}_p = -\frac{\varepsilon' + \varepsilon_G + \zeta}{eT} \nabla T + \nabla \left(\frac{\zeta}{e} - \varphi\right), \tag{2.23a}$$

and

$$\gamma_p = e/m_p c. \tag{2.23b}$$

Since for the holes the integral (2.15) changes its sign, but the relationship between $\chi_p$ and $\chi_p^*$ (2.22) does not contain a minus sign as was the case with (2.16), the current density

$$\mathbf{j}_p = \frac{pe^2}{m_p} \langle \chi_p^* \rangle \tag{2.24}$$

is similar to (2.21). The total current carried by electrons and holes is

$$\mathbf{j} = \mathbf{j}_n + \mathbf{j}_p = \frac{ne^2}{m_n} \langle \chi_n^* \rangle + \frac{pe^2}{m_p} \langle \chi_p^* \rangle, \tag{2.25}$$

where $\chi_n^*$ and $\chi_p^*$ are equal to (2.17) and to (2.23), respectively.

## 9.3. Electrical Conductivity of Nondegenerate Semiconductors with a Simple Energy-Band Structure

Consider the electric current in a homogeneous $n$-type semiconductor in the absence of a temperature gradient and of a magnetic field ($\nabla T = 0$, $\nabla \zeta = 0$, $\mathbf{H} = 0$). In this case

$$\chi^* = \tau \mathbf{P} = \tau(-\nabla \varphi) = \tau \mathbf{E}, \tag{3.1}$$

where $\mathbf{E}$ is the electric field intensity.[2]

It follows from (2.21) and (3.1) that

$$\mathbf{j} = \frac{ne^2}{m} \langle \tau \rangle \mathbf{E} = \sigma \mathbf{E}, \tag{3.2}$$

where $\sigma$ is the conductivity. Hence, and from (6.7.7a) follows the expression for the electron mobility

$$\mu = \frac{\sigma}{en} = \frac{e}{m} \langle \tau \rangle. \tag{3.3}$$

If the relaxation time $\tau$ is independent of the energy $\varepsilon$, then it follows from (2.20a) that

$$\mu = e\tau/m. \tag{3.3a}$$

If we were to introduce the average carrier drift velocity $\mathbf{v}_{\mathrm{dr}}$ by the definition

$$\mathbf{j} = \sigma \mathbf{E} = en\mathbf{v}_{\mathrm{dr}},$$

we would obtain by comparison with (3.3)

$$\mu = v_{\mathrm{dr}}/E,$$

i.e., the mobility would be numerically equal to the drift velocity in an electric field $E$ of unit intensity (but the dimensionality of mobility does not coincide with that of velocity!).

In the preceding chapter we considered several electron scattering mechanisms in crystals for which the relaxation time was of the type

$$\tau = a\varepsilon^r = a(k_0T)^r x^r = \tau_0 x^r, \tag{3.4}$$

---

[2] To simplify notation we omit the suffix $n$ for the corresponding quantities in expressions (3.1)-(3.6).

where $x = \varepsilon/k_0T$ and $\tau_0 = a\,(k_0T)^r$ is independent of $x$. It follows from (3.4), (2.20) and from Appendix 7, i.2 that

$$\langle\tau\rangle = \tau_0 \frac{4}{3\sqrt{\pi}}\,\Gamma\left(\frac{5}{2}+r\right). \tag{3.4a}$$

It follows from (8.4.11), (8.7.14), (8.6.7), (8.6.13) that for electron scattering by:

(a) acoustic vibrations $\tau = a_A\,(k_0T)^{-3/2}x^{-1/2}$; (3.5a)

(b) impurity ions $\tau = a_I\,(k_0T)^{3/2}x^{3/2}$; (3.5b)

(c) optical vibrations in ionic crystals at temperatures above the Debye temperature $\tau = a_P\,(k_0T)^{-1/2}x^{1/2}$; (3.5c)

(d) and below the Debye temperature

$$\tau = a_{0P}\exp\left(\frac{\hbar\omega_0}{k_0T}\right). \tag{3.5d}$$

The dependence of the relaxation time $\tau$ on $T$ and on $x = \varepsilon/k_0T$ in the region of effective piezoelectric scattering is the same as in case (c). The values of $a_A$, $a_I$, $a_P$ and $a_{0P}$ can be determined by comparing expressions (3.5a)-(3.5d) with corresponding expressions in Chapter 8.

Substitute (3.5a)-(3.5d) into (3.3). Making use of Appendix 7, i.2, we obtain

$$\text{(a)}\quad \mu = \frac{4}{3\sqrt{\pi}}\,\frac{e}{m}\,\frac{a_A}{(k_0T)^{3/2}} \propto T^{-3/2}, \tag{3.6a}$$

$$\text{(b)}\quad \mu = \frac{8}{\sqrt{\pi}}\,\frac{e}{m}\,a_I\,(k_0T)^{3/2} \propto T^{3/2}, \tag{3.6b}$$

$$\text{(c)}\quad \mu = \frac{8}{3\sqrt{\pi}}\,\frac{e}{m}\,\frac{a_P}{(k_0T)^{1/2}} \propto T^{-1/2}, \tag{3.6c}$$

$$\text{(d)}\quad \mu = \frac{e}{m}\,a_{0P}\exp\left(\frac{\hbar\omega_0}{k_0T}\right) \propto \exp\left(\frac{\hbar\omega_0}{k_0T}\right). \tag{3.6d}$$

In the case of several scattering mechanisms acting simultaneously, the relaxation time to be substituted into (3.2) is determined from equation (8.7.17).

Comparing (8.4.11), with (3.6a), we obtain for the mobility due to scattering by acoustic vibrations

$$\mu = 3\,\frac{\sqrt{\pi}}{2}\,\frac{e\rho v_0^2\hbar^4}{C^2 m^{*5/2}\,(k_0T)^{3/2}}, \tag{3.7}$$

where the crystal's density is $\rho = M/\Omega_0$. Of course, $\rho$, $v_0$, $C$ and $m^*$ all depend on the temperature, but this dependence is a weak one, and, therefore, mainly $\mu \propto T^{-3/2}$.

The electric current carried by the holes is calculated from equation (2.24). The total current is

$$\mathbf{j} = e\,(n\mu_n + p\mu_p)\,\mathbf{E}, \tag{3.8}$$

where $n$ and $p$ are the electron and hole concentrations, and the hole mobility $\mu_p$ due, for instance, to their interaction with acoustic vibrations, is determined from expression (3.7), in which the interaction constant $C$ and the effective mass $m^*$ should be taken for holes.

A dependence closely resembling (3.7) is observed in $n$-Ge: $\mu_n \propto \propto T^{-1.66}$ $(100° < T \leqslant 280°)$. However, for $p$-Ge, $\mu_p \propto T^{-2.33}$ (in the same temperature interval). For $n$- and $p$-type silicon $\mu_n \propto \propto T^{-2.6}$ $(300° < T \leqslant 400°)$, and $\mu_p \propto T^{-2.3}$ $(150° < T < 400°)$. Various causes are responsible for such deviations from the theory, which for semiconductors with covalent bonds stipulates the law $\mu \propto T^{-3/2}$: additional scattering by optical lattice vibrations, intervalley transitions of scattered electrons, deviations of the band's shape from the parabolic, i.e., the dependence of the electron's effective mass on its energy.

As the temperature is decreased, the impurity ion scattering, which results in the dependence (3.6b): $\mu \propto T^{3/2}$ becomes the predominant scattering mechanism. The transition to a $T^{3/2}$ dependence is observed in experiment, for instance, in $n$-Ge.

In some cases various scattering mechanisms simultaneously play a vital part. For example, the mobility due to the interaction with acoustic vibrations decreases with the rise in temperature as $T^{-3/2}$, whereas the mobility due to impurity ion scattering increases as $T^{3/2}$. For this reason in the high- and low-temperature ranges we can neglect the impurity ion scattering and the interaction with the acoustic phonons, respectively, but both mechanisms will be vital in the intermediate range. The results obtained here are valid for any type of dependence of relaxation time $\tau$ on the energy, therefore, all the equations above remain in force, if the effective value (8.7.17) is substituted for $\tau$.

The relaxation time of an electron interacting simultaneously with acoustic vibrations and with impurity ions is, according to (7.19), equal to

$$\tau = \frac{3\sqrt{\pi}}{4}\,\frac{m^*}{e}\,\mu_A\,\frac{x^{3/2}}{x^2+\beta^2}, \tag{3.9}$$

where, as before, $x = \varepsilon/k_0T$, $\beta = \sqrt{6\mu_A/\mu_I}$, $\mu_A$ and $\mu_I$ are the mobilities (3.6a) and (3.6b), which we have introduced instead of $a_A$ and $a_I$. Note that $\beta^2 \propto T^{-3}$.

It follows from (3.3) and (3.9) that the mobility

$$\mu = \frac{3\sqrt{\pi}}{4}\,\mu_A \left\langle \frac{x^{3/2}}{x^2+\beta^2} \right\rangle, \tag{3.10}$$

i.e., that it is determined by the integral

$$\int_0^\infty \frac{x^3 e^{-x}\,dx}{x^2+\beta^2} = J(\beta) \tag{3.11}$$

dependent on the parameter β.

It can be demonstrated that [9.3]

$$J(\beta) = 1 + \beta^2 (\mathrm{Si}\,\beta \sin\beta + \mathrm{Ci}\,\beta \cos\beta), \tag{3.12}$$

where

$$\mathrm{Si}\,\beta = -\int_\beta^\infty \frac{\sin x}{x}\,dx, \quad \mathrm{Ci}\,\beta = -\int_\beta^\infty \frac{\cos x}{x}\,dx \tag{3.12a}$$

are the integral sine and cosine.[3]

Eventually the mobility μ (3.10) can be represented in the form of a function of temperature in the entire temperature range from high to low temperatures. In the limiting cases $\mu_I \to \infty$ and $\mu_A \to \infty$, the mobility μ, in accordance with (3.10), is equal to $\mu = \mu_A$ and $\mu = \mu_I$, respectively.

The effect of several scattering mechanisms acting at a time on other kinetic coefficients (thermoelectric power, galvanomagnetic coefficients, and thermomagnetic coefficients are discussed below in Sections 9.4, 9.5 and 9.6) can be considered in a similar way. The integrals appearing in such calculations are presented in the appropriate literature [9.5].

## 9.4. Thermoelectric Phenomena in Nondegenerate Semiconductors with a Simple Energy-Band Structure

**9.4.1** In the presence of a temperature gradient $\nabla T$ in a semiconductor several new phenomena take place in it known as *thermoelectric phenomena*. Since an increase in temperature is accompanied by an increase in the average energy and frequently also in the number of charge carriers, the temperature gradient causes a free charge carrier flux to flow in the direction of $-\nabla T$. In a broken circuit in the stationary state the current density is everywhere zero, which means that the resulting electric field **E** establishes at each point in the semiconductor a current that compensates the carrier flux proportional to $\nabla T$. The electromotive force appearing as the

---

[3] The tables are to be found, for example, in [9.4].

result in the circuit is termed *thermoelectric power* (*thermo-emf*). Since both the electrons and the holes diffuse in the semiconductor from the hot end to the cold end, the thermoelectric power of intrinsic semiconductors is generally smaller than that of extrinsic semiconductors. In metals the electron concentration and practically also the electron energy are independent of the temperature, and because of that the thermoelectric power of metals is smaller than this of semiconductors.

The thermoelectric power between points (1) and (2) cannot be calculated with the aid of a contour integral of the type

$$\int_{(1)}^{(2)} E_s\, ds = -\int_{(1)}^{(2)} \nabla_s \varphi\, dS = \varphi_1 - \varphi_2, \tag{4.1}$$

where $\varphi$ is the electrostatic potential, because there is a temperature-dependent contact potential difference on the boundary separating the semiconductor from the metal contact. Hence, an instrument will additionally measure the difference between the contact potentials of points (1) and (2), each of which are at different temperatures. To take account of this effect, the integration should be performed not with respect to $-\nabla\varphi$ as in (4.1), but with respect to $\nabla\left(\frac{\zeta}{e} - \varphi\right)$, where $\zeta$ is the chemical potential of the free charge carriers. Indeed, if it is assumed that the temperatures of the semiconductor and of the metal at the place of contact are equal and that in the vicinity of the boundary they are in a state of statistical equilibrium, then the total chemical potential $(\zeta - e\varphi)$, which includes the external field, should not change in the act of crossing the contact. The curve *a* in Fig. 9.1 depicts the course of the electrostatic potential $\varphi$ in a circuit consisting of the semiconductor *AB* and of the metal wires *OA* and *BC*, the voltage drop across which is almost zero. Since the temperatures of the contacts *A* and *B* ($T_1$ and $T_2$) are different, the contact potential jumps at *A* and *B* are also different, and because of this the potential difference $V_1 - V_2$ measured by a voltmeter will not be equal to the difference $\varphi_1 - \varphi_2$ calculated from (4.1).

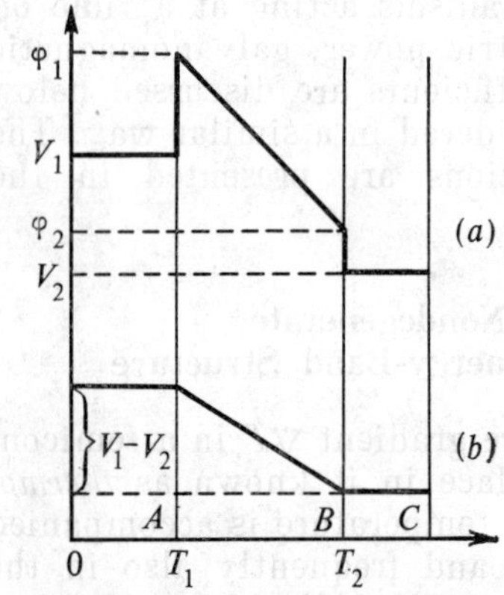

Fig. 9.1

Curve *b* depicts the course of the quantity $\varphi - \zeta/e$ which is continuous at points *A* and *B*. Since, on the other hand, $\zeta$ remains constant along each of the sections $OA$ and $BC$, the difference between the values of $\varphi - \zeta/e$ at points $A$ and $B$ is equal to $V_1 - V_2$ measured by the voltmeter.

Besides thermoelectric power there are two other thermoelectric phenomena of considerable interest.

In the presence of a temperature gradient in a current-carrying conductor, in addition to Joule's heat proportional to the square of the current $\mathbf{j}^2$, Thomson heat proportional to $(\mathbf{j}\cdot\nabla T)$ is liberated (or absorbed) in it.

Finally, the flow of current in an inhomogeneous conductor even in the absence of a temperature gradient is accompanied by the liberation or the absorption of heat (Peltier effect). This phenomenon takes place, for example, when current flows through a metal-semiconductor contact.

**9.4.2** In the absence of a magnetic field ($\mathbf{H} = 0$) and in the presence of a temperature gradient ($\nabla T \neq 0$, $\nabla\zeta \neq 0$) we obtain from (2.17)

$$\chi_n^* = \tau_n P_n = \tau_n \frac{\varepsilon - \zeta}{eT} \nabla T + \tau_n \nabla\left(\frac{\zeta}{e} - \varphi\right)$$
$$= \frac{k_0}{e}\left[x - \frac{\zeta}{k_0 T}\right]\tau_n \nabla T + \tau_n \nabla\left(\frac{\zeta}{e} - \varphi\right), \qquad (4.2)$$

where $x = \varepsilon/k_0T$. It follows from (2.21) that the electric current due to nonequilibrium carrier distribution (4.2) is equal to

$$\mathbf{j}_n = \frac{ne^2}{m_n}\langle\chi_n^*\rangle = \frac{ne^2}{m_n}\left\{\frac{k_0}{e}\left[\langle\tau_n x\rangle - \frac{\zeta}{k_0T}\langle\tau_n\rangle\right]\nabla T\right.$$
$$\left. + \nabla\left(\frac{\zeta}{e} - \varphi\right)\langle\tau_n\rangle\right\} = ne\mu_n\left\{\frac{k_0}{e}\left(g_n - \frac{\zeta}{k_0T}\right)\nabla T + \nabla\left(\frac{\zeta}{e} - \varphi\right)\right\}, \qquad (4.3)$$

where the electron mobility $\mu_n$ follows from expression (3.3) and

$$g_n = \frac{\langle\tau_n x\rangle}{\langle\tau_n\rangle}. \qquad (4.3a)$$

If the relaxation time $\tau_n \propto \varepsilon^r$, then

$$g_n = \frac{\Gamma(r+7/2)}{\Gamma(r+5/2)} = r + \frac{5}{2}. \qquad (4.3b)$$

For the four scattering mechanisms (3.5) the values of $g_n$ are:

(a) $g_n = 2$; (b) $g_n = 4$; (c) $g_n = 3$; (d) $g_n = 2.5$. (4.4)

The differential thermoelectric power $\alpha$ is defined as the ratio $\frac{\left|\nabla\left(\frac{\zeta}{e}-\varphi\right)\right|}{|\nabla T|}$ in an open circuit; from (4.3) we obtain for $\mathbf{j}_n = 0$

$$\alpha=\frac{\left|\nabla\left(\frac{\zeta}{e}-\varphi\right)\right|}{|\nabla T|}=\frac{k_0}{e}\left(g_n-\frac{\zeta}{k_0T}\right)$$
$$=\frac{k_0}{e}\left[g_n+\log\frac{2\,(2\pi m_n k_0T)^{3/2}}{nh^3}\right], \quad (4.5)$$

where $\zeta$ is taken from (6.2.15b) and $h = 2\pi\hbar$. Expression (4.5) for atomic semiconductors (with $g_n = 2$) was first obtained by N. L. Pisarenko in 1940.

For the holes we obtain from (2.23)

$$\chi_p^*=-\tau_p\frac{\varepsilon'+\varepsilon_G+\zeta}{eT}\nabla T+\tau_p\nabla\left(\frac{\zeta}{e}-\varphi\right). \quad (4.6)$$

Substituting this expression into (2.24), we obtain for the hole current

$$\mathbf{j}_p=-pe\mu_p\left\{\frac{k_0}{e}\left(g_p+\frac{\varepsilon_G+\zeta}{k_0T}\right)\nabla T-\nabla\left(\frac{\zeta}{e}-\varphi\right)\right\}, \quad (4.7)$$

where $g_p$ is similar to (4.3a).

The total current is

$$\mathbf{j}=\mathbf{j}_n+\mathbf{j}_p=ne\mu_n\left\{\nabla\left(\frac{\zeta}{e}-\varphi\right)+\frac{k_0}{e}\left(g_n-\frac{\zeta}{k_0T}\right)\nabla T\right\}$$
$$+pe\mu_p\left\{\nabla\left(\frac{\zeta}{e}-\varphi\right)-\frac{k_0}{e}\left(g_p+\frac{\varepsilon_G+\zeta}{k_0T}\right)\nabla T\right\}. \quad (4.8)$$

Making the total current $\mathbf{j} = 0$, we obtain for the differential thermoelectric power of an intrinsic semiconductor

$$\alpha=\frac{\left|\nabla\left(\frac{\zeta}{e}-\varphi\right)\right|}{|\nabla T|}=\frac{k_0}{e}\frac{1}{n\mu_n+p\mu_p}\left\{n\mu_n\left(g_n-\frac{\zeta}{k_0T}\right)\right.$$
$$\left.-p\mu_p\left(g_p+\frac{\varepsilon_G+\zeta}{k_0T}\right)\right\}$$
$$=\frac{k_0}{e}\frac{1}{n\mu_n+p\mu_p}\left\{n\mu_n\left[g_n+\log\frac{2\,(2\pi m_n k_0T)^{3/2}}{nh^3}\right]-\right.$$
$$\left.-p\mu_p\left(g_p+\log\frac{2\,(2\pi m_p k_0T)^{3/2}}{ph^3}\right)\right\}, \quad (4.9)$$

where $-\zeta$ and $\varepsilon_G + \zeta$ are taken from (6.2.15b), (6.2.15c). We see the contributions of electrons and holes to be of opposite signs; it follows, hence, that the thermoelectric power of an intrinsic semiconductor is, generally, less than that of an extrinsic one.

**9.4.3** Let us calculate the density of the energy flux transported by the electrons and holes:

$$\mathbf{w} = \int f_1^{(n)}(\mathbf{k}) \{\varepsilon - e\varphi\} \mathbf{v}_n \frac{d^3k}{4\pi^3} + \int f_1^{(p)}(k') \{\varepsilon' + \varepsilon_G + e\varphi\} \mathbf{v}_p \frac{d^3k'}{4\pi^3}. \quad (4.10)$$

The suffixes $n$ and $p$, as usual, mark the quantities relating to the electrons and holes. The expressions in the brackets in (4.10) take into account not only the kinetic energies of electrons and holes, but also their potential energies; the addend $\varepsilon_G$ means that the hole's energy $\varepsilon'$ is measured not from the bottom edge of the conduction band, but from the top edge of the valence band.

The $\mathbf{w}$ can be easily calculated from the nonequilibrium parts of the distribution functions of charge carriers in the flux $\mathbf{w}$ together with expressions (4.2) and (4.6) for $\chi^*(\varepsilon)$. The addends in the braces of (4.10) that are independent of $\varepsilon$ and $\varepsilon'$ contribute to $\mathbf{w}$ components proportional to $\mathbf{j}_n$ and $\mathbf{j}_p$. The addends $\varepsilon$ and $\varepsilon'$ require the computation of integrals with respect to $\tau x$ and $\tau x^2$ of the type (2.20). In the result we obtain[4]

$$\mathbf{w} = \varphi\mathbf{j} + \frac{\varepsilon_G}{e}\mathbf{j}_p - \frac{gk_0T}{e} n\mu_n \left[\nabla(\zeta - e\varphi) - \frac{\zeta - hk_0T}{T}\nabla T\right] + \frac{gk_0T}{e} p\mu_p \left[\nabla(\zeta - e\varphi) - \frac{\zeta + \varepsilon_G + hk_0T}{T}\nabla T\right], \quad (4.11)$$

where the equation for $\mathbf{j}_p$ is (4.7) and

$$h = \langle\tau x^2\rangle / \langle\tau_x\rangle. \quad (4.11a)$$

Substitute herein (and into $\mathbf{j}_p$) $\nabla(\zeta - e\varphi)$ from (4.8). Reducing the obtained addends to a common denominator, collecting similar terms associated with $n\mu_n$, $p\mu_p$, and their products, we obtain after lengthy although elementary computations

$$\mathbf{w} = \left(\varphi - \frac{\zeta}{e}\right)\mathbf{j} - \Pi\mathbf{j} - \varkappa\nabla T, \quad (4.12)$$

where the Peltier coefficient is

$$\Pi = T\frac{k_0}{e}\frac{1}{n\mu_n + p\mu_p}\left\{n\mu_n\left(g - \frac{\zeta}{k_0T}\right) - p\mu_p\left(g + \frac{\zeta + \varepsilon_G}{k_0T}\right)\right\}, \quad (4.12a)$$

and the heat conductivity coefficient is

$$\varkappa = n\mu_n(hg - g^2)\frac{k_0T}{e} + p\mu_p(hg - g^2)\frac{k_0^2T}{e} + \frac{n\mu_n p\mu_p}{n\mu_n + p\mu_p}\frac{(\varepsilon_G + 2gk_0T)^2}{eT}. \quad (4.12b)$$

[4] We make the assumption that $g_n = g_p \equiv g$ and $h_n = h_p \equiv h$ which is usually true.

If the relaxation time $\tau \propto \varepsilon^r$, the numerical value of the coefficient in (4.12b) is

$$hg - g^2 = g(h - g) = r + \frac{5}{2}, \tag{4.12c}$$

i.e., it coincides with $g$ from (4.3b). The value of this coefficient for various scattering mechanisms (3.5) is (4.4).

It follows from (4.12a) and (4.9) that

$$\Pi = \alpha T. \tag{4.13}$$

This is the so-called *Thomson relation* which, as we shall demonstrate now, is a corollary of the symmetry of the kinetic coefficients (1.5) and for this reason its validity is not limited to the case of the specific model of the semiconductor discussed above.

Making in (1.1) the current $\mathbf{j} = 0$, we obtain for the differential thermoelectric power (4.5)

$$\alpha = \frac{\left| \nabla \left( \frac{\zeta}{e} - \varphi \right) \right|}{|\nabla T|} = \frac{\beta}{\sigma}. \tag{4.14}$$

Substituting $\nabla \left( \frac{\zeta}{e} - \varphi \right)$ from (1.1) into (1.2), we obtain

$$\mathbf{w} = \left( \varphi - \frac{\zeta}{e} \right) \mathbf{j} + \frac{\gamma}{\sigma} \mathbf{j} + \left( \gamma \frac{\beta}{\sigma} - \varkappa \right) \nabla T.$$

Comparing this expression with (4.12), we obtain for the Peltier coefficient

$$\Pi = \gamma / \sigma. \tag{4.15}$$

The Thomson relation (4.13) follows immediately from (1.5) written for the scalar coefficients $\gamma$ and $\beta$ satisfying relations (4.14) and (4.15).

It follows from (4.12b) that in the case of bipolar conductivity, apart from the heat transported separately by electrons and holes, a fraction of the heat is due to the processes of generation—recombination of electron-hole pairs [the addend proportional to $np$ in (4.12b)]. In this case heat transport takes place as a result of the liberation of energy in the process of recombination of electrons and holes at the cold end, where their equilibrium concentrations are lower than in the hot part of the semiconductor. If the concentrations and the mobilities of electrons and holes are of the same order of magnitude, and if $\varepsilon_G \gg k_0 T$, then the heat conductivity due to electron-hole pairs will be $(\varepsilon_G / k_0 T)^2$ greater than the heat conductivity of electrons (holes).

In semiconductors the heat conductivity of the crystal lattice $\varkappa_0$ is frequently of the same order as the heat conductivity due to electrons (holes). In this case the member $-\varkappa_0 \nabla T$ should be added to the right-hand side of (4.12).

To obtain some interesting results from (4.12), let us take the divergence of both sides of (4.12) and take into account that in the stationary state div $\mathbf{w}$ = div $\mathbf{j}$ = 0; then

$$\operatorname{div}(-\varkappa\nabla T) = \mathbf{j}\cdot\left(\nabla\Pi + \nabla\left(\frac{\zeta}{e} - \varphi\right)\right). \tag{4.16}$$

Here we made use of the identity $\operatorname{div}(\psi\mathbf{a}) = \psi \operatorname{div}\mathbf{a} + \mathbf{a}\cdot\nabla\psi$. Applying expression (4.16) to an infinitesimal cylinder (Fig. 9.2) with bases parallel to the boundary separating two conductors and making use of Gauss' theorem and of an integral transformation of the gradient [9.6, Chap. II], we obtain

$$-\varkappa_1\left(\frac{\partial T}{\partial n}\right)_1 + \varkappa_2\left(\frac{\partial T}{\partial n}\right)_2 = (\Pi_1 - \Pi_2)\,j. \tag{4.17}$$

Here $\mathbf{n}$ is the normal to the boundary pointing in the direction of the second semiconductor; $\mathbf{j}$ is supposed to be parallel to $\mathbf{n}$.

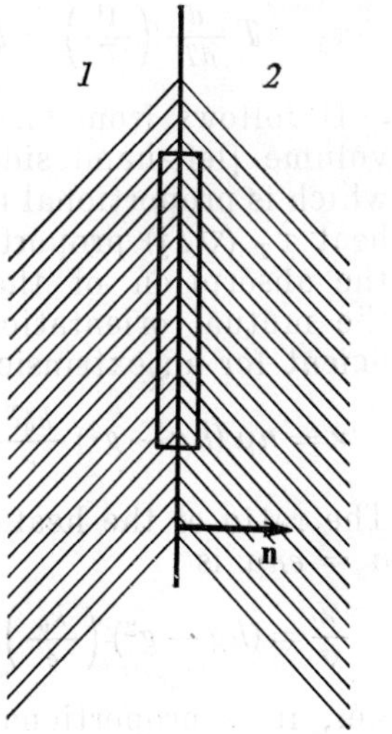


**Fig. 9.2**

In the derivation of (4.17) $j$ and $\left(\frac{\zeta}{e} - \varphi\right)$ on the boundary should be treated as continuous.

If $\left(\frac{\partial T}{\partial n}\right)_1$ and $\left(\frac{\partial T}{\partial n}\right)_2$ are of opposite signs, then the left-hand side of (4.17) will contain the sum of two heat fluxes flowing in opposite directions from the boundary (or to the boundary), and the right-hand side will contain the amount of heat liberated (or absorbed) per 1 s per 1 cm$^2$ of the boundary.

In the case of a semiconductor-metal contact the Peltier coefficient of the latter may be neglected.

Hence, the expression $(\Pi_1 - \Pi_2)\,j$ determines the liberation (absorption) of heat per 1 s per 1 cm$^2$ of the boundary between two conductors when a current $j$ flows through it.

From (4.8) and (4.12a) we obtain

$$\mathbf{j} = \sigma\nabla\left(\frac{\zeta}{e} - \varphi\right) + \frac{\sigma\Pi}{T}\nabla T. \tag{4.18}$$

Substituting $\nabla\left(\frac{\zeta}{e}-\varphi\right)$ from (4.18) into (4.16), we obtain

$$\operatorname{div}(-\varkappa\nabla T)=\frac{j^2}{\sigma}+\mathbf{j}\left(\nabla\Pi-\frac{\Pi}{T}\nabla T\right)=\frac{\mathbf{j}^2}{\sigma}+\tau_T(\nabla T\mathbf{j}), \tag{4.19}$$

where the *Thomson coefficient* $\tau_T$ can with the aid of relationship (4.13) be represented in the form

$$\tau_T=T\frac{d}{dT}\left(\frac{\Pi}{T}\right)=T\frac{d\alpha}{dT}. \tag{4.19a}$$

It follows from (4.19) that the heat liberated per 1 s in a unit volume [left-hand side of (4.19)] is equal to the Joule heat $j^2/\sigma$, which is proportional to the square of the current and to the Thomson heat $\tau_T$ $(\nabla T\mathbf{j})$ proportional to the current, both the liberation and the absorption of the Thomson heat are possible, depending on the mutual orientation of $\nabla T$ and $\mathbf{j}$. The heat conductivity coefficient for an extrinsic ($n$- or $p$-type) semiconductor is

$$\varkappa=n\mu(hg-g^2)\frac{k_0^2T}{e}. \tag{4.20}$$

The ratio of the heat conductivity $\varkappa$ to the electrical conductivity $\sigma=en\mu$ is

$$\frac{\varkappa}{\sigma}=(hg-g^2)\left(\frac{k_0}{e}\right)^2T=LT, \tag{4.21}$$

i.e., it is proportional to the absolute temperature $T$ (*the Wiedemann-Franz law*). In the case of a powerdependence of $\tau$ on the energy ($\tau\propto\varepsilon^r$) *the Lorenz number* is

$$L=(hg-g^2)\left(\frac{k_0}{e}\right)^2=\left(r+\frac{5}{2}\right)\left(\frac{k_0}{e}\right)^2. \tag{4.21a}$$

For various scattering mechanisms the factor $r+5/2$ in the Lorenz number assumes the values of (4.4).

Making use of (4.12a), we can easily write the expression for the Peltier coefficient $\Pi$ for an extrinsic (electron- or hole-type) semiconductor.

**9.4.4** Thermoelectric phenomena create the technical possibility of a direct conversion of thermal energy into electrical energy (thermoelectric power) and the possibility of cooling by means of electric current flowing through a contact between two conductors (Peltier effect).

The thermoelectric power of metals is small (at best, it is several ten μV/deg), and for this reason the technical uses of thermoelectric cells as effective generators of thermoelectric current became possible, only when it was discovered that the thermoelectric power of semiconductors reaches the value of several hundred μV/deg.

A. F. Ioffe in 1929 was the first to point out the prospects for using semiconductors in technical thermoelectric generators.

In the prewar years A. F. Ioffe and his staff succeeded in producing thermoelectric cells with an efficiency of 3%.

In the postwar years the work on the technical uses of thermoelectricity continued under A. F. Ioffe's guidance on a wider scale. At present such work is in progress throughout the world, but especially in the USSR, the USA and Great Britain.

We do not intend to discuss the problems of the technical uses of thermoelectric phenomena. Readers interested in the problem are referred to appropriate literature [9.7], [9.8], and [9.9].

## 9.5. Galvanomagnetic Phenomena in Nondegenerate Semiconductors with a Simple Energy-Band Structure

The term galvanomagnetic effects applies to phenomena involving currents flowing in the presence of an electric field **E** and a magnetic field **H**.

A theory can be developed for an arbitrary magnetic field, but in this case the kinetic coefficients cannot be expressed by elementary functions. For this reason we shall consider here the limiting cases of a weak and a strong magnetic field. We shall call a magnetic field weak if the dimensionless parameter in (2.11), (2.12) (for a simple band, when $u = \hbar/m^*$) is

$$\frac{e\tau u H}{\hbar c} = \frac{e\tau H}{cm^*} \approx \frac{\mu H}{c} \ll 1 \qquad (5.1)$$

Here the mobility $\mu \approx e\tau/m^*$, where $\tau$ is an appropriately averaged relaxation time (or just the relaxation time, if it is independent of the energy).

The criterion (5.1) can be expressed in a more visual way. Since $eH/cm^* = \omega_c$ is the cyclotron frequency (6.6.1a), $v\tau = l$ is the mean free path, and $cm^*v/eH = R$ is the radius of the circular electron orbit in a magnetic field, we obtain

$$\omega_c\tau = \frac{l}{R} \ll 1. \qquad (5.1a)$$

Hence, we shall regard a magnetic field as weak, if the electron's mean free path $l$ is much less than the radius of its circular orbit in the magnetic field, or if the free transit time $\tau$ is much less than $1/\omega_c = T_c/2\pi$, where $T_c$ is the period of the electron's revolution in this orbit.

**9.5.1** Consider first an extrinsic semiconductor with charge carriers of one type, for instance, with electrons. To simplify notation, we omit the suffix $n$ everywhere in expressions (5.2)-(5.18). In isothermic conditions

$$\mathbf{P}=\nabla\left(\frac{\zeta}{e}-\varphi\right)=\mathbf{E}, \tag{5.2}$$

as in the case of (3.1).

Let the magnetic field $\mathbf{H}$ be perpendicular to the electric field $\mathbf{E}$, then we obtain, in accordance with (2.17),

$$\chi^*=\frac{\tau\mathbf{E}+\gamma\tau^2\mathbf{H}\times\mathbf{E}}{1+(\gamma\tau H)^2}. \tag{5.3}$$

Applying the magnetic field $\mathbf{H}$ in the direction of the $z$ axis, and the electric field in the $xy$ plane, we obtain from (5.3) and (2.21)

$$\begin{aligned} j_x &= a_1E_x - a_2E_y,\\ j_y &= a_2E_x + a_1E_y, \end{aligned} \tag{5.4}$$

where

$$a_1=\frac{ne^2}{m}\left\langle\frac{\tau}{1+(\gamma\tau H)^2}\right\rangle, \tag{5.4a}$$

$$a_2=\frac{ne^2}{m}(\gamma H)\left\langle\frac{\tau^2}{1+(\gamma\tau H)^2}\right\rangle. \tag{5.4b}$$

The Hall effect is the appearance of an electric field perpendicular both to the current and to the magnetic field, which itself is perpendicular to the current. If the current flows in the $x$-axis direction so that $j_x = j$ and $j_y = 0$, then excluding $E_x$ from the equations (5.4), we obtain

$$E_y=-\frac{a_2}{a_1^2+a_2^2}j=RHj, \tag{5.5}$$

where by definition $R$ is the Hall constant. Hence,

$$R=-\frac{a_2}{(a_1^2+a_2^2)H}. \tag{5.5a}$$

In the absence of a magnetic field the current $\mathbf{j} = \sigma\mathbf{E} = \mathbf{E}/\rho$, where $\sigma$ and $\rho$ are, respectively, the conductivity and the resistivity. When a magnetic field $\mathbf{H} \parallel z$ is applied, the direction of the current $\mathbf{j} \parallel x$ no longer coincides with that of the electric field $\mathbf{E}$; therefore, the resistivity in the magnetic field is $\rho_{\mathbf{H}} = E_x/j = E_x/j_x$. The variation of resistivity in a magnetic field or *magnetoresistance* is

$$\frac{\rho_H-\rho}{\rho}\equiv\frac{\Delta\rho}{\rho}=\frac{E_x}{\rho j_x}-1=\frac{\sigma E_x}{j_x}-1. \tag{5.6}$$

Making $j_y=0$ in equations (5.4) and solving them for $E_x$, we obtain from (5.6)

$$\frac{\Delta\rho}{\rho}=\frac{\sigma a_1}{a_1^2+a_2^2}-1. \tag{5.6a}$$

If $\tau$ is independent of the energy, as is the case, for example, with ionic semiconductors in the low-temperature range, the averaging symbol $\langle\ \rangle$ in (5.4a) and (5.4b) can be dropped. It can easily be demonstrated that in this case in any magnetic field

$$\left.\begin{aligned} R &= -1/cen, \\ \Delta\rho/\rho &= 0. \end{aligned}\right\} \tag{5.7}$$

Consider the cases of weak and strong magnetic fields dependent on the electron's energy.

**A. Weak magnetic fields:** $\gamma\tau H \ll 1$ [condition (5.1)]. In the linear approximation in the magnetic field, in which only the terms of power not higher than the first are retained in the expansion in $H$, we obtain

$$a_1 = \frac{ne^2}{m}\langle\tau\rangle, \quad a_2 = \frac{ne^2}{m}(\gamma H)\langle\tau\rangle^2. \tag{5.8}$$

In the quadratic approximation, in which terms of the order of $H^2$ are retained, we have

$$a_1 = \frac{ne^2}{m}\{\langle\tau\rangle - (\gamma H)^2\langle\tau^3\rangle\}, \quad a_2 = \frac{ne^2}{m}(\gamma H)\langle\tau^2\rangle. \tag{5.9}$$

We obtain from (5.5a) and (5.6a) in the linear approximation in $H$ for $R$ and in the quadratic[5] approximation for $\Delta\rho/\rho$ $\left(\text{in the linear approximation in } H\ \frac{\Delta\rho}{\rho} = 0\right)$

$$R = -\frac{\langle\tau^2\rangle}{\langle\tau\rangle^2}\frac{1}{ecn}, \tag{5.10}$$

$$\frac{\Delta\rho}{\rho} = \left(\frac{eH}{mc}\right)^2 \frac{\langle\tau^3\rangle\langle\tau\rangle - \langle\tau^2\rangle^2}{\langle\tau\rangle^2}. \tag{5.11}$$

It follows immediately from those expressions that $R$ and $\Delta\rho/\rho$ for an energy-independent $\tau$ are equal to expressions (5.7).

If $\tau$ is a power function of energy (3.4), we can easily find the numerical coefficients in (5.10) and (5.11) making use of the definition (2.20). For various scattering mechanisms (3.5) we obtain

$$\text{(a)}\ R = -\frac{3\pi}{8}\frac{1}{ecn}; \quad \frac{\Delta\rho}{\rho} = \left(\frac{eH}{mc}\right)^2 \frac{\tau_0^2 L}{(k_0 T)^3}\left[1 - \frac{\pi}{4}\right]$$
$$= \frac{9\pi}{16}\left[1 - \frac{\pi}{4}\right]\left(\frac{\mu H}{c}\right)^2; \tag{5.12a}$$

---

[5] When calculating $\Delta\rho/\rho$, we should make use of the expansion $\frac{1}{a+bH^2} = \frac{1}{a}\frac{1}{1+\frac{b}{a}H^2} \approx \frac{1}{a}\left(1 - \frac{b}{a}H^2\right)$.

(b) $R = -\frac{315\pi}{512}\frac{1}{ecn}$; $\frac{\Delta\rho}{\rho} = \left(\frac{eH}{mc}\right)^2 \tau_{0I}^2 (k_0T)^3$

$$\times 120\left[\left(1-\frac{6615\pi}{32\,768}\right)\right] = \frac{15\pi}{8}\left[1-\frac{6615\pi}{32\,768}\right]\left(\frac{\mu H}{c}\right)^2, \quad (5.12b)$$

(c) $R = -\frac{45\pi}{128}\frac{1}{ecn}$; $\frac{\Delta\rho}{\rho} = \left(\frac{eH}{mc}\right)^2 \frac{\tau_{0p}^2}{(k_0T)^3}\left[1-\frac{75\pi}{256}\right]$

$$= \frac{27\pi}{64}\left[1-\frac{75\pi}{256}\right] = \left(\frac{\mu H}{c}\right)^2, \quad (5.12c)$$

(d) see (5.7).

We expressed the ratios $\Delta\rho/\rho$ in terms of mobilities (3.6a)-(3.6c). We see that in weak fields $\Delta\rho/\rho \propto H^2$. It follows from (5.10) that

$$|R|\sigma = |R|ne\mu = \frac{\langle\tau^2\rangle}{\langle\tau\rangle^2}\frac{1}{c}\mu, \quad (5.13)$$

where $\sigma$ is the conductivity, and $\mu$ is the mobility. Hence, in the case of charge carriers of one type we can find their concentration by measuring $R$ and their mobility, by measuring $R$ and $\sigma$.

Introduce the concept of the *Hall angle* $\theta$ between the directions of the current **j** and the field **E**:

$$\tan\theta \approx \theta = \left|\frac{E_y}{E_x}\right| = \left|\frac{a_2}{a_1}\right| = \frac{\langle\tau^2\rangle eH}{\langle\tau\rangle^2 mc} = \frac{\langle\tau^2\rangle}{\langle\tau\rangle^2}\frac{\mu H}{c}, \quad (5.14)$$

whence $\theta \approx \frac{\mu H}{c} \ll 1$.

**B. Strong magnetic fields:** $\gamma\tau H \gg 1$. Neglecting unity in comparison with $(\gamma\tau H)^2$, we obtain from (5.4a), (5.4b)

$$a_1 = \frac{ne^2}{m}\frac{1}{(\gamma H^2)}\left\langle\frac{1}{\tau}\right\rangle; \quad a_2 = \frac{ne^2}{m}\frac{1}{\gamma H}. \quad (5.15)$$

Here $\frac{a_1}{a_2} = \left\langle\frac{1}{\tau}\right\rangle\frac{1}{\gamma H} \ll 1$, so that $a_1^2 + a_2^2 \approx a_2^2$. We obtain from (5.5a) and (5.6a) making use of (5.15)

$$R = -1/ecn \quad (5.16)$$

$$\frac{\Delta\rho}{\rho} = \langle\tau\rangle\left\langle\frac{1}{\tau}\right\rangle - 1. \quad (5.17)$$

We see that in the case of strong magnetic fields the Hall constant is independent of the scattering mechanism, and the magnetoresistance $\Delta\rho/\rho$ reaches saturation.

For various scattering mechanisms (3.5) we have

(a) $\frac{\Delta\rho}{\rho} = \frac{32}{9\pi} - 1$, (b) $\frac{\Delta\rho}{\rho} = \frac{32}{3\pi} - 1$,

(c) $\frac{\Delta\rho}{\rho} = \frac{32}{9\pi} - 1$, (d) $\frac{\Delta\rho}{\rho} = 0$. (5.18)

Note that a magnetic field **H** applied parallel to the electric field **E** does not affect transport phenomena in a semiconductor with a standard band.

Indeed, in the case **H** ∥ **E** (**E** = **P**) it follows from (2.17) that

$$\chi^* = \frac{\tau\mathbf{E} + \gamma^2\tau^3 H^2 \mathbf{E}}{1+(\gamma\tau H)^2} = \tau\mathbf{E},$$

whence the magnetic field is seen not to affect the current.

Hence, in this case there is no Hall effect and no magnetoresistance.

**9.5.2** Consider now the galvanomagnetic effects in semiconductors with a mixed conductivity involving both electrons and holes.

In isothermic conditions ($\nabla T = 0$) $\mathbf{P}_n$ and $\mathbf{P}_p$ are equal to **E**. If **H** is as before perpendicular to **E**, $\chi_n^*$ is equal to (5.3)

$$\chi_n^* = \frac{\tau_n \mathbf{E} + \gamma_n \tau_n^2 \mathbf{H}\times\mathbf{E}}{1+(\gamma_n\tau_n H)^2}, \tag{5.19}$$

and $\chi_p^*$, following (2.23), is equal to

$$\chi_p^* = \frac{\tau_p \mathbf{E} - \gamma_p \tau_p^2 \mathbf{H}\times\mathbf{E}}{1+(\gamma_p\tau_p H)^2}. \tag{5.19a}$$

Here $\gamma_n$ and $\gamma_p$ are equal, respectively, to (2.17b) and (2.23b). The electric current density is given by (2.25)

$$\mathbf{j} = \mathbf{j}_n + \mathbf{j}_p = \frac{ne^2}{m_n}\langle\chi_n^*\rangle + \frac{pe^2}{m_p}\langle\chi_p^*\rangle. \tag{5.20}$$

Applying the magnetic field **H** in the direction of the $z$ axis and the electric field in the $xy$ plane, we obtain from (5.20), (5.19), and (5.19a) equations (5.4) in which

$$a_1 = \frac{ne^2}{m_n}\left\langle \frac{\tau_n}{1+(\gamma_n\tau_n H)^2}\right\rangle + \frac{pe^2}{m_p}\left\langle \frac{\tau_p}{1+(\gamma_p\tau_p H)^2}\right\rangle, \tag{5.21}$$

$$a_2 = \frac{ne^2}{m_n}(\gamma_n H)\left\langle \frac{\tau_n^2}{1+(\gamma_n\tau_n H)^2}\right\rangle - \frac{pe^2}{m_p}(\gamma_p H)\left\langle \frac{\tau_p^2}{1+(\gamma_p\tau_p H)^2}\right\rangle. \tag{5.21a}$$

We can, by analogy with the preceding item, consider the cases of a weak and a strong magnetic fields.

**A. Weak magnetic field** ($\gamma_n\tau_n H \ll 1$ and $\gamma_p\tau_p H \ll 1$). Determining $a_1$, $a_2$ and the ratio $a_2/(a_1^2+a_2^2)$ up to the terms of the first power in the magnetic field $H$, we obtain from (5.5a) for the Hall constant

$$R = -\frac{a_2}{(a_1^2+a_2^2)H} = -\frac{1}{ce}\,\frac{n\dfrac{e^2}{m_n^2}\langle\tau_n^2\rangle - p\dfrac{e^2}{m_p^2}\langle\tau_p^2\rangle}{\left[n\dfrac{e}{m_n}\langle\tau_n\rangle + p\dfrac{e}{m_p}\langle\tau_p\rangle\right]^2}. \tag{5.22}$$

In the case of an extrinsic semiconductor $p = 0$ (or $n = 0$), and (5.22) reduces to (5.10).

Making use of (3.3), we can express $R$ in terms of the mobilities

$$R = -\frac{1}{ce}\,\frac{\dfrac{\langle\tau_n^2\rangle}{\langle\tau_n\rangle^2}\,n\mu_n^2 - \dfrac{\langle\tau_p^2\rangle}{\langle\tau_p\rangle^2}\,p\mu_p^2}{[n\mu_n + p\mu_p]^2}. \tag{5.23}$$

If, for example, electrons and holes are scattered by acoustic vibrations (3.5a), then

$$\frac{\langle\tau_n^2\rangle}{\langle\tau_n\rangle^2} = \frac{\langle\tau_p^2\rangle}{\langle\tau_p\rangle^2} = \frac{8\pi}{3}, \tag{5.23a}$$

and

$$R = -\frac{8\pi}{3ce}\,\frac{n\mu_n^2 - p\mu_p^2}{(n\mu_n + p\mu_p)^2}. \tag{5.23b}$$

Since in an intrinsic semiconductor $n = p$, then

$$R = -\frac{8\pi}{3cen}\,\frac{\mu_n^2 - \mu_p^2}{(\mu_n + \mu_p)^2}. \tag{5.23c}$$

Hence, it follows that the Hall constant is small if $\mu_n$ is close to $\mu_p$.

Since the magnetoresistance $\Delta\rho/\rho$ in the approximation linear in the magnetic field is zero, we should determine $a_1$ and $a_2$ and calculate $\Delta\rho/\rho$ (5.6a) in the approximation quadratic in $H$, and obtain as a result of rather lengthy although elementary computations

$$\frac{\Delta\rho}{\rho} = \frac{1}{\sigma^2}\Big\{\Big(\frac{ne^2}{m_n}\Big)^2 (\gamma_n H)^2 [\langle\tau_n\rangle\langle\tau_n^3\rangle - \langle\tau_n^2\rangle^2]$$
$$+ \Big(\frac{pe^2}{m_p}\Big)^2 [(\gamma_p H)^2 \langle\tau_p\rangle\langle\tau_p^3\rangle - \langle\tau_p^2\rangle^2]$$
$$+ \Big(\frac{ne^2}{m_n}\Big)\Big(\frac{pe^2}{m_p}\Big)[(\gamma_p H)^2\langle\tau_n\rangle\langle\tau_p^3\rangle + (\gamma_n H)^2\langle\tau_p\rangle\langle\tau_n^3\rangle]\Big\}. \tag{5.24}$$

Note that in a weak magnetic field $\Delta\rho/\rho \propto H^2$. For an extrinsic semiconductor ($p = 0$) we obtain (5.11) from (5.24).

Making use of (3.3), we can express $\Delta\rho/\rho$ in terms of mobilities $\mu_n$ and $\mu_p$, as was done in (5.23). We shall not write down the general equation, but we would like to point out that in the case of scattering by acoustic vibrations

$$\frac{\Delta\rho}{\rho} = \frac{9\pi e}{16c^2}\,\frac{H^2}{\sigma}\Big[(n\mu_n^3 + p\mu_p^3) - \frac{\pi}{4}\,\frac{(n\mu_n^2 - p\mu_p^2)^2}{n\mu_n + p\mu_p}\Big]. \tag{5.25}$$

For an extrinsic semiconductor ($p = 0$, $\sigma = en\mu_n$) this expression reduces to (5.12a).

**B. Strong magnetic field** ($\gamma_n \tau_n H \gg 1$ and $\gamma_p \tau_p H \gg 1$).

In this case the appropriate equations should be expanded in the inverse powers of the magnetic field $1/H$. In the $1/H^2$ approximation we obtain from (5.21) and (5.21a)

$$a_1 = nm_n \frac{c^2}{H^2} \left\langle \frac{1}{\tau_n} \right\rangle + pm_p \frac{c^2}{H^2} \left\langle \frac{1}{\tau_p} \right\rangle, \tag{5.26}$$

$$a_2 = (n-p) \frac{ec}{H}. \tag{5.26a}$$

If $n \neq p$, then $\frac{a_1}{a_2} \approx \frac{1}{\gamma H} \left\langle \frac{1}{\tau} \right\rangle \ll 1$; therefore, $a_1^2 + a_2^2 \approx a_2^2$.

The Hall constant (5.5a) is

$$R = -\frac{a_2}{(a_1^2 + a_2^2) H} = -\frac{1}{a_2 H} = -\frac{1}{ec(n-p)}. \tag{5.27}$$

The magnetoresistance (5.6a) is

$$\frac{\Delta\rho}{\rho} = \frac{\sigma a_1}{a_2} - 1 = \frac{\sigma \left[ nm_n \left\langle \frac{1}{\tau_n} \right\rangle + pm_p \left\langle \frac{1}{\tau_p} \right\rangle \right]}{e^2 (n-p)^2} - 1, \tag{5.28}$$

i.e., it reaches saturation. For $p = 0$ and $\sigma = (e^2 n/m_n) \langle \tau_n \rangle$ (5.28) reduces to (5.17).

The case $n = p$ merits special consideration. It would appear at first glance that here $a_2 = 0$ (5.26a), $a_1 \neq 0$ (5.26), and the Hall constant $R = 0$. However, in the $1/H^2$ approximation not only $a_2 = 0$, but $(a_1^2 + a_2^2) H \propto 1/H^3 \approx 0$, as well; therefore in order to determine $R$ we have to compute $a_2$ and $(a_1^2 + a_2^2) H$ with an accuracy up to $1/H^3$. In this case

$$a_2 = -\frac{nc^3}{eH^3} \left[ m_n^2 \left\langle \frac{1}{\tau_n^2} \right\rangle - m_p^2 \left\langle \frac{1}{\tau_p^2} \right\rangle \right]. \tag{5.29}$$

Making use of this expression for $a_2$ and $a_1$ from (5.26), we obtain for the Hall constant in the $1/H^3$ approximation

$$R = -\frac{a_2}{(a_1^2 + a_2^2) H} = -\frac{1}{ecn} \frac{m_p^2 \langle 1/\tau_p^2 \rangle - m_n^2 \langle 1/\tau_n^2 \rangle}{[m_p \langle 1/\tau_p \rangle + m_n \langle 1/\tau_n \rangle]^2}. \tag{5.30}$$

Express $m_n$ and $m_p$ from the relationship $\mu = \frac{e}{m} \langle \tau \rangle$ in terms of the mobilities $\mu_n$ and $\mu_p$. If the scattering laws are the same for electrons and holes, $\langle \tau_n \rangle \langle 1/\tau_n \rangle = \langle \tau_p \rangle \langle 1/\tau_p \rangle$ and $\langle \tau_n \rangle^2 \langle 1/\tau_n^2 \rangle = \langle \tau_p \rangle^2 \langle 1/\tau_p^2 \rangle$, then

$$R = -\frac{\langle 1/\tau_n^2 \rangle}{\langle 1/\tau_n \rangle^2} \frac{1}{ecn} \frac{1-b}{1+b}, \tag{5.30a}$$

where $b = \mu_p/\mu_n$. If $\tau_n$ and $\tau_p$ are independent of the energy, the numerical coefficient in (5.30a) will be equal to unity. For scattering by acoustic vibrations $\langle 1/\tau_n^2\rangle \langle 1/\tau_n\rangle^2 = 45\pi/128$. The magnetoresistance (5.6a) in a strong magnetic field for $n = p$ can be calculated in the $1/H^2$ approximation in which $a_2 = 0$ and $a_1$ is equal to (5.26)

$$\frac{\Delta\rho}{\rho} = \frac{\sigma}{a_1} - 1 = \frac{\sigma H^2}{c^2 n\,[m_n \langle 1/\tau_n\rangle + m_p \langle 1/\tau_p\rangle]} - 1$$

$$= \left(\frac{eH}{c}\right)^2 \frac{\frac{1}{m_n}\langle\tau_n\rangle + \frac{1}{m_p}\langle\tau_p\rangle}{m_n \langle 1/\tau_n\rangle + m_p \langle 1/\tau_p\rangle}. \tag{5.31}$$

Here we neglected unity as compared with the first addend and made $\sigma = ne^2 \left[\frac{1}{m_n}\langle\tau_n\rangle + \frac{1}{m_p}\langle\tau_p\rangle\right]$.

Note that in a strong magnetic field for $n = p$ the magnetoresistance $\Delta\rho/\rho \propto H^2$, the same as in a weak magnetic field (5.24). If the scattering laws are the same for electrons and holes, we can proceed in the same way, as when we derived (5.30a) from (5.30), we obtain

$$\frac{\Delta\rho}{\rho} = \frac{1}{\langle\tau_n\rangle \langle 1/\tau_n\rangle} \left(\frac{\sqrt{\mu_n \mu_p}\, H}{c}\right)^2. \tag{5.31a}$$

When electrons and holes are scattered by acoustic vibrations, the numerical coefficient in (5.31a) is $9\pi/32$.

## 9.6. Thermomagnetic Phenomena in Nondegenerate Semiconductors with a Simple Energy-Band Structure

**9.6.1** If in addition to electric and magnetic fields there is also a temperature gradient in a semiconductor, transport phenomena taking place in it will be called *thermomagnetic*.

Consider first an extrinsic semiconductor with carriers of one type, for example, with electrons. In this case it follows from (2.17a) (the suffix $n$ has been omitted to simplify notation) that

$$\mathbf{P} = \frac{\varepsilon - \zeta}{eT} \nabla T + \nabla\left(\frac{\zeta}{e} - \varphi\right). \tag{6.1}$$

If the magnetic field $\mathbf{H}$ is perpendicular to the vector $\mathbf{P}$, we obtain from (2.17)

$$\chi^* = \frac{\tau \mathbf{P} + \gamma\tau^2 \mathbf{H} \times \mathbf{P}}{1 + (\gamma\tau H)^2}. \tag{6.2}$$

Apply the magnetic field **H** in the direction of the $z$ axis and the vectors $\nabla T$ and $\nabla\left(\frac{\zeta}{e}-\varphi\right)$ in the $xy$ plane. The $x$ and $y$ components of the current will be, by analogy with (5.4),

$$j_x = a_1\nabla_x\left(\frac{\zeta}{e}-\varphi\right)+b_1\nabla_x T - a_2\nabla_y\left(\frac{\zeta}{e}-\varphi\right)-b_2\nabla_y T,$$

$$j_y = a_2\nabla_x\left(\frac{\zeta}{e}-\varphi\right)+b_2\nabla_x T + a_1\nabla_y\left(\frac{\zeta}{e}-\varphi\right)+b_1\nabla_y T, \tag{6.3}$$

where $a_1$ and $a_2$ are determined via (5.4a) and (5.4b), respectively, and

$$b_1 = \frac{ne^2}{m}\left\langle\frac{\left[\left(\frac{k_0}{e}\right)x-\frac{\zeta}{eT}\right]\tau}{1+(\gamma\tau H)^2}\right\rangle, \tag{6.3a}$$

$$b_2 = \frac{ne^2}{m}(\gamma H)\left\langle\frac{\left[\left(\frac{k_0}{e}\right)x-\frac{\zeta}{eT}\right]\tau^2}{1+(\gamma\tau H)^2}\right\rangle. \tag{6.3b}$$

Here, as usual, $x = \varepsilon/k_0 T$.

The first example of thermomagnetic phenomena to be considered here will be the *Nernst-Ettingshausen effect*[6] consisting in the appearance of a transverse electric field $\nabla_y\left(\frac{\zeta}{e}-\varphi\right)$ in the absence of a current, $\mathbf{j}=0$, but in the presence of a temperature gradient, $\nabla_x T \neq 0$. Consider first the isothermic Nernst effect for which $\nabla_y T = 0$.

Making in equations (6.3) $j_x = j_y = \nabla_y T = 0$ and excluding from them $\nabla_x\left(\frac{\zeta}{e}-\varphi\right)$, we obtain

$$\nabla_y\left(\frac{\zeta}{e}-\varphi\right) = E_y = -\frac{a_1 b_2 - a_2 b_1}{a_1^2+a_2^2}\nabla_x T = -QH\nabla_x T, \tag{6.4}$$

where the *Nernst constant* is

$$Q = \frac{a_1 b_2 - a_2 b_1}{(a_1^2+a_2^2)H}. \tag{6.4a}$$

It follows from the expressions for the coefficients $a_i$ and $b_i$ (5.4a), (5.4b) and (6.3a), (6.3b), that if the relaxation time $\tau$ is independent of the electron energy $\varepsilon$, then $Q = 0$. As we shall see below, the sign of $Q$ depends on the type of the $\tau$ versus energy $\varepsilon$ dependence.

Consider now the Nernst effect in weak and strong magnetic fields.

---

[6] For the sake of brevity in the future it will be termed the *Nernst effect*.

**A. Weak fields:** $\gamma\tau H \ll 1$. In an approximation linear in the magnetic field the coefficients $a_1$ and $a_2$ are given by equations (5.8), and the coefficients $b_1$ and $b_2$ are, according to (6.3a), (6.3b) equal to

$$b_1 = \frac{ne^2}{m}\left\{\left(\frac{k_0}{e}\right)\langle\tau x\rangle - \frac{\zeta}{eT}\langle\tau\rangle\right\},$$

$$b_2 = \frac{ne^2}{m}(\gamma H)\left\{\left(\frac{k_0}{e}\right)\langle\tau^2 x\rangle - \frac{\zeta}{eT}\langle\tau^2\rangle\right\}, \tag{6.5}$$

whence

$$Q = \left(\frac{k_0}{e}\right)\frac{e}{mc}\frac{\langle\tau\rangle\langle\tau^2 x\rangle - \langle\tau^2\rangle\langle\tau x\rangle}{\langle\tau\rangle^2}. \tag{6.6}$$

If $\tau$ is of the form (3.4), then (see Appendix 7)

$$Q = \left(\frac{k_0}{e}\right)\frac{e}{mc}\tau_0\frac{\Gamma\left(2r+\frac{5}{2}\right)}{\Gamma\left(r+\frac{5}{2}\right)} = \frac{3\sqrt{\pi}}{4c} = \frac{\Gamma\left(2r+\frac{5}{2}\right)}{\Gamma^2\left(r+\frac{5}{2}\right)}r\left(\frac{k_0}{e}\right)\mu. \tag{6.6a}$$

Now we can determine the Nernst constant for various scattering mechanisms (3.5):

$$\text{(a)}\ Q = -\frac{\sqrt{\pi}}{4}\left(\frac{k_0}{e}\right)\frac{e}{mc}\frac{a_A}{(k_0T)^{3/2}} = -\frac{3\pi}{16c}\left(\frac{k_0}{e}\right)\mu, \tag{6.7a}$$

$$\text{(b)}\ Q = \frac{945\pi}{128}\left(\frac{k_0}{e}\right)\frac{e}{mc}a_I(k_0T)^{3/2} = \frac{945\pi}{1024c}\left(\frac{k_0}{e}\right)\mu, \tag{6.7b}$$

$$\text{(c)}\ Q = \frac{15\sqrt{\pi}}{32}\left(\frac{k_0}{e}\right)\frac{e}{mc}\frac{a_P}{(k_0T)^{1/2}} = \frac{45\pi}{256c}\left(\frac{k_0}{e}\right)\mu, \tag{6.7c}$$

$$\text{(d)}\ Q = 0. \tag{6.7d}$$

The sign of $Q$ coincides with that of the power of $r$ in (3.4). Therefore, $Q$ can change signs, if, as the temperature changes, one scattering mechanism is replaced by another.

**B. Strong fields:** $\gamma\tau H \gg 1$. Neglecting unity as compared with $(\gamma\tau H)^2$, we obtain from (6.3a), (6.3b)

$$b_1 = \frac{ne^2}{m}\frac{1}{(\gamma H)^2}\left\{\left(\frac{k_0}{e}\right)\left\langle\frac{x}{\tau}\right\rangle - \frac{\zeta}{eT}\left\langle\frac{1}{\tau}\right\rangle\right\},$$

$$b_2 = \frac{ne^2}{m}\frac{1}{\gamma H}\left\{\left(\frac{k_0}{e}\right)\langle x\rangle - \frac{\zeta}{eT}\right\}. \tag{6.8}$$

Substituting (6.8) and (5.15) into (6.4a), we obtain in the $1/H^2$ approximation

$$Q = \frac{1}{\gamma H^2}\left(\frac{k_0}{e}\right)\left\{\langle x\rangle\left\langle\frac{1}{\tau}\right\rangle - \left\langle\frac{x}{\tau}\right\rangle\right\}. \tag{6.9}$$

Note, that (3.4a) leads to $\langle x\rangle = 5/2$.

It can easily be demonstrated that in the case of electron scattering by acoustic vibrations (3.5a)

$$Q=-\frac{4}{3\sqrt{\pi}}\left(\frac{k_0}{e}\right)\frac{mc}{eH^2}\frac{(k_0T)^{3/2}}{a_A}=-\frac{16}{9\pi}\left(\frac{k_0}{e}\right)\frac{c}{\mu H^2}. \quad (6.10)$$

$Q$ for other scattering mechanisms is determined in a similar way.

Consider now the variation of the thermoelectric power in a magnetic field. Making again in equations (6.3) $j_x = j_y = 0$, $\nabla_y T = 0$, but this time excluding $\nabla_y\left(\frac{\zeta}{e}-\varphi\right)$, we obtain for the differential thermoelectric power in a magnetic field

$$\alpha_H=\frac{\left|\nabla_x\left(\varphi-\frac{\zeta}{e}\right)\right|}{|\nabla_x T|}=\frac{a_1b_1+a_2b_2}{a_1^2+a_2^2}. \quad (6.11)$$

For $H = 0$, $a_2 = b_2 = 0$, and the thermoelectric power

$$\alpha_{H=0}=\left.\frac{b_1}{a_1}\right|_{H=0}=\left(\frac{k_0}{e}\right)\left\{\frac{\langle\tau x\rangle}{\langle\tau\rangle}-\frac{\zeta}{k_0T}\right\}=\alpha \quad (6.11a)$$

coincides with (4.5).

It can easily be demonstrated that in an approximation linear in the magnetic field the variation of the thermoelectric power $\Delta\alpha = \alpha_H - \alpha = 0$. Hence, to compute $\Delta\alpha$, we have to determine $b_1$ and $b_2$ in an approximation quadratic in the magnetic field. It follows from (6.3a), (6.3b) that

$$b_1=\frac{ne^2}{m}\left\{\left(\frac{k_0}{e}\right)\langle\tau x\rangle-\frac{\zeta}{eT}\langle\tau\rangle-(\gamma H)^2\left[\left(\frac{k_0}{e}\right)\langle\tau^3x\rangle-\frac{\zeta}{eT}\langle\tau^3\rangle\right]\right\}, \quad (6.12)$$

$$b_2=\frac{ne^2}{m}(\gamma H)\left\{\left(\frac{k_0}{e}\right)\langle\tau^2x\rangle-\frac{\zeta}{eT}\langle\tau^2\rangle\right\}. \quad (6.12a)$$

Making use of the footnote on p. 519, we obtain from (6.12), (6.12a), (5.9), (6.11) in the approximation quadratic in the magnetic field

$$\Delta\alpha=\left(\frac{k_0}{e}\right)(\gamma H)^2\frac{\langle\tau\rangle\langle\tau^3\rangle\langle\tau x\rangle+\langle\tau\rangle\langle\tau^2\rangle\langle\tau^2x\rangle-\langle\tau^2\rangle\langle\tau^3x\rangle-\langle\tau^2\rangle^2\langle\tau x\rangle}{\langle\tau\rangle^3} \quad (6.13)$$

For $\tau$ independent of the electron energy $\Delta\alpha = 0$ (this can also be demonstrated for the case of an arbitrary magnetic field).

For interaction with the acoustic vibrations

$$\Delta\alpha=\left(\frac{k_0}{e}\right)\left(\frac{eH}{mc}\right)^2\frac{\tau_A^2}{(k_0T)^3}\left(1-\frac{\pi}{8}\right)=\frac{9\pi}{16}\left(1-\frac{\pi}{8}\right)\left(\frac{k_0}{e}\right)\left(\frac{\mu H}{c}\right)^2. \quad (6.14)$$

For scattering by impurity ions

$$\Delta\alpha = -\left(\frac{k_0}{e}\right)\left(\frac{eH}{mc}\right)^2 \tau_I^2 (k_0T)^3\, 340\left[1-\frac{297\,675\pi}{2\,785\,280}\right]$$
$$\approx -50.0\left(\frac{k_0}{e}\right)\left(\frac{\mu H}{c}\right)^2. \qquad (6.15)$$

It is worth noting that in the case (6.14) $\Delta\alpha$ rises and in the case (6.15) drops with an increase in the magnetic field.

In the case of strong magnetic fields, when unity can be neglected in comparison with $(\gamma\tau H)^2$, the coefficients $a_i$ and $b_i$ are determined via (5.15) and (6.8), respectively. Substituting these values into (6.11) and neglecting the quantities $1/H^4$ in comparison with $1/H^2$ and the quantities $1/H^2$ in comparison with the members independent of $H$, we observe $\alpha$ to reach saturation as $H \to \infty$:

$$\alpha_\infty = \left(\frac{k_0}{e}\right)\left\{\frac{5}{2}-\frac{\zeta}{k_0T}\right\}. \qquad (6.16)$$

The most interesting characteristic of $\alpha_\infty$ is that it is independent of the scattering mechanism. Hence, and from (4.5), we obtain

$$\Delta\alpha = \alpha_\infty - \alpha = \left(\frac{k_0}{e}\right)\left\{\frac{5}{2}-\frac{\langle\tau x\rangle}{\langle\tau\rangle}\right\}. \qquad (6.17)$$

The maximum variations of $\alpha$ in the cases of scattering by acoustic vibrations and by impurity ions are, respectively,

$$\Delta\alpha = \frac{1}{2}\left(\frac{k_0}{e}\right) \quad \text{and} \quad \Delta\alpha = -\frac{3}{2}\left(\frac{k_0}{e}\right). \qquad (6.17a)$$

Find additionally the variation of the electron heat conductivity in a transverse magnetic field.

The thermal energy flux is, by analogy with (4.10), equal to

$$\mathbf{w} = -\frac{en}{m}\langle\chi^*(\varepsilon - e\varphi)\rangle = \varphi\frac{ne^2}{m}\langle\chi^*\rangle$$
$$-\frac{en}{m}\langle\chi^*\varepsilon\rangle = \varphi\mathbf{j} - \frac{en}{m}k_0T\langle\chi^* x\rangle, \qquad (6.18)$$

where $\chi^*$ is equal to (6.2); here we made use of the fact that the electron flux is equal to $\mathbf{j}/(-e)$, where the electric current $\mathbf{j}$ is equal to (2.21).

If the temperature gradient is directed along the $x$ axis, then $\nabla_y T = 0$; but because of the magnetic field not only $w_x$ but also $w_y$ are nonzero.

Making in equations (6.3) $j_x = j_y = \nabla_y T = 0$, we can find from them $\nabla_x\left(\frac{\zeta}{e} - \varphi\right)$ and $\nabla_y\left(\frac{\zeta}{e} - \varphi\right)$ in terms of $\nabla_x T$, and substitute the results into the heat flux component $w_x$ which then will be proportional to $\nabla_x T$. The ratio $w_x/(-\nabla_x T)$ is equal to the heat conductivity coefficient of electrons in a magnetic field $\varkappa(H)$. In cases of weak and strong magnetic fields the heat conductivity can be expressed by elementary functions. We shall not write out these elementary, but lengthy computations, citing only the result for the case of a weak field, when the relaxation time is expressed by (3.4):

$$\varkappa(H)=n\mu\left(r+\frac{5}{2}\right)\frac{k_0^2T}{e}\left\{1-a_r\left(\frac{\mu H}{c}\right)^2\right\}, \tag{6.19}$$

where

$$a_r=\frac{9\pi}{16\left(r+\frac{5}{2}\right)}\left[\left(4r^2+3r+\frac{5}{2}\right)\frac{\Gamma\left(3r+\frac{5}{2}\right)}{\Gamma^3\left(r+\frac{5}{2}\right)}-r^2\frac{\Gamma^2\left(2r+\frac{5}{2}\right)}{\Gamma^4\left(r+\frac{5}{2}\right)}\right]. \tag{6.19a}$$

We see the correction to the heat conductivity in a weak magnetic field to be proportional to $H^2$.

In the case of scattering by acoustic vibrations $r=-1/2$, and, therefore,

$$\varkappa(H)=n\mu\frac{2k_0^2T}{e}\left\{1-\frac{9\pi}{16}\left(1-\frac{\pi}{32}\right)\left(\frac{\mu H}{c}\right)^2\right\}. \tag{6.19b}$$

The correction is equal to about $1.6\left(\frac{\mu H}{C}\right)^2$. $\varkappa(H)$ for other scattering mechanisms can be found in a similar way.

**9.6.2** We have found the Hall constant $R$ (5.5a), the magnetoresistance $\Delta\rho/\rho$ (5.6a), the Nernst constant $Q$ (6.4a), the variation of the thermoelectric power in a magnetic field $\Delta\alpha$ (6.13), and the heat conductivity coefficient $\varkappa(H)$ on the condition that $\nabla_y T=0$, i.e., when the temperature in the transverse direction is constant. The appropriate coefficients for this case have been called *isothermic*. However, in some cases the experimental conditions correspond rather to the assumption on the absence of a transverse heat flux $w_y=0$; the coefficient in this case is called *adiabatic*.

The calculation of the adiabatic coefficients should be based on equations (6.3) and equation (6.18) for

$$w_y=0=\varphi j_y-\frac{en}{m}k_0T\langle\chi_y^*x\rangle. \tag{6.20}$$

For instance, to find the adiabatic Hall constant $R_{ad}$ we should make $j_x=j$, $j_y=0$, and $\nabla_x T=0$; next we should express $\nabla_x\left(\frac{\zeta}{e}-\varphi\right)$ and $\nabla_y T$ in terms of $\nabla_y\left(\frac{\zeta}{e}-\varphi\right)$ from equation (6.20) and equation $j_y=0$ (6.3). Substituting them into the first equation (6.3) with

$j_x = j$, we obtain from it in the approximation linear in the magnetic field

$$\frac{\nabla_y\left(\frac{\zeta}{e}-\varphi\right)}{Hj} = R_{\text{ad}} = R_{\text{is}}\left(1+\frac{1}{4}\frac{e}{k_0}\alpha\right), \tag{6.21}$$

where $\alpha$ is the differential thermoelectric power and the isothermic Hall constant $R_{\text{is}}$ is equal to (5.5a). If the term $-\varkappa_0\nabla T$ connected with lattice heat conductivity is added to the heat flux **w**, equation (6.21) will, as can easily be checked, assume the form

$$R_{\text{ad}} = R_{\text{is}}\left(1+\frac{\frac{1}{4}\frac{e}{k_0}\alpha}{1+\frac{\varkappa_0}{\varkappa_e}}\right), \tag{6.21a}$$

where $\varkappa_e$ is the electron heat conductivity coefficient. Making, for instance, for tellurium $\varkappa_e = (2k_0^2T/e)\,n\mu$, $n = 4\times 10^{16}$ cm$^{-3}$, $\mu = 10^3$ cm$^2$/V·s, $\varkappa_0 = 4\times 10^{-3}$ cal/deg, we find $R_{\text{ad}}$ to be 0.4% greater than $R_{\text{is}}$, i.e., the correction for the adiabatic case to be small. In the same way we can determine the adiabatic Nernst constant ($w_y = 0$, $\nabla_y T \neq 0$). For a weak magnetic field it turns out to be

$$Q_{\text{ad}} = Q_{\text{is}}\left(1+\frac{\frac{7}{4}\frac{e}{k_0}\alpha}{1+\frac{\varkappa_0}{\varkappa_e}}\right), \tag{6.22}$$

i.e., the correction is in this case seven times as great as in the preceding case.

A similar procedure can be employed to calculate the adiabatic coefficients of other galvano- and thermomagnetic phenomena both for weak and strong magnetic fields.

**9.6.3** The general expressions for the Hall constant $R$ (5.5a), the magnetoresistance $\Delta\rho/\rho$ (5.6a), the Nernst constant $Q$ (6.4a) and the thermoelectric power in a magnetic field $\alpha_H$ (6.11) are valid for an arbitrary magnetic field. Hence, in the case of an arbitrary magnetic field the problem reduces to the computation of the coefficients $a_1$, $a_2$, $b_1$ and $b_2$ from the general equations (5.4a), (5.4b) and (6.3a), (6.3b). The calculation of, for example, the coefficients $a_1$ and $a_2$ from (5.4a), (5.4b) in the case of scattering by acoustic vibrations (3.5a) reduces to the computation of the integrals

$$\int_0^\infty \frac{x^2e^{-x}\,dx}{h+x}, \quad \int_0^\infty \frac{x^{3/2}e^{-x}\,dx}{h+x}, \tag{6.23}$$

functions of the parameter

$$h=\frac{\tau_A^2}{(k_0T)^3}\left(\frac{eH}{mc}\right)^2, \tag{6.23a}$$

which itself is a function of the magnetic field $H$ and of the temperature $T$.

The integrals (6.23) can be expressed in terms of the transcendental functions of $h$: of the probability integral and of the integral exponential function. We can also easily plot a graph of the integrals' dependence on $h$ from numerical data (or draw up a table). Similar integrals appear also in the computation of the coefficients $b_1$ and $b_2$, as well as in cases of other scattering mechanisms [9.10]. If we know the dependence of the coefficients $a_i$ and $b_i$ on $H$ and $T$, we can determine the dependence of the quantities $R$, $\Delta\rho/\rho$, $Q$, $\alpha_H$ and $\varkappa(H)$ on the magnetic field and the temperature.

**9.6.4** For mixed-conductivity semiconductors consider only the Nernst effect in a weak magnetic field.

In the approximation linear in $H$ the coefficients $a_1$, $a_2$, $b_1$, $b_2$ are, according to (5.8) and (6.5), equal to:

$$a_1=\frac{ne^2}{m_n}\langle\tau_n\rangle+\frac{pe^2}{m_p}\langle\tau_p\rangle, \tag{6.24a}$$

$$a_2=\frac{ne^2}{m_n}(\gamma_nH)\langle\tau_n^2\rangle-\frac{pe^2}{m_p}(\gamma_pH)\langle\tau_p^2\rangle, \tag{6.24b}$$

$$b_1=\frac{ne^2}{m_n}\left[\left(\frac{k_0}{e}\right)\langle\tau_nx\rangle-\frac{\zeta}{eT}\langle\tau_n\rangle\right]-\frac{pe^2}{m_p}\left[\left(\frac{k_0}{e}\right)\langle\tau_px\rangle+\right.$$
$$\left.+\frac{\varepsilon_G+\zeta}{eT}\langle\tau_p\rangle\right], \tag{6.24c}$$

$$b_2=\frac{ne^2}{m_n}(\gamma_nH)\left[\left(\frac{k_0}{e}\right)\langle\tau_n^2x\rangle-\frac{\zeta}{eT}\langle\tau_n^2\rangle\right]$$
$$+\frac{pe^2}{m_p}(\gamma_pH)\left[\left(\frac{k_0}{e}\right)\langle\tau_p^2x\rangle+\frac{\varepsilon_G+\zeta}{eT}\langle\tau_p^2\rangle\right]. \tag{6.24d}$$

In the same approximation in the magnetic field the Nernst constant is

$$Q=\frac{a_1b_1-a_2b_2}{(a_1^2+a_2^2)H}=\left[\frac{ne^2}{m_n}\langle\tau_n\rangle+\frac{pe^2}{m_p}\langle\tau_p\rangle\right]^{-2}$$
$$\times\left\{\left(\frac{ne^2}{m_n}\right)^2\gamma_n\left(\frac{k_0}{e}\right)[\langle\tau_n^2x\rangle\langle\tau_n\rangle-\langle\tau_nx\rangle\langle\tau_n^2\rangle]\right.$$
$$+\left(\frac{pe^2}{m_p}\right)^2\gamma_p\left(\frac{k_0}{e}\right)[\langle\tau_p^2x\rangle\langle\tau_p\rangle-\langle\tau_px\rangle\langle\tau_p^2\rangle]$$
$$+\left(\frac{ne^2}{m_n}\right)\left(\frac{pe^2}{m_p}\right)\left[\gamma_p\left(\frac{k_0}{e}\right)(\langle\tau_p^2x\rangle\langle\tau_n\rangle+\langle\tau_p^2\rangle\langle\tau_nx\rangle)\right.$$
$$+\gamma_n\left(\frac{k_0}{e}\right)(\langle\tau_n^2x\rangle\langle\tau_p\rangle+\langle\tau_n^2\rangle\langle\tau_px\rangle)$$
$$\left.\left.+\frac{\varepsilon_G}{eT}(\gamma_n\langle\tau_n^2\rangle\langle\tau_p\rangle+\gamma_p\langle\tau_p^2\rangle\langle\tau_n\rangle)\right]\right\}. \tag{6.25}$$

The braces in this expression contain three addends proportional to $n^2$, $p^2$ and $np$, respectively. Making the hole concentration $p = 0$, we obtain from (6.25) expression (6.6).

If the dependence of $\tau_n$ and of $\tau_p$ on the energy is described by a power law (3.4), we can, making use of (3.4a), express all the angle brackets in (6.25) in terms of the Γ-function. Next, making use of (3.3), we can, instead of $e/m_n$ and $e/m_p$, introduce the corresponding mobilities $\mu_n$ and $\mu_p$. To avoid making the discussion too extensive, we shall present the result only for the case of electron and hole scattering by acoustic vibrations

$$Q=-\frac{3\pi}{8ce}\left[\frac{k_0}{2}\frac{n\mu_n^3+p\mu_p^3}{(n\mu_n+p\mu_p)^2}-\frac{\varepsilon_G+\frac{7}{2}k_0T}{T}\frac{n\mu_n p\mu_p(\mu_n+\mu_p)}{(n\mu_n+p\mu_p)^2}\right]. \tag{6.25a}$$

## 9.7. Transport Phenomena in Semiconductors with a Simple Energy Band in the Case of Arbitrary Degeneracy

**9.7.1** Generalize some of the results in Sections 9.3-9.6 for the case of an arbitrary degeneracy of the charge carriers. Substitute in the expression (2.15) for the electron current density (2.13a) for $u_n$ and (2.16) for $\chi_n$ and transform the integral with the aid of (2.13) from the variable $k$ to $\varepsilon$. Then[7]

$$\mathbf{j}=\frac{2\sqrt{2}\,m^{1/2}e^2}{3\pi^2\hbar^3}\int_0^\infty \chi^*(\varepsilon)\left(-\frac{\partial f_0}{\partial\varepsilon}\right)\varepsilon^{3/2}\,d\varepsilon. \tag{7.1}$$

Write this expression in the form of (2.21)

$$\mathbf{j}=\frac{ne^2}{m}\langle\chi^*\rangle, \tag{7.1a}$$

where the meaning of the averaging symbol $\langle\ldots\rangle$ this time is as follows:

$$\langle\chi^*\rangle=\frac{1}{\zeta_0^{3/2}}\int_0^\infty \chi^*(\varepsilon)\left(-\frac{\partial f_0}{\partial\varepsilon}\right)\varepsilon^{3/2}\,d\varepsilon=\frac{1}{z_0^{3/2}}\int_0^\infty \chi^*(x)\left(-\frac{\partial f_0}{\partial x}\right)x^{3/2}\,dx. \tag{7.1b}$$

Here $\zeta_0$ is the Fermi energy (6.2.8a), $z_0 = \zeta_0/k_0T$ and $x = \varepsilon/k_0T$. In the case of arbitrary degeneracy the equilibrium distribution

[7] In all the expressions in this section the suffix $n$ will be omitted.

function is

$$f_0 = \frac{1}{\exp\frac{\varepsilon-\zeta}{k_0T}+1} = \frac{1}{e^{x-z}+1}, \tag{7.1c}$$

where $z = \zeta/k_0T$. In this sense (7.1b) is a generalization of expression (2.19) to which (7.1b) reduces, if the Boltzmann distribution (6.2.11a) is substituted for $f_0$.

It follows from (7.1b) for $\chi^* = \mathbf{a} = \text{const.}$ that

$$\langle \mathbf{a} \rangle = \mathbf{a}. \tag{7.1d}$$

We can easily demonstrate it, if we calculate the integral (7.1b) by parts and make use of the normalizing condition (6.2.5). If $\tau$ depends on the energy $\varepsilon$ of the type (3.4), the averaging procedure in equality (7.1a) in the limiting cases of weak and strong magnetic fields always reduces to the averaging of the energy powers, i.e.,

$$\langle \varepsilon^s \rangle = \zeta_0^{-3/2} \int_0^\infty \varepsilon^{s+3/2} \left(-\frac{\partial f_0}{\partial \varepsilon}\right) d\varepsilon, \tag{7.2}$$

$$\langle x^s \rangle = z_0^{-3/2} \int_0^\infty x^{s+3/2} \left(-\frac{\partial f_0}{\partial x}\right) dx. \tag{7.2a}$$

If $k_0T/\zeta_0 \ll 1$, (7.2) in the first ($\delta$-function) and second approximations in the degree of degeneracy is equal, respectively, to

$$\langle \varepsilon^s \rangle = \zeta_0^s, \tag{7.2b}$$

$$\langle \varepsilon^s \rangle = \zeta_0^s \left\{1 + \frac{\pi^2}{6} s \left(s+\frac{3}{2}\right) \left(\frac{k_0T}{\zeta_0}\right)^2\right\}, \tag{7.2c}$$

where we made use of equation (A.17.5) and of the expression for $\zeta$ (6.2.9).

Integrating (7.2a) by parts, we obtain

$$\langle x^s \rangle = z_0^{-3/2} (s + {}^3/_2)\, \mathscr{F}_{s+1/2}(z), \tag{7.2d}$$

where $z = \zeta/k_0T$, and the Fermi integral $\mathscr{F}_{s+1/2}$ is determined by (6.2.6).

The energy flux density can be obtained from (7.1) if the total electron energy $\varepsilon - e\varphi$, where $\varphi$ is the potential of the electric field acting on the electrons, is substituted for its charge $-e$ (under the integral sign). Hence, the energy flux density is

$$\mathbf{w} = -\frac{ne}{m} \langle \chi^* (\varepsilon - e\varphi) \rangle = -\frac{ne}{m} \langle \chi^* \varepsilon \rangle + \varphi \mathbf{j}. \tag{7.3}$$

If in the conditions of measurement the electric current is zero, $\mathbf{w}$ is equal to the first term on the right-hand side of (7.3).

9.7.2 The electric current in the absence of a temperature gradient and of a magnetic field is determined from (7.1a), where $\chi^*$, according to (3.1), is equal to $\tau\mathbf{E}$. Hence,

$$\mathbf{j}=\frac{ne^2}{m^*}\langle\tau\rangle\,\mathbf{E}, \tag{7.4}$$

whence the conductivity

$$\sigma=\frac{ne^2}{m^*}\langle\tau\rangle. \tag{7.5}$$

If $\tau$ is equal to (3.4), then

$$\langle\tau\rangle=a\,\langle\varepsilon^r\rangle=a\,(k_0T)^r\,\langle x^r\rangle. \tag{7.6}$$

Making use of (7.2d), we obtain in the case of arbitrary degeneracy

$$\sigma=\frac{2\sqrt{2}}{3\pi^2}\,\frac{e^2m^{1/2}}{\hbar^3}\,a\,(k_0T)^{r+3/2}\,(r+3/2)\,\mathcal{F}_{r+1/2}(z), \tag{7.7}$$

where $z=\zeta/k_0T$ for $n=$ const. is determined from (6.2.5), and in the presence of both donors and acceptors, from (6.2.15).

In the case of strong degeneracy (7.2b) is an adequate approximation, and

$$\sigma=\frac{ne^2}{m}\,\tau\,(\zeta_0), \tag{7.8}$$

where $\tau\,(\zeta_0)=\tau\,(\varepsilon)\,|_{\varepsilon=\zeta_0}$.

It will be seen from (7.1) that due to the presence in the integrand of the factor $\left(-\frac{\partial f_0}{\partial\varepsilon}\right)\approx\delta\,(\varepsilon-\zeta_0)$, only the electrons belonging to the layer about $k_0T$ wide near the Fermi surface $\varepsilon=\zeta_0$ take part in the current. Indeed, for other electrons $\delta\,(\varepsilon-\zeta_0)\approx 0$. For this reason the concept of electron mobility $\mu=\sigma/ne$ is not introduced in the case of degeneracy.

9.7.3 The expression for the differential thermoelectric power (4.5) can be written in the form

$$\alpha=\left(\frac{k_0}{e}\right)\left\{\frac{\langle\tau x\rangle}{\langle\tau\rangle}-z\right\}=\left(\frac{k_0}{e}\right)\frac{1}{k_0T}\left\{\frac{\langle\tau\varepsilon\rangle}{\langle\tau\rangle}-\zeta\right\}. \tag{7.9}$$

If $\tau$ is given by (3.4), for the case of an arbitrary degeneracy

$$\alpha=\left(\frac{k_0}{e}\right)\left\{\frac{(r+5/2)\,\mathcal{F}_{r+3/2}(z)}{(r+3/2)\,\mathcal{F}_{r+1/2}(z)}-z\right\}. \tag{7.10}$$

It follows immediately from (7.9) that in the first approximation with respect to degeneracy $\alpha=0$. Indeed, the first term in the braces of (7.9) in the $\delta$-function approximation is

$$\frac{\langle\tau\varepsilon\rangle}{\tau}=\frac{\tau\,(\zeta_0)\,\zeta_0}{\tau\,(\zeta_0)}=\zeta_0$$

and therefore cancels with the second term. In the second approximation with respect to degeneracy, we obtain, making use of (7.2d) and (6.2.9)[8],

$$\alpha=\frac{\pi^2}{3}\left(r+\frac{3}{2}\right)\left(\frac{k_0}{e}\right)\left(\frac{k_0T}{\zeta_0}\right). \tag{7.11}$$

The values of $r$ are different for different scattering mechanisms (3.5). For good metals, according to (8.4.14), $r = 3/2$.

It will be seen immediately from (7.11) that $\alpha$ for metals is substantially smaller than for nondegenerate semiconductors, since in the case of a strong degeneracy an additional factor $(k_0T/\zeta_0)$ appears in $\alpha$, equal for metals at room temperatures to $5 \times 10^{-3}$.

**9.7.4** Substituting (3.3), (4.3a) and (4.11a) into (4.20), we obtain for the heat conductivity coefficient

$$\varkappa=\frac{n}{m}k_0^2T\,\frac{\langle\tau\rangle\langle\tau x^2\rangle-\langle\tau x\rangle^2}{\langle\tau\rangle}=\frac{n}{m}\,\frac{1}{T}\,\frac{\langle\tau\rangle\langle\tau\varepsilon^2\rangle-\langle\tau\varepsilon\rangle^2}{\langle\tau\rangle}. \tag{7.12}$$

It can easily be shown in the same way as it has been done for $\alpha$ that $\varkappa$ in the first approximation with respect to degeneracy is zero. For $\tau$ given by (3.4) in the second approximation with respect to degeneracy we have

$$\varkappa=\frac{\pi^2}{3}k_0^2T\,\frac{n}{m}\,\tau(\zeta_0), \tag{7.13}$$

where use was made of expansion (7.2c). Hence, and from (7.8) follows the Wiedemann-Franz law:

$$\varkappa/\sigma = LT, \tag{7.14}$$

where the Lorenz number in the case of strong degeneracy

$$L=\frac{\pi^2}{3}\left(\frac{k_0}{e}\right)^2 \tag{7.14a}$$

is independent of the scattering mechanism.

For an intermediate degree of degeneracy $\varkappa$ can be expressed by (7.2d) in terms of the Fermi integrals. In the general case the Lorenz number is

$$L=\frac{(r+3/2)(r+7/2)\mathscr{F}_{r+1/2}\mathscr{F}_{r+5/2}-(r+5/2)^2\mathscr{F}^2_{r+3/2}}{(r+3/2)^2\mathscr{F}^2_{r+3/2}}\left(\frac{k_0}{e}\right)^2. \tag{7.15}$$

---

[8] Here and below we use the expansion ($x \ll 1$)

$$\frac{1+ax^2}{1+bx^2}\approx(1+ax^2)(1-bx^2)\approx 1+(a-b)\,x^2$$

to within the quantities of the order of $x^2$.

**9.7.5** Consider the principal galvanomagnetic phenomena. Equations (5.4), (5.4a), (5.4b) are valid when the magnetic field is perpendicular to the vector $\mathbf{P} = \tau\mathbf{E}$. Of course, here the symbol $\langle\ \rangle$ must be understood in the sense of the averaging procedure (7.1b). Hence, the equations for the Hall constant $R$ and for the magnetoresistance $\Delta\rho/\rho$ obtained in Section 9.5 for the cases of weak and strong magnetic fields remain valid, provided $\langle\ \rangle$ is the symbol for the averaging procedure (7.1b).

**A. Weak fields:** $\gamma\tau H \ll 1$. In this case we have, according to (5.10),

$$R = -\frac{\langle\tau^2\rangle}{\langle\tau\rangle^2}\frac{1}{ecn}. \tag{7.16}$$

In the first approximation with respect to degeneracy

$$\frac{\langle\tau^2\rangle}{\langle\tau\rangle^2} = \frac{\tau^2(\zeta_0)}{[\tau(\zeta_0)]^2} = 1,$$

therefore,

$$R = -1/ecn. \tag{7.17}$$

If (3.4) is valid and degeneracy is intermediate, the factor $\langle\tau^2\rangle/\langle\tau\rangle^2$ can be expressed in terms of the Fermi integrals.

The magnetoresistance is, according to (5.11),

$$\frac{\Delta\rho}{\rho} = \left(\frac{eH}{mc}\right)^2 \frac{\langle\tau^3\rangle\langle\tau\rangle - \langle\tau^2\rangle^2}{\langle\tau\rangle^2} \tag{7.18}$$

and vanishes in the first approximation. This agrees with the general conception that the magnetoresistance appears only when there are groups of electrons with different energies and different scattering mechanisms.

For a power dependence of the relaxation time on the energy (3.4) we obtain in the second approximation with respect to degeneracy

$$\Delta\rho/\rho = B_0 H^2, \tag{7.19}$$

where

$$B_0 = \frac{\pi^2}{3} r^2 \left[\frac{e\tau(\zeta_0)}{mc}\right]^2 \left(\frac{k_0 T}{\zeta_0}\right)^2. \tag{7.19a}$$

In the case of intermediate degeneracy $\Delta\rho/\rho$ (7.18) can be expressed in terms of the Fermi integrals.

Note that in metals in the cases of a weak field or strong degeneracy $\Delta\rho/\rho$ is temperature-independent, since $\tau(\zeta_0)$ is inversely proportional to the temperature (8.4.14).

**B. Strong fields:** $\gamma\tau H \gg 1$. For strong fields the Hall constant $R$ is, according to (5.16), equal to (7.17), no matter what the degree of degeneracy is. According to (5.17), the magnetoresistance is

$$\frac{\Delta\rho}{\rho} = \langle\tau\rangle \left\langle \frac{1}{\tau} \right\rangle - 1. \tag{7.20}$$

In the first approximation with respect to degeneracy $\Delta\rho/\rho = 0$ since $\langle\tau\rangle \left\langle \frac{1}{\tau} \right\rangle = 1$. When (3.4) is valid in the second approximation with respect to degeneracy, we obtain

$$\frac{\Delta\rho}{\rho} = \frac{\pi^2}{3} r^2 \left( \frac{k_0 T}{\zeta_0} \right)^2, \tag{7.21}$$

i.e., saturation for $H \to \infty$. For strong fields and in the case of strong degeneracy $\Delta\rho/\rho \propto T^2$.

**C. Arbitrary fields:** strong degeneracy. The value of $\Delta\rho/\rho$ can be calculated for an arbitrary magnetic field. In the second approximation with respect to degeneracy

$$\frac{\Delta\rho}{\rho} = \frac{B_0 H^2}{1 + (R\sigma H)^2}. \tag{7.22}$$

Here $B_0$ is given by (7.19a), and $R$ and $\sigma$ are the Hall constant (7.17) and the conductivity (7.8). From (7.22) it follows for strong magnetic fields that

$$\left( \frac{\Delta\rho}{\rho} \right)_{H\to\infty} = \frac{B_0}{(R\sigma)^2}, \tag{7.22a}$$

which coincides with (7.21).

The experimental data on magnetoresistance of metals are in poor agreement with equation (7.22). For very weak fields $\Delta\rho/\rho$ is, indeed, proportional to $H^2$, but soon the increase in $\Delta\rho/\rho$ becomes proportional to $H$ and does not approach real saturation. Moreover, the values of $B_0$ observed in experiment are several orders of magnitude greater than theoretical values. The temperature dependence of $\Delta\rho/\rho$ predicted by the theory is not confirmed in experiments. The main reason for such discrepancies lies, probably, in the complex electron energy-band pattern.

I. M. Lifshitz with his coworkers [9.11] has developed a theory of galvanomagnetic effects in metals, which takes into account the complex shape of the Fermi surface ($\varepsilon = \zeta_0$). He has demonstrated the close interrelation between the asymptotic behaviour of the components of the galvanomagnetic tensor in strong magnetic fields and the topology of the Fermi surface, the factor of foremost importance is the closed or open nature of the Fermi surface, i.e., whether it covers continuously the entire **k**-space or splits up into closed surfaces repeated in the reciprocal lattice space.

**9.7.6** Out of the thermomagnetic phenomena we shall consider only the transverse Nernst effect. The general form of equations (6.3) and the expressions for the coefficients $b_1$ and $b_2$ (6.3a), (6.3b) remain unaltered for all degrees of degeneracy of charge carriers, provided the symbol $\langle\ \rangle$ is understood as the averaging procedure (7.1b). This means that we can use (6.4a) for the Nernst constant $Q$ and the expressions for the cases of weak and strong fields derived from it.

**A. Weak fields:** $\gamma\tau H \ll 1$. It follows from (6.6a) that

$$Q=\left(\frac{k_0}{e}\right)\frac{e}{mc}\,\frac{\langle\tau\rangle\langle\tau^2x\rangle-\langle\tau^2\rangle\langle\tau x\rangle}{\langle\tau\rangle^2}. \tag{7.23}$$

Hence, it follows immediately that in the first approximation with respect to degeneracy $Q = 0$. This reflects the general principle, according to which the Nernst effect is only possible if there are different groups of electrons with different scattering mechanisms. $Q$ is also zero if $\tau$ is independent of $\varepsilon$. In case (3.4) in the second approximation with respect to degeneracy

$$Q=\frac{\pi^2}{3}r\left(\frac{k_0}{e}\right)\left(\frac{e\tau(\zeta_0)}{mc}\right)\left(\frac{k_0T}{\zeta_0}\right), \tag{7.24}$$

where we made use of expansion (7.2c).

Specifically, it follows from here that the sign of the effect, as in the nondegenerate case, coincides with the sign of $r$. The presence of the factor $k_0T/\zeta_0$ makes the effect a small one. In the range of intermediate degeneracies $Q$ can be expressed in terms of the Fermi integrals.

**B. Strong fields:** $\gamma\tau H \gg 1$. It will be seen from (6.9) that in the first approximation with respect to degeneracy $Q = 0$. In the second approximation with respect to degeneracy in the case of a power dependence of the relaxation time on the energy (3.4), we obtain

$$Q=\frac{\pi^2}{3}r\left(\frac{k_0}{e}\right)\frac{mc}{e\tau(\zeta_0)H^2}\left(\frac{k_0T}{\zeta_0}\right). \tag{7.25}$$

All the remarks made in connection with (7.24) are appropriate for this expression.

The variation of the thermoelectric power $\alpha$ in a magnetic field can be considered in exactly the same way as in the preceding case.

## 9.8. Transport Phenomena in Silicon- and Germanium-Type Semiconductors

**9.8.1** To begin with, consider transport phenomena in an electron-type semiconductor in which the energy minimum is not located in the centre of the Brillouin zone. In this case the constant-energy surfaces will generally have the shape not of spheres, but of ellipsoids. Near the energy minimum the dependence of the electron's energy $\varepsilon(k_x, k_y, k_z)$ on the wave vector components will, as in the case of a simple band, be quadratic.

As we saw in Section 4.15, the conduction band for silicon has six energy minima symmetrically arranged at points on the $\Delta$ axes, i.e., in the directions [100]; the conduction band of germanium has eight energy minima located on the Brillouin zone boundary in the [111] directions. The constant-energy surfaces $\varepsilon(\mathbf{k}) = \text{const.}$ near these minima are ellipsoids of revolution (4.15.1). Such energy bands became known as *multiellipsoid* or *multivalley* bands. The complications arising from the existence of a multiellipsoid energy-band pattern involve not only a more complex form of the kinetic coefficients, but also the transformation of the relaxation time $\tau$, if it can be introduced at all, from a scalar into a tensor.

However, it seems likely that even in the cases of an intricate energy-band pattern the electron relaxation time can be treated as an energy-dependent scalar. In this section we intend to limit ourselves to this approximation.

Since the distance between the equivalent minima in the **k**-space is of the order of the reciprocal lattice constant, the electron in the act of phonon absorption can pass from the region of one minimum into the region of another only if it absorbs a phonon with a wave vector $q \approx 1/a$, where $a$ is the lattice constant (wave vector conservation law).

At low temperatures such *interellipsoid transitions* will be comparatively rare, but at higher temperatures they can, as was demonstrated by C. Herring, in some cases explain a more pronounced dependence of mobility on the temperature ($\mu \propto T^{-n}$, where $n > 1.5$). The theory of interellipsoid transitions is as yet in a rudimentary state, and for this reason it shall not be discussed here. In the case of *intraellipsoid* electron scattering the currents belonging to the different ellipsoids add up independently.

**9.8.2** If the electron's energy $\varepsilon$ is measured from the value corresponding to one of the equivalent energy minima, and the electron's wave vector **k** from the point of the $i$-th minimum in the **k**-space, the expression for $\varepsilon$ in the major axes of the $i$-th energy ellipsoid will be

$$\varepsilon = \frac{\hbar^2}{2}\left[\frac{k_1^2}{m_1} + \frac{k_2^2}{m_2} + \frac{k_3^2}{m_3}\right] = \frac{\hbar^2}{2}\sum_{\alpha=1}^{3}\frac{k_\alpha^2}{m_\alpha}, \tag{8.1}$$

where $k_\alpha$ are the rectangular coordinate components of the wave vector **k**, and $m_\alpha$ are the components of the effective mass tensor (4.15.1). In the case of germanium and silicon, pointing the $z$ axis ($\alpha = 3$) in the direction of the ellipsoid's axis of rotation and making $m_3 = m_{\|}$ the *longitudinal mass* and making $m_1 = m_2 = m_{\perp}$ the *transverse mass*, we obtain

$$\varepsilon = \frac{\hbar^2}{2}\left[\frac{k_1^2 + k_2^2}{m_{\perp}} + \frac{k_3^2}{m_{\|}}\right]. \tag{8.1a}$$

It will now be convenient to introduce the variables

$$w_\alpha = \frac{\hbar}{\sqrt{2m_\alpha}} k_\alpha, \tag{8.2}$$

in which

$$\varepsilon = w_1^2 + w_2^2 + w_3^2. \tag{8.2a}$$

The components of the electron's velocity are

$$v_\alpha = \frac{1}{\hbar}\frac{\partial \varepsilon}{\partial k_\alpha} = \frac{\hbar}{m_\alpha} k_\alpha = \sqrt{\frac{2}{m_\alpha}}\, w_\alpha. \tag{8.3}$$

The kinetic equation (8.2.6) also remains valid in the case of a complex energy-band pattern; therefore,

$$\mathbf{v} \cdot \nabla_{\mathbf{r}} f - \frac{e}{\hbar}\left\{\mathbf{E} + \frac{1}{c}\mathbf{v} \times \mathbf{H}\right\} \cdot \nabla_{\mathbf{k}} f = -\frac{f - f_0}{\tau(\varepsilon)}. \tag{8.4}$$

Here, as has already been stated above, we assume that there is an energy-dependent relaxation time, so that the collision term is

$$\left(\frac{\partial f}{\partial t}\right)_c = -\frac{f - f_0}{\tau(\varepsilon)}. \tag{8.4a}$$

**9.8.3** Consider the electric conductivity due to one energy ellipsoid in the presence of an electric field only ($\nabla T = 0$, $\mathbf{H} = 0$). In this case the kinetic equation (8.4) takes the form:

$$\frac{e\tau}{\hbar}\mathbf{E} \cdot \nabla_{\mathbf{k}} f = f - f_0. \tag{8.5}$$

Make the nonequilibrium distribution function

$$f = f_0 + \mathbf{f}^{(10)} \cdot \mathbf{E} = f_0 + \sum_{\mu=1}^{3} f_\mu^{(10)} E_\mu. \tag{8.6}$$

This can be regarded as the expansion of $f$ into a power series in the electric field $E_\mu$ in which only the first-order terms have been retained. Since the nonequilibrium correction to the distribution function is a scalar, $\mathbf{f}^{(10)}$ is a vector, i.e., $f_\mu^{(10)}$ are components of a rank 1 tensor (Appendix 11). Substituting (8.6) into (8.5) and dropping the members of the order of $E^2$, we obtain

$$\frac{e\tau}{\hbar}\mathbf{E} \cdot \nabla_{\mathbf{k}} f_0 = \mathbf{E} \cdot \mathbf{f}^{(10)}, \tag{8.7}$$

whence (because **E** is arbitrary)

$$\mathbf{f}^{(10)} = \frac{e\tau}{\hbar} \nabla_{\mathbf{k}} f_0. \tag{8.8}$$

Since the equilibrium distribution function $f_0$ depends on **k** only through the energy $\varepsilon(\mathbf{k})$, it follows that

$$\nabla_{\mathbf{k}} f_0 = \frac{\partial f_0}{\partial \varepsilon} \nabla_{\mathbf{k}} \varepsilon = \frac{\partial f_0}{\partial \varepsilon} \hbar \mathbf{v}, \tag{8.9}$$

whence

$$\mathbf{f}^{(10)} = e\tau(\varepsilon) \frac{\partial f_0}{\partial \varepsilon} \mathbf{v}, \tag{8.10}$$

or

$$f_\mu^{(10)} = e\tau(\varepsilon) \frac{\partial f_0}{\partial \varepsilon} v_\mu. \tag{8.10a}$$

The electric current established by the electrons of the $i$-th ellipsoid is

$$\mathbf{j}^{(i)} = -e \sum_{(\mathbf{k})} f\mathbf{v} = e^2 \sum_{(\mathbf{k})} (\mathbf{E} \cdot \mathbf{v})\, \mathbf{v}\tau(\varepsilon) \left( -\frac{\partial f_0}{\partial \varepsilon} \right), \tag{8.11}$$

since the equilibrium distribution function $f_0$ does not contribute to the current. Here

$$\sum_{(\mathbf{k})} \equiv \frac{1}{4\pi^3} \int d^3k \tag{8.11a}$$

denotes summation (integration) over the wave vector **k**. A current component is, according to (8.11),

$$j_\alpha^{(i)} = e^2 \sum_{(\mathbf{k})} \sum_{\beta} \tau(\varepsilon) \left( -\frac{\partial f_0}{\partial \varepsilon} \right) v_\alpha v_\beta E_\beta = \sum_{\beta} \sigma_{\alpha\beta}^{(i)} E_\beta, \tag{8.12}$$

where

$$\sigma_{\alpha\beta}^{(i)} = e^2 \sum_{(\mathbf{k})} \tau(\varepsilon) \left( -\frac{\partial f_0}{\partial \varepsilon} \right) v_\alpha v_\beta \tag{8.12a}$$

is a component of the electric conductivity tensor of the $i$-th ellipsoid. Since $v_\alpha \propto k_\alpha$, and $\varepsilon$ is an even function of $k_\alpha$, $\sigma_{\alpha\beta} = 0$ if $\alpha \neq \beta$, i.e., the electric conductivity tensor is diagonal in the major axes of the energy ellipsoid. Hence,

$$\sigma_{\alpha\alpha}^{(i)} = e^2 \sum_{(\mathbf{k})} \tau(\varepsilon) \left( -\frac{\partial f_0}{\partial \varepsilon} \right) v_\alpha^2 = e^2 \sum_{(\mathbf{w})} \tau(\varepsilon) \left( -\frac{\partial f_0}{\partial \varepsilon} \right) \frac{2}{m_\alpha} w_\alpha^2, \tag{8.12b}$$

where the summation is now performed over **w**. Taking into account the isotropic relationship between $\varepsilon$ and $w_\alpha$ (8.2a), we can easily demonstrate that summation over **w** reduces to the summation over $\varepsilon$ with $1/3\varepsilon$ being substituted for $w_\alpha^2$. Then

$$\sigma_{\alpha\alpha}^{(i)}=\frac{e^2}{m_\alpha}\frac{2}{3}\sum_{(\varepsilon)}\tau(\varepsilon)\left(-\frac{\partial f_0}{\partial\varepsilon}\right)\varepsilon. \tag{8.12c}$$

Making use of the expression for the density of states (6.2.22), we obtain

$$\frac{2}{3}\sum_{(\varepsilon)}\tau(\varepsilon)\left(-\frac{\partial f_0}{\partial e}\right)\varepsilon$$

$$=\frac{2\sqrt{2}}{3\pi^2}\frac{(m_1m_2m_3)^{1/2}}{\hbar^3}\int_0^\infty\tau(\varepsilon)\left(-\frac{\partial f_0}{\partial\varepsilon}\right)\varepsilon^{3/2}\,d\varepsilon=n^{(i)}\langle\tau\rangle, \tag{8.13}$$

where $n^{(i)}$ is the number of electrons in the $i$-th ellipsoid, and the symbol $\langle\ \rangle$ is a direct generalization of (7.1b) to which it reduces, when $m_1 = m_2 = m_3 = m$. Hence,

$$\sigma_{\alpha\alpha}^{(i)}=\frac{e^2n^{(i)}}{m_\alpha}\langle\tau\rangle, \tag{8.14}$$

this being similar to (7.5).

Choose now a rectangular coordinate system with the axes pointing in the direction of the edges of the crystal's unit cell. The components of the total (of all the ellipsoids) electric conductivity tensor in this coordinate system will be

$$\sigma_{\lambda\mu}=\sum_i\sigma_{\lambda\mu}^{(i)}, \tag{8.15}$$

where $\sigma_{\lambda\mu}^{(i)}$ are the components of tensor (8.14), transformed to the coordinate system connected with the crystal's axes.

In a cubic crystal the electric conductivity tensor $\sigma_{\lambda\mu}$ is a scalar $\sigma$.

Consider now the simple case of silicon in which the energy minima are arranged along the $\langle 100\rangle$ directions. The axes of rotation of the energy ellipsoids (effective masses $m_{\|}$) are parallel to the same directions.

Hence, in the case of silicon the major axes of all six energy ellipsoids are parallel to the coordinate axes coinciding with the cube's edges of a unit cell. It follows from (8.15) and (8.14) that

$$\sigma=\sigma_{xx}=2\frac{e^2n^{(i)}}{m_{\|}}\langle\tau\rangle+4\frac{e^2n^{(i)}}{m_{\perp}}\langle\tau\rangle=\frac{e^2n}{m'}\langle\tau\rangle, \tag{8.16}$$

where the electron concentration is

$$n=N_cn^{(i)}=6n^{(i)} \tag{8.16a}$$

($N_c$ is the number of equivalent minima), and

$$\frac{1}{m'}=\frac{1}{3}\left(\frac{1}{m_{\|}}+\frac{2}{m_{\perp}}\right). \tag{8.16b}$$

Making use of the equations for the transformation of the tensor components (Appendix 11), we can demonstrate that equations (8.16a), (8.16b) are also valid in the case of germanium, when the major axes of the energy ellipsoids are not parallel to the cube's edges.

Note that (8.16b) follows from the requirement that $1/m'$ be proportional to the scalar invariant of the inverse effective mass tensor, i.e., to its trace $\sum_{\alpha} 1/m_\alpha$ (the factor 1/3 is obviously needed in the case of a scalar effective mass).

The mobility is

$$\mu=\frac{\sigma}{en}=\frac{e\langle\tau\rangle}{m'}, \tag{8.17}$$

this being similar to (3.3).

**9.8.4** Consider the Hall effect in a weak magnetic field. To this end we must solve the kinetic equation (8.4) (with $\nabla_{\mathbf{r}} f=0$) to within the members of the first order in $H$. Again considering an ellipsoid, we make the distribution function expressed in terms of the ellipsoid's major axes

$$f=f_0+\sum_{\mu} f_{\mu}^{(10)} E_{\mu}+\sum_{\mu\nu} f_{\mu\nu}^{(11)} E_{\mu} H_{\nu}, \tag{8.18}$$

where $f_{\mu\nu}^{(11)}$ is a rank 2 tensor (in this case, however, the correction proportional to $H$ is a scalar). Substitute (8.18) into the left-hand and the right-hand sides of the kinetic equation (2.4)

$$\frac{e\tau}{\hbar}\left\{\mathbf{E}+\frac{1}{c}\mathbf{v}\times\mathbf{H}\right\}\nabla_{\mathbf{k}}\left(f_0+\sum_{\mu} f_{\mu}^{(10)} E_{\mu}+\sum_{\mu\nu} f_{\mu\nu}^{(11)} E_{\mu} H_{\nu}\right.$$

$$=\sum_{\mu} f_{\mu}^{(10)} E_{\mu}+\sum_{\mu\nu} f_{\mu\nu}^{(11)} E_{\mu} H_{\nu}. \tag{8.19}$$

In accordance with the aforesaid, the expression $(e\tau/\hbar)\,\mathbf{E}\nabla_k f_0$ cancels with the first sum on the right-hand side.

Since

$$(\mathbf{v}\times\mathbf{H})\cdot\nabla_{\mathbf{k}} f_0=\mathbf{v}\times\mathbf{H}\,\frac{\partial f_0}{\partial\varepsilon}\,\nabla_{\mathbf{k}}\varepsilon \propto (\mathbf{v}\times\mathbf{H})\cdot\mathbf{v}=0,$$

it follows that the term on the left-hand side linear in **E** and **H** is

$$\frac{e\tau}{\hbar c}(\mathbf{v}\times\mathbf{H})\cdot\nabla_{\mathbf{k}}\left\{\sum_{\mu} f_{\mu}^{(10)}E_{\mu}\right\}=\frac{e\tau}{\hbar c}\,\mathbf{v}\times\mathbf{H}\sum_{\mu}E_{\mu}\nabla_{\mathbf{k}}\left(e\tau\frac{\partial f_0}{\partial\varepsilon}v_{\mu}\right)$$

$$=\frac{e^2\tau}{\hbar c}\,\mathbf{v}\times\mathbf{H}\sum_{\mu}E_{\mu}\left\{v_{\mu}\nabla_{\mathbf{k}}\left(\tau\frac{\partial f_0}{\partial\varepsilon}\right)+\tau\frac{\partial f_0}{\partial\varepsilon}\nabla_{\mathbf{k}}v_{\mu}\right\}. \tag{8.20}$$

Since $\tau$ depends only on $\varepsilon$, it follows that

$$\nabla_{\mathbf{k}}\left(\tau\frac{\partial f_0}{\partial\varepsilon}\right)=\frac{\partial}{\partial\varepsilon}\left(\tau\frac{\partial f_0}{\partial\varepsilon}\right)\nabla_{\mathbf{k}}\varepsilon=\frac{\partial}{\partial\varepsilon}\left(\tau\frac{\partial f_0}{\partial\varepsilon}\right)\hbar\mathbf{v},$$

and the product of $\mathbf{v}\times\mathbf{H}$ by the first term in the braces of (8.20) is zero. On the other hand, according to (8.3),

$$\nabla_{\mathbf{k}}v_{\mu}=\frac{\hbar}{m_{\mu}}\,\mathbf{e}_{\mu}. \tag{8.21}$$

Here $\mathbf{e}_{\mu}$ is a unit vector along the $\mu$ axis. Hence, (8.20) includes the scalar product

$$(\mathbf{v}\times\mathbf{H})\cdot\mathbf{e}_{\mu}=\mathbf{v}\times\mathbf{H}_{\mu}=\sum_{\alpha\nu}\delta_{\mu\alpha\nu}v_{\alpha}H_{\nu}, \tag{8.22}$$

where $\delta_{\mu\alpha\nu}$ is a symbol equal to $+1$ when $\mu\alpha\nu$ is an even transmutation of the indices 1, 2, 3, to $-1$ when $\mu\alpha\nu$ is an odd transmutation of 1, 2, 3, and zero in all other cases, when there are identical indices among $\mu\alpha\nu$.[9]

Finally, it follows from (8.19) that

$$\frac{e^2\tau^2}{c}\frac{\partial f_0}{\partial\varepsilon}\sum_{\mu\alpha\nu}\delta_{\mu\alpha\nu}\frac{v_{\alpha}}{m_{\mu}}E_{\mu}H_{\nu}=\sum_{\mu\nu}f_{\mu\nu}^{(11)}E_{\mu}H_{\nu}. \tag{8.23}$$

Comparing the coefficients of $E_{\mu}H_{\nu}$, we obtain

$$f_{\mu\nu}^{(11)}=\frac{e^2\tau^2}{c}\frac{\partial f_0}{\partial\varepsilon}\frac{1}{m_{\mu}}\sum_{\alpha}\delta_{\mu\alpha\nu}v_{\alpha}=\frac{e^2\tau^2}{c}\left(-\frac{\partial f_0}{\partial\varepsilon}\right)\frac{1}{m_{\mu}}\sum_{\alpha}\delta_{\mu\alpha\nu}v_{\alpha}. \tag{8.24}$$

The current due to the correction (8.23) to the distribution function is

$$j_{\lambda}^{(i)}=-e\sum_{(\mathbf{k})}\left\{v_{\lambda}\sum_{\mu\nu}f_{\mu\nu}^{(11)}E_{\mu}H_{\nu}\right\}=\frac{e^3}{c}\sum_{\mu\nu\alpha}\delta_{\mu\alpha\nu}E_{\mu}H_{\nu}\sum_{(\mathbf{k})}\tau^2\frac{\partial f_0}{\partial\varepsilon}\frac{v_{\alpha}v_{\lambda}}{m_{\mu}}. \tag{8.25}$$

---

9 See Appendix 24. Some authors use the notation $\varepsilon_{\mu\alpha\nu}$ for $\delta_{\mu\alpha\nu}$.

Expression (8.25), like (8.12a), is nonzero when $\alpha = \lambda$. In this case the integration with respect to **k** can be reduced, as in (8.12c), to the integration with respect to $\varepsilon$. As a result, we obtain

$$j_\lambda^{(i)} = -\frac{e^3}{c} \sum_{\mu\nu} \delta_{\mu\nu\lambda} n^{(i)} \langle \tau^2 \rangle \frac{1}{m_\mu m_\lambda} E_\mu H_\nu = \sum_{\mu\nu} \sigma_{\lambda\mu\nu}^{(i)} E_\mu H_\nu, \tag{8.26}$$

where the symbol $\langle\ \ \rangle$ was defined in (8.13), and the rank 3 tensor

$$\sigma_{\lambda\mu\nu}^{(i)} = -\frac{e^3}{c} n^{(i)} \langle \tau^2 \rangle \frac{\delta_{\lambda\mu\nu}}{m_\lambda m_\mu} \tag{8.26a}$$

determines the additional current in the $\lambda$-direction, proportional to $E_\mu H_\nu$[10]. With no account taken of the interellipsoid transitions the total current in the coordinate system connected with the crystal's axes is

$$j_\lambda = \sum_i j_\lambda^{(i)} = \sum_{\mu\nu} \sigma_{\lambda\mu\nu} E_\mu H_\nu, \tag{8.27}$$

where

$$\sigma_{\lambda\mu\nu} = \sum_i \sigma_{\lambda\mu\nu}^{(i)} \tag{8.27a}$$

and $\sigma_{\lambda\mu\nu}^{(i)}$ are the components of tensor (8.26a), transformed to the coordinate system connected with the crystal. In a cubic crystal a cyclic transmutation of the indices $\lambda\mu\nu$ does not change $\sigma_{\lambda\mu\nu}$. Besides, it seems natural to suppose that in a cubic crystal the Hall current should be proportional to $\mathbf{E} \times \mathbf{H}$. This means that

$$\sigma_{\lambda\mu\nu} = \eta \delta_{\lambda\mu\nu}, \tag{8.28}$$

where $\eta$ is a scalar in terms of which all the 27 components of the tensor $\sigma_{\lambda\mu\nu}$ are expressed. It follows from (8.28) and (8.27) that

$$\mathbf{j} = \eta \mathbf{E} \times \mathbf{H}. \tag{8.29}$$

Write an invariant (scalar)

$$\sum_{\lambda\mu\nu} \sigma_{\lambda\mu\nu} \delta_{\lambda\mu\nu} = \sum_{\lambda\mu\nu} \eta \delta_{\lambda\mu\nu}^2 = 6\eta \tag{8.30}$$

and express it in terms of $\sigma_{\lambda\mu\nu}^{(i)}$

$$6\eta = \sum_{\lambda\mu\nu} \sum_i \sigma_{\lambda\mu\nu}^{(i)} \delta_{\lambda\mu\nu} = \sum_i \sum_{\lambda\mu\nu} \sigma_{\lambda\mu\nu}^{(i)} \delta_{\lambda\mu\nu}. \tag{8.30a}$$

Since each sum belonging to the $i$-th ellipsoid is also an invariant, it can be computed in any coordinate system, including that of

[10] Of course, the dimension of $\sigma_{\lambda\mu\nu}^{i}$ is not that of conductivity.

the major axes of the tensor $\sigma^{(i)}_{\lambda\mu\nu}$. Making use of (8.26a), we obtain

$$\sum_{\lambda\mu\nu} \sigma^{(i)}_{\lambda\mu\nu}\delta_{\lambda\mu\nu} = -\frac{e^3}{c} n^{(i)} \langle\tau^2\rangle \sum_{\lambda\mu\nu} \frac{\delta^2_{\lambda\mu\nu}}{m_\lambda m_\mu} =$$

$$= -\frac{e^3}{c} n^{(i)} \langle\tau^2\rangle\, 2\left[\frac{1}{m_1 m_2} + \frac{1}{m_1 m_3} + \frac{1}{m_2 m_3}\right], \qquad (8.30b)$$

this expression being identical for all ellipsoids.

From (8.30a), (8.30b) we obtain

$$\eta = -\frac{e^3}{c} n \langle\tau^2\rangle \frac{1}{m''^2}, \qquad (8.31)$$

where

$$\frac{1}{m''^2} = \frac{1}{3}\left[\frac{1}{m_1 m_2} + \frac{1}{m_1 m_3} + \frac{1}{m_2 m_3}\right]. \qquad (8.31a)$$

For Ge and Si (8.31a) has the form

$$\frac{1}{m''^2} = \frac{1}{3}\left[\frac{2}{m_\perp m_\parallel} + \frac{1}{m_\perp^2}\right]. \qquad (8.31b)$$

It follows from (8.16) and (8.29) that the total current is

$$\mathbf{j} = \sigma\mathbf{E} + \eta\mathbf{E}\times\mathbf{H}. \qquad (8.32)$$

Making use of the definition of the Hall constant $R$ (5.5), we obtain from (8.32) in the approximation linear in the magnetic field

$$R = \frac{E_y}{jH} = \frac{\eta}{\sigma^2} = -\frac{1}{ecn}\frac{\langle\tau^2\rangle}{\langle\tau\rangle^2}\left(\frac{m'}{m''}\right)^2, \qquad (8.33)$$

the only difference from (6.10) being the presence of the factor $(m'/m'')^2$.

**9.8.5** Consider the magnetoconductivity in weak magnetic fields. With this in view we expand the nonequilibrium distribution function into a series in the magnetic field intensity up to the terms of the second order in $H$:

$$f = f_0 + \sum_\mu f^{(10)}_\mu E_\mu + \sum_{\mu\nu} f^{(11)}_{\mu\nu} E_\mu H_\nu + \sum_{\mu\nu\rho} f^{(12)}_{\mu\nu\rho} E_\mu H_\nu H_\rho, \qquad (8.34)$$

where $f^{(12)}_{\mu\nu\rho}$ is a rank 3 tensor.

If we take into account the values of $f^{(10)}_\mu$ and of $f^{(11)}_{\mu\nu}$, obviously we must substitute

$$\sum_{\mu\nu} f^{(11)}_{\mu\nu} E_\mu H_\nu$$

into the left-hand side of the kinetic equation and obtain a value proportional to

$$(\mathbf{v}\times\mathbf{H})\cdot\nabla_\mathbf{k} f^{(11)}_{\mu\nu} = -\frac{e^2}{cm_\mu}(\mathbf{v}\times\mathbf{H})\cdot\nabla_\mathbf{k}\left(\tau^2\frac{\partial f_0}{\partial\varepsilon}\sum_\alpha \delta_{\mu\nu\alpha}v_\alpha\right). \qquad (8.35)$$

Carrying out transformations similar to those of (8.20)-(8.22), we obtain from the kinetic equation

$$-\frac{e^3}{c^2}\tau^3\frac{\partial f_0}{\partial\varepsilon}\sum_{\mu\nu}\sum_{\alpha\beta\rho}\frac{\delta_{\mu\nu\alpha}\delta_{\alpha\beta\rho}}{m_\alpha m_\mu}v_\beta E_\mu H_\nu H_\rho=\sum_{\mu\nu\rho}f^{(12)}_{\mu\nu\rho}E_\mu H_\nu H_\rho, \tag{8.36}$$

whence

$$f^{(12)}_{\mu\nu\rho}=-\frac{e^3}{c^2}\sum_{\alpha\beta}\delta_{\mu\nu\alpha}\delta_{\alpha\beta\rho}\tau^3\frac{\partial f_0}{\partial\varepsilon}\frac{v_\beta}{m_\alpha m_\mu}. \tag{8.37}$$

Determine now the current from the $i$-th ellipsoid due to this correction to the distribution function:

$$j^{(i)}_\lambda=-e\sum_{(\mathbf{k})}v_\lambda\sum_{\mu\nu\rho}f^{(12)}_{\mu\nu\rho}E_\mu H_\nu H_\rho=\sum_{\mu\nu\rho}\sigma^{(i)}_{\lambda\mu\nu\rho}E_\mu H_\nu H_\rho. \tag{8.38}$$

Here $\sigma^{(i)}_{\lambda\mu\nu\rho}$ is a rank 4 magnetoconductivity tensor with $3\times3\times3\times3=81$ components. Substituting (8.36) into (8.38), we obtain

$$\sigma^{(i)}_{\lambda\mu\nu\rho}=-\frac{e^4}{c^2}\sum_{(\mathbf{k})}\sum_{\alpha\beta}\delta_{\mu\nu\alpha}\delta_{\alpha\beta\rho}\tau^3\left(-\frac{\partial f_0}{\partial\varepsilon}\right)\frac{v_\lambda v_\beta}{m_\alpha m_\mu}. \tag{8.39}$$

As in the preceding cases, integration with respect to **k** yields zero if $\beta\neq\lambda$. We can again change the integration variables from **k** to $\varepsilon$ and make use of definition (8.13) to obtain

$$\sigma^{(i)}_{\lambda\mu\nu\rho}=-\frac{e^4}{c^2}n^{(i)}\langle\tau^3\rangle\sum_\alpha\frac{\delta_{\alpha\mu\nu}\delta_{\alpha\lambda\rho}}{m_\alpha m_\mu m_\lambda}. \tag{8.40}$$

Since the substitution $\nu\rightleftarrows\rho$ does not change the term $E_\mu H_\nu H_\rho$, it will be convenient to rewrite $\sigma^{(i)}_{\lambda\mu\nu\rho}$ in the sum of (8.38) in a symmetrical form, making

$$(\sigma^{(i)}_{\lambda\mu\nu\rho})_{\text{sym}}={}^1/_2\,(\sigma^{(i)}_{\lambda\mu\nu\rho}+\sigma^{(i)}_{\lambda\mu\nu\rho}). \tag{8.41}$$

In this case there will be one product $E_\mu H_\nu H_\rho$ corresponding to each of the coefficients $(\sigma^{(i)}_{\lambda\mu\nu\rho})_{\text{sym}}$.

Below we shall assume such symmetrization terms in (8.38). Then dropping the symbol $(\ )_{\text{sym}}$, we obtain instead of (8.40)

$$\sigma^{(i)}_{\lambda\mu\nu\rho}=\frac{e^4}{c^2}n^{(i)}\langle\tau^3\rangle M^{(i)}_{\lambda\mu\nu\rho}, \tag{8.42}$$

where

$$M^{(i)}_{\lambda\mu\nu\rho}=-\frac{1}{2}\sum_\alpha\frac{1}{m_\alpha m_\lambda m_\mu}[\delta_{\alpha\mu\nu}\delta_{\alpha\lambda\rho}+\delta_{\alpha\mu\rho}\delta_{\alpha\lambda\nu}]. \tag{8.42a}$$

Hence, we obtain

$$M^{(i)}_{\lambda\lambda\lambda\lambda}=0, \tag{8.43}$$

$$M^{(i)}_{\lambda\lambda\nu\nu}=-\frac{1}{m_\lambda^2 m_\alpha} \quad (\alpha\neq\lambda,\ \nu) \tag{8.43a}$$

$$M^{(i)}_{\lambda\mu\lambda\mu}=M^{(i)}_{\lambda\mu\mu\lambda}=\frac{1}{2m_1m_2m_3} \tag{8.43b}$$

The coefficients $\sigma^{(i)}_{\lambda\mu\nu\rho}$ take into account the variation of current caused by the variation of electrical resistance (magnetoresistance). It follows from (8.42) and (8.43) that $\sigma^{(i)}_{\lambda\lambda\lambda\lambda}=0$, i.e., that the magnetoresistance along one of the energy ellipsoid's major axes is zero when the magnetic field coincides in direction with the current. This is a generalization of the result establishing the absence of the longitudinal magnetoresistance in the case of spherically symmetric constant energy surfaces (Section 9.5.1).

The different nonzero coefficients $M^{(i)}_{\lambda\mu\nu\rho}$ for germanium and silicon are

$$M^{(i)}_{1122}=-\frac{1}{m_\perp^2 m_{||}},\qquad M^{(i)}_{1133}=M^{(i)}_{2233}=-\frac{1}{m_\perp^3},$$

$$M^{(i)}_{3311}=M^{(i)}_{3322}=-\frac{1}{m_{||}^2 m_\perp},\qquad M^{(i)}_{1212}=M^{(i)}_{1313}=M^{(i)}_{2323}=\frac{1}{2m_{||}^2 m_\perp}. \tag{8.44}$$

Of course, all relationships of the type $M^{(i)}_{1122}=M^{(i)}_{2211}$ or $M^{(i)}_{1212}=M^{(i)}_{1221}$, etc. are also valid.

In silicon and germanium the totality of all the equivalent minima in the first Brillouin zone obey the cubic symmetry of the crystals. We distinguish between the two cases:

A. Six energy ellipsoids are arranged along the ⟨100⟩ directions (silicon).

B. Four energy ellipsoids are arranged along the ⟨111⟩ directions (germanium).

Denote by $x_i$ $(i=1,\ 2,\ 3)$ the axes of a rectangular coordinate system, pointing along the unit cell cube's edges and by $\xi_i$ $(i=1,\ 2,\ 3)$ the major axes of one of the energy ellipsoids.

In the case A for one pair of ellipsoids $\xi_1 \parallel x_1$, $\xi_2 \parallel x_2$, $\xi_3 \parallel x_3$, for the second pair, $\xi_1 \parallel x_1$, $\xi_2 \parallel x_3$, $\xi_3 \parallel x_2$ and for the third, $\xi_1 \parallel x_3$, $\xi_2 \parallel x_2$, $\xi_3 \parallel x_1$.

If the currents of the individual ellipsoids are simply added up, then, for example,

$$\sigma_{1122}=\sum_i \sigma^{(i)}_{\lambda\mu\nu\rho}=2\sigma^{(i)}_{1122}+2\sigma^{(i)}_{1133}+2\sigma^{(i)}_{3322}=n\frac{e^4}{c^2}\langle\tau^3\rangle M_{1122}, \tag{8.45}$$

where the electron concentration is $n = N_c n^{(i)}$, $N_c = 6$ is the number of equivalent minima, and

$$M_{1122} = \frac{1}{N_c} \sum_i M^{(i)}_{\lambda\mu\nu\rho} = \frac{1}{3} \left[ M^{(i)}_{1122} + M^{(i)}_{1133} + M^{(i)}_{3322} \right]$$

$$= -\frac{1}{3} \frac{m_\perp^2 + m_\perp m_\parallel + m_\parallel^2}{m_\perp^3 m_\parallel}. \qquad (8.45a)$$

All the other nonzero coefficients $M_{\lambda\mu\nu\rho}$ for both cases A and B can be calculated in a similar way. In case B the problem becomes somewhat more difficult because the major axes of the energy ellipsoids are not parallel to the cube's edges of the unit cell, and because of this in order to determine the components of the tensor $M_{\lambda\mu\nu\rho}$ in terms of the components of the tensors $M^{(i)}_{\lambda\mu\nu\rho}$, first we have to transform the latter to the coordinate system $O$ connected with the crystal (see Appendix 11).

In comparing theory with experiment it appears more convenient to introduce dimensionless coefficients

$$F^{(i)}_{\lambda\mu\nu\rho} = \frac{m''^4}{m'} M^{(i)}_{\lambda\mu\nu\rho} \qquad (8.46)$$

dependent only on the ratio of the effective mass tensor components instead of the quantities $M^{(i)}_{\lambda\mu\nu\rho}$. Making use of (8.16b), (8.31b) and (8.44), we can easily compile Table 9.1. The components of the magnetoconductivity tensor $\sigma_{\lambda\mu\nu\rho}$ in the coordinate system connected with the crystal's axes are

$$\sigma_{\lambda\mu\nu\rho} = \sum_i \sigma^{(i)}_{\lambda\mu\nu\rho} = \frac{e^4}{c^2} \langle \tau^3 \rangle \frac{m'}{m''^4} \sum_i n^{(i)} F^{(i)}_{\lambda\mu\nu\rho}$$

$$= \frac{e^4}{c^2} \langle \tau^3 \rangle \frac{m'}{m''^4} n \frac{1}{N_c} \sum_i F^{(i)}_{\lambda\mu\nu\rho}$$

$$= \frac{\langle \tau^3 \rangle}{\langle \tau \rangle^3} \left( \sigma \frac{\mu^2}{c^2} \right) \left( \frac{m'}{m''} \right)^4 F_{\lambda\mu\nu\rho}, \qquad (8.47)$$

where the summation is performed over all the equivalent minima in the Brillouin zone. Here $\sigma$ and $\mu$ are the electric conductivity and the mobility for $H = 0$ (8.16) and (8.17), respectively. $N_c$ is the number of equivalent minima in the Brillouin zone and

$$F_{\lambda\mu\nu\rho} = \frac{1}{N_c} \sum_i F^{(i)}_{\lambda\mu\nu\rho}. \qquad (8.47a)$$

When computing $F_{\lambda\mu\nu\rho}$, as in the case of $M_{\lambda\mu\nu\rho}$, we must transform the components of the tensors $F^{(i)}_{\lambda\mu\nu\rho}$ belonging to different

Table 9.1

| | |
|---|---|
| $F^{(i)}_{1122}$ | $-\dfrac{3(m_\perp+2m_\parallel)m_\perp}{(m_\parallel+2m_\perp)^2}$ |
| $F^{(i)}_{1133}=F^{(i)}_{2233}$ | $-\dfrac{3(m_\perp+2m_\parallel)m_\parallel}{(m_\parallel+2m_\perp)^2}$ |
| $F^{(i)}_{3311}=F^{(i)}_{3322}$ | $-\dfrac{3(m_\perp+2m_\parallel)m_\perp^2}{(m_\parallel+2m_\perp)^2 m_\parallel}$ |
| $F^{(i)}_{1212}=F^{(i)}_{1313}=F^{(i)}_{2323}$ | $\dfrac{3(m_\perp+2m_\parallel)m_\perp}{2(m_\parallel+2m_\perp)^2}$ |

minima from the main axes of the energy ellipsoids to a coordinate system connected with the crystal's axes. In case A this is an elementary operation; in case B we must compile a table of directional cosines between the main axes of the energy ellipsoids and the axes coinciding with the cube's edges of the unit cell. The results of such computations are presented in Table 9.2.

Write the electric current corresponding to the tensor component $\sigma_{\lambda\mu\nu\rho}$ (8.47) in the form

$$\frac{\langle\tau^3\rangle}{\langle\tau\rangle^3}\times\left(\frac{m'}{m''}\right)^4 F_{\lambda\mu\nu\rho}\times\left(\sigma E_\mu\frac{\mu H_\nu}{c}\frac{\mu H_\rho}{c}\right). \tag{8.48}$$

Here the first dimensionless multiplier of the order of unity reflects the dependence of $\tau$ on the energy $\varepsilon$, if $\tau$ is independent of $\varepsilon$, it is equal to 1. The second dimensionless multiplier is connected only with the ratio of the effective mass tensor components and with the shape of the energy ellipsoids in the Brillouin zone. Finally, the third multiplier is equal to the product of the ohmic current $\sigma E_\mu$ by two small dimensionless multipliers proportional to the magnetic field.

It will be seen from Table 9.2 that generally the symmetrical tensor $\sigma_{\lambda\mu\nu\rho}$ in a cubic crystal is described by three independent quantities: $\sigma_{\lambda\lambda\lambda\lambda}$, $\sigma_{\lambda\lambda\mu\mu}$, $\sigma_{\lambda\mu\lambda\mu}$ and $\sigma_{\lambda\mu\mu\lambda}$. The current corresponding to this tensor, for example in the direction of axis 1, is equal to

$$j_1=\sum_{\mu\nu\rho}\sigma_{1\mu\nu\rho}E_\mu H_\nu H_\rho=\sigma_{1111}E_1H_1^2+\sigma_{1122}E_1H_2^2+\sigma_{1133}E_1H_3^2$$
$$+(\sigma_{1212}+\sigma_{1221})E_2H_1H_2+(\sigma_{1313}+\sigma_{1331})E_3H_1H_3. \tag{8.49}$$

Writing the coefficients in a symmetrical form similar to (8.41) and using the same notation for them, we obtain

$$j_1=\sigma_{1111}E_1H_1^2+\sigma_{1122}E_1H_2^2+\sigma_{1133}E_1H_3^2+2\sigma_{1212}E_2H_1H_2+$$
$$+2\sigma_{1313}E_3H_1H_3 \tag{8.49a}$$

Table 9.2

| | $F_{\lambda\lambda\lambda\lambda}$ | $F_{\lambda\lambda\nu\nu}$ | $F_{\lambda\mu\lambda\mu} = F_{\lambda\mu\mu\lambda}$ |
|---|---|---|---|
| $\langle 100 \rangle$ | $0$ | $-\dfrac{\left(1+2\dfrac{m_\parallel}{m_\perp}\right)\left[1+\dfrac{m_\parallel}{m_\perp}+\left(\dfrac{m_\parallel}{m_\perp}\right)^2\right]}{\dfrac{m_\parallel}{m_\perp}\left(2+\dfrac{m_\parallel}{m_\perp}\right)^2}$ | $\dfrac{3\left(1+2\dfrac{m_\parallel}{m_\perp}\right)}{2\left(2+\dfrac{m_\parallel}{m_\perp}\right)^2}$ |
| $\langle 111 \rangle$ | $-\dfrac{2\left(\dfrac{m_\parallel}{m_\perp}-1\right)^2\left(1+2\dfrac{m_\parallel}{m_\perp}\right)}{3\dfrac{m_\parallel}{m_\perp}\left(2+\dfrac{m_\parallel}{m_\perp}\right)^2}$ | $-\dfrac{\left(1+2\dfrac{m_\parallel}{m_\perp}\right)\left[2+5\dfrac{m_\parallel}{m_\perp}+2\left(\dfrac{m_\parallel}{m_\perp}\right)^2\right]}{3\dfrac{m_\parallel}{m_\perp}\left(2+\dfrac{m_\parallel}{m_\perp}\right)^2}$ | $\dfrac{\left(1+2\dfrac{m_\parallel}{m_\perp}\right)\left[2+5\dfrac{m_\parallel}{m_\perp}+2\left(\dfrac{m_\parallel}{m_\perp}\right)^2\right]}{6\dfrac{m_\parallel}{m_\perp}\left(2+\dfrac{m_\parallel}{m_\perp}\right)^2}$ |
| | $\sigma_{\lambda\lambda\lambda\lambda}/\sigma$ | $\sigma_{\lambda\lambda\nu\nu}/\sigma$ | $\sigma_{\lambda\mu\lambda\mu}/\sigma = \sigma_{\lambda\mu\mu\lambda}/\sigma$ |
| | $-(a+b+c)$ | $-a-(R\sigma)^2$ | $-\dfrac{1}{2}b+\dfrac{1}{2}(R\sigma)^2$ |

Denoting

$$\sigma_{1111} = \gamma', \quad \sigma_{1212} = \sigma_{1313} = \alpha, \quad 2\sigma_{1212} = 2\sigma_{1313} = \beta, \tag{8.50}$$

write (8.49a) in the form

$$j_1 = \alpha E_1 (H_1^2 + H_2^2 + H_3^2) + \beta H_1 (E_1H_1 + E_2H_2 + E_3H_3) + \gamma E_1 H_1^2 \tag{8.51}$$

where

$$\gamma = \gamma' - \alpha - \beta. \tag{8.51a}$$

It follows from expression (8.51) that the density of the appropriate current is

$$\mathbf{j} = \alpha H^2 \mathbf{E} + \beta(\mathbf{E}\cdot\mathbf{H})\,\mathbf{H} + \gamma\mathbf{F}, \tag{8.52}$$

where the vector **F** is

$$\mathbf{F}\,(E_1H_1^2,\ E_2H_2^2,\ E_3H_3^2), \tag{8.52a}$$

i.e., its $\mu$-axis component is $E_\mu H_\mu^2$.

Taking into account (8.32), we obtain for the total current

$$\mathbf{j} = \sigma\mathbf{E} + \eta\mathbf{E}\times\mathbf{H} + \alpha H^2\mathbf{E} + \beta\,(\mathbf{E}\cdot\mathbf{H})\,\mathbf{H} + \gamma\mathbf{F}, \tag{8.53}$$

i.e., the current is characterized by five constants $\sigma$, $\eta$, $\alpha$, $\beta$ and $\gamma$. In the isotropic case, i.e., in the case of a spherical constant-energy surface, because of symmetry

$$\alpha = \beta \quad \text{and} \quad \gamma = 0. \tag{8.53a}$$

The general form of (8.53) with indefinite constants $\sigma$, $\eta$, $\alpha$, $\beta$, and $\gamma$ can be obtained for cubic crystals solely from considerations of symmetry. Hence it follows that any correct electron theory for the current **j** in the $H^2$ approximation must in the case of cubic crystals lead to an expression of the form (8.53).

The parameter measured in the experimental studies of magnetoresistance is not the dependence of the current **j** on the fields **E** and **H**, but the voltage (i.e., the electric field **E**) as a function of **j** and **H**; therefore, it would be desirable to solve expression (8.53) for **E**. It can, however, be demonstrated that for a cubic crystal one can, as in the case of (8.53), obtain the expression

$$\mathbf{E} = \rho\mathbf{j} + g\mathbf{j}\times\mathbf{H} + aH^2\mathbf{j} + b\,(\mathbf{j}\cdot\mathbf{H})\,\mathbf{H} + c\mathbf{L}, \tag{8.54}$$

where $\rho$, $g$, $b$, $c$ are constants and the components of the vector **L** are $j_1H_1^2$, $j_2H_2^2$, and $j_3H_3^2$, solely from the considerations of symmetry. The composition of expressions (8.53) and (8.54) is similar.

Substituting (8.54) into (8.53), leaving only the members of the order not exceeding $H^2$ and comparing the coefficients in front of **j**,

$\mathbf{j} \times \mathbf{H}$, $H^2\mathbf{j}$, $(\mathbf{j}\cdot\mathbf{H})$ and $\mathbf{L}$, we obtain a simple system of equations for $\rho$, $g$, $a$, $b$, $c$ whence

$$\left.\begin{aligned} &\rho=\frac{1}{\sigma}, \quad g=-\frac{\eta}{\sigma}, \quad a=-\frac{(\alpha+\eta^2/\sigma)}{\sigma}, \\ &b=-\frac{(\beta-\eta^2/\sigma)}{\sigma}, \quad c=-\frac{\gamma}{\sigma}. \end{aligned}\right\} \tag{8.55}$$

In the absence of a magnetic field, the electric field $\mathbf{E}$ coincides in direction with the current $\mathbf{j}$; therefore, the conductivity is $\rho = 1/\sigma = E/j$. In the presence of a magnetic field, when $\mathbf{E}$ is not parallel to $\mathbf{j}$, the magnetoresistance is $\rho_H = E_j/j = (\mathbf{E}\cdot\mathbf{j})/j^2$, where $E_j$ is the projection of the electric field on the direction of the current. Making use of (8.54), we obtain

$$\rho_H=\frac{\mathbf{E}\cdot\mathbf{j}}{j^2}=\rho+\rho a H^2+\rho b\,\frac{(\mathbf{j}\cdot\mathbf{H})^2}{j^2}+\rho c\,\frac{\mathbf{L}\cdot\mathbf{j}}{j^2}. \tag{8.56}$$

Hence

$$\frac{\Delta\rho}{\rho}=\frac{\rho_H-\rho}{\rho}=aH^2+\frac{b\,(j_1H_1+j_2H_2+j_3H_3)^2}{j^2}+c\,\frac{j_1^2H_1^2+j_2^2H_2^2+j_3^2H_3^2}{j^2}. \tag{8.56a}$$

The quantities $a$, $b$, $c$ can be determined from experiment. They are connected with $\sigma$, $\eta$ (or $R$), $\alpha$, $\beta$, $\gamma$ calculated from the theory, as demonstrated above, via expression (8.55). The ratios $\sigma_{\lambda\mu\nu\rho}/\sigma$ in Table 9.2 are expressed in terms of the constants $a$, $b$, $c$ and $R\sigma$ and can be measured in experiment.

It follows from the aforesaid that the Hall effect and magnetoresistance measurements in semiconductor single crystals for different current and magnetic field directions provide valuable information on the effective masses of charge carriers and on the location of constant-energy surfaces in the Brillouin zone. Such experiments, carried out on $n$-germanium and $n$-silicon before cyclotron resonance in them had been studied, resulted in a correct picture of conduction electron energy spectra.

As was remarked above, it follows from (8.42) and (8.43) that if the current and magnetic field are parallel to an energy ellipsoid's main axis, the longitudinal magnetoresistance is zero, as in the case of spherical constant-energy surfaces. Studies of longitudinal magnetoresistance in silicon led to the discovery that when the current and the magnetic field parallel to it point in the ⟨100⟩ direction, the magnetoresistance drops almost to zero. At the same time the investigation of longitudinal magnetoresistance in germanium showed that there is no such direction in a crystal in which it decreased appreciably. A conclusion was drawn from this fact that the equivalent minima in silicon are arranged along the ⟨100⟩

directions, whereas in germanium they are arranged along the ⟨111⟩ directions.

We shall not consider here the case of strong magnetic fields for a multiellipsoid model discussed in special literature [9.12], [9.13]. We would only like to point out that in strong magnetic fields the Hall constant for a multiellipsoid model is also equal to

$$R = -1/cen,$$

as in all cases involving the isotropic model (5.16). The explanation is that in strong magnetic fields the Hall current is not connected with the scattering of electrons, but is determined by their free drift with the velocity $(E/H)\,c$ in the direction perpendicular both to **E** and **H**.

**9.8.6** In order to determine the differential thermoelectric power, we have to start from equation (8.4) in which we must make $\mathbf{H} = 0$. Then

$$\mathbf{v}\nabla_r f - \frac{e}{\hbar}\mathbf{E}\nabla_k f = -\frac{f-f_0}{\tau(\varepsilon)}. \tag{8.57}$$

Up to members linear in $\mathbf{E} = -\nabla\varphi$ ($\varphi$ is the electrostatic potential) and in $\nabla T$ $f = f_0(\varepsilon)$ can be substituted into the left-hand side of (8.57). Then

$$\nabla_r f \approx \nabla_r f_0 = \left(-\frac{\partial f_0}{\partial\varepsilon}\right)\left\{\frac{\varepsilon-\zeta}{T}\nabla T + \nabla\zeta\right\}, \tag{8.58}$$

since in the presence of a temperature gradient both $T$ and $\zeta(T)$ are temperature-dependent.

Taking into account (8.58) and (8.9), we obtain from (8.57)

$$\left(-\frac{\partial f_0}{\partial\varepsilon}\right)\mathbf{v}\left[\frac{\varepsilon-\zeta}{T}\nabla T - e\nabla\left(\varphi-\frac{\zeta}{e}\right)\right] = -\frac{f-f_0}{\tau}. \tag{8.59}$$

The electric current is

$$\mathbf{j} = -e\sum_{(k)}(f-f_0)\,\mathbf{v}$$
$$= e\sum_{(k)}\left(-\frac{\partial f_0}{\partial\varepsilon}\right)\tau(\varepsilon)\left\{\mathbf{v}\left[\frac{\varepsilon-\zeta}{T}\nabla T - e\nabla\left(\varphi-\frac{\zeta}{e}\right)\right]\right\}\mathbf{v}, \tag{8.60}$$

where the summation $\sum_{(k)}$ (integration) is performed over the wave vector **k**. For the current $\mathbf{j} = 0$, we obtain from (8.60)

$$k_0\sum_{(k)}\left(-\frac{\partial f_0}{\partial\varepsilon}\right)\tau(\varepsilon)\frac{\varepsilon-\zeta}{k_0T}(\mathbf{v}\nabla T)\,\mathbf{v}$$
$$= e\sum_{(k)}\left(-\frac{\partial f_0}{\partial\varepsilon}\right)\tau(\varepsilon)\left(\mathbf{v}\nabla\left(\varphi-\frac{\zeta}{e}\right)\right)\mathbf{v}. \tag{8.61}$$

If $\nabla T$ points in the direction of the $\mu$ axis, then $\mathbf{v}\nabla T = \mathbf{v}_\mu \nabla_\mu T$, and the expression on the left-hand side of (8.61) is equal to

$$\nabla_\mu T k_0 \sum_{(k)} \left(-\frac{\partial f_0}{\partial \varepsilon}\right) \tau \frac{\varepsilon-\zeta}{k_0 T} v_\mu^2$$

$$= \nabla_\mu T k_0 \frac{1}{m_\mu} \frac{2}{3} \sum_{(\varepsilon)} \left(-\frac{\partial f_0}{\partial \varepsilon}\right) \tau \frac{\varepsilon-\zeta}{k_0 T} \varepsilon$$

$$= \nabla_\mu T k_0 \frac{1}{m_\mu} \left\{\frac{\langle \tau\varepsilon \rangle}{k_0 T} - \frac{\langle \tau\zeta \rangle}{k_0 T}\right\}, \tag{8.62}$$

where we made use of the considerations discussed above when we derived expressions (8.12b), (8.12c) and (8.13). The vector in the right-hand side of (8.61) should obviously be parallel to $\nabla_\mu T$, so that the right-hand side should be equal to

$$\frac{e}{m_\mu} \nabla_\mu \left(\varphi - \frac{\zeta}{e}\right) \langle \tau \rangle. \tag{8.63}$$

It follows from (8.62) and (8.63) that the differential thermoelectric power is

$$\alpha = \frac{\nabla_\mu \left(\varphi - \frac{\zeta}{e}\right)}{\nabla_\mu T} = \left(\frac{k_0}{e}\right) \left\{\frac{\langle \tau\varepsilon \rangle}{k_0 T \langle \tau \rangle} - \frac{\zeta}{k_0 T}\right\}. \tag{8.64}$$

We see that the first member in the braces coincides with the result obtained for the isotropic model (spherical constant-energy surfaces). The value of the chemical potential is equal to

$$\frac{\zeta}{k_0 T} = \ln \frac{4\pi^3 \hbar^3 n}{(2\pi k_0 T)^{3/2} (m_1 m_2 m_3)^{1/2} N_c}, \tag{8.64a}$$

in accordance with (6.2.12b) and (6.2.22a).

**9.8.7** Finally we shall deal briefly with a simplified theory of transport phenomena in semiconductors of the *p*-germanium type.

In Section 4.15 we described the structure of the valence bands of *p*-germanium and *p*-silicon. The most important point about it is the double degeneracy of the energy spectrum of holes for $\mathbf{k} = 0$ and the ensuing presence of two kinds of holes: the heavy and the light. It was pointed out that on isotropic averaging over the $\mathbf{k}$-space the hole energy becomes proportional to $|\mathbf{k}|^2$. The ratio of scalar effective masses of the heavy and light holes in germanium is $m_1/m_2 = 8.0$.

In such an isotropic approximation for the energies and in the assumption that the relaxation times $\tau_1$ and $\tau_2$ of the heavy and light holes depend on their energies we can easily calculate all the kinetic coefficients.

The comparison between the theory and experiment demonstrates that in germanium the ratio of the light and heavy hole mobilities is approximately equal to the inverse ratio of their effective masses [9.14]

$$\mu_2/\mu_1 \approx m_1/m_2 = 8.0. \tag{8.65}$$

Should $\tau_1$ and $\tau_2$ be much less than the transition times from the state of heavy holes to the state of light holes (we shall term such transitions interband), we could calculate the relaxation times with the aid of the usual method, i.e., without taking into account the interband transitions, and in this case, according to (8.4.11), we would obtain for the interaction with acoustic vibrations $\frac{\langle\tau_1\rangle}{\langle\tau_2\rangle} = \left(\frac{m_2}{m_1}\right)^{3/2}$. Here we assume that the deformation potential constant $\mathscr{E}_1$ (8.5.15) (or the interaction constant $C$) is the same for both heavy and light holes. A possible explanation for this assumption is that $\mathscr{E}_1$, according to (8.5.4), is equal to the shift of the valence band edge per relative variation of the crystal's volume caused by uniform compression. Since uniform compression does not remove the degeneracy, it follows that $\mathscr{E}_1$ should be the same for both types of holes.

Making use of the usual equation for the mobility (3.3), we obtain $\frac{\mu_2}{\mu_1} = \left(\frac{m_1}{m_2}\right)^{5/2}$, this being in contradiction with (8.65). This demonstrates that an important part in scattering should be played by interband transitions. It follows from (8.65) that the relaxation times of both heavy and light holes should be equal.

Determine the relaxation time of, for example, the heavy holes; take into account interband transitions and make the simplifying assumption that the transition probability in the act of scattering by acoustic vibrations has a simple isotropic form (8.3.11a), (8.3.11b) [9.15] even in the case of a degenerate valence band.

Assuming the collisions with phonons to be elastic, substituting $k_0T/\hbar\omega_q$ for $N_q$ and $N_q + 1$ and putting for the acoustic branch $\omega_q = v_0 q$ ($v_0$ is the velocity of longitudinal waves), we obtain an equal probability

$$W(\mathbf{k}, \mathbf{k}') = W_0 \delta(\varepsilon_{\mathbf{k}'} - \varepsilon_{\mathbf{k}}), \tag{8.66}$$

where

$$W_0 = \frac{4\pi C^2 k_0 T}{9NM\hbar v_0^2} \tag{8.66a}$$

both for the emission and for the absorption of a phonon.

Denote the distribution functions for the heavy and light holes by $f_1(\mathbf{k})$ and $f_2(\mathbf{k})$, respectively. Then the variation of $f_1$ due to collisions will, according to (8.2.8), be equal to

$$\left(\frac{\partial f_1}{\partial t}\right)_c = -W_0 \sum_{\mathbf{k}'} \{[f_1(\mathbf{k}) - f_1(\mathbf{k}')]\,\delta[\varepsilon_1(\mathbf{k}) - \varepsilon_1(\mathbf{k}')] + [f_1(\mathbf{k}) - f_2(\mathbf{k}')]\,\delta[\varepsilon_1(\mathbf{k}) - \varepsilon_2(\mathbf{k}')]\}, \tag{8.67}$$

where

$$\varepsilon_1(\mathbf{k}) = \hbar^2 k^2/2m_1, \quad \varepsilon_2(\mathbf{k}) = \hbar^2 k^2/2m_2 \tag{8.67a}$$

are the energies of the heavy and light holes.

The first expression in square brackets in (8.67) accounts for the transitions of the heavy holes to the heavy hole states (from state $\mathbf{k}$ to state $\mathbf{k}'$), and the second expression accounts for the transitions of the heavy holes to light hole states. Make, as usual,

$$f_1(\mathbf{k}) = f_{10}(\varepsilon) + \mathbf{g}_1(\varepsilon)\cdot\mathbf{k},$$
$$f_2(\mathbf{k}) = f_{20}(\varepsilon) + \mathbf{g}_2(\varepsilon)\cdot\mathbf{k}. \tag{8.68}$$

Substituting $f_1(\mathbf{k})$ into the first expression in square brackets of (8.67) and integrating with respect to $\mathbf{k}'$, we obtain a result coinciding with (Section 8.4)

$$\left(\frac{\partial f_1}{\partial t}\right)_c^{(11)} = -\frac{W_0(2\varepsilon)^{1/2}}{\pi^2\hbar^3} m_1^{3/2}(\mathbf{g}_1\cdot\mathbf{k}) = -\frac{(\mathbf{g}_1\cdot\mathbf{k})}{\tau_{11}}, \tag{8.69}$$

where $\tau_{11}$ is the heavy hole relaxation time due to their transitions inside the band.

Note that

$$f_{10}(\varepsilon) = f_{20}(\varepsilon) = e^{\frac{\zeta-\varepsilon}{k_0 T}}, \tag{8.70}$$

since in equilibrium the chemical potentials of the heavy and light holes are identical.

Substituting $f_1(\mathbf{k})$ and $f_2(\mathbf{k})$ into the second expression in square brackets of (8.67) and making use of (8.70), we easily see that the member proportional to $\mathbf{g}_2(\varepsilon)\,\mathbf{k}'$ vanishes as a result of integration with respect to $\mathbf{k}'$, and, therefore,

$$\left(\frac{\partial f_1}{\partial t}\right)_c^{1/2} = -\frac{W_0(2\varepsilon)^{1/2}}{\pi^2\hbar^3} m_2^{3/2}(\mathbf{g}_1\cdot\mathbf{k}) = -\frac{(\mathbf{g}_1\cdot\mathbf{k})}{\tau_{12}}, \tag{8.71}$$

where $\tau_{12}$ is the relaxation time for the transition of the heavy holes to the light hole state. Eventually the total relaxation time of the heavy holes is

$$\frac{1}{\tau_1} = \frac{1}{\tau_{11}} + \frac{1}{\tau_{12}} = \frac{W_0(2\varepsilon)^{1/2}}{\pi^2\hbar^3}(m_1^{3/2} + m_2^{3/2}). \tag{8.72}$$

A similar calculation for the light holes will easily be seen to yield a result for the inverse relaxation time $1/\tau_2$ coinciding with (8.72).

Hence, our approximation of isotropic scattering probability (8.66) and of a spherical constant-energy surface (8.67a) results in identical relaxation times for heavy and light holes, i.e.,

$$\tau_1 = \tau_2, \tag{8.73}$$

whence (8.65) follows immediately.

Conductivity due to holes of both types is, according to (3.3), equal to

$$\sigma = \frac{n_1 e^2}{m_1}\langle\tau_1\rangle + \frac{n_2 e^2}{m_2}\langle\tau_2\rangle = n_1 e\mu_1 + n_2 e\mu_2. \tag{8.74}$$

Here $n_i$ and $\mu_i$ are concentrations and mobilities of the light and heavy holes.

In order to find the Hall constant and the magnetoresistance make use of equations (5.4), where both the heavy and light hole currents determine $a_1$ and $a_2$:

$$a_1 = \frac{n_1 e^2}{m_1}\left\langle\frac{\tau_1}{1+(\gamma_1\tau_1 H)^2}\right\rangle + \frac{n_2 e^2}{m_2}\left\langle\frac{\tau_2}{1+(\gamma_2\tau_2 H)^2}\right\rangle,$$

$$a_2 = -\frac{n_1 e^2}{m_1}(\gamma_1 H)\left\langle\frac{\tau_1^2}{1+(\gamma_1\tau_1 H)^2}\right\rangle - \frac{n_2 e^2}{m_2}(\gamma_2 H)\left\langle\frac{\tau_2^2}{1+(\gamma_2\tau_2 H)^2}\right\rangle. \tag{8.75}$$

Here both addends in $a_2$ are negative, since, as before, $\gamma_1 = \frac{e}{m_{1c}} > 0$ and $\gamma_2 = \frac{e}{m_{2c}} > 0$, and $a_2$ for electrons is positive.

Should we take for $a_1$ and $a_2$ in (8.75) an approximation linear in $H$, we would obtain from equation (5.5a) for the Hall constant

$$R = \frac{1}{cen_1}\,\frac{\langle\tau_1^2\rangle + \left(\frac{n_2}{n_1}\right)\left(\frac{m_1}{m_2}\right)^2\langle\tau_2^2\rangle}{\left[\langle\tau_1\rangle + \left(\frac{n_2}{n_1}\right)\left(\frac{m_1}{m_2}\right)\langle\tau_2\rangle\right]^2}. \tag{8.76}$$

Assuming the energy dependence of $\tau_1$ and $\tau_2$ to be determined by expression (3.5a)[11], we can easily demonstrate that

$$\langle\tau^2\rangle = \frac{3\pi}{8}\left(\frac{m}{e}\right)\mu^2, \tag{8.77}$$

where $\mu$ is the mobility.

---

[11] Although germanium is a typical atomic crystal, this assumption cannot be regarded as a well-founded one, since the temperature dependence of mobility does not obey the law (6.9).

From (8.76), (8.77) and the expression for mobility (3.3), we obtain

$$R=\frac{3\pi}{8}\frac{1}{cen_1}\frac{1+(n_2/n_1)(\mu_2/\mu_1)^2}{[1+(n_2/n_1)(\mu_2/\mu_1)]^2}. \qquad (8.76a)$$

Making use of the quadratic approximation for $a_1$ and $a_2$ of the form (5.9), we can calculate the magnetoresistance in weak fields. Assuming that the hole relaxation time is expressed by (3.5a), we obtain in the result of a similar calculation

$$\frac{\Delta\rho}{\rho}=\frac{9\pi}{16}\left(\frac{\mu_1 H}{c}\right)^2\left[\frac{1+(n_2/n_1)(\mu_2/\mu_1)^3}{1+(n_2/n_1)(\mu_2/\mu_1)} - \frac{\pi}{4}\left(\frac{1+(n_2/n_1)(\mu_2/\mu_1)^2}{1+(n_2/n_1)(\mu_2/\mu_1)}\right)^2\right]. \qquad (8.78)$$

Making the concentration of light holes $n_2 = 0$, we obtain from (8.76a) and (8.78) expressions (5.12a)[12].

The ratio of the light and heavy hole mobilities can to a good approximation be assumed to be equal to the inverse ratio of their effective masses (8.65). We see the three expressions (8.74), (8.76a) and (8.78) to contain three unknowns $n_1$, $\mu_1$ and $n_2/n_1$.

That is why measurements of conductivity, Hall effect and magnetoresistance are, generally, adequate for the determination of the concentrations and mobilities of the light and heavy holes.

Actually, experimental data and theoretical equations used for comparing theory with experiment [9.14] are not limited to the case of weak fields, but include the cases of strong and intermediate fields as well.

It turns out that the agreement cetween theory and experiment is best when the concentration ratio is $n_2/n_1 = 00.2$; this is in contradiction with the data on effective masses obtained from cyclotron resonance experiments, since, according to (6.2.12),

$$\frac{n_2}{n_1}=\left(\frac{m_2}{m_1}\right)^{3/2}=(8.0)^{-3/2}=0.04$$

This contradiction is probably due to some simplifying assumptions employed in the theory and to the questionable assumption about the energy dependence of the relaxation time (see footnote 11).

## 9.9 Transport Phenomena in Semiconductors with a Spherical Nonparabolic Band

**9.9.1** Consider transport phenomena in semiconductors with charge carriers of a type whose energy depends on the magnitude of the wave vector $|\mathbf{k}| = k$, but is arbitrary in other respects, i.e., we

[12] Of course, the same result can be obtained if we make the heavy hole concenration $n_1 = 0$, but to do this one should transform (8.76a) and (8.78).

are dropping the assumption that the energy $\varepsilon$ is a homogeneous quadratic function of the wave vector components $k_x$, $k_y$ and $k_z$. Such a situation may occur in a semiconductor with a narrow forbidden band and is realized, for example, in indium antimonide (InSb).

The expression for the electron current (2.15)[13]

$$\mathbf{j} = -\frac{e}{3\pi^2}\int_0^\infty \chi(k)\left(-\frac{\partial f_0}{\partial \varepsilon}\right) u(k)\, k^4\, dk \tag{9.1}$$

is also valid in the case of a nonstandard spherical band. Here, in accordance with (2.2a)

$$u(k) = \frac{1}{\hbar k}\frac{\partial \varepsilon(k)}{\partial k}. \tag{9.1a}$$

Consider first electric conductivity in isothermic conditions ($\nabla T = 0$) and in the absence of a magnetic field ($H = 0$). In this case it follows from (2.11) that

$$\chi(k) = -eu(k)\,\tau(k)\,\mathbf{E} = -\frac{e}{\hbar k}\frac{\partial \varepsilon(k)}{\partial k}\,\tau(k)\,\mathbf{E}. \tag{9.1b}$$

Substituting (9.1a), (9.1b) into (9.1), we obtain for the conductivity

$$\sigma = \frac{j}{E} = \frac{e^2}{3\pi^2\hbar^2}\int_0^\infty \left(-\frac{\partial f_0}{\partial \varepsilon}\right)\left(\frac{\partial \varepsilon}{\partial k}\right)^2 \tau(k)\, k^4\, dk. \tag{9.2}$$

Make

$$\frac{\partial \varepsilon(k)}{\partial k} = \frac{\hbar^2 k}{m(\varepsilon)}, \tag{9.2a}$$

where $m(\varepsilon)$ is the effective mass dependent on the electron energy $\varepsilon$. In the case of a standard band (2.13), $m(\varepsilon) = m$.

Substitute (9.2a) into (9.2) and change over from the integration variable $k$ to $\varepsilon$:

$$\sigma = \frac{e^2}{3\pi^2}\int_0^\infty \left(-\frac{\partial f_0}{\partial \varepsilon}\right)\tau(\varepsilon)\frac{k^3(\varepsilon)}{m(\varepsilon)}\, d\varepsilon. \tag{9.3}$$

For conduction electrons in InSb the energy is, according to (4.13.36b), equal to

$$\varepsilon = \frac{\hbar^2 k^2}{2m_0} + \frac{1}{2}\varepsilon_G + \frac{(\varepsilon_G^2 + 8P^2k^2/3)^{1/2}}{2}.$$

[13] We omit the suffix $n$.

Here $m_0$ is the electron mass in vacuum, $P$ is a constant characterizing the "interaction" between the valence and conduction bands equal to (4.13.30b), $\varepsilon_G$ is the forbidden bandwidth. In the above expression the energy $\varepsilon$ is measured from the top edge of the valence band. In the future it will be more convenient to measure the energy from the bottom edge of the valence band; to this end we must substitute $\varepsilon + \varepsilon_G$ for $\varepsilon$. Then

$$\varepsilon = \frac{\hbar^2 k^2}{2m_0} + \frac{\varepsilon_G}{2}\left(\sqrt{1 - \frac{8P^2k^2}{3\varepsilon_G^2}} - 1\right). \tag{9.4}$$

For $Pk \ll \varepsilon_G$, i.e., for small $k$, the root can be expanded into a series. Then

$$\varepsilon = \frac{\hbar^2 k^2}{2m_0} + \frac{2P^2k^2}{3\varepsilon_G} = \frac{\hbar^2 k^2}{2m(0)}. \tag{9.5}$$

Here $m(0)$ is the electron effective mass at the bottom of the conduction band. From cyclotron resonance experiments we know that $m(0) = 0.013\, m_0$, i.e., that $m(0) \ll m_0$. In this approximation we obtain from (9.4)

$$\varepsilon = \frac{\varepsilon_G}{2}\left(\sqrt{1 + \frac{2\hbar^2 k^2}{m(0)\,\varepsilon_G}} - 1\right). \tag{9.6}$$

It follows hence, that

$$k(\varepsilon) = \frac{(2m(0)\,\varepsilon)^{1/2}}{\hbar}\left(1 + \frac{\varepsilon}{\varepsilon_G}\right)^{1/2}. \tag{9.6a}$$

From (9.2a) and (9.6) we obtain

$$m(\varepsilon) = m(0)\left(1 + \frac{2\varepsilon}{\varepsilon_G}\right). \tag{9.7}$$

The multiplier $(1 + 2\varepsilon/\varepsilon_G)$ on the right-hand side describes the increase in $m(\varepsilon)$ with $\varepsilon$ for $\varepsilon_G \to \infty$ $m(\varepsilon) = m(0)$, and the band becomes parabolic in shape.

Investigate the variation of the relaxation time $\tau(\varepsilon)$ in the case of a nonstandard band.

Consider the relaxation time due to the interaction of the electron with acoustic vibrations. It was estimated that in the case of a nonstandard spherical band the interaction of the electrons with the acoustic vibrations is also almost elastic. This means that the part of the phonon energy $\hbar\omega_q = \hbar v_0 q$ ($v_0$ is the speed of sound) in the energy conservation law can be neglected. It will be easily seen that in this case (8.4.1d), $0 \leqslant q \leqslant 2k$.

Expression (8.4.7) assumes the form

$$\frac{1}{\tau}=-\frac{V}{(2\pi)^3}\int_0^{2k} q^2\,dq\int_0^{\pi}\sin\vartheta\,d\vartheta\int_0^{2\pi}d\varphi\Big\{w(q)\,N_q\delta\,[\varepsilon((\mathbf{k}+\mathbf{q})^2)-\varepsilon(\mathbf{k}^2)]\frac{q\cos\beta}{k\cos\beta}-w(q)(N_q+1)\,\delta\,[\varepsilon(\mathbf{k}-\mathbf{q})^2-\varepsilon(\mathbf{k}^2)]\frac{q\cos\alpha}{k\cos\beta}\Big\}. \quad (9.8)$$

Here $q\cos\alpha$ and $k\cos\beta$ are projections of $\mathbf{q}$ and $\mathbf{k}$ on the vector $\chi$. The $\delta$-functions are obviously equal to

$$\delta\,[\varepsilon((\mathbf{k}\pm\mathbf{q})^2)-\varepsilon(\mathbf{k}^2)] = \delta\,[\varepsilon(k^2+q^2\pm 2kq\cos\vartheta)-\varepsilon(\mathbf{k}^2)] = \frac{1}{\left[\pm\frac{\partial\varepsilon}{\partial(k^2)}\,2kq\right]}\,\delta\left(\cos\vartheta\mp\frac{q}{2k}\right), \quad (9.9)$$

where $\mp q/2k$ are the roots of the equation

$$\varepsilon(k^2+q^2\pm 2kq\cos\vartheta)-\varepsilon(k^2)=0$$

or of the equation

$$k^2+q^2\pm 2kq\cos\vartheta-k^2=0.$$

Proceeding in the same way as we did in Section 8.4, we obtain

$$\frac{1}{\tau}=\frac{V}{8\pi^2}\frac{1}{k^2}\left(\frac{\partial\varepsilon}{\partial k}\right)^{-1}\int_0^{2k} w(q)\,(2N_q+1)\,q^3\,dq, \quad (9.10)$$

whence

$$\tau=\frac{9\pi}{4}\,\frac{Mv_0^2\hbar}{\Omega_0 c^2 k_0 T}\left(\frac{\partial\varepsilon}{\partial k}\right)\frac{1}{k^2}. \quad (9.11)$$

In the case of a standard band this immediately reduces to (8.4.11). Substituting (9.2a), (9.7) and (9.6a) into (9.11), we obtain

$$\tau=\frac{9\pi}{2}\,\frac{\rho v_0^2\hbar^4}{C^2\,[2m\,(0)\,k_0T]^{3/2}}\left(\frac{\varepsilon}{k_0T}\right)^{-1/2}\frac{\left(1+\frac{\varepsilon}{\varepsilon_G}\right)^{-1/2}}{1+\frac{2\varepsilon}{\varepsilon_G}}. \quad (9.12)$$

Here the crystal's density is $\rho=M/\Omega_0$, the second multiplier, which depends on $\varepsilon_G$, accounts for the deviation of the spherical band from the parabolic shape. For $\varepsilon_G\to\infty$ (9.12) reduces to (8.4.11).

Write the relaxation time (9.12) in the form

$$\tau=\tau_0 x^r\,\frac{(1+\beta x)^r}{1+2\beta x}, \quad (9.13)$$

where $x = \varepsilon/k_0T$, $\beta = k_0T/\varepsilon_G$ is a parameter characterizing the deviation of the shape of the band from parabolic, and $r = -1/2$. It can easily be demonstrated that this expression for the relaxation time is also applicable in the case of other scattering mechanisms having different values of $\tau_0$ and $r$ (3.5b)-(3.5d). Making use of (9.13), (9.6a) and (9.7), we obtain for the conductivity (9.3)

$$\sigma = \frac{e^2}{3\pi^2}\frac{[2m(0)\,k_0T]^{3/2}}{\hbar^3 m(0)}\tau_0 \int_0^\infty \left(-\frac{\partial f_0}{\partial x}\right)\frac{(x+\beta x^2)^{r+3/2}}{(1+2\beta x)^2}\,dx, \tag{9.14}$$

where $f_0 = [1 + \exp(x - z)]^{-1}$, $z = \zeta/k_0T$.

The coefficients of the other kinetic effects (the thermoelectric, galvanomagnetic and thermomagnetic) can be calculated in a similar way [9.16, Chap. V]. These coefficients are expressed in terms of integrals similar to (9.14) and are of the form

$$\mathcal{J}^m_{n,\,k}(z,\beta) = \int_0^\infty \left(-\frac{\partial f_0}{\partial x}\right)\frac{x^m(x+\beta x^2)^n}{(1+2\beta x)^k}\,dx. \tag{9.15}$$

They are sometimes called generalized or biparametric Fermi integrals. For $\beta = 0$ they are expressed in terms of the Fermi integrals (6.2.6). Indeed,

$$\mathcal{J}^m_{n,\,k}(z,0) = -\int_0^\infty x^{m+n}\frac{\partial f_0}{\partial x}\,dx = -\int_0^\infty x^{m+n}\,df_0$$

$$= -x^{m+n}f_0\Big|_0^\infty (m+n)\int_0^\infty f_0(x)\,x^{m+n-1}\,dx$$

$$= (m+n)\,\mathcal{F}_{m+n-1}(z). \tag{9.15a}$$

The tabulated generalized Fermi integrals (9.15) for parameters $-5 \leqslant z \leqslant 20$ and $0 \leqslant \beta \leqslant 1$ are given in Appendix Д in [9.16].

Express the electron concentration $n$ in terms of the chemical potential $z = \zeta/k_0T$ and the parameter $\beta = k_0T/\varepsilon_G$

$$n = 2\sum_k f_0(\varepsilon) = 2\int f_0(\varepsilon)\frac{d^3k}{(2\pi)^3} = \frac{1}{\pi^2}\int f_0(\varepsilon)\,k^2\,dk$$

$$= \frac{1}{3\pi^2}\int f_0(\varepsilon)\,dk^3 = \frac{1}{3\pi^2}\int\left(-\frac{\partial f_0}{\partial\varepsilon}\right)k^3(\varepsilon)\,d\varepsilon. \tag{9.16}$$

Substituting herein $k(\varepsilon)$ from (9.6a), we obtain

$$n = \frac{[2m(0)\,k_0T]^{3/2}}{3\pi^2\hbar^3}\int\left(-\frac{\partial f_0}{\partial x}\right)(x+\beta x^2)^{3/2}\,dx$$

$$= \frac{[2m(0)\,k_0T]^{3/2}}{3\pi^2\hbar^3}\mathcal{J}^0_{3/2,\,0}(z,\beta). \tag{9.16a}$$

It follows hence and from (9.15a) that for a parabolic band ($\beta = 0$)

$$n = \frac{(2mk_0T)^{3/2}}{2\pi^2\hbar^3}\mathcal{F}_{1/2}(z), \tag{9.16b}$$

this coinciding with (6.2.5)

In the case of a strongly degenerate semiconductor $-\partial f_0/\partial\varepsilon = \delta(\varepsilon - \zeta)$; therefore, it follows from (9.16) that

$$k(\zeta) = (3\pi^2 n)^{1/3}. \tag{9.17}$$

Hence and from (9.6) it follows that

$$\zeta = \frac{\varepsilon_G}{2}\left[\sqrt{1 + \frac{2\hbar^2(3\pi^2 n)^{2/3}}{m(0)\varepsilon_G}} - 1\right]. \tag{9.18}$$

Substituting this expression into (9.7), we obtain

$$m(\zeta) = m(0)\sqrt{1 + \frac{2\hbar^2(3\pi^2 n)^{2/3}}{m(0)\varepsilon_G}}. \tag{9.19}$$

It follows that $m(\zeta)$ in indium antimonide increases almost four-fold, when the concentration $n$ changes from $10^{16}$ to $5 \times 10^{18}$ cm$^{-3}$.

A paper by J. Kolodziejczak [9.17] contains the dependence of mobility $\sigma/en$ on concentration for $n$-InSb, which in the case of a parabolic band does not exist at all.

From (9.18) and (9.19) we obtain a simple relationship between the Fermi level $\zeta$ and the corresponding effective mass

$$\zeta = \frac{\varepsilon_G}{2}\left[\frac{m(\zeta)}{m(0)} - 1\right]. \tag{9.20}$$

## 9.10 Phonon Drag Effect in Semiconductors

**9.10.1** In all the preceding paragraphs of this chapter we assumed the phonons to be in a state of a statistical equilibrium, i.e., the phonon occupancy numbers $N_q$ are determined by Planck's function (8.4.10). At the same time it is obvious that in the presence of an electric current and electron scattering by phonons the directional momenta of the electrons must be transmitted to the phonons, and because of this their distribution cannot remain isotropic.

This circumstance has not been taken into account in the preceding equations, but still it is clear that the reciprocal effect of the deviation of the phonon distribution from its equilibrium value on the deviation of the electron distribution from its equilibrium value is an effect of the second order and can practically be neglected. However, the deviation of the phonon distribution function from its equilibrium value can be the result of a temperature gradient in the crystal.

As was first shown by L. E. Gurevich (1945), the deviation of the phonon distribution from its equilibrium value caused by a temperature gradient can under certain conditions play a substantial part in thermoelectric phenomena in metals. This effect became known as *phonon drag.*

The most interesting phenomena involving phonon drag are observed in semiconductors. In 1953 H. P. R. Frederikse and independently of him T. H. Geballe, observed a considerable increase in the thermoelectric power $\alpha$ in $p$-germanium at low temperatures, not predicted by the usual theory.

Herring and Frederikse correctly interpreted this phenomenon as the effect of phonon drag of holes in germanium. Indeed, the temperature gradient causes a flux of phonons to flow from the hot end of the specimen to the cold end. This flux drags charge carriers with it, thereby causing an increase in the thermoelectric current. As we shall see below, for a correct estimate of the magnitude of the drag effect we have to take into account that due to the conservation laws, the electrons scattered by acoustic vibrations can interact only with phonons whose wave vector $\mathbf{q}$ is of the order of the electron wave vector $\mathbf{k}$. For this reason the relaxation time of phonons in the equations for phonon drag does not coincide with the relaxation time of phonons determined from lattice heat conductivity. We shall consider the theory of the phonon drag effect in semiconductors with one type of charge carriers in the absence of degeneracy and on the assumption of a simple energy-band structure (2.13).

**9.10.2** The quantitative theory of the phonon drag effect is based on the kinetic equation.[14] As we shall see below, the phonon drag effect is in a sense connected with the inelastic character of electron scattering by acoustic vibrations, which is usually ignored. Hence, the determination of $(\partial f/\partial t)_c$ should be based on the general expression (8.2.7)

$$\left[\frac{\partial f(\mathbf{k})}{\partial t}\right]_c = \sum_{\mathbf{k}'} \{W(\mathbf{k}', \mathbf{k}) f(\mathbf{k}') - W(\mathbf{k}, \mathbf{k}') f(\mathbf{k})\}. \qquad (10.1)$$

We do not take into account the Pauli principle, since we consider electrons in a nondegenerate state. $W(\mathbf{k}, \mathbf{k}')$ is the transition probability from state $\mathbf{k}$ to state $\mathbf{k}'$.

It follows from the wave vector conservation law for electron-phonon interaction in the absence of interband transitions that

$$\mathbf{k}' = \mathbf{k} \pm \mathbf{q}. \qquad (10.2)$$

[14] Here we follow the line of presentation suggested by Yu. N. Obraztsov.

Here the plus sign corresponds to the absorption and the minus sign corresponds to the emission of a phonon. Accordingly, we obtain four addends for the sum (10.1)

$$\left[\frac{\partial f(\mathbf{k})}{\partial t}\right]_c = \sum_{\mathbf{q}} \{W^-(\mathbf{k}+\mathbf{q}, \mathbf{k}) f(\mathbf{k}+\mathbf{q}) + W^+(\mathbf{k}-\mathbf{q}, \mathbf{k}) f(\mathbf{k}-\mathbf{q}) - W^+(\mathbf{k}, \mathbf{k}+\mathbf{q}) f(\mathbf{k}) - W^-(\mathbf{k}, \mathbf{k}-\mathbf{q}) f(\mathbf{k})\}, \tag{10.3}$$

where $W^+$ and $W^-$ are the acoustic phonon absorption and emission probabilities, so that, for instance, $W^-(\mathbf{k}+\mathbf{q}, \mathbf{k})$ is the probability of the transition $\mathbf{k}+\mathbf{q} \rightarrow \mathbf{k}$ involving the emission of a phonon with the wave vector $\mathbf{q}$.

Making use of (8.3.11a)-(8.3.11c), write (10.3) in the form

$$\left(\frac{\partial f}{\partial t}\right)_c = \sum_{\mathbf{q}} w(q) \{f(\mathbf{k}+\mathbf{q})(N_{\mathbf{q}}+1)\,\delta(\varepsilon_{\mathbf{k}} - \varepsilon_{\mathbf{k}+\mathbf{q}} + \hbar\omega_{\mathbf{q}}) + f(\mathbf{k}-\mathbf{q}) N_{\mathbf{q}} \delta(\varepsilon_{\mathbf{k}} - \varepsilon_{\mathbf{k}-\mathbf{q}} - \hbar\omega_{\mathbf{q}}) - f(\mathbf{k}) N_{\mathbf{q}} \delta(\varepsilon_{\mathbf{k}+\mathbf{q}} - \varepsilon_{\mathbf{k}} - \hbar\omega_{\mathbf{q}}) - f(\mathbf{k})(N_{\mathbf{q}}+1)\,\delta(\varepsilon_{\mathbf{k}-\mathbf{q}} - \varepsilon_{\mathbf{k}} + \hbar\omega_{\mathbf{q}})\}. \tag{10.4}$$

Substituting in the second and fourth members of the sum $-\mathbf{q}$ for $\mathbf{q}$ and making use of the even character of the $\delta$-function, we obtain

$$\left(\frac{\partial f}{\partial t}\right)_c = \sum_{\mathbf{q}} w(q) \{[f(\mathbf{k}+\mathbf{q})(N_{\mathbf{q}}+1) - f(\mathbf{k}) N_{\mathbf{q}}]\,\delta(\varepsilon_{\mathbf{k}+\mathbf{q}} - \varepsilon_{\mathbf{k}} - \hbar\omega_{\mathbf{q}}) + [f(\mathbf{k}+\mathbf{q}) N_{-\mathbf{q}} - f(\mathbf{k})(N_{-\mathbf{q}}+1)]\,\delta(\varepsilon_{\mathbf{k}+\mathbf{q}} - \varepsilon_{\mathbf{k}} + \hbar\omega_{\mathbf{q}})\}. \tag{10.5}$$

Up to now we have always assumed the phonon distribution to be equilibrium in which $N_{-\mathbf{q}} = N_{\mathbf{q}}$.

Make in the nonequilibrium case

$$N_{\mathbf{q}} = N_{\mathbf{q}}^{(0)} + N'_{\mathbf{q}}, \tag{10.6}$$

where $N_{\mathbf{q}}^{(0)}$ is the equilibrium phonon distribution function.

As we shall demonstrate now, the nonequilibrium correction to the phonon distribution function satisfies the condition

$$N'_{-\mathbf{q}} = -N'_{\mathbf{q}}. \tag{10.6a}$$

The nonequilibrium correction $N'_{\mathbf{q}}$ can be determined from the equations

$$\left(\frac{\partial N'_{\mathbf{q}}}{\partial t}\right)_c = -\frac{N'_{\mathbf{q}}}{\tau_{\mathrm{ph}}} \tag{10.7}$$

and

$$\left(\frac{\partial N'_{\mathbf{q}}}{\partial t}\right)_{\mathrm{f}}+\left(\frac{\partial N'_{\mathbf{q}}}{\partial t}\right)_{\mathrm{c}}=0 \tag{10.7a}$$

where $\tau_{\mathrm{ph}}(q)$ is the relaxation time of long-wave longitudinal acoustic phonons, and the field member is

$$\left(\frac{\partial N'_{\mathbf{q}}}{\partial t}\right)_{\mathrm{f}}=\mathbf{v}_{\mathrm{gr}}\cdot\nabla_r N_{\mathbf{q}}^{(0)}. \tag{10.7b}$$

Here $\mathbf{v}_{\mathrm{gr}}$ is the group velocity of long-wave phonons equal to $v_0\frac{\mathbf{q}}{q}$, where $v_0$ is the velocity of the longitudinal sound waves and

$$\nabla_r N_{\mathbf{q}}^{(0)}=\frac{dN_{\mathbf{q}}^{(0)}}{dT}\nabla T=\frac{k_0}{\hbar v_0 q}\nabla T, \tag{10.7c}$$

since for the long-wave phonons $N_{\mathbf{q}}^{(0)}\approx\frac{k_0T}{\hbar v_0 q}$.

From (10.7a)-(10.7c) we obtain

$$N'_{\mathbf{q}}=\tau_{\mathrm{ph}}\left(\frac{\partial N'_{\mathbf{q}}}{\partial t}\right)_{\mathrm{f}}=\frac{k_0\tau_{\mathrm{ph}}}{\hbar q}\left(\frac{\mathbf{q}}{q}\nabla T\right), \tag{10.8}$$

whence follows (10.6a).

Herring's studies of the heat conductivity of crystals, which we are unable to discuss here, have shown that if we take into account the elastic anisotropy of crystals to avoid divergence of heat conductivity for $q\to 0$, we can obtain the expression for the relaxation time of longitudinal long-wave phonons interacting with other phonons in crystals of cubic symmetry

$$\tau_{\mathrm{ph}}(q)=1/A_lT^3q^2. \tag{10.9}$$

If the elastic anisotropy is taken into account, the consideration can be limited to three-phonon collisions.

Making in (10.5) $f(\mathbf{k})$ and $N_{\mathbf{q}}$ equal to their equilibrium values, we obtain $\sum_{\mathbf{q}}\{\ \}=0$. Previously it has always been assumed that $N_{\mathbf{q}}=N_{\mathbf{q}}^{(0)}$, but that

$$f(\mathbf{k})=f_0(\varepsilon_{\mathbf{k}})+f_1(\mathbf{k}), \tag{10.10}$$

where $f_1(\mathbf{k})\neq 0$, and $(\partial f/\partial t)_{\mathrm{c}}$ has been determined, which, when the relaxation time existed, turned out to be proportional to $f_1(\mathbf{k})$. Substituting expressions (10.10) and (10.6) into (10.5), we obtain, apart from the usual term proportional to $f_1(\mathbf{k})$, an additional term due to $N'_{\mathbf{q}}\neq 0$, this term being responsible for the phonon drag effect. To retain the same order of magnitude of all the terms, we can make the electron distribution functions in the term responsible for the phonon drag equal to their equilibrium values.

Hence, the collision term connected with the phonon drag is

$$\left(\frac{\partial f}{\partial t}\right)_c^{ph} = \sum_{(\mathbf{q})} w(q)\, N'_{\mathbf{q}} \{[f_0(\varepsilon_{\mathbf{k}+\mathbf{q}}) - f_0(\varepsilon_{\mathbf{k}})]\, \delta(\varepsilon_{\mathbf{k}+\mathbf{q}} - \varepsilon_{\mathbf{k}} - \hbar\omega_{\mathbf{q}})$$
$$- [f_0(\varepsilon_{\mathbf{k}+\mathbf{q}} - f_0(\varepsilon_{\mathbf{k}})]\, \delta(\varepsilon_{\mathbf{k}+\mathbf{q}} - \varepsilon_{\mathbf{k}} + \hbar\omega_{\mathbf{q}})\}, \tag{10.11}$$

where we made use of (10.6a). If we were to ignore the fact that the scattering is inelastic, i.e., drop the terms $\hbar\omega_{\mathbf{q}}$ in the arguments of the $\delta$-functions, we would obtain zero for the whole expression. Because of this, when we consider phonon drag, we must take into account inelastic effects due to the absorption and emission of phonons.

Since the change in the electron energy following the absorption or emission of a phonon is small, it follows that

$$f_0(\varepsilon_{\mathbf{k}+\mathbf{q}}) - f_0(\varepsilon_{\mathbf{k}}) = \frac{\partial f_0}{\partial \varepsilon}(\varepsilon_{\mathbf{k}+\mathbf{q}} - \varepsilon_{\mathbf{k}}), \tag{10.12}$$

where the difference $\varepsilon_{\mathbf{k}+\mathbf{q}} - \varepsilon_{\mathbf{k}}$ for the first square brackets in (10.11) is equal to $\hbar\omega_{\mathbf{q}}$ and for the second it is equal to $\hbar\omega_{\mathbf{q}}$. As a result, we obtain

$$\left(\frac{\partial f}{\partial t}\right)_c^{ph} = \sum_{(\mathbf{q})} 2w(q)\, N'_{\mathbf{q}} \frac{\partial f_0}{\partial \varepsilon} \hbar\omega_{\mathbf{q}} \delta(\varepsilon_{\mathbf{k}+\mathbf{q}} - \varepsilon_{\mathbf{k}}), \tag{10.13}$$

where the inelasticity in the argument of the $\delta$-function in the sum (the value of which is of the order of $\hbar\omega_{\mathbf{q}} = \hbar v_0 q$) is consistently ignored.

Choosing a polar axis coinciding in direction with **k** in the **q**-space and replacing summation by integration, we obtain

$$\left(\frac{\partial f}{\partial t}\right)_c^{ph} = \frac{V}{(2\pi)^3} \int_{q_{min}}^{q_{max}} q^2\, dq \int_0^{\pi} \sin\vartheta\, d\vartheta$$
$$\times \int_0^{2\pi} d\varphi \left\{2w(q)\, N'_{\mathbf{q}} \frac{\partial f_0}{\partial \varepsilon} \hbar v_0 q \delta(\varepsilon_{\mathbf{k}+\mathbf{q}} - \varepsilon_{\mathbf{k}})\right\}, \tag{10.14}$$

where $V$ is the volume of the crystal's principal region.

Substituting herein (8.3.11c) for $w(q)$ and (10.8) for $N'_{\mathbf{q}}$ and integrating precisely in the same way as was done in Section 8.4.1, we obtain

$$\left(\frac{\partial f}{\partial t}\right)_c^{ph} = \frac{C^2\Omega_0 k_0 m^*}{9\pi M(\hbar k)^3} \frac{\partial f_0}{\partial \varepsilon} (\hbar\mathbf{k}\cdot\nabla T) \int_0^{2k} \tau_{ph}(q)\, q^3\, dq, \tag{10.15}$$

where $\Omega_0 = V/N$ is the volume of a crystal cell.

Expressing $C^2$ from (8.4.11) in terms of the relaxation time $\tau'$ due to the electron's interaction with acoustic vibrations and introducing the averaging procedure of the type

$$\tilde{\tau}_{\text{ph}} = \frac{1}{4k^4} \int_0^{2k} \tau_{\text{ph}}(q)\, q^3\, dq \tag{10.16}$$

we obtain

$$\left(\frac{\partial f}{\partial t}\right)_{\text{c}}^{\text{ph}} = \frac{v_0^2}{T} \frac{\tilde{\tau}_{\text{ph}}}{\tau'} (\hbar \mathbf{k} \cdot \nabla T) \frac{\partial f_0}{\partial \varepsilon}. \tag{10.17}$$

In case (10.9)

$$\tilde{\tau}_{\text{ph}} = \frac{1}{2A_l T^3 k^2} = \frac{\hbar^2}{4A_l m^* T^3 \varepsilon}, \tag{10.18}$$

where $\varepsilon = \hbar^2 k^2/2m^*$ is the electron energy.

With the phonon drag effect taken into account, the kinetic equation in the approximation of a scalar effective mass assumes the form

$$\frac{\hbar \mathbf{k}}{m^*} \nabla_r f + \frac{1}{\hbar} \mathbf{F} \cdot \nabla_{\mathbf{k}} f = \left(\frac{\partial f}{\partial t}\right)_{\text{c}} + \left(\frac{\partial f}{\partial t}\right)_{\text{c}}^{\text{ph}}, \tag{10.19}$$

where $\hbar \mathbf{k}/m^* = \mathbf{v}$ is the electron's velocity, $\mathbf{F}$ is the force acting on it, and $(\partial f/\partial t)_{\text{c}}$ is the usual collision term connected with the deviation of the electron distribution function $f$ from its equilibrium value.

We can substitute the equilibrium distribution function into the left-hand side of (10.19) (with the exception of the addend proportional to the magnetic field) and obtain, by analogy with (2.5),

$$\nabla_r f = \nabla_r f_0 = \frac{\partial f_0}{\partial T} \nabla T = -\frac{\partial f_0}{\partial \varepsilon} \left[\frac{\varepsilon - \zeta}{T} + \frac{\partial \zeta}{\partial T}\right] \nabla T. \tag{10.20}$$

Substituting (10.17) and (10.20) into (10.19), we have

$$-\frac{\partial f_0}{\partial \varepsilon} \left[\frac{\varepsilon + m^* v_0^2 (\tilde{\tau}_{\text{ph}}/\tau') - \zeta}{T} + \frac{\partial \zeta}{\partial T}\right] \left(\frac{\hbar \mathbf{k}}{m^*} \nabla T\right) + \frac{1}{\hbar} \mathbf{F} \cdot \nabla_{\mathbf{k}} f$$
$$= \left(\frac{\partial f}{\partial t}\right)_{\text{c}} = -\frac{f - f_0}{\tau}, \tag{10.21}$$

where $\tau$ is the total electron relaxation time, which may be due not only to the electron's interaction with acoustic vibrations but also to other scattering mechanisms, such as scattering by impurity ions, etc.

It will be seen from (10.21) that in our assumptions the inclusion of the phonon drag effect reduces to the inclusion of an additional energy-dependent term in the energy sum equal to $m^* v_0^2 (\tilde{\tau}_{\text{ph}}/\tau')$.

This enables the additive terms in the kinetic phenomena responsible for the phonon drag to be easily identified.

**9.10.3** Determine the contribution of the phonon drag to the thermoelectric power and the Nernst effect.

It will be seen from (4.2) that the phonon drag determines an additional vector

$$\chi_{\text{ph}}^{*}=\frac{k_0}{e}\,\tau\,\frac{m^{*}v_0^2}{k_0T}\,\frac{\tilde{\tau}_{\text{ph}}}{\tau'}\,\nabla T, \tag{10.22}$$

which, according to (2.21), results in an additional current

$$\mathbf{j}_{\text{ph}}=\frac{ne^2}{m^{*}}\langle\chi_{\text{ph}}^{*}\rangle=\frac{ne^2}{m^{*}}\left(\frac{k_0}{e}\right)\frac{m^{*}v_0^2}{k_0T}\left\langle\frac{\tau\tilde{\tau}_{\text{ph}}}{\tau'}\right\rangle\nabla T. \tag{10.23}$$

Hence, and from (4.5) it follows that the phonon part of the thermoelectric power is

$$\alpha_{\text{ph}}=\left(\frac{k_0}{e}\right)\frac{m^{*}v_0^2}{k_0T}\,\frac{\langle\tau\tilde{\tau}_{\text{ph}}/\tau'\rangle}{\langle\tau\rangle}. \tag{10.24}$$

If the sole electron scattering mechanism is acoustic vibrations, then $\tau=\tau'$, and

$$\alpha_{\text{ph}}=\left(\frac{k_0}{e}\right)\frac{m^{*}v_0^2}{k_0T}\,\frac{\langle\tilde{\tau}_{\text{ph}}\rangle}{\langle\tau\rangle}. \tag{10.24a}$$

It follows from (10.18) that $\langle\tilde{\tau}_{\text{ph}}\rangle\propto T^{-4}$, and since $\langle\tau\rangle\propto T^{-3/2}$, we obtain

$$\alpha_{\text{ph}}\propto T^{-7/2}, \tag{10.24b}$$

i.e., it rises rapidly with the fall in the temperature.

The behaviour of $\alpha_{\text{ph}}$ in the case of other scattering mechanisms, when $\tau\neq\tau'$, can be studied in a similar way.

In order to determine the part played by the phonon drag in the Nernst effect, consider the structure of the constant $Q$ (6.4a). To be definite, take the case of a weak magnetic field, in which the coefficients $a_i$ and $b_i$ are in the approximation linear in $H$ equal, respectively, to (5.8) and (6.5).

In accordance with the aforesaid, the correction $m^{*}v_0^2\,(\tilde{\tau}_{\text{ph}}/\tau')$ due to the phonon drag is an additive part of the energy $\varepsilon$ in $b_1$ and $b_2$, since those coefficients stand in front of $\nabla_x T$ ($\nabla_y T=0!$) in (6.3). Hence, in the approximation linear in $H$

$$Q_{\text{ph}}=\frac{a_1b_2^{\text{ph}}-a_2b_1^{\text{ph}}}{a_1^2H}=\left(\frac{k_0}{e}\right)\frac{m^{*}v_0^2}{k_0T}\,\frac{e}{m^{*}c}$$

$$\times\frac{\langle\tau\rangle\left\langle\frac{\tilde{\tau}_{\text{ph}}}{\tau'}\tau^2\right\rangle-\langle\tau^2\rangle\left\langle\frac{\tilde{\tau}_{\text{ph}}}{\tau'}\tau\right\rangle}{\langle\tau\rangle^2}. \tag{10.25}$$

If the electrons are scattered exclusively by longitudinal vibrations, then $\tau = \tau'$, and

$$Q_{\rm ph} = \left(\frac{k_0}{e}\right) \frac{m^* v_0^2}{k_0 T} \frac{e}{m^* c} \frac{\langle\tau\rangle \langle\tilde{\tau}_{\rm ph}\tau\rangle - \langle\tau^2\rangle \langle\tilde{\tau}_{\rm ph}\rangle}{\langle\tau\rangle^2}. \tag{10.25a}$$

If $\tilde{\tau}_{\rm ph}$ is determined by expression (10.18), so that $\langle\tilde{\tau}_{\rm ph}\rangle \propto T^{-4}$, and $\langle\tau\rangle \propto T^{-3/2}$, then

$$Q_{\rm ph} \propto T^{-5}, \tag{10.25b}$$

i.e., rises rapidly with the fall in temperature. It is interesting to note that if $\tau_{\rm ph} \propto 1/q$ (and not $1/q^2$), then $Q_{\rm ph} = 0$.

It follows from (10.24a), (4.5), (10.25a), and (6.6) that the order of magnitude of the ratio

$$\frac{Q_{\rm ph}/Q}{\alpha_{\rm ph}/\alpha} \approx \frac{\langle\tau x\rangle}{\langle\tau\rangle} - \frac{\zeta}{k_0 T}. \tag{10.26}$$

For nondegenerate semiconductors $(-\zeta/k_0T) \gg 1$; therefore, in them the phonon drag plays a relatively greater part in the Nernst effect than in the thermoelectric power.

The phonon drag effect in thermomagnetic phenomena was first experimentally observed in hole-type germanium by I. V. Mochan, Yu. N. Obraztsov, and T. V. Krylova [9.18] and independently of them by Herring and Geballe [9.19].

Here we presented only the fundamentals of the theory of phonon drag of electrons in semiconductors. Several circumstances complicate the phenomena observed in experiment.

First, with the fall in temperature the phonon's mean free path $l_{\rm ph} = v_0\tau_{\rm ph}$ [see (10.18)] increases rapidly and becomes comparable to the linear dimensions $L$ of the specimen being studied; then $\tau_{\rm ph} = L/v_0$ and no longer depends on $T$ and $q$. This will obviously change the temperature dependences of $\alpha_{\rm ph}$ and $Q_{\rm ph}$ obtained above.

Second, we assumed that it was possible to neglect the reciprocal effect of the free electrons on the nonequilibrium (because of the presence of a temperature gradient) phonon distribution function. If there is a notable reduction in the phonon heat conductivity due to the scattering of phonons by free electrons, the phonon distribution function becomes less nonequilibral, and the phonon drag effect decreases. This phenomenon is observed both, when the carrier concentration is increased, and when the temperature is reduced. Because of this, as the temperature is reduced, the phonon drag exhibits saturation.

Third, new and interesting phenomena are observed in the studies of the phonon drag effect in semiconductors with a complex energy-band pattern, especially of the $p$-germanium type in which both light and heavy holes are present [9.19].

All these subjects are to be found in specialized literature, in particular in [9.20].

The effect of the phonon drag on other nonisothermic phenomena is calculated in a similar fashion.

## 9.11 Quantum Mechanics Theory of Galvano- and Thermomagnetic Phenomena in Semiconductors[15]

Consider first the problem on the limits of validity of the kinetic equation used in this chapter for calculating the kinetic coefficients.

We shall consider two factors that place limitations on the applicability of the kinetic equation.

The first involves the process of scattering of conduction electrons, and the second involves the quantization of electron motion in a magnetic field.

**9.11.1** It follows from the Heisenberg principle that the energy of a conduction electron can be determined up to $\Delta\varepsilon > h/\Delta t$, where $\Delta t$ is the lifetime of the electron's quasi-stationary state, which is of the order of the relaxation time $\tau$. On the other hand, for the kinetic equation to be applicable the uncertainty in the electron's energy $\Delta\varepsilon$ must be much less than the energy interval $k_0T$ in which the distribution function $f(\mathbf{k})$ contained in the kinetic equation varies appreciably. Hence

$$h/\tau < \Delta\varepsilon \ll k_0T.$$

This yields the following criterion of applicability of the kinetic equation:

$$\tau \gg h/k_0T. \tag{11.1}$$

This criterion can also be derived from other considerations.

The essential foundations for the applicability of the kinetic equation are those discussed in Appendix 20.2. The transition probability from the state $i$ to the state $f$ is proportional to the time $t$ and is given by (A.20.7). The function of $\omega_{fi}$ in the square brackets of this expression has a sharp maximum the half-width of which is $\Delta\omega \approx 2\pi/t$ or, in energy units, $\hbar\Delta\omega = \Delta\varepsilon \approx h/t$. Since for a sufficiently long time interval $t$, the limit of the expression in the square brackets, transforms into a $\delta$-function of the difference in the energies of the final and initial states of the system, the quantity $\Delta\varepsilon \approx$ $\approx h/t$ may be said to determine the degree of accuracy with which

[15] The presentation in this section is of a more concise nature than in other sections of the book. For a more detailed acquaintance with the problem we recommend the book [9.16. Chaps. VII and VIII]. See also [9.21].

the energy conservation law is being fulfilled. In this sense, the relationship $\Delta\varepsilon t \approx h$ can be regarded as the Heisenberg uncertainty relation for the energy and the time.

Since the time of measurement $t$ must be less than the relaxation time $\tau$, and $\Delta\varepsilon$, as we have seen above, must be much less than $k_0T$, we arrive again at the criterion of applicability of the kinetic equation (11.1).

The order of magnitude of the average energy of a conduction electron with the effective mass $m$ is $m\overline{v}^2 = k_0T$, where $\overline{v}$ is the average electron velocity, the corresponding de Broiglie wavelength is $\lambda = h/m\overline{v}$, and the mean free path is $l = \overline{v}\tau$. Making use of these relations, we can write (11.1) in the form

$$l \gg \lambda. \tag{11.2}$$

For a better understanding of criterion (11.2) note that the greater the electron's mean free path $l$, the weaker is its interaction with the scattering agent, and the less the de Broglie wavelength $\lambda$, the greater the electron's speed (energy).

Since the electron's mobility is $\mu = (e/m) \langle\tau\rangle$ (3.3), criterion (1.1) can be written in the form

$$\mu \gg \frac{eh}{mk_0T}. \tag{11.3}$$

Making $m = 10^{-27}$ g and $T = 300$ K, we obtain that the mobility at room temperature

$$\mu \gg 40 \text{ cm}^2/\text{V s}. \tag{11.3a}$$

The kinetic equation cannot be used to describe the behavior of charge carriers whose mobility is less than that on the right-hand side of (11.3a).

Criterion (11.1) [or the equivalent criteria (11.2) and (11.3)] is valid in the case of nondegenerate electrons. It can be shown that in the case of degenerate electrons the criterion of applicability of the kinetic equation for kinetic effects operating in the zeroth approximation in the chemical potential $\zeta$ [for instance, for the electric conductivity (7.8)] is of the form

$$\tau \gg h/\zeta. \tag{11.4}$$

It is not quite clear what form the criterion of applicability of the kinetic equation for degenerate electrons should take in the case of kinetic effects operating only in the $k_0T/\zeta$ approximation [for instance, in the case of thermoelectric power (7.11)].

**9.11.2** It was pointed out in Section 9.5 that if the dimensionless parameter $\gamma\tau H \approx \mu H/c \gg 1$, the magnetic field is called *strong*; in the case of a reverse inequality it is termed *weak*. The kinetic equation is applicable both in the case of a weak and of a strong field.

The quantum mechanics treatment of the motion of a free electron in a magnetic field shows that the energy corresponding to its motion in a plane perpendicular to the magnetic field is quantized (6.5.19b). The difference between the neighboring energy levels is $2\mu^* H = \hbar\omega_c$, where $\mu^* = e\hbar/2mc$ is the effective Bohr magneton, and $\omega_c = eH/mc$ is the cyclotron frequency. Obviously, if

$$k_0 T \ll 2\mu^* H = \hbar\omega_c, \tag{11.5}$$

the discrete nature of the levels cannot be neglected, since the only effect of thermal motion is that it slightly widens the electron's quantum levels.

The magnetic fields satisfying inequality (11.5) are called *quantizing*.

The ratio of the weakest magnetic fields satisfying the condition of strong fields $\frac{\mu H_{st}}{c} = 1$ to the weakest quantizing magnetic fields $H_q$ (11.5) is equal to

$$\frac{H_{st}}{H_q} = \frac{c}{\mu}\frac{2\mu^*}{k_0 T} = \frac{1}{2\pi}\frac{eh}{\mu m_0 k_0 T} \ll 1. \tag{11.6}$$

If (11.3) is fulfilled, then $H_q > H_{st}$, and in this case there is a region of strong magnetic fields that are not quantizing fields, i.e., for which the kinetic equation is valid. In quantizing magnetic fields the electron's motion cannot be described in terms of continuous values of energy, momentum, coordinate, and because of that the kinetic equation is inapplicable.

**9.11.3** Consider an electron-type conductor in isothermic conditions ($\nabla T = 0$), placed in a magnetic field **H** pointing in the direction of the $z$ axis and in an electric field **E** lying in the $xy$ plane.

The phenomenological equations (1.1a) assume the form

$$j_x = \sigma_{xx}E_x - \sigma_{yx}E_y, \quad j_y = \sigma_{yx}E_x + \sigma_{xx}E_y, \tag{11.7}$$

since $\sigma_{xx} = \sigma_{yy}$ and $\sigma_{xy} = -\sigma_{yx}$. The tensor components $\sigma_{xx}$ and $\sigma_{yx}$ coincide with $a_1$ and $a_2$ in (5.4). In a strong magnetic field, in accordance with (5.15),

$$\sigma_{yx} = \frac{ne^2}{m}\frac{1}{\gamma H} = \frac{ecn}{H}. \tag{11.8}$$

This value of $\sigma_{yx}$ independent of the scattering mechanism is also obtained in quantum mechanics.

It follows from (5.15) that in the case of a strong magnetic field

$$\frac{\sigma_{xx}}{\sigma_{yx}}=\frac{a_1}{a_2}=\left\langle\frac{1}{\tau}\right\rangle\frac{mc}{eH}\approx\frac{1}{\gamma\tau H}\ll 1, \tag{11.9}$$

i.e., that the dissipative component of the electric conductivity tensor $\sigma_{xx}$ is much smaller than its nondissipative component $\sigma_{yx}$. Relation (11.9) holds for quantizing magnetic fields as well. The parameter usually measured in experiment is the transverse magnetoresistance $\rho_H$, which, according to (5.6), is

$$\rho_H=\frac{\sigma_{xx}}{\sigma_{xx}^2+\sigma_{yx}^2}\approx\frac{\sigma_{xx}}{\sigma_{yx}^2}. \tag{11.10}$$

The dissipative component of the electric conductivity tensor $\sigma_{xx}$ in quantizing magnetic fields was first calculated by the Roumanian physicist S. Titeica (1935). Titeica's theory is based on some self-evident semiclassical conceptions; however, it has been subsequently rigorously confirmed by E. N. Adams and T. D. Holstein in 1959 who considered the motion equations for the density matrix.

Titeica bases his calculations on the quantum mechanics equations of motion for an electron in crossed electric and magnetic fields (Section 7.7.2). In the case of a simple band the electron's energy (7.7.23) in an approximation linear in the electric field $E$ is

$$\varepsilon=\varepsilon(0)+(2N+1)\mu^*H+\frac{\hbar^2k_z^2}{2m}-\lambda^2eEk_y, \tag{11.11}$$

where the subscript c has been dropped in the notation of appropriate quantities. It follows from (7.7.21b) that in the same approximation in the electric field the energy is

$$\varepsilon=\varepsilon(0)+(2N+1)\mu^*H+\frac{\hbar^2k_z^2}{2m}-eEx_0,$$

where $x_0\equiv x_c$ is the coordinate of the oscillator's equilibrium position. Hence, we can assume that the electron's energy $\varepsilon$ depends on the quantum numbers $N$, $k_z$, and $x_0$ and on the electric field $E$, i.e., $\varepsilon(N,\ k_z,\ x_0;\ E)\equiv\varepsilon_\nu^E$, where $\nu\equiv\{N,\ k_z,\ x_0\}$. Assuming that the electron in the state $\nu$ is "located" at point $x_0$, we make the current

$$j_x=-e\sum_{N,\,N',\,k_z,\,k_z'}\ \sum_{\substack{x_0<0\\ x_0'>0}}\{f_0(\varepsilon_\nu^E)[1-f_0(\varepsilon_{\nu'}^E)]W_{\nu\nu'}^E$$
$$-f_0(\varepsilon_{\nu'}^E)[1-f_0(\varepsilon_\nu^E)]W_{\nu'\nu}^E\}. \tag{11.12}$$

Here $f_0(\varepsilon_\nu^E)$ is the equilibrium distribution function for the electrons with the energy $\varepsilon_\nu^E$, and $W_{\nu\nu'}^E$ is the transition probability per unit time from the state $\nu$ (with $x_0<0$) to the state $\nu'$ (with $x_0'>0$) due to scattering, i.e., the probability of such a scattering of the electron in the result of which it crosses the plane $x = 0$ from left to right; the factor $[1 - f_0(\varepsilon^E_{\nu'})]$, in accordance with the Pauli exclusion principle, is the probability that the state $\nu'$ is unoccupied. The second addend in the braces obviously determines the electron flux across the plane $x = 0$ in the opposite direction.

Expanding the expression in braces in (11.12) in the powers of the electric field and retaining only the first addend proportional to $E$, we obtain, comparing the result with (11.7),

$$\sigma_{xx} = e^2 \sum_{\nu\nu'} \left[ -\frac{\partial f_0(\varepsilon_\nu)}{\partial \varepsilon_\nu} \right] \frac{(x_0' - x_0)^2}{2} W_{\nu\nu'}, \tag{11.13}$$

where all the quantities under the summation sign are for the field $E = 0$. Adams and Holstein, making use of (11.13), have determined the dependence of $\rho_H$ (11.10) on the magnetic field $H$ and on the temperature $T$ for different scattering mechanisms in the quantum limit $N = N' = 0$ $(k_0T \ll \hbar\omega_c)$. This dependence turned out to be different for nondegenerate and degenerate semiconductors.

In the case of a strong degeneracy of charge carriers $-\partial f_0/\partial\varepsilon = \delta(\zeta - \varepsilon)$, where $\zeta$ is their chemical potential. It follows from (6.5.23a) that in this case the density of states in (11.13) is

$$g(\varepsilon) \propto \sum_N \frac{1}{\sqrt{\varepsilon - (2N+1)\mu^* H}} \delta(\zeta - \varepsilon) = \sum_N \frac{1}{\sqrt{\zeta - (2N+1)\mu^* H}}. \tag{11.14}$$

Hence, $g(\varepsilon)$ and, consequently, $\sigma_{xx}$ as well, increase sharply every time the Landau level crosses the level of the chemical potential. Such oscillations of $\sigma_{xx}$, and, consequently, of $\rho_H$ (11.10), are periodic with a period determined by the condition

$$2\Delta N \mu^* H + (2N + 1)\mu^* \Delta H = 0$$

for $\Delta N = 1$. Hence

$$-\frac{\Delta H}{H^2} = \Delta\left(\frac{1}{H}\right) = \frac{2\mu^*}{\zeta} = \frac{e\hbar}{cm\zeta}. \tag{11.15}$$

Such oscillations became known as the *Shubnikov-de Haas oscillations* after the scientists who discovered them for bismuth in 1930. They can be observed only in the case of a strong charge carrier degeneracy, their period $\Delta(1/H)$ being dependent on the conduction electron concentration (since $\zeta$ depends on it).

**9.11.4** In 1964 V. L. Gurevich and Yu. A. Firsov theoretically predicted a new type of oscillations of $\rho_H$, which became known as *magnetophonon oscillations*. They are caused by inelastic scattering of electrons by optical phonons.

Consider the magnetophonon resonance from a qualitative point of view. In the absence of scattering the conduction electrons placed in crossed electric ($\mathbf{E} \parallel x$) and magnetic ($\mathbf{H} \parallel z$) fields oscillate in the $xy$ plane and drift with a constant velocity $v_y = c\,(E/H)$ in the direction of the $y$ axis [see (11.8)]. The current $j_x$ is entirely due to electron scattering. The contribution to the current $j_x$ of the electrons with the energy $\varepsilon$ and the magnetic quantum number $N$ will be proportional to their number $f_0(\varepsilon)\, g(\varepsilon, N)$, where $f_0(\varepsilon)$ is the Fermi distribution function, and $g(\varepsilon, N)$ is the density of states for a fixed $N$ [see (6.5.23a)]; $j_x$ will obviously also be proportional to the probability of the transition $(\varepsilon, N) \to (\varepsilon', N')$:

$$W_{NN'}(\varepsilon, \varepsilon')\,\delta(\varepsilon + \hbar\omega_0 - \varepsilon')\,[1 - f_0(\varepsilon')]\, g(\varepsilon', N'),$$

where $W_{NN'}(\varepsilon, \varepsilon')$ is a smooth function of its indices and arguments, $\delta$-function expresses the energy conservation law in the case of the absorption[16] of an optical phonon $\hbar\omega_0$, and the expression in square brackets determines the number of vacant states with the energy $\varepsilon'$.

The total current $j_x$ will be proportional to the product of all those multipliers added up over $N$ and $N'$ and integrated with respect to $\varepsilon$ and $\varepsilon'$. Integrating with respect to $\varepsilon'$ with the aid of the $\delta$-function and making use of expression (6.5.23a) for the density of states, we obtain

$$j_x \propto \sum_{N,\,N'} \int d\varepsilon\, \frac{W_{NN'}(\varepsilon,\ \varepsilon + \hbar\omega_0)\, f_0(\varepsilon)\,[1 - f_0(\varepsilon + \hbar\omega_0)]}{\sqrt{\varepsilon - (N + 1/2)\,\hbar\omega_c}\,\sqrt{\varepsilon - [(N' + 1/2)\,\hbar\omega_c - \hbar\omega]}}, \qquad (11.16)$$

where we made $2\mu^* H = \hbar\omega_c$ ($\omega_c = eH/mc$ is the cyclotron frequency). If the quantities subtracted from $\varepsilon$ in the expressions under the radical signs in (11.16) are different, the integral with respect to $\varepsilon$ will not have any particular points. If, on the other hand, these subtrahends become equal as the magnetic field is varied, the integral will diverge logarithmically, the result being a sharp increase in $j_x$.

The subtrahends under the roots in (11.16) coincide, if

$$(N' - N)\,\omega_c = \omega_0, \qquad (11.17)$$

i.e., if $\omega_0$ is a multiple of $\omega_c$. Taking into account that $\omega_c = eH/mc$ we obtain for the period of oscillations of $j_x$ by analogy with (11.15)

$$\Delta\left(\frac{1}{H}\right) = \frac{e}{mc\omega_0}. \qquad (11.18)$$

An essential distinction of the magnetophonon resonance from the Shubnikov-de Haas oscillations is that the latter are observed only

[16] The result remains valid for the case of phonon emission.

in the case of charge carrier degeneracy, and that the period of oscillations (11.15) depends on their concentration. The magnetophonon resonance can be observed both in degenerate and in nondegenerate semiconductors, the period of oscillations (11.18) being independent of the carrier concentration. From (11.18) we can obtain the effective mass $m$, provided we know the maximum frequency of the optical phonons $\omega_0$. The theory of magnetophonon resonance agrees well with experiment.[17]

**9.11.5** Consider briefly the theory of thermomagnetic phenomena in quantizing magnetic fields.

The phenomenological equations (1.1a) in the presence of a temperature gradient $\nabla_x T$ in the direction of the $x$ axis take the form

$$j_x = \sigma_{xx}\nabla_x\left(\frac{\zeta}{e}-\varphi\right)+\sigma_{xy}\nabla_y\left(\frac{\zeta}{e}-\varphi\right)-\beta_{xx}\nabla_x T,$$

$$j_y = \sigma_{yx}\nabla_x\left(\frac{\zeta}{e}-\varphi\right)+\sigma_{yy}\nabla_y\left(\frac{\zeta}{e}-\varphi\right)-\beta_{yx}\nabla_x T. \qquad (11.19)$$

Since the chemical potential $\zeta$ depends on the temperature whose gradient points in the $x$-direction, it follows that $\nabla_y\left(\frac{\zeta}{e}-\varphi\right)=E_y$, $\nabla_x\left(\frac{\zeta}{e}-\varphi\right)\equiv E_x^*\neq E_x$. In an isotropic conductor $\sigma_{xx}=\sigma_{yy}$, and $\sigma_{xy}=-\sigma_{yx}$. In the zeroth scattering approximation only $\sigma_{yx}$ and $\beta_{yx}$ are nonzero. Besides, in addition to (11.9) we obtain from (6.8) in the case of strong fields ($\gamma\tau H\gg 1$)

$$\frac{\beta_{xx}}{\beta_{yx}}=\frac{b_1}{b_2}\approx\frac{1}{\gamma\tau H}\ll 1. \qquad (11.20)$$

Finally we obtain for the thermoelectric power and the Nernst constant in strong magnetic fields in the roughest scattering approximation

$$\alpha_H=\frac{|E_x^*|}{|\nabla_x T|}=\frac{\beta_{yx}}{\sigma_{yx}}, \qquad (11.21)$$

$$Q=-\frac{E_y}{\nabla_x T H}=\frac{\sigma_{yx}\beta_{xx}-\sigma_{xx}\beta_{yx}}{\sigma_{yx}^2 H}, \qquad (11.21a)$$

which agrees with (6.11) and (6.4a).

In order to determine $\alpha_H$ and $Q$, we must in addition to (11.8) and (11.13) compute $\beta_{yx}$ and $\beta_{xx}$. This entails the following principal difficulty. In the classical case in the presence of a temperature gradient the temperature $T$ must be regarded as a function of the coordinate $x$, i.e., $T=T(x)$. In the quantum mechanics case this does not make sense, because the electron's coordinate $x$ in a quantiz-

[17] For detailed information on the problem see [9.22].

ing magnetic field is not a quantum number, i.e., has no definite value. Applying the principle of local thermodynamical equilibrium A. I. Anselm and B. M. Askerov (1960) suggested that in the presence of a temperature gradient in the $x$ direction the temperature is a function of the equilibrium position of the magnetic oscillator $x_0$ (6.5.18a). Using the same argument as was used in the derivation of $\sigma_{xx}$ (11.13), we can demonstrate that

$$\beta_{xx}=\frac{e}{2T}\sum_{\nu\nu'}\left[\frac{\partial f_0(\varepsilon_\nu)}{\partial\varepsilon_\nu}\right](\varepsilon_\nu-\zeta)(x_0-x_0')^2 W_{\nu\nu'}. \qquad (11.22)$$

Since we are interested in effects linear in the electric field $\mathbf{E}$ and the temperature gradient $\nabla_x T$, we can in (11.22) take all the quantities for $E=0$. Note that the arguments in support of this expression contained in the paper by A. I. Anselm, Yu. N. Obraztsov and R. G. Tarkhanyan (1965) were similar to those put forwad by Adams and Holstein in support of (11.13).

If we were to use the quantum mechanics expression for the current density $j_y^{(\nu)}$ of the electron in the state $\nu=\{N, x_0, k_z\}$ and the assumption that $T=T(x_0)$ as the starting point, we would obtain for the current density

$$j_y=\sum_\nu j_y^{(\nu)} f_0\left(\frac{\varepsilon_\nu-\zeta(x_0)}{k_0T(x_0)}\right). \qquad (11.23)$$

Expanding the Fermi function $f_0$ in the powers of the increment $\Delta x_0$ and identifying $\Delta T/\Delta x_0=\nabla_x T$, we obtain

$$\beta_{yx}=\frac{(2m)^{1/2}e\omega_c}{4\pi^2\hbar}\sum_N\left(N+\frac{1}{2}\right)\int_{\varepsilon_N}^{\infty}\left(\frac{\varepsilon-\zeta}{T}\right)\frac{\partial f_0}{\partial\varepsilon}\frac{d\varepsilon}{\sqrt{\varepsilon-\varepsilon_N}}, \qquad (11.23a)$$

where $\varepsilon_N=(N+1/2)\hbar\omega_c$. It turned out, however, that this expression for $\beta_{yx}$ did not meet the Onsager principle (1.5a). A correct expression for $\beta_{yx}$ was obtained by Yu. N. Obraztsov (1963). He drew attention to the fact that when calculating the nondissipative current $j_y$, we should add to the volume current determined by (11.23a) the surface currents flowing along the specimen's faces perpendicular to the $x$ axis. In the absence of a temperature gradient ($\nabla_x T=0$) those currents compensate each other and are responsible only for the diamagnetism of the conduction electrons (Section 6.5). On the other hand, when $\nabla_x T\neq 0$ there is no such compensation, and a current appears in the $y$ direction proportional to $\nabla_x T$. If we take that current into account, we obtain the following simple expression for $\beta_{yx}$:

$$\beta_{yx}=-cS/H, \qquad (11.24)$$

where $S$ is the electron entropy per 1 $cm^3$, and $c$ is the velocity of light. Hence, and from (11.21) and (11.8) we obtain

$$\alpha_H = s/c, \tag{11.25}$$

where $s = S/n$ is the entropy per electron.

Obraztsov has demonstrated that in the case of nondegenerate electrons

$$\alpha_H = \frac{k_0}{e}\left[\frac{3}{2} + \frac{\hbar\omega_c}{2k_0T}\coth\left(\frac{\hbar\omega_c}{2k_0T}\right) - \frac{\zeta(H)}{k_0T}\right]. \tag{11.25a}$$

In the quasi-classical case $(\hbar\omega_c/2k_0T) \ll 1$; then $(\hbar\omega_c/2k_0T) \times \times \coth(\hbar\omega_c/2k_0T) \approx 1$, and

$$\alpha_H = \frac{k_0}{e}\left[\frac{5}{2} - \frac{\zeta}{k_0T}\right],$$

this coinciding with (9.6.16).

Equation (11.25a) was confirmed by experiments with InSb (I. L. Drichko and I. V. Mochan, 1964; S. Poorie and H. Geballe, 1964).[18]

Making use of (11.8), (11.13), (11.22) and (11.24), we can find the Nernst constant $Q$ (11.21a) for various scattering mechanisms and compare the results with experiment. Unfortunately, there are no such experiments that could be unambiguously compared with the theory.

---

[18] The discrepancy between experiment and theory for $(\hbar\omega_c/2k_0T) > 1$ can be explained neither by the spin splitting of Landau levels nor by the deviation of the band's shape from the parabolic. The origin of this discrepancy remains unclear.

# Appendices

### Appendix 1

Consider oblique coordinates in a plane. It is easy to generalize the results to a three-dimensional case.

Figure A.1 depicts the $X_1$ and $X_2$ axes of a system of rectangular coordinates and the $\Xi_1$ and $\Xi_2$ axes of an oblique system, both systems having a common origin at $O$.

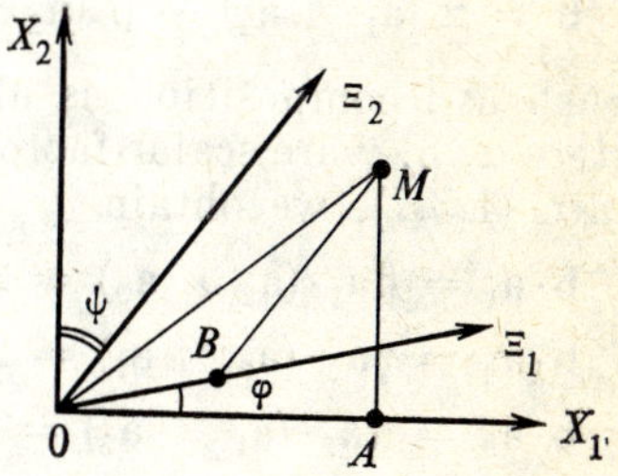

Fig. A.1

The position of a certain point $M$ is determined by the position vector $\overline{OM}$, by the rectangular coordinates $OA = x_1$ and $\overline{AM} = x_2$, or by oblique coordinates $\overline{OB} = \xi_1$ and $\overline{BM} = \xi_2$ ($\overline{BM} \parallel \Xi_2$). The vector $\overline{OM}$ may be regarded as the sum of the vectors $\overline{OB}$ and $\overline{BM}$, i.e.,

$$\overline{OM} = \overline{OB} + \overline{BM}. \tag{A.1.1}$$

Projecting both sides of this equality on the $X_1$ and $X_2$ axes, we obtain

$$x_1 = \xi_1 \cos\varphi + \xi_2 \sin\psi, \quad x_2 = \xi_1 \sin\varphi + \xi_2 \cos\psi.$$

Solving this system for $\xi_1$ and $\xi_2$, we obtain

$$\xi_1 = \alpha_{11}x_1 + \alpha_{12}x_2, \quad \xi_2 = \alpha_{21}x_1 + \alpha_{22}x_2, \tag{A.1.2}$$

where

$$\alpha_{11} = \frac{\cos\psi}{\cos(\varphi+\psi)}, \quad \alpha_{12} = -\frac{\sin\psi}{\cos(\varphi+\psi)}, \quad \alpha_{21} = -\frac{\sin\varphi}{\cos(\varphi+\psi)},$$

$$\alpha_{22} = \frac{\cos\varphi}{\cos(\varphi+\psi)}. \tag{A.1.3}$$

Similarly, in the case of a three-dimensional oblique system of coordinates whose origin coincides with that of a Cartesian system, the coordinates $\xi_i$ are homogeneous linear functions of the Cartesian coordinates $x_k$:

$$\xi_i = \sum_{k=1}^{3} \alpha_{ik} x_k \quad (i = 1,\ 2,\ 3), \tag{A.1.3a}$$

where $\alpha_{ik}$ are functions of the angles which the $\Xi_i$ axes make with the $X_k$ axes.

**Appendix 2**

To solve equation (1.3.7), suppose that the unknown vector has been decomposed into the vectors $\mathbf{a}_1 \times \mathbf{a}_2$, $\mathbf{a}_2 \times \mathbf{a}_3$, $\mathbf{a}_3 \times \mathbf{a}_1$ not lying in one plane

$$\mathbf{b} = \alpha\,(\mathbf{a}_1 \times \mathbf{a}_2) + \beta\,(\mathbf{a}_2 \times \mathbf{a}_3) + \gamma\,(\mathbf{a}_3 \times \mathbf{a}_1). \tag{A.2.1}$$

Such a decomposition is always possible and unique [A.1, § 102]. Here $\alpha$, $\beta$, $\gamma$ are scalar factors to be determined. Substituting (A.2.1) into (1.3.7), we obtain

$$\mathbf{b}\cdot\mathbf{a}_1 = \beta\mathbf{a}_1\cdot(\mathbf{a}_2 \times \mathbf{a}_3) = 2\pi g_1,$$

$$\mathbf{b}\cdot\mathbf{a}_2 = \gamma\mathbf{a}_2\cdot(\mathbf{a}_3 \times \mathbf{a}_1) = 2\pi g_2,$$

$$\mathbf{b}\cdot\mathbf{a}_3 = \alpha\mathbf{a}_3\cdot(\mathbf{a}_1 \times \mathbf{a}_2) = 2\pi g_3,$$

whence

$$\alpha = \frac{2\pi g_3}{\Omega_0}, \quad \beta = \frac{2\pi g_1}{\Omega_0}, \quad \gamma = \frac{2\pi g_2}{\Omega_0}. \tag{A.2.2}$$

(1.38) follows from (A.2.1) and (A.2.2).

**Appendix 3**

A.3.1 Consider two rectangular coordinate frames $(x_1,\ x_2,\ x_3)$ and $(x_1',\ x_2',\ x_3')$ with a common origin at $O$. Denote the cosines of the angles between the axes of the two systems by $\alpha_{ik} = \cos\,(\widehat{x_i\, x_k'})$, where $i$ and $k$ independently assume the values 1, 2 and 3. It is an established fact that

$$\sum_{i=1}^{3} \alpha_{ik}\alpha_{il} = \sum_{i=1}^{3} \alpha_{ki}\alpha_{li} = \delta_{kl}. \tag{A.3.1}$$

For $k = l$ ($\delta_{kl} = 1$) those equalities mean that the sum of the squares of direction cosines is equal to unity, and for $k \neq l$, $\delta_{kl} = 0$ is the condition of orthogonality of the $x_k$ and $x_l$ axes (or the $x_k'$ and $x_l'$ axes). The projection of the geometrical sum (of the position vector

**r**) on a given direction (e.g., on $x_1'$) is known to be equal to the sum of the projections of its components; therefore,

$$\left.\begin{array}{l} x_1' = \alpha_{11}x_1 + \alpha_{12}x_2 + \alpha_{13}x_3, \\ \text{and, similarly,} \\ x_2' = \alpha_{21}x_1 + \alpha_{22}x_2 + \alpha_{23}x_3, \\ x_3' = \alpha_{31}x_1 + \alpha_{32}x_2 + \alpha_{33}x_3, \end{array}\right\} \quad (A.3.2)$$

or

$$x_i' = \sum_{k=1}^{3} \alpha_{ik}x_k. \quad (A.3.3)$$

Write (A.3.2) in a more compact form:

$$\mathbf{r}' = \boldsymbol{\alpha}\mathbf{r}, \quad (A.3.4)$$

where

$$\boldsymbol{\alpha} = \begin{pmatrix} \alpha_{11} & \alpha_{12} & \alpha_{13} \\ \alpha_{21} & \alpha_{22} & \alpha_{23} \\ \alpha_{31} & \alpha_{32} & \alpha_{33} \end{pmatrix} \quad (A.3.5)$$

is the *matrix* of the linear transformation (A.3.2). The matrix $\boldsymbol{\alpha}$ (A.3.5) is of *rank three*, i.e., it is a square array consisting of three rows and three columns and of $3^2 = 9$ elements $\alpha_{ik}$. If the elements of the matrix $\alpha_{ik}$ meet the condition (A.3.1), the linear transformation (A.3.2) is termed *orthogonal*. The inverse transformation from the primed coordinates $x_k'$ to the unprimed $x_i$ can be obtained by analogy with (A.3.2)

$$\begin{aligned} x_1 &= \alpha_{11}x_1' + \alpha_{21}x_2' + \alpha_{31}x_3', \\ x_2 &= \alpha_{12}x_1' + \alpha_{22}x_2' + \alpha_{32}x_3', \\ x_3 &= \alpha_{13}x_1' + \alpha_{23}x_2' + \alpha_{33}x_3', \end{aligned} \quad (A.3.6)$$

which can be written in a form similar to (A.3.4)

$$\mathbf{r} = \boldsymbol{\alpha}^{-1}\mathbf{r}', \quad (A.3.6a)$$

where $\boldsymbol{\alpha}^{-1}$ is the matrix of *inverse transformation* (A.3.6).

We see that the matrix of the inverse orthogonal transformation $\boldsymbol{\alpha}^{-1}$ (A.3.6) can be obtained from the matrix of the direct orthogonal transformation $\boldsymbol{\alpha}$ (A.3.2) by interchanging the rows and the columns, i.e.,

$$(\boldsymbol{\alpha}^{-1})_{ik} = (\boldsymbol{\alpha})_{ki}. \quad (A.3.7)$$

The scalar product of the vectors **a** and **b** is equal to

$$\mathbf{a}\cdot\mathbf{b} = ab\cos\widehat{(\mathbf{a}, \mathbf{b})} = \sum_{i=1}^{3} a_i b_i = \sum_{i=1}^{3} a_i' b_i', \quad (A.3.8)$$

where $a_i'$, $b_i'$ are the components of the vectors **a** and **b** in the coordinate system $(x_1', x_2', x_3')$. The last of these equalities follows both from the geometrical meaning of a scalar product and from formal considerations if $a_i$ and $b_i$ are expressed in terms of $a_i'$ and $b_i'$ with the aid of equations (A.3.6) and if use is made of the orthogonality conditions (A.3.1).

**A.3.2** As a generalization of (A.3.2) we write the expression for a homogeneous linear transformation of $n$ variables in the form

$$\begin{aligned} x_1' &= \alpha_{11}x_1 + \alpha_{12}x_2 + \ldots + \alpha_{1n}x_n, \\ x_2' &= \alpha_{21}x_1 + \alpha_{22}x_2 + \ldots + \alpha_{2n}x_n, \\ &\ldots\ldots\ldots\ldots\ldots\ldots\ldots\ldots \\ x_n' &= \alpha_{n1}x_1 + \alpha_{n2}x_2 + \ldots + \alpha_{nn}x_n, \end{aligned} \tag{A.3.9}$$

or

$$x_i' = \sum_{k=1}^{n} \alpha_{ik}x_k \ (i = 1, 2, \ldots, n), \tag{A.3.10}$$

which can be written in a form similar to (A.3.4):

$$\mathbf{r}' = \alpha\mathbf{r}. \tag{A.3.11}$$

Here $\mathbf{r} \equiv \{x_1, x_2, \ldots, x_n\}$ and $\mathbf{r}' \equiv \{x_1', x_2', \ldots, x_n'\}$ are position vectors in the $n$-dimensional space, and $\alpha$ is the matrix of the linear transformation (A.3.9) of rank $n$, i.e.,

$$\alpha = \begin{pmatrix} \alpha_{11} & \alpha_{12} & \ldots & \alpha_{1n} \\ \alpha_{21} & \alpha_{22} & \ldots & \alpha_{2n} \\ \ldots & \ldots & \ldots & \ldots \\ \alpha_{n1} & \alpha_{n2} & \ldots & \alpha_{nn} \end{pmatrix}. \tag{A.3.12}$$

In the general case the variables $x_i'$, $x_k$ and the matrix elements $\alpha_{ik}$ are complex numbers.

The matrix (A.3.12) of rank $n$ is a square array consisting of $n$ rows, $n$ columns, and $n^2$ elements $\alpha_{ik}$. The elements $\alpha_{ii}$ are said to be occupying positions on the *principal diagonal* of the matrix (i.e., on the diagonal passing from the top left corner of the array to its right lower corner).

A consecutive application of two linear transformations of the type (A.3.11) with matrices $\alpha$ and $\beta$ yields

$$\mathbf{r}' = \alpha\mathbf{r}, \ \mathbf{r}'' = \beta\mathbf{r}',$$

whence

$$\mathbf{r}'' = \beta\mathbf{r}' = \beta\alpha\mathbf{r}. \tag{A.3.13}$$

In the expanded form
$$x_l'' = \sum_{i=1}^{n} \beta_{li} x_i' = \sum_{i=1}^{n} \sum_{k=1}^{n} \beta_{li} \alpha_{ik} x_k = \sum_{k=1}^{n} (\beta\alpha)_{lk} x_k, \qquad (A.3.14)$$
where
$$(\beta\alpha)_{lk} = \sum_{i=1}^{n} \beta_{li} \alpha_{ik}. \qquad (A.3.15)$$

The latter relationship formulates the *matrix multiplication rule*. To obtain the $lk$-th element of a matrix equal to the product of the matrices $\beta\alpha$, we must multiply the elements of the $l$-th row of matrix $\beta$ by the elements of the $k$-th column of matrix $\alpha$ and add the products up. This rule can be depicted schematically as follows:

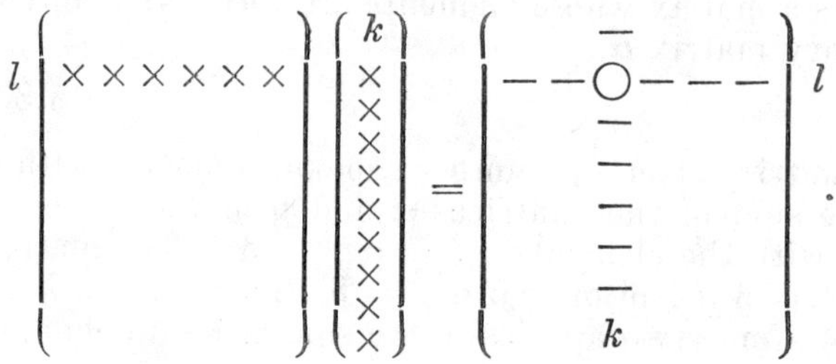

$$l \begin{Bmatrix} \times \times \times \times \times \times \end{Bmatrix} \begin{Bmatrix} k \\ \times \\ \times \\ \times \\ \times \\ \times \\ \times \\ \times \\ \times \end{Bmatrix} = \begin{Bmatrix} - \\ - \\ ---O--- \\ - \\ - \\ - \\ k \end{Bmatrix} l \qquad (A.3.16)$$

In some cases it is expedient to generalize the concept of a matrix (A.3.5) to include the case of rectangular arrays in which the number of rows is not equal to the number of columns. The matrix multiplication rule (A.3.15) can be directly generalized to the case in which the number $n$ of columns in $\beta$ is equal to the number $n$ of rows in $\alpha$ (in this case the index $i$ in (A.3.15) runs through the values from 1 to $n$). Then the product $\beta\alpha$ is a rectangular matrix, the number of whose rows is equal to the number of rows in $\beta$ and the number of columns is equal to the number of columns in $\alpha$. In the important case of multiplication of a square matrix by a single-column matrix we obtain a singlecolumn matrix.

It is obvious that in general a matrix product is *not commutative* (see A.3.16):

$$\beta\alpha \neq \alpha\beta; \qquad (A.3.17)$$

at the same time it satisfies the associative law

$$\gamma(\beta\alpha) = (\gamma\beta)\alpha \qquad (A.3.18)$$

which can be easily verified through the use of the matrix multiplication rule (A.3.15).

**A.3.3** If only the elements on the principal diagonal of a square matrix are nonzero, the matrix is called *diagonal* and its elements are

$$D_{ik} = A_i \delta_{ik}, \tag{A.3.19}$$

where $\delta_{ik}$ is the Kronecker delta.

The term *unit matrix* **E** applies to a diagonal matrix whose elements are equal to unity:

$$E_{ik} = \delta_{ik}. \tag{A.3.20}$$

Making use of the matrix multiplication rule (A.3.15), we can easily demonstrate that for any matrix

$$\mathbf{E}\boldsymbol{\alpha} = \boldsymbol{\alpha}\mathbf{E} = \boldsymbol{\alpha}. \tag{A.3.21}$$

The *zero matrix* **0** is a matrix whose elements are zeros. It is obvious that for an arbitrary matrix $\boldsymbol{\alpha}$,

$$\mathbf{0}\boldsymbol{\alpha} = \boldsymbol{\alpha}\mathbf{0} = \mathbf{0}. \tag{A.3.22}$$

The product of a matrix $\boldsymbol{\alpha}$ times a number $c$ means a matrix with the elements $c\alpha_{ik}$. The sum of the matrices $\boldsymbol{\alpha}$ and $\boldsymbol{\beta}$ of the same rank means a matrix $\boldsymbol{\gamma}$ with the elements $\gamma_{ik} = \alpha_{ik} + \beta_{ik}$. The matrix $\boldsymbol{\alpha}$ is equal to the matrix $\boldsymbol{\beta}$ if and only if $\alpha_{ik} = \beta_{ik}$ for all $i$ and $k$.

The term *trace* of a matrix $\boldsymbol{\alpha}$ applies to the sum of its diagonal elements

$$\operatorname{Tr} \boldsymbol{\alpha} = \sum_i \alpha_{ii}. \tag{A.3.23}$$

The trace of a product of matrices is independent of their order in the product; indeed

$$\begin{aligned}\operatorname{Tr}(\boldsymbol{\beta\alpha}) &= \sum_i (\boldsymbol{\beta\alpha})_{ii} = \sum_i \sum_k \beta_{ik}\alpha_{ki} = \sum_k \Big(\sum_i \alpha_{ki}\beta_{ik}\Big) \\ &= \sum_k (\boldsymbol{\alpha\beta})_{kk} = \operatorname{Tr}(\boldsymbol{\alpha\beta}).\end{aligned} \tag{A.3.24}$$

The matrix *inverse* to $\boldsymbol{\alpha}$, denoted by $\boldsymbol{\alpha}^{-1}$ performs the transformation inverse to the transformation (A.3.11):

$$\mathbf{r} = \boldsymbol{\alpha}^{-1}\mathbf{r}'. \tag{A.3.25}$$

Substituting herein (A.3.11), we obtain

$$\mathbf{r} = \boldsymbol{\alpha}^{-1}\boldsymbol{\alpha}\mathbf{r}, \tag{A.3.26}$$

whence

$$\left.\begin{array}{l}\boldsymbol{\alpha}^{-1}\boldsymbol{\alpha} = \mathbf{E} \text{ is a unit matrix,} \\ \text{similarly} \\ \boldsymbol{\alpha}\boldsymbol{\alpha}^{-1} = \mathbf{E}.\end{array}\right\} \tag{A.3.27}$$

These relations are obviously valid for the orthogonal transformations (A.3.4) and (A.3.6a) as well.

We have seen (see A.3.7) that the matrix inverse to the orthogonal transformation $\boldsymbol{\alpha}^{-1}$ is the matrix obtained by the interchange in $\boldsymbol{\alpha}$ of the rows and the columns (this is equivalent to the "reflection" of all its elements in the principal diagonal), this is called the *transpose* of a matrix. Denoting the transpose $\boldsymbol{\alpha}$ by $\tilde{\boldsymbol{\alpha}}$ (i.e., $\tilde{\alpha}_{ik} = \alpha_{ki}$), we obtain for an orthogonal transformation

$$\boldsymbol{\alpha}^{-1} = \tilde{\boldsymbol{\alpha}} \qquad \text{(A.3.28)}$$

in agreement with (A.3.7).

If $\boldsymbol{\alpha}$ is a matrix of rank $n$, the matrix equation (A.3.27) is equivalent to $n^2$ first-order algebraic equations in $n^2$ unknowns $\alpha_{ik}^{-1}$. Since the right-hand sides in $n$ of these equations are equal to unity (in the other equations they are equal to zero), the solution of $n^2$ inhomogeneous equations in the unknowns $\alpha_{ik}^{-1}$ exists if and only if det $[\alpha_{ik}]$ made up of the elements of matrix $\boldsymbol{\alpha}$ does not vanish. Hence, a matrix $\boldsymbol{\alpha}$ has an inverse matrix $\boldsymbol{\alpha}^{-1}$ only if $|\alpha_{ik}| \neq 0$, i.e., if $\boldsymbol{\alpha}$ is a *nonsingular* matrix.

It can be easily demonstrated that

$$(\boldsymbol{\beta\alpha})^{-1} = \boldsymbol{\alpha}^{-1}\boldsymbol{\beta}^{-1}, \qquad \text{(A.3.29)}$$

and

$$\widetilde{(\boldsymbol{\beta\alpha})} = \tilde{\boldsymbol{\alpha}}\tilde{\boldsymbol{\beta}}. \qquad \text{(A.3.30)}$$

It follows from (A.3.27) and (A.3.28) that for the orthogonal transformation

$$\tilde{\boldsymbol{\alpha}}\boldsymbol{\alpha} = \mathbf{E}. \qquad \text{(A.3.31)}$$

Since the matrix multiplication rule (A.3.15) coincides with the multiplication rule for determinants [A.2, p. 250], it follows that

$$\det \tilde{\boldsymbol{\alpha}} \det \boldsymbol{\alpha} = \det \mathbf{E} = 1. \qquad \text{(A.3.32)}$$

Since the replacement of the rows by the columns does not change the determinant [A.2, p. 250], it follows that $\det \tilde{\boldsymbol{\alpha}} = \det \boldsymbol{\alpha}$ and that, consequently,

$$(\det \boldsymbol{\alpha})^2 = 1. \qquad \text{(A.3.33)}$$

Therefore, the determinant of the matrix of an orthogonal transformation (A.3.2) is

$$\det \boldsymbol{\alpha} = \pm 1. \qquad \text{(A.3.33a)}$$

It can be easily demonstrated that $+1$ corresponds to a simple rotation of the system and $-1$ corresponds to a rotation accompanied by inversion, i.e., to the transition from a right-handed coordinate system to a left-handed (or vice versa). Indeed, a simple rotation may be imagined as a continuous process of rotation from the initial state in which both systems coincide and which is described by the unity transformation with the matrix $E_{ik} = \delta_{ik}$. Since the determinant corresponding to this transformation is det $[E_{ik}] =$ det $[\delta_{ik}] = 1$, and since it cannot change abruptly in the course of a continuous rotation, the determinant corresponding to a simple rotation is equal to $+1$.

In the case of inversion $J$: $x_i' = -x_i$ $(i = 1, 2, 3)$, and the corresponding determinant is $| J | = [-\delta_{ik}] = -1$. Hence, it follows immediately that the determinant corresponding to an arbitrary rotation accompanied by inversion is also equal to $-1$. Orthogonal transformations with the determinant equal to $+1$ became known as *proper rotations* (or simply rotations), whereas orthogonal transformations with the determinant equal to $-1$ become known as *improper rotations*.

**A.3.4** Consider some matrices important for applications.

Denote by $\boldsymbol{\alpha}^*$ a matrix *complex conjugate* to $\alpha$, so that $(\alpha^*)_{ik} = \alpha_{ik}^*$.

The term *hermitian conjugate* matrix $\boldsymbol{\alpha}^+$ applies to the complex-conjugate transpose of $\alpha$, i.e., to $\boldsymbol{\alpha}^+ = \widetilde{\boldsymbol{\alpha}^*}$ in which $\alpha_{ik}^+ = \alpha_{ki}^*$.

A *self-conjugate*, or *hermitian*, matrix $\boldsymbol{\alpha}$ exhibits the property $\boldsymbol{\alpha}^+ = \boldsymbol{\alpha}$; the elements of a hermitian matrix symmetrical about the principal diagonal are complex-conjugate.

The term *unitary* matrix applies to a matrix, the complex-conjugate of which is equal to the inverse matrix:

$$\mathbf{U}^+ = \mathbf{U}^{-1}, \text{ or } \mathbf{U}\mathbf{U}^+ = \mathbf{U}^+\mathbf{U} = \mathbf{E}. \tag{A.3.34}$$

A unitary matrix performs a unitary transformation, which is a generalization of an orthogonal transformation. To demonstrate this point, we shall introduce the concept of the *inner* (or *scalar*) *product* of two $n$-dimensional complex vectors $\mathbf{a}$ $(a_1, a_2, \ldots, a_n)$ and $\mathbf{b}$ $(b_1, b_2, \ldots, b_n)$, which by definition is

$$\mathbf{a}\cdot\mathbf{b} = \sum_{i=1}^{h} a_i^* b_i. \tag{A.3.35}$$

If $\mathbf{a}$ and $\mathbf{b}$ are real three-dimensional vectors, the definition (A.3.35) reduces to an ordinary scalar product (A.3.8). Demonstrate next that

$$\mathbf{a}\cdot\boldsymbol{\alpha}\mathbf{b} = \boldsymbol{\alpha}^+\mathbf{a}\cdot\mathbf{b}, \tag{A.3.36}$$

where $\boldsymbol{\alpha}$ is a linear transformation matrix of rank $n$. Indeed, the left-hand side of (A.3.36) is equal to

$$\sum_i a_i^* (\boldsymbol{\alpha}\mathbf{b})_i = \sum_{ik} a_i^* \alpha_{ik} b_k,$$

and the right-hand side is equal to

$$\sum_i (\boldsymbol{\alpha}^+\mathbf{a})_i^* b_i = \sum_{ik} a_k^* \alpha_{ki} b_i,$$

the only difference is in the designation of the dummy summation indices.

Let us now apply unitary transformation $\mathbf{U}$ to the vectors $\mathbf{a}$ and $\mathbf{b}$ in (A.3.35):

$$\mathbf{Ua} \cdot \mathbf{Ub} = \mathbf{U}^+\mathbf{Ua} \cdot \mathbf{b} = \mathbf{Ea} \cdot \mathbf{b} = \mathbf{a} \cdot \mathbf{b}, \tag{A.3.37}$$

where we made use of (A.3.36) and (A.3.34).

To conclude this section we shall introduce the important concept of a *similarity transformation* of the matrix $\boldsymbol{\alpha}$:

$$\boldsymbol{\alpha}' = \mathbf{s}^{-1}\boldsymbol{\alpha}\mathbf{s}, \tag{A.3.38}$$

where $\mathbf{s}$ is an arbitrary nonsingular matrix (i.e., a matrix that has an inverse).

Premultiplying (A.3.38) by $\mathbf{s}$, multiplying the result by $\mathbf{s}^{-1}$, and taking into account that $\mathbf{s}^{-1}\mathbf{s} = \mathbf{E}$, we obtain

$$\boldsymbol{\alpha} = \mathbf{s}\boldsymbol{\alpha}'\mathbf{s}^{-1}. \tag{A.3.38a}$$

It can be demonstrated that any matrix equation is invariant under a similarity transformation. For example, the matrix equation

$$\boldsymbol{\beta}\boldsymbol{\alpha} + \boldsymbol{\gamma} = \boldsymbol{\varepsilon} \tag{A.3.39}$$

under the similarity transformation (A.3.38a) takes the form

$$\mathbf{s}\boldsymbol{\beta}'\mathbf{s}^{-1}\mathbf{s}\boldsymbol{\alpha}'\mathbf{s}^{-1} + \mathbf{s}\boldsymbol{\gamma}'\mathbf{s}^{-1} = \mathbf{s}\boldsymbol{\varepsilon}'\mathbf{s}^{-1}.$$

Premultiplying this expression by $\mathbf{s}^{-1}$ and multiplying the result by $\mathbf{s}$, we obtain

$$\boldsymbol{\beta}'\boldsymbol{\alpha}' + \boldsymbol{\gamma}' = \boldsymbol{\varepsilon}', \tag{A.3.39a}$$

i.e., (A.3.39) is invariant under a similarity transformation.

Demonstrate in addition that the trace of a matrix is invariant under a similarity transformation:

$$\mathrm{Tr}\,\boldsymbol{\alpha} = \mathrm{Tr}\,(\mathbf{s}\boldsymbol{\alpha}'\mathbf{s}^{-1}) = \mathrm{Tr}\,(\mathbf{s}^{-1}\mathbf{s}\boldsymbol{\alpha}') = \mathrm{Tr}\,(\mathbf{E}\boldsymbol{\alpha}') = \mathrm{Tr}\boldsymbol{\alpha}', \tag{A.3.40}$$

where we made use of (A.3.24).

**A.3.5** The vector $\mathbf{r}'$ in (A.3.11) is in general not proportional to the vector $\mathbf{r}$. On the other hand, if for some vector $\mathbf{r} = \mathbf{x}\,(x_1, x_2, \ldots, x_n)$,

$$\boldsymbol{\alpha}\mathbf{x} = \lambda\mathbf{x}, \tag{A.3.41}$$

where $\lambda$ is a scalar, then $\mathbf{x}$ is termed *eigenvector*, and $\lambda$ *eigenvalue* of matrix $\boldsymbol{\alpha}$.

Equation (A.3.41), when projected on the axes, takes the form

$$\sum_{k=1}^{n} \alpha_{ik} x_k = \lambda x_i ,$$

or

$$\sum_{k=1}^{n} (\alpha_{ik} - \lambda \delta_{ik}) x_k = 0. \tag{A.3.42}$$

We see that the eigenvector components $x_k$ are determined from the system of homogeneous linear equations (A.3.42), which has nonzero solutions only if its determinant is equal to zero:

$$\begin{vmatrix} \alpha_{11} - \lambda & \alpha_{12} & \ldots & \alpha_{1n} \\ \alpha_{21} & \alpha_{22} - \lambda & \ldots & \alpha_{2n} \\ \ldots & \ldots & \ldots & \ldots \\ \alpha_{n1} & \alpha_{n2} & \ldots & \alpha_{nn} - \lambda \end{vmatrix} = 0, \tag{A.3.43}$$

or

$$\det [\alpha_{ik} - \lambda \delta_{ik}] = 0. \tag{A.3.43a}$$

The $n$-th degree equation (A.3.43) in $\lambda$ is termed a *characteristic* or *secular* equation; $n$ roots of this equation $\lambda_1, \lambda_2, \ldots, \lambda_n$ (some of which may coincide) determine $n$ eigenvalues of the matrix $\boldsymbol{\alpha}$. Each eigenvalue $\lambda_i$ is associated with an eigenvector $\mathbf{x}^{(i)}$ (or several eigenvectors, in which case the eigenvalue is said to be *degenerate*). Let us prove an important theorem concerning the eigenvalues and the eigenvectors of a hermitian matrix **H**. Let

$$\mathbf{H}\mathbf{x}^{(1)} = \lambda_1 \mathbf{x}^{(1)}, \tag{A.3.44}$$

$$\mathbf{H}\mathbf{x}^{(2)} = \lambda_2 \mathbf{x}^{(2)}, \tag{A.3.44a}$$

where $\mathbf{x}^{(1)}$, $\mathbf{x}^{(2)}$ and $\lambda_1$, $\lambda_2$ are two eigenvectors and two eigenvalues of **H**.

In terms of projections on the coordinate axes,

$$\sum_{k} H_{ik} x_k^{(1)} = \lambda_1 x_i^{(1)}, \tag{A.3.45}$$

$$\sum_{k} H_{ik} x_k^{(2)} = \lambda_2 x_i^{(2)}. \tag{A.3.45a}$$

Multiply both sides of (A.3.45) by $x_i^{(2)*}$ and sum over $i$ to obtain

$$\sum_{k,\,i} H_{ik} x_k^{(1)} x_i^{(2)*} = \lambda_1 \sum_{i} x_i^{(1)} x_i^{(2)*} .$$

Take the complex-conjugate of (A.3.45a), multiply it by $x_i^{(1)}$ and sum over $i$:

$$\sum_{k,i} H_{ik}^* x_k^{(2)*} x_i^{(1)} = \lambda_2^* \sum_i x_i^{(2)*} x_i^{(1)}.$$

Since **H** is hermitian, it follows that $H_{ik}^* = H_{ki}$, but in that case the left-hand sides of the last two equations are equal (in the second equation the dummy indices, $i$ and $k$, can be interchanged) and, therefore,

$$(\lambda_1 - \lambda_2^*) \sum_i x_i^{(2)*} x_i^{(1)} = 0. \tag{A.3.46}$$

Suppose first that $\lambda_1 = \lambda_2$ and that $\mathbf{x}^{(1)} = \mathbf{x}^{(2)} \neq 0$; then it follows from (A.3.46) that

$$\lambda_1 = \lambda_2^* = \lambda_1^* \tag{A.3.47}$$

(the last equality follows from $\lambda_1 = \lambda_2$), i.e., that the *eigenvalues* of hermitian matrices are real. Suppose on the contrary that $\lambda_1 \neq \neq \lambda_2 = \lambda_2^*$, then it follows from (A.3.46) that

$$\sum_i x_i^{(2)*} x_i^{(1)} = \mathbf{x}^{(2)} \cdot \mathbf{x}^{(1)} = 0, \tag{A.3.48}$$

i.e., that the eigenvectors of hermitian matrices (associated with different eigenvalues) are mutually orthogonal. If the eigenvalue $\lambda$ is *degenerate* and corresponds, for example, to three eigenvectors $\mathbf{x}^{(1)}$, $\mathbf{x}^{(2)}$, $\mathbf{x}^{(3)}$, the eigenvectors associated with $\lambda$ can also be made mutually orthogonal by means of the *Gram-Schmidt orthogonalization process.*

Any linear combination of the vectors $\mathbf{x}^{(1)}$, $\mathbf{x}^{(2)}$, and $\mathbf{x}^{(3)}$ will obviously be also an eigenvector corresponding to $\lambda$. Make $\mathbf{y}^{(1)} = \mathbf{x}^{(1)}$, $\mathbf{y}^{(2)} = \mathbf{x}^{(2)} + p_1\mathbf{y}^{(1)}$. Choose $p_1$ such that $\mathbf{y}^{(1)}$ is orthogonal to $\mathbf{y}^{(2)}$:

$$\mathbf{y}^{(1)} \cdot \mathbf{y}^{(2)} = 0 = \mathbf{y}^{(1)} \cdot \mathbf{x}^{(2)} + p_1 \mathbf{y}^{(1)} \cdot \mathbf{y}^{(1)},$$

whence

$$p_1 = \frac{\mathbf{y}^{(1)} \cdot \mathbf{x}^{(2)}}{\mathbf{y}^{(1)} \cdot \mathbf{y}^{(1)}}.$$

Now make $\mathbf{y}^{(3)} = \mathbf{x}^{(3)} + p_2\mathbf{y}^{(1)} + p_3\mathbf{y}^{(2)}$ and choose $p_2$ and $p_3$ such that $\mathbf{y}^{(3)}$ is orthogonal to $\mathbf{y}^{(1)}$ and $\mathbf{y}^{(2)}$. Eventually we shall have three mutually orthogonal eigenvectors $\mathbf{y}^{(i)}$ ($i = 1, 2, 3$) corresponding to $\lambda$. This process can obviously be extended to cover the case of an arbitrary degeneracy of the eigenvalue $\lambda$.

**A.3.6** We shall now introduce an important generalization of the product of matrices of different ranks.

Let the square matrix $\boldsymbol{\alpha}$ be of rank $n$ and the square matrix $\boldsymbol{\beta}$, be of rank $m$ ($n \neq m$). The term *direct product* of $\boldsymbol{\alpha}$ and $\boldsymbol{\beta}$, denoted $\boldsymbol{\alpha} \times \boldsymbol{\beta}$ applies to a matrix with the elements

$$(\boldsymbol{\alpha} \times \boldsymbol{\beta})_{ij,\,kl} = \alpha_{ik}\beta_{jl}. \tag{A.3.49}$$

Here the double index $ij$, in which $i$ runs through the values from 1 to $n$ and $j$ runs through the values from 1 to $m$, numbers the rows of the matrix $\boldsymbol{\alpha} \times \boldsymbol{\beta}$ in accordance with a certain arbitrary convention. For instance, the values 11, 12, 13, . . ., $1m$, 21, 22, 23, . . ., $2m$, $n1$, $n2$, . . ., $nm$ of the index $ij$ may be chosen to correspond to the rows 1, 2, 3, . . ., $nm$. The same rule must be adhered to in the numeration of the columns (E. Wigner noted that the marking of the rows and the columns should be identical). Every change in the marking reduces to an interchange of the rows accompanied by a simultaneous interchange of the columns. This can be demonstrated to be of no importance for the properties to be proved below.

**Theorem.** *Let there be two matrices $\boldsymbol{\alpha}$ and $\boldsymbol{\alpha}'$ of rank n and two matrices $\boldsymbol{\beta}$ and $\boldsymbol{\beta}'$ of rank m. Then*

$$(\boldsymbol{\alpha} \times \boldsymbol{\beta})\,(\boldsymbol{\alpha}' \times \boldsymbol{\beta}') = \boldsymbol{\alpha}\boldsymbol{\alpha}' \times \boldsymbol{\beta}\boldsymbol{\beta}'. \tag{A.3.50}$$

*Here in the left-hand side we first compute the direct products $\boldsymbol{\alpha} \times \boldsymbol{\beta}$ and $\boldsymbol{\alpha}' \times \boldsymbol{\beta}'$ and then the ordinary product of the obtained matrices of rank $n \cdot m$. On the right-hand side we first compute ordinary matrix products $\boldsymbol{\alpha}\boldsymbol{\alpha}'$ and $\boldsymbol{\beta}\boldsymbol{\beta}'$ and then their direct product.*

*Proof.* The element with the index $(ij, kl)$ on the left-hand side is equal to

$$\{(\boldsymbol{\alpha} \times \boldsymbol{\beta})\,(\boldsymbol{\alpha}' \times \boldsymbol{\beta}')\}_{ij,\,kl} = \sum_{rs} (\boldsymbol{\alpha} \times \boldsymbol{\beta})_{ij,\,rs}\,(\boldsymbol{\alpha}' \times \boldsymbol{\beta}')_{rs,\,kl}$$

$$= \sum_{rs} \alpha_{ir}\beta_{js}\alpha'_{rk}\beta'_{sl}, \tag{A.3.51}$$

where the first equality expresses the conventional matrix multiplication rule (A.3.15) (i.e., the summation over the "inner" index $rs$) and the second equality is based on (A.3.49).

The same element with the index $(ij, kl)$ on the right-hand side of (A.3.50) is equal to

$$\{\boldsymbol{\alpha}\boldsymbol{\alpha}' \times \boldsymbol{\beta}\boldsymbol{\beta}'\}_{ij,\,kl} = (\boldsymbol{\alpha}\boldsymbol{\alpha}')_{ik}\,(\boldsymbol{\beta}\boldsymbol{\beta}')_{jl} = \sum_{rs} \alpha_{ir}\alpha'_{rk}\beta_{js}\beta'_{sl},$$

which coincides with (A.3.51). Hence, (A.3.50) has been proved.

In addition, we compute the trace of the matrix equal to the direct product of two matrices:

$$\mathrm{Tr}\,(\alpha\times\beta)=\sum_{ij}(\alpha\times\beta)_{ij,\,ij}$$
$$=\sum_{ij}\alpha_{ii}\beta_{jj}=\sum_{i}\alpha_{ii}\sum_{j}\beta_{jj}=\mathrm{Tr}\,\alpha\times\mathrm{Tr}\,\beta \qquad \text{(A.3.52)}$$

i.e., the trace of the direct product of two matrices is equal to the product of the traces of the matrices.

## Appendix 4

Consider the properties of an ideal Fermi gas at absolute zero (Ya. I. Frenkel). Let the volume $\Delta V$ contain $\Delta N$ electrons in the lowest energy state. The Pauli principle requires an elementary cell of the phase space, $(\Delta x\Delta y\Delta z)\,(\Delta p_x\Delta p_y\Delta p_z)=(2\pi\hbar)^3$, to contain not more than two electrons (with opposite spins). In the lowest energy state the $\Delta N$ electrons will fill in momentum space a sphere of the radius $p_0$ determined from the condition

$$2\,\frac{\text{total phase space}}{(2\pi\hbar)^3}=2\,\frac{\Delta V\,(4\pi/3)\,p_0^3}{(2\pi\hbar)^3}=\Delta N, \qquad \text{(A.4.1)}$$

whence

$$p_0=2\pi\hbar\left(\frac{3n}{8\pi}\right)^{1/3}, \qquad \text{(A.4.2)}$$

where $n=\Delta N/\Delta V$ is the electron concentration. The maximum kinetic energy of the electrons at the point with a concentration $n$ is equal to

$$\varepsilon_0=\frac{1}{2m}p_0^2=\frac{(2\pi\hbar)^2}{2m}\left(\frac{3n}{8\pi}\right)^{2/3}. \qquad \text{(A.4.3)}$$

The number of quantum states (the statistical weight) per cubic centimetre in the momentum interval $[p,\ p+dp]$ or in the interval of kinetic energy $[\varepsilon,\ \varepsilon+d\varepsilon]$ is equal to

$$d_pn=2\,\frac{1\ \text{cm}^3\times\text{volume of a spherical layer }dp}{(2\pi\hbar)^3}=\frac{8\pi p^2\,dp}{(2\pi\hbar)^3}$$
$$d_\varepsilon n=\frac{8\sqrt{2}\pi\,m^{3/2}}{(2\pi\hbar)^3}\sqrt{\varepsilon}\,d\varepsilon, \qquad \text{(A.4.4)}$$

where $\varepsilon=p^2/2m$. Figure A.2 depicts the electron distribution (quantum states) densities per unit intervals of momentum and energy.

Calculate the density of the kinetic energy of the electrons (i.e., the kinetic energy per cubic centimetre):

$$\mathscr{E}_k = \int \frac{1}{2m} p^2 \, d_p n = \frac{8\pi}{2m(2\pi\hbar)^3} \int_0^{p_0} p^4 \, dp = \frac{3^{5/3}\pi^{4/3}\hbar^2}{10m} n^{5/3}. \quad \text{(A.4.5)}$$

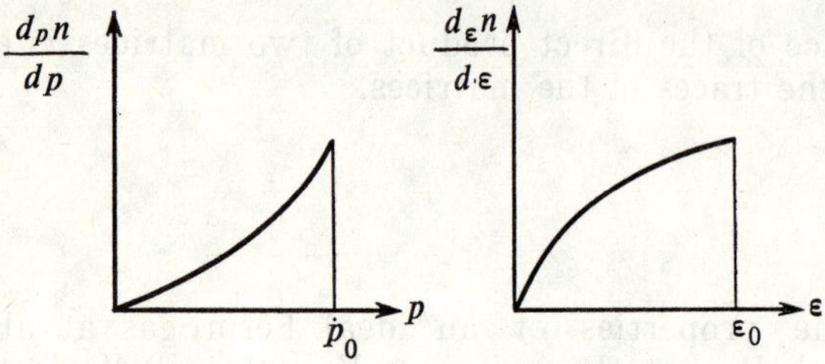


**Fig. A.2**

## Appendix 5

**The concept of normal (principal) coordinates of a mechanical system.** Consider a simple mechanical system with two degrees of freedom that can be used to illustrate the concept of *normal*, or *principal, coordinates.* Imagine two particles with identical (for the sake of simplicity) masses $m$ capable of moving along the $x$ axis. The quasi-elastic forces $-\beta u_1$ and $-\beta u_2$ act on the first and the second particles, respectively, attracting them to their respective equilibrium positions $O_1$ and $O_2$ ($u_1$ and $u_2$ are displacements of the particles from their centres $O_1$ and $O_2$). In addition, the particles interact with a quasi-elastic force equal to $\pm\gamma(u_1 - u_2)$. The equations of motion are of the form

$$m\ddot{u}_1 = -\beta u_1 - \gamma(u_1 - u_2),$$
$$m\ddot{u}_2 = -\beta u_2 - \gamma(u_2 - u_1). \quad \text{(A.5.1)}$$

We shall look for the solution of this system of differential equations in a complex form

$$u_1 = A_1 e^{i\omega t}, \quad u_2 = A_2 e^{i\omega t}. \quad \text{(A.5.2)}$$

Substituting (A.5.2) into (A.5.1) and canceling out the multiplier $e^{i\omega t}$, we obtain a homogeneous linear system of algebraic equations for the amplitudes $A_1$ and $A_2$

$$(\omega_A^2 - \omega^2) A_1 - \omega_B^2 A_2 = 0,$$
$$-\omega_B^2 A_1 + (\omega_A^2 - \omega^2) A_2 = 0, \quad \text{(A.5.3)}$$

where $m\omega_A^2 = \beta + \gamma$ and $m\omega_B^2 = \gamma$.

As we have seen in a similar case in Section 3.3.2, the system (A.5.3) determines only the ratio $A_1/A_2$. If follows from (A.5.3) that

$$\frac{A_1}{A_2} = \frac{\omega_B^2}{\omega_A^2 - \omega^2} = \frac{\omega_A^2 - \omega^2}{\omega_B^2}. \tag{A.5.4}$$

Solving this *quadratic characteristic equation* for $\omega^2$, we obtain two roots

$$\omega_1^2 = \omega_A^2 + \omega_B^2 = \frac{\beta + 2\gamma}{m}, \quad \omega_2^2 = \omega_A^2 - \omega_B^2 = \frac{\beta}{m}. \tag{A.5.5}$$

It follows from (A.5.4) that each of these roots yields for the ratio $A_1/A_2$

$$\left(\frac{A_1}{A_2}\right)_{\omega=\omega_1} = -1, \text{ i.e., } A_1^{(1)} = -A_2^{(1)} \equiv c_1,$$

$$\left(\frac{A_1}{A_2}\right)_{\omega=\omega_2} = +1, \text{ i.e., } A_1^{(2)} = +A_2^{(2)} \equiv c_2.$$

Hence, and from the linearity of equations (A.5.1) it follows that the general solution is of the form

$$u_1 = c_1 e^{i\omega_1 t} + c_2 e^{i\omega_2 t}, \quad u_2 = -c_1 e^{i\omega_1 t} + c_2 e^{i\omega_2 t}. \tag{A.5.6}$$

We see that the coordinates $u_1$ and $u_2$ consist of sums each addend of which varies in time harmonically with its own frequency. At this point we may ask whether such (*normal*) coordinates could be chosen in the system that would vary harmonically with time. In the above case this is quite simple. It is seen immediately from (A.5.6) that

$$(u_1 + u_2) \sim e^{i\omega_2 t}; \quad (u_1 - u_2) \sim e^{i\omega_1 t}.$$

Define the normal coordinates thus:

$$q_1 = \frac{u_1 - u_2}{\sqrt{2}}, \quad q_2 = \frac{u_1 + u_2}{\sqrt{2}} \tag{A.5.7}$$

(the constant factor $\sqrt{2}$ was chosen for the sake of convenience). Solving (A.5.7) for $u_1$ and $u_2$, we obtain

$$u_1 = \frac{\sqrt{2}}{2}(q_1 + q_2), \quad u_2 = \frac{\sqrt{2}}{2}(q_2 - q_1). \tag{A.5.8}$$

Express now the kinetic energy $\mathcal{T}$ and the potential energy $\Phi$ in terms of the normal coordinates $q_1$ and $q_2$ and the generalized velocities $\dot{q}_1$ and $\dot{q}_2$. The kinetic energy is

$$\mathcal{T} = \frac{m}{2}(\dot{u}_1^2 + \dot{u}_2^2) = \frac{m}{2}(\dot{q}_1^2 + \dot{q}_2^2). \tag{A.5.9}$$

The potential energy is

$$\Phi=\frac{1}{2}\{\beta(u_1^2+u_2^2)+\gamma(u_1-u_2)^2\}=\frac{1}{2}\{(\beta+\gamma)(u_1^2+u_2^2)-2\gamma u_1u_2\}.$$

This can be easily checked if we compute the forces $-\partial\Phi/\partial u_1$ and $-\partial\Phi/\partial u_2$ acting on the particles. Substituting the normal coordinates from equations (A.5.8) into $\Phi$, we obtain

$$\Phi=\frac{m}{2}(\omega_1^2q_1^2+\omega_2^2q_2^2). \tag{A.5.10}$$

Hence, the Lagrange function is

$$\mathscr{L}=\mathscr{T}-\Phi=\frac{m}{2}(\dot{q}_1^2+\dot{q}_2^2)-\frac{m}{2}(\omega_1^2q_1^2+\omega_2^2q_2^2). \tag{A.5.11}$$

We see $\mathscr{L}$ in normal coordinates $q_i$ and the corresponding generalized velocities $\dot{q}_i$ to reduce to a sum of quadratic terms (it does not include mixed members, for instance, proportional to $q_1q_2$). This condition is easily seen to be necessary and sufficient for each coordinate to be a harmonic function of time. Write the Lagrange equations of the second kind $\frac{d}{dt}\frac{\partial\mathscr{L}}{\partial\dot{q}_i}-\frac{\partial\mathscr{L}}{\partial q_i}=0$ for the Lagrange function (A.5.11). We have

$$\frac{d}{dt}\frac{\partial\mathscr{L}}{\partial\dot{q}_i}=m_i\ddot{q}_i \quad\text{and}\quad \frac{\partial\mathscr{L}}{\partial q_i}=-m\omega_i^2q_i,$$

whence

$$\ddot{q}_i+\omega_i^2q_i=0 \quad (i=1,\ 2). \tag{A.5.12}$$

We see that $q_i\propto e^{iw_it}$, in accordance with the above. Since $\Phi$ and, consequently, the Lagrange function $\mathscr{L}=\mathscr{T}-\Phi$, expressed in the variables $u_i$, contain a mixed member $-2\gamma u_1u_2$, the variables $u_1$ and $u_2$ in the equation of motion (A.5.1) cannot be separated. For this reason the time-dependence of each coordinate $u_i$ is not harmonic but more complex.

Let us introduce generalized momenta corresponding to the normal coordinates $q_i$, $p_i=\partial\mathscr{L}/\partial\dot{q}_i=m\dot{q}_i$.

The Hamiltonian function for our system is

$$\mathscr{H}(q,\ p)=\mathscr{T}+\Phi=\frac{1}{2m}(p_1^2+p_2^2)+\frac{1}{2m}(\omega_1^2q_1^2+\omega_2^2q_2^2), \tag{A.5.13}$$

i.e., it also contains only the squares of the canonical variables $q$ and $p$. The problem of determining the normal coordinates of a system may be said to consist in the reduction of the Hamiltonian function $\mathscr{H}$ to the sum of the squares of $q$ and $p$.

**Appendix 6**

Compute these two sums:

$$L=\sum_{\mathbf{a_n}} e^{i\mathbf{q}\cdot\mathbf{a_n}} \quad \text{and} \quad M=\sum_{\mathbf{q}} e^{i\mathbf{q}\cdot\mathbf{a_n}}, \tag{A.6.1}$$

where the vector of the direct lattice is

$$\mathbf{a_n}=n_1\mathbf{a}_1+n_2\mathbf{a}_2+n_3\mathbf{a}_3=\sum_{k=1}^{3} n_k\mathbf{a}_k\,(n_k=1, 2, 3, \ldots, G), \tag{A.6.1a}$$

and the wave vector is[1]

$$\mathbf{q}=\frac{1}{G}(g_1\mathbf{b}_1+\ _2\mathbf{b}_2+g_3\mathbf{b}_3)=\sum_{i=1}^{3}\frac{g_i}{G}\mathbf{b}_i\left(-\frac{G}{2}<g_i<\frac{G}{2}\right), \tag{A.6.1b}$$

as stipulated by the Born-von Karman cyclic conditions (Section 3.5.6). The sum $L$ is a three-dimensional generalization of expression (3.4.5). The first sum is

$$L=\sum_{n_1n_2n_3}\exp\left[i\frac{1}{G}\sum_{i=1}^{3} g_i\mathbf{b}_i\sum_{k=1}^{3} n_k\mathbf{a}_k\right]$$

$$=\sum_{n_1n_2n_3}\exp\left[\frac{2\pi i}{G}\sum_{i=1}^{3} g_i n_i\right], \tag{A.6.2}$$

where we made use of the condition $\mathbf{b}_i\cdot\mathbf{a}_k=2\pi\delta_{ik}$ [see (1.3.9)].

1) $\mathbf{q}\neq 0$; in this case at least one $g_i\neq 0$ and

$$l_i\equiv\exp\left(\frac{2\pi i}{G}g_i\right)\neq 1. \tag{A.6.3}$$

In this case

$$L=\sum_{n_1n_2n_3} l_1^{n_1}l_2^{n_2}l_3^{n_3}\propto\sum_{n_i} l_i^{n_i},$$

but

$$\sum_{n_i=1}^{G} l_i^{n_i}=l_i+l_i^2+\ \ldots\ +l_i^G=\frac{l_i\,(1-l_i^G)}{1-l_i}=0,$$

since all the $l_i^G=\exp(2\pi i g_i)=1$ and the denominator, according to (A.6.3), is nonzero. Hence, in this case $L=0$.

2) $\mathbf{q}=0$, i.e., all the $g_i=0$, all the $l_i=1$, and $L=\sum_{n_1n_2n_3} 1^{n_1}1^{n_2}1^{n_3}=\sum_{n_1n_2n_3} 1=G^3=N=$ number of cells in the crystal's principal region.

[1] $G$ is conveniently assumed to be a large odd number.

The case $\mathbf{q} = b_{\mathbf{g}}$ is the vector of the reciprocal lattice is obviously equivalent to the case $\mathbf{q} = 0$.

Hence,

$$L = \sum_{\mathbf{a_n}} e^{i\mathbf{q}\cdot\mathbf{a_n}} = N\delta_{\mathbf{q}\mathbf{b_g}}, \tag{A.6.4}$$

where, in particular, $\mathbf{b_g} = 0$.

The second sum is equal to

$$M \equiv \sum_{g_1 g_2 g_3} \exp\left[\frac{2\pi i}{G}\sum_{i=1}^{3} g_i n_i\right]. \tag{A.6.5}$$

Denote

$$m_i \equiv \exp\left(\frac{2\pi i}{G} n_i\right), \tag{A.6.6}$$

then

$$M = \sum_{g_1 g_2 g_3} m_1^{g_1} m_2^{g_2} m_3^{g_3}. \tag{A.6.7}$$

Consider two more cases:

1) $\mathbf{a_n} \neq 0$; then at least one $n_i \neq 0$ corresponds to $m_i \neq 1$. In this case the sum $M$ has a factor

$$\sum_{g_i = -\frac{G-1}{2}}^{\frac{G-1}{2}} m_i^{g_i} = m_i^{-\frac{G-1}{2}} + m_i^{-\frac{G-3}{2}} + \ldots + m_i^{\frac{G-1}{2}}$$

$$= m_i^{-\frac{G-1}{2}} (1 + m_i + \ldots + m_i^{G-1}) = m_i^{-\frac{G-1}{2}} \frac{1 - m_i^G}{1 - m_i} = 0,$$

since the numerator $1 - m_i^G = 1 - \exp(2\pi i n_i) = 1 - 1 = 0$, and the denominator is nonzero.

2) $\mathbf{a_n} = 0$; then all the $n_i = 0$, and, therefore, all $m_i = 1$. In this case, as in the preceding one, the sum $M = N$. Hence,

$$M = \sum_{\mathbf{q}} e^{i\mathbf{q}\cdot\mathbf{a_n}} = N\delta_{\mathbf{a_n}0}. \tag{A.6.8}$$

## Appendix 7

**A.7.1** We present here some definite integrals commonly used in semiconductor theory (e.g., see [A.1, p. 258]).

$$\int_{-\infty}^{\infty} e^{-\alpha x^2}\,dx = 2\int_{0}^{\infty} e^{-\alpha x^2}\,dx = \frac{\sqrt{\pi}}{\alpha^{1/2}}, \tag{A.7.1}$$

$$\int_{-\infty}^{\infty} e^{-\alpha x^2} x^2\,dx = 2\int_{0}^{\infty} e^{-\alpha x^2} x^2\,dx = \frac{1}{2}\frac{\sqrt{\pi}}{\alpha^{3/2}}, \tag{A.7.2}$$

$$\int_{-\infty}^{\infty} e^{-\alpha x^2} x^4\, dx = 2\int_{0}^{\infty} e^{-\alpha x^2} x^4\, dx = \frac{3}{4}\,\frac{\sqrt{\pi}}{\alpha^{5/2}}, \tag{A.7.3}$$

$$\int_{0}^{\infty} e^{-\alpha x^2} x\, dx = \frac{1}{2\alpha}, \tag{A.7.4}$$

$$\int_{0}^{\infty} e^{-\alpha x^2} x^3\, dx = \frac{1}{2\alpha^2}, \tag{A.7.5}$$

$$\int_{0}^{\infty} e^{-\alpha x^2} x^5\, dx = \frac{1}{\alpha^3}. \tag{A.7.6}$$

A.7.2 Define the gamma function $\Gamma(z)$ by means of the equality

$$\Gamma(z) = \int_{0}^{\infty} x^{z-1} e^{-x}\, dx. \tag{A.7.7}$$

The definite integral on the right-hand side with respect to the real variable $x$ depends on the parameter $z$. The equality (A.7.7) determines the gamma function for any complex values of $z$ with a positive real part.

Deduce an important recurrence equation

$$\Gamma(z+1) = z\Gamma(z). \tag{A.7.8}$$

Integrating by parts, we obtain

$$\Gamma(z+1) = \int_{0}^{\infty} x^{z} e^{-x}\, dx = -\int_{0}^{\infty} x^{z}\, de^{-x} = -x^{z}e^{-x}\Big|_{0}^{\infty} + \int_{0}^{\infty} z x^{z-1} e^{-x}\, dx$$

$$= z\int_{0}^{\infty} x^{z-1} e^{-x}\, dx = z\Gamma(z).$$

To determine the value of $\Gamma(z)$ for integral and half odd integral values of $z$, compute

$$\Gamma(1) = \int_{0}^{\infty} e^{-x}\, dx = 1 \tag{A.7.9}$$

and

$$\Gamma\left(\frac{1}{2}\right) = \int_{0}^{\infty} x^{1/2} e^{-x}\, dx = \sqrt{\pi}, \tag{A.7.10}$$

where we introduced the integration variable $t = x^{1/2}$.

Making use of the recurrence equation (A.7.8) and of the values of $\Gamma$ (1) and $\Gamma$ (1/2), we obtain

$$\Gamma(2) = 1,\quad \Gamma(3) = 1\cdot 2,\quad \Gamma(4) = 1\cdot 2\cdot 3,\ \ldots,\ \Gamma(n) = (n-1)! \qquad \text{(A.7.11)}$$

$$\Gamma\left(\frac{3}{2}\right)=\frac{1}{2}\sqrt{\pi},\quad \Gamma\left(\frac{5}{2}\right)=\frac{3}{2}\,\frac{1}{2}\sqrt{\pi},\quad \Gamma\left(\frac{7}{2}\right)=\frac{5}{2}\,\frac{3}{2}\,\frac{1}{2}\sqrt{\pi},$$

$$\Gamma\left(\frac{2n+1}{2}\right)=\frac{(2n-1)!!}{2^n}\sqrt{\pi}, \qquad \text{(A.7.12)}$$

where $(2n-1)!!$ is a product of successive odd numbers from 1 to $(2n-1)$.

**Appendix 8**

**A.8.1** To simplify calculations deduce formula (4.2.9), i.e., calculate the energy $\mathscr{E}$ of a system consisting of $N = 2$ electrons. The generalization to the case of $N$ electrons does not present any principal difficulties, the only thing we have to do is to make use of well-known properties of determinants.

Making use of the wave function (4.2.8) and of the Hamiltonian (4.2.1a) for a system of two electrons, we obtain

$$\mathscr{E} = \int \Phi^* \hat{\mathscr{H}} \Phi \, d\tau_1\, d\tau_2 = \frac{1}{2}\int [\varphi_1^*(1)\,\varphi_2^*(2) - \varphi_1^*(2)\,\varphi_2^*(1)]$$
$$\times \left\{\hat{\mathscr{H}}_1 + \hat{\mathscr{H}}_2 + \frac{e^2}{r_{12}}\right\} [\varphi_1(1)\,\varphi_2(2) - \varphi_1(2)\,\varphi_2(1)]\, d\tau_1\, d\tau_2, \qquad \text{(A.8.1)}$$

where $\hat{\mathscr{H}}_i$ $(i = 1, 2)$ are given by expression (4.2.1b).

Opening the brackets in the integrand we obtain

$$\mathscr{E} = \frac{1}{2}\left\{\int \varphi_1^*(1)\, \hat{\mathscr{H}}_1 \varphi_1(1)\, d\tau_1 + \int \varphi_1^*(2)\, \hat{\mathscr{H}}_2 \varphi_1(2)\, d\tau_2\right.$$
$$+ \int \varphi_2^*(1)\, \hat{\mathscr{H}}_1 \varphi_2(1)\, d\tau_1 + \int \varphi_2^*(2)\, \hat{\mathscr{H}}_2 \varphi_2(2)\, d\tau_2$$
$$+ \int |\varphi_1(1)|^2 \frac{e^2}{r_{12}} |\varphi_2(2)|^2\, d\tau_1\, d\tau_2$$
$$+ \int |\varphi_1(2)|^2 \frac{e^2}{r_{12}} |\varphi_2(1)|^2\, d\tau_1\, d\tau_2$$
$$- \int \varphi_1^*(1)\,\varphi_2(1) \frac{e^2}{r_{12}} \varphi_1(2)\,\varphi_2^*(2)\, d\tau_1\, d\tau_2$$
$$\left. - \int \varphi_1^*(2)\,\varphi_2(2) \frac{e^2}{r_{12}} \varphi_1(1)\,\varphi_2^*(1)\, d\tau_1\, d\tau_2\right\}.$$

Here the first integral is equal to the second, the third to the fourth, the fifth to the sixth and the seventh to the eighth, since the only difference between the integrals in those pairs is in the integration variables, and this does not affect the value of a definite integral. Hence,

$$\mathscr{E} = \sum_{i}^{1,\,2} \int \varphi_i^* (1)\, \hat{\mathscr{H}}_1 \varphi_i (1)\, d\tau_1$$

$$+ \frac{1}{2} \sum_{i \neq j}^{1,\,2} \int |\, \varphi_i (1)\,|^2 \frac{e^2}{z_{12}} \,|\, \varphi_j (2)\,|^2 \, d\tau_1\, d\tau_2$$

$$- \frac{1}{2} \sum_{i \neq j}^{1,\,2} \int \varphi_i^* (1)\, \varphi_j (1) \frac{e^2}{r_{12}} \varphi_i (2)\, \varphi_j^* (2)\, d\tau_1\, d\tau_2. \qquad \text{(A.8.2)}$$

For $N = 2$ this coincides with (4.2.9).

**A.8.2** It follows from (4.2.9a) that

$$\delta\mathscr{E} = \sum_{i}^{1,\,N} \int d\mathbf{r}_1 \delta\psi_{n_i}^* (\mathbf{r}_1) \Big\{ \hat{\mathscr{H}}_1 \psi_{n_i} (\mathbf{r}_1)$$

$$+ \Big[ \sum_{j}{}' \int \frac{e^2 \,|\, \psi_{n_j} (\mathbf{r}_2)\,|^2\, d\mathbf{r}_2}{r_{12}} \psi_{n_i} (\mathbf{r}_1)$$

$$- \sum_{\substack{\text{parallel} \\ \text{spins}}}{}' \Big( \int \frac{e^2 \psi_{n_i} (\mathbf{r}_2)\, \psi_{n_j}^* (\mathbf{r}_2) d\mathbf{r}_2}{r_{12}} \Big) \psi_{n_j} (\mathbf{r}_1) \Big] \Big\}. \qquad \text{(A.8.3)}$$

The factor 1/2 in front of the double sums in (4.2.9a) vanishes, since, when the functions $\psi_{n_i}^*$ are varied, the function with given $n_i$ occurs twice: the first time in the sum over $i$ and the second time in the sum over $j$. Take into account the additional condition (4.2.10a) with the aid of the method of indefinite Lagrange multipliers. To this end, multiply (4.2.10a) by $-\lambda_{ij}$, sum up over $i$ and $j$, add to (A.8.3), and equate to zero:

$$\sum_{i}^{1,\,N} \int d\mathbf{r}_1 \delta\psi_{n_i}^* (\mathbf{r}_1) \Big\{ \hat{\mathscr{H}}_1 \psi_{n_i} (\mathbf{r}_1) + \ldots - \sum_{j} \lambda_{ij} \psi_{n_j} (\mathbf{r}_1) \Big\} = 0. \qquad \text{(A.8.4)}$$

Since the variations $\delta\psi_{n_i}^*$ are arbitrary, the expression in braces in (A.8.4) is zero for all values of $i$. It can be demonstrated that we can always choose the solutions so as to make the matrix $\lambda_{ij}$ diagonal. Denoting $\lambda_{ii} = \mathscr{E}_{n_i}$, we obtain equation (4.2.11).

**Appendix 9**

Substitute the Bloch functions (4.3.1) into the expression (4.2.11b) for $\mathcal{U}_{eff}$ making $n_i \equiv \mathbf{k}$ and $n_j \equiv \mathbf{k}'$:

$$\mathcal{U}_{eff}(\mathbf{r}_1) = \sum_{\mathbf{k}'}{}' \int \frac{e^2 \mid u_{\mathbf{k}'}(\mathbf{r}_2) \mid^2}{\mid \mathbf{r}_2 - \mathbf{r}_1 \mid} d\mathbf{r}_2$$

$$- \sum_{\mathbf{k}'}{}' \frac{u_{\mathbf{k}'}(\mathbf{r}_1) e^{i\mathbf{k}'\cdot\mathbf{r}_1}}{u_{\mathbf{k}}(\mathbf{r}_1) e^{i\mathbf{k}\cdot\mathbf{r}_1}} \int \frac{e^2 u_{\mathbf{k}}(\mathbf{r}_2) e^{i\mathbf{k}\cdot\mathbf{r}_2} u^*_{\mathbf{k}'}(\mathbf{r}_2) e^{-i\mathbf{k}'\cdot\mathbf{r}_2}}{\mid \mathbf{r}_2 - \mathbf{r}_1 \mid} d\mathbf{r}_2.$$

Substitute in this expression $\mathbf{r}_1 + \mathbf{a}_n$ for $\mathbf{r}_1$ and $\mathbf{r}_2 + \mathbf{a}_n$ for $\mathbf{r}_2$; the latter is the change of the integration variable. Making use of the properties of periodicity (4.3.1a) of the functions $u_{\mathbf{k}}$ and $u_{\mathbf{k}'}$ and canceling out the factor $e^{i(\mathbf{k}'-\mathbf{k})\mathbf{a}_n}$ on the right-hand side of the second sum, we see that $\mathcal{U}_{eff}(\mathbf{r}_1 + \mathbf{a}_n)$ is equal to the initial expression for $\mathcal{U}_{eff}(\mathbf{r}_1)$, i.e., we have thus proved that $\mathcal{U}_{eff}$ is periodic.

**Appendix 10**

When we substitute (4.3.1) into the Schrödinger equation (4.3.7), we have to calculate the result of applying the Laplace operator $\nabla^2 \equiv \equiv \text{div grad} \equiv \partial^2/\partial x^2 + \partial^2/\partial y^2 + \partial^2/\partial z^2$ to the product of two functions of $\mathbf{r}$ ($\mathbf{r} \equiv x, y, z$): $u_{\mathbf{k}}(\mathbf{r})\, e^{i\cdot\mathbf{kr}}$. We can easily check the validity of the following equation:

$$\nabla^2 \{f(\mathbf{r})\, \varphi(\mathbf{r})\} = \text{div grad}\,(f\varphi) = f\nabla^2\varphi + \varphi\nabla^2 f + 2\,(\text{grad}\, f\cdot \text{grad}\, \varphi). \tag{A.10.1}$$

Making $f = u_{\mathbf{k}}(\mathbf{r})$ and $\varphi = e^{i\mathbf{k}\cdot\mathbf{r}}$, we obtain

$$\text{grad}\, f = \text{grad}\, u_{\mathbf{k}},\quad \text{grad}\, \varphi = i\mathbf{k}e^{i\mathbf{k}\cdot\mathbf{r}},$$

$$\nabla^2 f = \nabla^2 u_{\mathbf{k}},\quad \nabla^2\varphi = -k^2 e^{i\mathbf{k}\cdot\mathbf{r}}.$$

Substituting these values into (A.10.1), we obtain

$$\nabla^2(f\varphi) = \nabla^2(u_{\mathbf{k}} e^{i\mathbf{k}\cdot\mathbf{r}})$$

$$= \{\nabla^2 u_{\mathbf{k}} + 2i\mathbf{k}\cdot\text{grad}\, u_{\mathbf{k}} - k^2 u_{\mathbf{k}}\}\, e^{i\mathbf{k}\cdot\mathbf{r}}. \tag{A.10.2}$$

Making use of this expression, we can easily obtain (4.3.8).

**Appendix 11**

**A.11.1** Consider the simplest properties connected with the concept of a tensor. A vector is usually defined as a quantity that, in contrast to a scalar, is characterized not only by its numerical value, but also by its direction in space. The simplest examples of a scalar and a vector are, for instance, the mass of a particle $m$ and its position vector $\mathbf{r}$ determining its position in space. As we shall now demonstrate, the definition of a vector requires a clarification that at the same time will enable us to define a tensor.

Consider two rectangular coordinate systems $(x_1, x_2, x_3)$ and $(x_1', x_2', x_3')$ with a common origin $O$. Denote the cosines of the angles between the axes of both systems by $\alpha_{ik} = \cos(\widehat{x_i, x_k'})$, where $i$ and $k$ independently run through the values 1, 2, 3.

Let us ask the following question. Is the projection of the position vector **r** on a certain axis $x_i$ a scalar? On one hand, the projection is characterized only by its numerical value, on the other hand, in the primed coordinate system $x_i' \neq x_i$, whereas the scalar mass $m$ remains the same in both systems.

It is known from analytical geometry[2] that the projections of a position vector **r** in both coordinate systems are related thus:

$$x_1' = x_1\alpha_{11} + x_2\alpha_{12} + x_3\alpha_{13},$$
$$x_2' = x_1\alpha_{21} + x_2\alpha_{22} + x_3\alpha_{23},$$
$$x_3' = x_1\alpha_{31} + x_2\alpha_{32} + x_3\alpha_{33}, \qquad \text{(A.11.1)}$$

or briefly

$$x_i' = \sum_{k=1}^{3} \alpha_{ik}x_k \quad (i = 1, 2, 3). \qquad \text{(A.11.2a)}$$

Similarly,

$$x_i = \sum_{k=1}^{3} \alpha_{ki}x_k' \quad (i = 1, 2, 3). \qquad \text{(A.11.2b)}$$

Let us now give a rigorous definition of the concept of a vector. We shall understand a vector to be a set of three quantities $A_i$ $(i = 1, 2, 3)$, which, in the process of transition from one system to another, undergo transformations obeying the rules (A.11.2a) and (A.11.2b):

$$A_i' = \sum_k \alpha_{ik}A_k, \quad A_i = \sum_k \alpha_{ki}A_k'. \qquad \text{(A.11.2c)}$$

**A.11.2** Consider the rules governing the transformation of products of the type $A_iB_k = T_{ik}$ consisting of the components of two vectors **A** and **B** in the process of transition from one coordinate system to another.

It is obvious that

$$T_{ik}' = A_i'B_k' = \sum_l \alpha_{il}A_l \sum_m \alpha_{km}B_m = \sum_{l,\,m} \alpha_{il}\alpha_{km}A_lB_m,$$

---

[2] See Section A.3.1.

i.e.,

$$T'_{ik} = \sum_{l,\, m} \alpha_{il}\alpha_{km}T_{lm}. \tag{A.11.3}$$

Any set of nine quantities

$$(T_{ik}) = \begin{pmatrix} T_{11} & T_{12} & T_{13} \\ T_{21} & T_{22} & T_{23} \\ T_{31} & T_{32} & T_{33} \end{pmatrix}, \tag{A.11.4}$$

that is transformed in accordance with the rule (A.11.3) is termed a rank 2 tensor. From this point of view it is natural to term a vector and a scalar tensors of rank 1 and rank 0. Accordingly, the term rank 3 tensor applies to a set of twenty-seven quantities $T_{ikl}$, which can be transformed in accordance with the rule

$$T'_{ikl} = \sum_{m,\, n,\, p} \alpha_{im}\alpha_{kn}\alpha_{lp}T_{mnp}. \tag{A.11.5}$$

**A.11.3** Thus, a tensor is not simply a collection of scalar quantities remaining constant in the process of transition to a new coordinate system, but a set of quantities whose transformation obeys a definite rule.

In physics tensors usually appear as coefficients in relations connecting the components of various vectors and scalars with vectors. Let us consider some examples of tensors. The differential Ohm's law in an isotropic medium is of the form

$$\mathbf{j} = \sigma\mathbf{E}, \tag{A.11.6}$$

where $\mathbf{j}$ is the vector of current density, $\mathbf{E}$ is the vector of electric field intensity, and $\sigma$ is the electric conductivity of the material at some point. In projections on rectangular coordinate axes we have

$$j_x = \sigma E_x, \quad j_y = \sigma E_y, \quad j_z = \sigma E_z. \tag{A.11.6a}$$

The generalization of (A.11.6a) for the case of an anisotropic medium is based on a natural assumption that each current density component is a homogeneous linear function of all the electric field intensity components:

$$\begin{aligned} j_x &= \sigma_{xx}E_x + \sigma_{xy}E_y + \sigma_{xz}E_z, \\ j_y &= \sigma_{yx}E_x + \sigma_{yy}E_y + \sigma_{yz}E_z, \\ j_z &= \sigma_{zx}E_x + \sigma_{zy}E_y + \sigma_{zz}E_z, \end{aligned} \tag{A.11.7}$$

or briefly,

$$j_i = \sum_k \sigma_{ik}E_k, \tag{A.11.7a}$$

where the indices $i$ and $k$ assume the values $x$, $y$ and $z$ or, respectively, 1, 2 and 3.

In the course of a transition to the primed coordinate system the vector components $j_i$ and $E_k$ are transformed in accordance with the rule (A.11.2c). It can be demonstrated that the transformation rule for the quantities $\sigma_{ik}$ is (A.11.3), i.e., that they form a rank 2 tensor.

It follows from (A.11.7a) and (A.11.2c) that

$$\sum_l \alpha_{li} j'_l = \sum_k \sigma_{ik} \sum_m \sigma_{mk} E'_m.$$

Multiplying both sides by $\alpha_{ni}$ and summing over $i$, we obtain

$$\sum_l \Big(\sum_i \alpha_{li}\alpha_{ni}\Big) j'_l = \sum_m \Big(\sum_{i,k} \alpha_{ni}\alpha_{mk}\sigma_{ik}\Big) E'_m.$$

Making use of (A.3.1) for the left-hand side, we obtain

$$j'_n = \sum_m \sigma'_{nm} E'_m, \tag{A.11.7b}$$

where

$$\sigma'_{nm} = \sum_{i,k} \alpha_{ni}\alpha_{mk}\sigma_{ik}. \tag{A.11.7c}$$

Comparing (A.11.7c) with (A.11.3), we see that the coefficients in (A.11.7) are the components of a rank 2 tensor.

It can be demonstrated that the electric conductivity tensor

$$(\sigma_{ik}) = \begin{pmatrix} \sigma_{xx} & \sigma_{xy} & \sigma_{xz} \\ \sigma_{yx} & \sigma_{yy} & \sigma_{yz} \\ \sigma_{zx} & \sigma_{zy} & \sigma_{zz} \end{pmatrix} \tag{A.11.8}$$

is a symmetrical tensor:

$$\sigma_{ik} = \sigma_{ki}. \tag{A.11.9}$$

Thus, we see that the components symmetrical about the *principal diagonal* of the tensor $\sigma_{ii}$ are equal. In isotropic dielectrics the following relationship between the electric field intensity **E** and the electric induction vector **D** exists

$$\mathbf{D} = \varepsilon \mathbf{E}, \tag{A.11.10}$$

where $\varepsilon$ is the dielectric constant. The generalization of this relation to anisotropic media (crystals) is as follows:

$$D_i = \sum_k \varepsilon_{ik} E_k, \tag{A.11.10a}$$

where $\varepsilon_{ik}$ is a symmetrical tensor of the dielectric constant.

The energy of an electron in a crystal close to its extremum value is, according to (4.3.18), equal to

$$\varepsilon(\mathbf{k}) - \varepsilon(\mathbf{k}_0) = \Delta\varepsilon = 1/2 \sum_{i,l} m_{il}^{-1} p_i p_l, \tag{A.11.11}$$

where $p_i$ and $p_l$ are the components of the electron's quasi-momentum vector, $m_{il}^{-1}$ are quantities determined by equality (4.3.19) and $i$ and $l$ are indices that assume the values $x$, $y$ and $z$ (or 1, 2 and 3). In a new (primed) coordinate system the value of the energy $\Delta\varepsilon$ is invariant, and the components of the vector **p** are transformed in accordance with the equations (A.11.2c):

$$p_i = \sum_n \alpha_{ni} p'_n, \quad p_l = \sum_s \alpha_{sl} p'_s.$$

Hence,

$$(\Delta\varepsilon) = \frac{1}{2} \sum_{n,s} (m_{ns}^{-1})' \, p'_n p'_s, \tag{A.11.11a}$$

where

$$(m_{ns}^{-1})' = \sum_{i,l} \alpha_{ni} \alpha_{sl} m_{il}^{-1}. \tag{A.11.11b}$$

In the process of transition to a new coordinate system the quantities $m_{il}^{-1}$ will be seen from the latter equality to be transformed in accordance with the rule (A.11.13), i.e., like the components of a rank 2 tensor. From (4.3.19) the inverse effective mass tensor $(m_{il}^{-1})$ is seen to be symmetric, i.e., $m_{il}^{-1} = m_{li}^{-1}$.

Just as a vector, i.e., a rank 1 tensor, can be represented as a directed segment (an arrow), a symmetrical rank 2 tensor $(T_{ik})$ can be represented as a second-order surface

$$\sum_{i,k} T_{ik} x_i x_k = 1, \tag{A.11.12}$$

or, in an expanded form,

$$T_{11}x_1^2 + T_{22}x_2^2 + T_{33}x_3^2 + 2T_{12}x_1x_2 + 2T_{13}x_1x_3 + 2T_{23}x_2x_3 = 1. \tag{A.11.12a}$$

It can be demonstrated that if all $T_{ii} > 0$, the surface is an ellipsoid termed a *tensor ellipsoid.*

**A.11.4** The tensor components change in the process of transition from one coordinate system to another. It can be proved that for a symmetrical rank 2 tensor we can always choose a coordinate system in which the tensor assumes a diagonal form, i.e., only the tensor components $T_{ii}$ located on its principal diagonal are nonzero. In this case the tensor ellipsoid takes the form

$$T_{11}x_1^2 + T_{22}x_2^2 + T_{33}x_3^2 = 1. \tag{A.11.12b}$$

We see that from the point of view of analytical geometry the choice of such a coordinate system corresponds to the reduction of the tensor ellipsoid to its principal axes.

In the principal axes the electric conductivity and inverse effective mass tensors take the form

$$(\sigma_{ik})=\begin{pmatrix}\sigma_1 & 0 & 0\\ 0 & \sigma_2 & 0\\ 0 & 0 & \sigma_3\end{pmatrix},\quad (m_{ik}^{-1})=\begin{pmatrix}m_1^{-1} & 0 & 0\\ 0 & m_2^{-1} & 0\\ 0 & 0 & m_3^{-1}\end{pmatrix}. \tag{A.11.13}$$

The quantity $1 \div m_i^{-1}$ denoted here by $m_i$ is not a tensor component, i.e., it is not transformed in accordance with the rule (A.11.3). It has the dimensionality of mass and is termed an electron effective mass in a crystal.

Since the electric conductivity and inverse effective mass tensors are material constants of a crystal, the directions of their principal axes have a definite orientation with respect to the crystal's symmetry axes (e.g., in the case of a crystal belonging to the orthorhombic system the principal axes are parallel to the edges of a unit cell). In general, the principal axes of various material tensors, e.g., of the dielectric constant and of the inverse effective mass, need not coincide. Only the microscopic theory of such tensors can provide an answer to this question. If the crystal's symmetry is such that the principal axes $x_1$ and $x_2$ of a tensor are physically equivalent, then obviously $\sigma_1 = \sigma_2$, and $m_1 = m_2$. In a cubic crystal all three principal axes of a tensor (parallel to the cube's edges) are equivalent: e.g., for this reason the electric conductivity tensor is of the form

$$(\sigma_{ik})=\begin{pmatrix}\sigma & 0 & 0\\ 0 & \sigma & 0\\ 0 & 0 & \sigma\end{pmatrix}, \tag{A.11.13a}$$

and, consequently, the rule (A.11.7) reduces to the isotropic relation (A.11.6a). The rank 2 tensors in crystals of cubic symmetry may be said to reduce to scalars.

**Appendix 12**

The Bloch electron function $\psi_k(\mathbf{r}) = u_k(\mathbf{r})\, e^{i\mathbf{k}\cdot\mathbf{r}}$ in a periodic crystal field $V(\mathbf{r})$ satisfies the Schrödinger equation

$$\nabla^2\psi(\mathbf{r}) + \frac{2m}{\hbar^2}[\varepsilon(\mathbf{k}) - V(\mathbf{r})]\,\psi(\mathbf{r}) = 0. \tag{A.12.1}$$

Differentiating both sides with respect to $k_x$, we obtain

$$\nabla^2\frac{\partial\psi}{\partial k_x} + \frac{2m}{\hbar^2}\left[\frac{\partial\varepsilon}{\partial k_x}\psi - (\varepsilon - V)\frac{\partial\psi}{\partial k_x}\right] = 0. \tag{A.12.2}$$

We have

$$\frac{\partial\psi}{\partial k_x} = \frac{\partial}{\partial k_x}(ue^{i\mathbf{k}\cdot\mathbf{r}}) = ix\psi + e^{i\mathbf{k}\cdot\mathbf{r}}\frac{\partial u}{\partial k_x}, \tag{A.12.3}$$

and

$$\nabla^2 \frac{\partial \psi}{\partial k_x} = \nabla^2 (ix\psi) + \nabla^2 \left( e^{i\mathbf{k}\cdot\mathbf{r}} \frac{\partial u}{\partial k_x} \right)$$

$$= 2i \frac{\partial \psi}{\partial x} + ix\nabla^2\psi + \nabla^2 \left( e^{i\mathbf{k}\cdot\mathbf{r}} \frac{\partial u}{\partial k_x} \right). \quad \text{(A.12.4)}$$

Substituting (A.12.4) and (A.12.3) into (A.12.2), we obtain

$$2i \frac{\partial \psi}{\partial x} + \frac{2m}{\hbar^2} \frac{\partial \varepsilon}{\partial k_x} \psi + ix \left\{ \nabla^2\psi + \frac{2m}{\hbar^2} (\varepsilon - V) \psi \right\}$$

$$+ \left[ \nabla^2 + \frac{2m}{\hbar^2} (\varepsilon - V) \right] e^{i\mathbf{k}\cdot\mathbf{r}} \frac{\partial u}{\partial k_x} = 0, \quad \text{(A.12.5)}$$

where the expression in braces vanishes because of (A.12.1).

Premultiplying both sides of equation (A.12.5) by $\psi^*$ and integrating over the crystal's principal region, we obtain

$$2i \int \psi^* \frac{\partial \psi}{\partial x} d\tau + \frac{2m}{\hbar^2} \frac{\partial \varepsilon}{\partial k_x} \int \psi^* \psi \, d\tau$$

$$+ \int e^{i\mathbf{k}\cdot\mathbf{r}} \frac{\partial u}{\partial k_x} \left[ \nabla^2 + \frac{2m}{\hbar^2} (\varepsilon - V) \right] \psi^* \, d\tau = 0, \quad \text{(A.12.6)}$$

where in the transformation of the last integral we made use of the fact that the operator $\nabla^2$ is self-conjugate. The last integral is zero, since $\psi^*$ also satisfies (A.12.1). Hence, if we take into account that $\psi$ is normalized to the principal region, it will follow from (A.12.6) that

$$\frac{\hbar}{mi} \int \psi^* \frac{\partial \psi}{\partial x} d\tau = \frac{1}{\hbar} \frac{\partial \varepsilon}{\partial k_x}. \quad \text{(A.12.7)}$$

Comparing this expression with (4.3.28), we obtain (4.3.32).

**Appendix 13**

Figure A.3 depicts a cube and a rectangular coordinate system with the axes $x$, $y$ and $z$ directed along its edges. The black circle depicts an atom in the cube's centre, and the white circles depict atoms on its faces. If the cube's edge length is $a$, and if $\mathbf{i}_0$, $\mathbf{j}_0$, $\mathbf{k}_0$ are the unit vectors of the coordinate system, the basis vectors of a face-centred cubic lattice can be chosen as follows:

$$\mathbf{a}_1 = \frac{a}{2} (\mathbf{i}_0 + \mathbf{k}_0),$$

$$\mathbf{a}_2 = \frac{a}{2} (\mathbf{i}_0 + \mathbf{j}_0),$$

$$\mathbf{a}_3 = \frac{a}{2} (\mathbf{j}_0 + \mathbf{k}_0). \quad \text{(A.13.1)}$$

The vectors $\mathbf{a}_1$, $\mathbf{a}_2$ and $\mathbf{a}_3$ connect the origin with the nearest O-atoms.

It follows directly from (A.13.1) that

$$\mathbf{a}_2 \times \mathbf{a}_3 = \frac{a^2}{4}(\mathbf{i}_0 - \mathbf{j}_0 + \mathbf{k}_0), \tag{A.13.1a}$$

and, therefore, the volume of a unit cell is

$$\Omega_0 = |\mathbf{a}_1 \cdot (\mathbf{a}_2 \times \mathbf{a}_3)| = a^3/4, \tag{A.13.1b}$$

i.e., it is equal to 1/4 of the cube's volume.

Basis vectors of a body-centred cubic lattice can be conveniently chosen as follows:

$$\mathbf{a}_1' = \frac{a}{2}(\mathbf{i}_0 - \mathbf{j}_0 + \mathbf{k}_0), \quad \mathbf{a}_2' = \frac{a}{2}(\mathbf{i}_0 + \mathbf{j}_0 - \mathbf{k}_0),$$

$$\mathbf{a}_3' = \frac{a}{2}(-\mathbf{i}_0 + \mathbf{j}_0 + \mathbf{k}_0), \tag{A.13.2}$$

i.e., they should connect the origin $O$ with the central ●-atoms of the cubes located in front, below, and to the left of the one depicted in Fig. A.3. The basis vectors in Fig. 1.6*b* were chosen similarly.

It follows from (A.13.2) that

$$\mathbf{a}_2' \times \mathbf{a}_3' = \frac{a^2}{2}(\mathbf{i}_0 + \mathbf{k}_0). \tag{A.13.2a}$$

The volume of a unit cell is

$$\Omega_0' = |\mathbf{a}_1' \cdot (\mathbf{a}_2' \times \mathbf{a}_3')| = a^3/2, \tag{A.13.2b}$$

i.e., it is equal to 1/2 of the cube's volume.

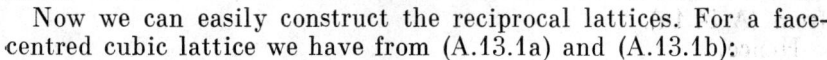

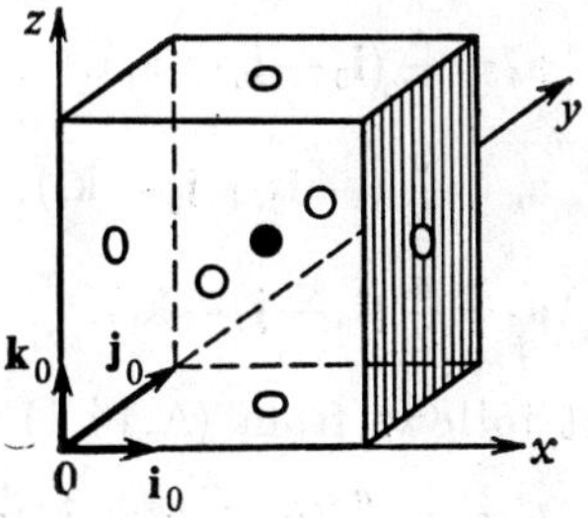

**Fig. A.3**

Now we can easily construct the reciprocal lattices. For a face-centred cubic lattice we have from (A.13.1a) and (A.13.1b):

$$\mathbf{b}_1 = \frac{1}{\Omega_0}\mathbf{a}_2 \times \mathbf{a}_3 = \frac{1}{a}(\mathbf{i}_0 - \mathbf{j}_0 + \mathbf{k}_0),$$

and, similarly,

$$\mathbf{b}_2 = \frac{1}{\Omega_0}\mathbf{a}_3 \times \mathbf{a}_1 = \frac{1}{a}(\mathbf{i}_0 + \mathbf{j}_0 - \mathbf{k}_0),$$

$$\mathbf{b}_3 = \frac{1}{\Omega_0}\mathbf{a}_1 \times \mathbf{a}_2 = \frac{1}{a}(-\mathbf{i}_0 + \mathbf{j}_0 + \mathbf{k}_0). \tag{A.13.3}$$

For a body-centred cubic lattice we obtain similarly from (A.13.2a) and (A.13.2b)

$$\mathbf{b}_1' = \frac{1}{\Omega_0'}\mathbf{a}_2' \times \mathbf{a}_3' = \frac{1}{a}(\mathbf{i}_0 + \mathbf{k}_0),$$

$$\mathbf{b}_2' = \frac{1}{\Omega_0'}\mathbf{a}_3' \times \mathbf{a}_1' = \frac{1}{a}(\mathbf{i}_0 + \mathbf{j}_0),$$

$$\mathbf{b}_3' = \frac{1}{\Omega_0'}\mathbf{a}_1' \times \mathbf{a}_2' = \frac{1}{a}(\mathbf{j}_0 + \mathbf{k}_0). \tag{A.13.4}$$

Comparing (A.13.3) with (A.13.2), and (A.13.4) with (A.13.1), we see that the reciprocal lattice of a face-centred cube is body-centred, and vice versa.

## Appendix 14

To compute the sum $\sum_{\mathbf{n}_0}^{1,8} e^{i\mathbf{k}\cdot\mathbf{a}_{n_0}}$ in the case of a body-centred cubic lattice, turn to Fig. A.3. The vector $\mathbf{a}_{\mathbf{n}_0} = \mathbf{a}_1$. For the ●-atom in the centre of the cube

$$\mathbf{a}_1 = \frac{a}{2}(\mathbf{i}_0 + \mathbf{j}_0 + \mathbf{k}_0). \tag{A.14.1}$$

Similarly for the other seven atoms

$$\mathbf{a}_2 = \frac{a}{2}(-\mathbf{i}_0 + \mathbf{j}_0 + \mathbf{k}_0), \quad \mathbf{a}_3 = \frac{a}{2}(-\mathbf{i}_0 - \mathbf{j}_0 + \mathbf{k}_0),$$

$$\mathbf{a}_4 = \frac{a}{2}(\mathbf{i}_0 - \mathbf{j}_0 + \mathbf{k}_0), \quad \mathbf{a}_5 = \frac{a}{2}(\mathbf{i}_0 + \mathbf{j}_0 - \mathbf{k}_0),$$

$$\mathbf{a}_6 = \frac{a}{2}(-\mathbf{i}_0 + \mathbf{j}_0 - \mathbf{k}_0), \quad \mathbf{a}_7 = \frac{a}{2}(-\mathbf{i}_0 - \mathbf{j}_0 - \mathbf{k}_0),$$

$$\mathbf{a}_8 = \frac{a}{2}(\mathbf{i}_0 - \mathbf{j}_0 - \mathbf{k}_0). \tag{A.14.1a}$$

It follows from (A.14.1) that

$$\mathbf{k}\cdot\mathbf{a}_1 = \frac{a}{2}(k_x + k_y + k_z);$$

similar results are obtained for the products $\mathbf{k}\cdot\mathbf{a}_i$ $(i = 2, 3, \ldots, 8)$ from (A.14.1a).

Hence,

$$\sum_{\mathbf{n}_0}^{1,8} e^{i\mathbf{k}\cdot\mathbf{a}_{\mathbf{n}_0}} = e^{\frac{ia}{2}(k_x+k_y+k_z)} + e^{\frac{ia}{2}(-k_x+k_y+k_z)}$$

$$+ e^{\frac{ia}{2}(-k_x-k_y+k_z)} + e^{\frac{ia}{2}(k_x-k_y+k_z)} + e^{\frac{ia}{2}(k_x+k_y-k_z)}$$

$$+ e^{\frac{ia}{2}(-k_x+k_y-k_z)} + e^{\frac{ia}{2}(-k_x-k_y-k_z)} + e^{\frac{ia}{2}(k_x-k_y-k_z)}.$$

Carrying the factor $e^{ia(k_x+k_y)/2}$ out of the first and fifth addends, we obtain for these addends:

$$e^{\frac{ia}{2}(k_x+k_y)}\left[e^{\frac{iak_z}{2}} + e^{-\frac{iak_z}{2}}\right] = 2e^{\frac{ia}{2}(k_x+k_y)}\cos\frac{ak_z}{2}.$$

Performing a similar operation for the second and sixth addends, for the third and seventh addends; and for the fourth and eighth addends, we shall obtain in each case an expression containing the same factor 2cos $(ak_z/2)$. Taking it out of the common bracket, transform the remaining expression in a similar way. The result will be equation (4.7.10b).

**Appendix 15**

We substitute the electron wave function in a crystal (4.7.17) into the Schrödinger equation (4.7.3) to obtain

$$\sum_{\mathbf{n}} e^{i\mathbf{k}\cdot\mathbf{a_n}}\left[-\frac{\hbar^2}{2m}\nabla^2+V(\mathbf{r})-\varepsilon\right]$$

$$\times\{\alpha\psi_x(\mathbf{r}-\mathbf{a_n})+\beta\psi_y(\mathbf{r}-\mathbf{a_n})+\gamma\psi_z(\mathbf{r}-\mathbf{a_n})\}=0. \quad \text{(A.15.1)}$$

Since $\psi_x$, $\psi_y$, $\psi_z$ are electron wave functions in an isolated atom, they satisfy the equation

$$-\frac{\hbar^2}{2m}\nabla^2\psi_\mu(\mathbf{r}-\mathbf{a_n})+[\mathcal{U}(|\mathbf{r}-\mathbf{a_n}|)-\varepsilon_0]\,\psi_\mu(\mathbf{r}-\mathbf{a_n})=0, \quad \text{(A.15.2)}$$

where $\mu = x$, $y$, or $z$, $\varepsilon_0$ is the *p*-electron's energy, and $\mathcal{U}(|\mathbf{r}-\mathbf{a_n}|)$ is its potential energy in the field of the isolated **n**-th site.

Substituting the quantities $-\frac{\hbar^2}{2m}\nabla^2\psi_\mu$ in (A.15.1), in accordance with (A.15.2), we obtain

$$\sum_{\mathbf{n}} e^{i\mathbf{k}\cdot\mathbf{a_n}}(\varepsilon-\varepsilon_0)\,[\alpha\psi_x(\mathbf{r}-\mathbf{a_n})+\beta\psi_y(\mathbf{r}-\mathbf{a_n})+\gamma\psi_z(\mathbf{r}-\mathbf{a_n})]$$

$$=\sum_{\mathbf{n}} e^{i\mathbf{k}\cdot\mathbf{a_n}}[V(\mathbf{r})-\mathcal{U}(|\mathbf{r}-\mathbf{a_n}|)]$$

$$\times\{\alpha\psi_x(\mathbf{r}-\mathbf{a_n})+\beta\psi_y(\mathbf{r}-\mathbf{a_n})+\gamma\psi_z(\mathbf{r}-\mathbf{a_n})\}. \quad \text{(A.15.3)}$$

Multiply both sides of this equality by the wave function of the zeroth site $\psi_x(\mathbf{r})$ and integrate over the volume of the crystal's principal region.[3] Assuming the wave functions $\psi_\mu$ to be orthonormal, i.e.,

$$\int \psi_\mu(\mathbf{r})\,\psi_\nu(\mathbf{r})\,d\tau=\delta_{\mu\nu} \quad (\mu \text{ and } \nu = x,\ y,\ z), \quad \text{(A.15.4)}$$

and neglecting the overlap integral of wave functions belonging to different sites, we obtain for the left-hand side of (A.15.3)

$$\alpha\,(\varepsilon-\varepsilon_0). \quad \text{(A.15.5)}$$

---

[3] It is obvious that with respect to the sum $\sum_{\mathbf{n}}$ embracing all crystal sites in (A.15.3), the zeroth site is equivalent to any other site $\mathbf{n}'$, we could have multiplied (A.15.3) by $\psi(\mathbf{r}-\mathbf{a_{n'}})$.

The right-hand side of (A.15.3) for $\mathbf{n} = 0$ will contain integrals of the form

$$\mathcal{T}_{x\nu} = \int \psi_x(\mathbf{r})\,[V(\mathbf{r}) - \mathcal{U}(\mathbf{r})]\,\psi_\nu(\mathbf{r})\,d\tau \quad (\nu = x, y, z). \qquad \text{(A.15.6)}$$

For $\nu = x$ the integral

$$\mathcal{T}_{xx} = \int \psi_x^2(\mathbf{r})\,[V(\mathbf{r}) - \mathcal{U}(\mathbf{r})]\,d\tau = -C < 0 \qquad \text{(A.15.7)}$$

is similar to (4.7.7). For $\nu = y$ or $\nu = z$

$$\mathcal{T}_{xy} = \mathcal{T}_{xz} = 0. \qquad \text{(A.15.8)}$$

This can be verified if the coordinate system is rotated about the $z$ axis until the $x$ axis coincides with the $y$ axis, and the $y$ axis coincides with the $-x$ axis. This, according to (4.7.16), will transform the function $\psi_x$ into $\psi_y$ and $\psi_y$ into $-\psi_x$. Since such a transformation of the coordinates leaves $[V(\mathbf{r}) - \mathcal{U}(\mathbf{r})]$ invariant, because of the cubic symmetry of the crystal field, the integral as a whole will change sign. On the other hand, any coordinate transformation leaves the value of a definite integral unchanged, hence $\mathcal{T}_{xy} = -\mathcal{T}_{xy}$, whence (A.15.8) follows directly. It follows from the same considerations of the crystal field's cubic symmetry that

$$\mathcal{T}_{xx} = \mathcal{T}_{yy} = \mathcal{T}_{zz} = -C. \qquad \text{(A.15.7a)}$$

On the right-hand side of (A.15.3) take into account the nearest neighbours of the zeroth site, i.e., the members with $\mathbf{a_n} = \pm a\mathbf{i}_0$, $\pm a\mathbf{j}_0$, $\pm a\mathbf{k}_0$, where $\mathbf{i}_0$, $\mathbf{j}_0$ and $\mathbf{k}_0$ are unit vectors of rectangular coordinate axes $x$, $y$ and $z$. The corresponding integrals will be of the form

$$\mathcal{T}_{x\nu}(\mathbf{a}_{\mathbf{n}_0}) = \int \psi_x(\mathbf{r})\,[V(\mathbf{r}) - \mathcal{U}(|\mathbf{r} - \mathbf{a}_{\mathbf{n}_0}|)]\,\psi_\nu(\mathbf{r} - \mathbf{a}_{\mathbf{n}_0})\,d\tau, \quad \text{(A.15.9)}$$

where $\nu = x$, $y$ or $z$.

Consider first the case $\nu = y$ or $z$. Take a neighbouring atom[4] $\mathbf{n}_0$ on the $z$ axis, i.e., make $\mathbf{a}_{\mathbf{n}_0} = a\mathbf{k}_0$. If only the fields of the zeroth atom and of the neighbouring atom we chose on the $z$ axis are taken into account (this is permissible in the region where the values of the wave functions (A.15.9) differ appreciably from zero), then $[V(\mathbf{r}) - \mathcal{U}(|\mathbf{r} - a\mathbf{k}_0|)]$ is independent of the azimuthal angle $\varphi$. Since $\psi_x \propto \cos\varphi$, $\psi_y \propto \sin\varphi$, and $\psi_z$ is independent of $\varphi$, it is easily seen that the integral (A.15.9) with respect to $\varphi$ in the cases $\nu = y$ or $\nu = z$ is equal to zero. It should, however, be noted that in the case $\nu = x$, when the integral (A.15.9) is nonzero, its value will be different depending on whether the neighbouring atom is located on

[4] The only reason for choosing a neighbouring atom on the $z$ axis is to retain the customary orientation of the spherical coordinate frame.

the $x$ axis or on the $y$ or $z$ axis. In the first case the wave functions $\psi_x(\mathbf{r})$ and $\psi_x(\mathbf{r} - a\mathbf{i}_0)$, which change signs when crossing the planes perpendicular to the $x$ axis (as shown in Fig. A.4), will be of different

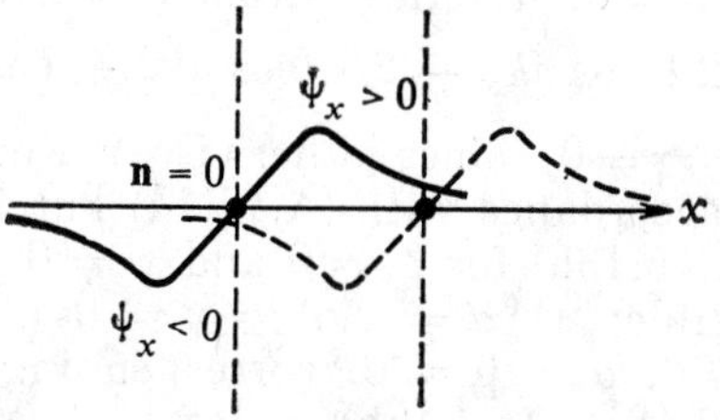


**Fig. A.4**

signs in the region where they overlap, and since $[V(\mathbf{r}) - \mathcal{U}(|\mathbf{r} \pm a\mathbf{i}_0|)] < 0$, we can expect the integral (A.15.9) to be positive in this case:

$$\mathcal{T}_{xx}(\pm a\mathbf{i}_0) = \int \psi_x(\mathbf{r})\,[V(\mathbf{r}) - \mathcal{U}(|\mathbf{r} \mp a\mathbf{i}_0|)]\,\psi_x(\mathbf{r} \mp a\mathbf{i}_0)\,d\tau = A > 0. \quad \text{(A.15.10)}$$

In the second case the value of the integral will be different, and it will be still more difficult to judge its sign. We denote

$$\mathcal{T}_{xx}(\pm a\mathbf{j}_0) = \mathcal{T}_{xx}(\pm a\mathbf{k}_0)$$
$$= \int \psi_x(\mathbf{r})\,[V(\mathbf{r}) - \mathcal{U}(|\mathbf{r} \mp a\mathbf{j}_0|)]\,\psi_x(\mathbf{r} \mp a\mathbf{j}_0)\,d\tau = -B. \quad \text{(A.15.11)}$$

Taking into account the six nearest neighbours in a simple cubic lattice, we obtain for the right-hand side of (A.15.3)

$$\alpha\,[-C + A\,(e^{ik_xa} + e^{-ik_xa}) - B\,(e^{ik_ya} + e^{-ik_ya} + e^{ik_za} + e^{-ik_za})]$$
$$= \alpha\,[-C + 2A\cos ak_x - 2B\,(\cos ak_y + \cos ak_z)]. \quad \text{(A.15.12)}$$

It follows from (A.15.3), (A.15.5) and (A.15.12) that

$$\alpha\,[\varepsilon - \varepsilon_0 + C - 2A\cos ak_x + 2B\,(\cos ak_y + \cos ak_z)] = 0. \quad \text{(A.15.13)}$$

Should we multiply (A.15.3) by $\psi_y(\mathbf{r})$ or $\psi_z(\mathbf{r})$ and integrate over the volume of the crystal's principal region, we would obtain instead of (A.15.13) two other equalities which can be obtained as a result of a cyclic permutation of $x$, $y$ and $z$ and of $\alpha$, $\beta$ and $\gamma$:

$$\beta\,[\varepsilon - \varepsilon_0 + C - 2A\cos ak_y + 2B\,(\cos ak_z + \cos ak_x)] = 0, \quad \text{(A.15.13a)}$$

$$\gamma\,[\varepsilon - \varepsilon_0 + C - 2A\cos ak_z + 2B\,(\cos ak_x + \cos ak_y)] = 0. \quad \text{(A.15.13b)}$$

The equalities (A.15.13), (A.15.13a), (A.15.13b) may simultaneously have a trivial solution if we make $\alpha = \beta = \gamma = 0$, which is of no interest since in this case the wave function (4.7.17) is identically equal to zero. If $\alpha \neq 0$, it follows from (A.15.13) that

$$\varepsilon = \varepsilon_0 - C + 2A \cos ak_x - 2B (\cos ak_y + \cos ak_z), \qquad \text{(A.15.14)}$$

and, therefore, $\beta = \gamma = 0$, since otherwise $\varepsilon$ would depend on $k_x$, $k_y$, and $k_z$ not in accordance with (A.15.14) but in accordance with (A.15.13a) and (A.15.13b) for $\beta \neq 0$ and $\gamma \neq 0$. Thus, we see that three cases are possible: (1) $\alpha \neq 0, \beta = \gamma = 0$; (2) $\beta \neq 0, \alpha = \gamma = 0$, and (3) $\gamma \neq 0$, $\alpha = \beta = 0$, corresponding to the dispersion rules (4.7.18), (4.7.18a), (4.7.18b).

## Appendix 16

Denote the projections of the vectors $\mathbf{r}_n$, $\mathbf{r}_p$, $\mathbf{R}$, and $\mathbf{r}$ on the $x$ axis by the same letters but not in bold type. Since (5.3.2) also hold for projections, they will be valid for $r_n$, $r_p$, $R$, and $r$.

Compute $\partial^2\psi/\partial r_n^2$:

$$\frac{\partial\psi}{\partial r_n} = \frac{\partial\psi}{\partial R}\frac{\partial R}{\partial r_n} + \frac{\partial\psi}{\partial r}\frac{\partial\rho}{\partial r_n} = \frac{m_n}{M}\frac{\partial\psi}{\partial R} + \frac{\partial\psi}{\partial r},$$

$$\frac{\partial^2\psi}{\partial r_n^2} = \left(\frac{m_n}{M}\frac{\partial^2\psi}{\partial R^2} + \frac{\partial^2\psi}{\partial r\,\partial R}\right)\frac{m_n}{M} + \left(\frac{m_n}{M}\frac{\partial^2\psi}{\partial R\,\partial r} + \frac{\partial^2\psi}{\partial r^2}\right)$$

$$= \left(\frac{m_n}{M}\right)^2 \frac{\partial^2\psi}{\partial R^2} + 2\frac{m_n}{M}\frac{\partial^2\psi}{\partial r\,\partial R} + \frac{\partial^2\psi}{\partial r^2}.$$

Similarly,

$$\frac{\partial^2\psi}{\partial r_p^2} = \left(\frac{m_p}{M}\right)^2 \frac{\partial^2\psi}{\partial R^2} - 2\frac{m_p}{M}\frac{\partial^2\psi}{\partial r\,\partial R} + \frac{\partial^2\psi}{\partial r^2},$$

whence

$$\frac{1}{m_n}\frac{\partial^2\psi}{\partial r_n^2} + \frac{1}{m_p}\frac{\partial^2\psi}{\partial r_p^2} = \frac{1}{M}\frac{\partial^2\psi}{\partial R^2} + \frac{1}{\mu}\frac{\partial^2\psi}{\partial r^2}.$$

Since similar equalities hold for the projections on the $y$ and $z$ axes, (5.3.3) follows directly.

## Appendix 17

In order to calculate the integral $I$ (6.2.7) in the next approximation, we shall introduce the variable $\eta = (\varepsilon - \zeta)/k_0T$ and expand the function $\varphi(\varepsilon) = \psi(\eta)$ into a power series in $\eta$. Since the function $-\partial f_0/\partial\varepsilon$ is of the delta-type, only values of $\varepsilon$ close to $\zeta$ (i.e., small $\eta$'s) will play any part in the integral:

$$\psi(\eta) = \psi(0) + \psi'(0)\,\eta + \frac{1}{2}\psi''(0)\,\eta^2 + \dots \qquad \text{(A.17.1)}$$

Further,

$$-\frac{\partial f_0}{\partial \varepsilon}\, d\varepsilon = -\frac{\partial f_0}{\partial \eta}\, d\eta = -\frac{\partial}{\partial \eta}\left(\frac{1}{e^{\eta}+1}\right) d\eta = \frac{e^{-\eta}}{(1+e^{-\eta})^2}\, d\eta. \qquad (A.17.2)$$

Substituting $-\infty$ for the lower limit in the integral $I$ equal to $-z$ (for the variable $\eta$) (this is permissible since $z \gg 1$), we obtain

$$I = \int_{-\infty}^{+\infty} \psi(\eta)\left(-\frac{\partial f_0}{\partial \eta}\right) d\eta$$

$$= \psi(0) \int_{-\infty}^{+\infty} \left(-\frac{\partial f_0}{\partial \eta}\right) d\eta + \psi'(0) \int_{-\infty}^{+\infty} \frac{\eta e^{-\eta}}{(1+e^{-\eta})^2}\, d\eta$$

$$+ \frac{1}{2}\psi''(0) \int_{-\infty}^{+\infty} \frac{\eta^2 e^{-\eta}}{(1+e^{-\eta})^2}\, d\eta. \qquad (A.17.3)$$

The first integral on the right-hand side is unity, the second is zero (because the integrand is an odd function), the third is

$$\frac{1}{2}\int_{-\infty}^{+\infty} \frac{\eta^2 e^{-\eta}}{(1+e^{-\eta})^2}\, d\eta = \int_{0}^{\infty} \frac{\eta^2 e^{-\eta}}{(1+e^{-\eta})^2}\, d\eta$$

$$= \int_{0}^{\infty} \eta^2 \left[e^{-\eta} - 2e^{-2\eta} + 3e^{-3\eta} - \ldots\right] d\eta$$

$$= 2\left(1 - \frac{1}{2^2} + \frac{1}{3^2} - \frac{1}{4^2} + \ldots\right)$$

$$= 2\,\frac{\pi^2}{12} = \frac{\pi^2}{6}. \qquad (A.17.4)$$

When we computed the last integral, we expanded the factor $(1+e^{-\eta})^{-2}$ in the integrand into a series in $e^{-\eta}$, and then we made use of the equation for the sum of the alternating series of inverse squares of natural numbers [A.1, p. 156]

Finally

$$I = \psi(0) + \frac{\pi^2}{6}\psi''(0). \qquad (A.17.5)$$

In the case (6.2.5)

$$\psi(\eta) = \frac{\sqrt{2}}{\pi^2}\,\frac{m^{*3/2}}{\hbar^3}\,\frac{2}{3}\,\zeta + k_0 T \eta^{3/2}. \qquad (A.17.6)$$

Making use of (A.17.5) and (A.17.6), we obtain for (6.2.5) the following equation:

$$n = \frac{(2m^*)^{3/2}}{3\pi^2\hbar^3}\,\zeta^{3/2}\left[1 + \frac{\pi^2}{8}\left(\frac{k_0 T}{\zeta}\right)^2\right].$$

Assuming that it is possible to put in the square brackets $\zeta = \zeta_0$ and solving the equality thus obtained for $\zeta$, which stands in front of the square brackets, we obtain

$$\zeta = \frac{\hbar^2}{2m^*}(3\pi^2 n)^{2/3}\left[1+\frac{\pi^2}{8}\left(\frac{k_0T}{\zeta_0}\right)^2\right]^{-2/3}.$$

Expanding the expression in square brackets into a series in $(k_0T/\zeta_0)^2$ and retaining only the first two members, we obtain (6.2.9).

## Appendix 18

In accordance with Poisson's equation [A.3, p. 77],

$$\sum_{N=-\infty}^{\infty} \varphi(2\pi N+t) = \frac{1}{2\pi}\sum_{l=-\infty}^{\infty} e^{ilt}\int_{-\infty}^{+\infty}\varphi(\tau)\, e^{-il\tau}\, d\tau. \tag{A.18.1}$$

In our case (6.5.25b),

$$\varphi(2\pi N+t) = \frac{1}{(2\pi)^{3/2}}(2\pi N+t)^{3/2}, \tag{A.18.2}$$

if we substitute $-N$ for the summation index $N$ and make

$$t = 2\pi\varepsilon - \pi. \tag{A.18.3}$$

Since $\varphi(2\pi N + t)$ is real, we have in our case

$$-[\varepsilon - 1/2] \leqslant N \leqslant 0, \tag{A.18.4}$$

where $[x]$ denotes the largest integer contained in $x$. The variation limits of the argument of the function $\varphi(2\pi N + t)$ are from $2\pi(\varepsilon - 1/2) - 2\pi[\varepsilon - 1/2]$ (for large $\varepsilon$ this is close to zero, which will be assumed in the future) to $t$. This determines the limits of the integral in (A.18.1) since outside these limits $\varphi(\tau) \equiv 0$.

Hence, $\Phi(\varepsilon)$ in (5.25a) for large $\varepsilon$ is equal to

$$\Phi(\varepsilon) = \frac{1}{(2\pi)^{5/2}}\sum_{l=-\infty}^{\infty} e^{ilt}\int_0^t \tau^{3/2} e^{-il\tau}\, d\tau. \tag{A.18.5}$$

The addend for $l = 0$ is equal to

$$\int_0^t \tau^{3/2}\, d\tau = \frac{2}{5}t^{5/2}. \tag{A.18.6}$$

For the sum of the two addends $+l$ and $-l$ we obtain

$$(-1)^l\left\{[\cos(2\pi l\varepsilon) + i\sin(2\pi l\varepsilon)]\int_0^t \tau^{3/2}[\cos(l\tau) - i\sin(l\tau)]\, d\tau\right.$$

$$\left. + [\cos(2\pi l\varepsilon) - i\sin(2\pi l\varepsilon)]\int_0^t \tau^{3/2}[\cos(l\tau) + i\sin(l\tau)]\, d\tau\right\}$$

$$= (-1)^l 2 \left[ \cos(2\pi l\varepsilon) \int_0^t \tau^{3/2} \cos(l\tau)\, d\tau \right.$$

$$\left. + \sin(2\pi l\varepsilon) \int_0^t \tau^{3/2} \sin(l\tau)\, d\tau \right]. \qquad (A.18.7)$$

Here we took into account that $e^{\pm il\pi} = (-1)^l$ and made use of the relation $e^{\pm i\alpha} = \cos\alpha \pm i \sin\alpha$. Integrating by parts in (A.18.7) with the aim of lowering the power of $\tau$ from 3/2 to —1/2 and substituting $(\pi/2l)\, x^2$ for the variable $\tau$ in the last integral, we obtain

$$\Phi(\varepsilon) = \frac{1}{(2\pi)^{5/2}} \left[ \frac{2}{5} (2\pi\varepsilon)^{5/2} + 3 \sum_{l=1}^{\infty} \frac{(-1)^l}{l^{5/2}} \left\{ (2\pi l\varepsilon)^{1/2} \right.\right.$$

$$\left.\left. - \sqrt{\frac{\pi}{2}} [\sin(2\pi l\varepsilon)\, S(\sqrt{4l\varepsilon}) + \cos(2\pi l\varepsilon)\, C(\sqrt{4l\varepsilon})] \right\} \right], \qquad (A.18.8)$$

where

$$S(u) = \int_0^u \sin\left(\frac{\pi}{2} x^2\right) dx, \qquad (A.18.8a)$$

$$C(u) = \int_0^u \cos\left(\frac{\pi}{2} x^2\right) dx \qquad (A.18.8b)$$

are Fresnel integrals [A.4, p. 73]; the functions $S(u)$ and $C(u)$ oscillate with a period of the order of 1, same as the trigonometric functions in (A.18.8) for $l = 1$ but with an attenuating amplitude; $C(\infty) = S(\infty) = 0.5$.

The sum in (A.18.8) is equal to [A.1, i. 156]

$$\sum_{l=1}^{\infty} \frac{(-1)^l}{l^{5/2}} (2\pi l\varepsilon)^{1/2} = (2\pi\varepsilon)^{1/2} \sum_{l=1}^{\infty} \frac{(-1)^l}{l^2}$$

$$= (2\pi\varepsilon)^{1/2} \left( -\frac{\pi^2}{12} \right); \qquad (A.18.9)$$

(6.5.26) follows directly from (A.18.8) and (A.18.9).

**Appendix 19**

With the aim of deducing the Kramers-Kronig relations (7.1.9a) continue the function $\varepsilon(\omega)$ analytically to the upper complex half-plane, making

$$\omega = \omega_1 + i\omega_2 \quad (\omega_2 > 0). \qquad (A.19.1)$$

Then (7.1.7a) assumes the form

$$\varepsilon(\omega)-1=\varepsilon(\omega_1+i\omega_2)-1=\int_0^\infty f(t)\,e^{i\omega_1 t}e^{-i\omega_2 t}\,dt. \qquad (A.19.2)$$

It follows from the properties of the function $f(t)$ and from the presence in the integrand of a decaying exponent $\exp(-\omega_2 t)$ that the integral converges, and, therefore, the function $\varepsilon(\omega)-1$ has no singular points in the upper half-plane.

If the function $\chi(\omega)=\chi(\omega_1+i\omega_2)$ has no singular points inside a closed contour $C$, it follows from the Cauchy theorem [A.2, p. 525] that

$$\oint_C \chi(\omega)\,d\omega=0. \qquad (A.19.3)$$

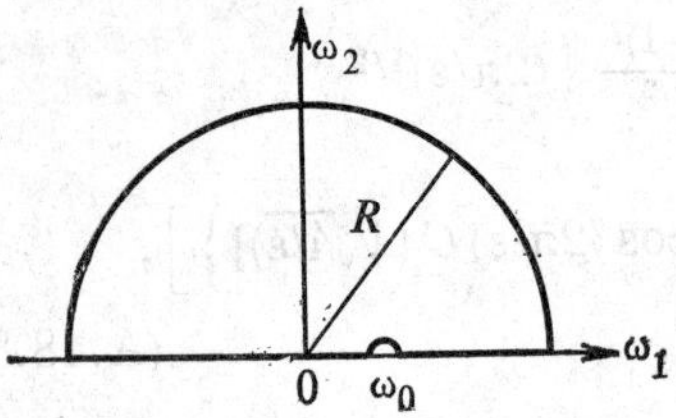


Fig. A.5

Make

$$\chi(\omega)=\frac{\varepsilon(\omega)-1}{\omega-\omega_0}, \qquad (A.19.4)$$

where $\omega_0$ is some fixed positive real value of $\omega$. Choose the contour $C$, as shown in Fig. A.5, i.e., in the shape of a semicircle of a large radius $R$, of two segments of the real axis, and of a semicircle of a small radius $\rho$ with its centre at point $\omega_0$; in this case the integral (A.19.3) with the value of $\chi(\omega)$ equal to (A.19.4) will be

$$\int_{-R}^{\omega_0-\rho}\frac{\varepsilon(\omega_1)-1}{\omega_1-\omega_0}\,d\omega_1+\int_{(\rho)}\frac{\varepsilon(\omega)-1}{\omega-\omega_0}\,d\omega$$

$$+\int_{\omega_0+\rho}^{R}\frac{\varepsilon(\omega_1)-1}{\omega_1-\omega_0}\,d\omega_1+\int_{(R)}\frac{\varepsilon(\omega)-1}{\omega-\omega_0}\,d\omega=0. \qquad (A.19.5)$$

In the limit $R\to\infty$ and $\rho\to 0$ the last integral will be zero, since $\varepsilon(\omega)-1$ tends exponentially to zero for $R\to\infty$. To compute the second integral, take into account that $\omega-\omega_0=\rho e^{i\varphi}$, so that for a specified $\rho$ we have $d\omega=i\rho e^{i\varphi}\,d\varphi$. Then for $\rho\to 0$

$$\int_{\rho\to 0}\frac{\varepsilon(\omega)-1}{\omega-\omega_0}\,d\omega=\int_{\rho\to 0}\frac{\varepsilon(\omega)-1}{\rho e^{i\varphi}}\,i\rho e^{i\varphi}\,d\varphi$$

$$=i\,[\varepsilon(\omega_0)-1]\,(-\pi). \qquad (A.19.5a)$$

Finally, the sum of the first and third integrals for $\rho\to 0$ yields the principal value of the integral. Hence, we obtain from (A.19.5a)

$$\int_{-\infty}^{\infty}\frac{\varepsilon(\omega_1)-1}{\omega_1-\omega_0}\,d\omega_1-i\pi\,[\varepsilon(\omega_0)-1]=0,$$

and we can rewrite it in the form

$$\varepsilon_1(\omega)+i\varepsilon_2(\omega)-1=\frac{1}{i\pi}\int\frac{\varepsilon_1(x)+i\varepsilon_2(x)-1}{x-\omega}\,dx, \qquad (A.19.6)$$

where we introduced the notation $\omega_0 \equiv \omega$ and $\omega_1 \equiv x$.

Separating in the last equality the real parts from the imaginary parts, we obtain

$$\varepsilon_1(\omega)-1=\frac{1}{\pi}\int\limits_{-\infty}^{\infty}\frac{\varepsilon_2(x)}{x-\omega}\,dx, \qquad (A.19.7)$$

$$\varepsilon_2(\omega)=-\frac{1}{\pi}\int\limits_{-\infty}^{\infty}\frac{\varepsilon_1(x)-1}{x-\omega}\,dx, \qquad (A.19.7a)$$

this coinciding with the first equalities in (7.1.7) and (7.1.7a).

To obtain the second equality in (7.1.7), we transform the integral in (A.19.7):

$$\int\limits_{-\infty}^{\infty}\frac{\varepsilon_2(x)}{x-\omega}\,dx=\int\limits_{-\infty}^{0}\frac{\varepsilon_2(x)}{x-\omega}\,dx+\int\limits_{0}^{\infty}\frac{\varepsilon_2(x)}{x-\omega}\,dx.$$

Substitute in the first integral on the right-hand side $-x$ for $x$ and make use of the fact that the function $\varepsilon_2(x)$ in (7.1.6a) is odd. The right-hand side will take the form

$$\int\limits_{0}^{\infty}\frac{\varepsilon_2(x)}{x+\omega}\,dx+\int\limits_{0}^{\infty}\frac{\varepsilon_2(x)}{x-\omega}\,dx=2\int\limits_{0}^{\infty}\frac{x\varepsilon_2(x)}{x^2-\omega^2}\,dx,$$

which coincides with (7.1.7).

The second equality in (7.1.7a) is obtained in a similar way.

**Appendix 20**

**A.20.1** Consider certain points of the theory of quantum transitions of importance to us.

Let the full Hamiltonian of a system be

$$\hat{\mathscr{H}}(t)=\hat{\mathscr{H}}_0+\hat{\mathscr{H}}'(t), \qquad (A.20.1)$$

where $\hat{\mathscr{H}}'(t)$ is a small time-dependent perturbation. The time-dependent Schrödinger equation with the Hamiltonian (A.20.1) is of the form

$$i\hbar\frac{\partial\Psi(t)}{\partial t}=\hat{\mathscr{H}}(t)\,\Psi(t). \qquad (A.20.1a)$$

Expand the solution of this equation $\Psi(t)$ in a closed system of eigenfunctions $\psi_n = u_n \exp\left(-\frac{i\varepsilon_n t}{\hbar}\right)$ of the unperturbed Hamiltonian $\hat{\mathscr{H}}_0$

$$\Psi(t) = \sum_n a_n(t) u_n \exp\left(-\frac{i\varepsilon_n t}{\hbar}\right) \tag{A.20.2}$$

with

$$\hat{\mathscr{H}}_0 u_n = \varepsilon_n u_n. \tag{A.20.2a}$$

Making $a_n(t) = a_n^{(0)} + a_n^{(1)}(t) + a_n^{(2)}(t) + \ldots$, where $a_n^{(0)}$) is the unperturbed (initial) value of $a_n(t)$, and $a_n^{(1)}(t)$, $a_n^{(2)}(t)$ are corrections of the first and second orders of smallness in $\hat{\mathscr{H}}'(t)$, we can demonstrate that [A.5, Sec. 29]

$$\frac{da_k^{(1)}}{dt} = \frac{1}{i\hbar} \sum_n \mathscr{H}'_{kn} a_n^{(0)} \exp\left[\frac{i}{\hbar}(\varepsilon_k - \varepsilon_n) t\right], \tag{A.20.3}$$

$$\frac{da_k^{(2)}}{dt} = \frac{1}{i\hbar} \sum_n \mathscr{H}'_{kn} a_n^{(1)} \exp\left[\frac{i}{\hbar}(\varepsilon_k - \varepsilon_n) t\right], \tag{A.20.3a}$$

where the matrix element is

$$\mathscr{H}'_{kn} = \int u_k^* \hat{\mathscr{H}}' u_n \, d\tau, \tag{A.20.4}$$

and $\sum_n$ means the summation over the discrete and integration with respect to the continuous states of the unperturbed system.

**A.20.2** Suppose that at the initial instant ($t = 0$) the system is in the $i$-th quantum state. Then $a_i^{(0)} = 1$, and all other $a_n^{(0)} = 0$ ($n \neq i$). We are interested in the amplitude $a_f^{(1)}(t)$ of the final state $f$ at the time $t$, if the perturbation $\hat{\mathscr{H}}'(t)$ is "switched on" at the instant $t = 0$. It is obvious that $a_f^{(1)}(0) = 0$; therefore, from (A.20.3) we have

$$a_f^{(1)}(t) = \frac{1}{i\hbar} \int_0^t \mathscr{H}'_{fi}(t') \exp\left[\frac{i}{\hbar}(\varepsilon_f - \varepsilon_i) t'\right] dt'. \tag{A.20.5}$$

If $\mathscr{H}'$ is independent of time,[5] we obtain

$$a_f^{(1)}(t) = -\frac{1}{\hbar} \mathscr{H}'_{fi} \frac{\exp(i\omega_{fi} t) - 1}{\omega_{fi}}, \tag{A.20.6}$$

where $\mathscr{H}'_{fi} = \text{const.}$, and $\omega_{fi} = (\varepsilon_f - \varepsilon_i)/\hbar$.

It is known from quantum mechanics that the squares of the moduli of the coefficients in expansion (A.20.2) determine the relative

[5] This statement is inaccurate, since it is assumed that $\hat{\mathscr{H}}' = \text{const.}$ for $t > 0$ but $\hat{\mathscr{H}}' = 0$ for $t' < 0$.

probabilities of the corresponding states; therefore, the probability of finding the system in the state $f$ at the instant $t$ is

$$|a_f^{(1)}(t)|^2 = |\mathcal{H}'_{fi}|^2 \frac{4\sin^2\left(\frac{\omega_{fi}t}{2}\right)}{\hbar^2\omega_{fi}^2}$$

$$= \frac{2\pi t}{\hbar^2}|\mathcal{H}'_{fi}|^2 \left[\frac{t}{2\pi}\frac{\sin^2\left(\frac{\omega_{fi}t}{2}\right)}{\left(\frac{\omega_{fi}t}{2}\right)^2}\right]. \qquad (A.20.7)$$

Since at the initial instant $t = 0$ we have $a_i(0) = 1$ and $a_f^{(1)} = 0$, it follows that $|a_f^{(1)}(t)|^2$ can be regarded as the probability of transition of the system from the state $i$ to the state $f$ in the time $t$.

Let us demonstrate that the expression in square brackets on the right-hand side of this expression for $t \gg 1/\omega_{fi}$ behaves like a $\delta$-function of $\omega_{fi}$, i.e., for large $t$'s

$$\left[\frac{t\sin^2(\omega_{fi}t/2)}{2\pi(\omega_{fi}t/2)^2}\right]_{t\gg 1/\omega_{fi}} \to \delta(\omega_{fi}). \qquad (A.20.8)$$

Figure A.6 depicts the factor $\sin^2(\omega_{fi}t/2)/\omega_{fi}^2$ as a function of $\omega_{fi}$. For $\omega_{fi} = 0$ the expression in square brackets in (A.20.8) is equal to $t/2\pi$, i.e., in units of $1/\omega_{fi}$ it is very large. On the other hand,

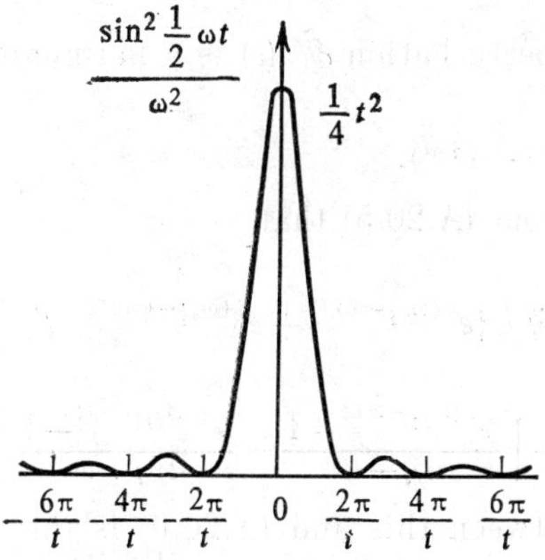


**Fig. A.6**

the half-width of the central maximum, outside which the expression in the square brackets is practically zero, is quite small, since it is equal to $2\pi/t$. It remains to be proved that the integral of the expression in square brackets with respect to $d\omega_{fi}$ is unity:

$$\int_{-\infty}^{\infty}\left[\frac{t}{2\pi}\frac{\sin^2(\omega_{fi}t/2)}{(\omega_{fi}t/2)^2}\right]d\omega_{fi}=\frac{1}{\pi}\int_{-\infty}^{\infty}\frac{\sin^2 x}{x^2}\,dx=1,$$

since the last integral is equal to $\pi$ [A.4, p. 58].

Hence, the $i\to f$ transition probability per unit time is

$$w_{if}=\frac{|a_f^{(1)}(t)|^2}{t}=\frac{2\pi}{\hbar^2}|\mathscr{H}'_{fi}|^2\,\delta(\omega_{fi}). \qquad (A.20.9)$$

Since [A.5, p. 56]

$$\delta(\omega_{fi})=\delta\left(\frac{\varepsilon_f-\varepsilon_i}{\hbar}\right)=\hbar\delta(\varepsilon_f-\varepsilon_i),$$

it follows that

$$w_{if}=\frac{2\pi}{\hbar}|\mathscr{H}'_{fi}|^2\,\delta(\varepsilon_f-\varepsilon_i). \qquad (A.20.10)$$

Note that owing to the structure of the expression in square brackets in (A.20.7), which can be reduced to a $\delta$-function, $|a_f^{(1)}(t)|^2$ is proportional to $t$, i.e., $w_{if}$ is independent of time.

To determine the total number of transitions per unit time, $W$ we must add up all the $w_{if}$'s over the states $i$ and $f$:

$$W=\sum_{i,f}w_{if}=\frac{2n}{\hbar}\sum_{i,f}|\mathscr{H}'_{fi}|^2\,\delta(\varepsilon_f-\varepsilon_i). \qquad (A.20.10a)$$

**A.20.3** If the perturbation $\hat{\mathscr{H}}'(t)$ is a harmonic function of time, i.e., if

$$\hat{\mathscr{H}}'_{(t)}=\hat{\mathscr{H}}^0(e^{-i\omega t}+e^{i\omega t}), \qquad (A.20.11)$$

then it follows from (A.20.5) that

$$a_f^{(1)}(t)=\frac{1}{i\hbar}\int_0^t\mathscr{H}^0_{fi}\,[e^{i(\omega_{fi}-\omega)t'}+e^{i(\omega_{fi}+\omega)t'}]\,dt'$$

$$=-\frac{\mathscr{H}^0_{fi}}{\hbar}\left[\frac{e^{i(\omega_{fi}-\omega)t}-1}{\omega_{fi}-\omega}+\frac{e^{i(\omega_{fi}+\omega)t}-1}{\omega_{fi}+\omega}\right]. \qquad (A.20.12)$$

The difference between this and (A.20.6) is that now we have two addends in the expression for the amplitude $a_f^{(1)}(t)$, the difference between the first and (A.20.6) is that $\omega_{fi}-\omega$ is substituted for $\omega_{fi}$, and between the second and (A.20.6) is that $\omega_{fi}+\omega$ is substituted for $\omega_{fi}$. The expression corresponding to the first addend is $\delta(\omega_{fi}-\omega)$, i.e., the energy conservation law $\varepsilon_f=\varepsilon_i+\hbar\omega$ for the *absorption* of a quantum $\hbar\omega$ by the system; the expression corresponding to the second addend is $\delta(\omega_{fi}+\omega)$, i.e. the energy conservation law $\varepsilon_f = \varepsilon_i - \hbar\omega$ for the *emission* of a quantum by the system. The two physical situations cannot occur simultaneously; therefore, either the first or the second addend in the square brackets (A.20.12) should be retained. For example, in the case of the absorption of a quantum $\hbar\omega$ we obtain instead of (A.20.10)

$$w_{if} = \frac{2\pi}{\hbar} |\mathcal{H}^0_{fi}|^2 \delta(\varepsilon_f - \varepsilon_i - \hbar\omega). \qquad (A.20.13)$$

In the case of emission of a quantum $\hbar\omega$ we obtain the same equation only there will be a plus sign in front of $\hbar\omega$.

The total number of quanta $\hbar\omega$ (photons) absorbed per second per one cubic centimetre of the crystal's volume in the course of electron transitions from the valence ($v$) to the conduction ($c$) band is

$$W_{vc} = \frac{2\pi}{\hbar} \sum_{i,f} |\mathcal{H}^0_{fi}|^2 \delta(\varepsilon_f - \varepsilon_i - \hbar\omega). \qquad (A.20.13a)$$

Here the summation is performed over the occupied states (per 1 cm³) of the valence band ($i \equiv v$) and over the vacant states in the conduction band ($f \equiv c$), which satisfy the energy conservation law.

**A.20.4** Consider the second approximation of the theory of quantum transitions (A.20.3a); the need for it arises, for instance, in the case where an electron goes over from the valence to the conduction band absorbing two quanta, a photon and a phonon.

Let the perturbation

$$\hat{\mathcal{H}}'(t) = \sum_k \hat{\mathcal{B}}^k e^{-i\omega_k t} \qquad (A.20.14)$$

consist of a sum of members each of which is a harmonic function of frequency $\omega_k$. We consider the case when the system absorbs a quanta $\hbar\omega_k$, this placing no limitation on the generality.

From (A.20.3a) and (A.20.14) we obtain for the transition $i \to$ via the intermediate state $m$

$$\frac{da_f^{(2)}}{dt} = \frac{1}{i\hbar} \sum_k \sum_m \mathcal{H}^k_{fm} a_m^{(1)} e^{i(\omega_{fm} - \omega_k)t}$$

$$= -\frac{1}{i\hbar^2} \sum_{k,k'} \sum_m \mathcal{H}^k_{fm} \mathcal{H}^{k'}_{mi} e^{i(\omega_{fm} - \omega_k)t} \left\{ \frac{e^{i(\omega_{mi} - \omega_{k'})t} - 1}{\omega_{mi} - \omega_{k'}} \right\}$$

$$= -\frac{1}{i\hbar^2} \sum_{k,k'} \sum_m \frac{\mathcal{H}^k_{fm} \mathcal{H}^{k'}_{mi}}{\omega_{mi} - \omega_{k'}} \{ e^{i(\omega_{fi} - \omega_k - \omega_{k'})t} - e^{i(\omega_{fm} - \omega_k)t} \}. \qquad (A.20.15)$$

Integrating, we obtain

$$a_f^{(2)}(t) = -\frac{1}{\hbar^2} \sum_{k,\,k'} \sum_m \frac{\mathcal{H}_{fm}^{k} \mathcal{H}_{mi}^{k'}}{\omega_{mi} - \omega_{k'}} \times \left\{ \frac{e^{i(\omega_{fi} - \omega_k - \omega_{k'})t} - 1}{\omega_{fi} - \omega_k - \omega_{k'}} - \frac{e^{i(\omega_{fm} - \omega_k)t} - 1}{\omega_{fm} - \omega_k} \right\}. \qquad \text{(A.20.16)}$$

This expression consists of a sum of members that depend on $m$, $k$ and $k'$. It is seen from (A.20.6) that only the squares of the moduli of separate addends entering into the expression in braces in (A.20.16) result in the proportionality of $|a_f^{(2)}(t)|^2$ to time and in the respective energy conservation law. The square of the modulus of the first addend in braces in (A.20.16) leads to the energy conservation law: $\varepsilon_f = \varepsilon_i + \hbar\omega_k + \hbar\omega_{k'}$; in such conditions the second "parasitic" addend in the braces, which is the result of a nonphysical situation of an instantaneous appearance of a perturbation at the instant $t = 0$ [A.5, Sec. 29], can be neglected. We arrive at the conclusion that to calculate $w_{if} = |a_f^{(2)}(t)|^2/t$, we must take the squares of the moduli of the individual addends on the right-hand side of (A.20.16) (dropping the second addend in the braces), since only they provide for the proportionality of $|a_f^{(2)}(t)|^2$ to the time $t$ and for the fulfillment of the law of energy conservation between the initial and final state.

Hence, we obtain for the probability of an $i \to f$ transition via an intermediate state $m$

$$w_{if} = \frac{2\pi}{\hbar} \sum_{k,\,k'} \sum_m \frac{|\mathcal{H}_{fm}^{k}|^2 \, |\mathcal{H}_{mi}^{k'}|^2}{(\varepsilon_m - \varepsilon_i - \hbar\omega_{k'})^2} \, \delta(\varepsilon_f - \varepsilon_i - \hbar\omega_k - h\omega_{k'}). \qquad \text{(A.20.17)}$$

For a definite process this expression assumes a more concrete form. In Section 7.3 we considered indirect interband transitions. Here the following two-stage transition is possible: a valence electron near $\mathbf{k} \approx 0$ first absorbs a photon, the result being its direct transition, and subsequently it absorbs a phonon $\hbar\omega_{\mathrm{q}}$ and goes over to another point of the Brillouin zone. Thus, the transition $i \to m$ takes place as a result of the electron's interaction with a photon $\hbar\omega$, which means that $\mathcal{H}_{mi}^{k'} = \mathcal{H}_{mi}^{\mathrm{ph}}$; on the other hand, the transition $m \to f$ involves the electron's interaction with a phonon $\hbar\omega_{\mathrm{q}}$; therefore, $\mathcal{H}_{fm}^{k} = \mathcal{H}_{fm}^{\mathrm{ph}}$.

Finally (A.20.17) for this transition takes the form

$$w_{if} = \frac{2\pi}{\hbar} \sum_m \frac{|\mathcal{H}_{fm}^{\mathrm{ph}}|^2 \, |\mathcal{H}_{mi}^{\mathrm{ph}}|^2}{(\varepsilon_m - \varepsilon_i - \hbar\omega)^2} \, \delta(\varepsilon_f - \varepsilon_i - \hbar\omega_{\mathrm{q}} - \hbar\omega). \qquad \text{(A.20.18)}$$

The expression for the total number of transitions $W$ can be obtained from (A.20.18) in the same way as in (A.20.13a) by summation over the initial state $i$ and the final state $f$.

**Appendix 21**

Write out the matrix element (8.3.7a) making use of (8.3.1) and (8.3.3):

$$\langle \mathbf{k}',\ N'_{\mathbf{q}j} \mid \Delta V \mid \mathbf{k},\ N_{\mathbf{q}j}\rangle$$
$$= -\frac{1}{\sqrt{NM}} \int \Big[ \psi^*_{\mathbf{k}'}(\mathbf{r}) \prod_{\mathbf{q}j} \psi^*_{N'_{\mathbf{q}j}}(Q_{\mathbf{q}j}) \operatorname{grad} V \cdot \sum_{\mathbf{q}j}{}' \mathbf{e}_{\mathbf{q}j} (a_{\mathbf{q}j} e^{i\mathbf{q}\cdot\mathbf{r}}$$
$$+ a^*_{\mathbf{q}j} e^{-i\mathbf{q}\cdot\mathbf{r}}) \psi_{\mathbf{k}}(\mathbf{r}) \prod_{\mathbf{q}j} \psi_{N_{\mathbf{q}j}}(Q_{\mathbf{q}j}) \Big] d\tau \prod_{\mathbf{q}j} dQ_{\mathbf{q}j}. \qquad \text{(A.21.1)}$$

Change the orders of integration and summation and separate the integrals with respect to the electron coordinates and to the normal lattice vibrations:

$$-\frac{1}{\sqrt{NM}} \sum_{\mathbf{q}j}{}' \Big\{ \Big[ \mathbf{e}_{\mathbf{q}j} \frac{1}{N} \int \operatorname{grad} V e^{i(\mathbf{k}+\mathbf{q}-\mathbf{k}')\cdot\mathbf{r}} u_{\mathbf{k}}(\mathbf{r})\, u^*_{\mathbf{k}'}(\mathbf{r})\, d\tau \Big]$$
$$\times \Big[ \int \prod_{\mathbf{q}j} \psi^*_{N'_{\mathbf{q}j}}(Q_{\mathbf{q}j})\, a_{\mathbf{q}j} \prod_{\mathbf{q}j} \psi_{N_{\mathbf{q}j}}(Q_{\mathbf{q}j})(dQ_{\mathbf{q}j}) \Big]$$
$$+ \Big[ \mathbf{e}_{\mathbf{q}j} \cdot \frac{1}{N} \int \operatorname{grad} V e^{i(\mathbf{k}-\mathbf{q}-\mathbf{k}')\cdot\mathbf{r}} u_{\mathbf{k}}(\mathbf{r})\, u^*_{\mathbf{k}'}(\mathbf{r})\, d\tau \Big]$$
$$\times \Big[ \int \prod_{\mathbf{q}j} \psi^*_{N'_{\mathbf{q}j}}(Q_{\mathbf{q}j})\, a^*_{\mathbf{q}j} \prod_{\mathbf{q}j} \psi_{N_{\mathbf{q}j}}(Q_{\mathbf{q}j})\,(dQ_{\mathbf{q}j}) \Big] \Big\}, \qquad \text{(A.21.2)}$$

where $(dQ_{\mathbf{q}j}) \equiv \prod_{\mathbf{q}j} dQ_{\mathbf{q}j}$.

Denote the first, second, third and fourth square brackets in (A.21.2) successively by $K^+$, $L$, $K^-$ and $L^*$. Consider first $L$. The integrals of all the pairs $\psi^*_{N'_{\mathbf{q}j}}(Q_{\mathbf{q}j})\, \psi_{N_{\mathbf{q}j}}(Q_{\mathbf{q}j})$ for $Q_{\mathbf{q}j}$ not corresponding to $a_{\mathbf{q}j}$ are equal to unity if $N'_{\mathbf{q}j} = N_{\mathbf{q}j}$ and to zero in all other cases. The matrix element of $a_{\mathbf{q}j}$, according to (3.10.25), is nonzero only for $N'_{\mathbf{q}j} = N_{\mathbf{q}j} - 1$, and in this case

$$L = \langle N_{\mathbf{q}j} - 1 \mid a_{\mathbf{q}j} \mid N_{\mathbf{q}j}\rangle = \sqrt{\frac{\hbar N_{\mathbf{q}j}}{2\omega_{\mathbf{q}j}}} \qquad \text{(A.21.3)}$$

with $L^* = 0$. Using the same arguments, we find $L^*$ to be nonzero only if $N'_{\mathbf{q}j} = N_{\mathbf{q}j} + 1$, and in this case, according to (3.10.26),

$$L^* = \langle N_{\mathbf{q}j} + 1 \mid a^*_{\mathbf{q}j} \mid N_{\mathbf{q}j}\rangle = \sqrt{\frac{\hbar (N_{\mathbf{q}j} + 1)}{2\omega_{\mathbf{q}j}}}. \qquad \text{(A.21.3a)}$$

Hence, the interaction of electrons with lattice vibrations results only in such transitions in the course of which the number of phonons of a definite kind $\mathbf{q}j$ either decreases or increases by unity, the number of all the other phonons remaining unchanged.

Consider now the integral $K^+$ with respect to the electron coordinates $\mathbf{r}$. In $K^+$ substitute $a_n + \mathbf{r}'$ for $\mathbf{r}$, where $\mathbf{r}'$ varies within one unit cell. Taking into account the three-dimensional periodicity (with lattice periods) of $V(\mathbf{r})$, $u_{\mathbf{k}}(\mathbf{r})$ and $u^*_{\mathbf{k}'}(\mathbf{r})$, we obtain

$$K^+ = \frac{1}{N} \sum_n e^{i(\mathbf{k}+\mathbf{q}-\mathbf{k}')\cdot \mathbf{a_n}} \int e^{i(\mathbf{k}+\mathbf{q}-\mathbf{k}')\cdot \mathbf{r}} \mathbf{e}_{\mathbf{q}j} \cdot \operatorname{grad} V(\mathbf{r})\, u_{\mathbf{k}}(\mathbf{r})\, u^*_{\mathbf{k}'}(\mathbf{r})\, d\tau_0, \tag{A.21.4}$$

where the integration is performed inside the unit cell, and where we dropped the prime at $\mathbf{r}'$ in the integrand. According to (A.6.4), the sum over $\mathbf{n}$ will be nonzero and equal to $N$ if

$$\mathbf{k}' = \mathbf{k} + \mathbf{q}. \tag{A.21.4a}$$

However, in this case the following possibility should be reckoned with. The sum of the vectors $\mathbf{k} + \mathbf{q}$ may fail to correspond to a point in the first Brillouin zone. In this case in order to reduce the vector $\mathbf{k}'$ to the first Brillouin zone we must make

$$\mathbf{k}' = \mathbf{k} + \mathbf{q} + \mathbf{b}_g, \tag{A.21.4b}$$

where $\mathbf{b}_g$ is a vector of the reciprocal lattice. In this case the sum over $\mathbf{n}$ in (A.21.4) will also be nonzero.

The term for the scattering corresponding to (A.21.4b) is *Umklapp process*, and it plays an important part in the establishment of thermal equilibrium in lattices (Peierls).

Umklapp processes play no part in the kinetic phenomena considered below, and, accordingly, they will be ignored in the future.

In the case (A.21.4a) we have

$$K^+ = \int \mathbf{e}_{\mathbf{q}j} \cdot \operatorname{grad} V u_{\mathbf{k}}(\mathbf{r})\, u^*_{\mathbf{k}'}(\mathbf{r})\, d\tau_0 = \int u_{\mathbf{k}} u^*_{\mathbf{k}'} \frac{\partial V}{\partial s}\, d\tau_0$$
$$= \int \frac{\partial}{\partial s} (u_{\mathbf{k}} u^*_{\mathbf{k}'} V)\, d\tau_0 - \int V \frac{\partial}{\partial s} (u_{\mathbf{k}} u^*_{\mathbf{k}'})\, d\tau_0, \tag{A.21.5}$$

where $\mathbf{k}' = \mathbf{k} + \mathbf{q}$, and $\partial/\partial s$ means differentiation with respect to the direction $\mathbf{e}_{\mathbf{q}j}$. The first integral on the right-hand side of (A.21.5) can be reduced to a surface integral [A.1, §11]:

$$\int \frac{\partial}{\partial s} (u_{\mathbf{k}} u^*_{\mathbf{k}'} V)\, d\tau_0 = \mathbf{e}_{\mathbf{q}j} \cdot \int \operatorname{grad} (u_{\mathbf{k}} u^*_{\mathbf{k}'} V)\, d\tau_0 = \mathbf{e}_{\mathbf{q}j} \oint \boldsymbol{\nu} u_{\mathbf{k}} u^*_{\mathbf{k}'} V\, d\sigma_0, \tag{A.21.5a}$$

where $\boldsymbol{\nu}$ is the external normal to the surface of the unit cell. Since $u_{\mathbf{k}} u^*_{\mathbf{k}'} V$ at the corresponding points of the opposite faces of the unit cell is the same, and $\mathbf{e}_{\mathbf{q}j}\cdot\mathbf{v} = \cos\widehat{(\mathbf{v}, \mathbf{e}_{\mathbf{q}j})}$ is equal in magnitude but opposite in sign, the integral (A.21.5a) as a whole is zero. Hence,

$$K^{+} = -\int V \frac{\partial}{\partial s}(u_{\mathbf{k}}u_{\mathbf{k}'}^{*})\,d\tau_0. \qquad \text{(A.21.5b)}$$

The functions $u_{\mathbf{k}}$ and $u_{\mathbf{k}'}^{*}$ satisfy (4.3.8):

$$-\frac{\hbar^2}{2m}\nabla^2 u_{\mathbf{k}} + V(\mathbf{r})\,u_{\mathbf{k}} - \frac{i\hbar^2}{m}(\mathbf{k}\cdot\operatorname{grad} u_{\mathbf{k}}) = \left(\varepsilon_{\mathbf{k}} - \frac{\hbar^2 k^2}{2m}\right)u_{\mathbf{k}},$$

$$-\frac{\hbar^2}{2m}\nabla^2 u_{\mathbf{k}'}^{*} + V(\mathbf{r})\,u_{\mathbf{k}'}^{*} + \frac{i\hbar^2}{m}(\mathbf{k}'\cdot\operatorname{grad} u_{\mathbf{k}'}^{*}) = \left(\varepsilon_{\mathbf{k}'} - \frac{\hbar^2 k'^2}{2m}\right)u_{\mathbf{k}'}^{*}.$$

Multiplying the first equation by $\partial u_{\mathbf{k}'}^{*}/\partial s$ and the second by $\partial u_{\mathbf{k}}/\partial s$, adding up, and integrating over the unit cell, we obtain

$$\begin{aligned}
&-\int V \frac{\partial}{\partial s}(u_{\mathbf{k}}u_{\mathbf{k}'}^{*})\,d\tau_0 \\
&= \int \Big\{ -\frac{\hbar^2}{2m}\left[\frac{\partial u_{\mathbf{k}'}^{*}}{\partial s}\nabla^2 u_{\mathbf{k}} + \frac{\partial u_{\mathbf{k}}}{\partial s}\nabla^2 u_{\mathbf{k}'}^{*}\right] \\
&- \frac{i\hbar^2}{m}\left[\frac{\partial u_{\mathbf{k}'}^{*}}{\partial s}(\mathbf{k}\cdot\operatorname{grad} u_{\mathbf{k}}) - \frac{\partial u_{\mathbf{k}}}{\partial s}(\mathbf{k}'\cdot\operatorname{grad} u_{\mathbf{k}'}^{*})\right] \\
&- \left[\left(\varepsilon_{\mathbf{k}} - \frac{\hbar^2 k^2}{2m}\right)u_{\mathbf{k}}\frac{\partial u_{\mathbf{k}'}^{*}}{\partial s} + \left(\varepsilon_{\mathbf{k}'} - \frac{\hbar^2 k'^2}{2m}\right)u_{\mathbf{k}'}^{*}\frac{\partial u_{\mathbf{k}}}{\partial s}\right]\Big\}\,d\tau_0.
\end{aligned} \qquad \text{(A.21.6)}$$

Transform the second addends in each of the square brackets, making use of the facts that the operators $\nabla^2$ and $i\nabla$ are self-conjugate and that the integrals of the type of (A.21.5a) are equal to zero:

$$\begin{aligned}
1)\ \int \frac{\partial u_{\mathbf{k}}}{\partial s}\nabla^2 u_{\mathbf{k}'}^{*}\,d\tau_0 &= \int u_{\mathbf{k}'}^{*}\nabla^2\left(\frac{\partial u_{\mathbf{k}}}{\partial s}\right)d\tau_0 = \int u_{\mathbf{k}'}^{*}\frac{\partial}{\partial s}(\nabla^2 u_{\mathbf{k}})\,d\tau_0 \\
&= \int \frac{\partial}{\partial s}(u_{\mathbf{k}'}^{*}\nabla^2 u_{\mathbf{k}})\,d\tau_0 - \int \frac{\partial u_{\mathbf{k}'}^{*}}{\partial s}\nabla^2 u_{\mathbf{k}}\,d\tau_0 \\
&= -\int \frac{\partial u_{\mathbf{k}'}^{*}}{\partial s}\nabla^2 u_{\mathbf{k}}\,d\tau_0;
\end{aligned}$$

$$\begin{aligned}
2)\ -\int \frac{\partial u_{\mathbf{k}}}{\partial s}\,i\nabla u_{\mathbf{k}'}^{*}\,d\tau_0 &= \int u_{\mathbf{k}'}^{*}(+i)\nabla\frac{\partial u_{\mathbf{k}}}{\partial s}\,d\tau_0 \\
&= \int u_{\mathbf{k}'}^{*}\frac{\partial}{\partial s}(+i\nabla u_{\mathbf{k}})\,d\tau_0 \\
&= \int \frac{\partial}{\partial s}[u_{\mathbf{k}'}^{*}(+i)\nabla u_{\mathbf{k}}]\,d\tau_0 \\
&\quad - \int \frac{\partial u_{\mathbf{k}'}^{*}}{\partial s}\,i\nabla u_{\mathbf{k}}\,d\tau_0 = -\int \frac{\partial u_{\mathbf{k}'}^{*}}{\partial s}\,i\nabla u_{\mathbf{k}}\,d\tau_0;
\end{aligned}$$

$$\begin{aligned}
(3)\ \int u_{\mathbf{k}'}^{*}\frac{\partial u_{\mathbf{k}}}{\partial s}\,d\tau_0 &= \int \frac{\partial}{\partial s}(u_{\mathbf{k}'}^{*}u_{\mathbf{k}})\,d\tau_0 \\
&\quad - \int u_{\mathbf{k}}\frac{\partial u_{\mathbf{k}'}^{*}}{\partial s}\,d\tau_0 = -\int u_{\mathbf{k}}\frac{\partial u_{\mathbf{k}'}^{*}}{\partial s}\,d\tau_0.
\end{aligned}$$

Making use of these three equalities, we obtain from (A.21.5b) and (A.21.6)

$$K^+ = -\frac{\hbar^2 i}{m}\left[(\mathbf{k}-\mathbf{k}')\cdot\int \operatorname{grad} u_{\mathbf{k}} \frac{\partial u^*_{\mathbf{k}'}}{\partial s}\,d\tau_0\right]$$

$$-\left[\varepsilon_{\mathbf{k}}-\varepsilon_{\mathbf{k}'}-\left(\frac{\hbar^2 k^2}{2m}-\frac{\hbar^2 k'^2}{2m}\right)\right]\int u_{\mathbf{k}} \frac{\partial u^*_{\mathbf{k}'}}{\partial s}\,d\tau_0. \qquad (A.21.7)$$

If the effective mass $m^*$ is equal to the electron mass $m$, the expression in square brackets in the second addend is equal to zero, since $\varepsilon_{\mathbf{k}} = \hbar^2k^2/2m$ and $\varepsilon_{\mathbf{k}'} = \hbar^2k'^2/2m$. If the effective mass is of the order of magnitude of $m$, the order of magnitude of the expression in square brackets is $\hbar^2k^2/m$. The integral in square brackets in the first addend is of the order of magnitude of

$$\int \operatorname{grad} u_{\mathbf{k}} \frac{\partial u^*_{\mathbf{k}}}{\partial s}\,d\tau_0 \approx \frac{1}{a}\int u_{\mathbf{k}} \frac{\partial u^*_{\mathbf{k}'}}{\partial s}\,d\tau_0,$$

where $a$ is the lattice constant. Accordingly, the ratio of the second addend to the first will be of the order of magnitude of

$$\frac{\hbar^2k^2}{m} \div \frac{\hbar^2 k}{ma} = ak,$$

this in the case of semiconductors is much less than unity.

This estimate is true if the electron's energy extremum is at the centre of the Brillouin zone, at $\mathbf{k} = 0$.

Neglecting the second addend in (A.21.7) and making use of (A.21.4a), we obtain

$$K^+ = \frac{\hbar^2 i}{m}\mathbf{q}\cdot\int \operatorname{grad} u_{\mathbf{k}}\int \frac{\partial u^*_{\mathbf{k}'}}{\partial s}\,d\tau_0 = \frac{\hbar^2 i}{m}\sum_{\alpha,\beta} q_\alpha e_\beta$$

$$\times\int \frac{\partial u_{\mathbf{k}}}{\partial x_\alpha}\frac{\partial u^*_{\mathbf{k}'}}{\partial x_\beta}\,d\tau_0, \qquad (A.21.8)$$

where $e_\beta$ is a rectangular component of the polarization vector $\mathbf{e}_{\mathbf{q}j}$. Here use is made of the relation

$$\frac{\partial u^*_{\mathbf{k}'}}{\partial s} = \mathbf{e}_{\mathbf{q}j}\cdot\operatorname{grad} u^*_{\mathbf{k}'}.$$

In a cubic crystal $u_{\mathbf{k}}$ is either an even or an odd function of $x_\alpha$; therefore, $\partial u_{\mathbf{k}}/\partial x_\alpha$ is in the first case an odd function and in the second an even function of $x_\alpha$. In integrating in (A.21.8) with respect to $x_\alpha$ from $-a/2$ to $+a/2$ ($a$ is the lattice constant), the integral for $\beta \neq \alpha$ is equal to zero; therefore,

$$K^+ = \frac{\hbar^2 i}{m}\sum_{\alpha} q_\alpha e_\alpha \int \frac{\partial u_{\mathbf{k}}}{\partial x_\alpha}\frac{\partial u^*_{\mathbf{k}'}}{\partial x_\alpha}\,d\tau_0.$$

It can be demonstrated that $u_{\mathbf{k}}$ depends little on $\mathbf{k}$, so that $u_{\mathbf{k}} \approx \approx u_{\mathbf{k}'} \equiv u$. Since all three rectangular axes in a cubic crystal are equivalent, it follows that

$$\int \frac{\partial u_{\mathbf{k}}}{\partial x_\alpha} \frac{\partial u_{\mathbf{k}'}}{\partial x_\alpha} d\tau_0 \approx \int \left| \frac{\partial u}{\partial x_\alpha} \right|^2 d\tau_0 = \frac{1}{3} \int |\operatorname{grad} u|^2 d\tau_0$$

for all $\alpha$.

Thus, we have

$$K^+ = \frac{\hbar^2 i}{m} \frac{1}{3} \int |\operatorname{grad} u|^2 d\tau_0 (\mathbf{q} \cdot \mathbf{e}_{\mathbf{q}j}). \tag{A.21.8a}$$

If we consider three vibration branches, one longitudinal ($j = 1$) $\mathbf{e}_{\mathbf{q}1} \| \mathbf{q}$ and two transverse ($j = 2, 3$) $\mathbf{e}_{\mathbf{q}2}$ and $\mathbf{e}_{\mathbf{q}3} \perp \mathbf{q}$, we see that in our approximations only the longitudinal vibrations interact with the electron.

Finally, we obtain for (A.21.8a)

$$K^+ = i \frac{2}{3} qC, \tag{A.21.8b}$$

where

$$C = \frac{\hbar^2}{2m} \int |\operatorname{grad} u|^2 d\tau_0. \tag{A.21.8c}$$

Combining (A.21.3) with (A.21.8b), we obtain the expression (8.3.10) for the matrix element describing the absorption of a phonon.

Considering the case when $L^* \neq 0$ and is equal to (A.21.3a), i.e., the case of an electron emitting a phonon, we can easily see that the only difference between the computation of $K^-$ and $K^+$ is that in the former case

$$\mathbf{k}' = \mathbf{k} - \mathbf{q} \tag{A.21.9}$$

takes the place of (A.21.4a) with the result that the only difference between $K^-$ and $K^+$ is in the sign. Combining (A.21.3a) with the expression for $K^-$, we obtain (8.3.10a).

### Appendix 22

Prove relation (8.5.15) by direct quantum mechanics calculation of the variation of energy of an electron occupying a level near the bottom of the conduction band caused by a deformation of the crystal [A.6].

The lattice constant of a cubic crystal in the case of homogeneous extension is

$$a = a_0 (1 + \varepsilon), \tag{A.22.1}$$

where $a_0$ is the lattice constant of an undeformed crystal and $\varepsilon = \varepsilon_{ii} = \partial u_i / \partial x_i$.

If the decrease in the level of the bottom edge of the conduction band caused by this extension is $\Delta\mathscr{E}$, then by definition (8.5.4),

$$\mathscr{E}_1 = -\frac{\Delta\mathscr{E}}{3\varepsilon}. \tag{A.22.2}$$

In a homogeneous deformed crystal, same as in an undeformed one, the conduction electron is described by the Bloch function

$$\psi_{\mathbf{k}}(\mathbf{r}) = u_{\mathbf{k}}(\mathbf{r})\, e^{i\mathbf{k}\cdot\mathbf{r}}, \tag{A.22.3}$$

where $u_{\mathbf{k}}(\mathbf{r})$ is a function periodic in three dimensions with the period $a$. The modulating function $u_{\mathbf{k}=0} \equiv u_0$ near the bottom of the conduction band $\mathbf{k} = 0$ satisfies (4.3.8)

$$-\frac{\hbar^2}{2m}\nabla^2 u_0 + V(\mathbf{r})\, u_0 = \mathscr{E} u_0, \tag{A.22.4}$$

where $V(\mathbf{r})$ is the periodic potential acting on the electron, and $\mathscr{E}$ is the energy of the bottom edge of the conduction band in a deformed crystal.

For perturbation theory to be applicable, both the perturbed and the unperturbed wave functions must satisfy the same boundary conditions, which for the function $u_0$ are replaced by conditions of periodicity. To make $u_0$ in a deformed and in an undeformed crystal satisfy the same periodicity conditions, we introduce dimensionless coordinates

$$x' = \frac{x}{a} = \frac{x}{a_0(1+\varepsilon)} \tag{A.22.5}$$

with similar expressions for $y'$ and $z'$. An increase in $x$ by $a = a_0(1+\varepsilon)$ both in a deformed ($\varepsilon \neq 0$) and in an undeformed ($\varepsilon = 0$) crystal will obviously cause a change in the coordinate $x'$ by 1, and, therefore, the period of $u_0$ in the dimensionless coordinates $x'$ will in both cases be the same and equal to 1. Introducing dimensionless coordinates $\mathbf{r}' = \mathbf{r}/a$, we obtain instead of (A.22.4)

$$-\frac{\hbar^2}{2m} a^{-2}\nabla^2_{\mathbf{r}'} u_0 + V(a\mathbf{r}')\, u_0 = \mathscr{E} u_0. \tag{A.22.6}$$

Denoting the periodic potential and the electron's energy (for $\mathbf{k} = 0$) in an undeformed crystal by $V_0$ and $\mathscr{E}_0$, respectively, we obtain by analogy with (A.22.6)

$$-\frac{\hbar^2}{2m} a_0^{-2}\nabla^2_{\mathbf{r}'} u_0 + V_0(a_0\mathbf{r}')\, u_0 = \mathscr{E}_0 u_0, \tag{A.22.7}$$

where $\mathbf{r}' = \mathbf{r}/a_0$.

In all real cases the deformation $\varepsilon \ll 1$; therefore, we calculate the displacement of the bottom edge of the conduction band in terms of the perturbation theory in the first approximation in $\varepsilon$, i.e., neglecting corrections of the order of $\varepsilon^2$ and higher.

It is established in quantum mechanics that to determine the perturbation energy in the first approximation we can take the wave functions in the zeroth approximation. This means that $u_0$ in (A.22.6) and (A.22.7) may be assumed to be identical. Premultiplying (A.22.6) and (A.22.7) by $u_0^*(\mathbf{r}/a_0)$, subtracting the equalities term-by-term and integrating with respect to $d\tau' = d\tau/a_0^3$, we obtain

$$(\mathscr{E}-\mathscr{E}_0)\int u_0^* u_0\,d\tau' = \int u_0^*\Big\{-\frac{\hbar^2}{2m}a_0^{-2}(1+\varepsilon)^{-2}\nabla_{\mathbf{r}'}^2 + V[a_0(1+\varepsilon)\mathbf{r}'] + \frac{\hbar^2}{2m}a_0^{-2}\nabla_{\mathbf{r}'}^2 - V_0(a_0\mathbf{r}')\Big\}u_0\,d\tau'. \quad \text{(A.22.8)}$$

If we normalize the Bloch function (A.22.3) in accordance with (4.3.6), we obtain

$$\int u_0^* u_0\,d\tau' = \frac{1}{a_0^3}\int |u_0|^2\,d\tau = \frac{1}{a_0^3}. \quad \text{(A.22.9)}$$

We expand the first addend in the braces in (A.22.8) into a series in $\varepsilon$ leaving only members of the zeroth and first orders, cancel out the zeroth-order member with the third addend in the braces, and return to the variable $\mathbf{r} = \mathbf{r}'a_0$; we then obtain from (A.22.8), making use of (A.22.9),

$$\mathscr{E}-\mathscr{E}_0 = \Delta\mathscr{E} = \varepsilon\frac{\hbar^2}{m}\int u_0^*\nabla^2 u_0\,d\tau + \int u_0^*\{V[(1+\varepsilon)\mathbf{r}] - V_0(\mathbf{r})\}u_0\,d\tau. \quad \text{(A.22.10)}$$

Applying Green's equation [A.2, p. 361, equation (9)], we obtain

$$\int u_0^*\nabla^2 u_0\,d\tau = -\int |\nabla u_0|^2\,d\tau + \oint u_0^*(\nabla u_0\cdot d\sigma), \quad \text{(A.22.11)}$$

where the last integral over the surface of the unit cell is zero.

According to the hypothesis of deformable ions (8.3.3), $V[(1+\varepsilon)\mathbf{r}] = V_0(\mathbf{r})$; it then follows from (A.22.10), (A.22.11) and (A.22.2), if use is made of the definition of $C$ (8.3.10b), that

$$\mathscr{E}_1 = -\frac{\Delta\mathscr{E}}{3\varepsilon} = \frac{\hbar^2}{3m}\int |\nabla u_0|^2\,d\tau = \frac{2}{3}C, \quad \text{(A.22.12)}$$

this coinciding with (8.5.15).

**Appendix 23**

The calculation of $\chi_n$ (or $\chi_p$) from equation (9.2.8) [or (9.2.10)] reduces to the solution of the vector equation

$$\mathbf{x} = \mathbf{a} + \mathbf{b}\times\mathbf{x} \quad \text{(A.23.1)}$$

for the unknown vector $\mathbf{x}$.

Note that from (A.23.1) follows

$$\mathbf{b}\cdot\mathbf{x}=\mathbf{b}\cdot\mathbf{a}$$

since

$$\mathbf{b}\cdot(\mathbf{b}\times\mathbf{x})=0.$$

Substituting into the right-hand side of (A.23.1) the expression $\mathbf{a}+\mathbf{b}\times\mathbf{x}$ for $\mathbf{x}$, we obtain

$$\mathbf{x}=\mathbf{a}+\mathbf{b}\times\mathbf{a}+\mathbf{b}\cdot(\mathbf{b}\times\mathbf{x}). \tag{A.23.2}$$

Making use of the identity transformation [A.2, pp. 215-6]

$$\mathbf{b}\times(\mathbf{b}\times\mathbf{x})=\mathbf{b}\cdot(\mathbf{b}\cdot\mathbf{x})-\mathbf{x}b^2$$

and of (A.23.2), we obtain

$$\mathbf{x}=\mathbf{a}+\mathbf{b}\times\mathbf{a}+\mathbf{b}\cdot(\mathbf{b}\cdot\mathbf{a})-\mathbf{x}b^2,$$

whence

$$\mathbf{x}=\frac{\mathbf{a}+\mathbf{b}\times\mathbf{a}+(\mathbf{a}\cdot\mathbf{b})\cdot\mathbf{b}}{1+b^2}. \tag{A.23.3}$$

Using this expression, we obtain (9.2.11) and (9.2.12).

## Appendix 24

The indices $\mu$, $\alpha$ and $\nu$ in the symbol $\delta_{\mu\alpha\nu}$ independently assume the values 1, 2, 3. Consider all possible *permutations* of the numbers 1, 2, 3:

$$(123),\ (231),\ (312),\ (132),\ (213),\ (321). \tag{A.24.1}$$

Their number is equal to $3! = 6$.

Apply the term *disorder* to the permutation when a larger number precedes a smaller one, and count the number of disorders in the permutations (A.24.1). It is easily seen to be equal to

$$0,\ 2,\ 2,\ 1,\ 1,\ 3, \tag{A.24.2}$$

i.e., the number of disorders in the first three permutations is even and in the last three it is odd.

If we make

$\delta_{\mu\alpha\nu}=+1$ when $(\mu\alpha\nu)$ forms an even number of disorders,

$\delta_{\mu\alpha\nu}=-1$ when $(\mu\alpha\nu)$ forms an odd number of disorders,

$\delta_{\mu\alpha\nu}=0$ when there are identical indices among $(\mu\alpha\nu)$,

we find immediately that the vector product $\mathbf{v}\times\mathbf{H}$ can be written in the form (9.8.22) with the aid of the symbol $\delta_{\mu\alpha\nu}$.

Apply the term *transposition* to the operation in which two elements (indices) change places in the course of a permutation.

A transposition is easily seen to change the number of disorders by an odd number. Hence, transpositions transform permutations with an even number of disorders into those with an odd number of disorders, and vice versa, so that

$$\delta_{\mu\alpha\nu} = -\delta_{\alpha\mu\nu}, \tag{A.24.3}$$

etc.

A cyclic permutation of the indices $\mu \to \alpha$, $\alpha \to \nu$, $\nu \to \mu$ does not change the symbol $\delta_{\mu\alpha\nu}$. It is obvious that $\delta^2_{\mu\alpha\nu} = 1$ or 0.

It can be demonstrated that $\delta_{\mu\alpha\nu}$ is a rank 3 tensor in which only 6 out of 27 components are nonzero.

The above concepts and results can easily be generalized to the case of permutations of $n$ elements [A.1, i. 117].

# References

## Chapter 1


1.1 E. Kreyszig: *Advanced Engineering Mathematics*, 3rd ed. (J. Wiley & Sons, New York 1972)

1.2 R. O'Connor: *Fundamentals of Chemistry*, 2nd ed. (Harper and Row, New York 1977)

1.3 G. A. Korn, T. M. Korn: *Mathematical Handbook for Scientists and Engineers*, 2nd ed. (McGraw-Hill, New York 1968)

1.4 J. M. Ziman: *Principles of the Theory of Solids*, (Cambridge University Press, Cambridge 1964)


## Chapter 2


2.1 *Solid State Theory: Methods and Applications*, ed. by P. T. Lansberg (J. Wiley & Sons, London 1969)

2.2 M. Tinkham: *Group Theory and Quantum Mechanics* (McGraw-Hill, New York 1964)

2.3 G. L. Bir, G. E. Pikus: *Symmetry and Strain-Induced Effects in Semiconductors* (J. Wiley & Sons, New York 1974)

2.4 A. S. Davydov: *Quantum Mechanics*, 2nd ed. (Pergamon Press, Oxford 1976)

2.5 E. P. Wigner: *Group Theory and Its Application to the Quantum Mechanics of Atomic Spectra* (Academic Press, New York 1957)

2.6 L. D. Landau, E. M. Lifshitz: *Quantum Mechanics: Non-Relativistic Theory*, 3rd ed. (Pergamon Press, Oxford 1977)

2.7 A. Sommerfeld: *Lectures on Theoretical Physics*, Vol. 1: *Mechanics* (Academic Press, New York 1964)

2.8 G. Ya. Ljubarskii: *Group Theory and Its Application to Physics* (Pergamon Press, Oxford 1960)

2.9 H. Jones: *The Theory of Brillouin Zones and Electronic States in Crystals* (North-Holland, Amsterdam 1960)

2.10 E. S. Fedorov: *Symmetry of Crystals* (American Crystallographic Association 1971)

2.11 A. Schönflies: *Theorie der Kristallstrukturen* (Berlin 1923)

2.12 M. I. Petrashen', E. D. Trifonov: *Applications of Group Theory in Quantum Mechanics* (MIT Press, Cambridge, Mass. 1969)

2.13 M. Hamermesh: *Group Theory and Its Application to Physical Problems* (Addison-Wesley/Pergamon Press, New York 1962)

2.14 L. D. Landau, E. M. Lifshitz: *Mechanics*, 3rd ed. (Pergamon Press, Oxford 1976)

2.15 M. Born, K. H. Huang: *Dynamical Theory of Crystal Lattices* (Oxford University Press, Oxford 1954)

2.16 E. Kreyszig: *Advanced Engineering Mathematics*, 3rd ed. (J. Wiley & Sons, New York 1972)

**Chapter 3**

3.1 L. D. Landau, E. M. Lifshitz: *Theory of Elasticity*, 2nd ed. (Pergamon Press, Oxford 1970)

3.2 J. F. Nye: *Physical Properties of Crystals* (Clarendon Press, Oxford 1957)

3.3 A. S. Davydov: *Quantum Mechanics*, 2nd ed. (Pergamon Press, Oxford 1976)

3.4 W. Heitler: *Elementary Wave Mechanics* (Clarendon Press, Oxford 1947)

3.5 L. D. Landau, E. M. Lifshitz: *Quantum Mechanics: Non-Relativistic Theory*, 3rd ed. (Pergamon Press, Oxford 1977)

3.6 Ya. I. Frenkel': *Vvedenie v teoriyu metallov* (Introduction to the Theory of Metals) (Moscow 1958)

3.7. E. Kreyszig: *Advanced Engineering Mathematics*, 3rd ed. (J. Wiley & Sons, New York 1972)

3.8 R. Resnik, D. Halliday: *Physics*: Part 1, 3rd ed. (J. Wiley & Sons, New York 1977)

3.9 M. A. Morrison, T. L. Estle, N. F. Lane: *Quantum States of Atoms, Molecules, and Solids* (Prentice-Hall, Englewood Cliffs, N. J. 1976)

3.10. L.D. Landau, E. M. Lifshitz: *Mechanics*, 3rd ed. (Pergamon Press, Oxford 1976)

3.11. M. Born, K. H. Huang: *Dynamical Theory of Crystal Lattices* (Oxford University Press, Oxford 1954)

3.12 G. A. Korn, T. M. Korn: *Mathematical Handbook for Scientists and Engineers*, 2nd ed. (McGraw-Hill, New York 1968)

3.13 L. P. Bouckaert, R. Smoluchowski, E. P. Wigner: "Theory of Brillouin zones and symmetry properties of wave functions in crystals" in *Symmetry in the Solid State*, ed. by R. S. Knox, A. Gold (W. A. Benjamin, New York 1964)

3.14 H. Jones: *The Theory of Brillouin Zones and Electronic States in Crystals* (North-Holland, Amsterdam 1960)

3.15 I. E. Tamm: *Fundamentals of the Theory of Electricity* (Mir Publishers, Moscow 1979)

3.16 N. E. Kochin: *Vektornoe ischislenie i nachala tenzornogo ischisleniya* (Vector and Tensor Analysis), 9th ed. (Nauka Publishers, Moscow 1965)

3.17 A. I. Anselm: *Osnovy statisticheskoi fiziki i termodinamiki* (The Fundamentals of Statistical Physics and Thermodynamics) (Moscow 1973)

3.18 L. D. Landau, E. M. Lifshitz: *Statistical Physics*: Part 1, 3rd ed. (Pergamon Press, Oxford 1979)

3.19 E. W. Kellermann: *Phil. Trans. Roy. Soc.* **238** (1940), 3

3.20 E. W. Kellermann: *Proc. Roy. Soc.* **A178** (1941), 17

3.21 F. A. Johnson, W. Cochran: in *Proc. 6th Intern. Conf. Phys. Semicond. Exeter* (The Institute of Physics and The Physical Society, London 1962), p. 498

3.22 P. W. Bridgman: *Dimensional Analysis* (Yale University press, New Haven, Conn. 1932)

## Chapter 4

4.1 A. S. Davydov: *Quantum Mechanics*, 2nd ed. (Pergamon Press, Oxford 1976)
4.2 E. Spenke: *Electronische Halbleiter* (Springer, Berlin 1956)
4.3 G. G. Darwin: *Proc. Roy. Soc.* London **A154** (1936), 61
4.4 I. V. Savelyev: *Physics: A General Course*, Vol. II: *Electricity and Magnetism, Waves, Optics* (Mir Publishers, Moscow 1980)
4.5 S. Brown, S. J. Barnett: *Phys. Rev.* **87** (1952), 601
4.6 N. Mott, W. Jones: *The Theory of Properties of Metals and Alloys* (Oxford University Press, Oxford 1958)
4.7 A. H. Wilson: *The Theory of Metals* (Cambridge University Press, Cambridge 1958)
4.8 *Symmetry in the Solid State*, ed. by R. S. Knox, A. Gold (W. A. Benjamin, New York 1964)
4.9 L. D. Landau, E. M. Lifshitz: *The Classical Theory of Fields*, 4th ed. (Pergamon Press, Oxford 1975)
4.10 I. E. Tamm: *Fundamentals of the Theory of Electricity* (Mir Publishers, Moscow 1979)
4.11 R. Courant, F. John: *Introduction to Calculus and Analysis*, Vol. 1 (Wiley-Interscience, New York 1965)
4.12 J. Mathews, R. L. Walker: *Mathematical Methods of Physics* (W. A. Benjamin, New York 1964)
4.13 R. J. Elliott: *Phys. Rev.* **96** (1954), 280.
4.14 G. Dresselhous: *Phys. Rev.* **100** (1955), 580
4.15 L. D. Landau, E. M. Lifshitz: *Quantum Mechanics: Non-Relativistic Theory*, 3rd ed. (Pergamon Press, Oxford 1977)
4.16 G. A. Korn, T. M. Korn: *Mathematical Handbook for Scientists and Engineers*, 2nd ed. (McGraw-Hill, New York 1968)
4.17 I. M. Tsidil'kovskii: *Elektrony i dyrki v poluprovodnikakh* (Electrons and Holes in Semiconductors) (Moscow 1972)
4.18 O. Madelung: *Physics of III-V Compounds* (New York 1964)

## Chapter 5

5.1 L. I. Schiff: *Quantum Mechanics*, 3rd ed. (McGraw-Hill, New York 1968)
5.2 P. T. Matthews: *Introduction to Quantum Mechanics*, 3rd ed. (McGraw-Hill, London 1974)
5.3 V. I. Smirnov: *Kurs vysshei matematiki* (A Course of Higher Mathematics), 10th ed., Vol. 3, Part 1 (Nauka Publishers, Moscow 1974)
5.4 S. I. Pekar: *Issledovaniya po elektronnoi teorii kristallov* (Studies in the Electron Theory of Crystals) (Gostekhizdat, Moscow 1951)
5.5 *Polarons and Excitons*, ed. by C. G. Kuper, G. D. Whitfield (Scottish Universities' Summer School 1963) (Oliver and Boyd, Edinburgh 1963)
5.6 I. E. Tamm: *Fundamentals of the Theory of Electricity* (Mir Publishers, Moscow 1979)
5.7 L. D. Landau, E. M. Lifshitz: *Quantum Mechanics: Non-Relativistic Theory*, 3rd ed. (Pergamon Press, Oxford 1977)
5.8 G. A. Korn, T. M. Korn: *Mathematical Handbook for Scientists and Engineers*, 2nd ed. (McGraw-Hill, New York 1968)
5.9 Yu. A. Firsov: in *Polyarony* (Polarons) (Moscow 1975)

## Chapter 6


6.1 A. I. Anselm: *Osnovy statisticheskoi fiziki i termodinamiki* (The Fundamentals of Statistical Physics and Thermodynamics) (Moscow 1973)

6.2 T. McDougall, E. C. Stoner: *Phil. Trans. Roy. Soc.* **A237** (1938), 350

6.3 A. C. Beer, M. N. Chase, P. F. Choquard: *Helv. Phys. Acta* 28 (1955), 529

6.4 P. Rhodes: *Proc. Roy. Soc.* **A204** (1950), 396

6.5 S. V. Vonsovskii: *Magnetism*, 2 Vols. (Halsted Press, New York 1975)

6.6 I. E. Tamm: *Fundamentals of the Theory of Electricity* (Mir Publishers, Moscow 1979)

6.7 M. A. Morrison, T. L. Estle, N. F. Lane: *Quantum States of Atoms, Molecules, and Solids* (Prentice-Hall, Englewood Cliffs, N.J. 1976)

6.8 A. S. Davydov: *Quantum Mechanics*, 2nd ed. (Pergamon Press, Oxford 1976)

6.9 I. V. Savelyev: *Physics: A General Course*, Vol. II: *Electricity and Magnetism, Waves, Optics* (Mir Publishers, Moscow 1980)

6.10 L. D. Landau, E. M. Lifshitz: *The Classical Theory of Fields*, 4th ed. (Pergamon Press, Oxford 1975)

6.11 L. D. Landau, E. M. Lifshitz: *Quantum Mechanics: Non-Relativistic Theory*, 3rd ed. (Pergamon Press, Oxford 1977)

6.12 G. Busch, U. Winkler: "Bestimmung der characteristischen Grössen eines Halbleiters aus elektrischen, optischen und magnetischen Messungen" in *Ergebnisse der exacten Naturwissenschaften*, Bd. 29 (Zurich 1956)

6.13 E. Kreyszig: *Advanced Engineering Mathematics*, 3rd ed. (J. Willey & Sons, New York 1972)

6.14 T. M. Luttinger: *Phys. Rev.* **102** (1955), 1030

6.15 V. I. Smirnov: *Kurs vysshei matematiki* (*A Course of Higher Mathematics*), 21st ed., Vol. 2 (Nauka Publishers, Moscow 1974)

6.16 G. E. Pikus: *Osnovy teorii poluprovodnikovykh priborov* (The Fundamentals of the Theory of Semiconductor Devices) (Nauka Publishers, Moscow 1965)

6.17 E. V. Shpol'skii: *Atomnaya fizika* (Atomic Physics), 4th ed., Vol. 2 (Nauka Publishers, Moscow 1974)

6.18 R. Paul: *Hableiterphysik* (VEB Verlag Technik, Berlin 1974)


## Chapter 7


7.1 V. I. Smirnov: *Kurs vysshei matematiki* (*A Course of Higher Mathematics*), 8th ed., Vol. 3, Part 2 (Nauka Publishers, Moscow 1969)

7.2 B. Velicky: *Czech. J. Phys.* **B11** (1961), 787

7.3 I. E. Tamm: *Fundamentals of the Theory of Electricity* (Mir Publishers, Moscow 1979)

7.4 D. I. Blokhintsev: *Osnovy kvantovoi mekhaniki* (Principles of Quantum Mechanics), 5th ed. (Nauka Publishers, Moscow 1976)

7.5 R. J. Elliot: *Phys. Rev.* **108**. (1957); 1384

7.6 L. D. Landau, E. M. Lifshitz: *Quantum Mechanics: Non-Relativistic Theory*, 3rd ed. (Pergamon Press, Oxford 1977)

7.7 G. L. Bir, G. N. Pikus: *Simmetriya i deformatsionnye effecty v poluprovodnikakh* (Symmetry and Strain-Induced Effects in Semiconductors) (Nauka Publishers, Moscow 1972)

7.8 *The Optical Properties of Solids* (International School of Physics "Enrico Fermi", Course 34), ed. by J. Tauc (Academic Press, New York 1966)

7.9 I. S. Gradshteyn, I. M. Ryzhik: *Tables of Integrals, Series, and Products*, 4th ed. (Academic Press, New York 1965)

7.10 W. A. Blanpied: *Modern Physics: An Introduction to Its Mathematical Language* (Holt, Rinehart and Winston, New York 1971)

7.11 *Semiconductors and Semimetals*, Vol. 3: *Optical Properties of III-V Compounds*, ed. by R. K. Willardson, A. C. Beer (Academic Press, New York 1967)

7.12 K. Huang: *Proc. Roy. Soc.* **A208** (1951), 352

7.13 *Optical Properties of Solids*, ed. by F. Abelès (North-Holland, Amsterdam 1972)

7.14 N. N. Lebedev: *Spetsial'nye funktsii i ikh primeneniya* (Special Functions and Their Applications) (Moscow-Leningrad 1953)

7.15 V. I. Smirnov: *Kurs vysshei matematiki* (A Course of Higher Mathematics), 21st ed., Vol. 2 (Nauka Publishers, Moscow 1974)

7.16 H. Jeffreys, B. Jeffreys: *Methods of Mathematical Physics*, 3rd ed. (Cambridge University Press, London 1962)

**Chapter 8**

8.1 N. E. Kochin: *Vektornoe ischislenie i nachala tenzornogo ischisleniya* (Vector and Tensor Analysis), 9th ed. (Nauka Publishers, Moscow 1965)

8.2 M. Born: *Atomic Physics* (Blackie and Son, Glasgow 1957)

8.3 C. Herring, E. Vogt: *Phys. Rev.* **101** (1956), 944

8.4 A. G. Samoilovich, I. Ya. Korenblat, I. V. Dakhovskii, V. D. Iskra: *Fizika Tverdogo Tela*, No. 3 (1961), 3285

8.5 L. I. Schiff: *Quantum Mechanics*, 3rd ed. (McGraw-Hill, New York 1968)

8.6 G. E. Pikus: *ZhTF* **28** (1958), 2390

8.7 A. Sommerfeld: *Lectures on Theoretical Physics*, Vol. 2: *Mechanics of Deformable Bodies* (Academic Press, New York 1950)

8.8 H. Meijer, D. Polder: *Physica* **19** (1953), 255

8.9 I. E. Tamm: *Fundamentals of the Theory of Electricity* (Mir Publishers, Moscow 1979)

8.10 A. S. Davydov: *Quantum Mechanics*, 2nd ed. (Pergamon Press, Oxford 1975)

8.11 V. I. Kogan, V. M. Galitskiy: *Problems in Quantum Mechanics* (Prentice-Hall, Englewood Cliffs, N.J. 1963)

8.12 C. Erginsoy: *Phys. Rev.* **79** (1950), 1013

**Chapter 9**

9.1 A. I. Anselm: *Osnovy statisticheskoi fiziki i termodinamiki* (The Fundamentals of Statistical Physics and Thermodynamics) (Moscow 1973)

9.2 E. Kreyszig: *Advanced Engineering Mathematics*, 3rd ed. (J. Wiley & Sons, New York 1972)

9.3 A. I. Anselm, V. I. Klyachkin: *ZhETF* **22** (1952), 297

9.4 E. Jahnke, F. Emde: *Tables of Higher Functions*, 6th ed. (McGraw-Hill, New York 1960)

9.5 R. B. Dingle, A. Doreen, S. K. Roy: *Applied Scientific Research* (Hague) **B6** (1956), 155

9.6 N. E. Kochin: *Vektornoe ischislenie i nachala tenzornogo ischisleniya* (Vector and Tensor Analysis), 9th ed. (Nauka Publishers, Moscow 1965)
9.7 A. F. Ioffe: *Poluprovodnikovye termoelementy* (Semiconductor Thermocouples) (Izd Akad. Nauk SSSR, Moscow-Leningrad 1960)
9.8 A. F. Ioffe, L. S. Stil'bans, E. K. Iordanishvili, T. S. Stavitskaya: *Termoelektricheskoe okhlazhdenie* (Thermoelectric Cooling) (Izd. Akad. Nauk SSSR, Moscow-Leningrad 1959)
9.9 A. S. Okhotin, A. S. Pushkarskii, H. P. Borovikova, V. A. Simonov: *Metody izmereniya kharakteristik termoelektricheskikh materialov i preobrazovatelei* (Methods of Measuring the Characteristics of Thermoelectric Materials and Transducers) (Nauka Publishers, Moscow 1974)
9.10 V. A. Johnson, W. J. Whitesell, *Phys. Rev.* 89 (1953), 941
9.11 I. M. Lifshitz, M. Ya. Azbel', M. I. Kaganov: *Elektronnaya teoriya metallov* (Electron Theory of Metals) (Nauka Publishers, Moscow 1971)
9.12 B. Abeles, S. Meiboom: *Phys. Rev.* 95 (1954), 31
9.13 M. Shibuya: *Phys. Rev.* 95 (1954), 1385
9.14 R. K. Willardson, T. C. Harman, A. C. Beer: *Phys. Rev.* 96 (1954), 1512
9.15 G. E. Pikus: *ZhTF* 27 (1957), 1606
9.16 B. M. Askerov: *Kineticheskie effecty v poluprovodnikakh* (Kinetic Effects in Semiconductors) (Nauka Publishers, Moscow 1970)
9.17 J. Kolodziejczak: *Acta Phys. Polon.* 20 (1961), 289
9.18 I. V. Mochan, Yu. N. Obraztsov, T. V. Krylova: *ZhTF* 27 (1957), 242
9.19 C. Herring, T. H. Geballe: *Bull. of the Amer. Phys. Soc.*, Series II, 1 (1956), 117
9.20 C. Herring: in *Halbleiter und Phosphore*, ed. by M. Schön, H. Welker (Vieweg., Braunschweig 1958), p. 184
9.21 P. C. Zyryanov, M. I. Klinger: *Kvantovaya teoriya yavlenii electronnogo perenosa v kristallicheskikh poluprovodnikakh* (Quantum Theory of Electron Transport in Crystalline Semiconductors) (Nauka Publishers, Moscow 1976)
9.22 P. V. Parfen'ev, G. I. Kharus, I. M. Tsidil'kovskii, S. S. Shalyt: *UFN* 112 (1974), 3

**Appendices**

A.1 V. I. Smirnov: *Kurs vysshei matematiki* (A Course of Higher Mathematics), 21st ed., Vol. 2 (Nauka Publishers, Moscow 1974)
A.2 E. Kreyszig: *Advanced Engineering Mathematics*, 3rd ed. (J. Wiley & Sons, New York 1972)
A.3 R. Courant, D. Hilbert: *Methods of Mathematical Physics*, Vol. 1 (Interscience, New York 1953)
A.4 J. Mathews, R. L. Walker: *Mathematical Methods of Physics* (W. A. Benjamin, New York 1964)
A.5 L. I. Schiff: *Quantum Mechanics*, 3rd ed. (McGraw-Hill, New York 1968)
A.6 G. E. Pikus: *ZhTF* 28 (1958), 2390

# Name Index

# Subject Index